World Checklist of

Palms

World Checklist of

Palms

Rafaël Govaerts and John Dransfield

PLANTS PEOPLE
POSSIBILITIES

First published in 2005 by
Royal Botanic Gardens, Kew
Richmond, Surrey, TW9 3AB, UK
www.kew.org

ISBN 1-84246-084-6

Production editor: Ruth Linklater
Typesetting and page layout: Christine Beard
Cover design: Jeff Eden

Information Services Department,
Royal Botanic Gardens, Kew.

For information or to purchase all Kew titles please visit
www.kewbooks.com or email publishing@kew.org

Contents

Introduction

The palms are widely recognised as one of the most economically significant families of flowering plants. Not only does the family provide major industrial crops such as the African oil palm, the coconut and the date palm, but also produces a wealth of locally significant crops, some actively planted, others harvested from the wild. These locally important crops often form the basis of rural life in the tropics and subtropics. There is also a rapidly growing horticultural trade in palms world-wide. As with so many groups of plants, there has been a tendency towards over-description of the variation encountered, some of this due to the difficulty of making good representative herbarium collections of palms. Narrow species concepts have lead to the description of new taxa based on variation in characters later shown to display continuous variation. Political fragmentation of contiguous land masses such as continental South America and Indochina has also lead to the description of political species that are later shown to be conspecific across the broad geographical area. There is thus a real need for a nomenclator that provides a reliable guide to accepted names. This is particularly important as a basic framework for exchange of information in economically important but highly species-rich groups such as the rattans, and in providing a framework for addressing conservation status and needs.

In the last fifteen years there has been a great surge in palm research that has resulted in the publication of many new monographs and floristic accounts and although much more research is required particularly in the Old World tropics before a world palm monograph can be produced, there is enough published information for the production of a meaningful world checklist.

As work goes on, names inevitably change — new taxonomic research provides new insights that result in name changes and overlooked names also emerge. We do not pretend that this list is complete or entirely correct. We solicit input and constructive criticism. Please send comments by email to r.govaerts@kew.org and j.dransfield@kew.org or by regular post to the authors at Herbarium, Royal Botanic Gardens, Kew, Richmond, Surrey, TW9 3AB, U.K.

The data are also available on the World Wide Web as a searchable database (www.kew.org/monocotchecklist/). The database is live and comments can be made.

We are aware that as soon as this checklist is printed it will become out of date. The database that underpins the checklist will constantly be updated. Any changes that are made will be apparent in the on-line, fully searchable checklist, available at www.kew.org

Currently, the checklist includes accepted 2364 species in 190 accepted genera.

How to use this Checklist

Structure

The checklist is derived from a database encompassing 24 fields and complying with the data standards proposed by the Taxonomic Databases Working Group (TDWG). Its compilation was effected using Microsoft Access version 3.0 for Windows, within which editing was also carried out. Accepted and synonymous names respectively appear in identical, linked tables. In output, these are combined with the bibliographic citation and genus tables, the arrangement of genera being alphabetical with accepted and synonymous genera intercalated. The genus table furnishes the generic heading, which for accepted genera contains the number of published species and distribution.

Names

Names of genera and their species and infraspecific taxa are listed alphabetically. For each accepted taxon, associated synonyms are listed chronologically if heterotypic, with any homotypic synonyms following in a given lead. Doubtful and excluded taxa are likewise summarised in an appendix following the last genus. Place and date of publication of all names are given. Citation of authors follows *Authors of Plant Names* (Brummitt & Powell 1992), an updated version of which can be found on the internet (http://www. ipni.org/ipni/query_author.html); for book abbreviations, the standard is *Taxonomic Literature*, 2nd edn. (Stafleu & Cowan 1976–88; supplements, 1992–2000); and periodicals are abbreviated according to *Botanico-Periodicum-Huntianum/ Supplementum* (Bridson 1991). A question mark (?) following a name and author indicates that a place of publication has yet to be established. Names of nothospecies ('hybrids') are preceded by a multiplication sign (×), with the place of publication being followed by the names of the parents if known. Basionyms or replaced synonyms of accepted names are designated by an asterisk (*). For genera, the number of accepted species and the geographical distribution are furnished. Generic types are not indicated, but reference may be made to Farr *et al.* (1979, 1986), Greuter *et al.* (1993) or Uhl & Dransfield (1987).

Acceptance of taxa

Acceptance of species and infraspecific taxa is based not only on assessments of literature and common practice but also, where possible, by reference to specialist advice and (where necessary) to the herbarium or living collections. Generic limits follow Uhl & Dransfield (1987), modified in Dransfield & Uhl (1998), and with further modifications that will be published in the Second Edition of Genera Palmarum (Dransfield *et al.*, in prep.). Names of hybrids of horticultural origin are not accepted but full references can be found in the list of unplaced names and parentage given where known.

New names and combinations

A few necessary new names and combinations have been made in the text to bring known taxa into line with current practice.

Geographical Distribution

Distributions of **species** and taxa of lower rank are furnished in two ways: firstly by a generalised statement in narrative form, and secondly as TDWG geographical codes (Brummitt, 2001 or http://www.tdwg.org/geogrphy.html) expressed to that system's third level. Examples of the former include:

E. & C. U.S.A.
Texas to C. America
Mexico (Veracruz)
Europe to Iran
E. Himalaya, Tibet, China (W. Yunnan)
Philippines (Luzon)
S. Trop. America

When the presence of a taxon in a given region or location is not certainly known, a question mark is used, e.g. New Ireland ?; when an exact location within a country is not known, a question mark within brackets is used, e.g. Mexico (?). Distributions of **genera** are furnished in a relatively simplified form, any special features being given within brackets.

With respect to the TDWG codes, the **region** is indicated by the two-digit number (representative of the first two levels), the first digit also indicating the continent. The letter codes following the digits, when given, represent the third-level **unit** (a country, state or other comparable area). They usually are the first three letters of a given unit's name, but sometimes are contractions. 'ALL' is used if the species is known to occur in every area of a given region. If the country code is not known, '+' is used. For taxa that are known or appear to be extinct in a given region, '†' is used after the country code. Naturalisation is expressed by putting the third-level codes in lower case and, if in a second-level region all occurrences are the result of naturalisation, the code number for the region is placed in brackets. The application of question marks is as indicated above for geographical regions. Examples include:

12 SPA	[SW. Europe: Spain]
32 +	[C. Asia (more exact distribution not known)]
36 CHN? 38 JAP KOR	[Doubtful in China and Eastern Asia; China, Japan, Korea]
51 NZN NZS	[New Zealand: North and South Islands]
6 ARI 77 NWM TEX	[SW. & SC. U.S.A., Mexico & C. America:
79 ALL 80 GUA HON	Arizona, New Mexico, Texas, Mexico, Guatemala and Honduras]
77 TEX†	[SC. U.S.A.: Texas, where extinct]
(10)(11) 42 50 51 60 85	[of a genus: New Guinea, Australia, SW. Pacific Islands, New Zealand, and S. America; naturalised in NW. and C. Europe]

References

Bridson, G., comp. & ed. (1991). *Botanico-Periodicum Huntianum/Supplementum.* Hunt Institute for Botanical Documentation, Pittsburgh.

Burnett, J. (1994). IOPI and the Global Plant Checklist project. *Biology International*, **29**: 40–44.

Brummitt, R.K. (1992). *Vascular Plant Families and Genera.* 804 pp. Royal Botanic Gardens, Kew.

Brummitt, R.K. & Powell, C.E. (1992). *Authors of Plant Names.* 732 pp. Royal Botanic Gardens, Kew.

Brummitt, R.K. (2001). *World Geographical Scheme for Recording Plant Distributions edition 2.* xv, 137 pp. Hunt Institute for Botanical Documentation, Carnegie-Mellon University, Pittsburgh, Penna. (for the International Working Group on Taxonomic Databases for Plant Sciences). (Plant Taxonomic Database Standards, 2: version 1.0.)

Dransfield, J. & Uhl, N.W. (1998). *Palmae.* In Kubitzki, K. (ed.). *Families and Genera of Vascular Plants. Volume IV. Flowering Plants. Monocotyledons.* pp. 306–389. Springer, Berlin.

Dransfield, J., Uhl, N.W., Asmussen, C., Baker, W.J., Harley, M.M. & Lewis, C. (in prep.). *Genera Palmarum Ed. 2.*

Farr, E.R., Leussink, J.A. & Stafleu, F.A. (eds), 1979. *Index Nominum Genericorum (Plantarum).* 3 vols. Bohn, Scheltema & Holkema, Utrecht. (Regnum Vegetabile, 100–102).

Farr, E.R., Leussink, J.A. & Zijlstra, G., eds. (1986). *Index Nominum Genericorum (Plantarum): Supplementum I.* xv, 126 pp. Bohn, Scheltema & Holkema, The Hague. (Regnum Vegetabile, 113.)

Frodin, D.G. & Govaerts, R. (1996). *World Checklist and Bibliography of Magnoliaceae.* vii, 72 pp. Royal Botanic Gardens, Kew.

Govaerts, R. (1995–). *World Checklist of Seed Plants.* Antwerp: MIM (letter 'A'); Continental Publishing (letter 'B', 1996, letter 'C', 1999). [In continuation]

Greuter, W. *et al.* (1993). *Names in Current Use for Extant Plant Genera* (Names in current use, 3). xxvii, 1464 pp. Koeltz, Koenigstein. (Regnum Vegetabile)

Stafleu, F. & Cowan, R.S. (1976–88). *Taxonomic Literature: A Selective Guide to Botanical Publications and Collections with Dates, Commentaries and Types.* 2nd edn. 7 vols. Utrecht: Bohn, Scheltema & Holkema. (Regnum Vegetabile 94, 98, 105, 110, 112, 115, 116.) Continued as Stafleu, F. *et al.* (1992–). *Taxonomic Literature, Supplement.* Vols. 1– . Koeltz, Koenigstein. (Regnum Vegetabile 125, *passim.* As of 2000 six volumes published.)

Uhl, N.W. & Dransfield, J. (1987). *Genera Palmarum.* xxi, 610 pp. Allen Press, Lawrence, Kansas.

Abbreviations

al., alii **others**
arch. **archipelago**
auct., auctorum **of authors**
C. **central**
cham. **chamaephyte** [life-form]
cit. **citations; cited**
Co., comitas **county** *or* (China) *hsien*
comb., combinatio **combination**
cons., conservandus **conserved**
cult., cultus **cultivated**
descr., descriptio **description**
Distr. **District**
E. **east(ern)**
etc., et cetera **and the rest**
e.g., exempli gratia **for example**
hort., hortorum **of gardens** *or*
hortulanorum **of horticulturists**
I./Is **island/islands**
ICBN **International Code of Botanical Nomenclature**
i.e., id est **that is**
ign., ignotus **unknown**
in litt., in litteris **in correspondence**
ined., ineditus **unpublished** [provisional name, not accepted by the author]
inq., inquilinus **naturalised**
i.q., idem quod **the same as**
Medit. **Mediterranean**
Mt./Mts **mountain/mountains**
N. **north(ern)**
nom. cons., nomen conservandum **name to be conserved** [ICBN]
nom. illeg., nomen illegitimum **illegitimate name** [ICBN]
nom. inval., nomen invalidum **invalidly published name** [ICBN]
nom. nud., nomen nudum **name without a description** [ICBN]
nom. provis., nomen provisorum **provisional name, not accepted by the author(s)**
nom. rejic., nomen rejiciendum **rejected name** [ICBN]
nom. superfl., nomen superfluum **name superfluous when published** [ICBN]
nov., novus **new**
Pen. **peninsula(r)**
p.p., pro parte **in part**
pro syn., pro synonymo **as a synonym**
prov. **province**
q.e., quod est **which is**
q.v., quod vide **which see**
reg., regio **region**

Rep. **republic**
S. **south(ern)**
seq., sequens **following**
sine descr. lat., sine descriptione latine **without description in Latin**
s.l., sensu lato **in the broad sense**
sp., spp. **species** (both singular and plural)
sphalm., sphalmate **by mistake**
s.s. sensu stricto **in the narrow sense**
st. **status**
subtrop. **subtropical**
syn., synonymon **synonym**
temp. **temperate**
trop. **tropical**
viz, videlicet **namely**
W. **west(ern)**
? **not known, doubtful** (all contexts)
* **basionym or replaced synonym** (before a name)
× **nothotaxon** (before a genus or species name)
+ **range more than as indicated but not certainly known**
† **extinct** (after a TDWG code)

Acknowledgements

Several palm specialists kindly provided their opinions concerning whether taxa should be accepted or not or have pointed out errors. We have tried to follow their suggestions as far as possible; any remaining errors are our own. We especially wish to thank Bill Baker, Anders Barfod, Ross Bayton, Finn Borchsenius, Nilce Costa, John Dowe, Tom Evans, Andrew Henderson, Don Hodel, Francis Kahn, Carl Lewis, Don Munro, Carlo Morici, Saw Leng Guan and Scott Zona. Scott Zona, in particular, generously shared his views on the taxonomy and nomenclature of *Ptychosperma*, *Ptychococcus* and their relatives. Sasha Barrow, Alison Hoare and Helen Sanderson at various times worked on a palm database at Kew, with much expert advice from Bob Allkin; this database provided geographical distribution and some taxonomic data that have been incorporated in the Checklist. Numerous correspondents and users of the prototype have suggested corrections. We thank them all. Finally we thank Phil Cribb for his support and encouragement of the Monocot Checklist project, of which the Palm Checklist is an output.

Acanthococos

Acanthococos Barb.Rodr. = **Acrocomia** Mart.

Acanthococos emensis Toledo = **Acrocomia hassleri** (Barb.Rodr.) W.J.Hahn

Acanthococos hassleri Barb.Rodr. = **Acrocomia hassleri** (Barb.Rodr.) W.J.Hahn

Acanthococos sericea Burret = **Acrocomia hassleri** (Barb.Rodr.) W.J.Hahn

Acanthophoenix

Acanthophoenix H.Wendl., Ann. Gén. Hort. 6: 181 (1866).
1 species Mascarenes. 29.

Acanthophoenix crinita (Bory) H.Wendl. = **Acanthophoenix rubra** (Bory) H.Wendl.

Acanthophoenix grandis auct. = **?**

Acanthophoenix rubra (Bory) H.Wendl., Fl. Serres Jard. Paris 2(6): 181 (1866).
Mascarenes. 29 MAU REU.
Areca crinita Bory, Voy. îles Afrique 1: 307 (1804).
Acanthophoenix crinita (Bory) H.Wendl., Ann. Gén. Hort. 6: 181 (1866).
*Areca rubra Bory, Voy. îles Afrique 1: 306 (1804).
Sublimia centennina Comm. ex Mart., Hist. Nat. Palm. 3: 174 (1838), nom. inval.
Areca cincta Walp., Ann. Bot. Syst. 5: 808 (1860).
Calamus dealbatus H.Wendl. in O.C.E.de Kerchove de Denterghem, Palmiers: 236 (1878).
Calamus verschaffeltii H.Wendl. in O.C.E.de Kerchove de Denterghem, Palmiers: 238 (1878).
Areca herbstii W.Watson, Gard. Chron., n.s., 22: 426 (1884).

Acoelorrhaphe

Acoelorrhaphe H.Wendl., Bot. Zeitung (Berlin) 37: 148 (1879).
1 species, S. Florida, Caribbean, Mexico to Colombia. 78 79 80 81.
Paurotis O.F.Cook, Mem. Torrey Bot. Club 12: 21 (1902).
Acanthosabal Prosch., Gard. Chron., III, 77: 91 (1925).

Acoelorrhaphe arborescens (Sarg.) Becc. = **Acoelorrhaphe wrightii** (Griseb. & H.Wendl.) H.Wendl. ex Becc.

Acoelorrhaphe cookii Bartlett = **Brahea dulcis** (Kunth) Mart.

Acoelorrhaphe pimo (Becc.) Bartlett = **Brahea pimo** Becc.

Acoelorrhaphe pinetorum Bartlett = **Acoelorrhaphe wrightii** (Griseb. & H.Wendl.) H.Wendl. ex Becc.

Acoelorrhaphe salvadorensis (H.Wendl. ex Becc.) Bartlett = **Brahea dulcis** (Kunth) Mart.

Acoelorrhaphe schippii (Burret) Dahlgren = **Brahea dulcis** (Kunth) Mart.

Acoelorrhaphe wrightii (Griseb. & H.Wendl.) H.Wendl. ex Becc., Webbia 2: 109 (1908).
S. Florida, Caribbean, S. Mexivo to Colombia. 78 FLA 79 MXG MXT 80 BLZ COS GUA HON
**Copernicia wrightii* Griseb. & H.Wendl., Cat. Pl. Cub.: 220 (1866). *Paurotis wrightii* (Griseb. & H.Wendl.) Britton in N.L.Britton & J.A.Shafer, N. Amer. Trees: 141 (1908).
Serenoa arborescens Sarg., Bot. Gaz. 27: 90 (1899).

Paurotis arborescens (Sarg.) O.F.Cook, Mem. Torrey Bot. Club 12: 22 (1902). *Acoelorrhaphe arborescens* (Sarg.) Becc., Webbia 2: 113 (1908).
Paurotis androsana O.F.Cook, Mem. Torrey Bot. Club 12: 22 (1902).
Acanthosabal caespitosa Prosch., Gard. Chron., III, 77: 92 (1925).
Brahea psilocalyx Burret, Notizbl. Bot. Gart. Berlin-Dahlem 11: 1037 (1934). *Paurotis psilocalyx* (Burret) Lundell, Wrightia 2: 116 (1961).
Acoelorrhaphe pinetorum Bartlett, Publ. Carnegie Inst. Wash. 461: 33 (1935).
Paurotis schippii Burret, Notizbl. Bot. Gart. Berlin-Dahlem 12: 303 (1935).

Acanthorrhiza

Acanthorrhiza H.Wendl. = **Cryosophila** Blume

Acanthorrhiza aculeata (Liebm. ex Mart.) H.Wendl. = **Cryosophila nana** (Kunth) Blume

Acanthorrhiza arborea Drude ex Hook.f. = **Trinax arborea**

Acanthorrhiza argentea (Lodd. ex Schult. & Schult.f.) O.F.Cook = **Coccothrinax argentea** (Lodd. ex Schult. & Schult.f.) Sarg. ex Becc.

Acanthorrhiza chuco (Mart.) Drude = **Chelyocarpus chuco** (Mart.) H.E.Moore

Acanthorrhiza collinsii O.F.Cook = **Cryosophila stauracantha** (Heynh.) R.J.Evans

Acanthorrhiza kalbreyeri Dammer ex Burret = **Cryosophila kalbreyeri** (Dammer ex Burret) Dahlgren

Acanthorrhiza mocinoi (Kunth) Benth. & Hook.f. = **Cryosophila nana** (Kunth) Blume

Acanthorrhiza stauracantha (Heynh.) H.Wendl. ex Linden = **Cryosophila stauracantha** (Heynh.) R.J.Evans

Acanthorrhiza wallisii H.Wendl. = **?**

Acanthorrhiza warscewiczii H.Wendl. = **Cryosophila warscewiczii** (H.Wendl.) Bartlett

Acanthosabal

Acanthosabal Prosch. = **Acoelorrhaphe** H.Wendl.

Acanthosabal caespitosa Prosch. = **Acoelorrhaphe wrightii** (Griseb. & H.Wendl.) H.Wendl. ex Becc.

Acrista

Acrista O.F.Cook = **Prestoea** Hook.f.

Acrista monticola O.F.Cook = **Prestoea acuminata** var. **montana** (Graham) A.J.Hend. & Galeano

Acrocomia

2 species, Mexico to Trop. America.

Acrocomia Mart., Hist. Nat. Palm. 2: 66 (1824).
Mexico to Trop. America. 79 80 81 82 83 84 85.
Acanthococos Barb.Rodr., Palm. Hassler.: 1 (1900).

Acrocomia aculeata (Jacq.) Lodd. ex Mart., Hist. Nat. Palm. 3: 286 (1845).
Mexico to Trop. America. 79 MXG MXS MXT 80 BLZ COS ELS GUA HON NIC PAN 81 CUB DOM HAI JAM LEE PUE TRT WIN 82 FRG GUY SUR VEN 83 BOL CLM 84 BZC BZE BZL BZN BZS 85 AGE PAR.

Cocos aculeata Jacq., Select. Stirp. Amer. Hist.: 278 (1763). *Acrocomia sclerocarpa* Mart., Hist. Nat. Palm. 2: 66 (1824), nom. illeg.

Palma spinosa Mill., Gard. Dict. ed. 8: 3 (1768). *Acrocomia spinosa* (Mill.) N.E.Moore, Gentes Herb. 9: 238 (1963).

Bactris globosa Gaertn., Fruct. Sem. Pl. 1: 22 (1788). *Acrocomia globosa* (Gaertn.) Lodd. ex Mart., Hist. Nat. Palm. 3: 286 (1845).

Bactris minor Gaertn., Fruct. Sem. Pl. 1: 9 (1788), nom. illeg.

Cocos fusiformis Sw., Fl. Ind. Occid. 1: 616 (1797). *Acrocomia fusiformis* (Sw.) Sweet, Hort. Brit.: 432 (1826).

Acrocomia sphaerocarpa Desf., Tabl. École Bot., ed. 3: 30 (1829).

Acrocomia guianensis Lodd. ex G.Don in J.C.Loudon, Hort. Brit.: 382 (1830).

Acrocomia minor Lodd. ex G.Don in J.C.Loudon, Hort. Brit.: 382 (1830).

Acrocomia lasiospatha Mart. in A.D.d'Orbigny, Voy. Amér. Mér. 7(3): 81 (1844).

Acrocomia totai Mart. in A.D.d'Orbigny, Voy. Amér. Mér. 7(3): 78 (1844).

Bactris pavoniana Mart. in A.D.d'Orbigny, Voy. Amér. Mér. 7(3): 70 (1844).

Acrocomia horrida Lodd. ex Mart., Hist. Nat. Palm. 3: 286 (1845).

Acrocomia mexicana Karw. ex Mart., Hist. Nat. Palm. 3: 285 (1845).

Acrocomia tenuifrons Lodd. ex Mart., Hist. Nat. Palm. 3: 286 (1845).

Acrocomia cubensis Lodd. ex H.Wendl., Index Palm.: 1 (1854).

Acrocomia vinifera Oerst., Vidensk. Meddel. Dansk Naturhist. Foren. Kjøbenhavn 1858: 47 (1858).

Acrocomia antioquiensis Posada-Ar., Bull. Soc. Bot. France 25: 184 (1878).

Acrocomia zapotecis Karw. ex H.Wendl. in O.C.E.de Kerchove de Denterghem, Palmiers: 230 (1878).

Astrocaryum sclerocarpum H.Wendl. in O.C.E.de Kerchove de Denterghem, Palmiers: 232 (1878).

Acrocomia glaucophylla Drude in C.F.P.von Martius & auct. suc. (eds.), Fl. Bras. 3(2): 392 (1881).

Acrocomia intumescens Drude in C.F.P.von Martius & auct. suc. (eds.), Fl. Bras. 3(2): 391 (1881).

Acrocomia sclerocarpa var. *wallaceana* Drude in C.F.P.von Martius & auct. suc. (eds.), Fl. Bras. 3(2): 391 (1881). *Acrocomia wallaceana* (Drude) Becc., Pomona Coll. J. Econ. Bot. 2: 362 (1912).

Acrocomia microcarpa Barb.Rodr., Vellosia, ed. 2, 1: 107 (1891).

Acrocomia mokayayba Barb.Rodr., Pl. Jard. Rio de Janeiro 5: 11 (1896).

Acrocomia odorata Barb.Rodr., Palm. Mattogross.: 48 (1898).

Acrocomia media O.F.Cook, Bull. Torrey Bot. Club 28: 566 (1901).

Acrocomia erioacantha Barb.Rodr., Contr. Jard. Bot. Rio de Janeiro 2: 85 (1902).

Acrocomia ulei Dammer, Notizbl. Königl. Bot. Gart. Berlin 6: 266 (1915).

Acrocomia pilosa Léon, Mem. Soc. Cub. Hist. Nat. "Felipe Poey" 14: 52 (1940).

Acrocomia belizensis L.H.Bailey, Gentes Herb. 4: 445 (1941).

Acrocomia chunta Covas & Ragonese, Revista Argent. Agron. 8: 2 (1941).

Acrocomia hospes L.H.Bailey, Gentes Herb. 4: 449 (1941).

Acrocomia ierensis L.H.Bailey, Gentes Herb. 4: 473 (1941).

Acrocomia karukerana L.H.Bailey, Gentes Herb. 4: 466 (1941).

Acrocomia panamensis L.H.Bailey, Gentes Herb. 4: 444 (1941).

Acrocomia quisqueyana L.H.Bailey, Gentes Herb. 4: 471 (1941).

Acrocomia subinermis Léon ex L.H.Bailey, Gentes Herb. 4: 474 (1941).

Acrocomia antiguana L.H.Bailey, Gentes Herb. 8: 142 (1949).

Acrocomia christopherensis L.H.Bailey, Gentes Herb. 8: 140 (1949).

Acrocomia grenadana L.H.Bailey, Gentes Herb. 8: 144 (1949).

Acrocomia viegasii L.H.Bailey, Gentes Herb. 8: 139 (1949).

Acrocomia antiguana L.H.Bailey = **Acrocomia aculeata** (Jacq.) Lodd. ex Mart.

Acrocomia antioquiensis Posada-Ar. = **Acrocomia aculeata** (Jacq.) Lodd. ex Mart.

Acrocomia armentalis (Morales) L.H.Bailey & E.Z.Bailey = **Gastrococos crispa** (Kunth) H.E.Moore

Acrocomia belizensis L.H.Bailey = **Acrocomia aculeata** (Jacq.) Lodd. ex Mart.

Acrocomia christopherensis L.H.Bailey = **Acrocomia aculeata** (Jacq.) Lodd. ex Mart.

Acrocomia chunta Covas & Ragonese = **Acrocomia aculeata** (Jacq.) Lodd. ex Mart.

Acrocomia crispa (Kunth) C.F.Baker ex Becc. = **Gastrococos crispa** (Kunth) H.E.Moore

Acrocomia cubensis Lodd. ex H.Wendl. = **Acrocomia aculeata** (Jacq.) Lodd. ex Mart.

Acrocomia erioacantha Barb.Rodr. = **Acrocomia aculeata** (Jacq.) Lodd. ex Mart.

Acrocomia fusiformis (Sw.) Sweet = **Acrocomia aculeata** (Jacq.) Lodd. ex Mart.

Acrocomia glaucophylla Drude = **Acrocomia aculeata** (Jacq.) Lodd. ex Mart.

Acrocomia globosa (Gaertn.) Lodd. ex Mart. = **Acrocomia aculeata** (Jacq.) Lodd. ex Mart.

Acrocomia grenadana L.H.Bailey = **Acrocomia aculeata** (Jacq.) Lodd. ex Mart.

Acrocomia guianensis Lodd. ex G.Don = **Acrocomia aculeata** (Jacq.) Lodd. ex Mart.

Acrocomia hassleri (Barb.Rodr.) W.J.Hahn, Principes 35: 170 (1991).
Brazil to Paraguay. 84 BZC BZE BZL BZS 85 PAR.
Acanthococos hassleri Barb.Rodr., Palm. Hassler.: 2 (1900).

Acanthococos sericea Burret, Notizbl. Bot. Gart. Berlin-Dahlem 15: 109 (1940).

Acanthococos emensis Toledo, Arq. Bot. Estado São Paulo, n.s., f.m., 3: 4 (1952).

Acrocomia horrida Lodd. ex Mart. = **Acrocomia aculeata** (Jacq.) Lodd. ex Mart.

Acrocomia hospes L.H.Bailey = **Acrocomia aculeata** (Jacq.) Lodd. ex Mart.

Acrocomia ierensis L.H.Bailey = **Acrocomia aculeata** (Jacq.) Lodd. ex Mart.

Acrocomia intumescens Drude = **Acrocomia aculeata** (Jacq.) Lodd. ex Mart.

Acrocomia karukerana L.H.Bailey = **Acrocomia aculeata** (Jacq.) Lodd. ex Mart.

Acrocomia lasiospatha Mart. = **Acrocomia aculeata** (Jacq.) Lodd. ex Mart.

Acrocomia media O.F.Cook = **Acrocomia aculeata** (Jacq.) Lodd. ex Mart.

Acrocomia mexicana Karw. ex Mart. = **Acrocomia aculeata** (Jacq.) Lodd. ex Mart.

Acrocomia microcarpa Barb.Rodr. = **Acrocomia aculeata** (Jacq.) Lodd. ex Mart.

Acrocomia minor Lodd. ex G.Don = **Acrocomia aculeata** (Jacq.) Lodd. ex Mart.

Acrocomia mokayayba Barb.Rodr. = **Acrocomia aculeata** (Jacq.) Lodd. ex Mart.

Acrocomia odorata Barb.Rodr. = **Acrocomia aculeata** (Jacq.) Lodd. ex Mart.

Acrocomia panamensis L.H.Bailey = **Acrocomia aculeata** (Jacq.) Lodd. ex Mart.

Acrocomia pilosa Léon = **Acrocomia aculeata** (Jacq.) Lodd. ex Mart.

Acrocomia quisqueyana L.H.Bailey = **Acrocomia aculeata** (Jacq.) Lodd. ex Mart.

Acrocomia sclerocarpa Mart. = **Acrocomia aculeata** (Jacq.) Lodd. ex Mart.

Acrocomia sclerocarpa var. *wallaceana* Drude = **Acrocomia aculeata** (Jacq.) Lodd. ex Mart.

Acrocomia sphaerocarpa Desf. = **Acrocomia aculeata** (Jacq.) Lodd. ex Mart.

Acrocomia spinosa (Mill.) N.E.Moore = **Acrocomia aculeata** (Jacq.) Lodd. ex Mart.

Acrocomia subinermis Léon ex L.H.Bailey = **Acrocomia aculeata** (Jacq.) Lodd. ex Mart.

Acrocomia tenuifrons Lodd. ex Mart. = **Acrocomia aculeata** (Jacq.) Lodd. ex Mart.

Acrocomia totai Mart. = **Acrocomia aculeata** (Jacq.) Lodd. ex Mart.

Acrocomia ulei Dammer = **Acrocomia aculeata** (Jacq.) Lodd. ex Mart.

Acrocomia viegasii L.H.Bailey = **Acrocomia aculeata** (Jacq.) Lodd. ex Mart.

Acrocomia vinifera Oerst. = **Acrocomia aculeata** (Jacq.) Lodd. ex Mart.

Acrocomia wallaceana (Drude) Becc. = **Acrocomia aculeata** (Jacq.) Lodd. ex Mart.

Acrocomia zapotecis Karw. ex H.Wendl. = **Acrocomia aculeata** (Jacq.) Lodd. ex Mart.

Acrostigma

Acrostigma O.F.Cook & Doyle = **Wettinia** Poepp. ex Endl.

Acrostigma aequale O.F.Cook & Doyle = **Wettinia aequalis** (O.F.Cook & Doyle) R.Bernal

Actinokentia

Actinokentia Dammer, Bot. Jahrb. Syst. 39: 20 (1906). 2 species, New Caledonia. 60.

Actinokentia divaricata (Brongn.) Dammer, Bot. Jahrb. Syst. 39: 21 (1906). C. & SE. New Caledonia. 60 NWC.

Kentia divaricata Planch. ex Brongn., Compt. Rend. Hebd. Séances Acad. Sci. 77: 398 (1873), pro syn.

**Kentiopsis divaricata* Brongn., Compt. Rend. Hebd. Séances Acad. Sci. 77: 398 (1873). *Drymophloeus divaricatus* (Brongn.) Benth. & Hook.f. ex Becc., Ann. Jard. Bot. Buitenzorg 2: 168 (1885).

Kentia polystemon Pancher ex H.Wendl. in O.C.E.de Kerchove de Denterghem, Palmiers: 248 (1878).

Actinokentia schlechteri Dammer, Bot. Jahrb. Syst. 39: 21 (1906).

Actinokentia huerlimannii H.E.Moore, Gentes Herb. 12: 17 (1980). SE. New Caledonia. 60 NWC.

Actinokentia schlechteri Dammer = **Actinokentia divaricata** (Brongn.) Dammer

Actinophloeus

Actinophloeus (Becc.) Becc. = **Ptychosperma** Labill.

Actinophloeus ambiguus (Becc.) Becc. = **Ptychosperma ambiguum** (Becc.) Becc. ex Martelli

Actinophloeus angustifolius (Blume) L.H.Bailey = **Drymophloeus litigiosus** (Becc.) H.E.Moore

Actinophloeus bleeseri Burret = **Ptychosperma bleeseri** Burret

Actinophloeus capitis-yorki (H.Wendl. & Drude) Burret = **Ptychosperma elegans** (R.Br.) Blume

Actinophloeus cuneatus Burret = **Ptychosperma cuneatum** (Burret) Burret

Actinophloeus furcatus Becc. = **Ptychosperma furcatum** (Becc.) Becc. ex Martelli

Actinophloeus guppyanus Becc. = **Ptychococcus guppyanus** (Becc.) Burret

Actinophloeus hospitus Burret = **Ptychosperma macarthurii** (H.Wendl. ex H.J.Veitch) H.Wendl. ex Hook.f.

Actinophloeus kerstenianus (Sander) Burret = **Balaka seemannii** (H.Wendl.) Becc.

Actinophloeus kraemerianus Becc. = **Ptychococcus kraemerianus** (Becc.) Burret

Actinophloeus linearis Burret = **Ptychosperma lineare** (Burret) Burret

Actinophloeus macarthurii (H.Wendl. ex H.J.Veitch) Becc. ex Raderm. = **Ptychosperma macarthurii** (H.Wendl. ex H.J.Veitch) H.Wendl. ex Hook.f.

Actinophloeus macrospadix Burret = **Ptychosperma microcarpum** (Burret) Burret

Actinophloeus microcarpus Burret = **Ptychosperma microcarpum** (Burret) Burret

Actinophloeus montanus (K.Schum. & Lauterb.) Burret = **Ptychosperma caryotoides** Ridl.

Actinophloeus nicolai (Sander ex André) Burret = **Ptychosperma nicolai** (Sander ex André) Burret

Actinophloeus propinquus (Becc.) Becc. = **Ptychosperma propinquum** (Becc.) Becc. ex Martelli

Actinophloeus punctulatus Becc. = **Ptychosperma lauterbachii** Becc.

Actinophloeus sanderianus (Ridl.) Burret = **Ptychosperma sanderianum** Ridl.

Actinophloeus schumannii Becc. = **Brassiophoenix schumannii** (Becc.) Essig

Actinorhytis

Actinorhytis H.Wendl. & Drude, Linnaea 39: 184 (1875).
1 species, Papuasia. (41) (42) 43.

Actinorhytis calapparia (Blume) H.Wendl. & Drude ex
Scheff., Ann. Jard. Bot. Buitenzorg 1: 156 (1876).
Papuasia. (41) tha (42) mly sum 43 NWG SOL.
*Areca calapparia Blume, Rumphia 2: t. 100 (1843).
Seaforthia calapparia (Blume) Mart., Hist. Nat.
Palm. 3: 313 (1849). *Ptychosperma calapparia*
(Blume) Miq., Fl. Ned. Ind. 3: 20 (1855). *Pinanga
calapparia* (Blume) H.Wendl. in O.C.E.de
Kerchove de Denterghem, Palmiers: 253 (1878),
nom. illeg.
Areca cocoides Griff., Calcutta J. Nat. Hist. 4: 454
(1845).
Actinorhytis poamau Becc., Webbia 4: 274 (1914).

Actinorhytis poamau Becc. = **Actinorhytis calapparia**
(Blume) H.Wendl. & Drude ex Scheff.

Adelodypsis

Adelodypsis Becc. = **Dypsis** Noronha ex Mart.

Adelodypsis boiviniana (Baill.) Becc. = **Dypsis
boiviniana** Baill.

Adelodypsis gracilis (Bory ex Mart.) Becc. = **Dypsis
pinnatifrons** Mart.

Adelodypsis sambiranensis (Jum. & H.Perrier)
H.P.Guérin = **Dypsis pinnatifrons** Mart.

Adelonenga

Adelonenga (Becc.) Hook.f. = **Hydriastele** H.Wendl. &
Drude

Adelonenga geelvinkiana (Becc.) Becc. = **Hydriastele
geelvinkiana** (Becc.) Burret

Adelonenga kasesa (Lauterb.) Becc. = **Hydriastele
kasesa** (Lauterb.) Burret

Adelonenga microspadix (Warb. ex K.Schum. &
Lauterb.) Becc. = **Hydriastele microspadix** (Warb.
ex K.Schum. & Lauterb.) Burret

Adelonenga variabilis (Becc.) Becc. = **Hydriastele
variabilis** (Becc.) Burret

Adonidia

Adonidia Becc., Philipp. J. Sci. 14: 329 (1919).
1 species, Philippines. 42.

Adonidia merrillii (Becc.) Becc., Philipp. J. Sci. 14: 329
(1919).
Philippines (Palawan). 42 PHI.
*Normanbya merrillii Becc., Philipp. J. Sci., C 4: 606
(1909). *Veitchia merrillii* (Becc.) H.E.Moore,
Gentes Herb. 8: 501 (1957).

Aeria

Aeria O.F.Cook = **Gaussia** H.Wendl.

Aeria attenuata O.F.Cook = **Gaussia attenuata**
(O.F.Cook) Becc.

Aeria vinifera (Mart.) O.F.Cook = **Pseudophoenix
vinifera** (Mart.) Becc.

Aiphanes

Aiphanes Willd., Samml. Deutsch. Abh. Koningl. Akad.
Wiss. Berlin 1803: 250 (1806).
23 species, Trop. America. 80 81 82 83 84.

Marara H.Karst., Linnaea 28: 389 (1856).
Curima O.F.Cook, Bull. Torrey Bot. Club 28: 561
(1901).
Tilmia O.F.Cook, Bull. Torrey Bot. Club 28: 565
(1901).

Aiphanes acanthophylla (Mart.) Burret = **Aiphanes
minima** (Gaertn.) Burret

Aiphanes acaulis Galeano & R.Bernal, Principes 29: 20
(1985).
W. Colombia. 83 CLM.

Aiphanes aculeata Willd. = **Aiphanes horrida** (Jacq.)
Burret

Aiphanes caryotifolia (Kunth) H.Wendl. = **Aiphanes
horrida** (Jacq.) Burret

Aiphanes chiribogensis Borchs. & Balslev, Nordic J.
Bot. 9: 386 (1989 publ. 1990).
W. Ecuador. 83 ECU.

Aiphanes chocoensis Gentry = **Aiphanes macroloba**
Burret

Aiphanes concinna H.E.Moore = **Aiphanes lindeniana**
(H.Wendl.) H.Wendl.

Aiphanes corallina (Mart.) H.Wendl. = **Aiphanes
minima** (Gaertn.) Burret

Aiphanes deltoidea Burret, Notizbl. Bot. Gart. Berlin-
Dahlem 11: 568 (1932).
Colombia to Peru and N. Brazil. 83 CLM PER 84
BZN.

Aiphanes disticha (Linden) Burret = **Martinezia
disticha**

Aiphanes duquei Burret, Notizbl. Bot. Gart. Berlin-
Dahlem 13: 492 (1937).
Colombia. 83 CLM.

Aiphanes echinocarpa Dugand = **Aiphanes linearis**
Burret

Aiphanes eggersii Burret, Notizbl. Bot. Gart. Berlin-
Dahlem 11: 563 (1932).
W. Ecuador. 83 ECU.

Aiphanes elegans (Linden & H.Wendl.) H.Wendl. =
Aiphanes horrida (Jacq.) Burret

Aiphanes ernestii (Burret) Burret = **Aiphanes horrida**
(Jacq.) Burret

Aiphanes erinacea (H.Karst.) H.Wendl. in O.C.E.de
Kerchove de Denterghem, Palmiers: 230 (1878).
Colombia to Ecuador. 83 CLM ECU.
*Marara erinacea H.Karst., Linnaea 28: 391 (1856).

Aiphanes erosa (Mart.) Burret = **Aiphanes minima**
(Gaertn.) Burret

Aiphanes fosteriorum H.E.Moore = **Aiphanes hirsuta**
subsp. **fosteriorum** (H.E.Moore) Brochs. & R.Bernal

Aiphanes fuscopubens L.H.Bailey = **Aiphanes hirsuta**
Burret

Aiphanes gelatinosa H.E.Moore, Gentes Herb. 8: 227
(1951).
S. Colombia to N. Ecuador. 83 CLM ECU.

Aiphanes gracilis Burret = **Aiphanes weberbaueri**
Burret

Aiphanes grandis Borchs. & Balslev, Nordic J. Bot. 9:
388 (1989 publ. 1990).
SW. Ecuador. 83 ECU.

Aiphanes hirsuta Burret, Notizbl. Bot. Gart. Berlin-
Dahlem 11: 573 (1932).
Costa Rica to Ecuador. 80 COS PAN 83 CLM ECU.

Aiphanes pachyclada Burret, Notizbl. Bot. Gart. Berlin-Dahlem 11: 574 (1932).

Aiphanes fuscopubens L.H.Bailey, Gentes Herb. 6: 209 (1943).

subsp. **fosteriorum** (H.E.Moore) Brochs. & R.Bernal, Fl. Neotrop. Monogr. 70: 65 (1996).
SW. Colombia to NW. Ecuador. 83 CLM ECU.
*Aiphanes fosteriorum H.E.Moore, Gentes Herb. 8: 225 (1951).

subsp. **hirsuta**
Costa Rica to Colombia. 80 COS PAN 83 CLM.

subsp. **intermedia** Brochs. & R.Bernal, Fl. Neotrop. Monogr. 70: 64 (1996).
W. Colombia. 83 CLM.

subsp. **kalbreyeri** (Burret) Brochs. & R.Bernal, Fl. Neotrop. Monogr. 70: 63 (1996).
W. Colombia. 83 CLM.
*Aiphanes kalbreyeri Burret, Notizbl. Bot. Gart. Berlin-Dahlem 11: 572 (1932).

Aiphanes horrida (Jacq.) Burret, Notizbl. Bot. Gart. Berlin-Dahlem 11: 575 (1932).
Trinidad, S. Trop. America. 81 TRT 82 VEN 83 BOL CLM PER 84 BZN.
*Caryota horrida Jacq., Fragm. Bot.: 20 (1801).
Aiphanes aculeata Willd., Samml. Deutsch. Abh. Koningl. Akad. Wiss. Berlin 1803: 251 (1806). *Euterpe aculeata* (Willd.) Spreng., Syst. Veg. 2: 140 (1825). *Martinezia aculeata* (Willd.) Klotzsch, Linnaea 20: 455 (1847). *Marara aculeata* (Willd.) H.Karst. ex H.Wendl. in O.C.E.de Kerchove de Denterghem, Palmiers: 250 (1878).
Martinezia caryotifolia Kunth in F.W.H.von Humboldt, A.J.A.Bonpland & C.S.Kunth, Nov. Gen. Sp. 1: 305 (1816). *Marara caryotifolia* (Kunth) H.Karst. ex H.Wendl. in O.C.E.de Kerchove de Denterghem, Palmiers: 250 (1878). *Aiphanes caryotifolia* (Kunth) H.Wendl. in O.C.E.de Kerchove de Denterghem, Palmiers: 230 (1878). *Tilmia caryotifolia* (Kunth) O.F.Cook, Bull. Torrey Bot. Club 28: 565 (1901).
Bactris praemorsa Poepp. ex Mart. in A.D.d'Orbigny, Voy. Amér. Mér. 7(3): 66 (1844). *Aiphanes praemorsa* (Poepp. ex Mart.) Burret, Notizbl. Bot. Gart. Berlin-Dahlem 11: 575 (1932).
Martinezia aiphanes Mart. in A.D.d'Orbigny, Voy. Amér. Mér. 7(3): 77 (1844).
Martinezia truncata Brongn. ex Mart. in A.D. d'Orbigny, Voy. Amér. Mér. 7(3): 75 (1844). *Aiphanes truncata* (Brongn. ex Mart.) H.Wendl. in O.C.E.de Kerchove de Denterghem, Palmiers: 230 (1878).
Marara bicuspidata H.Karst., Linnaea 28: 390 (1856).
Martinezia elegans Linden & H.Wendl., Linnaea 28: 351 (1856). *Aiphanes elegans* (Linden & H.Wendl.) H.Wendl. in O.C.E.de Kerchove de Denterghem, Palmiers: 230 (1878).
Martinezia ulei Dammer, Notizbl. Königl. Bot. Gart. Berlin 6: 266 (1915), nom. illeg. *Martinezia ernestii* Burret, Notizbl. Bot. Gart. Berlin-Dahlem 11: 327 (1932). *Aiphanes ernestii* (Burret) Burret, Notizbl. Bot. Gart. Berlin-Dahlem 11: 560 (1932).
Aiphanes killipii Burret, Notizbl. Bot. Gart. Berlin-Dahlem 11: 562 (1932).
Aiphanes orinocensis Burret, Notizbl. Bot. Gart. Berlin-Dahlem 11: 561 (1932).
Martinezia killipii Burret, Notizbl. Bot. Gart. Berlin-Dahlem 11: 326 (1932).

Aiphanes kalbreyeri Burret = **Aiphanes hirsuta** subsp. **kalbreyeri** (Burret) Brochs. & R.Bernal

Aiphanes killipii Burret = **Aiphanes horrida** (Jacq.) Burret

Aiphanes leiospatha Burret = **?** [83 CLM]

Aiphanes leiostachys Burret, Notizbl. Bot. Gart. Berlin-Dahlem 11: 570 (1932).
Colombia (Antioquia). 83 CLM.

Aiphanes lindeniana (H.Wendl.) H.Wendl. in O.C.E.de Kerchove de Denterghem, Palmiers: 230 (1878).
Colombia. 83 CLM.
*Martinezia lindeniana H.Wendl., Linnaea 28: 349 (1856).
Aiphanes concinna H.E.Moore, Gentes Herb. 8: 223 (1951).

Aiphanes linearis Burret, Notizbl. Bot. Gart. Berlin-Dahlem 11: 577 (1932).
Colombia (Antioquia, Valle). 83 CLM.
Aiphanes echinocarpa Dugand, Caldasia 2: 455 (1944).

Aiphanes luciana L.H.Bailey = **Aiphanes minima** (Gaertn.) Burret

Aiphanes macroloba Burret, Notizbl. Bot. Gart. Berlin-Dahlem 11: 576 (1932).
Colombia to NW. Ecuador. 83 CLM ECU.
Aiphanes monostachys Burret, Notizbl. Bot. Gart. Berlin-Dahlem 11: 574 (1932).
Aiphanes chocoensis Gentry, Ann. Missouri Bot. Gard. 68: 112 (1981).

Aiphanes minima (Gaertn.) Burret, Notizbl. Bot. Gart. Berlin-Dahlem 11: 558 (1932).
Hispaniola to Windward Is. 81 DOM PUE WIN.
*Bactris minima Gaertn., Fruct. Sem. Pl. 2: 269 (1791).
Bactris acanthophylla Mart. in A.D.d'Orbigny, Voy. Amér. Mér. 7(3): 70 (1844). *Martinezia acanthophylla* (Mart.) Becc. in I.Urban, Symb. Antill. 8: 79 (1920). *Aiphanes acanthophylla* (Mart.) Burret, Notizbl. Bot. Gart. Berlin-Dahlem 11: 558 (1932).
Bactris erosa Mart., Hist. Nat. Palm. 3: 281 (1845). *Martinezia erosa* (Mart.) Linden, Cat. Gén. 87: 5 (1871). *Aiphanes erosa* (Mart.) Burret, Notizbl. Bot. Gart. Berlin-Dahlem 11: 558 (1932).
Martinezia corallina Mart., Hist. Nat. Palm. 3: 284 (1845). *Aiphanes corallina* (Mart.) H.Wendl. in O.C.E.de Kerchove de Denterghem, Palmiers: 230 (1878). *Curima corallina* (Mart.) O.F.Cook, Bull. Torrey Bot. Club 28: 563 (1901).
Bactris martineziifolia H.Wendl. in O.C.E.de Kerchove de Denterghem, Palmiers: 234 (1878), nom. inval.
Curima colophylla O.F.Cook, Bull. Torrey Bot. Club 28: 561 (1901).
Aiphanes luciana L.H.Bailey, Gentes Herb. 8: 166 (1949).
Aiphanes vincentiana L.H.Bailey, Gentes Herb. 8: 170 (1949).

Aiphanes monostachys Burret = **Aiphanes macroloba** Burret

Aiphanes orinocensis Burret = **Aiphanes horrida** (Jacq.) Burret

Aiphanes pachyclada Burret = **Aiphanes hirsuta** Burret

Aiphanes parvifolia Burret, Notizbl. Bot. Gart. Berlin-Dahlem 11: 569 (1932).
Colombia. 83 CLM.

Aiphanes pilaris R.Bernal, Caldasia 23: 163 (2001).
Colombia. 83 CLM.

Aiphanes praemorsa (Poepp. ex Mart.) Burret = **Aiphanes horrida** (Jacq.) Burret

Aiphanes praga Kunth = **Prestoea acuminata** var. **acuminata**

Aiphanes schultzeana Burret = **Aiphanes ulei** (Dammer) Burret

Aiphanes simplex Burret, Notizbl. Bot. Gart. Berlin-Dahlem 11: 567 (1932).
Colombia. 83 CLM.

Aiphanes spicata Brochs. & R.Bernal, Fl. Neotrop. Monogr. 70: 78 (1996).
N. Peru. 83 PER.

Aiphanes tessmannii Burret = **Aiphanes weberbaueri** Burret

Aiphanes tricuspidata Borchs., M.Ruíz & Bernal, Brittonia 41: 156 (1989).
Colombia to Ecuador. 83 CLM ECU.

Aiphanes truncata (Brongn. ex Mart.) H.Wendl. = **Aiphanes horrida** (Jacq.) Burret

Aiphanes ulei (Dammer) Burret, Notizbl. Bot. Gart. Berlin-Dahlem 11: 568 (1932).
W. South America to N. Brazil. 83 CLM ECU PER 84 BZN.
Martinezia ulei Dammer, Verh. Bot. Vereins Prov. Brandenburg 48: 127 (1906 publ. 1907). *Martinezia ernestii* Burret, Notizbl. Bot. Gart. Berlin-Dahlem 11: 327 (1932). *Aiphanes ernestii* (Burret) Burret, Notizbl. Bot. Gart. Berlin-Dahlem 11: 560 (1932). *Aiphanes schultzeana* Burret, Notizbl. Bot. Gart. Berlin-Dahlem 15: 36 (1940).

Aiphanes verrucosa Borchs. & Balslev, Nordic J. Bot. 9: 389 (1989 publ. 1990).
Ecuador. 83 ECU.

Aiphanes vincentiana L.H.Bailey = **Aiphanes minima** (Gaertn.) Burret

Aiphanes weberbaueri Burret, Notizbl. Bot. Gart. Berlin-Dahlem 11: 565.
S. Ecuador to Peru. 83 ECU PER.
Aiphanes gracilis Burret, Notizbl. Bot. Gart. Berlin-Dahlem 11: 566 (1932).
Aiphanes tessmannii Burret, Notizbl. Bot. Gart. Berlin-Dahlem 11: 564 (1932).

Alfonsia

Alfonsia Kunth = **Elaeis** Jacq.

Alfonsia oleifera Kunth = **Elaeis oleifera** (Kunth) Cortés

Allagoptera

Allagoptera Nees in M.A.P.Wied-Neuwied, Reise Bras. 2: 335 (1821).
4 species, Brazil to Bolivia and Argentina (Misiones). 83 84 85.
Diplothemium Mart., Hist. Nat. Palm. 2: 107 (1826).
Diplothenium Voigt, Syll. Pl. Nov. 2: 51 (1828), orth. var.

Allagoptera arenaria (M.Gómez) Kuntze, Revis. Gen. Pl. 2: 726 (1891).
E. Brazil. 84 BZE BZL.
Cocos arenaria M.Gómez, Mem. Acad. Real Sci. Lisboa 3(Mem.): 61 (1812). *Diplothemium arenarium* (M.Gómez) Vasc. & Franco, Portugaliae Act. Biol., Sér. B, Sist. 2: 412 (1948).

Allagoptera pumila Nees in M.A.P.Wied-Neuwied, Reise Bras. 2: 335 (1821).
Diplothemium littorale Mart., Hist. Nat. Palm. 2: 110 (1826).
Diplothemium maritimum Mart., Hist. Nat. Palm. 2: 108 (1826).

Allagoptera anisitsii (Barb.Rodr.) H.E.Moore = **Allagoptera leucocalyx** (Drude) Kuntze

Allagoptera brevicalyx M.Moraes, Brittonia 45: 21 (1993).
Brazil (Bahia). 84 BZE.

Allagoptera campestris (Mart.) Kuntze, Revis. Gen. Pl. 2: 726 (1891).
Brazil to Argentina (Misiones). 84 BZC BZE BZL BZS 85 AGE PAR.
Diplothemium campestre Mart., Hist. Nat. Palm. 2: 109 (1826).

Allagoptera caudescens (Mart.) Kuntze = **Polyandrococos caudescens** (Mart.) Barb.Rodr.

Allagoptera hassleriana (Barb.Rodr.) H.E.Moore = **Allagoptera leucocalyx** (Drude) Kuntze

Allagoptera leucocalyx (Drude) Kuntze, Revis. Gen. Pl. 2: 726 (1891).
Brazil to Bolivia and Argentina (Misiones). 83 BOL 84 BZC BZE BZL BZS 85 AGE PAR.
Diplothemium leucocalyx Drude in C.F.P.von Martius & auct. suc. (eds.), Fl. Bras. 3(2): 431 (1881).
Diplothemium jangadense S.Moore, Trans. Linn. Soc. London, Bot. 4: 499 (1895).
Diplothemium anisitsii Barb.Rodr., Palm. Paraguay.: 16 (1899). *Allagoptera anisitsii* (Barb.Rodr.) H.E.Moore, Principes 6: 38 (1962).
Diplothemium hasslerianum Barb.Rodr., Palm. Hassler.: 10 (1900). *Allagoptera hassleriana* (Barb.Rodr.) H.E.Moore, Principes 6: 39 (1962).

Allagoptera pumila Nees = **Allagoptera arenaria** (M.Gómez) Kuntze

Allagoptera torallyi (Mart.) Kuntze = **Parajubaea torallyi** (Mart.) Burret

Alloschmidia

Alloschmidia H.E.Moore, Gentes Herb. 11: 293 (1978).
1 species, New Caledonia. 60.

Alloschmidia glabrata (Becc.) H.E.Moore, Gentes Herb. 11: 294 (1978).
NE. New Caledonia. 60 NWC.
Basselinia glabrata Becc., Webbia 5: 145 (1921).

Alsmithia

Alsmithia H.E.Moore, Principes 26: 124 (1982).
1 species, Fiji. 60.

Alsmithia longipes H.E.Moore, Principes 26: 124 (1982).
Fiji. 60 FIJ.

Ammandra

Ammandra O.F.Cook, J. Wash. Acad. Sci. 17: 220 (1927).
1 species, W. South America. 83.

Ammandra decasperma O.F.Cook, J. Wash. Acad. Sci. 17: 220 (1927). *Phytelephas decasperma* (O.F.Cook) Dahlgren, Field Mus. Nat. Hist., Bot. Ser. 14: 231 (1936).
Colombia to N. Ecuador. 83 CLM ECU.

Phytelephas dasyneura Burret, Notizbl. Bot. Gart. Berlin-Dahlem 11: 5 (1930). *Ammandra dasyneura* (Burret) Barfod, Opera Bot. 105: 43 (1991), without exact basionym page ref.

Ammandra dasyneura (Burret) Barfod = **Ammandra decasperma** O.F.Cook

Ammandra natalia Balslev & A.J.Hend. = **Aphandra natalia** (Balslev & A.Hend.) Barfod

Amylocarpus

Amylocarpus Barb.Rodr. = **Bactris** Jacq. ex Scop.

Amylocarpus acanthocnemis (Mart.) Barb.Rodr. = **Bactris simplicifrons** Mart.

Amylocarpus angustifolius Huber = **Bactris simplicifrons** Mart.

Amylocarpus arenarius (Barb.Rodr.) Barb.Rodr. = **Bactris simplicifrons** Mart.

Amylocarpus cuspidatus (Mart.) Barb.Rodr. = **Bactris cuspidata** Mart.

Amylocarpus ericetinus (Barb.Rodr.) Barb.Rodr. = **Bactris hirta** var. **hirta**

Amylocarpus formosus (Barb.Rodr.) Barb.Rodr. = **Bactris hirta** var. **pectinata** (Mart.) Govaerts

Amylocarpus geonomoides (Drude) Barb.Rodr. = **Bactris hirta** var. **pectinata** (Mart.) Govaerts

Amylocarpus hirtus (Mart.) Barb.Rodr. = **Bactris hirta** Mart.

Amylocarpus hylophilus (Spruce) Barb.Rodr. = **Bactris hirta** var. **pectinata** (Mart.) Govaerts

Amylocarpus inermis (Trail ex Barb.Rodr.) Barb.Rodr. = **Bactris simplicifrons** Mart.

Amylocarpus linearifolius (Barb.Rodr.) Barb.Rodr. = **Bactris hirta** var. **pectinata** (Mart.) Govaerts

Amylocarpus luetzelburgii Burret = **Bactris simplicifrons** Mart.

Amylocarpus marajay (Barb.Rodr.) Barb.Rodr. = **Bactris cuspidata** Mart.

Amylocarpus microspathus (Barb.Rodr.) Barb.Rodr. = **Bactris simplicifrons** Mart.

Amylocarpus mitis (Mart.) Barb.Rodr. = **Bactris cuspidata** Mart.

Amylocarpus obovatus Burret = **Bactris simplicifrons** Mart.

Amylocarpus pectinatus (Mart.) Barb.Rodr. = **Bactris hirta** var. **pectinata** (Mart.) Govaerts

Amylocarpus platyspinus Barb.Rodr. = **Bactris hirta** var. **hirta**

Amylocarpus pulcher (Trail) Barb.Rodr. = **Bactris hirta** var. **hirta**

Amylocarpus setipinnatus (Barb.Rodr.) Barb.Rodr. = **Bactris hirta** var. **pectinata** (Mart.) Govaerts

Amylocarpus simplicifrons (Mart.) Barb.Rodr. = **Bactris simplicifrons** Mart.

Amylocarpus syagroides (Barb.Rodr. & Trail) Barb.Rodr. = **Bactris syagroides** Barb.Rodr. & Trail

Amylocarpus tenuissimus (Barb.Rodr.) Barb.Rodr. = **Bactris simplicifrons** Mart.

Amylocarpus xanthocarpus (Barb.Rodr.) Barb.Rodr. = **Bactris simplicifrons** Mart.

Ancistrophyllum

Ancistrophyllum (G.Mann & H.Wendl.) H.Wendl. = **Laccosperma** Drude

Ancistrophyllum acutiflorum Becc. = **Laccosperma acutiflorum** (Becc.) J.Dransf.

Ancistrophyllum laeve (G.Mann & H.Wendl.) Drude = **Laccosperma laeve** (G.Mann & H.Wendl.) Kuntze

Ancistrophyllum laurentii De Wild. = **Laccosperma secundiflorum** (P.Beauv.) Kuntze

Ancistrophyllum majus Burret = **Laccosperma secundiflorum** (P.Beauv.) Kuntze

Ancistrophyllum opacum (Drude) Drude = **Laccosperma opacum** Drude

Ancistrophyllum robustum Burret = **Laccosperma robustum** (Burret) J.Dransf.

Ancistrophyllum secundiflorum (P.Beauv.) G.Mann & H.Wendl. = **Laccosperma secundiflorum** (P.Beauv.) Kuntze

Anothea

Anothea O.F.Cook = **Chamaedorea** Willd.

Anothea scandens (Liebm.) O.F.Cook = **Chamaedorea elatior** Mart.

Antia

Antia O.F.Cook = **Coccothrinax** Sarg.

Antia crinita (Becc.) O.F.Cook = **Coccothrinax crinita** (Griseb. & H.Wendl. ex C.H.Wright) Becc.

Antongilia

Antongilia Jum. = **Dypsis** Noronha ex Mart.

Antongilia perrieri Jum. = **Dypsis perrieri** (Jum.) Beentje & J.Dransf.

Aphandra

Aphandra Barfod, Opera Bot. 105: 44 (1991).
  1 species, W. South America to N. Brazil. 83 84.

Aphandra natalia (Balslev & A.Hend.) Barfod, Opera Bot. 105: 46 (1991).
  E. Ecuador to N. Peru and N. Brazil. 83 ECU PER 84 BZN.
    Ammandra natalia Balslev & A.J.Hend., Syst. Bot. 12: 501 (1987).

Arausiaca

Arausiaca Blume = **Oania**

Arausiaca excelsa Blume = **Oania regalis**

Archontophoenix

Archontophoenix H.Wendl. & Drude, Linnaea 39: 182 (1875).
  6 species, NE. & E. Australia. 50 (63).
    Loroma O.F.Cook, J. Wash. Acad. Sci. 5: 117 (1915).

Archontophoenix alexandrae (F.Muell.) H.Wendl. & Drude, Linnaea 39: 212 (1875).
  N. & NE. Queensland. 50 QLD (63) haw.
    Ptychosperma alexandrae F.Muell., Fragm. 5: 47 (1865).
    Archontophoenix alexandrae var. *schizanthera* H.Wendl. & Drude, Linnaea 39: 212 (1875).

Archontophoenix veitchii H.Wendl. & Drude, Linnaea 39: 214 (1875).

Ptychosperma beatriceae F.Muell., Melbourne Chem. Druggist 4(Suppl.): 77 (1882). *Archontophoenix beatriceae* (F.Muell.) F.M.Bailey, Queensl. Fl. 5: 1675 (1902).

Ptychosperma drudei H.Wendl. in G.Bentham & J.D.Hooker, Gen. Pl. 3: 892 (1883). *Saguaster drudei* (H.Wendl.) Kuntze, Revis. Gen. Pl. 2: 735 (1891).

Jessenia glazioviana Dammer, Bot. Jahrb. Syst. 31(70): 21 (1901).

Archontophoenix alexandrae var. *schizanthera* H.Wendl. & Drude = **Archontophoenix alexandrae** (F.Muell.) H.Wendl. & Drude

Archontophoenix beatriceae (F.Muell.) F.M.Bailey = **Archontophoenix alexandrae** (F.Muell.) H.Wendl. & Drude

Archontophoenix cunninghamiana (H.Wendl.) H.Wendl. & Drude, Linnaea 39: 214 (1875).
E. Australia. 50 NSW QLD.

Seaforthia elegans Hook., Bot. Mag. 83: t. 4961 (1857), nom. illeg.

*Ptychosperma cunninghamianum H.Wendl., Bot. Zeitung (Berlin) 16: 346 (1858). *Loroma cunninghamiana* (H.Wendl.) O.F.Cook, J. Wash. Acad. Sci. 5: 118 (1915).

Jessenia amazonum Drude in C.F.P.von Martius & auct. suc. (eds.), Fl. Bras. 3(2): 474 (1882).

Loroma amethystina O.F.Cook, J. Wash. Acad. Sci. 5: 118 (1915).

Archontophoenix jardinei F.M.Bailey = **Ptychosperma elegans** (R.Br.) Blume

Archontophoenix maxima Dowe, Austrobaileya 4: 235 (1994).
NE. Queensland. 50 QLD.

Archontophoenix myolensis Dowe, Austrobaileya 4: 237 (1994).
NE. Queensland (Myola Reg.). 50 QLD.

Archontophoenix purpurea Hodel & Dowe, Austrobaileya 4: 238 (1994).
NE. Queensland. 50 QLD.

Archontophoenix tuckeri Dowe, Austrobaileya 4: 240 (1994).
N. Queensland. 50 QLD.

Archontophoenix veitchii H.Wendl. & Drude = **Archontophoenix alexandrae** (F.Muell.) H.Wendl. & Drude

Areca

Areca L., Sp. Pl.: 1189 (1753).
48 species, Trop. & Subtrop. Asia. 36 (38) 40 41 42 43 (62).

Mischophloeus Scheff., Ann. Jard. Bot. Buitenzorg 2: 115 (1876).

Gigliolia Becc., Malesia 1: 171 (1877).

Pichisermollia H.C.Monteiro, Rodriguésia 28: 195 (1976).

Areca abdulrahmanii J.Dransf., Bot. J. Linn. Soc. 81: 33 (1980).
Borneo (Sarawak). 42 BOR.

Areca ahmadii J.Dransf., Kew Bull. 39: 4 (1984).
Borneo (Sarawak). 42 BOR.

Areca alba Bory = **Dictyosperma album** (Bory) Scheff.

Areca aliceae W.Hill ex F.Muell. = **Areca triandra** Roxb. ex Buch.-Ham.

Areca amdjahii Furtado = **Areca minuta** Scheff.

Areca andersonii J.Dransf., Kew Bull. 39: 6 (1984).
Borneo (Sarawak). 42 BOR.

Areca angulosa Giseke = **?** [40 IND]

Areca appendiculata F.M.Bailey = **Oraniopsis appendiculata** (F.M.Bailey) J.Dransf. A.K.Irvine & N.W.Uhl

Areca arundinacea Becc., Malesia 1: 23 (1877).
Borneo (Sarawak). 42 BOR.

Areca augusta Kurz = **Rhopaloblaste augusta** (Kurz) H.E.Moore

Areca aurea Van Houtte = **Dictyosperma album** var. **aureum** Balf.f.

Areca bacaba Arruda = **Oenocarpus bacaba** Mart.

Areca banaensis (Magalon) Burret = **Nenga banaensis** (Magalon) Burret

Areca banksii A.Cunn. ex Kunth = **Rhopalostylis sapida** (Sol. ex G.Forst.) H.Wendl. & Drude

Areca baueri Hook.f. ex Lem. = **Rhopalostylis baueri** (Hook.f. ex Lem.) H.Wendl. & Drude

Areca bifaria Hodel = **Areca tunku** J.Dransf. & C.K.Lim

Areca bongayensis Becc. ex Furtado = **Areca minuta** Scheff.

Areca borbonica Kunth = **Dictyosperma album** (Bory) Scheff.

Areca borneensis Becc. = **Areca triandra** Roxb. ex Buch.-Ham.

Areca brachypoda J.Dransf., Kew Bull. 39: 8 (1984).
Borneo (Sarawak). 42 BOR.

Areca calapparia Blume = **Actinorhytis calapparia** (Blume) H.Wendl. & Drude ex Scheff.

Areca caliso Becc., Leafl. Philipp. Bot. 8: 2998 (1919).
Philippines. 42 PHI.

Areca camarinensis Becc., Philipp. J. Sci. 14: 309 (1919).
Philippines (Luzon). 42 PHI.

Areca cathechu Burm.f. = **Areca catechu** L.

Areca catechu L., Sp. Pl.: 1189 (1753).
Cultigen from Malesia. (36) chc chh (38) tai (40) ban ind srl (41) cbd tha lao vie 42 mly PHI (43) nwg sol van (62) crl mrn.

Areca cathechu Burm.f., Fl. Indica: 241 (1768).

Areca faufel Gaertn., Fruct. Sem. Pl. 1: 19 (1788).

Areca hortensis Lour., Fl. Cochinch.: 568 (1790).

Sublimia areca Comm. ex Mart., Hist. Nat. Palm. 3: 169 (1838), nom. inval.

Areca himalayana Griff. ex H.Wendl. in O.C.E.de Kerchove de Denterghem, Palmiers: 231 (1878).

Areca nigra Giseke ex H.Wendl. in O.C.E.de Kerchove de Denterghem, Palmiers: 231 (1878).

Areca celebica Burret, Repert. Spec. Nov. Regni Veg. 32: 115 (1933).
Sulawesi. 42 SUL.

Areca chaiana J.Dransf., Kew Bull. 39: 10 (1984).
Borneo (Sarawak). 42 BOR.

Areca cincta Walp. = **Acanthophoenix rubra** (Bory) H.Wendl.

Areca cocoides Griff. = **Actinorhytis calapparia** (Blume) H.Wendl. & Drude ex Scheff.

Areca communis Zipp. ex Blume = **Drymophloeus litigiosus** (Becc.) H.E.Moore

Areca concinna Thwaites, Enum. Pl. Zeyl.: 328 (1864).
SW. Sri Lanka. 40 SRL.

Areca congesta Becc., Bot. Jahrb. Syst. 58: 441 (1923).
New Guinea. 43 NWG.

Areca cornuta Giseke = **?** [**40 IND**]

Areca coronata Blume ex Mart. = **Pinanga coronata**
(Blume ex Mart.) Blume

Areca costata (Blume) Kurz = **Pinanga coronata**
(Blume ex Mart.) Blume

Areca costulata Becc., Philipp. J. Sci. 14: 310 (1919).
Philippines (Leyte). 42 PHI.

Areca crinita Bory = **Acanthophoenix rubra** (Bory)
H.Wendl.

Areca cuneifolia Stokes = **? Drymophloeus sp.**

Areca curvata Griff. = **Pinanga disticha** (Roxb.)
H.Wendl.

Areca curvata Griff. = **Pinanga paradoxa** (Griff.) Scheff.

Areca dayung J.Dransf., Bot. J. Linn. Soc. 81: 30 (1980).
Borneo (Sarawak). 42 BOR.

Areca dicksonii Roxb. = **Pinanga dicksonii** (Roxb.)
Blume

Areca disticha Roxb. = **Pinanga disticha** (Roxb.)
H.Wendl.

Areca elaeocarpa Reinw. ex Kunth = **Drymophloeus
oliviformis** (Giseke) *Areca erythrocarpa* H.Wendl. =
Cyrtostachys renda Blume

Areca erythropoda Miq. = **Cyrtostachys renda** Blume

Areca faufel Gaertn. = **Areca catechu** L.

Areca flavescens Voss = **Dypsis lutescens** (H.Wendl.)
Beentje & J.Dransf.

Areca furcata Becc., Malesia 1: 23 (1877).
Borneo (Sarawak). 42 BOR.

Areca furfuracea H.Wendl. = **Dictyosperma album**
(Bory) Scheff.

Areca gigantea H.Wendl. = **Pinanga rumphiana** (Mart.)
J.Dransf. & Govaerts

Areca glandiformis Lam. = **?** [**42 MOL**]

Areca globulifera Lam. = **Pinanga globulifera** (Lam.)
Blume

Areca gracilis Thouars ex Kunth = **Dypsis pinnatifrons**
Mart.

Areca gracilis Roxb. = **Pinanga gracilis** Blume

Areca gracilis Buch.-Ham. = **?** [**40 IND**]

Areca guppyana Becc., Webbia 4: 258 (1914).
Solomon Is. (Shortland Is.). 43 SOL.

Areca haematocarpon Griff. = **Pinanga malaiana**
(Mart.) Scheff.

Areca hallieriana Becc. = **Areca kinabaluensis** Furtado

Areca henrici Furtado = **Areca vestiaria** Giseke

Areca herbstii W.Watson = **Acanthophoenix rubra**
(Bory) H.Wendl.

Areca hewittii Furtado = **Areca minuta** Scheff.

Areca hexasticha Kurz = **Pinanga hexasticha** (Kurz)
Scheff.

Areca himalayana Griff. ex H.Wendl. = **Areca catechu** L.

Areca horrida Griff. = **Oncosperma horridum** (Griff.)
Scheff.

Areca hortensis Lour. = **Areca catechu** L.

Areca hullettii Furtado = **Areca minuta** Scheff.

Areca humilis Roxb. ex H.Wendl. = **Pinanga disticha**
(Roxb.) H.Wendl.

Areca humilis Blanco ex H.Wendl. = **Areca triandra**
Roxb. ex Buch.-Ham.

Areca humilis Willd. = **Areca oryziformis var. saxatilis**

Areca hutchinsoniana Becc., Philipp. J. Sci. 14: 312
(1919).
Philippines (Mindanao). 42 PHI.
Areca mammillata var. *mindanaoensis* Becc., Philipp.
J. Sci., C 4: 62 (1909).

Areca insignis (Becc.) J.Dransf., Kew Bull. 39: 13
(1984).
NW. Borneo. 42 BOR.
Gigliolia insignis Becc., Malesia 1: 172 (1877).
Pichisermollia insignis (Becc.) H.C.Monteiro,
Rodriguésia 28: 196 (1976).

var. **insignis**
NW. Borneo. 42 BOR.

var. **moorei** (J.Dransf.) J.Dransf., Kew Bull. 39: 13
(1984).
NW. Borneo. 42 BOR.
Pichisermollia insignis var. *moorei* J.Dransf., Bot.
J. Linn. Soc. 81: 40 (1980).

Areca ilsemannii André = **?**

Areca ipot Becc., Leafl. Philipp. Bot. 2: 639 (1909).
Philippines. 42 PHI.

Areca jobiensis Becc., Malesia 1: 21 (1877).
New Guinea. 43 NWG.

Areca jugahpunya J.Dransf., Kew Bull. 39: 13 (1984).
Borneo (Sarawak). 42 BOR.

Areca kinabaluensis Furtado, Repert. Spec. Nov. Regni
Veg. 33: 228 (1933).
Borneo. 42 BOR.
Areca hallieriana Becc., Atti Soc. Tosc. Sci. Nat. Pisa
Processi Verbali 44: 114 (1934).

Areca klingkangensis J.Dransf., Kew Bull. 39: 15 (1984).
Borneo (Sarawak). 42 BOR.

Areca lactea Miq. = **Dictyosperma album** (Bory)
Scheff.

Areca langloisiana Potztal = **Areca vestiaria** Giseke

Areca lansiformis Giseke = **?** [**40 IND**]

Areca laosensis Becc., Webbia 3: 191 (1910).
Indo-China. 41 LAO THA.

Areca latiloba Ridl. = **Areca montana** Ridl.

Areca latisecta (Blume) Scheff. = **Pinanga latisecta**
Blume

Areca laxa Buch.-Ham. = **Areca triandra** Roxb. ex
Buch.-Ham.

Areca ledermanniana Becc., Bot. Jahrb. Syst. 58: 441
(1923).
New Guinea. 43 NWG.

Areca leptopetala Burret = **Areca vestiaria** Giseke

Areca litoralis Blume = **Drymophloeus litigiosus**
(Becc.) H.E.Moore

Areca lutescens Bory = **Hyophorbe indica** Gaertn.

Areca macrocalyx Zipp. ex Blume, Rumphia 2: 75 (1839).
New Guinea. 43 NWG.

Areca macrocarpa Becc., Philipp. J. Sci., C 4: 601 (1909).
Philippines (Mindanao). 42 PHI.

Areca madagascariensis Mart. = **?** [29 MDG]

Areca malaiana (Mart.) Griff. = **Pinanga malaiana** (Mart.) Scheff.

Areca mammillata Becc. = **Areca vidaliana** Becc.

Areca mammillata var. *mindanaoensis* Becc. = **Areca hutchinsoniana** Becc.

Areca micholitzii Sander = **?** [43 NWG]

Areca microspadix Burret = **Nenga banaensis** (Magalon) Burret

Areca minor W.Hill = **Linospadix minor** (F.Muell.) F.Muell.

Areca minuta Scheff., Ann. Jard. Bot. Buitenzorg 1: 146 (1876).
Borneo. 42 BOR.
  Areca tenella Becc., Malesia 1: 22 (1877).
  Areca amdjahii Furtado, Repert. Spec. Nov. Regni Veg. 33: 231 (1933).
  Areca bongayensis Becc. ex Furtado, Repert. Spec. Nov. Regni Veg. 33: 231 (1933).
  Areca hewittii Furtado, Repert. Spec. Nov. Regni Veg. 33: 231 (1933).
  Areca hullettii Furtado, Repert. Spec. Nov. Regni Veg. 33: 231 (1933).

Areca monostachya Mart. = **Linospadix monostachya** (Mart.) H.Wendl. & Drude

Areca montana Ridl., Mat. Fl. Malay. Penins. 2: 136 (1907).
Thailand to W. Malesia. 41 THA 42 JAW MLY SUM.
  Areca latiloba Ridl., J. Straits Branch Roy. Asiat. Soc. 86: 310 (1922).
  Areca recurvata Hodel, Palm J. 134: 28 (1997).

Areca multifida Burret, Notizbl. Bot. Gart. Berlin-Dahlem 13: 331 (1936).
New Guinea. 43 NWG.

Areca nagensis Griff. = **Areca triandra** Roxb. ex Buch.-Ham.

Areca nannospadix Burret, J. Arnold Arbor. 12: 265 (1931).
New Guinea (Kep. Aru). 43 NWG.

Areca nenga Blume ex Mart. = **Nenga pumila** (Blume) H.Wendl.

Areca nibung Mart. = **Oncosperma horridum** (Griff.) Scheff.

Areca nibung Griff. ex H.Wendl. = **Oncosperma tigillarium** (Jack) Ridl.

Areca nigasolu Becc., Webbia 4: 256 (1914).
Solomon Is. 43 SOL.

Areca nigra Giseke ex H.Wendl. = **Areca catechu** L.

Areca nobilis auct. = **Nephrosperma van-houtteanum** (H.Wendl. ex Van Houtt.) Balf.f.

Areca normanbyi F.Muell. = **Normanbya normanbyi** (F.Muell.) L.H.Bailey

Areca novohibernica (Lauterb.) Becc., Bot. Jahrb. Syst. 52: 23 (1914).
Bismarck Arch. 43 BIS.
  Nenga novohibernica Lauterb., Bot. Jahrb. Syst. 45: 357 (1911).

Areca oleracea Jacq. = **Roystonea oleracea** (Jacq.) O.F.Cook

Areca oliviformis Giseke = **Drymophloeus oliviformis** (Giseke) Mart.

Areca oliviformis var. *gracilis* Giseke = **Drymophloeus oliviformis** (Giseke) Mart.

Areca oriziformis Gaertn. = **Pinanga globulifera** (Lam.) Blume

Areca oriziformis var. *gracilis* Giseke = **Pinanga coronata** (Blume ex Mart.) Blume

Areca oriziformis var. *saxatilis* Burm.f. ex Giseke = **?** [42 MOL] Pinanga **?**

Areca oviformis Giseke = **?** [40 IND]

Areca oxycarpa Miq., Verh. Kon. Ned. Akad. Wetensch., Afd. Natuurk. 11(5): 1 (1868).
Sulawesi. 42 SUL.

Areca paniculata (Miq.) Scheff. = **Areca vestiaria** Giseke

Areca paradoxa Griff. = **Pinanga paradoxa** (Griff.) Scheff.

Areca parens Becc., Philipp. J. Sci. 14: 307 (1919).
Philippines (Luzon). 42 PHI.

Areca passalacquae Kunth = **Medemia argun** (Mart.) Wurttenb. ex H.Wendl.

Areca pisifera Lodd. ex Hook.f. = **Dictyosperma album** (Bory) Scheff.

Areca polystachya (Miq.) H.Wendl. = **Areca triandra** Roxb. ex Buch.-Ham.

Areca procera Zipp. ex Blume = **Gronophyllum procerum** (Blume) H.E.Moore

Areca propria Miq. = **Dictyosperma album** (Bory) Scheff.

Areca pumila Blume = **Nenga pumila** (Blume) H.Wendl.

Areca punicea Zipp. ex Blume = **Pinanga rumphiana** (Mart.) J. Dransf. & Govaerts

Areca purpurea auct. = **?** [29 MAU]

Areca rechingeriana Becc., Webbia 3: 163 (1910).
Solomon Is. 43 SOL.

Areca recurvata Hodel = **Areca montana** Ridl.

Areca rheophytica J.Dransf., Kew Bull. 39: 18 (1984).
Borneo (Sabah). 42 BOR.

Areca ridleyana Becc. ex Furtado, Repert. Spec. Nov. Regni Veg. 33: 236 (1933).
Pen. Malaysia. 42 MLY.

Areca rostrata Burret, Notizbl. Bot. Gart. Berlin-Dahlem 12: 322 (1935).
New Guinea. 43 NWG.

Areca rubra H.Wendl. = **Dictyosperma album** (Bory) Scheff.

Areca rubra Bory = **Acanthophoenix rubra** (Bory) H.Wendl.

Areca salomonensis Burret, Notizbl. Bot. Gart. Berlin-Dahlem 13: 70 (1936).
Solomon Is. 43 SOL.

Areca sanguinea Zipp. ex Blume = **Pinanga rumphiana** (Mart.) J. Dransf. & Govaerts

Areca sapida Sol. ex G.Forst. = **Rhopalostylis sapida** (Sol. ex G.Forst.) H.Wendl. & Drude

Areca sechellarum (H.Wendl.) Baill. = **Phoenicophorium borsigianum** (K.Koch) Stuntz

Areca speciosa Lem. = **Hyophorbe amaricaulis** Mart.

Areca spicata Lam. = **Calyptrocalyx spicatus** (Lam.) Blume

Areca spinosa Hasselt & Kunth = **Oncosperma tigillarium** (Jack) Ridl.

Areca subacaulis (Becc.) J.Dransf., Kew Bull. 29: 20 (1984).
Borneo (Sarawak). 42 BOR.
*Gigliolia subacaulis Becc., Malesia 1: 174 (1877). *Pichisermollia subacaulis* (Becc.) H.C.Monteiro, Rodriguésia 28: 196 (1976).

Areca sylvestris Lour. = **Pinanga sylvestris** (Lour.) Hodel

Areca tenella Becc. = **Areca minuta** Scheff.

Areca tigillaria Jack = **Oncosperma tigillarium** (Jack) Ridl.

Areca torulo Becc., Webbia 4: 253 (1914).
Solomon Is. 43 SOL.

Areca triandra Roxb. ex Buch.-Ham., Mem. Wern. Nat. Hist. Soc. 5: 310 (1826).
Trop. & Subtrop. Asia. 36 CHC CHS 40 ASS BAN 41 AND CBD LAO MYA THA VIE 42 BOR MLY PHI SUM.
Areca laxa Buch.-Ham., Mem. Wern. Nat. Hist. Soc. 5(2): 309 (1826).
Areca nagensis Griff., Calcutta J. Nat. Hist. 5: 453 (1845). *Nenga nagensis* (Griff.) Scheff., Ann. Jard. Bot. Buitenzorg 1: 120 (1876).
Ptychosperma polystachyum Miq., Fl. Ned. Ind., Eerste Bijv.: 590 (1861). *Areca polystachya* (Miq.) H.Wendl. in O.C.E.de Kerchove de Denterghem, Palmiers: 232 (1878).
Areca triandra var. *bancana* Scheff., Natuurk. Tijdschr. Ned.-Indië 32: 165 (1873).
Areca borneensis Becc., Malesia 1: 22 (1877).
Areca humilis Blanco ex H.Wendl. in O.C.E.de Kerchove de Denterghem, Palmiers: 231 (1878).
Areca aliceae W.Hill ex F.Muell., Gartenflora 28: 199 (1879).

Areca triandra var. *bancana* Scheff. = **Areca triandra** Roxb. ex Buch.-Ham.

Areca tunku J.Dransf. & C.K.Lim, Principes 36: 81 (1992).
Pen. Thailand to Sumatera. 41 THA 42 MLY SUM.
Areca bifaria Hodel, Palm J. 136: 7 (1997).

Areca vaginata Giseke = **Drymophloeus oliviformis** (Giseke) Mart.

Areca verschaffeltii Lem. = **Hyophorbe verschaffeltii** H.Wendl.

Areca vestiaria Giseke, Prael. Ord. Nat. Pl.: 78 (1792).
Pinanga vestiaria (Giseke) Blume, Rumphia 2: 77 (1839). *Seaforthia vestiaria* (Giseke) Mart., Hist. Nat. Palm. 3: 313 (1849). *Ptychosperma vestiarium* (Giseke) Miq., Fl. Ned. Ind. 3: 31 (1855). *Mischophloeus vestiarius* (Giseke) Merr., Interpr. Herb. Amboin.: 121 (1917).
Sulawesi to Maluku. 42 MOL SUL.
Drymophloeus vestiarius Miq., Palm. Archip. Ind.: 24 (1868), nom. inval.
Ptychosperma paniculatum Miq., Verh. Kon. Ned. Akad. Wetensch., Afd. Natuurk. 11(5): 3 (1868). *Areca paniculata* (Miq.) Scheff., Tijdschr. Ned.-Indië 32: 179 (1871). *Mischophloeus paniculatus* (Miq.) Scheff., Ann. Jard. Bot. Buitenzorg 2: 152 (1876).
Areca henrici Furtado, Repert. Spec. Nov. Regni Veg. 33: 232 (1933).
Areca leptopetala Burret, Notizbl. Bot. Gart. Berlin-Dahlem 13: 199 (1936).
Areca langloisiana Potztal, Willdenowia 2: 628 (1960).

Areca vidaliana Becc., Philipp. J. Sci., C 2: 222 (1907).
Philippines (Palawan to N. Borneo (Banggi, Balambangan). 42 BOR PHI.
Areca mammillata Becc., Philipp. J. Sci., C 2: 220 (1907).

Areca wallichiana Mart. = **Iguanura wallichiana** (Mart.) Becc.

Areca warburgiana Becc., Bot. Jahrb. Syst. 52: 24 (1914).
New Guinea. 43 NWG.

Areca wendlandiana (Scheff.) H.Wendl. = **Nenga pumila** (Blume) H.Wendl.

Areca whitfordii Becc., Philipp. J. Sci., C 2: 219 (1907).
Philippines. 42 PHI.

Arecastrum

Arecastrum (Druce) Becc. = **Syagrus** Mart.

Arecastrum × *campos-portoanum* (Bondar) A.D.Hawkes = **Syagrus** × **campos-portoana** (Bondar) Glassman

Arecastrum romanzofianum (Cham.) Becc. = **Syagrus romanzoffiana** (Cham.) Glassman

Arenga

Arenga Labill. ex DC., Bull. Sci. Soc. Philom. Paris 2: 162 (1800), nom. cons.
20 species, Trop. & Subtrop. Asia to N. Australia. 36 38 40 41 42 50.
Saguerus Steck, Sagu: 15 (1757).
Gomutus Corrêa, Ann. Mus. Natl. Hist. Nat. 9: 288 (1807).
Blancoa Blume, Rumphia 2: 128 (1843), nom. illeg.
Didymosperma H.Wendl. & Drude ex Hook.f. in G.Bentham & J.D.Hooker, Gen. Pl. 3: 917 (1883).

Arenga ambong Becc. = **Arenga undulatifolia** Becc.

Arenga australasica (H.Wendl. & Drude) T.S.Blake ex H.E.Moore, Gentes Herb. 9: 268 (1963).
N. & NE. Queensland. 50 QLD.
*Saguerus australasicus H.Wendl. & Drude, Linnaea 39: 219 (1875). *Normanbya australasicus* (H.Wendl. & Drude) Baill., Hist. Pl. 13: 364 (1895).

Arenga bonnetti Hook.f. = **?** [40 IND]

Arenga borneensis (Becc.) J.Dransf. = **Arenga hastata** (Becc.) Whitmore

Arenga brevipes Becc., Malesia 3: 95 (1889). *Saguerus brevipes* (Becc.) Kuntze, Revis. Gen. Pl. 2: 736 (1891).
Sumatera, Borneo. 42 BOR SUM.

Arenga caudata (Lour.) H.E.Moore, Principes 4: 114 (1960).
Hainan to Indo-China and N. Pen. Malaysia. 36 CHH 41 CBD LAO MYA THA VIE 42 MLY.
*Borassus caudatus Lour., Fl. Cochinch. 2: 760 (1790). *Wallichia caudata* (Lour.) Mart., Hist. Nat. Palm. 3: 315 (1853). *Blancoa caudata* (Lour.) Kuntze, Revis. Gen. Pl. 2: 727 (1891). *Didymosperma caudatum* (Lour.) H.Wendl. & Drude ex B.D.Jacks., Index Kew. 1: 756 (1893).
Didymosperma caudatum var. *tonkinense* Becc., Webbia 3: 208 (1910). *Didymosperma tonkinense* (Becc.) Becc. ex Gagnep. in H.Lecomte, Fl. Indo-Chine 6: 966 (1937).

Arenga engleri Becc., Malesia 3: 184 (1889).
Didymosperma engleri (Becc.) Warb., Monsunia 1: t. 2, f. 1 (1900). *Arenga tremula* var. *engleri* (Becc.) Hatus., Fl. Ryukyus: 754 (1971), no basionym ref.
Nansei-shoto to Taiwan. 38 jap NNS TAI.

Arenga gamuto (Houtt.) Merr. = **Arenga pinnata** (Wurmb) Merr.

Arenga gracilicaulis F.M.Bailey = **Arenga microcarpa** Becc.

Arenga griffithii Seem. ex H.Wendl. = **Arenga pinnata** (Wurmb) Merr.

Arenga hastata (Becc.) Whitmore, Principes 14: 124 (1970).
W. Malesia. 42 BOR MLY SUM.
Didymosperma borneense Becc., Malesia 3: 197 (1889). *Blancoa borneensis* (Becc.) Kuntze, Revis. Gen. Pl. 2: 727 (1891). *Arenga borneensis* (Becc.) J.Dransf., Bot. J. Linn. Soc. 81: 40 (1980).
*Didymosperma hastatum Becc., Malesia 3: 199 (1889). *Blancoa hastata* (Becc.) Kuntze, Revis. Gen. Pl. 2: 727 (1891).

Arenga hookeriana (Becc.) Whitmore, Principes 14: 124 (1970).
Pen. Thailand to N. Pen. Malaysia. 41 THA 42 MLY.
*Didymosperma hookerianum Becc., Malesia 3: 186 (1889).

Arenga javanica H.Wendl. = **?** [42 JAW] **Korthalsia sp.**

Arenga listeri Becc., Hooker's Icon. Pl. 20: 1985 (1891).
Christmas I. 42 XMS.

Arenga longicarpa C.F.Wei, Acta Bot. Austro Sin. 4: 7 (1989).
S. China. 36 CHS.

Arenga manillensis (H.Lodd) H.Wendl. = **?** [42 PHI]

Arenga micrantha C.F.Wei, Acta Phytotax. Sin. 36: 404 (1988).
SE. Tibet to E. Himalaya. 36 CHT 40 EHM.

Arenga microcarpa Becc. in K.M.Schumann & U.M.Hollrung, Fl. Kais. Wilh. Land: 16 (1889). *Didymosperma microcarpum* (Becc.) Warb. ex K.Schum. & Lauterb., Fl. Schutzgeb. Südsee: 204 (1900).
New Guinea. 43 NWG.
Arenga gracilicaulis F.M.Bailey, Queensland Agric. J. 3: 203 (1898).

Arenga mindorensis Becc. in G.H.Perkins & al., Fragm. Fl. Philipp.: 48 (1904). *Saguerus mindorensis* (Becc.) O.F.Cook, U.S.D.A. Bur. Pl. Industr. Invent. Seeds 33: 53 (1915).
Philippines (Mindoro). 42 PHI.

Arenga nana (Griff.) H.E.Moore, Principes 4: 114 (1960).
Assam. 40 ASS.
*Wallichia nana Griff., Calcutta J. Nat. Hist. 5: 488 (1845). *Harina nana* (Griff.) Griff., Palms Brit. E. Ind.: 176 (1850). *Didymosperma nanum* (Griff.) H.Wendl. & Drude in O.C.E.de Kerchove de Denterghem, Palmiers: 243 (1878). *Blancoa nana* (Griff.) Kuntze, Revis. Gen. Pl. 2: 727 (1891).

Arenga obtusifolia Mart., Hist. Nat. Palm. 3: 191 (1838).
Pen. Thailand to Jawa. 41 THA 42 JAW MLY SUM.
Gomutus obtusifolius Blume, Rumphia 2: 131 (1843), nom. inval.

Arenga pinnata (Wurmb) Merr., Interpr. Herb. Amboin.: 119 (1917).
S. China to C. Malesia. 36 CHC CHH CHS 40 ASS ind 41 MYA THA 42 MLY PHI.
Saguerus gamuto Houtt., Handl. Pl.-Kruidk. 1: 410 (1773), nom. inval. *Arenga gamuto* (Houtt.) Merr., Philipp. J. Sci., C 9: 63 (1914).
*Saguerus pinnatus Wurmb, Verh. Batav. Genootsch. Kunsten 1: 351 (1779).

Borassus gomutus Lour., Fl. Cochinch. 2: 618 (1790). *Sagus gomutus* (Lour.) Perr., Mém. Soc. Linn. Paris 3: 142 (1824).
Arenga saccharifera Labill. ex DC., Bull. Sci. Soc. Philom. Paris 2: 162 (1800). *Gomutus saccharifer* (Labill. ex DC.) Spreng., Syst. Veg. 2: 624 (1825). *Saguerus saccharifer* (Labill. ex DC.) Blume, Rumphia 2: 128 (1843).
Gomutus rumphii Corrêa, Ann. Mus. Natl. Hist. Nat. 9: 288 (1807). *Saguerus rumphii* (Corrêa) Roxb. ex Ainslie, Mat. ind. 2: 225 (1826).
Caryota onusta Blanco, Fl. Filip.: 741 (1837).
Gomutus vulgaris Oken, Allg. Naturgesch. 3(1): 675 (1841).
Arenga griffithii Seem. ex H.Wendl. in O.C.E.de Kerchove de Denterghem, Palmiers: 232 (1878).

Arenga porphyrocarpa (Blume) H.E.Moore, Principes 4: 114 (1960).
Sumatera, Jawa. 42 JAW SUM.
*Orania porphyrocarpa Blume in C.F.P.von Martius, Hist. Nat. Palm. 3: 187 (1838). *Wallichia porphyrocarpa* (Blume) Mart., Hist. Nat. Palm. 3: 190 (1845). *Didymosperma porphyrocarpum* (Blume) H.Wendl. & Drude ex Hook.f., Rep. Progr. Condition Roy. Bot. Gard. Kew 1882: 61 (1884). *Blancoa porphyrocarpa* (Blume) Kuntze, Revis. Gen. Pl. 2: 727 (1891).
Caryota humilis Reinw. ex Kunth, Enum. Pl. 3: 193 (1841), pro syn.
Wallichia horsfieldii Blume, Rumphia 2: 112 (1843). *Didymosperma horsfieldii* (Blume) H.Wendl. & Drude in O.C.E.de Kerchove de Denterghem, Palmiers: 243 (1878). *Blancoa horsfieldii* (Blume) Kuntze, Revis. Gen. Pl. 2: 727 (1891).
Wallichia orania Blume, Rumphia 2: 113 (1843).
Wallichia reinwardtiana Miq., Pl. Jungh.: 157 (1852). *Blancoa reinwardtiana* (Miq.) Kuntze, Revis. Gen. Pl. 2: 727 (1891). *Didymosperma reinwardtianum* (Miq.) H.Wendl. & Drude ex B.D.Jacks., Index Kew. 1: 756 (1893).

Arenga retroflorescens H.E.Moore & Meijer, Principes 9: 100 (1965).
Borneo (Sabah). 42 BOR.

Arenga saccharifera Labill. ex DC. = **Arenga pinnata** (Wurmb) Merr.

Arenga tremula (Blanco) Becc., Philipp. J. Sci., C 4: 612 (1909).
Philippines. 42 PHI.
*Caryota tremula Blanco, Fl. Filip.: 744 (1837). *Wallichia tremula* (Blanco) Mart., Hist. Nat. Palm. 3: 315 (1853). *Didymosperma tremulum* (Blanco) H.Wendl. & Drude ex B.D.Jacks., Index Kew. 1: 756 (1893).

Arenga tremula var. *engleri* (Becc.) Hatus. = **Arenga engleri** Becc.

Arenga undulatifolia Becc., Malesia 3: 92 (1886). *Saguerus undulatifolius* (Becc.) Kuntze, Revis. Gen. Pl. 2: 736 (1891).
Borneo, Philippines (Palawan), Sulawesi. 42 BOR PHI SUL.
Arenga ambong Becc., Philipp. J. Sci., C 2: 229 (1907).

Arenga westerhoutii Griff., Calcutta J. Nat. Hist. 5: 474 (1845). *Saguerus westerhoutii* (Griff.) H.Wendl. & Drude in O.C.E.de Kerchove de Denterghem, Palmiers: 256 (1878).
Bhutan to Pen. Malaysia. 40 ASS EHM 41 CBD LAO MYA THA 42 MLY.

Arenga wightii Griff., Calcutta J. Nat. Hist. 5: 475 (1845). *Saguerus wightii* (Griff.) H.Wendl. & Drude in O.C.E.de Kerchove de Denterghem, Palmiers: 256 (1878).
SW. India. 40 IND.

Arikury

Arikury Becc. = **Syagrus** Mart.

Arikury schizophylla (Mart.) Becc. = **Syagrus schizophylla** (Mart.) Glassman

Arikuryroba

Arikuryroba Barb.Rodr. = **Syagrus** Mart.

Arikuryroba capanemae Barb.Rodr. = **Syagrus schizophylla** (Mart.) Glassman

Arikuryroba ruschiana (Bondar) Toledo = **Syagrus ruschiana** (Bondar) Glassman

Arikuryroba schizophylla (Mart.) L.H.Bailey = **Syagrus schizophylla** (Mart.) Glassman

Arikuryroba × tostana (Bondar) A.D.Hawkes = **Syagrus × tostana** (Bondar) Glassman

Aristeyera

Aristeyera H.E.Moore = **Asterogyne** H.Wendl. ex Hook.f.

Aristeyera ramosa H.E.Moore = **Asterogyne ramosa** (H.E.Moore) Wess.Boer

Aristeyera spicata H.E.Moore = **Asterogyne spicata** (H.E.Moore) Wess.Boer

Arrudaria

Arrudaria Macedo = **Copernicia** Mart. ex Endl.

Arrudaria cerifera (Arruda) Macedo = **Copernicia prunifera** (Mill.) H.E.Moore

Asraoa

Asraoa J.Joseph = **Wallichia** Roxb.

Asraoa triandra J.Joseph = **Wallichia triandra** (J.Joseph) S.K.Basu

Asterogyne

Asterogyne H.Wendl. ex Hook.f. in G.Bentham & J.D.Hooker, Gen. Pl. 3: 914 (1883).
5 species, C. & S. Trop. America. 80 82 83.
Aristeyera H.E.Moore, J. Arnold Arbor. 47: 3 (1966).

Asterogyne guianensis Granv. & A.J.Hend., Brittonia 40: 76 (1988).
SE. French Guiana. 82 FRG.

Asterogyne martiana (H.Wendl.) H.Wendl. ex Drude in H.G.A.Engler & K.A.E.Prantl (eds.), Nat. Pflanzenfam. 2(3): 59 (1889).
C. America to NW. Ecuador. 80 BLZ COS GUA HON NIC PAN 83 CLM ECU.
*Geonoma martiana H.Wendl., Linnaea 28: 342 (1856).
Geonoma trifurcata Oerst., Vidensk. Meddel. Dansk Naturhist. Foren. Kjøbenhavn 1858: 84 (1858).
Asterogyne minor Burret, Bot. Jahrb. Syst. 63: 141 (1930).

Asterogyne minor Burret = **Asterogyne martiana** (H.Wendl.) H.Wendl. ex Drude

Asterogyne ramosa (H.E.Moore) Wess.Boer, Verh. Kon. Ned. Akad. Wetensch., Afd. Natuurk., Tweede Sect. 58(1): 81 (1968).
Venezuela (Pen. de Paria). 82 VEN.
*Aristeyera ramosa H.E.Moore, J. Arnold Arbor. 48: 144 (1967).

Asterogyne spicata (H.E.Moore) Wess.Boer, Verh. Kon. Ned. Akad. Wetensch., Afd. Natuurk., Tweede Sect. 58(1): 82 (1968).
Venezuela (Miranda). 82 VEN.
*Aristeyera spicata H.E.Moore, J. Arnold Arbor. 47: 5 (1966).

Asterogyne yaracuyense A.J.Hend. & Steyerm., Brittonia 38: 309 (1986).
Venezuela (Cerro La Chapa). 82 VEN.

Astrocaryum

Astrocaryum G.Mey., Prim. Fl. Esseq.: 265 (1818).
36 species, Mexico to Trop. America. 79 80 81 82 83 84.
Avoira Giseke, Prael. Ord. Nat. Pl.: 38 (1792).
Toxophoenix Schott, Tagebücher, Anh. 2: 12 (1820).
Hexopetion Burret, Notizbl. Bot. Gart. Berlin-Dahlem 12: 156 (1934).

Astrocaryum acanthopodium Barb.Rodr. = **Astrocaryum paramaca** Mart.

Astrocaryum acaule Mart., Hist. Nat. Palm. 2: 78 (1824).
S. Trop. America. 82 GUY VEN 83 CLM 84 BZN.
Astrocaryum luetzelburgii Burret, Notizbl. Bot. Gart. Berlin-Dahlem 10: 1021 (1930).
Astrocaryum huebneri Burret, Repert. Spec. Nov. Regni Veg. 35: 128 (1934).

Astrocaryum aculeatissimum (Schott) Burret, Repert. Spec. Nov. Regni Veg. 35: 152 (1934).
E. & S. Brazil. 84 BZE BZL BZS.
*Toxophoenix aculeatissima Schott, Tagebücher, Anh. 2: 12 (1820).
Astrocaryum ayri Mart., Hist. Nat. Palm. 2: 71 (1824).

Astrocaryum aculeatum G.Mey., Prim. Fl. Esseq.: 266 (1818).
Trinidad, S. Trop. America. 81 TRT 82 FRG GUY SUR VEN 83 BOL CLM 84 BZN.
Astrocaryum tucuma Mart., Hist. Nat. Palm. 2: 77 (1824).
Astrocaryum aureum Griseb. & H.Wendl. in A.H.R.Grisebach, Fl. Brit. W. I.: 521 (1864).
Astrocaryum candescens Barb.Rodr., Enum. Palm. Nov.: 22 (1875).
Astrocaryum princeps Barb.Rodr., Enum. Palm. Nov.: 22 (1875).
Astrocaryum jucuma Linden, Ill. Hort. 28: 15 (1881).
Astrocaryum manaoense Barb.Rodr., Vellosia, ed. 2, 1: 105 (1891).
Astrocaryum macrocarpum Huber, Bull. Herb. Boissier, II, 4: 271 (1906).

Astrocaryum aculeatum Barb.Rodr. = **Astrocaryum rodriguesii** Trail

Astrocaryum aculeatum Wallace = **Bactris balanophora** Spruce

Astrocaryum alatum Loomis, J. Wash. Acad. Sci. 29: 142 (1939).
C. America. 80 COS NIC PAN.

Astrocaryum arenarium Barb.Rodr. = **?** [84 BZC]

Astrocaryum argenteum auct. = **?**

Astrocaryum aureum Griseb. & H.Wendl. = **Astrocaryum aculeatum** G.Mey.

Astrocaryum awarra de Vriese = **Astrocaryum vulgare** Mart.

Astrocaryum ayri Mart. = **Astrocaryum aculeatissimum** (Schott) Burret

Astrocaryum borsigianum K.Koch = **Phoenicophorium borsigianum** (K.Koch) Stuntz

Astrocaryum burity Barb.Rodr. = **?** [84]

Astrocaryum campestre Mart., Hist. Nat. Palm. 2: 79 (1824).
Brazil to Bolivia. 83 BOL 84 BZC BZE BZN.
Bactris bradei Burret, Notizbl. Bot. Gart. Berlin-Dahlem 14: 259 (1938).

Astrocaryum candescens Barb.Rodr. = **Astrocaryum aculeatum** G.Mey.

Astrocaryum carnosum F.Kahn & B.Millán, Bull. Inst. Franç. Études Andines 21: 504 (1992).
Peru. 83 PER.

Astrocaryum chambira Burret, Repert. Spec. Nov. Regni Veg. 35: 122 (1934).
S. Trop. America. 82 VEN 83 CLM ECU PER 84 BZN.

Astrocaryum chichon Linden = **Astrocaryum mexicanum** Liebm. ex Mart.

Astrocaryum chonta Mart. in A.D.d'Orbigny, Voy. Amér. Mér. 7(3): 84 (1844).
Peru to Bolivia. 83 BOL PER.

Astrocaryum ciliatum F.Kahn & B.Millán, Bull. Inst. Franç. Études Andines 21: 500 (1992). *Astrocaryum murumuru* var. *ciliatum* (F.Kahn & B.Millán) A.J.Hend., Palms Amazon: 245 (1995).
SE. Colombia. 83 CLM.

Astrocaryum cohune (S.Watson) Standl. = **Astrocaryum mexicanum** Liebm. ex Mart.

Astrocaryum confertum H.Wendl. ex Burret, Repert. Spec. Nov. Regni Veg. 35: 136 (1934).
Costa Rica to Panama. 80 COS PAN.
Astrocaryum polystachyum H.Wendl. ex Hemsl., Biol. Cent.-Amer., Bot. 3: 414 (1885), nom. nud.

Astrocaryum crispum (Kunth) M.Gómez & Roig = **Gastrococos crispa** (Kunth) H.E.Moore

Astrocaryum cuatrecasanum Dugand = **Astrocaryum urostachys** Burret

Astrocaryum dasychaetum Burret = **Astrocaryum gynacanthum** Mart.

Astrocaryum echinatum Barb.Rodr. = **?** [84 BZC]

Astrocaryum faranae F.Kahn & E.Ferreira, Candollea 50: 321 (1995).
Brazil (Acre). 84 BZN.

Astrocaryum farinosum Barb.Rodr., Enum. Palm. Nov.: 21 (1875).
S. Guyana to N. Brazil. 82 GUY 84 BZN.

Astrocaryum ferrugineum F.Kahn & B.Millán, Bull. Inst. Franç. Études Andines 21: 467 (1992). *Astrocaryum murumuru* var. *ferrugineum* (F.Kahn & B.Millán) A.J.Hend., Palms Amazon: 245 (1995).
Brazil (Amazonas). 84 BZN.

Astrocaryum filare W.Bull. ex Hook.f. = **?**

Astrocaryum flexuosum H.Wendl. = **?**

Astrocaryum giganteum Barb.Rodr., Contr. Jard. Bot. Rio de Janeiro 2: 82 (1902).
N. Brazil. 84 BZN.

Astrocaryum gratum F.Kahn & B.Millán, Bull. Inst. Franç. Études Andines 21: 520 (1992).
Peru to Bolivia. 83 BOL PER.

Astrocaryum guara Burret = **Astrocaryum jauari** Mart.

Astrocaryum guianense Splitg. ex Mart. = **Astrocaryum vulgare** Mart.

Astrocaryum gymnopus Burret = **Astrocaryum gynacanthum** Mart.

Astrocaryum gynacanthum Mart., Hist. Nat. Palm. 2: 71 (1824).
S. Trop. America. 82 FRG GUY SUR VEN 83 BOL CLM PER 84 BZN.
Astrocaryum munbaca Mart., Hist. Nat. Palm. 2: 74 (1824).
Astrocaryum gynacanthum var. *munbaca* Trail, J. Bot. 6: 78 (1877).
Astrocaryum gymnopus Burret, Notizbl. Bot. Gart. Berlin-Dahlem 10: 1020 (1930).
Astrocaryum dasychaetum Burret, Repert. Spec. Nov. Regni Veg. 35: 141 (1934).
Astrocaryum gynacanthum var. *dasychaetum* Burret, Repert. Spec. Nov. Regni Veg. 35: 141 (1934).
Astrocaryum dasychaetum Burret, Repert. Spec. Nov. Regni Veg. 35: 141 (1934), pro syn.

Astrocaryum gynacanthum var. *dasychaetum* Burret = **Astrocaryum gynacanthum** Mart.

Astrocaryum gynacanthum var. *munbaca* Trail = **Astrocaryum gynacanthum** Mart.

Astrocaryum horridum Barb.Rodr. = **Astrocaryum javarense** (Trail) Drude

Astrocaryum huaimi Mart. in A.D.d'Orbigny, Voy. Amér. Mér. 7(3): 86 (1844).
WC. Brazil to Peru and Bolivia. 83 BOL PER 84 BZC.
Astrocaryum leiospatha Barb.Rodr., Palm. Mattogross.: 56 (1898).

Astrocaryum huebneri Burret = **Astrocaryum acaule** Mart.

Astrocaryum huicungo Dammer ex Burret, Repert. Spec. Nov. Regni Veg. 35: 146 (1934). *Astrocaryum murumuru* var. *huicungo* (Dammer ex Burret) A.J.Hend., Palms Amazon: 245 (1995).
N. Peru. 83 PER.

Astrocaryum humile Wallace = **Bactris acanthocarpa** var. **exscapa** Barb.Rodr.

Astrocaryum iriartoides Willis ex Regel. = **?**

Astrocaryum jucuma Linden = **Astrocaryum aculeatum** G.Mey.

Astrocaryum jauari Mart., Hist. Nat. Palm. 2: 76 (1824).
S. Trop. America. 82 FRG GUY SUR VEN 83 CLM ECU PER 84 BZN.
Astrocaryum guara Burret, Notizbl. Bot. Gart. Berlin-Dahlem 11: 15 (1930).

Astrocaryum javarense (Trail) Drude in C.F.P.von Martius & auct. suc. (eds.), Fl. Bras. 3(2): 372 (1881).
N. Peru. 83 PER.
Astrocaryum paramaca var. *javarense* Trail, J. Bot. 15: 77 (1877). *Astrocaryum murumuru* var. *javarense* (Trail) A.J.Hend., Palms Amazon: 246 (1995).
Astrocaryum horridum Barb.Rodr., Vellosia, ed. 2, 1: 104 (1891).

Astrocaryum kewense Barb.Rodr. = **?** [84]

Astrocaryum leiospatha Barb.Rodr. = **Astrocaryum huaimi** Mart.

Astrocaryum luetzelburgii Burret = **Astrocaryum acaule** Mart.

Astrocaryum macrocalyx Burret, Repert. Spec. Nov. Regni Veg. 35: 150 (1934). *Astrocaryum murumuru* var. *macrocalyx* (Burret) A.J.Hend., Palms Amazon: 246 (1995). N. Peru. 83 PER.

Astrocaryum macrocarpum Huber = **Astrocaryum aculeatum** G.Mey.

Astrocaryum malybo H.Karst., Linnaea 28: 245 (1856). Colombia. 83 CLM.

Astrocaryum manaoense Barb.Rodr. = **Astrocaryum aculeatum** G.Mey.

Astrocaryum mexicanum Liebm. ex Mart., Hist. Nat. Palm. 3: 323 (1853). *Hexopetion mexicanum* (Liebm. ex Mart.) Burret, Notizbl. Bot. Gart. Berlin-Dahlem 12: 156 (1934).
Mexico to C. America. 79 MXG MXS MXT 80 BLZ ELS GUA HON NIC.
Astrocaryum rostratum Hook.f., Bot. Mag. 80: t. 4773 (1854).
Astrocaryum chichon Linden, Ill. Hort. 28: 15 (1881).
Bactris cohune S.Watson, Proc. Amer. Acad. Arts 21: 467 (1886). *Astrocaryum cohune* (S.Watson) Standl., Trop. Woods 21: 25 (1930).

Astrocaryum minus Trail, J. Bot. 15: 78 (1877). French Guiana to N. Brazil. 82 FRG 84 BZN.

Astrocaryum munbaca Mart. = **Astrocaryum gynacanthum** Mart.

Astrocaryum murumuru Mart., Hist. Nat. Palm. 2: 70 (1824).
N. South America to N. Brazil. 82 FRG GUY SUR VEN 84 BZN.
Astrocaryum yauaperyense Barb.Rodr., Vellosia, ed. 2, 1: 103 (1886).

Astrocaryum murumuru var. *ciliatum* (F.Kahn & B.Millán) A.Hend. = **Astrocaryum ciliatum** F.Kahn & B.Millán

Astrocaryum murumuru var. *ferrugineum* (F.Kahn & B.Millán) A.Hend. = **Astrocaryum ferrugineum** F.Kahn & B.Millán

Astrocaryum murumuru var. *huicungo* (Dammer ex Burret) A.Hend. = **Astrocaryum huicungo** Dammer ex Burret

Astrocaryum murumuru var. *javarense* (Trail) A.Hend. = **Astrocaryum javarense** (Trail) Drude

Astrocaryum murumuru var. *macrocalyx* (Burret) A.Hend. = **Astrocaryum macrocalyx** Burret

Astrocaryum murumuru var. *perangustatum* (F.Kahn & B.Millán) A.Hend. = **Astrocaryum perangustatum** F.Kahn & B.Millán

Astrocaryum murumuru var. *urostachys* (Burret) A.Hend. = **Astrocaryum urostachys** Burret

Astrocaryum panamense Linden = ? [83 CLM]

Astrocaryum paramaca Mart. in A.D.d'Orbigny, Voy. Amér. Mér. 7(3): 88 (1844).
Guianas to N. Brazil. 82 FRG SUR 84 BZN.
Bactris paraensis Splitg. ex de Vriese, Jaarb. Kon. Ned. Maatsch. Aanm. Tuinb. 1848: 10 (1848).
Astrocaryum acanthopodium Barb.Rodr., Enum. Palm. Nov.: 20 (1875).

Astrocaryum perangustatum F.Kahn & B.Millán, Bull. Inst. Franç. Études Andines 21: 517 (1992).

Astrocaryum murumuru var. *perangustatum* (F.Kahn & B.Millán) A.J.Hend., Palms Amazon: 246 (1995). Peru (Pasco). 83 PER.

Astrocaryum pictum Balf.f. = **Phoenicophorium borsigianum** (K.Koch) Stuntz

Astrocaryum plicatum Drude = **Astrocaryum sciophilum** (Miq.) Pulle

Astrocaryum polystachyum H.Wendl. ex Hemsl. = **Astrocaryum confertum** H.Wendl. ex Burret

Astrocaryum princeps Barb.Rodr. = **Astrocaryum aculeatum** G.Mey.

Astrocaryum pumilum H.Wendl. = ?

Astrocaryum pygmaeum Drude = ? [84]

Astrocaryum rodriguesii Trail, J. Bot. 15: 79 (1877). French Guiana to N. Brazil. 82 FRG 84 BZN.
Astrocaryum aculeatum Barb.Rodr., Enum. Palm. Nov.: 20 (1875), nom. illeg.

Astrocaryum rostratum Hook.f. = **Astrocaryum mexicanum** Liebm. ex Mart.

Astrocaryum sciophilum (Miq.) Pulle, Enum. Vasc. Pl. Surinam: 73 (1906).
Guianas to N. Brazil. 82 FRG GUY SUR 84 BZN.
Bactris sciophila Miq., Verh. Nat. Wet. Haarlem 7: 208 (1851).
Astrocaryum plicatum Drude in C.F.P.von Martius & auct. suc. (eds.), Fl. Bras. 3(2): 375 (1881).

Astrocaryum sclerocarpum H.Wendl. = **Acrocomia aculeata** (Jacq.) Lodd. ex Mart.

Astrocaryum sclerophyllum Drude = ? [84]

Astrocaryum scopatum F.Kahn & B.Millán, Bull. Inst. Franç. Études Andines 21: 503 (1992).
N. Peru. 83 PER.

Astrocaryum sechellarum (H.Wendl.) Baill. = **Phoenicophorium borsigianum** (K.Koch) Stuntz

Astrocaryum segregatum Drude = **Astrocaryum vulgare** Mart.

Astrocaryum sociale Barb.Rodr., Vellosia, ed. 2, 1: 103 (1891).
N. Brazil (Near Manaus). 84 BZN.

Astrocaryum standleyanum L.H.Bailey, Gentes Herb. 3: 88 (1933).
Costa Rica to Ecuador. 80 COS PAN 83 CLM ECU.
Astrocaryum trachycarpum Burret, Repert. Spec. Nov. Regni Veg. 35: 138 (1934).

Astrocaryum tenuifolium Linden = ? [84]

Astrocaryum trachycarpum Burret = **Astrocaryum standleyanum** L.H.Bailey

Astrocaryum triandrum Galeano-Garces, R.Bernal & F.Kahn, Candollea 43: 279 (1988).
Colombia (Antioquia, Caldas). 83 CLM.

Astrocaryum tucuma Mart. = **Astrocaryum aculeatum** G.Mey.

Astrocaryum tucumoides Drude = **Astrocaryum vulgare** Mart.

Astrocaryum ulei Burret, Repert. Spec. Nov. Regni Veg. 35: 147 (1934).
Brazil (Acre) to N. Bolivia. 83 BOL 84 BZN.

Astrocaryum urostachys Burret, Repert. Spec. Nov. Regni Veg. 35: 151 (1934). *Astrocaryum murumuru* var. *urostachys* (Burret) A.J.Hend., Palms Amazon: 247 (1995).

Colombia to N. Peru. 83 CLM ECU PER.
Astrocaryum cuatrecasanum Dugand, Caldasia 1(1): 18 (1940).

Astrocaryum vulgare Mart., Hist. Nat. Palm. 2: 74 (1824).
Guianas to N. Brazil. 82 FRG GUY SUR 84 BZN.
Astrocaryum awarra de Vriese, Jaarb. Kon. Ned. Maatsch. Aanm. Tuinb. 1848: 12 (1848).
Astrocaryum guianense Splitg. ex Mart., Hist. Nat. Palm. 3: 323 (1853).
Astrocaryum segregatum Drude in C.F.P.von Martius & auct. suc. (eds.), Fl. Bras. 3(2): 382 (1881).
Astrocaryum tucumoides Drude in C.F.P.von Martius & auct. suc. (eds.), Fl. Bras. 3(2): 381 (1881).

Astrocaryum warszewiczii H.Karst. = **?**

Astrocaryum weddellii Drude = **? [84]**

Astrocaryum yauaperyense Barb.Rodr. = **Astrocaryum murumuru** Mart.

Atitara

Atitara Barrère ex Kuntze = **Desmoncus** Mart.

Atitara aerea (Drude) Barb.Rodr. = **Desmoncus polyacanthos** var. **polyacanthos**

Atitara ataxacantha (Barb.Rodr.) Kuntze = **Desmoncus orthacanthos** Mart.

Atitara caespitosa (Barb.Rodr.) Barb.Rodr. = **Desmoncus polyacanthos** var. **polyacanthos**

Atitara chinantlensis (Liebm. ex Mart.) Kuntze = **Desmoncus orthacanthos** Mart.

Atitara costaricensis Kuntze = **Desmoncus costaricensis** (Kuntze) Burret

Atitara cuyabaensis (Barb.Rodr.) Barb.Rodr. = **Desmoncus orthacanthos** Mart.

Atitara drudeana Kuntze = **Desmoncus orthacanthos** Mart.

Atitara dubia Kuntze = **Desmoncus polyacanthos** var. **polyacanthos**

Atitara granatensis (W.Bull) Kuntze = **Desmoncus granatensis**

Atitara horrida (Splitg. ex Mart.) Kuntze = **Desmoncus orthacanthos** Mart.

Atitara inermis (Barb.Rodr.) Barb.Rodr. = **Desmoncus polyacanthos** var. **polyacanthos**

Atitara latifrons (W.Bull) Kuntze = **Desmoncus latifrons**

Atitara leptoclona (Drude) Barb.Rodr. = **Desmoncus polyacanthos** var. **polyacanthos**

Atitara leptospadix (Mart.) Kuntze = **Desmoncus mitis** var. **leptospadix** (Mart.) A.J.Hend.

Atitara lophacantha (Mart.) Barb.Rodr. = **Desmoncus orthacanthos** Mart.

Atitara macroacantha (Mart.) Kuntze = **Desmoncus pycnacanthos**

Atitara macrocarpa (Barb.Rodr.) Barb.Rodr. = **Desmoncus orthacanthos** Mart.

Atitara macrodon (Barb.Rodr.) Barb.Rodr. = **Desmoncus phoenicocarpus** Barb.Rodr.

Atitara major (Crueg. ex Griseb.) Kuntze = **Desmoncus orthacanthos** Mart.

Atitara mitis (Mart.) Kuntze = **Desmoncus mitis** Mart.

Atitara nemorosa (Barb.Rodr.) Barb.Rodr. = **Desmoncus phoenicocarpus** Barb.Rodr.

Atitara oligacantha (Barb.Rodr.) Kuntze = **Desmoncus polyacanthos** var. **polyacanthos**

Atitara orthacantha (Mart.) Kuntze = **Desmoncus orthacanthos** Mart.

Atitara oxyacantha (Mart.) Kuntze = **Desmoncus polyacanthos** var. **polyacanthos**

Atitara palustris (Trail) Kuntze = **Desmoncus orthacanthos** Mart.

Atitara paraensis Barb.Rodr. = **Desmoncus polyacanthos** var. **polyacanthos**

Atitara phengophylla (Drude) Kuntze = **Desmoncus polyacanthos** var. **polyacanthos**

Atitara philippiana (Barb.Rodr.) Barb.Rodr. = **Desmoncus polyacanthos** var. **polyacanthos**

Atitara phoenicocarpa (Barb.Rodr.) Kuntze = **Desmoncus phoenicocarpus** Barb.Rodr.

Atitara polyacantha (Mart.) Kuntze = **Desmoncus polyacanthos** Mart.

Atitara prostrata (Lindm.) Barb.Rodr. = **Desmoncus orthacanthos** Mart.

Atitara prunifera (Poepp. ex Mart.) Kuntze = **Desmoncus polyacanthos** var. **prunifer** (Poepp. ex Mart.) A.J.Hend.

Atitara pumila (Trail) Kuntze = **Desmoncus mitis** var. **mitis**

Atitara pycnacantha (Mart.) Kuntze = **Desmoncus polyacanthos** var. **polyacanthos**

Atitara riparia (Spruce) Kuntze = **Desmoncus polyacanthos** var. **polyacanthos**

Atitara rudenta (Mart.) Barb.Rodr. = **Desmoncus orthacanthos** Mart.

Atitara setosa (Mart.) Kuntze = **Desmoncus polyacanthos** var. **polyacanthos**

× Attabignya

× *Attabignya* Balick, A.B.Anderson & Med.-Costa = **Attalea** Kunth

× *Attabignya minarum* Balick, A.B.Anderson & Med.-Costa = **Attalea × minarum** (Balick, A.B.Anderson & Med.-Costa) Zona

Attalea

Attalea Kunth in F.W.H.von Humboldt, A.J.A.Bonpland & C.S.Kunth, Nov. Gen. Sp. 1: 309 (1816).
69 species, Mexico to Trop. America. 79 80 81 82 83 84 85.
Maximiliana Mart., Hist. Nat. Palm. 2: 131 (1826).
Orbignya Mart. ex Endl., Gen. Pl.: 257 (1837).
Lithocarpos O.Targ.Tozz. ex Steud., Nomencl. Bot., ed. 2, 2: 56 (1841).
Scheelea H.Karst., Linnaea 28: 264 (1856).
Englerophoenix Kuntze, Revis. Gen. Pl. 2: 728 (1891).
Pindarea Barb.Rodr., Pl. Jard. Rio de Janeiro 5: 17 (1896).
Bornoa O.F.Cook, Natl. Hort. Mag. 18: 264 (1939), no latin descr.
Heptantra O.F.Cook, Natl. Hort. Mag. 18: 277 (1939).
Temenia O.F.Cook, Natl. Hort. Mag. 18: 276 (1939).
Parascheelea Dugand, Caldasia 1(1): 10 (1940).
Sarinia O.F.Cook, Natl. Hort. Mag. 21: 68 (1942).
Ynesa O.F.Cook, Natl. Hort. Mag. 21: 71 (1942).
Markleya Bondar, Arch. Jard. Bot. Rio de Janeiro 15: 50 (1957).

× *Attabignya* Balick, A.B.Anderson & Med.-Costa, Brittonia 39: 27 (1987).
Maximbignya Glassman, Illinois Biol. Monogr. 59: 199 (1999).

Attalea acaulis Burret = **Attalea funifera** Mart.

Attalea acaulis H.Wendl. = **Attalea funifera** Mart.

Attalea agrestis Barb.Rodr. = **Attalea microcarpa** Mart.

Attalea allenii H.E.Moore, Gentes Herb. 8: 191 (1949).
Panama to NW. Colombia. 80 PAN 83 CLM.

Attalea amygdalina Kunth in F.W.H.von Humboldt, A.J.A.Bonpland & C.S.Kunth, Nov. Gen. Sp. 1: 310 (1816).
Colombia (Río Cauca valley). 83 CLM.
Attalea uberrima Dugand, Mutisia 18: 4 (1953).
Attalea victoriana Dugand, Mutisia 18: 9 (1953).

Attalea amylacea (Barb.Rodr.) Zona, Palms 46: 132 (2002).
Brazil. 84 BZL.
*Scheelea amylacea Barb.Rodr., Pl. Jard. Rio de Janeiro 1: 17 (1891).

Attalea anisitsiana (Barb.Rodr.) Zona, Palms 46: 132 (2002).
Brazil (Mato Grosso) to Paraguay. 84 BZC 85 PAR.
*Scheelea anisitsiana Barb.Rodr., Palm. Mattogross.: 63 (1898).
Scheelea quadrisperma Barb.Rodr., Palm. Paraguay.: 23 (1899).
Attalea parviflora Barb.Rodr., Bull. Herb. Boissier, II, 3: 625 (1903). *Scheelea parviflora* (Barb.Rodr.) Barb.Rodr., Sert. Palm. Brasil. 1: 53 (1903).
Scheelea quadrisulcata Barb.Rodr., Contr. Jard. Bot. Rio de Janeiro 4: 107 (1907).

Attalea apoda Burret, Repert. Spec. Nov. Regni Veg. 32: 105 (1933).
Brazil (S. Minas Gerais to Goiás). 84 BZC BZL.
Attalea camposportoana Burret, Notizbl. Bot. Gart. Berlin-Dahlem 14: 257 (1938).

Attalea attaleoides (Barb.Rodr.) Wess.Boer, Indig. Palms Surin.: 157 (1965).
Guianas to N. Brazil. 82 FRG SUR 84 BZN.
*Maximiliana attaleoides Barb.Rodr., Enum. Palm. Nov.: 41 (1875). *Englerophoenix attaleoides* (Barb. Rodr.) Barb.Rodr., Sert. Palm. Brasil. 1: 76 (1903).
Attalea transitiva Barb.Rodr., Prot.-App. Enum. Palm. Nov.: 49 (1879).

Attalea barreirensis Glassman, Illinois Biol. Monogr. 59: 25 (1999).
Brazil (Bahia). 84 BZE.

Attalea bassleriana (Burret) Zona, Palms 46: 132 (2002).
N. Peru. 83 PER.
*Scheelea bassleriana Burret, Notizbl. Bot. Gart. Berlin-Dahlem 10: 655 (1929).
Scheelea brachyclada Burret, Notizbl. Bot. Gart. Berlin-Dahlem 10: 680 (1929).
Scheelea stenorhyncha Burret, Notizbl. Bot. Gart. Berlin-Dahlem 10: 675 (1929).

Attalea blepharopus Mart. in A.D.d'Orbigny, Voy. Amér. Mér. 7(3): 116 (1844). *Scheelea blepharopus* (Mart.) Burret, Notizbl. Bot. Gart. Berlin-Dahlem 10: 541 (1929).
Bolivia. 83 BOL. — Provisionally accepted.

Attalea boehmii Drude = **?** [2]

Attalea borgesiana Bondar = **Attalea humilis** Mart. ex Spreng.

Attalea brasiliensis Glassman, Illinois Biol. Monogr. 59: 65 (1999).
Brazil (Goiás). 84 BZC.

Attalea brejinhoensis (Glassman) Zona, Palms 46: 132 (2002).
NE. Brazil (Bahia). 84 BZE.
*Orbignya brejinhoensis Glassman, Illinois Biol. Monogr. 59: 84 (1999).

Attalea burretiana Bondar = **Attalea oleifera** Barb.Rodr.

Attalea butyracea (Mutis ex L.f.) Wess.Boer, Pittieria 17: 312 (1988).
S. Venezuela to W. South America. 82 VEN 83 BOL CLM ECU PER.
*Cocos butyracea Mutis ex L.f., Suppl. Pl.: 454 (1782). *Scheelea butyracea* (Mutis ex L.f.) H.Karst. ex H.Wendl. in O.C.E.de Kerchove de Denterghem, Palmiers: 256 (1878).
Attalea gomphococca Mart., Hist. Nat. Palm. 3: 301 (1845). *Scheelea gomphococca* (Mart.) Burret, Notizbl. Bot. Gart. Berlin-Dahlem 10: 541 (1929).
Scheelea excelsa H.Karst., Linnaea 28: 267 (1856).
Scheelea regia H.Karst., Linnaea 28: 266 (1856).
Attalea humboldtiana Spruce, J. Linn. Soc., Bot. 11: 163 (1869). *Scheelea humboldtiana* (Spruce) Burret, Notizbl. Bot. Gart. Berlin-Dahlem 10: 541 (1929).
Scheelea excelsa Barb.Rodr., Pl. Jard. Rio de Janeiro 1: 30 (1891), nom. illeg.
Attalea wallisii Huber, Bull. Herb. Boissier, II, 6: 267 (1906). *Scheelea wallisii* (Huber) Burret, Notizbl. Bot. Gart. Berlin-Dahlem 10: 542 (1929).
Scheelea dryanderae Burret, Notizbl. Bot. Gart. Berlin-Dahlem 11: 1049 (1934).
Attalea pycnocarpa Wess.Boer, Pittieria 17: 299 (1988).

Attalea burretiana Bondar = **Attalea oleifera** Barb.Rodr.

Attalea camopiensis (Glassman) Zona, Palms 46: 132 (2002).
French Guiana. 82 FRG.
*Scheelea camopiensis Glassman, Illinois Biol. Monogr. 59: 138 (1999).

Attalea camposportoana Burret = **Attalea apoda** Burret

Attalea cephalotus Poepp. ex Mart. in A.D.d'Orbigny, Voy. Amér. Mér. 7(3): 119 (1844). *Scheelea cephalotes* (Poepp. ex Mart.) H.Karst., Linnaea 28: 269 (1856).
N. Peru. 83 PER.

Attalea ceraensis Barb.Rodr. = **?** [84]

Attalea cohune Mart. in A.D.d'Orbigny, Voy. Amér. Mér. 7(3): 121 (1844). *Orbignya cohune* (Mart.) Dahlgren ex Standl., Trop. Woods 30: 3 (1932).
SE. Mexico to C. America. 79 MXT 80 BLZ ELS GUA HON NIC.
Orbignya dammeriana Barb.Rodr., Sert. Palm. Brasil. 1: 62 (1903).

Attalea colenda (O.F.Cook) Balslev & A.J.Hend., Brittonia 39: 1 (1987).
SW. Colombia to W. Ecuador. 83 CLM ECU.
*Ynesa colenda O.F.Cook, Natl. Hort. Mag. 21: 71 (1942).

Attalea compta Mart., Hist. Nat. Palm. 2: 137 (1826).
Brazil (Minas Gerais). 84 BZL.

Attalea compta var. *acaulis* Mart. = **Attalea humilis** Mart. ex Spreng.

Attalea concentrista Bondar = **Attalea oleifera** Barb. Rodr.

Attalea concinna (Barb.Rodr.) Burret = **Attalea dubia** (Mart.) Burret

Attalea coronata Lodd. ex H.Wendl. = **?**

Attalea crassispatha (Mart.) Burret, Kongl. Svenska Vetenskapsakad. Handl., III, 6(7): 23 (1929).
SW. Haiti. 81 HAI.
  *Maximiliana crassispatha Mart. in A.D.d'Orbigny, Voy. Amér. Mér. 7(3): 110 (1844). *Bornoa crassispatha* (Mart.) O.F.Cook, Natl. Hort. Mag. 18: 266 (1939). *Orbignya crassispatha* (Mart.) Glassman, Illinois Biol. Monogr. 59: 94 (1999).

Attalea cryptanthera Wess.Boer = **Attalea maripa** (Aubl.) Mart.

Attalea cuatrecasana (Dugand) A.J.Hend., Galeano & R.Bernal, Field Guide to the Palms of the Americas: 265 (1995).
W. Colombia. 83 CLM.
  *Orbignya cuatrecasana Dugand, Caldasia 2: 285 (1943).

Attalea dahlgreniana (Bondar) Wess.Boer, Indig. Palms Surin.: 158 (1965).
Suriname to N. Brazil. 82 SUR 84 BZN.
  *Markleya dahlgreniana Bondar, Arch. Jard. Bot. Rio de Janeiro 15: 50 (1957). *Maximbignya dahlgreniana* (Bondar) Glassman, Illinois Biol. Monogr. 59: 199 (1999).

Attalea degranvillei (Glassman) Zona, Palms 46: 132 (2002).
French Guiana. 82 FRG.
  *Scheelea degranvillei Glassman, Illinois Biol. Monogr. 59: 139 (1999).

Attalea dubia (Mart.) Burret, Notizbl. Bot. Gart. Berlin-Dahlem 10: 516 (1929).
SE. & S. Brazil. 84 BZL BZS.
  *Orbignya dubia Mart., Hist. Nat. Palm. 3: 304 (1845). *Scheelea dubia* (Mart.) Burret, Bot. Jahrb. Syst. 63: 73 (1929). *Pindarea dubia* (Mart.) A.D.Hawkes, Arq. Bot. Estado São Paulo, n.s., f.m., 2: 191 (1952). *Pindarea concinna* Barb.Rodr., Pl. Jard. Rio de Janeiro 5: 17 (1896). *Attalea concinna* (Barb.Rodr.) Burret, Notizbl. Bot. Gart. Berlin-Dahlem 10: 537 (1929). *Pindarea fastuosa* Barb.Rodr., Pl. Jard. Rio de Janeiro 5: 23 (1896).

Attalea eichleri (Drude) A.J.Hend., Palms Amazon: 143 (1995).
Brazil to Bolivia. 83 BOL 84 BZC BZE BZN.
  Orbignya humilis Mart. in A.D.d'Orbigny, Voy. Amér. Mér. 7(3): 129 (1847).
  *Orbignya eichleri Drude in C.F.P.von Martius & auct. suc. (eds.), Fl. Bras. 3(2): 449 (1881). *Orbignya campestris* Barb.Rodr., Palm. Mattogross.: 78 (1898). *Orbignya longibracteata* Barb.Rodr., Palm. Mattogross.: 79 (1898). *Orbignya macrocarpa* Barb.Rodr., Palm. Mattogross.: 74 (1898). *Orbignya urbaniana* Dammer, Bot. Jahrb. Syst. 31(70): 23 (1901).

Attalea exigua Drude in C.F.P.von Martius & auct. suc. (eds.), Fl. Bras. 3(2): 439 (1881).
Brazil. 84 BZC BZE BZL.

Attalea excelsa Mart. = **Attalea phalerata** Mart. ex Spreng.

Attalea fairchildensis (Glassman) Zona, Palms 46: 132 (2002).
Colombia. 83 CLM.

*Scheelea fairchildensis Glassman, Illinois Biol. Monogr. 59: 163 (1999).

Attalea ferruginea Burret = **Attalea racemosa** Spruce

Attalea funifera Mart., Hist. Nat. Palm. 2: 136 (1826).
Sarinia funifera (Mart.) O.F.Cook, Natl. Hort. Mag. 21: 78 (1942).
NE. Brazil. 84 BZE.
  Lithocarpos cocciformis O.Targ.Tozz. ex Steud., Nomencl. Bot., ed. 2, 2: 56 (1841). *Attalea acaulis* H.Wendl., Index Palm.: 4 (1854). *Attalea acaulis* Burret, Repert. Spec. Nov. Regni Veg. 32: 103 (1933).

Attalea geraensis Barb.Rodr., Pl. Jard. Rio de Janeiro 6: 22 (1898).
Brazil to Paraguay. 84 BZC BZE BZL 85 PAR.
  Attalea guaranitica Barb.Rodr., Palm. Paraguay.: 27 (1899).

Attalea glassmanii Zona = **Attalea speciosa** Mart.

Attalea goeldiana Huber = **Attalea insignis** (Mart.) Drude

Attalea gomphococca Mart. = **Attalea butyracea** (Mutis ex L.f.) Wess.Boer

Attalea grandis H.Wendl. = **?**

Attalea guacuyule (Liebm. ex Mart.) Zona, Palms 46: 133 (2002).
SW. Mexico. 79 MXS.
  Cocos cocoyule Mart., Hist. Nat. Palm. 3: 324 (1853). **Cocos guacuyule* Liebm. ex Mart., Hist. Nat. Palm. 3: 323 (1853). *Orbignya guacuyule* (Liebm. ex Mart.) Hern.-Xol., Bol. Soc. Bot. México 9: 17 (1949).

Attalea guaranitica Barb.Rodr. = **Attalea geraensis** Barb.Rodr.

Attalea guianensis (Glassman) Zona, Palms 46: 133 (2002).
French Guiana. 82 FRG.
  *Scheelea guianensis Glassman, Illinois Biol. Monogr. 59: 137 (1999).

Attalea hoehnei Burret, Notizbl. Bot. Gart. Berlin-Dahlem 10: 522 (1929).
Brazil (Mato Grosso). 84 BZC.

Attalea huebneri (Burret) Zona, Palms 46: 133 (2002).
N. Brazil. 84 BZN.
  *Scheelea huebneri Burret, Notizbl. Bot. Gart. Berlin-Dahlem 10: 663 (1929).

Attalea humboldtiana Spruce = **Attalea butyracea** (Mutis ex L.f.) Wess.Boer

Attalea humilis Mart. ex Spreng., Syst. Veg. 2: 624 (1825).
E. Brazil. 84 BZE BZL.
  Attalea compta var. *acaulis* Mart., Hist. Nat. Palm. 2: t. 75 (1826). *Attalea butyrosa* Lodd. ex H.Wendl. in O.C.E.de Kerchove de Denterghem, Palmiers: 232 (1878). *Cocos butyrosa* H.Wendl. in O.C.E.de Kerchove de Denterghem, Palmiers: 140 (1878). *Attalea borgesiana* Bondar, Bol. Inst. Centr. Fomento Econ. Bahia 3: 16 (1939).

Attalea iguadummat de Nevers, Ann. Missouri Bot. Gard. 74: 506 (1987).
Panama. 80 PAN.

Attalea insignis (Mart.) Drude in H.G.A.Engler & K.A.E. Prantl (eds.), Nat. Pflanzenfam., Nachtr. 1: 56 (1897).
Colombia to N. Peru. 83 CLM ECU PER.
  *Maximiliana insignis Mart., Hist. Nat. Palm. 2: 133 (1826). *Englerophoenix insignis* (Mart.) Kuntze,

Revis. Gen. Pl. 3(2): 322 (1898). *Scheelea insignis* (Mart.) H.Karst., Linnaea 28: 269 (1956).

Scheelea attaleoides H.Karst., Linnaea 38: 265 (1856).

Attalea goeldiana Huber, Bull. Herb. Boissier, II, 6: 268 (1906). *Scheelea goeldiana* (Huber) Burret, Notizbl. Bot. Gart. Berlin-Dahlem 10: 541 (1929).

Attalea kewensis (Hook.f.) Zona, Palms 46: 133 (2002).
N. Peru. 83 PER.
*Scheelea kewensis Hook.f., Bot. Mag. 123: t. 7552 (1897).

Attalea lapidea (Gaertn.) Burret = **Cocos lapidea**

Attalea lauromuelleriana (Barb.Rodr.) Zona, Palms 46: 133 (2002).
SE. Brazil. 84 BZL.
*Scheelea lauromuelleriana Barb.Rodr., Contr. Jard. Bot. Rio de Janeiro 4: 108 (1907).

Attalea leandroana (Barb.Rodr.) Zona, Palms 46: 133 (2002).
Brazil. 84 +.
*Scheelea leandroana Barb.Rodr., Pl. Jard. Rio de Janeiro 1: 19 (1891).

Attalea liebmannii (Becc.) Zona, Palms 46: 133 (2002).
S. Mexico (to Veracruz). 79 MXG MXS MXT.
Cocos regia Liebm. in C.F.P.von Martius, Hist. Nat. Palm. 3: 323 (1853). *Scheelea liebmannii* Becc., Bibliot. Agr. Colon. Firenze 1916: 113 (1916).

Attalea limbata Seem. ex H.Wendl. = ?

Attalea luetzelburgii (Burret) Wess.Boer, Pittieria 17: 303 (1988).
SE. Colombia to SW. Venezuela and NW. Brazil. 82 VEN 83 CLM 84 BZN.
*Orbignya luetzelburgii Burret, Notizbl. Bot. Gart. Berlin-Dahlem 10: 1019 (1930). *Parascheelea luetzelburgii* (Burret) Dugand, Caldasia 1(3): 24 (1941).
Parascheelea anchistropetala Dugand, Caldasia 1(1): 12 (1940).

Attalea lundellii (Bartlett) Zona, Palms 46: 133 (2002).
Guatemala. 80 GUA.
*Scheelea lundellii Bartlett, Publ. Carnegie Inst. Wash. 461: 46 (1935).

Attalea lydiae (Drude) Barb.Rodr. = **Attalea speciosa** Mart.

Attalea macoupi Sagot ex Drude = **Attalea spectabilis** Mart.

Attalea macrocarpa (H.Karst.) Linden, Ill. Hort. 28: 15 (1881).
Venezuela. 82 VEN.
*Scheelea macrocarpa H.Karst., Linnaea 28: 268 (1856).
Scheelea passargei Burret, Notizbl. Bot. Gart. Berlin-Dahlem 10: 671 (1929).

Attalea macrolepis (Burret) Wess.Boer, Pittieria 17: 311 (1988).
S. Venezuela. 82 VEN.
*Scheelea macrolepis Burret, Notizbl. Bot. Gart. Berlin-Dahlem 10: 688 (1929).

Attalea macropetala (Burret) Wess.Boer = **Attalea maripa** (Aubl.) Mart.

Attalea magdalenae Linden = ? [83 CLM]

Attalea magdalenica (Dugand) Zona, Palms 46: 133 (2002).
Colombia. 83 CLM.
*Scheelea magdalenica Dugand, Mutisia 26: 1 (1959).

Attalea manaca Linden = ? [83 CLM]

Attalea maracaibensis Mart. in A.D.d'Orbigny, Voy. Amér. Mér. 7(3): 124 (1844). *Scheelea maracaibensis* (Mart.) Burret, Notizbl. Bot. Gart. Berlin-Dahlem 10: 541 (1929).
Colombia to Venezuela. 82 VEN 83 CLM.

Attalea maripa (Aubl.) Mart. in A.D.d'Orbigny, Voy. Amér. Mér. 7(3): 123 (1844). *Ethnora maripa* (Mart.) O.F.Cook, J. Wash. Acad. Sci. 30: 297 (1940).
Trinidad, S. Trop. America. 81 TRT 82 FRG GUY SUR VEN 83 BOL CLM ECU PER 84 BZC BZN.
*Palma maripa Aubl., Hist. Pl. Guiane 2: 974 (1775). *Scheelea maripa* (Aubl.) H.Wendl. in O.C.E.de Kerchove de Denterghem, Palmiers: 256 (1878). *Maximiliana maripa* (Aubl.) Drude in C.F.P.von Martius & auct. suc. (eds.), Fl. Bras. 3(2): 452 (1881). *Englerophoenix maripa* (Aubl.) Kuntze, Revis. Gen. Pl. 2: 728 (1891).
Maximiliana regia Mart., Hist. Nat. Palm. 2: 132 (1826). *Englerophoenix regia* (Mart.) Kuntze, Revis. Gen. Pl. 2: 728 (1891). *Temenia regia* (Mart.) O.F.Cook, Natl. Hort. Mag. 18: 276 (1939). *Attalea regia* (Mart.) Wess.Boer, Indig. Palms Surin.: 150 (1965).
Maximiliana elegans H.Karst., Linnaea 28: 271 (1856).
Maximiliana martiana H.Karst., Linnaea 28: 273 (1856).
Maximiliana caribaea Griseb. & H.Wendl. in A.H.R.Grisebach, Fl. Brit. W. I.: 522 (1864). *Englerophoenix caribaeum* (Griseb. & H.Wendl.) Kuntze, Revis. Gen. Pl. 2: 728 (1891).
Maximiliana tetrasticha Drude in C.F.P.von Martius & auct. suc. (eds.), Fl. Bras. 3(2): 455 (1881). *Scheelea tetrasticha* (Drude) Burret, Notizbl. Bot. Gart. Berlin-Dahlem 10: 667 (1929).
Maximiliana longirostrata Barb.Rodr., Vellosia, ed. 2, 1: 112 (1891). *Englerophoenix longirostrata* (Barb.Rodr.) Barb.Rodr., Sert. Palm. Brasil. 1: 77 (1903).
Maximiliana macrogyne Burret, Notizbl. Bot. Gart. Berlin-Dahlem 10: 692 (1929).
Maximiliana macropetala Burret, Notizbl. Bot. Gart. Berlin-Dahlem 10: 699 (1929). *Attalea macropetala* (Burret) Wess.Boer, Indig. Palms Surin.: 155 (1965).
Maximiliana stenocarpa Burret, Notizbl. Bot. Gart. Berlin-Dahlem 10: 696 (1929).
Attalea cryptanthera Wess.Boer, Pittieria 17: 310 (1988).

Attalea maripensis (Glassman) Zona, Palms 46: 133 (2002).
French Guiana. 82 FRG.
*Scheelea maripensis Glassman, Illinois Biol. Monogr. 59: 140 (1999).

Attalea microcarpa Mart. in A.D.d'Orbigny, Voy. Amér. Mér. 7(3): 125 (1844). *Orbignya microcarpa* (Mart.) Burret, Notizbl. Bot. Gart. Berlin-Dahlem 10: 507 (1929).
S. Trop. America. 82 FRG GUY SUR VEN 83 CLM PER 84 BZN.
Attalea agrestis Barb.Rodr., Enum. Palm. Nov.: 42 (1875). *Orbignya agrestis* (Barb.Rodr.) Burret, Notizbl. Bot. Gart. Berlin-Dahlem 10: 511 (1929).
Orbignya sagotii Trail ex Thurn, Timehri 3: 276 (1884). *Attalea sagotii* (Trail ex Thurn) Wess.Boer, Indig. Palms Surin.: 162 (1965).
Orbignya sabulosa Barb.Rodr., Vellosia 1: 54 (1888).
Orbignya polysticha Burret, Notizbl. Bot. Gart. Berlin-Dahlem 11: 324 (1932). *Attalea polysticha* (Burret) Wess.Boer, Pittieria 17: 301 (1988).

Attalea × minarum (Balick, A.B.Anderson & Med.-Costa) Zona, Palms 46: 133 (2002). A. compta × A. vitrivir.
Brazil (Minas Gerais). 84 BZL.
**× Attabignya minarum* Balick, A.B.Anderson & Med.-Costa, Brittonia 39: 27 (1987).

Attalea monosperma Barb.Rodr. = **Attalea spectabilis** Mart.

Attalea moorei (Glassman) Zona, Palms 46: 133 (2002).
N. Peru. 83 PER.
*Scheelea moorei Glassman, Illinois Biol. Monogr. 59: 127 (1999).

Attalea nucifera H.Karst., Linnaea 28: 255 (1856).
Colombia (Río Magdalena valley). 83 CLM.

Attalea oleifera Barb.Rodr., Revista Brasil. (1879-81) 7: 123 (1881).
NE. Brazil. 84 BZE.
Attalea burretiana Bondar, Bol. Inst. Centr. Fomento Econ. Bahia 12: 30, 63 (1942).
Attalea concentrista Bondar, Bol. Inst. Centr. Fomento Econ. Bahia 13: 63 (1942).

Attalea osmantha (Barb.Rodr.) Wess.Boer, Pittieria 17: 318 (1988).
Trinidad, Tobago, N. Venezuela. 81 TRT 82 VEN.
**Scheelea osmantha* Barb.Rodr., Pl. Jard. Rio de Janeiro 4: 24 (1894).
Scheelea urbaniana Burret, Notizbl. Bot. Gart. Berlin-Dahlem 10: 672 (1929).
Scheelea curvifrons L.H.Bailey, Gentes Herb. 7: 443 (1947).

Attalea parviflora Barb.Rodr. = **Attalea anisitsiana** (Barb.Rodr.) Zona

Attalea peruviana Zona, Palms 46: 133 (2002).
N. Peru. 83 PER.
**Scheelea tessmannii* Burret, Notizbl. Bot. Gart. Berlin-Dahlem 10: 682 (1929).

Attalea phalerata Mart. ex Spreng., Syst. Veg. 2: 624 (1825). *Scheelea phalerata* (Mart. ex Spreng.) Burret, Notizbl. Bot. Gart. Berlin-Dahlem 10: 541 (1929).
WC. Brazil. 84 BZC.

var. **phalerata**
WC. Brazil. 84 BZC.
Attalea excelsa Mart., Hist. Nat. Palm. 2: 138 (1826).
Maximiliana princeps Mart. in A.D.d'Orbigny, Voy. Amér. Mér. 8: t. 4 (1842), nom. nud.
Scheelea princeps var. *corumbaensis* Barb.Rodr., Palm. Mattogross.: 66 (1898). *Scheelea corumbaensis* (Barb.Rodr.) Barb.Rodr., Sert. Palm. Brasil. 1: 54 (1903).
Scheelea martiana Burret, Notizbl. Bot. Gart. Berlin-Dahlem 10: 541 (1929).
Scheelea microspadix Burret, Notizbl. Bot. Gart. Berlin-Dahlem 15: 104 (1940).

var. **concinna** L.R.Moreno & O.I.Moreno, Revista Soc. Boliv. Bot. 4: 68 (2003).
Bolivia. 83 BOL.

Attalea × piassabossu Bondar, Bol. Inst. Centr. Fomento Econ. Bahia 13: 29, 61 (1942). A. funifera × A. oleifera.
Brazil (Bahia). 84 BZE.

Attalea pindobassu Bondar, Bol. Inst. Centr. Fomento Econ. Bahia 13: 30, 62 (1942).
Brazil (Bahia). 84 BZE.

Attalea pixuna Barb.Rodr. = **Attalea spectabilis** Mart.

Attalea plowmanii (Glassman) Zona, Palms 46: 133 (2002).
N. Peru. 83 PER.
**Scheelea plowmanii* Glassman, Illinois Biol. Monogr. 59: 144 (1999).

Attalea polysticha (Burret) Wess.Boer = **Attalea microcarpa** Mart.

Attalea princeps Mart. in A.D.d'Orbigny, Voy. Amér. Mér. 7(3): 113 (1844). *Scheelea princeps* (Mart.) H.Karst., Linnaea 28: 269 (1856).
Bolivia. 83 BOL.

Attalea purpurea Linden = ? [83 CLM]

Attalea puruensis Linden = ? [84 BZN]

Attalea pycnocarpa Wess.Boer = **Attalea butyracea** (Mutis ex L.f.) Wess.Boer

Attalea racemosa Spruce, J. Linn. Soc., Bot. 11: 166 (1869). *Orbignya racemosa* (Spruce) Drude in C.F.P.von Martius & auct. suc. (eds.), Fl. Bras. 3(2): 448 (1881).
S. Trop. America. 82 VEN 83 CLM PER 84 BZN.
Attalea ferruginea Burret, Notizbl. Bot. Gart. Berlin-Dahlem 11: 1044 (1934).

Attalea regia (Mart.) Wess.Boer = **Attalea maripa** (Aubl.) Mart.

Attalea rhynchocarpa Burret, Notizbl. Bot. Gart. Berlin-Dahlem 12: 617 (1935).
Colombia. 83 CLM. — Provisionally accepted.

Attalea rossii Lodd. ex Loudon = ? [84]

Attalea rostrata Oerst., Vidensk. Meddel. Dansk Naturhist. Foren. Kjøbenhavn 1858: 50 (1858). *Scheelea rostrata* (Oerst.) Burret, Notizbl. Bot. Gart. Berlin-Dahlem 10: 541 (1929).
C. America. 79 COS GUA HON NIC PAN.
Scheelea costaricensis Burret, Notizbl. Bot. Gart. Berlin-Dahlem 10: 684 (1929).
Scheelea preussii Burret, Notizbl. Bot. Gart. Berlin-Dahlem 10: 678 (1929).
Scheelea zonensis L.H.Bailey, Gentes Herb. 3: 36 (1933).

Attalea sagotii (Trail ex Thurn) Wess.Boer = **Attalea microcarpa** Mart.

Attalea salazarii (Glassman) Zona, Palms 46: 133 (2002).
N. Peru. 83 PER.
**Scheelea salazarii* Glassman, Illinois Biol. Monogr. 59: 146 (1999).

Attalea salvadorensis Glassman, Illinois Biol. Monogr. 59: 63 (1999).
Brazil (Bahia). 84 BZE.

Attalea seabrensis Glassman, Illinois Biol. Monogr. 59: 50 (1999).
Brazil (Bahia). 84 BZE.

Attalea septuagenata Dugand, Mutisia 18: 3 (1953).
Colombia (Amazonas). 83 CLM.

Attalea speciosa Mart., Hist. Nat. Palm. 2: 138 (1826). *Orbignya speciosa* (Mart.) Barb.Rodr., Sert. Palm. Brasil. 1: 60 (1903).
Guianas to Bolivia. 82 GUY SUR 83 BOL 84 BZE BZL BZN.
Orbignya phalerata Mart. in A.D.d'Orbigny, Voy. Amér. Mér. 7(3): 126 (1844). *Heptantra phalerata* (Mart.) O.F.Cook, Natl. Hort. Mag. 18: 277 (1939).
Attalea glassmanii Zona, Palms 46: 132 (2002).
Orbignya cuci Kunth ex H.Wendl. in O.C.E.de Kerchove de Denterghem, Palmiers: 252 (1878).

Orbignya lydiae Drude in C.F.P.von Martius & auct. suc. (eds.), Fl. Bras. 3(2): 448 (1881). *Attalea lydiae* (Drude) Barb.Rodr., Sert. Palm. Brasil. 1: 65 (1903).

Orbignya martiana Barb.Rodr., Palm. Mattogross.: 68 (1898).

Orbignya huebneri Burret, Notizbl. Bot. Gart. Berlin-Dahlem 10: 501 (1929).

Orbignya macropetala Burret, Notizbl. Bot. Gart. Berlin-Dahlem 10: 507 (1929).

Orbignya barbosiana Burret, Notizbl. Bot. Gart. Berlin-Dahlem 11: 690 (1932).

Attalea spectabilis Mart., Hist. Nat. Palm. 2: 136 (1826). *Orbignya spectabilis* (Mart.) Burret, Notizbl. Bot. Gart. Berlin-Dahlem 10: 508 (1929). Brazil (Pará). 84 BZN.
> *Attalea monosperma* Barb.Rodr., Enum. Palm. Nov.: 42 (1875).
> *Attalea pixuna* Barb.Rodr., Enum. Palm. Nov.: 43 (1875).
> *Orbignya pixuna* (Barb.Rodr.) Barb.Rodr., Prot.-App. Enum. Palm. Nov.: 49 (1879).
> *Attalea macoupi* Sagot ex Drude in C.F.P.von Martius & auct. suc. (eds.), Fl. Bras. 3(2): 441 (1881).

Attalea spinosa Meyen = ? **[83 PER]**

Attalea × teixeirana (Bondar) Zona, Palms 46: 133 (2002). A. eichleri × A. speciosa. NE. Brazil. 84 BZE.
> *Orbignya × teixeirana* Bondar, Arch. Jard. Bot. Rio de Janeiro 13: 58 (1954).

Attalea tessmannii Burret, Notizbl. Bot. Gart. Berlin-Dahlem 10: 538 (1929). Brazil (Acre) to N. Peru. 83 PER 84 BZN.

Attalea tiasse Linden = ? **[84]**

Attalea transitiva Barb.Rodr. = **Attalea attaleoides** (Barb.Rodr.) Wess.Boer

Attalea uberrima Dugand = **Attalea amygdalina** Kunth

Attalea venatorum Mart. = ?

Attalea victoriana Dugand = **Attalea amygdalina** Kunth

Attalea vitrivir Zona, Palms 46: 133 (2002). Brazil (Bahia, Minas Gerais). 84 BZE BZL.
> **Orbignya oleifera* Burret, Notizbl. Bot. Gart. Berlin-Dahlem 14: 240 (1938).

Attalea × voeksii Noblick ex Glassman, Illinois Biol. Monogr. 59: 191 (1999). A. funifera × A. humilis. Brazil (Bahia). 84 BZE.

Attalea wallisii Huber = **Attalea butyracea** (Mutis ex L.f.) Wess.Boer

Attalea weberbaueri (Burret) Zona, Palms 46: 133 (2002). Peru (Junín). 83 PER.
> **Scheelea weberbaueri* Burret, Notizbl. Bot. Gart. Berlin-Dahlem 10: 659 (1929).

Attalea wesselsboeri (Glassman) Zona, Palms 46: 133 (2002). Venezuela (Barinas). 82 VEN.
> **Scheelea wesselsboeri* Glassman, Illinois Biol. Monogr. 59: 170 (1999).

Augustinea

Augustinea H.Karst. = **Bactris** Jacq. ex Scop.

Augustinea balanoidea Oerst. = **Bactris major** var. **major**

Augustinea major (Jacq.) H.Karst. = **Bactris major** Jacq.

Augustinea ovata Oerst. = **Bactris major** var. **major**

Avoira

Avoira Giseke = **Astrocaryum** G.Mey.

Avoira conanam Giseke = ? **[82 FRG]**

Avoira nodosa Giseke = ? **[82 FRG]**

Avoira scandens Giseke = ? **[82 FRG]**

Avoira sylvestris Giseke = ? **[82 FRG]**

Avoira uliginosa Giseke = ? **[82 FRG]**

Avoira vulgaris Giseke = ? **[82 FRG] Astrocaryum sp.**

Bactris

Bactris Jacq. ex Scop., Intr. Hist. Nat.: 70 (1777). 76 species, Mexico to Trop. America. 79 80 81 82 83 84 85.
> *Guilielma* Mart., Palm. Fam.: 21 (1824).
> *Guilielma* Link, Handbuch 1: 259 (1829).
> *Augustinea* H.Karst., Linnaea 28: 395 (1856).
> *Pyrenoglyphis* H.Karst., Fl. Columb. 2: 141 (1869).
> *Amylocarpus* Barb.Rodr., Contr. Jard. Bot. Rio de Janeiro 3: 69 (1902).
> *Yuyba* L.H.Bailey, Gentes Herb. 7: 416 (1947).

Bactris acanthocarpa Mart., Hist. Nat. Palm. 2: 92 (1826). S. Trop. America. 82 FRG GUY SUR VEN 83 BOL CLM ECU PER 84 BZC BZE BZL BZN.

var. **acanthocarpa**
NE. Brazil (to Espírito Santo). 84 BZE BZL.
> *Bactris bicuspidata* Spruce, J. Linn. Soc., Bot. 11: 146 (1871). *Pyrenoglyphis bicuspidata* (Spruce) Burret, Repert. Spec. Nov. Regni Veg. 34: 253 (1934).
> *Bactris mindellii* Barb.Rodr., Pl. Jard. Rio de Janeiro 6: 19 (1898).

var. *crispata* Drude = **Bactris acanthocarpa** var. **exscapa** Barb.Rodr.

var. **exscapa** Barb.Rodr., Enum. Palm. Nov.: 31 (1875). *Bactris exscapa* (Barb.Rodr.) Barb.Rodr., Sert. Palm. Brasil. 2: 9 (1903). S. Trop. America. 82 GUY VEN 83 BOL CLM ECU PER 84 BZC BZN.
> *Astrocaryum humile* Wallace, Palm Trees Amazon: 115 (1853). *Bactris humilis* (Wallace) Burret, Repert. Spec. Nov. Regni Veg. 34: 194 (1934).
> *Bactris interruptepinnata* Barb.Rodr., Enum. Palm. Nov.: 37 (1875).
> *Bactris acanthocarpa* var. *crispata* Drude in C.F.P.von Martius & auct. suc. (eds.), Fl. Bras. 3(2): 350 (1881).
> *Bactris aculeifera* Drude in C.F.P.von Martius & auct. suc. (eds.), Fl. Bras. 3(2): 352 (1881).
> *Bactris tarumanensis* Barb.Rodr., Vellosia 1: 44 (1888).
> *Bactris fragae* Lindm., Bih. Kongl. Svenska Vetensk.-Akad. Handl. 26(5): 11 (1901).
> *Bactris macrocalyx* Burret, Repert. Spec. Nov. Regni Veg. 34: 194 (1934).
> *Bactris microcalyx* Burret, Repert. Spec. Nov. Regni Veg. 34: 195 (1934).
> *Bactris pinnatisecta* Burret, Repert. Spec. Nov. Regni Veg. 34: 191 (1934).
> *Bactris leptochaete* Burret, Notizbl. Bot. Gart. Berlin-Dahlem 12: 620 (1935).
> *Bactris devia* H.E.Moore, Gentes Herb. 8: 157 (1949).

var. **intermedia** A.J.Hend., Palms Amazon: 174 (1995). Guianas to N. Brazil. 82 FRG SUR 84 BZN.

var. **trailiana** (Barb.Rodr.) A.J.Hend., Fl. Neotrop. Monogr. 79: 28 (2000).
S. Trop. America. 82 VEN 83 BOL CLM 84 BZN.
Bactris trailiana Barb.Rodr., Enum. Palm. Nov.: 27 (1875). *Bactris acanthocarpa* subsp. *trailiana* (Barb.Rodr.) Trail, J. Bot. 15: 46 (1877).

subsp. *trailiana* (Barb.Rodr.) Trail = **Bactris acanthocarpa** var. **trailiana** (Barb.Rodr.) A.Hend.

Bactris acanthocarpoides Barb.Rodr., Enum. Palm. Nov.: 31 (1875).
Guianas to N. Brazil. 82 FRG GUY SUR 84 BZN.

Bactris acanthocnemis Mart. = **Bactris simplicifrons** Mart.

Bactris acanthophylla Mart. = **Aiphanes minima** (Gaertn.) Burret

Bactris acanthospatha (Trail) Trail ex Drude = **Bactris macroacantha** Mart.

Bactris actinoneura Drude & Trail ex Drude = **Bactris maraja** var. **trichospatha**(Trail) A.Hend.

Bactris aculeifera Drude = **Bactris acanthocarpa** var. **exscapa** Barb.Rodr.

Bactris acuminata Liebm. ex Mart. = **Bactris mexicana** Mart.

Bactris albonotata L.H.Bailey = **Bactris major** Jacq.

Bactris alleniana L.H.Bailey = **Bactris glandulosa** Oerst.

Bactris amoena Burret = **Bactris simplicifrons** Mart.

Bactris ana-juliae Cascante, Palms 44: 146 (2000).
Costa Rica. 80 COS.

Bactris angustifolia Dammer = **Bactris bifida** Mart.

Bactris anisitsii Barb.Rodr. = **Bactris glaucescens** Drude

Bactris arenaria Barb.Rodr. = **Bactris simplicifrons** Mart.

Bactris aristata Mart. = **Bactris fissifrons** Mart.

Bactris armata Barb.Rodr. = **Bactris maraja** Mart.

Bactris arundinacea (Trail) Drude = **Bactris tomentosa** Mart.

Bactris atrox Burret = **Bactris hirta** var. **pectinata** (Mart.) Govaerts

Bactris aubletiana Trail, J. Bot. 14: 372 (1876).
Guianas. 82 FRG SUR.

Bactris augustinea L.H.Bailey = **Bactris major** Jacq.

Bactris aureodrupa L.H.Bailey = **Bactris gracilior** Burret

Bactris bahiensis Noblick ex A.J.Hend., Fl. Neotrop. Monogr. 79: 32 (2000).
E. Brazil. 84 BZE BZL.

Bactris baileyana H.E.Moore = **Bactris glandulosa** var. **baileyana** (H.E.Moore) Nevers

Bactris balanoidea (Oerst.) H.Wendl. = **Bactris major** Jacq.

Bactris balanophora Spruce, J. Linn. Soc., Bot. 11: 146 (1871).
Colombia to N. Brazil. 82 VEN 83 CLM 84 BZN.
Astrocaryum aculeatum Wallace, Palm Trees Amazon: 111 (1853), nom. illeg.

Bactris barronis L.H.Bailey, Gentes Herb. 3: 101 (1933).
Panama to Colombia. 80 PAN 83 CLM.

Bactris beata L.H.Bailey = **Bactris major** Jacq.

Bactris bella Burret = **Bactris maraja** var. **juruensis** (Trail) A.Hend.

Bactris bergantina Steyerm. = **Bactris setulosa** H.Karst.

Bactris bicuspidata Spruce = **Bactris acanthocarpa** Mart.

Bactris bidentula Spruce, J. Linn. Soc., Bot. 11: 146 (1871).
S. Trop. America. 82 VEN 83 CLM PER 84 BZC BZN.
Bactris palustris Barb.Rodr., Enum. Palm. Nov.: 36 (1875).
Bactris nigrispina Barb.Rodr., Palm. Hassler.: 15 (1900).

Bactris bifida Mart., Hist. Nat. Palm. 2: 105 (1826). *Pyrenoglyphis bifida* (Mart.) Burret, Repert. Spec. Nov. Regni Veg. 34: 183, 242 (1933).
W. South America to N. Brazil. 83 CLM PER 84 BZN.
Bactris bifida var. *humaitensis* Trail, J. Bot. 15: 47 (1877).
Bactris bifida var. *puruensis* Trail, J. Bot. 15: 47 (1877).
Bactris angustifolia Dammer, Verh. Bot. Vereins Prov. Brandenburg 48: 128 (1906 publ. 1907).

Bactris bifida Oerst. = **Bactris glandulosa** Oerst.

Bactris bifida var. *humaitensis* Trail = **Bactris bifida** Mart.

Bactris bifida var. *puruensis* Trail = **Bactris bifida** Mart.

Bactris bijugata Burret = **Bactris maraja** Mart.

Bactris bradei Burret = **Astrocaryum campestre** Mart.

Bactris brevifolia Spruce = **Bactris simplicifrons** Mart.

Bactris broadwayi L.H.Bailey = **Bactris major** Jacq.

Bactris brongniartii Mart. in A.D.d'Orbigny, Voy. Amér. Mér. 7(3): 59 (1844). *Pyrenoglyphis brongniartii* (Mart.) Burret, Repert. Spec. Nov. Regni Veg. 34: 251 (1934).
S. Trop. America. 82 FRG GUY SUR VEN 83 BOL CLM PER 84 BZC BZN.
Bactris pallidispina Mart. in A.D.d'Orbigny, Voy. Amér. Mér. 7(3): 62 (1844). *Pyrenoglyphis pallidispina* (Mart.) Burret, Repert. Spec. Nov. Regni Veg. 34: 249 (1934).
Bactris flavispina Heynh., Alph. Aufz. Gew. 2: 57 (1846), nom. inval.
Guilielma tenera H.Karst., Linnaea 28: 399 (1856). *Bactris tenera* (H.Karst.) H.Wendl. in O.C.E.de Kerchove de Denterghem, Palmiers: 234 (1878). *Pyrenoglyphis tenera* (H.Karst.) Burret, Repert. Spec. Nov. Regni Veg. 34: 250 (1934).
Bactris maraya-acu Barb.Rodr., Enum. Palm. Nov.: 36 (1875).
Bactris rivularis Barb.Rodr., Enum. Palm. Nov.: 36 (1875). *Pyrenoglyphis rivularis* (Barb.Rodr.) Burret, Repert. Spec. Nov. Regni Veg. 34: 251 (1934).
Bactris piscatorum Wedd. ex Drude in C.F.P.von Martius & auct. suc. (eds.), Fl. Bras. 3(2): 354 (1881). *Pyrenoglyphis piscatorum* (Wedd. ex Drude) Burret, Repert. Spec. Nov. Regni Veg. 34: 251 (1934).
Bactris cuyabaensis Barb.Rodr., Palm. Mattogross.: 42 (1898).
Bactris stictacantha Burret, Notizbl. Bot. Gart. Berlin-Dahlem 11: 18 (1930).
Pyrenoglyphis microcarpa Burret, Repert. Spec. Nov. Regni Veg. 34: 250 (1934). *Bactris burretii* Glassman, Rhodora 65: 259 (1963).

Bactris burretii Glassman = **Bactris brongniartii** Mart.

Bactris campestris Poepp. in C.F.P.von Martius, Hist. Nat. Palm. 2: 146 (1837).
Trinidad, Tobago, S. Trop. America. 81 TRT 82 FRG GUY SUR VEN 83 CLM 84 BZE BZN.
Bactris leptocarpa Trail ex Thurn, Timehri 3: 253 (1884).
Bactris savannarum Britton, Bull. Torrey Bot. Club 50: 51 (1923).
Bactris lanceolata Burret, Notizbl. Bot. Gart. Berlin-Dahlem 10: 1023 (1930).

Bactris capillacea (Trail) Drude = **Bactris tomentosa** Mart.

Bactris capinensis Huber = **Bactris tomentosa** Mart.

Bactris caribaea H.Karst. = **Bactris gasipaes** var. **chichagui** (H.Karst.) A.Hend.

Bactris carolensis Spruce = **Bactris simplicifrons** Mart.

Bactris caryotifolia Mart., Hist. Nat. Palm. 2: 106 (1826).
Brazil (Bahia to Rio de Janeiro). 84 BZE BZL.

Bactris cateri L.H.Bailey = **Bactris major** Jacq.

Bactris caudata H.Wendl. ex Burret, Repert. Spec. Nov. Regni Veg. 34: 230 (1934).
C. America. 80 COS NIC PAN.
Bactris dasychaeta Burret, Repert. Spec. Nov. Regni Veg. 34: 215 (1934).

Bactris chaetochlamys Burret = **Bactris maraja** var. **trichospatha** (Trail) A.Hend.

Bactris chaetophylla Mart. = **Bactris plumeriana** Mart.

Bactris chaetorhachis Mart. = **Bactris major** Jacq.

Bactris chaetospatha Mart. = **Bactris maraja** var. **chaetospatha** (Mart.) A.Hend.

Bactris chapadensis Barb.Rodr. = **Bactris major** var. **infesta** (Mart.) Drude

Bactris chloracantha Poepp. = **Bactris maraja** Mart.

Bactris charnleyae Nevers, A.J.Hend. & Grayum, Proc. Calif. Acad. Sci. 49: 176 (1996).
Panama (San Blas). 80 PAN.

Bactris chaveziae A.J.Hend., Fl. Neotrop. Monogr. 79: 49 (2000).
Peru to Bolivia and Brazil (Acre). 83 BOL PER 84 BZN.
Bactris concinna var. *sigmoidea* A.J.Hend., Palms Amazon: 185 (1995).

Bactris chlorocarpa Burret = **Bactris maraja** var. **juruensis** (Trail) A.Hend.

Bactris ciliata (Ruiz & Pav.) Mart. = **Bactris gasipaes** Kunth

Bactris circularis L.H.Bailey = **Bactris setulosa** H.Karst.

Bactris coccinea Barb.Rodr. = **Bactris gasipaes** var. **chichagui** (H.Karst.) A.Hend.

Bactris cohune S.Watson = **Astrocaryum mexicanum** Liebm. ex Mart.

Bactris coloniata L.H.Bailey, Gentes Herb. 3: 106 (1933).
Panama to Peru. 80 PAN 83 CLM ECU PER.

Bactris coloradonis L.H.Bailey, Gentes Herb. 3: 104 (1933).
Costa Rica to W. Ecuador. 80 COS PAN 83 CLM ECU.
Bactris porschiana Burret, Ann. Naturhist. Mus. Wien 46: 229 (1933).

Bactris concinna Mart., Hist. Nat. Palm. 2: 99 (1826).
Pyrenoglyphis concinna (Mart.) Burret, Repert. Spec. Nov. Regni Veg. 34: 242 (1934).
W. South America to N. Brazil. 83 BOL CLM ECU PER 84 BZN.
Bactris concinna var. *inundata* Spruce, J. Linn. Soc., Bot. 11: 154 (1871).
Bactris concinna subsp. *depauperata* Trail, J. Bot. 6: 48 (1877).

Bactris concinna subsp. *depauperata* Trail = **Bactris concinna** Mart.

Bactris concinna var. *inundata* Spruce = **Bactris concinna** Mart.

Bactris concinna var. *sigmoidea* A.Hend. = **Bactris chaveziae** A.Hend.

Bactris confluens Linden & H.Wendl. = **?** [82 VEN]

Bactris confluens var. *acanthospatha* Trail = **Bactris macroacantha**

Bactris corazillo H.Wendl. = **Bactris corossilla** H.Karst.

Bactris constanciae Barb.Rodr., Enum. Palm. Nov.: 37 (1875).
Guianas to N. Brazil. 82 FRG SUR 84 BZN.

Bactris corossilla H.Karst., Linnaea 28: 407 (1856).
Venezuela to Peru. 82 VEN 83 CLM ECU PER 84 BZN.
Bactris corazillo H.Wendl., Index Palm.: 5 (1854), orth. var.
Bactris cuesco Engl., Linnaea 33: 665 (1865).
Bactris duplex H.E.Moore, Gentes Herb. 8: 160 (1949).
Bactris duidae Steyerm., Fieldiana, Bot. 28: 73 (1951).
Bactris venezuelensis Steyerm., Fieldiana, Bot. 28: 80 (1951).

Bactris cruegeriana Griseb. = **Bactris major** Jacq.

Bactris cubensis Burret, Kongl. Svenska Vetenskapsakad. Handl. 6(7): 25 (1929).
E. Cuba. 81 CUB.

Bactris cuesa Crueg. ex Griseb. = **Bactris setulosa** H.Karst.

Bactris cuesco Engl. = **Bactris corossilla** H.Karst.

Bactris curuena (Trail) Trail ex Drude = **Bactris major** var. **infesta** (Mart.) Drude

Bactris cuspidata Mart., Hist. Nat. Palm. 2: 101 (1826).
Amylocarpus cuspidatus (Mart.) Barb.Rodr., Contr. Jard. Bot. Rio de Janeiro 3: 72 (1902).
Guianas to N. Brazil. 82 FRG SUR 84 BZN.
Bactris mitis Mart., Hist. Nat. Palm. 2: 102 (1826).
Bactris cuspidata var. *mitis* (Mart.) Drude in C.F.P.von Martius & auct. suc. (eds.), Fl. Bras. 3(2): 329 (1881). *Amylocarpus mitis* (Mart.) Barb.Rodr., Contr. Jard. Bot. Rio de Janeiro 3: 72 (1902).
Bactris floccosa Spruce, J. Linn. Soc., Bot. 11: 146 (1871).
Bactris marajay Barb.Rodr., Enum. Palm. Nov.: 29 (1875). *Amylocarpus marajay* (Barb.Rodr.) Barb. Rodr., Contr. Jard. Bot. Rio de Janeiro 3: 72 (1902).
Bactris cuspidata var. *angustipinnata* Trail, J. Bot. 6: 4 (1877).
Bactris cuspidata var. *coriacea* Trail, J. Bot. 1877: 4 (1877).

Bactris cuspidata var. *angustipinnata* Trail = **Bactris cuspidata** Mart.

Bactris cuspidata var. *coriacea* Trail = **Bactris cuspidata** Mart.

Bactris cuspidata var. *mitis* (Mart.) Drude = **Bactris cuspidata** Mart.

Bactris cuspidata var. *tenuis* (Wallace) Drude = **Bactris simplicifrons** Mart.

Bactris cuvaro H.Karst. = **Bactris setulosa** H.Karst.

Bactris cuyabaensis Barb.Rodr. = **Bactris brongniartii** Mart.

Bactris cyagroides Barb.Rodr. = **Bactris syagroides** Barb.Rodr. & Trail

Bactris dahlgreniana Govaerts = **Bactris gasipaes** var. **chichagui** (H.Karst.) A.Hend.

Bactris dakamana (L.H.Bailey) Glassman = **Bactris simplicifrons** Mart.

Bactris dasychaeta Burret = **Bactris caudata** H.Wendl. ex Burret

Bactris demerarana L.H.Bailey = **Bactris major** Jacq.

Bactris devia H.E.Moore = **Bactris acanthocarpa** var. **exscapa***Bactris divisicupula* L.H.Bailey = **Bactris maraja** Mart.

Bactris dianeura Burret, Repert. Spec. Nov. Regni Veg. 34: 217 (1934).
C. America. 80 COS NIC PAN.

Bactris duidae Steyerm. = **Bactris corossilla** H.Karst.

Bactris duplex H.E.Moore = **Bactris corossilla** H.Karst.

Bactris elatior Wallace = **Bactris maraja** Mart.

Bactris elegans Barb.Rodr. & Trail in J.Barbosa Rodrigues, Enum. Palm. Nov.: 35 (1875).
S. Trop. America. 82 FRG GUY SUR 83 BOL CLM 84 BZN.

Bactris elegantissima Burret, Notizbl. Bot. Gart. Berlin-Dahlem 15: 4 (1940).

Bactris elegantissima Burret = **Bactris elegans** Barb. Rodr. & Trail

Bactris ellipsoidalis L.H.Bailey = **Bactris major** Jacq.

Bactris ericetina Barb.Rodr. = **Bactris hirta** Mart.

Bactris erosa Mart. = **Aiphanes minima** (Gaertn.) Burret

Bactris erostrata Burret = **Bactris maraja** Mart.

Bactris escragnollei Glaz. ex Burret = **Bactris setosa** Mart.

Bactris essequiboensis (L.H.Bailey) Glassman = **Bactris simplicifrons** Mart.

Bactris eumorpha Trail = **Bactris tomentosa** Mart.

Bactris eumorpha subsp. *arundinacea* Trail = **Bactris tomentosa** Mart.

Bactris exaltata Barb.Rodr. = **Bactris major** var. **infesta** (Mart.) Drude

Bactris exscapa (Barb.Rodr.) Barb.Rodr. = **Bactris acanthocarpa** var. **exscapa***Barb.Rodr.

Bactris falcata J.R.Johnst. = **Bactris setulosa** H.Karst.

Bactris faucium Mart. in A.D.d'Orbigny, Voy. Amér. Mér. 7(3): 60 (1844).
W. Bolivia. 83 BOL.

Bactris ferruginea Burret, Repert. Spec. Nov. Regni Veg. 34: 222 (1934).
E. Brazil. 84 BZE BZL.

Bactris fissifrons Mart., Hist. Nat. Palm. 2: 103 (1826).
W. South America to NW. Brazil. 83 CLM ECU PER 84 BZN.
Bactris aristata Mart., Hist. Nat. Palm. 2: 97 (1826).

Pyrenoglyphis aristata (Mart.) Burret, Repert. Spec. Nov. Regni Veg. 34: 242 (1934).
Bactris fissifrons var. *robusta* Trail, J. Bot. 6: 9 (1877).

Bactris fissifrons var. *robusta* Trail = **Bactris fissifrons** Mart.

Bactris flavispina Heynh. = **Bactris brongniartii** Mart.

Bactris floccosa Spruce = **Bactris cuspidata** Mart.

Bactris formosa Barb.Rodr. = **Bactris hirta** var. **pectinata** (Mart.) Govaerts

Bactris fragae Lindm. = **Bactris acanthocarpa** var. **exscapa** Barb.Rodr.

Bactris fusca Oerst. = **Bactris glandulosa** Oerst.

Bactris fuscospina L.H.Bailey = **Bactris maraja** Mart.

Bactris gasipaes Kunth in F.W.H.A.von Humboldt, A.J.A.Bonpland & C.S.Kunth, Nov. Gen. Sp. 1: 302 (1816). *Guilielma gasipaes* (Kunth) L.H.Bailey, Gentes Herb. 2: 187 (1930).
C. & S. Trop. America. 80 COS HON NIC PAN 82 FRG VEN 83 BOL CLM ECU PER 84 BZC BZN. — Cooked fruits edible.

var. **chichagui** (H.Karst.) A.J.Hend., Fl. Neotrop. Monogr. 79: 73 (2000).
S. Trop. America. 82 VEN 83 BOL CLM ECU PER 84 BZC BZN.
Guilielma macana Mart. in A.D.d'Orbigny, Voy. Amér. Mér. 7(3): 74 (1844). *Bactris macana* (Mart.) Pittier, Man. Pl. Usual. Venez.: 276 (1926).
Bactris caribaea H.Karst., Linnaea 28: 403 (1856). *Guilielma caribaea* (H.Karst.) H.Wendl. in O.C.E.de Kerchove de Denterghem, Palmiers: 246 (1878).
*Bactris speciosa var. *chichagui* H.Karst., Linnaea 28: 402 (1856).
Guilielma mattogrossensis Barb.Rodr., Palm. Mattogross.: 33 (1898). *Bactris coccinea* Barb.Rodr., Contr. Jard. Bot. Rio de Janeiro 4: 110 (1907).
Guilielma microcarpa Huber, Bol. Mus. Goeldi Paraense Hist. Nat. Ethnogr. 4: 476 (1904). *Bactris dahlgreniana* Govaerts, World Checklist Seed Pl. 2(1): 9 (1996).
Guilelma microcarpa Huber, Bull. Herb. Boissier, II, 6: 270 (1906).

var. **gasipaes**
C. & S. Trop. America. 80 COS HON NIC PAN 82 FRG VEN 83 BOL CLM ECU PER 84 BZN. — Cooked fruits edible.
Martinezia ciliata Ruiz & Pav., Syst. Veg. Fl. Peruv. Chil.: 295 (1798), nom. rejic. *Bactris ciliata* (Ruiz & Pav.) Mart., Hist. Nat. Palm. 2: 95 (1826). *Guilielma ciliata* (Ruiz & Pav.) H.Wendl. in O.C.E.de Kerchove de Denterghem, Palmiers: 246 (1878).
Guilielma speciosa Mart., Hist. Nat. Palm. 2: 82 (1824). *Bactris speciosa* (Mart.) H.Karst., Linnaea 28: 402 (1856).
Guilielma insignis Mart. in A.D.d'Orbigny, Voy. Amér. Mér. 7(3): 71 (1844). *Bactris insignis* (Mart.) Baill., Hist. Pl. 13: 305 (1895).
Guilielma chontaduro H.Karst. & Triana in J.J.Triana, Nuev. Jen. Esp.: 15 (1855).
Guilielma utilis Oerst., Vidensk. Meddel. Dansk Naturhist. Foren. Kjøbenhavn 1858: 46 (1858). *Bactris utilis* (Oerst.) Benth. & Hook.f. ex Hemsl., Biol. Cent.-Amer., Bot. 3: 413 (1885).

Guilielma speciosa var. *coccinea* Barb.Rodr., Enum. Palm. Nov.: 23 (1875).

Guilielma speciosa var. *flava* Barb.Rodr., Enum. Palm. Nov.: 23 (1875).

Guilielma speciosa var. *mitis* Drude in C.F.P.von Martius & auct. suc. (eds.), Fl. Bras. 3(2): 363 (1881).

Guilielma speciosa var. *ochracea* Barb.Rodr., Vellosia 1: 40 (1888).

Bactris gastoniana Barb.Rodr., Vellosia 1: 40 (1888).
Pyrenoglyphis gastoniana (Barb.Rodr.) Burret, Repert. Spec. Nov. Regni Veg. 34: 242 (1934).
Guianas to N. Brazil. 82 FRG SUR 84 BZN.

Bactris gaviona (Trail) Trail ex Drude = **Bactris major** var. **infesta** (Mart.) Drude

Bactris geonomoides Drude = **Bactris hirta** var. **pectinata** (Mart.) Govaerts

Bactris geonomoides var. *setosa* Drude = **Bactris hirta** var. **pectinata**(Mart.) Govaerts

Bactris glandulosa Oerst., Vidensk. Meddel. Dansk Naturhist. Foren. Kjøbenhavn 1858: 184 (1859).
Costa Rica to NW. Colombia. 80 COS PAN 83 CLM.
Bactris bifida Oerst., Vidensk. Meddel. Dansk Naturhist. Foren. Kjøbenhavn 1858: 44 (1858), nom. illeg. *Bactris oerstediana* Trail, J. Bot. 15: 43 (1877), nom. illeg.

var. **baileyana** (H.E.Moore) Nevers, Proc. Calif. Acad. Sci. 49: 183 (1996).
Costa Rica to Panama. 80 COS PAN.
Bactris baileyana H.E.Moore, Gentes Herb. 8: 155 (1949).

var. **glandulosa**
Costa Rica to NW. Colombia. 80 COS PAN 83 CLM.
Bactris fusca Oerst., Vidensk. Meddel. Dansk Naturhist. Foren. Kjøbenhavn 1858: 43 (1858).
Bactris macrotricha Burret, Repert. Spec. Nov. Regni Veg. 34: 232 (1934).
Bactris alleniana L.H.Bailey, Gentes Herb. 6: 228 (1943).

Bactris glassmanii Med.-Costa ex A.J.Hend., Fl. Neotrop. Monogr. 79: 78 (2000).
NE. Brazil. 84 BZE.

Bactris glaucescens Drude in C.F.P.von Martius & auct. suc. (eds.), Fl. Bras. 3(2): 345 (1881).
WC. Brazil (incl. Rondônia) to Paraguay. 83 BOL 84 BZC BZN 85 PAR.
Bactris glaucescens var. *malanacantha* Drude in C.F.P. von Martius & auct. suc. (eds.), Fl. Bras. 3(2): 345 (1881).
Bactris anisitsii Barb.Rodr., Palm. Paraguay.: 19 (1899).
Bactris tucum Burret, Repert. Spec. Nov. Regni Veg. 34: 225 (1934).

Bactris glaucescens var. *malanacantha* Drude = **Bactris glaucescens** Drude

Bactris glazioviana Drude = **Bactris vulgaris** Barb.Rodr.

Bactris gleasonii (L.H.Bailey) Glassman = **Bactris simplicifrons** Mart.

Bactris globosa Gaertn. = **Acrocomia aculeata** (Jacq.) Lodd. ex Mart.

Bactris gracilior Burret, Repert. Spec. Nov. Regni Veg. 34: 216 (1934).
C. America. 80 COS NIC PAN.

Bactris aureodrupa L.H.Bailey, Gentes Herb. 6: 232 (1943).

Bactris gracilis Barb.Rodr. = **Bactris simplicifrons** Mart.

Bactris granariuscarpa Barb.Rodr. = **Bactris maraja** Mart.

Bactris granatensis (H.Karst.) H.Wendl. = **Bactris pilosa** H.Karst.

Bactris grayumii Nevers & A.J.Hend., Proc. Calif. Acad. Sci. 49: 188 (1996).
C. America. 80 COS NIC.

Bactris guineensis (L.) H.E.Moore, Gentes Herb. 9: 251 (1963).
C. America to Venezuela. 80 COS NIC PAN 83 VEN 83 CLM.
Cocos guineensis L., Mant. Pl. 1: 137 (1767).
Bactris minor Jacq., Select. Stirp. Amer. Hist., ed. 2: 134 (1781). *Bactris rotunda* Stokes, Bot. Mat. Med. 4: 394 (1812), nom. illeg.
Cocos acicularis Sw., Prodr.: 58 (1788).
Guilielma piriti H.Karst., Linnaea 28: 397 (1856). *Bactris piriti* (H.Karst.) H.Wendl. in O.C.E.de Kerchove de Denterghem, Palmiers: 234 (1878).
Bactris horrida Oerst., Vidensk. Meddel. Dansk Naturhist. Foren. Kjøbenhavn 1858: 41 (1858).
Bactris oraria L.H.Bailey, Gentes Herb. 6: 232 (1943).

Bactris gymnospatha Burret = **Bactris maraja** Mart.

Bactris halmoorei A.J.Hend., Fl. Neotrop. Monogr. 79: 86 (2000).
N. Brazil to Peru. 83 PER 84 BZN.

Bactris hatschbachii Noblick ex A.J.Hend., Fl. Neotrop. Monogr. 79: 87 (2000).
Brazil (São Paulo, Paraná). 84 BZL BZS.

Bactris herrerana Cascante, Palms 44: 148 (2000).
Costa Rica. 80 COS.

Bactris hirsuta Burret = **Bactris pilosa** H.Karst.

Bactris hirta Mart., Hist. Nat. Palm. 2: 105 (1826).
Amylocarpus hirtus (Mart.) Barb.Rodr., Contr. Jard. Bot. Rio de Janeiro 3: 72 (1902).
S. Trop. America. 82 FRG GUY SUR VEN 83 BOL CLM PER 84 BZE BZL BZN.

var. **hirta**
Colombia to N. Peru and Brazil (Amazonas). 83 CLM PER 84 BZN.
Bactris ericetina Barb.Rodr., Enum. Palm. Nov.: 26 (1875). *Amylocarpus ericetinus* (Barb.Rodr.) Barb.Rodr., Contr. Jard. Bot. Rio de Janeiro 3: 71 (1902).
Bactris hirta subsp. *pulchra* Trail, J. Bot. 15: 4 (1877). *Bactris pulchra* (Trail) Drude in C.F.P.von Martius & auct. suc. (eds.), Fl. Bras. 3(2): 324 (1881). *Amylocarpus pulcher* (Trail) Barb.Rodr., Contr. Jard. Bot. Rio de Janeiro 3: 72 (1902). *Bactris hirta* var. *pulchra* (Trail) A.J.Hend., Palms Amazon: 196 (1995).
Bactris longipes var. *exilis* Trail, J. Bot. 15: 5 (1877).
Bactris unaensis Barb.Rodr., Palm. Hassler.: 14 (1900).
Amylocarpus platyspinus Barb.Rodr., Contr. Jard. Bot. Rio de Janeiro 3: 72 (1902). *Bactris platyspina* (Barb.Rodr.) Burret, Repert. Spec. Nov. Regni Veg. 34: 189 (1934).
Bactris mollis Dammer, Verh. Bot. Vereins Prov. Brandenburg 48: 129 (1906 publ. 1907). *Bactris hirta* var. *mollis* (Dammer) A.J.Hend., Palms Amazon: 196 (1995).

var. **jenmanii** A.J.Hend., Fl. Neotrop. Monogr. 79: 92 (2000).
Guyana. 82 GUY.

var. **lakoi** (Burret) A.J.Hend., Fl. Neotrop. Monogr. 79: 92 (2000).
N. Brazil to N. Peru. 83 PER 84 BZN.
Bactris lakoi Burret, Repert. Spec. Nov. Regni Veg. 34: 187 (1934).

var. **pectinata** (Mart.) Govaerts in R.H.A.Govaerts & J.Dransfield World Checklist Palms: (2004).
S. Trop. America. 82 FRG GUY SUR VEN 83 BOL CLM PER 84 BZE BZL BZN.
Bactris pectinata Mart., Hist. Nat. Palm. 2: 98 (1826). *Amylocarpus pectinatus* (Mart.) Barb. Rodr., Contr. Jard. Bot. Rio de Janeiro 3: 72 (1902).
Bactris longipes Poepp. in C.F.P.von Martius, Hist. Nat. Palm. 2: 145 (1837).
Bactris integrifolia Wallace, Palm Trees Amazon: 91 (1853).
Bactris simplicifrons Spruce, J. Linn. Soc., Bot. 11: 148 (1869), nom. illeg.
Bactris hylophila Spruce, J. Linn. Soc., Bot. 11: 146 (1871). *Bactris pectinata* subsp. *hylophila* (Spruce) Trail, J. Bot. 6: 6 (1877). *Amylocarpus hylophilus* (Spruce) Barb.Rodr., Contr. Jard. Bot. Rio de Janeiro 3: 72 (1902).
Bactris microcarpa Spruce, J. Linn. Soc., Bot. 11: 146 (1871). *Bactris pectinata* subsp. *microcarpa* (Spruce) Trail, J. Bot. 6: 6 (1877).
Bactris turbinata Spruce, J. Linn. Soc., Bot. 11: 146 (1871). *Bactris pectinata* subsp. *turbinata* (Spruce) Trail, J. Bot. 6: 7 (1877).
Bactris linearifolia Barb.Rodr., Enum. Palm. Nov.: 31 (1875). *Amylocarpus linearifolius* (Barb. Rodr.) Barb.Rodr., Contr. Jard. Bot. Rio de Janeiro 3: 72 (1902).
Bactris setipinnata Barb.Rodr., Enum. Palm. Nov.: 32 (1875). *Amylocarpus setipinnatus* (Barb. Rodr.) Barb.Rodr., Contr. Jard. Bot. Rio de Janeiro 3: 72 (1902).
Bactris pectinata var. *nana* Trail, J. Bot. 6: 6 (1877).
Bactris pectinata var. *setipinnata* (Barb.Rodr.) Trail, J. Bot. 6: 6 (1877).
Bactris pectinata var. *spruceana* Trail, J. Bot. 6: 7 (1877). *Bactris hirta* var. *spruceana* (Trail) A.J.Hend., Fl. Neotrop. Monogr. 79: 92 (2000).
Bactris pectinata var. *subintegrifolia* Trail, J. Bot. 6: 7 (1877).
Bactris geonomoides Drude in C.F.P.von Martius & auct. suc. (eds.), Fl. Bras. 3(2): 325 (1881). *Amylocarpus geonomoides* (Drude) Barb.Rodr., Contr. Jard. Bot. Rio de Janeiro 3: 72 (1902).
Bactris geonomoides var. *setosa* Drude in C.F.P.von Martius & auct. suc. (eds.), Fl. Bras. 3(2): 325 (1881).
Bactris hylophila var. *glabrescens* Drude in C.F.P.von Martius & auct. suc. (eds.), Fl. Bras. 3(2): 332 (1881).
Bactris hylophila var. *macrocarpa* Drude in C.F.P.von Martius & auct. suc. (eds.), Fl. Bras. 3(2): 332 (1881).
Bactris hylophila var. *nana* Drude in C.F.P.von Martius & auct. suc. (eds.), Fl. Bras. 3(2): 332 (1881).
Bactris formosa Barb.Rodr., Vellosia 1: 43 (1888). *Amylocarpus formosus* (Barb.Rodr.) Barb.Rodr., Contr. Jard. Bot. Rio de Janeiro 3: 72 (1902).

Bactris atrox Burret, Repert. Spec. Nov. Regni Veg. 34: 182 (1933).
Bactris hoppii Burret, Repert. Spec. Nov. Regni Veg. 34: 181 (1933).
Bactris huebneri Burret, Repert. Spec. Nov. Regni Veg. 34: 182 (1933).
Pyrenoglyphis hoppii Burret, Repert. Spec. Nov. Regni Veg. 34: 246 (1934). *Bactris pyrenoglyphoides* A.D.Hawkes, Arq. Bot. Estado São Paulo, n.s., f.m. 2: 184 (1952).

Bactris hirta var. *mollis* (Dammer) A.Hend. = **Bactris hirta** Mart.

Bactris hirta subsp. *pulchra* Trail = **Bactris hirta** Mart.

Bactris hirta var. *pulchra* (Trail) A.Hend. = **Bactris hirta** Mart.

Bactris hirta var. *spruceana* (Trail) A.Hend. = **Bactris hirta** var. **pectinata**(Mart.) Govaerts

Bactris hondurensis Standl., Publ. Field Columbian Mus., Bot. Ser. 8: 4 (1930).
C. America to E. Ecuador. 80 COS HON NIC PAN 83 CLM ECU.
Bactris obovata H.Wendl. in O.C.E.de Kerchove de Denterghem, Palmiers: 234 (1878), nom. nud.
Bactris wendlandiana Burret, Repert. Spec. Nov. Regni Veg. 34: 198 (1934).
Bactris villosa H.Wendl. ex Hemsl., Biol. Cent.-Amer., Bot. 3: 413 (1885), nom. nud.
Bactris pubescens Burret, Repert. Spec. Nov. Regni Veg. 34: 197 (1934).
Bactris standleyana Burret, Repert. Spec. Nov. Regni Veg. 34: 199 (1934).
Bactris paula L.H.Bailey, Gentes Herb. 6: 226 (1943). *Yuyba paula* (L.H.Bailey) L.H.Bailey, Gentes Herb. 8: 173 (1949).

Bactris hoppii Burret = **Bactris hirta** var. **pectinata** (Mart.) Govaerts

Bactris horrida Oerst. = **Bactris guineensis** (L.) H.E.Moore

Bactris horridispatha Noblick ex A.J.Hend., Fl. Neotrop. Monogr. 79: 98 (2000).
Brazil (Bahia). 84 BZE.

Bactris huberiana Burret = **Bactris simplicifrons** Mart.

Bactris huebneri Burret = **Bactris hirta** var. **pectinata** (Mart.) Govaerts

Bactris humilis (Wallace) Burret = **Bactris acanthocarpa** var. **exscapa**Barb.Rodr.

Bactris hylophila Spruce = **Bactris hirta** var. **pectinata** (Mart.) Govaerts

Bactris hylophila var. *glabrescens* Drude = **Bactris hirta** var. **pectinata**(Mart.) Govaerts

Bactris hylophila var. *macrocarpa* Drude = **Bactris hirta** var. **pectinata**(Mart.) Govaerts

Bactris hylophila var. *nana* Drude = **Bactris hirta** var. **pectinata** (Mart.) Govaerts

Bactris incommoda Trail = **Bactris maraja** var. **juruensis** (Trail) A.Hend.

Bactris inermis Trail ex Barb.Rodr. = **Bactris simplicifrons** Mart.

Bactris inermis var. *tenuissima* Barb.Rodr. = **Bactris simplicifrons** Mart.

Bactris infesta Mart. = **Bactris major** var. **infesta** (Mart.) Drude

Bactris insignis (Mart.) Baill. = **Bactris gasipaes** Kunth

Bactris integrifolia Wallace = **Bactris hirta** var. **pectinata** (Mart.) Govaerts

Bactris interruptepinnata Barb.Rodr. = **Bactris acanthocarpa** var. **exscapa**Barb.Rodr.

Bactris inundata Mart. = **Bactris riparia** Mart.

Bactris jamaicana L.H.Bailey, Gentes Herb. 4: 177 (1938).
Jamaica. 81 JAM.

Bactris juruensis Trail = **Bactris maraja** var. **juruensis** (Trail) A.Hend.

Bactris juruensis var. *lissospatha* Trail = **Bactris simplicifrons** Mart.

Bactris kalbreyeri Burret = **Bactris setulosa** H.Karst.

Bactris kamarupa Steyerm. = **Bactris maraja** var. **trichospatha** (Trail) A.Hend.

Bactris killipii Burret, Repert. Spec. Nov. Regni Veg. 34: 175 (1933).
Colombia to Peru and N. Brazil. 83 CLM PER 84 BZN.

Bactris krichana Barb.Rodr. = **Bactris maraja** var. **juruensis** (Trail) A.Hend.

Bactris kuhlmannii Burret = **Bactris simplicifrons** Mart.

Bactris kunorum Nevers & Grayum, Proc. Calif. Acad. Sci. 49: 195 (1996).
C. Panama. 80 PAN.

Bactris lakoi Burret = **Bactris hirta** var. **lakoi** (Burret) A.Hend.

Bactris lanceolata Burret = **Bactris campestris** Poepp.

Bactris leptocarpa Trail ex Thurn = **Bactris campestris** Poepp.

Bactris leptochaete Burret = **Bactris acanthocarpa** var. **exscapa** Barb.Rodr.

Bactris leptospadix Burret = **Bactris maraja** Mart.

Bactris leptotricha Burret = **Bactris maraja** Mart.

Bactris leucantha H.Wendl. = **Bactris maraja** Mart.

Bactris lindmanniana Drude ex Lindm. = **Bactris setosa** Mart.

Bactris linearifolia Barb.Rodr. = **Bactris hirta** var. **pectinata** (Mart.) Govaerts

Bactris littoralis Barb.Rodr. = **Bactris riparia** Mart.

Bactris longicuspis Burret = **Bactris maraja** var. **trichospatha** (Trail) A.Hend.

Bactris longifrons Mart. = **Bactris riparia** Mart.

Bactris longipes Poepp. = **Bactris hirta** var. **pectinata** (Mart.) Govaerts

Bactris longipes var. *exilis* Trail = **Bactris hirta** Mart.

Bactris longisecta Burret = **Bactris maraja** var. **trichospatha** (Trail) A.Hend.

Bactris longiseta H.Wendl. ex Burret, Repert. Nov. Regni Veg. 34: 213 (1934).
Costa Rica. 80 COS.
Bactris polystachya H. Wendl. ex Grayum, Phytologia 84: 308 (1998 publ. 1999).

Bactris luetzelburgii Burret = **Bactris simplicifrons** Mart.

Bactris luetzelburgii var. *anacantha* Burdet = **Bactris simplicifrons** Mart.

Bactris macana (Mart.) Pittier = **Bactris gasipaes** var. **chichagui** (H.Karst.) A.Hend.

Bactris macroacantha Mart., Hist. Nat. Palm. 2: 95 (1826).
Colombia to Peru and N. Brazil. 83 CLM PER 84 BZN.
Bactris confluens var. *acanthospatha* Trail, J. Bot. 15: 44 (1877). *Bactris acanthospatha* (Trail) Trail ex Drude in C.F.P.von Martius & auct. suc. (eds.), Fl. Bras. 3(2): 354 (1881).
Bactris platyacantha Burret, Notizbl. Bot. Gart. Berlin-Dahlem 14: 265 (1938).

Bactris macrocalyx Burret = **Bactris acanthocarpa** var. **exscapa**Barb.Rodr.

Bactris macrocarpa Wallace = **Bactris maraja** Mart.

Bactris macrotricha Burret = **Bactris glandulosa** Oerst.

Bactris maguirei (L.H.Bailey) Steyerm. = **Bactris simplicifrons** Mart.

Bactris major Jacq., Select. Stirp. Amer. Hist., ed. 2: 134 (1781). *Bactris ovata* Stokes, Bot. Mat. Med. 4: 394 (1812), nom. illeg. *Augustinea major* (Jacq.) H.Karst., Linnaea 28: 395 (1856). *Pyrenoglyphis major* (Jacq.) H.Karst., Fl. Columb. 2: 141 (1869).
Trinidad, Mexico to S. Trop. America. 79 MXG MXS MXT 80 BLZ COS ELS GUA HON NIC PAN 81 TRT 82 FRG GUY SUR VEN 83 BOL CLM ECU PER 84 BZE BZN.

var. **infesta** (Mart.) Drude in C.F.P.von Martius & auct. suc. (eds.), Fl. Bras. 3(2): 54 (1881).
S. Trop. America. 82 GUY SUR VEN 83 BOL ECU PER 84 BZE BZN.
Bactris infesta Mart. in A.D.d'Orbigny, Voy. Amér. Mér. 7(3): 54 (1844). *Pyrenoglyphis infesta* (Mart.) Burret, Repert. Spec. Nov. Regni Veg. 34: 248 (1934).
Bactris exaltata Barb.Rodr., Enum. Palm. Nov.: 32 (1875). *Pyrenoglyphis exaltata* (Barb.Rodr.) Burret, Repert. Spec. Nov. Regni Veg. 34: 246 (1934).
Bactris nemorosa Barb.Rodr., Enum. Palm. Nov.: 32 (1875). *Pyrenoglyphis nemorosa* (Barb.Rodr.) Burret, Repert. Spec. Nov. Regni Veg. 34: 247 (1934).
Bactris socialis subsp. *curuena* Trail, J. Bot. 15: 48 (1877). *Bactris curuena* (Trail) Trail ex Drude in C.F.P.von Martius & auct. suc. (eds.), Fl. Bras. 3(2): 359 (1881). *Pyrenoglyphis curuena* (Trail) Burret, Repert. Spec. Nov. Regni Veg. 34: 248 (1934).
Bactris socialis subsp. *gaviona* Trail, J. Bot. 15: 48 (1877). *Bactris gaviona* (Trail) Trail ex Drude in C.F.P.von Martius & auct. suc. (eds.), Fl. Bras. 3(2): 360 (1881). *Pyrenoglyphis gaviona* (Trail) Burret, Repert. Spec. Nov. Regni Veg. 34: 246 (1934).
Bactris chapadensis Barb.Rodr., Palm. Mattogross.: 41 (1898). *Pyrenoglyphis chapadensis* (Barb. Rodr.) Burret, Repert. Spec. Nov. Regni Veg. 34: 248 (1934).
Bactris mattogrossensis Barb.Rodr., Palm. Mattogross.: 38 (1898). *Pyrenoglyphis mattogrossensis* (Barb.Rodr.) Burret, Repert. Spec. Nov. Regni Veg. 34: 248 (1934).

var. **major**
Trinidad, Mexico to Brazil. 79 MXG MXS MXT 80 BLZ COS ELS GUA HON NIC PAN 81 TRT 82 FRG GUY SUR VEN 83 CLM 84 BZE BZN.
Bactris chaetorhachis Mart. in A.D.d'Orbigny, Voy. Amér. Mér. 7(3): 61 (1844). *Pyrenoglyphis*

chaetorhachis (Mart.) Burret, Repert. Spec. Nov. Regni Veg. 34: 245 (1934).

Bactris minax Miq., Verh. Nat. Wet. Haarlem 7: 207 (1851).

Augustinea balanoidea Oerst., Vidensk. Meddel. Dansk Naturhist. Foren. Kjøbenhavn 1858: 39 (1858). *Pyrenoglyphis balanoidea* (Oerst.) H.Karst., Fl. Columb. 2: 141 (1869). *Bactris balanoidea* (Oerst.) H.Wendl. in O.C.E.de Kerchove de Denterghem, Palmiers: 233 (1878).

Augustinea ovata Oerst., Vidensk. Meddel. Dansk Naturhist. Foren. Kjøbenhavn 1858: 38 (1858). *Pyrenoglyphis ovata* (Oerst.) H.Karst., Fl. Columb. 2: 142 (1869). *Bactris ovata* (Oerst.) H.Wendl. in O.C.E.de Kerchove de Denterghem, Palmiers: 234 (1878), nom. illeg. *Bactris augustinea* L.H.Bailey, Gentes Herb. 3: 95 (1933).

Bactris cruegeriana Griseb., Fl. Brit. W. I.: 520 (1864). *Pyrenoglyphis cruegeriana* (Griseb.) H.Karst., Fl. Columb. 2: 141 (1869).

Bactris megalocarpa Trail ex Thurn, Timehri 1: 242 (1882). *Bactris major* var. *megalocarpa* (Trail ex Thurn) A.J.Hend., Palms Amazon: 203 (1995).

Bactris ottostapfiana Barb.Rodr., Contr. Jard. Bot. Rio de Janeiro 4: 112 (1907). *Pyrenoglyphis ottostapfiana* (Barb.Rodr.) Burret, Notizbl. Bot. Gart. Berlin-Dahlem 12: 158 (1934).

Bactris superior L.H.Bailey, Gentes Herb. 3: 99 (1933). *Pyrenoglyphis superior* (L.H.Bailey) Burret, Repert. Spec. Nov. Regni Veg. 34: 246 (1934).

Bactris albonotata L.H.Bailey, Gentes Herb. 7: 396 (1947).

Bactris beata L.H.Bailey, Gentes Herb. 7: 399 (1947).

Bactris broadwayi L.H.Bailey, Gentes Herb. 7: 388 (1947).

Bactris cateri L.H.Bailey, Gentes Herb. 7: 396 (1947).

Bactris ellipsoidalis L.H.Bailey, Gentes Herb. 7: 389 (1947).

Bactris obovoidea L.H.Bailey, Gentes Herb. 7: 392 (1947).

Bactris planifolia L.H.Bailey, Gentes Herb. 7: 392 (1947).

Bactris swabeyi L.H.Bailey, Gentes Herb. 7: 399 (1947).

Bactris demerarana L.H.Bailey, Gentes Herb. 8: 162 (1949).

var. **socialis** (Mart.) Drude in C.F.P.von Martius & auct. suc. (eds.), Fl. Bras. 3(2): 359 (1881).
Bolivia. 83 BOL.
Bactris socialis Mart. in A.D.d'Orbigny, Voy. Amér. Mér. 7(3): 56 (1844). *Pyrenoglyphis socialis* (Mart.) Burret, Repert. Spec. Nov. Regni Veg. 34: 246 (1934).

Bactris major var. *megalocarpa* (Trail ex Thurn) A.Hend. = **Bactris major** Jacq.

Bactris maraja Mart., Hist. Nat. Palm. 2: 93 (1826). *Pyrenoglyphis maraja* (Mart.) Burret, Repert. Spec. Nov. Regni Veg. 34: 252 (1934).
C. & S. Trop. America. 80 COS PAN 82 FRG GUY SUR VEN 83 BOL CLM ECU PER 84 BZN.

var. **chaetospatha** (Mart.) A.J.Hend., Palms Amazon: 205 (1995).
N. Brazil to Peru. 83 PER 84 BZN.
Bactris chaetospatha Mart., Hist. Nat. Palm. 2: 147 (1837).

var. **juruensis** (Trail) A.J.Hend., Palms Amazon: 205 (1995).
W. South America to N. Brazil. 82 FRG? 83 BOL CLM PER 84 BZN.
Bactris incommoda Trail, J. Bot. 15: 42 (1877).
Bactris juruensis Trail, J. Bot. 15: 40 (1877).
Bactris piranga Trail, J. Bot. 15: 41 (1877).
Bactris krichana Barb.Rodr., Vellosia 1: 41 (1888).
Bactris penicillata Barb.Rodr., Vellosia 1: 42 (1888).
Bactris bella Burret, Notizbl. Bot. Gart. Berlin-Dahlem 12: 157 (1934).
Bactris chlorocarpa Burret, Notizbl. Bot. Gart. Berlin-Dahlem 12: 622 (1935).
Bactris pulchella Burret, Notizbl. Bot. Gart. Berlin-Dahlem 12: 621 (1935).
Bactris microspadix Burret, Notizbl. Bot. Gart. Berlin-Dahlem 14: 264 (1938).

var. **maraja**
C. & S. Trop. America. 80 COS PAN 82 GUY SUR VEN 83 BOL CLM ECU PER 84 BZN.
Bactris chloracantha Poepp. in C.F.P.von Martius, Hist. Nat. Palm. 2: 145 (1837).
Bactris elatior Wallace, Palm Trees Amazon: 81 (1853).
Bactris macrocarpa Wallace, Palm Trees Amazon: 85 (1853).
Bactris leucantha H.Wendl., Linnaea 28: 345 (1856). *Pyrenoglyphis leucantha* (H.Wendl.) Burret, Repert. Spec. Nov. Regni Veg. 34: 249 (1934).
Bactris sanctae-paulae Engl., Linnaea: 667 (1865).
Bactris granariuscarpa Barb.Rodr., Enum. Palm. Nov.: 37 (1875).
Bactris monticola Barb.Rodr., Enum. Palm. Nov.: 34 (1875).
Bactris paucijuga Barb.Rodr., Enum. Palm. Nov.: 34 (1875).
Bactris sylvatica Barb.Rodr., Enum. Palm. Nov.: 30 (1875).
Bactris umbraticola Barb.Rodr., Enum. Palm. Nov.: 34 (1875).
Bactris umbrosa Barb.Rodr., Enum. Palm. Nov.: 29 (1875).
Bactris maraja subsp. *limnaia* Trail, J. Bot. 6: 44 (1877). *Bactris maraja* var. *limnaia* (Trail) Drude in C.F.P.von Martius & auct. suc. (eds.), Fl. Bras. 3(2): 343 (1881). *Bactris sobralensis* var. *limnaia* (Trail) Barb.Rodr., Sert. Palm. Brasil. 2: 102 (1903).
Bactris maraja subsp. *sobralensis* Trail, J. Bot. 6: 44 (1877). *Bactris maraja* var. *sobralensis* (Trail) Drude in C.F.P.von Martius & auct. suc. (eds.), Fl. Bras. 3(2): 343 (1881). *Bactris sobralensis* (Trail) Barb.Rodr., Sert. Palm. Brasil. 2: 102 (1903).
Bactris armata Barb.Rodr., Enum. Palm. Nov.: 27 (1879).
Bactris gymnospatha Burret, Notizbl. Bot. Gart. Berlin-Dahlem 10: 1024 (1930).
Bactris erostrata Burret, Repert. Spec. Nov. Regni Veg. 34: 207 (1934).
Bactris leptospadix Burret, Repert. Spec. Nov. Regni Veg. 34: 210 (1934).
Bactris leptotricha Burret, Repert. Spec. Nov. Regni Veg. 34: 207 (1934).
Bactris sigmoidea Burret, Repert. Spec. Nov. Regni Veg. 34: 206 (1934).
Bactris bijugata Burret, Notizbl. Bot. Gart. Berlin-Dahlem 14: 264 (1938).
Bactris divisicupula L.H.Bailey, Gentes Herb. 6: 230 (1943).

Bactris fuscospina L.H.Bailey, Gentes Herb. 6: 228 (1943).

var. **trichospatha** (Trail) A.J.Hend., Fl. Neotrop. Monogr. 79: 119 (2000).
S. Trop. America. 82 FRG GUY SUR VEN 83 BOL CLM PER 84 BZN.
  Bactris trichospatha Trail, J. Bot. 15: 41 (1877).
  Bactris trichospatha var. *patens* Drude in C.F.P.von Martius & auct. suc. (eds.), Fl. Bras. 3(2): 339 (1881), nom. inval.
  Bactris trichospatha subsp. *jurutensis* Trail, J. Bot. 15: 42 (1877). *Bactris trichospatha* var. *jurutensis* (Trail) Drude in C.F.P.von Martius & auct. suc. (eds.), Fl. Bras. 3(2): 339 (1881).
  Bactris trichospatha var. *robusta* Trail, J. Bot. 15: 42 (1877).
  Bactris actinoneura Drude & Trail ex Drude in C.F.P.von Martius & auct. suc. (eds.), Fl. Bras. 3(2): 344 (1881).
  Bactris chaetochlamys Burret, Repert. Spec. Nov. Regni Veg. 34: 208 (1934).
  Bactris longisecta Burret, Repert. Spec. Nov. Regni Veg. 34: 205 (1934).
  Bactris longicuspis Burret, Notizbl. Bot. Gart. Berlin-Dahlem 15: 5 (1940).
  Bactris kamarupa Steyerm., Fieldiana, Bot. 28: 75 (1951).

Bactris maraja var. *limnaia* (Trail) Drude = **Bactris maraja** Mart.

Bactris maraja subsp. *limnaia* Trail = **Bactris maraja** Mart.

Bactris maraja var. *sobralensis* (Trail) Drude = **Bactris maraja** Mart.

Bactris maraja subsp. *sobralensis* Trail = **Bactris maraja** Mart.

Bactris marajay Barb.Rodr. = **Bactris cuspidata** Mart.

Bactris maraya-acu Barb.Rodr. = **Bactris brongniartii** Mart.

Bactris martiana A.J.Hend., Fl. Neotrop. Monogr. 79: 120 (2000).
W. South America to N. Brazil. 83 CLM ECU PER 84 BZN.

Bactris martineziifolia H.Wendl. = **Aiphanes minima** (Gaertn.) Burret

Bactris mattogrossensis Barb.Rodr. = **Bactris major** var. **infesta** (Mart.) Drude

Bactris megalocarpa Trail ex Thurn = **Bactris major** Jacq.

Bactris megistocarpa Burret = **?** [84 BZN] **Astrocaryum ?**

Bactris mexicana Mart. in A.D.d'Orbigny, Voy. Amér. Mér. 7(3): 65 (1844).
Mexico to C. America. 79 MXG MXS MXT 80 BLZ GUA HON NIC.

var. **mexicana**
Mexico. 79 MXG MXS MXT.
  Bactris acuminata Liebm. ex Mart., Hist. Nat. Palm. 3: 321 (1853).

var. **trichophylla** (Burret) A.J.Hend., Proc. Calif. Acad. Sci. 49: 204 (1996).
C. America. 80 BLZ GUA HON NIC.
  Bactris trichophylla Burret, Repert. Spec. Nov. Regni Veg. 32: 113 (1933).

Bactris microcalyx Burret = **Bactris acanthocarpa** var. **exscapa** Barb.Rodr.

Bactris microcarpa Spruce = **Bactris hirta** var. **pectinata** (Mart.) Govaerts

Bactris microspadix Burret = **Bactris maraja** var. **juruensis** (Trail) A.Hend.

Bactris microspatha Barb.Rodr. = **Bactris simplicifrons** Mart.

Bactris militaris H.E.Moore, Gentes Herb. 8: 229 (1951).
Costa Rica. 80 COS.

Bactris minax Miq. = **Bactris major** Jacq.

Bactris mindellii Barb.Rodr. = **Bactris acanthocarpa** Mart.

Bactris minima Gaertn. = **Aiphanes minima** (Gaertn.) Burret

Bactris minor Jacq. = **Bactris guineensis** (L.) H.E. Moore

Bactris minor Gaertn. = **Acrocomia aculeata** (Jacq.) Lodd. ex Mart.

Bactris mitis Mart. = **Bactris cuspidata** Mart.

Bactris mitis subsp. *inermis* Trail = **Bactris simplicifrons** Mart.

Bactris mitis subsp. *tenuis* (Wallace) Trail = **Bactris simplicifrons** Mart.

Bactris mitis subsp. *uaupensis* (Spruce) Trail = **Bactris simplicifrons** Mart.

Bactris mollis Dammer = **Bactris hirta** Mart.

Bactris monticola Barb.Rodr. = **Bactris maraja** Mart.

Bactris × moorei Wess.Boer, Acta Bot. Neerl. 20: 169 (1971). B. acanthocarpa × B. oligoclada.
Venezuela. 82 VEN.

Bactris multiramosa Burret = **Bactris syagroides** Barb.Rodr. & Trail

Bactris naevia Poepp. ex Burret = **Bactris simplicifrons** Mart.

Bactris negrensis Spruce = **Bactris simplicifrons** Mart.

Bactris negrensis var. *carolensis* (Spruce) Burret = **Bactris simplicifrons** Mart.

Bactris negrensis var. *minor* Spruce = **Bactris simplicifrons** Mart.

Bactris nemorosa Barb.Rodr. = **Bactris major** var. **infesta** (Mart.) Drude

Bactris neomilitaris Nevers & A.J.Hend., Brittonia 51: 77 (1999).
Costa Rica to Panama. 80 COS PAN.

Bactris nigrispina Barb.Rodr. = **Bactris bidentula** Spruce

Bactris obovata H.Wendl. = **Bactris hondurensis** Standl.

Bactris obovata Burret = **Bactris simplicifrons** Mart.

Bactris obovoidea L.H.Bailey = **Bactris major** Jacq.

Bactris oerstediana Trail = **Bactris glandulosa** Oerst.

Bactris oligocarpa Barb.Rodr., Enum. Palm. Nov.: 28 (1875). *Pyrenoglyphis oligocarpa* (Barb.Rodr.) Burret, Repert. Spec. Nov. Regni Veg. 34: 242 (1934).
Guianas to N. Brazil. 82 FRG SUR 84 BZN.
  Bactris oligocarpa var. *brachycaulis* Trail, J. Bot. 6: 47 (1877).

Bactris oligocarpa var. *brachycaulis* Trail = **Bactris oligocarpa** Barb.Rodr.

Bactris oligoclada Burret, Notizbl. Bot. Gart. Berlin-Dahlem 11: 325 (1932).
Venezuela, Guyana. 82 GUY VEN.

Bactris oraria L.H.Bailey = **Bactris guineensis** (L.) H.E.Moore

Bactris ottostaffiana Barb.Rodr. = **Bactris major** Jacq.

Bactris ovata Stokes = **Bactris major** Jacq.

Bactris ovata (Oerst.) H.Wendl. = **Bactris major** Jacq.

Bactris pallidispina Mart. = **Bactris brongniartii** Mart.

Bactris palustris Barb.Rodr. = **Bactris bidentula** Spruce

Bactris panamensis Nevers & Grayum, Proc. Calif. Acad. Sci. 49: 205 (1996).
Panama. 80 PAN.

Bactris paraensis Splitg. ex de Vriese = **Astrocaryum paramaca** Mart.

Bactris paucijuga Barb.Rodr. = **Bactris maraja** Mart.

Bactris paucisecta Burret = **Bactris simplicifrons** Mart.

Bactris paula L.H.Bailey = **Bactris hondurensis** Standl.

Bactris pavoniana Mart. = **Acrocomia aculeata** (Jacq.) Lodd. ex Mart.

Bactris pectinata Mart. = **Bactris hirta** var. **pectinata** (Mart.) Govaerts

Bactris pectinata subsp. *hylophila* (Spruce) Trail = **Bactris hirta** var. **pectinata** (Mart.) Govaerts

Bactris pectinata subsp. *microcarpa* (Spruce) Trail = **Bactris hirta** var. **pectinata** (Mart.) Govaerts

Bactris pectinata var. *nana* Trail = **Bactris hirta** var. **pectinata** (Mart.) Govaerts

Bactris pectinata var. *setipinnata* (Barb.Rodr.) Trail = **Bactris hirta** var. **pectinata** (Mart.) Govaerts

Bactris pectinata var. *spruceana* Trail = **Bactris hirta** var. **pectinata** (Mart.) Govaerts

Bactris pectinata var. *subintegrifolia* Trail = **Bactris hirta** var. **pectinata** (Mart.) Govaerts

Bactris pectinata subsp. *turbinata* (Spruce) Trail = **Bactris hirta** var. **pectinata** (Mart.) Govaerts

Bactris penicillata Barb.Rodr. = **Bactris maraja** var. **juruensis** (Trail) A.Hend.

Bactris pickelii Burret, Repert. Spec. Nov. Regni Veg. 34: 199 (1934).
NE. Brazil (to Espírito Santo). 84 BZE BZL.

Bactris pilosa H.Karst., Linnaea 28: 405 (1856).
Panama to Venezuela and Ecuador. 80 PAN 82 VEN 83 CLM ECU.
  Guilielma granatensis H.Karst., Linnaea 28: 400 (1856).
  Bactris granatensis (H.Karst.) H.Wendl. in O.C.E.de Kerchove de Denterghem, Palmiers: 234 (1878).
  Bactris hirsuta Burret, Notizbl. Bot. Gart. Berlin-Dahlem 11: 17 (1930).

Bactris pinnatisecta Burret = **Bactris acanthocarpa** var. **exscapa**Barb.Rodr.

Bactris piranga Trail = **Bactris maraja** var. **juruensis** (Trail) A.Hend.

Bactris piritu (H.Karst.) H.Wendl. = **Bactris guineensis** (L.) H.E.Moore

Bactris piscatorum Wedd. ex Drude = **Bactris brongniartii** Mart.

Bactris planifolia L.H.Bailey = **Bactris major** Jacq.

Bactris platyacantha Burret = **Bactris macroacantha** Mart.

Bactris platyspina (Barb.Rodr.) Burret = **Bactris hirta** Mart.

Bactris pliniana Granv. & A.J.Hend., Brittonia 46: 147 (1994).
Guianas to N. Brazil. 82 FRG GUY SUR 84 BZN.

Bactris plumeriana Mart. in A.D.d'Orbigny, Voy. Amér. Mér. 7(3): 64 (1844).
Hispaniola. 81 DOM HAI.
  Palma gracilis Mill., Gard. Dict. ed. 8: 5 (1768).
  Bactris chaetophylla Mart. in A.D.d'Orbigny, Voy. Amér. Mér. 7(3): 71 (1844).

Bactris polyclada Burret = **Bactris vulgaris** Barb.Rodr.

Bactris polystachya H. Wendl. ex Grayum = **Bactris longiseta** H.Wendl. ex Burret

Bactris porschiana Burret = **Bactris coloradonis** L.H. Bailey

Bactris praemorsa Poepp. ex Mart. = **Aiphanes aculeata** Willd.

Bactris ptariana Steyerm., Fieldiana, Bot. 28: 77 (1951).
SE. Venezuela to Guyana. 82 GUY VEN.

Bactris pubescens Burret = **Bactris hondurensis** Standl.

Bactris pulchella Burret = **Bactris maraja** var. **juruensis** (Trail) A.Hend.

Bactris pulchra (Trail) Drude = **Bactris hirta** Mart.

Bactris pulchra var. *inermis* Dammer = **Bactris simplicifrons** Mart.

Bactris pyrenoglyphoides (Burret) A.D.Hawkes = **Bactris major** var. **infesta** (Mart.) Drude

Bactris rhaphidacantha Wess.Boer, Indig. Palms Surin.: 93 (1965).
Suriname to Brazil (Amapá). 82 FRG SUR 84 BZN.

Bactris riparia Mart., Hist. Nat. Palm. 2: 97 (1826).
W. South America to N. Brazil. 83 BOL CLM ECU PER 84 BZN.

Bactris rivularis Barb.Rodr. = **Bactris brongniartii** Mart.

Bactris rostrata Galeano & R.Bernal, Caldasia 24: 280 (2002).
Colombia. 83 CLM.

Bactris rotunda Stokes = **Bactris guineensis** (L.) H.E.Moore

Bactris sanctae-paulae Engl. = **Bactris maraja** Mart.

Bactris savannarum Britton = **Bactris campestris** Poepp.

Bactris schultesii (L.H.Bailey) Glassman, Rhodora 65: 259 (1963).
Colombia to Peru. 83 CLM ECU PER.
  Yuyba schultesii L.H.Bailey, Gentes Herb. 8: 174 (1949).

Bactris sciophila Miq. = **Astrocaryum sciophilum** (Miq.) Pulle

Bactris setiflora Burret, Notizbl. Bot. Gart. Berlin-Dahlem 14: 328 (1939).
Ecuador (Pastaza). 83 ECU.

Bactris setipinnata Barb.Rodr. = **Bactris hirta** var. **pectinata** (Mart.) Govaerts

Bactris setosa var. *santensis* Barb.Rodr. = **Bactris setosa** Mart.

Bactris setosa Mart., Hist. Nat. Palm. 2: 94 (1826).
Brazil. 84 BZC BZE BZL BZS.
  Bactris lindmanniana Drude ex Lindm., Bih. Kongl. Svenska Vetensk.-Akad. Handl. 26(5): 12 (1901).
  Bactris setosa var. *santensis* Barb.Rodr., Sert. Palm. Brasil. 2: 25 (1903).

Bactris escragnollei Glaz. ex Burret, Repert. Spec. Nov. Regni Veg. 34: 223 (1934).

Bactris setulosa H.Karst., Linnaea 28: 408 (1856).
Colombia to Tobago and Peru. 81 TRT VNA 82 SUR VEN 83 CLM ECU PER.
Bactris cuvaro H.Karst., Linnaea 28: 406 (1856).
Bactris cuesa Crueg. ex Griseb., Fl. Brit. W. I.: 520 (1864).
Bactris falcata J.R.Johnst., Proc. Amer. Acad. Arts 40: 683 (1905).
Bactris sworderiana Becc., Repert. Spec. Nov. Regni Veg. 16: 437 (1920).
Bactris kalbreyeri Burret, Repert. Spec. Nov. Regni Veg. 34: 231 (1934).
Bactris circularis L.H.Bailey, Gentes Herb. 7: 388 (1947).
Bactris bergantina Steyerm., Fieldiana, Bot. 28: 71 (1951).

Bactris sigmoidea Burret = **Bactris maraja** Mart.

Bactris simplex Burret = **Bactris simplicifrons** Mart.

Bactris simplicifrons Mart., Hist. Nat. Palm. 2: 103 (1826). *Amylocarpus simplicifrons* (Mart.) Barb. Rodr., Contr. Jard. Bot. Rio de Janeiro 3: 71 (1902). *Yuyba simplicifrons* (Mart.) L.H.Bailey, Gentes Herb. 7: 416 (1947).
Trinidad, Tobago, S. Trop. America. 81 TRT 82 FRG GUY SUR 83 BOL CLM ECU PER 84 BZE BZN.
Bactris acanthocnemis Mart. in A.D.d'Orbigny, Voy. Amér. Mér. 7(3): 67 (1844). *Bactris simplicifrons* var. *acanthocnemis* (Mart.) Drude in C.F.P.von Martius & auct. suc. (eds.), Fl. Bras. 3(2): 321 (1881). *Amylocarpus acanthocnemis* (Mart.) Barb.Rodr., Contr. Jard. Bot. Rio de Janeiro 3: 71 (1902).
Bactris tenuis Wallace, Palm Trees Amazon: 87 (1853). *Bactris mitis* subsp. *tenuis* (Wallace) Trail, J. Bot. 15: 3 (1877). *Bactris cuspidata* var. *tenuis* (Wallace) Drude in C.F.P.von Martius & auct. suc. (eds.), Fl. Bras. 3(2): 329 (1881).
Bactris brevifolia Spruce, J. Linn. Soc., Bot. 11: 144 (1871). *Bactris simplicifrons* var. *brevifolia* (Spruce) Trail, J. Bot. 15: 1 (1877).
Bactris carolensis Spruce, J. Linn. Soc., Bot. 11: 145 (1871). *Bactris simplicifrons* var. *carolensis* (Spruce) Trail, J. Bot. 15: 1 (1877). *Bactris negrensis* var. *carolensis* (Spruce) Burret, Repert. Spec. Nov. Regni Veg. 34: 177 (1933).
Bactris negrensis Spruce, J. Linn. Soc., Bot. 11: 145 (1871). *Bactris simplicifrons* var. *negrensis* (Spruce) Trail, J. Bot. 15: 1 (1877).
Bactris negrensis var. *minor* Spruce, J. Linn. Soc., Bot. 11: 145 (1871).
Bactris uaupensis Spruce, J. Linn. Soc., Bot. 11: 145 (1871). *Bactris mitis* subsp. *uaupensis* (Spruce) Trail, J. Bot. 15: 3 (1877).
Bactris arenaria Barb.Rodr., Enum. Palm. Nov.: 29 (1875). *Amylocarpus arenarius* (Barb.Rodr.) Barb. Rodr., Contr. Jard. Bot. Rio de Janeiro 3: 72 (1902).
Bactris gracilis Barb.Rodr., Enum. Palm. Nov.: 27 (1875).
Bactris inermis Trail ex Barb.Rodr., Enum. Palm. Nov.: 30 (1875). *Amylocarpus inermis* (Trail ex Barb.Rodr.) Barb.Rodr., Sert. Palm. Brasil. 2: t. 45a (1903).
Bactris inermis var. *tenuissima* Barb.Rodr., Enum. Palm. Nov.: 30 (1875). *Amylocarpus tenuissimus* (Barb.Rodr.) Barb.Rodr., Contr. Jard. Bot. Rio de Janeiro 3: 72 (1902). *Bactris tenuissima* (Barb.Rodr.) Barb.Rodr., Sert. Palm. Brasil. 3: t. 10d (1903).

Bactris microspatha Barb.Rodr., Enum. Palm. Nov.: 26 (1875). *Amylocarpus microspathus* (Barb.Rodr.) Barb. Rodr., Contr. Jard. Bot. Rio de Janeiro 3: 72 (1902).
Bactris juruensis var. *lissospatha* Trail, J. Bot. 6: 40 (1877).
Bactris mitis subsp. *inermis* Trail, J. Bot. 15: 3 (1877).
Bactris simplicifrons var. *subpinnata* Trail, J. Bot. 6: 2 (1877).
Bactris xanthocarpa Barb.Rodr., Enum. Palm. Nov., App.: 30 (1879). *Amylocarpus xanthocarpus* (Barb. Rodr.) Barb.Rodr., Contr. Jard. Bot. Rio de Janeiro 3: 71 (1902).
Bactris pulchra var. *inermis* Dammer, Verh. Bot. Vereins Prov. Brandenburg 48: 128 (1906 publ. 1907). *Bactris ulei* Burret, Repert. Spec. Nov. Regni Veg. 34: 177 (1933).
Amylocarpus angustifolius Huber, Bol. Mus. Goeldi Paraense Hist. Nat. Ethnogr. 7: 285 (1913). *Bactris huberiana* Burret, Repert. Spec. Nov. Regni Veg. 34: 174 (1933).
Amylocarpus luetzelburgii Burret, Notizbl. Bot. Gart. Berlin-Dahlem 10: 1023 (1930).
Amylocarpus obovatus Burret, Notizbl. Bot. Gart. Berlin-Dahlem 11: 17 (1930).
Bactris luetzelburgii Burret, Notizbl. Bot. Gart. Berlin-Dahlem 10: 1022 (1930).
Bactris obovata Burret, Notizbl. Bot. Gart. Berlin-Dahlem 11: 16 (1930).
Bactris amoena Burret, Repert. Spec. Nov. Regni Veg. 34: 180 (1933).
Bactris luetzelburgii var. *anacantha* Burdet, Repert. Spec. Nov. Regni Veg. 34: 174 (1933).
Bactris naevia Poepp. ex Burret, Repert. Spec. Nov. Regni Veg. 34: 179 (1933).
Bactris paucisecta Burret, Repert. Spec. Nov. Regni Veg. 34: 171 (1933).
Bactris simplex Burret, Repert. Spec. Nov. Regni Veg. 34: 179 (1933).
Bactris kuhlmannii Burret, Notizbl. Bot. Gart. Berlin-Dahlem 14: 262 (1938).
Yuyba trinitensis L.H.Bailey, Gentes Herb. 7: 416 (1947). *Bactris trinitensis* (L.H.Bailey) Glassman, Rhodora 65: 259 (1963).
Yuyba dakamana L.H.Bailey, Bull. Torrey Bot. Club 75: 108 (1948). *Bactris dakamana* (L.H.Bailey) Glassman, Rhodora 65: 259 (1963).
Yuyba essequiboensis L.H.Bailey, Bull. Torrey Bot. Club 75: 108 (1948). *Bactris essequiboensis* (L.H.Bailey) Glassman, Rhodora 65: 259 (1963).
Yuyba maguirei L.H.Bailey, Bull. Torrey Bot. Club 75: 106 (1948). *Bactris maguirei* (L.H.Bailey) Steyerm., Fieldiana, Bot. 28: 80 (1951).
Yuyba stahelii L.H.Bailey, Bull. Torrey Bot. Club 75: 106 (1948). *Bactris stahelii* (L.H.Bailey) Glassman, Rhodora 65: 259 (1963).
Yuyba gleasonii L.H.Bailey, Gentes Herb. 8: 174 (1949). *Bactris gleasonii* (L.H.Bailey) Glassman, Rhodora 65: 259 (1963).
Bactris sororopanae Steyerm., Fieldiana, Bot. 28: 78 (1951).

Bactris simplicifrons Spruce = **Bactris hirta** var. **pectinata** (Mart.) Govaerts

Bactris simplicifrons var. *acanthocnemis* (Mart.) Drude = **Bactris simplicifrons** Mart.

Bactris simplicifrons var. *brevifolia* (Spruce) Trail = **Bactris simplicifrons** Mart.

Bactris simplicifrons var. *carolensis* (Spruce) Trail = **Bactris simplicifrons** Mart.

Bactris simplicifrons var. *negrensis* (Spruce) Trail = **Bactris simplicifrons** Mart.

Bactris simplicifrons var. *subpinnata* Trail = **Bactris simplicifrons** Mart.

Bactris sobralensis (Trail) Barb.Rodr. = **Bactris maraja** Mart.

Bactris sobralensis var. *limnaia* (Trail) Barb.Rodr. = **Bactris maraja** Mart.

Bactris socialis Mart. = **Bactris major** var. **socialis** (Mart.) Drude

Bactris socialis subsp. *curuena* Trail = **Bactris major** var. **infesta** (Mart.) Drude

Bactris socialis subsp. *gaviona* Trail = **Bactris major** var. **infesta** (Mart.) Drude

Bactris soeiroana Noblick ex A.J.Hend., Fl. Neotrop. Monogr. 79: 154 (2000). Brazil (Bahia). 84 BZE.

Bactris sororopanae Steyerm. = **Bactris simplicifrons** Mart.

Bactris speciosa (Mart.) H.Karst. = **Bactris gasipaes** Kunth

Bactris speciosa var. *chichagui* H.Karst. = **Bactris gasipaes** var. **chichagui**(H.Karst.) A.Hend.

Bactris sphaerocarpa Trail, J. Bot. 15: 8 (1877). *Bactris tomentosa* var. *sphaerocarpa* (Trail) A.J.Hend., Palms Amazon: 221 (1995). Colombia to Peru and N. Brazil. 83 CLM PER 84 BZN.
  Bactris sphaerocarpa subsp. *pinnatisecta* Trail, J. Bot. 6: 9 (1877).
  Bactris sphaerocarpa var. *platyphylla* Trail, J. Bot. 6: 8 (1877).
  Bactris sphaerocarpa var. *ensifolia* Trail ex Drude in C.F.P.von Martius & auct. suc. (eds.), Fl. Bras. 3(2): 325 (1881).
  Bactris sphaerocarpa var. *minor* Trail ex Drude in C.F.P.von Martius & auct. suc. (eds.), Fl. Bras. 3(2): 325 (1881).
  Bactris sphaerocarpa var. *schizophylla* Drude in C.F.P.von Martius & auct. suc. (eds.), Fl. Bras. 3(2): 325 (1881).

Bactris sphaerocarpa var. *ensifolia* Trail ex Drude = **Bactris sphaerocarpa** Trail

Bactris sphaerocarpa var. *minor* Trail ex Drude = **Bactris sphaerocarpa** Trail

Bactris sphaerocarpa subsp. *pinnatisecta* Trail = **Bactris sphaerocarpa** Trail

Bactris sphaerocarpa var. *platyphylla* Trail = **Bactris sphaerocarpa** Trail

Bactris sphaerocarpa var. *schizophylla* Drude = **Bactris sphaerocarpa** Trail

Bactris stahelii (L.H.Bailey) Glassman = **Bactris simplicifrons** Mart.

Bactris standleyana Burret = **Bactris hondurensis** Standl.

Bactris stictacantha Burret = **Bactris brongniartii** Mart.

Bactris superior L.H.Bailey = **Bactris major** Jacq.

Bactris swabeyi L.H.Bailey = **Bactris major** Jacq.

Bactris sworderiana Becc. = **Bactris setulosa** H.Karst.

Bactris syagroides Barb.Rodr. & Trail in J.Barbosa Rodrigues, Enum. Palm. Nov.: 33 (1875). *Amylocarpus syagroides* (Barb.Rodr. & Trail) Barb.Rodr., Contr. Jard. Bot. Rio de Janeiro 3: 72 (1902).

Brazil (Amazonas, Pará). 84 BZN.
  Bactris cyagroides Barb.Rodr., Enum. Palm. Nov.: 33 (1875), orth. var.
  Bactris multiramosa Burret, Notizbl. Bot. Gart. Berlin-Dahlem 14: 263 (1938).

Bactris sylvatica Barb.Rodr. = **Bactris maraja** Mart.

Bactris tarumanensis Barb.Rodr. = **Bactris acanthocarpa** var. **exscapa**Barb.Rodr.

Bactris tefensis A.J.Hend., Palms Amazon: 220 (1995). Brazil (Amazonas). 84 BZN.

Bactris tenera (H.Karst.) H.Wendl. = **Bactris brongniartii** Mart.

Bactris tenerrima Mart. ex Drude = **Desmoncus mitis** var. **tenerrimus** (Mart. ex Drude) A.Hend.

Bactris tenuis Wallace = **Bactris simplicifrons** Mart.

Bactris tenuissima (Barb.Rodr.) Barb.Rodr. = **Bactris simplicifrons** Mart.

Bactris timbuiensis H.Q.B.Fernald, Bol. Mus. Biol. Prof. Mello-Leitão. Sér. Bot. 5: 4 (1996). Brazil (Espírito Santo). 84 BZL.

Bactris tomentosa Mart., Hist. Nat. Palm. 2: 100 (1826). French Guiana to N. & NE. Brazil. 82 FRG 84 BZE BZN.
  Bactris eumorpha Trail, J. Bot. 15: 9 (1877).
  Bactris eumorpha subsp. *arundinacea* Trail, J. Bot. 15: 10 (1877). *Bactris arundinacea* (Trail) Drude in C.F.P.von Martius & auct. suc. (eds.), Fl. Bras. 3(2): 333 (1881).
  Bactris tomentosa subsp. *capillacea* Trail, J. Bot. 15: 5 (1877). *Bactris capillacea* (Trail) Drude in C.F.P.von Martius & auct. suc. (eds.), Fl. Bras. 3(2): 336 (1881).
  Bactris capinensis Huber, Bol. Mus. Goeldi Paraense Hist. Nat. Ethnogr. 6: 60 (1910).

Bactris tomentosa subsp. *capillacea* Trail = **Bactris tomentosa** Mart.

Bactris tomentosa var. *sphaerocarpa* (Trail) A.Hend. = **Bactris sphaerocarpa** Trail

Bactris trailiana Barb.Rodr. = **Bactris acanthocarpa** var. **trailiana** (Barb.Rodr.) A.Hend.

Bactris trichophylla Burret = **Bactris mexicana** var. **trichophylla** (Burret) A.Hend.

Bactris trichospatha Trail = **Bactris maraja** var. **trichospatha** (Trail) A.Hend.

Bactris trichospatha var. *jurutensis* (Trail) Drude = **Bactris maraja** var. **trichospatha** (Trail) A.Hend.

Bactris trichospatha subsp. *jurutensis* Trail = **Bactris maraja** var. **trichospatha** (Trail) A.Hend.

Bactris trichospatha var. *patens* Drude = **Bactris maraja** var. **trichospatha**(Trail) A.Hend.

Bactris trichospatha var. *robusta* Trail = **Bactris maraja** var. **trichospatha**(Trail) A.Hend.

Bactris trinitensis (L.H.Bailey) Glassman = **Bactris simplicifrons** Mart.

Bactris tucum Burret = **Bactris glaucescens** Drude

Bactris turbinata Spruce = **Bactris hirta** var. **pectinata** (Mart.) Govaerts

Bactris turbinocarpa Barb.Rodr., Enum. Palm. Nov.: 33 (1875). *Pyrenoglyphis turbinocarpa* (Barb.Rodr.) Burret, Repert. Spec. Nov. Regni Veg. 34: 248 (1934). Suriname to Brazil (Pará). 82 SUR 84 BZN.

Bactris uaupensis Spruce = **Bactris simplicifrons** Mart.

Bactris ulei Burret = **Bactris simplicifrons** Mart.

Bactris umbraticola Barb.Rodr. = **Bactris maraja** Mart.

Bactris umbrosa Barb.Rodr. = **Bactris maraja** Mart.

Bactris unaensis Barb.Rodr. = **Bactris hirta** Mart.

Bactris utilis (Oerst.) Benth. & Hook.f. ex Hemsl. = **Bactris gasipaes** Kunth

Bactris venezuelensis Steyerm. = **Bactris corossilla** H.Karst.

Bactris vexans Burret = **?** [84 BZN]

Bactris villosa H.Wendl. ex Hemsl. = **Bactris hondurensis** Standl.

Bactris vulgaris Barb.Rodr., Enum. Palm. Nov., App.: 42 (1879).
E. & S. Brazil. 84 BZE BZL BZS.
Bactris glazioviana Drude in C.F.P.von Martius & auct. suc. (eds.), Fl. Bras. 3(2): 348 (1881).
Bactris polyclada Burret, Repert. Spec. Nov. Regni Veg. 34: 226 (1934).

Bactris wendlandiana Burret = **Bactris hondurensis** Standl.

Bactris xanthocarpa Barb.Rodr. = **Bactris simplicifrons** Mart.

Bacularia

Bacularia F.Muell. ex Hook.f. = **Linospadix** H.Wendl.

Bacularia aequisegmentosa Domin = **Linospadix aequisegmentosa** (Domin) Burret

Bacularia albertisiana Becc. = **Linospadix albertisiana** (Becc.) Burret

Bacularia angustisecta Becc. = **Linospadix albertisiana** (Becc.) Burret

Bacularia arfakiana (Becc.) F.Muell. = **Calyptrocalyx arfakianus** (Becc.) Dowe & M.D.Ferrero

Bacularia canina Becc. = **Linospadix canina** (Becc.) Burret

Bacularia flabellata (Becc.) F.Muell. = **Calyptrocalyx flabellatus** (Becc.) Dowe & M.D.Ferrero

Bacularia intermedia C.T.White = **Linospadix minor** (F.Muell.) F.Muell.

Bacularia longicruris Becc. = **Linospadix albertisiana** (Becc.) Burret

Bacularia microcarya Domin = **Linospadix microcarya** (Domin) Burret

Bacularia minor (F.Muell.) F.Muell. = **Linospadix minor** (F.Muell.) F.Muell.

Bacularia monostachya (Mart.) F.Muell. = **Linospadix monostachya** (Mart.) H.Wendl. & Drude

Bacularia palmeriana F.M.Bailey = **Linospadix palmeriana** (F.M.Bailey) Burret

Bacularia sessilifolia Becc. = **Linospadix microcarya** (Domin) Burret

Balaka

Balaka Becc., Ann. Jard. Bot. Buitenzorg 2: 91 (1885).
11 species, SW. Pacific. 60.

Balaka brachychlamys Burret, Repert. Spec. Nov. Regni Veg. 24: 276 (1928).
Samoa. 60 SAM.

Balaka burretiana Christoph., Bernice P. Bishop Mus. Bull. 128: 32 (1935).

Balaka burretiana Christoph. = **Balaka brachychlamys** Burret

Balaka cuneata Burret = **Balaka seemannii** (H.Wendl.) Becc.

Balaka gracilis Burret = **Balaka seemannii** (H.Wendl.) Becc.

Balaka kersteniana (Sander) Becc. ex Martelli = **Balaka seemannii** (H.Wendl.) Becc.

Balaka leprosa A.C.Sm. = **Balaka longirostris** Becc.

Balaka longirostris Becc., Webbia 4: 270 (1914).
Fiji (Viti Levu). 60 FIJ.
Balaka leprosa A.C.Sm., J. Arnold Arbor. 31: 146 (1950).

Balaka macrocarpa Burret, Occas. Pap. Bernice Pauahi Bishop Mus. 11(4): 5 (1935).
Fiji (Viti Levu, Vanua Levu). 60 FIJ.

Balaka microcarpa Burret, Notizbl. Bot. Gart. Berlin-Dahlem 15: 89 (1940).
Fiji (SE. Viti Levu). 60 FIJ.
Balaka microcarpa var. *longicuspis* Burret, Notizbl. Bot. Gart. Berlin-Dahlem 15: 90 (1940).

Balaka microcarpa var. *longicuspis* Burret = **Balaka microcarpa** Burret

Balaka minuta Burret, Repert. Spec. Nov. Regni Veg. 24: 278 (1928).
Samoa (Savai'i). 60 SAM.
Drymophloeus minutus Rech., Denkschr. Kaiserl. Akad. Wiss., Wien. Math.-Naturwiss. Kl. 85: 237 (1910).
Vitiphoenix minuta (Rech.) Burret, Repert. Spec. Nov. Regni Veg. 24: 278 (1928).

Balaka pauciflora (H.Wendl.) H.E.Moore, Gentes Herb. 8: 535 (1957).
Fiji (Ovalau). 60 FIJ.
*Ptychosperma pauciflorum H.Wendl., Bonplandia 10: 193 (1862). Saguaster pauciflora (H.Wendl.) Kuntze, Revis. Gen. Pl. 2: 735 (1891). Vitiphoenix pauciflora (H.Wendl.) Burret, Repert. Spec. Nov. Regni Veg. 24: 270 (1928). Drymophloeus pauciflorus (H.Wendl.) Becc., Atti Soc. Tosc. Sci. Nat. Pisa Processi Verbali 44: 151 (1934).

Balaka perbrevis (H.Wendl.) Becc. = **Balaka seemannii** (H.Wendl.) Becc.

Balaka polyclada Burret = **Balaka tahitensis** (H.Wendl.) Becc.

Balaka rechingeriana Burret = **Balaka tahitensis** (H.Wendl.) Becc.

Balaka reineckei (Warb.) Burret = **Balaka tahitensis** (H.Wendl.) Becc.

Balaka samoensis Becc., Webbia 4: 267 (1914).
Vitiphoenix samoensis (Becc.) Burret, Notizbl. Bot. Gart. Berlin-Dahlem 13: 600 (1935).
Samoa. 60 SAM.
Balaka siliensis Christoph., Bernice P. Bishop Mus. Bull. 128 :34 (1935).

Balaka seemannii (H.Wendl.) Becc., Ann. Jard. Bot. Buitenzorg 2: 91 (1885).
Fiji (Vanua Levu, Taveuni). 60 FIJ.
Ptychosperma perbreve H.Wendl., Bonplandia 10: 193 (1862). *Balaka perbrevis* (H.Wendl.) Becc., Ann. Jard. Bot. Buitenzorg 2: 91 (1885). *Saguaster perbrevis* (H.Wendl.) Kuntze, Revis. Gen. Pl. 2: 735 (1891).

Ptychosperma seemannii H.Wendl., Bonplandia 10: 192 (1862). *Saguaster seemannii* (H.Wendl.) Kuntze, Revis. Gen. Pl. 2: 735 (1891). *Drymophloeus seemannii* (H.Wendl.) Becc. ex Martelli, Nuovo Giorn. Bot. Ital., n.s., 41: 711 (1935).
Kentia kersteniana Sander, Gard. Chron. 1898(2): 357 (1898). *Ptychosperma kerstenianum* (Sander) Burret, Repert. Spec. Nov. Regni Veg. 24: 263 (1928). *Actinophloeus kerstenianus* (Sander) Burret, Repert. Spec. Nov. Regni Veg. 24: 263 (1928). *Balaka kersteniana* (Sander) Becc. ex Martelli, Nuovo Giorn. Bot. Ital., n.s., 42: 30 (1935).
Balaka gracilis Burret, Repert. Spec. Nov. Regni Veg. 24: 274 (1928).
Drymophloeus kerstenianus Sander ex Burret, Repert. Spec. Nov. Regni Veg. 24: 263 (1928), pro syn.
Balaka cuneata Burret, Occas. Pap. Bernice Pauahi Bishop Mus. 11(4): 6 (1935).
Vitiphoenix seemannii Becc. ex Martelli, Nuovo Giorn. Bot. Ital., n.s., 42: 87 (1935), nom. inval.

Balaka siliensis Christoph. = **Balaka samoensis** Becc.

Balaka spectabilis Burret = **Veitchia vitiensis** (H.Wendl.) H.E.Moore

Balaka streptostachys D.Fuller & Dowe, Palms 43: 10 (1999).
Fiji. 60 FIJ.

Balaka tahitensis (H.Wendl.) Becc., Webbia 4: 271 (1914).
Samoa. 60 SAM.
Ptychosperma tahitensis H.Wendl., Bonplandia 10: 196 (1862). *Saguaster tahitensis* (H.Wendl.) Kuntze, Revis. Gen. Pl. 2: 735 (1891).
Drymophloeus reineckei Warb., Bot. Jahrb. Syst. 25: 590 (1898). *Balaka reineckei* (Warb.) Burret, Repert. Spec. Nov. Regni Veg. 24: 276 (1928).
Balaka polyclada Burret, Repert. Spec. Nov. Regni Veg. 24: 278 (1928).
Balaka rechingeriana Burret, Repert. Spec. Nov. Regni Veg. 24: 275 (1928).
Vitiphoenix polyclada Burret, Repert. Spec. Nov. Regni Veg. 24: 279 (1928).

Balaka tuasivica Christoph., Bernice P. Bishop Mus. Bull. 128: 36 (1935).
Samoa (Savai'i). 60 SAM.

Barbosa

Barbosa Becc. = **Syagrus** Mart.

Barbosa getuliana (Bondar) A.D.Hawkes = **Syagrus macrocarpa** Barb.Rodr.

Barbosa pseudococos Becc. = **Syagrus pseudococos** (Raddi) Glassman

Barcella

Barcella (Trail) Drude in C.F.P.von Martius & auct. suc. (eds.), Fl. Bras. 3(2): 459 (1881).
1 species, N. Brazil. 84.

Barcella odora (Trail) Drude in C.F.P.von Martius & auct. suc. (eds.), Fl. Bras. 3(2): 459 (1881).
Brazil (Amazonas). 84 BZN.
Elaeis odora Trail, J. Bot. 15: 81 (1877).

Barkerwebbia

Barkerwebbia Becc. = **Heterospathe** Scheff.

Barkerwebbia elegans Becc. = **Heterospathe elegans** (Becc.) Becc.

Barkerwebbia humilis (Becc.) Becc. ex Martelli = **Heterospathe humilis** Becc.

Basselinia

Basselinia Vieill., Bull. Soc. Linn. Normandie, II, 6: 230 (1872 publ. 1873).
11 species, New Caledonia. 60.
Microkentia H.Wendl. ex Hook.f. in G.Bentham & J.D.Hooker, Gen. Pl. 3: 895 (1883).
Nephrocarpus Dammer, Bot. Jahrb. Syst. 39: 21 (1906).

Basselinia billardieri (Brongn.) Becc. = **Basselinia gracilis** (Brongn. & Gris) Vieill.

Basselinia deplanchei (Brongn. & Gris) Vieill., Bull. Soc. Linn. Normandie, II, 6: 232 (1873).
EC. & SE. New Caledonia. 60 NWC.
Kentia deplanchei Brongn. & Gris, Bull. Soc. Bot. France 11: 314 (1864). *Cyphokentia deplanchei* (Brongn. & Gris) Brongn., Compt. Rend. Hebd. Séances Acad. Sci. 77: 401 (1873). *Clinostigma deplanchei* (Brongn. & Gris) Becc., Malesia 1: 41 (1877). *Microkentia deplanchei* (Brongn. & Gris) Hook.f. ex Salomon, Palmen: 88 (1887).
Cyphokentia surculosa Brongn., Compt. Rend. Hebd. Séances Acad. Sci. 77: 401 (1873). *Clinostigma surculosum* (Brongn.) Becc., Malesia 1: 41 (1877). *Microkentia surculosa* (Brongn.) Hook.f. ex Salomon, Palmen: 88 (1887). *Basselinia surculosa* (Brongn.) Becc., Webbia 5: 137 (1921).
Microkentia schlechteri Dammer, Bot. Jahrb. Syst. 39: 20 (1906).

Basselinia eriostachys (Brongn.) Becc. = **Basselinia gracilis** (Brongn. & Gris) Vieill.

Basselinia glabrata Becc. = **Alloschmidia glabrata** (Becc.) H.E.Moore

Basselinia favieri H.E.Moore, Allertonia 3: 363 (1984).
NE. New Caledonia (E. Mt. Panié). 60 NWC.

Basselinia gracilis (Brongn. & Gris) Vieill., Bull. Soc. Linn. Normandie, II, 6: 231 (1873).
New Caledonia. 60 NWC.
Kentia gracilis Brongn. & Gris, Bull. Soc. Bot. France 11: 315 (1864). *Cyphokentia gracilis* (Brongn. & Gris) Brongn., Compt. Rend. Hebd. Séances Acad. Sci. 77: 401 (1873). *Clinostigma gracile* (Brongn. & Gris) Becc., Malesia 1: 41 (1877). *Microkentia gracilis* (Brongn. & Gris) Hook.f. ex Salomon, Palmen: 88 (1887).
Cyphokentia billardieri Brongn., Compt. Rend. Hebd. Séances Acad. Sci. 77: 401 (1873). *Microkentia billardieri* (Brongn.) Hook.f. ex Salomon, Palmen: 88 (1887). *Basselinia billardieri* (Brongn.) Becc., Webbia 5: 145 (1921).
Cyphokentia eriostachys Brongn., Compt. Rend. Hebd. Séances Acad. Sci. 77: 401 (1873). *Microkentia eriostachys* (Brongn.) Hook.f. ex Salomon, Palmen: 88 (1887). *Basselinia eriostachys* (Brongn.) Becc. in F.Sarasin & J.Roux, Nova Caledonia, Bot. 1: 123 (1920).
Clinostigma billardieri Becc., Malesia 1: 41 (1877).
Clinostigma eriostachys Becc., Malesia 1: 41 (1877).
Basselinia heterophylla Becc., Webbia 5: 131 (1921).
Microkentia heterophylla Becc. ex Daniker, Vierteljahrschr. Naturf. Ges. Zürich 77(19): 87 (1932), nom. nud.

Basselinia heterophylla Becc. = **Basselinia gracilis** (Brongn. & Gris) Vieill.

Basselinia humboldtiana (Brongn.) H.E.Moore, Allertonia 3: 365 (1984).
EC. New Caledonia. 60 NWC.
Cyphokentia humboldtiana Brongn., Compt. Rend. Hebd. Séances Acad. Sci. 77: 400 (1873). *Clinostigma humboldtianum* (Brongn.) Becc., Malesia 1: 40 (1877). *Kentia humboldtiana* (Brongn.) Brongn. ex H.Wendl. in O.C.E.de Kerchove de Denterghem, Palmiers: 248 (1878).

Basselinia iterata H.E.Moore, Allertonia 3: 366 (1984).
NE. New Caledonia. 60 NWC.

Basselinia pancheri (Brongn. & Gris) Vieill., Bull. Soc. Linn. Normandie, II, 6: 232 (1873).
W. & S. New Caledonia. 60 NWC.
Kentia pancheri Brongn. & Gris, Bull. Soc. Bot. France 11: 316 (1864). *Cyphokentia pancheri* (Brongn. & Gris) Brongn., Compt. Rend. Hebd. Séances Acad. Sci. 77: 400 (1873). *Clinostigma pancheri* (Brongn. & Gris) Becc., Malesia 1: 40 (1877). *Microkentia pancheri* (Brongn. & Gris) Hook.f. ex Salomon, Palmen: 88 (1887).
Nephrocarpus schlechteri Dammer, Bot. Jahrb. Syst. 39: 22 (1906).

Basselinia porphyrea H.E.Moore, Allertonia 3: 367 (1984).
SW. New Caledonia. 60 NWC.

Basselinia sordida H.E.Moore, Allertonia 3: 361 (1984).
NW. & WC. New Caledonia. 60 NWC.

Basselinia surculosa (Brongn.) Becc. = **Basselinia deplanchei** (Brongn. & Gris) Vieill.

Basselinia tomentosa Becc., Webbia 5: 141 (1921).
SC. New Caledonia (Mt. Nakada, Mt. Nemara). 60 NWC.

Basselinia velutina Becc., Webbia 5: 143 (1921).
New Caledonia. 60 NWC.

Basselinia vestita H.E.Moore, Allertonia 3: 368 (1984).
C. New Caledonia (Mé Ori). 60 NWC.

Beata

Beata O.F.Cook = **Coccothrinax** Sarg.

Beata ekmanii (Burret) O.F.Cook = **Coccothrinax ekmannii**

Beccariophoenix

Beccariophoenix Jum. & H.Perrier, Ann. Fac. Sci. Marseille 23: 35 (1915).
1 species, Madagascar. 29.

Beccariophoenix madagascariensis Jum. & H.Perrier, Ann. Fac. Sci. Marseille 23: 35 (1915).
EC. & SE. Madagascar. 29 MDG.

Beethovenia

Beethovenia Engel = **Ceroxylon** Bonpl. ex DC.

Beethovenia cerifera Engl. = **Ceroxylon ceriferum** (H.Karst.) Pittier

Bejaudia

Bejaudia Gagnep. = **Myrialepis** Becc.

Bejaudia cambodiensis Gagnep. = **Myrialepis paradoxa** (Kurz) J.Dransf.

Bentinckia

Bentinckia Berry ex Roxb., Fl. Ind. ed. 1832, 3: 621 (1832).
2 species, Nicobar Is., India. 40 41.
Keppleria Mart. ex Endl., Gen. Pl.: 251 (1837).

Bentinckia condapanna Berry ex Roxb., Fl. Ind. ed. 1832, 3: 621 (1832).
S. India. 40 IND.

Bentinckia ceramica Miq. = **Rhopaloblaste ceramica** (Miq.) Burret

Bentinckia nicobarica (Kurz) Becc., Ann. Jard. Bot. Buitenzorg 2: 165 (1885).
Nicobar Is. 41 NCB.
Orania nicobarica Kurz, J. Bot. 13: 331 (1975).

Bentinckia renda (Blume) Mart. = **Cyrtostachys renda** Blume

Bentinckiopsis

Bentinckiopsis Becc. = **Clinostigma** H.Wendl.

Bentinckiopsis carolinensis (Becc.) Becc. = **Clinostigma carolinense** (Becc.) H.E.Moore & Fosberg

Bessia

Bessia Raf. = **Corypha** L.

Bessia sanguinolenta Raf. = **Corypha umbraculifera** L.

Bismarckia

Bismarckia Hildebr. & H.Wendl., Bot. Zeitung (Berlin) 39: 93 (1881).
1 species, Madagascar. 29.

Bismarckia nobilis Hildebr. & H.Wendl., Bot. Zeitung (Berlin) 39: 94 (1881). *Medemia nobilis* (Hildebr. & H.Wendl.) Gall., Compt. Rend. Hebd. Séances Acad. Sci. 138: 1120 (1904).
N. & W. Madagascar. 29 MDG.

Bisnicholsonia

Bisnicholsonia Kuntze = **Neonicholsonia** Dammer

Blancoa

Blancoa Blume = **Arenga** Labill. ex DC.

Blancoa borneensis (Becc.) Kuntze = **Arenga hastata** (Becc.) Whitmore

Blancoa caudata (Lour.) Kuntze = **Arenga caudata** (Lour.) H.E.Moore

Blancoa hastata (Becc.) Kuntze = **Arenga hastata** (Becc.) Whitmore

Blancoa horsfieldii (Blume) Kuntze = **Arenga porphyrocarpa** (Blume) H.E.Moore

Blancoa nana (Griff.) Kuntze = **Arenga nana** (Griff.) H.E.Moore

Blancoa porphyrocarpa (Blume) Kuntze = **Arenga porphyrocarpa** (Blume) H.E.Moore

Blancoa reinwardtiana (Miq.) Kuntze = **Arenga porphyrocarpa** (Blume) H.E.Moore

Borassodendron

Borassodendron Becc., Webbia 4: 359 (1914).
2 species, Pen. Thailand to W. Malesia. 41 42.

Borassodendron borneense J.Dransf., Reinwardtia 8: 355 (1972).
Borneo. 42 BOR.

Borassodendron machadonis (Ridl.) Becc., Webbia 4: 361 (1914).

Pen. Thailand to Pen. Malaysia. 41 THA 42 MLY.
Borassus machadonis Ridl., J. Straits Branch Roy.
Asiat. Soc. 44: 203 (1905).

Borassus

Borassus L., Sp. Pl.: 1187 (1753).
5 species, Trop. & S. Africa, Madagascar, Trop. & Subtrop.
Asia. 22 23 24 25 26 29 26 27 40 41 42 43 50?
Lontarus Adans., Fam. Pl. 2: 25 (1763).
Borassus aethiopum Mart., Hist. Nat. Palm. 3: 221
(1838). *Borassus flabellifer* var. *aethiopum* (Mart.)
Warb. in H.G.A.Engler (ed.), Pflanzenw. Ost-Afrikas,
B: 20 (1895).
Trop. & S. Africa. 22 BEN GHA IVO MLI NGA SEN
23 CAF CMN 24 CHA ETH soc SUD 25 KEN
TAN UGA 26 MLW MOZ ZIM 27 TVL.
Borassus aethiopum var. *bagamojense* Becc., Webbia
4: 337 (1914).
Borassus aethiopum var. *senegalense* Becc., Webbia 4:
334 (1914).
Borassus deleb Becc., Webbia 4: 339 (1914).

Borassus aethiopum var. *bagamojense* Becc. = **Borassus
aethiopum** Mart.

Borassus aethiopum var. *senegalense* Becc. = **Borassus
aethiopum** Mart.

Borassus caudatum Lour. = **Arenga caudata** (Lour.)
H.E.Moore

Borassus deleb Becc. = **Borassus aethiopum** Mart.

Borassus dichotomus White = **Hyphaene dichotoma**
(White) Furtado

Borassus flabellifer L., Sp. Pl.: 1187 (1753).
Trop. & Subtrop. Asia. 36 CHC 40 IND SRL 41
CBD LAO MLY THA VIE 42 JAW LSI mly 43
NWG 50 QLD?
Borassus flabelliformis L., Syst. Nat. ed. 13, 2: 827
(1770).
Lontarus domestica Gaertn., Fruct. Sem. Pl. 1: 21
(1788).
Borassus tunicatus Lour., Fl. Cochinch. 2: 618 (1790).
Pholidocarpus tunicatus (Lour.) H.Wendl. in
O.C.E. de Kerchove de Denterghem, Palmiers: 235
(1878).
Borassus sundaicus Becc., Webbia 4: 321 (1914).

Borassus flabellifer var. *aethiopum* (Mart.) Warb. =
Borassus aethiopum Mart.

Borassus flabellifer var. *madagascariense* Jum. &
H.Perrier = **Borassus madagascariense** (Jum. &
H.Perrier) Bojer ex Jum. & H.Perrier

Borassus flabelliforme L. = **Borassus flabellifer** L.

Borassus gomutus Lour. = **Arenga pinnata** (Wurmb)
Merr.

Borassus heineanus Becc., Webbia 4: 354 (1914).
New Guinea. 43 NWG.

Borassus ihur Giseke = **Pholidocarpus ihur** (Giseke)
Blume

Borassus machadonis Ridl. = **Borassodendron
machadonis** (Ridl.) Becc.

Borassus madagascariensis (Jum. & H.Perrier) Bojer
ex Jum. & H.Perrier, Ann. Inst. Bot.-Géol. Colon.
Marseille, III, 1(1): 61, t. 33-35 (1913).
W. Madagascar. 29 MDG.
Borassus flabellifer var. *madagascariensis* Jum. &
H.Perrier, Ann. Inst. Bot.-Géol. Colon. Marseille, II,
5: 389, f. 2–4 (1907).

Borassus pinnatifrons Jacq. = **Chamaedorea pinnatifrons**
(Jacq.) Oerst.

Borassus sambiranensis Jum. & H.Perrier, Ann. Inst.
Bot.-Géol. Colon. Marseille, III, 1(1): 67 (1913).
NW. Madagascar. 29 MDG.

Borassus sonneratii Giseke = **Lodoicea maldivica**
(J.F.Gmel.) Pers. ex H.Wendl.

Borassus sundaicum Becc. = **Borassus flabellifer** L.

Borassus tunicatum Lour. = **Borassus flabellifer** L.

Bornoa

Bornoa O.F.Cook = **Attalea** Kunth

Bornoa crassispatha (Mart.) O.F.Cook = **Attalea
crassispatha** (Mart.) Burret

Brahea

Brahea Mart., Hist. Nat. Palm. 3: 243 (1838).
10 species, Mexico to C. America. 79 80.
Erythea S.Watson, Bot. California 2: 211 (1880).
Glaucothea O.F.Cook, J. Wash. Acad. Sci. 5: 237 (1915).

Brahea aculeata (Brandegee) H.E.Moore, Principes 24:
91 (1980).
Mexico (Sonora, Sinaloa, Durango). 79 MXE MXN.
Erythea aculeata Brandegee, Zoe 5: 196 (1905).
Glaucothea aculeata (Brandegee) I.M.Johnst.,
Proc. Calif. Acad. Sci., IV, 12: 993 (1924).

Brahea armata S.Watson, Proc. Amer. Acad. Arts 11:
146 (1876). *Glaucothea armata* (S.Watson) O.F.Cook,
J. Wash. Acad. Sci. 5: 239 (1915).
Mexico (Baja California, Sonora). 79 MXN.
Erythea armata S.Watson, Bot. California 2: 212 (1880).
Brahea roezlii Linden, Ill. Hort. 28: 38 (1881). *Erythea
roezlii* (Linden) Becc. ex Martelli, Ann. Roy. Bot.
Gard. (Calcutta) 13: 320 (1931).
Brahea glauca Hook.f., Hooker's J. Bot. Kew Gard.
Misc. 1882: 64 (1884).
Brahea lucida Hook.f., Hooker's J. Bot. Kew Gard.
Misc. 1882: 64 (1884).
Brahea nobilis Hook.f., Hooker's J. Bot. Kew Gard.
Misc. 1882: 64 (1884).
Erythea elegans Franceschi ex Becc., Webbia 2: 138
(1908). *Glaucothea elegans* (Franceschi ex Becc.)
I.M.Johnst., Proc. Calif. Acad. Sci., IV, 12: 993
(1924). *Brahea elegans* (Franceschi ex Becc.)
H.E.Moore, Baileya 19: 168 (1975).
Erythea clara L.H.Bailey, Gentes Herb. 6: 197 (1943).
Brahea clara (L.H.Bailey) Espejo & López-Ferr.,
Sida 15: 617 (1993).

Brahea bella L.H.Bailey = **Brahea dulcis** (Kunth) Mart.

Brahea berlandieri Bartlett = **Brahea dulcis** (Kunth)
Mart.

Brahea brandegeei (Purpus) H.E.Moore, Baileya 19:
168 (1975).
Mexico (S. Baja California, Sonora). 79 MXN.
Erythea brandegeei Purpus, Gartenflora 1903: 11
(1903). *Glaucothea brandegeei* (Purpus) I.M.Johnst.,
Proc. Calif. Acad. Sci., IV, 12: 992 (1924).

Brahea calcarea Liebm. in C.F.P.von Martius, Hist. Nat.
Palm. 3: 319 (1853).
W. Mexico to Guatemala. 79 MXN MXS MXT? 80
GUA.
Brahea nitida André, Rev. Hort. 59: 344 (1887).
Brahea prominens L.H.Bailey, Gentes Herb. 6: 192
(1943).

Brahea clara (L.H.Bailey) Espejo & López-Ferr. = **Brahea armata** S.Watson

Brahea conduplicala Linden = **?**

Brahea conzattii Bartlett = **Brahea dulcis** (Kunth) Mart.

Brahea decumbens Rzed., Ciencia (Mexico) 15: 89 (1955).
Mexico (Tamaulipas, San Luis Potosí). 79 MXE.

Brahea dulcis (Kunth) Mart., Hist. Nat. Palm. 3: 244 (1838).
Mexico to C. America. 79 MXC MXE MXG MXN MXS 80 BLZ ELS GUA HON NIC.
> *Corypha dulcis* Kunth in F.W.H.von Humboldt, A.J.A.Bonpland & C.S.Kunth, Nov. Gen. Sp. 1: 300 (1816).
> *Corypha frigida* Mohl ex Mart., Hist. Nat. Palm. 3: 244 (1839).
> *Brahea frigida* Devansaye, Rev. Hort. 47: 32 (1875), pro syn.
> *Livistona occidentalis* Hook.f., Rep. Progr. Condition Roy. Bot. Gard. Kew 1887: 64 (1884).
> *Thrinax tunica* Hook.f., Rep. Progr. Condition Roy. Bot. Gard. Kew 1882: 64 (1884).
> *Brahea salvadorensis* H.Wendl. ex Becc., Webbia 2: 105 (1908). *Acoelorrhaphe salvadorensis* (H.Wendl. ex Becc.) Bartlett, Publ. Carnegie Inst. Wash. 461: 32 (1935). *Erythea salvadorensis* (H.Wendl. ex Becc.) H.E.Moore, Gentes Herb. 8: 217 (1951).
> *Brahea dulcis* var. *montereyensis* Becc., Ann. Roy. Bot. Gard. (Calcutta) 13: 403 (1931).
> *Acoelorrhaphe cookii* Bartlett, Publ. Carnegie Inst. Wash. 461: 32 (1935).
> *Brahea berlandieri* Bartlett, Publ. Carnegie Inst. Wash. 461: 31 (1935).
> *Brahea conzattii* Bartlett, Publ. Carnegie Inst. Wash. 461: 30 (1935).
> *Brahea schippii* Burret, Notizbl. Bot. Gart. Berlin-Dahlem 12: 304 (1935). *Acoelorrhaphe schippii* (Burret) Dahlgren, Field Mus. Nat. Hist., Bot. Ser. 14: 9 (1936).
> *Brahea bella* L.H.Bailey, Gentes Herb. 6: 194 (1943).
> *Copernicia depressa* Liebm. ex Dalgrem, Field Mus. Nat. Hist., Bot. Ser. 14: t. 170 (1959), nom. nud.

Brahea dulcis J.G.Cooper = **Washingtonia filifera** (Linden ex André) H.Wendl. ex de Bary

Brahea dulcis var. *montereyensis* Becc. = **Brahea dulcis** (Kunth) Mart.

Brahea edulis H.Wendl. ex S.Watson, Proc. Amer. Acad. Arts 11: 120 (1876). *Erythea edulis* (H.Wendl. ex S.Watson) S.Watson, Bot. California 2: 212 (1880). Guadalupe. 79 MXI.

Brahea elegans (Franceschi ex Becc.) H.E.Moore = **Brahea armata** S.Watson

Brahea filamentosa S.Watson = **Washingtonia filifera** (Linden ex André) H.Wendl. ex de Bary

Brahea filifera (Linden ex André) W.Watson = **Washingtonia filifera** (Linden ex André) H.Wendl. ex de Bary

Brahea frigida Devansaye = **Brahea dulcis** (Kunth) Mart.

Brahea glauca Hook.f. = **Brahea armata** S.Watson

Brahea lucida Hook.f. = **Brahea armata** S.Watson

Brahea minima (Nutt.) H.Wendl. = **Sabal minor** (Jacq.) Pers.

Brahea moorei L.H.Bailey ex H.E.Moore, Gentes Herb. 8: 219 (1951).
NE. Mexico. 79 MXE.

Brahea nitida André = **Brahea calcarea** Liebm.

Brahea nobilis Hook.f. = **Brahea armata** S.Watson

Brahea pimo Becc., Webbia 2: 103 (1908). *Acoelorrhaphe pimo* (Becc.) Bartlett, Publ. Carnegie Inst. Wash. 461: 32 (1935). *Erythea pimo* (Becc.) H.E.Moore, Gentes Herb. 8: 216 (1951).
SW. Mexico. 79 MXS.

Brahea prominens L.H.Bailey = **Brahea calcarea** Liebm.

Brahea psilocalyx Burret = **Acoelorrhaphe wrightii** (Griseb. & H.Wendl.) H.Wendl. ex Becc.

Brahea robusta Voss = **Washingtonia robusta** H.Wendl.

Brahea roezlii Linden = **Brahea armata** S.Watson

Brahea salvadorensis H.Wendl. ex Becc. = **Brahea dulcis** (Kunth) Mart.

Brahea sarukhanii H.J.Quero, Palms 44: 110 (2000).
Mexico (Nayarit, Jalisco). 79 MXS.

Brahea schippii Burret = **Brahea dulcis** (Kunth) Mart.

Brahea serrulata (Michx.) H.Wendl. = **Serenoa repens** (W.Bartram) Small

Brassiophoenix

Brassiophoenix Burret, Notizbl. Bot. Gart. Berlin-Dahlem 12: 345 (1935).
2 species, New Guinea. 43.

Brassiophoenix drymophloeoides Burret, Notizbl. Bot. Gart. Berlin-Dahlem 12: 346 (1935).
Papua New Guinea. 43 NWG.

Brassiophoenix schumannii (Becc.) Essig, Principes 19: 102 (1975).
Papua New Guinea. 43 NWG.
> *Actinophloeus schumannii* Becc. in K.M.Schumann & U.M.Hollrung, Fl. Kais. Wilh. Land: 15 (1889). *Drymophloeus schumannii* (Becc.) Warb. ex K.Schum. & Lauterb., Fl. Schutzgeb. Südsee: 207 (1900). *Ptychococcus schumannii* (Becc.) Burret, Repert. Spec. Nov. Regni Veg. 24: 262 (1928).

Brongniartikentia

Brongniartikentia Becc., Webbia 5: 116 (1921).
2 species, New Caledonia. 60.

Brongniartikentia lanuginosa H.E.Moore, Gentes Herb. 11(3): 154 (1976).
NE. New Caledonia. 60 NWC.

Brongniartikentia vaginata (Brongn.) Becc., Webbia 5: 117 (1921).
S. New Caledonia. 60 NWC.
> *Cyphokentia vaginata* Brongn., Compt. Rend. Hebd. Séances Acad. Sci. 77: 402 (1873). *Clinostigma vaginatum* (Brongn.) Becc., Malesia 1: 41 (1877).

Burretiokentia

Burretiokentia Pic.Serm., Webbia 11: 122 (1955).
5 species, New Caledonia. 60.
> *Rhynchocarpa* Becc., Palme Nuova Caledonia: 37 (1920), nom. illeg.

Burretiokentia dumasii Pintaud & Hodel, Principes 42: 160 (1998).
WC. New Caledonia (Mé Maoya massif). 60 NWC.

Burretiokentia grandiflora Pintaud & Hodel, Principes 42: 162 (1998).
SE. New Caledonia. 60 NWC.

Burretiokentia hapala H.E.Moore, Principes 13: 67 (1969).
N. New Caledonia. 60 NWC.

Burretiokentia koghiensis Pintaud & Hodel, Principes 42: 164 (1998).
SE. New Caledonia. 60 NWC.

Burretiokentia vieillardii (Brongn. & Gris) Pic.Serm., Webbia 11: 124 (1955).
New Caledonia. 60 NWC.
Kentia vieillardii Brongn. & Gris, Bull. Soc. Bot. France 11: 313 (1864). *Cyphosperma vieillardii* (Brongn. & Gris) H.Wendl. ex Salomon, Palmen: 87 (1887). *Rhynchocarpa viellardii* (Brongn. & Gris) Becc., Palme Nuova Caledonia: 37 (1920).

Butia

Butia (Becc.) Becc., Agric. Colon. 10: 489 (1916).
9 species, Brazil to S. South America. 84 85.

Butia amadelpha (Barb.Rodr.) Burret = **Butia paraguayensis** (Barb.Rodr.) L.H.Bailey

Butia archeri (Glassman) Glassman, Principes 23: 70 (1979).
Brazil (Goiás, Brasília D.F., Minas Gerais, São Paulo). 84 BZC BZL.
Syagrus archeri Glassman, Fieldiana, Bot. 31: 235 (1967).

Butia arenicola (Barb.Rodr.) Burret = **Butia paraguayensis** (Barb.Rodr.) L.H.Bailey

Butia argentea (Engel) Nehrl. = **Syagrus sancona** (Kunth) H.Karst.

Butia bonnetii Becc. = **Butia capitata** (Mart.) Becc.

Butia campicola (Barb.Rodr.) Noblick
Paraguay. 85 PAR.
Cocos campicola Barb.Rodr., Palm. Hassler.: 6 (1900). *Syagrus campicola* (Barb.Rodr.) Becc., Agric. Colon. 10: 465 (1916).

Butia capitata (Mart.) Becc., Agric. Colon. 10: 504 (1916).
Brazil to Uruguay. 84 BZC BZE BZL BZS 85 URU.
Cocos capitata Mart., Hist. Nat. Palm. 2: 114 (1826). *Calappa capitata* (Mart.) Kuntze, Revis. Gen. Pl. 2: 982 (1891). *Syagrus capitata* (Mart.) Glassman, Fieldiana, Bot. 32: 143 (1970).
Cocos leiospatha Barb.Rodr., Prot.-App. Enum. Palm. Nov.: 44 (1879). *Calappa leiospatha* (Barb.Rodr.) Kuntze, Revis. Gen. Pl. 2: 982 (1891). *Butia leiospatha* (Barb.Rodr.) Becc., Agric. Colon. 10: 520 (1916).
Cocos odorata Barb.Rodr., Pl. Jard. Rio de Janeiro 1: 11 (1891). *Butia capitata* var. *odorata* (Barb.Rodr.) Becc., Agric. Colon. 10: 513 (1916).
Cocos pulposa Barb.Rodr., Pl. Jard. Rio de Janeiro 1: 14 (1891). *Butia capitata* var. *pulposa* (Barb.Rodr.) Becc., Agric. Colon. 19: 516 (1916). *Butia pulposa* (Barb.Rodr.) Nehrl., Amer. Eag!e 24(17): 1 (5 Sept. 1929).
Cocos elegantissima Chabaud, Rev. Hort. 78: 144 (1906), nom. illeg. *Butia capitata* var. *elegantissima* (Chabaud) Becc., Agric. Colon. 10: 517 (1916).
Cocos erythrospatha Chabaud, Rev. Hort. 78: 144 (1906). *Butia capitata* var. *erythrospatha* (Chabaud) Becc., Agric. Colon. 10: 515 (1916).

Cocos lilaceiflora Chabaud, Rev. Hort. 78: 144 (1906). *Butia capitata* var. *lilaceiflora* (Chabaud) Becc., Agric. Colon. 10: 518 (1916).
Butia bonnetii Becc., Agric. Colon. 10: 504 (1916).
Butia capitata var. *subglobbosa* Becc., Agric. Colon. 10: 513 (1916).
Butia capitata var. *virescens* Becc., Agric. Colon. 10: 519 (1916).
Cocos nehrlingiana Abbott ex Nehrl., Amer. Eagle, 17 Feb.: (1927). *Butia capitata* var. *nehrlingiana* (Abbott ex Nehrl.) L.H.Bailey, Gentes Herb. 4: 33 (1936). *Butia nehrlingiana* (Abbott ex Nehrl.) Abbott ex Nehrl., Amer. Eagle 24(17): 1 (5 Sept. 1929).
Butia capitata var. *strictior* L.H.Bailey, Gentes Herb. 4: 32 (1936).
Butia capitata var. *rubra* Mattos, Loefgrenia 71: [1] (1977).

Butia capitata var. *elegantissima* (Chabaud) Becc. = **Butia capitata** (Mart.) Becc.

Butia capitata var. *erythrospatha* (Chabaud) Becc. = **Butia capitata** (Mart.) Becc.

Butia capitata var. *lilaceiflora* (Chabaud) Becc. = **Butia capitata** (Mart.) Becc.

Butia capitata var. *nehrlingiana* (Abbott ex Nehrl.) L.H.Bailey = **Butia capitata** (Mart.) Becc.

Butia capitata var. *odorata* (Barb.Rodr.) Becc. = **Butia capitata** (Mart.) Becc.

Butia capitata var. *pulposa* (Barb.Rodr.) Becc. = **Butia capitata** (Mart.) Becc.

Butia capitata var. *rubra* Mattos = **Butia capitata** (Mart.) Becc.

Butia capitata var. *strictior* L.H.Bailey = **Butia capitata** (Mart.) Becc.

Butia capitata var. *subglobbosa* Becc. = **Butia capitata** (Mart.) Becc.

Butia capitata var. *virescens* Becc. = **Butia capitata** (Mart.) Becc.

Butia dyeriana (Barb.Rodr.) Burret = **Butia paraguayensis** (Barb.Rodr.) L.H.Bailey

Butia eriospatha (Mart. ex Drude) Becc., Agric. Colon. 10: 496 (1916).
S. Brazil. 84 BZS.
Butia eriospatha subsp. *punctata* Bomhard in ?, .
Cocos eriospatha Mart. ex Drude in C.F.P.von Martius & auct. suc. (eds.), Fl. Bras. 3(2): 424 (1881). *Calappa eriospatha* (Mart. ex Drude) Kuntze, Revis. Gen. Pl. 2: 982 (1891). *Syagrus eriospatha* (Mart. ex Drude) Glassman, Fieldiana, Bot. 32: 145 (1970).

Butia eriospatha subsp. *punctata* Bomhard = **Butia eriospatha** (Mart. ex Drude) Becc.

Butia leiospatha (Barb.Rodr.) Becc. = **Butia capitata** (Mart.) Becc.

Butia microspadix Burret, Notizbl. Bot. Gart. Berlin-Dahlem 10: 1050 (1930).
Brazil (São Paulo, Paraná, Rio Grande do Sul). 84 BZL BZS.
Syagrus hatschbachii Glassman, Fieldiana, Bot. 31: 240 (1967).

Butia nehrlingiana (Abbott ex Nehrl.) Abbott ex Nehrl. = **Butia capitata** (Mart.) Becc.

Butia paraguayensis (Barb.Rodr.) L.H.Bailey, Gentes Herb. 4: 47 (1936).

Brazil, Paraguay, Uruguay, NE. Argentina. 84 BZC BZL BZS 85 AGE PAR URU.

Cocos paraguayensis Barb.Rodr., Palm. Paraguay.: 9 (1899). *Butia yatay* var. *paraguayensis* (Barb.Rodr.) Becc., Agric. Colon. 10: 503 (1916). *Syagrus paraguayensis* (Barb.Rodr.) Glassman, Fieldiana, Bot. 32: 151 (1970). *Butia yatay* subsp. *paraguayensis* (Barb.Rodr.) Xifreda & Sanso, Hickenia 2: 207 (1996).

Cocos amadelpha Barb.Rodr., Palm. Hassler.: 7 (1900). *Butia amadelpha* (Barb.Rodr.) Burret, Notizbl. Bot. Gart. Berlin-Dahlem 10: 1050 (1930). *Syagrus amadelpha* (Barb.Rodr.) Frambach ex Dahlgren, Field Mus. Nat. Hist., Bot. Ser. 14: 108 (1936).

Cocos arenicola Barb.Rodr., Sert. Palm. Brasil. 1: 100 (1903). *Butia arenicola* (Barb.Rodr.) Burret, Notizbl. Bot. Gart. Berlin-Dahlem 10: 1051 (1930). *Syagrus arenicola* (Barb.Rodr.) Frambach, Field Mus. Nat. Hist., Bot. Ser. 14: 109 (1936).

Cocos dyeriana Barb.Rodr., Sert. Palm. Brasil. 1: 95 (1903). *Syagrus dyeriana* (Barb.Rodr.) Becc., Agric. Colon. 10: 466 (1916). *Butia dyeriana* (Barb.Rodr.) Burret, Notizbl. Bot. Gart. Berlin-Dahlem 13: 696 (1937).

Cocos wildemaniana Barb.Rodr., Sert. Palm. Brasil. 1: 101 (1903). *Butia wildemaniana* (Barb.Rodr.) Burret, Notizbl. Bot. Gart. Berlin-Dahlem 10: 1050 (1930). *Syagrus wildemaniana* (Barb.Rodr.) Frambach ex Dahlgren, Field Mus. Nat. Hist., Bot. Ser. 14: 124 (1936).

Butia pungens Becc., Agric. Colon. 10: 523 (1916).

Butia poni (Hauman) Burret = **Butia yatay** (Mart.) Becc.

Butia pulposa (Barb.Rodr.) Nehrl. = **Butia capitata** (Mart.) Becc.

Butia pungens Becc. = **Butia paraguayensis** (Barb.Rodr.) L.H.Bailey

Butia purpurascens Glassman, Principes 23: 67 (1979). Brazil (Goiás). 84 BZC.

Butia stolonifera (Barb.Rodr.) Becc., Agric. Colon. 10: 492 (1916).
Uruguay (Pan d'Azucar). 85 URU. — Provisionally accepted.
Cocos stolonifera Barb.Rodr., Contr. Jard. Bot. Rio de Janeiro 1: 40 (1901).

Butia wildemaniana (Barb.Rodr.) Burret = **Butia paraguayensis** (Barb.Rodr.) L.H.Bailey

Butia yatay (Mart.) Becc., Agric. Colon. 10: 498 (1916). S. Brazil, Uruguay, NE. Argentina. 84 BZS 85 AGE URU.
Cocos yatay Mart. in A.D.d'Orbigny, Voy. Amér. Mér. 7(3): 93 (1844). *Calappa yatay* (Mart.) Kuntze, Revis. Gen. Pl. 2: 982 (1891). *Syagrus yatay* (Mart.) Glassman, Fieldiana, Bot. 32: 157 (1970).
Cocos poni Hauman, Physis (Buenos Aires) 4: 604 (1919). *Butia poni* (Hauman) Burret, Notizbl. Bot. Gart. Berlin-Dahlem 10: 1051 (1930).

Butia yatay subsp. *paraguayensis* (Barb.Rodr.) Xifreda & Sanso = **Butia paraguayensis** (Barb.Rodr.) L.H.Bailey

Butia yatay var. *paraguayensis* (Barb.Rodr.) Becc. = **Butia paraguayensis** (Barb.Rodr.) L.H.Bailey

× *Butiarecastrum*

× *Butiarecastrum* Prosch. = × **Butyagrus**

× *Butiarecastrum nabonnandii* Prosch. = × **Butyagrus nabonnandii**

× *Butyagrus*

× *Butyagrus* Vorster = **Butia** × **Syagrus**

× *Butyagrus nabonnandii* (Prosch.) Vorster = **Butia capitata** × **Syagrus romzanoffiana**

Calamosagus

Calamosagus Griff. = **Korthalsia** Blume

Calamosagus harinifolius Griff. = **Korthalsia laciniosa** (Griff.) Mart.

Calamosagus laciniosus Griff. = **Korthalsia laciniosa** (Griff.) Mart.

Calamosagus ochriger Griff. = **Korthalsia rigida** Blume

Calamosagus polystachys (Mart.) H.Wendl. = **Korthalsia rigida** Blume

Calamosagus scaphiger (Mart.) Griff. = **Korthalsia rostrata** Blume

Calamosagus wallichiifolius Griff. = **Korthalsia laciniosa** (Griff.) Mart.

Calamus

Calamus L., Sp. Pl.: 325 (1753). *Palmijuncus* Rumph. ex Kuntze, Revis. Gen. Pl. 2: 731 (1891), nom. illeg. 374 species, Trop. Africa, Trop. & Subtrop. Asia to SW. Pacific. 22 23 24 25 26 36 38 40 41 42 43 50 60.
Rotanga Boehm., Defin. Gen. Pl.: 395 (1760).
Rotang Adans., Fam. Pl. 2: 24 (1763).
Zalaccella Becc., Ann. Roy. Bot. Gard. (Calcutta) 11(1): 496 (1908).
Cornera Furtado, Gard. Bull. Singapore 14: 518 (1955).
Schizospatha Furtado, Gard. Bull. Singapore 14: 525 (1955).

Calamus acanthochlamys J.Dransf., Kew Bull. 45: 85 (1990).
Borneo (Brunei, Sarawak). 42 BOR.

Calamus acanthophyllus Becc., Webbia 3: 229 (1910). E. Thailand to Laos. 41 LAO THA.

Calamus acanthopis Griff. = **Daemonorops grandis** (Griff.) Mart.

Calamus acanthospathus Griff., Calcutta J. Nat. Hist. 5: 39 (1845). *Palmijuncus acanthospathus* (Griff.) Kuntze, Revis. Gen. Pl. 2: 733 (1891).
C. Himalaya to S. China and Indo-China. 36 CHC CHS CHT 40 ASS EHM IND NEP 41 LAO MYA THA.
Calamus montanus T.Anderson, J. Linn. Soc., Bot. 11: 9 (1871). *Palmijuncus montanus* (T.Anderson) Kuntze, Revis. Gen. Pl. 2: 733 (1891).
Calamus feanus Becc. in J.D.Hooker, Fl. Brit. India 6: 448 (1892).
Calamus feanus var. *medogensis* S.J.Pei & S.Y.Chen, Acta Phytotax. Sin. 27: 137 (1989).
Calamus yunnanensis var. *densiflorus* S.J.Pei & S.Y.Chen, Acta Phytotax. Sin. 27: 135 (1989), nom. inval.
Calamus yunnanensis var. *intermedius* S.J.Pei & S.Y.Chen, Acta Phytotax. Sin. 27: 137 (1989), nom. inval.
Calamus yunnanensis Govaerts, World Checklist Seed Pl. 3(1): 11 (1999).

Calamus accedens (Blume) Miq. = **Daemonorops rubra** (Reinw. ex Mart.) Blume

Calamus acidus Becc., Ann. Roy. Bot. Gard. (Calcutta) 11(1): 496 (1908).
Sulawesi. 42 SUL. — Provisionally accepted.

Calamus acuminatus Becc., Ann. Roy. Bot. Gard. (Calcutta) 11(App.): 16 (1913).
N. Borneo. 42 BOR.

Calamus adspersus (Blume) Blume, Rumphia 3: 40 (1847).
Jawa. 42 JAW.
*Daemonorops adspersa Blume, Rumphia 2: ii (1838). *Palmijuncus adspersus* (Blume) Kuntze, Revis. Gen. Pl. 2: 733 (1891).

Calamus aggregatus Burret, Notizbl. Bot. Gart. Berlin-Dahlem 15: 812 (1943).
Myanmar. 41 MYA.

Calamus aidae Fernando, Gard. Bull. Singapore 41: 49 (1988 publ. 1989).
Philippines. 42 PHI.

Calamus akimensis Becc. = **Calamus deerratus** H.Mann & H.Wendl.

Calamus albus Pers., Syn. Pl. 1: 383 (1805). *Rotang albus* (Pers.) Baill., Hist. Pl. 13: 299 (1895).
Maluku. 42 MOL. — Provisionally accepted.

Calamus altiscandens Burret, J. Arnold Arbor. 20: 196 (1939).
New Guinea. 43 NWG.

Calamus amarus Lour. = **Calamus tenuis** Roxb.

Calamus amboinensis Miq. = **Daemonorops calapparia** (Mart.) Blume

Calamus amischus Burret = **Calamus australis** Mart.

Calamus amplectens Becc. = **Calamus javensis** Blume

Calamus amplijugus J.Dransf., Kew Bull. 36: 787 (1982).
W. Borneo. 42 BOR.

Calamus anceps Blume = **Calamus melanoloma** Mart.

Calamus andamanicus Kurz, J. Asiat. Soc. Bengal, Pt. 2, Nat. Hist. 43(2): 211 (1874). *Palmijuncus andamanicus* (Kurz) Kuntze, Revis. Gen. Pl. 2: 733 (1891).
Andaman Is. (incl. Coco Is.), Nicobar Is. 41 AND NCB.

Calamus angustifolius Griff. = **Daemonorops angustifolia** (Griff.) Mart.

Calamus anomalus Burret, Notizbl. Bot. Gart. Berlin-Dahlem 12: 320 (1935).
New Guinea (W. Owen Standley Range). 43 NWG.
Calamus setiger Burret, Notizbl. Bot. Gart. Berlin-Dahlem 13: 320 (1936). *Schizospatha setigera* (Burret) Furtado, Gard. Bull. Singapore 14: 526 (1955).

Calamus aquatilis Ridl. = **Calamus erinaceus** (Becc.) J.Dransf.

Calamus arborescens Griff., Calcutta J. Nat. Hist. 5: 33 (1845). *Palmijuncus arborescens* (Griff.) Kuntze, Revis. Gen. Pl. 2: 733 (1891).
Myanmar to Pen. Malaysia. 41 MYA THA 42 MLY.
Calamus hostilis Wall. ex Voigt, Hort. Suburb. Calcutt.: 639 (1845), nom. nud.

Calamus arfakianus Becc. in L.S.Gibbs, Fl. Arfak Mts.: 95 (1917).
W. New Guinea. 43 NWG.

Calamus aruensis Becc., Malesia 3: 61 (1886). *Palmijuncus aruensis* (Becc.) Kuntze, Revis. Gen. Pl. 2: 733 (1891).
New Guinea to N. Queensland. 43 NWG 50 QLD.
Calamus hollrungii Becc. in K.M.Schumann &

U.M.Hollrung, Fl. Kais. Wilh. Land: 17 (1889).
Calamus latisectus Burret, Notizbl. Bot. Gart. Berlin-Dahlem 13: 319 (1936).

Calamus arugda Becc., Philipp. J. Sci., C 4: 622 (1909).
Philippines (Luzon). 42 PHI.

Calamus ashtonii J.Dransf., Bot. J. Linn. Soc. 81: 9 (1980).
Borneo (Brunei, Sarawak). 42 BOR.

Calamus asperrimus Blume in J.J.Roemer & J.A.Schultes, Syst. Veg. 7: 1327 (1830). *Palmijuncus asperrimus* (Blume) Kuntze, Revis. Gen. Pl. 2: 733 (1891). *Rotang asperrimus* (Blume) Baill., Hist. Pl. 13: 300 (1895).
Jawa. 42 JAW.

Calamus aureus Reinw. ex Mart. = **Calamus ornatus** Blume

Calamus australis Mart., Hist. Nat. Palm. 3: 342 (1853). *Palmijuncus australis* (Mart.) Kuntze, Revis. Gen. Pl. 2: 733 (1891).
NE. Queensland. 50 QLD.
Calamus obstruens F.Muell., Fragm. 5: 48 (1865).
Calamus jaboolum F.M.Bailey, Bull. Dept. Agric. Queensland 13: 14 (1899).
Calamus amischus Burret, Notizbl. Bot. Gart. Berlin-Dahlem 15: 800 (1943).

Calamus austroguangxiensis S.J.Pei & S.Y.Chen, Acta Phytotax. Sin. 27: 144 (1989).
China (Guangxi). 36 CHS.

Calamus axillaris Becc. in J.D.Hooker, Fl. Brit. India 6: 456 (1893).
Pen. Thailand to Pen. Malaysia, NW. Borneo. 41 THA 42 BOR MLY.
Calamus hendersonii Furtado, Gard. Bull. Singapore 15: 100 (1956).
Calamus riparius Furtado, Gard. Bull. Singapore 15: 103 (1956).

Calamus bacularis Becc., For. Borneo: 609 (1902).
Borneo (Sarawak). 42 BOR.

Calamus balansaeanus Becc. = **Calamus henryanus** Becc.

Calamus balansaeanus var. *castaneolepis* (C.F.Wei) S.J.Pei & S.Y.Chen = **Calamus henryanus** Becc.

Calamus balerensis Fernando, Gard. Bull. Singapore 41: 51 (1988 publ. 1989).
Philippines (Luzon). 42 PHI.

Calamus balingensis Furtado, Gard. Bull. Singapore 15: 240 (1956).
Pen. Thailand to Pen. Malaysia. 41 THA 42 MLY.

Calamus bankae W.J.Baker & J.Dransf., Kew Bull. 57: 860 (2002).
Papua New Guinea. 43 NWG.

Calamus baratangensis Renuka & Vij.Kumar, Rheedea 4: 141 (1994).
Andaman Is. 41 AND.

Calamus barbatus Zipp. ex Blume, Rumphia 3: 42 (1847). *Daemonorops barbata* (Zipp. ex Blume) Mart., Hist. Nat. Palm. 3: 330 (1853). *Palmijuncus barbatus* (Zipp. ex Blume) Kuntze, Revis. Gen. Pl. 2: 732 (1891). *Rotang barbatus* (Zipp. ex Blume) Baill., Hist. Pl. 13: 300 (1895).
New Guinea. 43 NWG.

Calamus basui Renuka & Vij.Kumar, Rheedea 4: 120 (1994).
Andaman Is. 41 AND.

Calamus batanensis (Becc.) Baja-Lapis, Sylvatrop 12: 73 (1987 publ. 1989).
N. Philippines. 42 PHI.
Calamus siphonospathus var. *batanensis* Becc., Philipp. J. Sci. 3: 342 (1908).

Calamus barteri Drude = **Calamus deerratus** H.Mann & H.Wendl.

Calamus belumutensis Furtado = **Calamus luridus** Becc.

Calamus benkulensis Becc., Ann. Roy. Bot. Gard. (Calcutta) 11(App.): 59 (1913).
Sumatera. 42 SUM.

Calamus benomensis Furtado = **Calamus viridispinus** Becc.

Calamus bicolor Becc., Ann. Roy. Bot. Gard. (Calcutta) 11(App.): 126 (1913).
Philippines (Mindanao). 42 PHI.

Calamus bifacialis Burret = **Daemonorops scapigera** Becc.

Calamus billitonensis Becc. ex K.Heyne, Nutt. Pl. Ned.-Ind., ed. 2, 1: 365 (1922).
Sumatera (Belitung). 42 SUM.

Calamus bimaniferus T.Evans & al., Kew Bull. 55: 936 (2000).
Laos. 41 LAO.

Calamus blancoi Kunth = **Calamus usitatus** Blanco

Calamus blumei Becc., Ann. Roy. Bot. Gard. (Calcutta) 11(1): 340 (1908).
Pen. Thailand to W. Malesia. 41 THA 42 BOR MLY SUM.
Calamus mawaiensis Furtado, Gard. Bull. Singapore 15: 75 (1956).
Calamus penibukanensis Furtado, Gard. Bull. Singapore 15: 79 (1956).
Calamus slootenii Furtado, Gard. Bull. Singapore 15: 79 (1956).

Calamus bonianus Becc. = **Calamus tetradactylus** Hance

Calamus boniensis Becc. ex K.Heyne, Nutt. Pl. Ned.-Ind., ed. 2, 1: 365 (1922).
Sulawesi. 42 SUL.

Calamus borneensis Miq. = **Calamus javensis** Blume

Calamus borneensis Becc. = **Calamus javensis** Blume

Calamus bousigonii Becc., Rec. Bot. Surv. India 2: 209 (1902).
Indo-China. 41 CBD THA VIE.

subsp. **bousigonii**
Indo-China. 41 CBD THA VIE.

subsp. **smitinandii** J.Dransf., Kew Bull. 55: 713 (2000).
Pen. Thailand. 41 THA.

Calamus brandisii Becc. in J.D.Hooker, Fl. Brit. India 6: 448 (1892).
S. India. 40 IND.

Calamus brachystachys Becc. = **Calamus conirostris** Becc.

Calamus brassii Burret, Notizbl. Bot. Gart. Berlin-Dahlem 12: 316 (1935).
New Guinea. 43 NWG.

Calamus brevifolius Becc., Bot. Jahrb. Syst. 58: 455 (1923).
New Guinea. 43 NWG.

Calamus brevifrons Mart. = **Calamus usitatus** Blanco

Calamus brevispadix Ridl. = **Calamus viridispinus** var. **viridispinus**

Calamus bubuensis Becc. = **Calamus viridispinus** var. **viridispinus**

Calamus burckianus Becc., Rec. Bot. Surv. India 2: 198 (1902).
Jawa. 42 JAW.

Calamus burkillianus Becc. ex Ridl., Fl. Malay Penins. 5: 56 (1925).
Pen. Thailand to Pen. Malaysia. 41 THA 42 MLY.
Calamus chibehensis Furtado, Gard. Bull. Singapore 15: 244 (1956).

Calamus buroensis Mart., Hist. Nat. Palm. 3: 336 (1853).
Maluku. 42 MOL.
Calamus viminalis subsp. *prostratus* Blume in J.J.Roemer & J.A.Schultes, Syst. Veg. 7(2): 1328 (1830).

Calamus cabrae De Wild. & T.Durand = **Eremospatha cabrae** (De Wild. & T.Durand) De Wild.

Calamus caesius Blume, Rumphia 3: 57 (1847).
Palmijuncus caesius (Blume) Kuntze, Revis. Gen. Pl. 2: 733 (1891). *Rotang caesius* (Blume) Baill., Hist. Pl. 13: 300 (1895).
Pen. Thailand to Philippines (Palawan). 41 THA 42 BOR MLY PHI SUM.
Calamus glaucescens Blume, Rumphia 3: 65 (1847), nom. illeg. *Palmijuncus glaucescens* Kuntze, Revis. Gen. Pl. 2: 733 (1891).

Calamus calapparius Mart. = **Daemonorops calapparia** (Mart.) Blume

Calamus calicarpus Griff. = **Daemonorops calicarpa** (Griff.) Mart.

Calamus calolepis Miq. = **Calamus melanoloma** Mart.

Calamus cambojensis Becc. = **Calamus tetradactylus** Hance

Calamus caryotoides A.Cunn. ex Mart., Hist. Nat. Palm. 3: 338 (1853). *Palmijuncus caryotoides* (A.Cunn. ex Mart.) Kuntze, Revis. Gen. Pl. 2: 733 (1891).
NE. Queensland. 50 QLD.

Calamus castaneus Griff., Calcutta J. Nat. Hist. 5: 28 (1845).
Pen. Thailand to Sumatera. 41 THA 42 MLY SUM.
Calamus griffithianus Mart., Hist. Nat. Palm. 3: 332 (1853). *Palmijuncus griffithianus* (Mart.) Kuntze, Revis. Gen. Pl. 2: 733 (1891). *Calamus castaneus* var. *griffithianus* (Mart.) Furtado, Gard. Bull. Singapore 15: 50 (1956).

Calamus castaneus var. *griffithianus* (Mart.) Furtado = **Calamus castaneus** Griff.

Calamus cawa Blume, Rumphia 3: 31 (1847). *Rotang cawa* (Blume) Baill., Hist. Pl. 13: 299 (1895).
Maluku. 42 MOL.

Calamus ceratophorus Conrard, Notul. Syst. (Paris) 7: 24 (1938).
S. Vietnam. 41 VIE.

Calamus chibehensis Furtado = **Calamus burkillianus** Becc. ex Ridl.

Calamus ciliaris Blume in J.J.Roemer & J.A.Schultes, Syst. Veg. 7: 1330 (1830). *Palmijuncus ciliaris* (Blume) Kuntze, Revis. Gen. Pl. 2: 733 (1891).
W. Sumatera to Jawa. 42 JAW SUM.

Calamus ciliaris var. *peninsularis* Furtado = **Calamus exilis** Griff.

Calamus cochinchinensis Hook.f. = ? [41 VIE]

Calamus cochleatus Miq. = **Daemonorops didymophylla** Becc.

Calamus cockburnii J.Dransf., Malaysian Forester 41: 338 (1978).
Pen. Malaysia (Pahang). 42 MLY.

Calamus collinus Griff. = **Calamus erectus** Roxb.

Calamus compsostachys Burret, Notizbl. Bot. Gart. Berlin-Dahlem 13: 598 (1937).
China (Guangdong). 36 CHS.

Calamus comptus J.Dransf., Kew Bull. 45: 98 (1990).
W. Borneo. 42 BOR.

Calamus concinnus Mart., Hist. Nat. Palm. 3: 332 (1853). *Palmijuncus concinnus* (Mart.) Kuntze, Revis. Gen. Pl. 2: 733 (1891).
Myanmar to Pen. Malaysia. 41 MYA THA 42 MLY.
    Plectocomiopsis ferox Ridl., Fl. Malay Penins. 5: 66 (1925).

Calamus congestiflorus J.Dransf., Kew Bull. 36: 785 (1982).
Borneo (Sabah). 42 BOR.

Calamus conirostris Becc. in J.D.Hooker, Fl. Brit. India 6: 461 (1893). *Cornera conirostris* (Becc.) Furtado, Gard. Bull. Singapore 14: 519 (1955).
W. Malesia. 42 BOR MLY SUM.
    Calamus brachystachys Becc., Rec. Bot. Surv. India 2: 215 (1902).

Calamus conjugatus Furtado, Gard. Bull. Straits Settlem. 8: 246 (1935).
Borneo (Sarawak). 42 BOR.

Calamus convallium J.Dransf., Kew Bull. 36: 800 (1982).
Borneo. 42 BOR.

Calamus corneri Furtado, Gard. Bull. Singapore 15: 219 (1956).
Pen. Malaysia. 42 MLY.

Calamus corrugatus Becc., Rec. Bot. Surv. India 2: 201 (1902).
Borneo (Sarawak). 42 BOR.

Calamus crassifolius J.Dransf., Kew Bull. 45: 94 (1990).
Borneo (Sarawak). 42 BOR.

Calamus crinitus (Blume) Miq. = **Daemonorops crinita** Blume

Calamus cumingianus Becc., Rec. Bot. Surv. India 2: 210 (1902).
Philippines. 42 PHI.

Calamus curag Blanco ex Becc., Ann. Roy. Bot. Gard. (Calcutta) 11: 498 (1902).
Philippines. 42 PHI. — Provisionally accepted.

Calamus cuthbertsonii Becc., Nuovo Giorn. Bot. Ital. 20: 179 (1888).
New Guinea. 43 NWG.

Calamus curtisii Ridl. = **Calamus exilis** Griff.

Calamus cuspidatus G.Mann & H.Wendl. = **Eremospatha cuspidata** (G.Mann & H.Wendl.) H.Wendl.

Calamus dachangensis Furtado = **Calamus gibbsianus** Becc.

Calamus dasyacanthus W.J.Baker & al., Kew Bull. 58: 364 (2003).
New Guinea. 43 NWG.

Calamus dealbatus H.Wendl. = **Acanthophoenix rubra** (Bory) H.Wendl.

Calamus deerratus G.Mann & H.Wendl., Trans. Linn. Soc. London 24: 429 (1864). *Palmijuncus deerratus* (G.Mann & H.Wendl.) Kuntze, Revis. Gen. Pl. 2: 733 (1891). *Eremospatha deerrata* (G.Mann & H.Wendl.) T.Durand & Schinz, Consp. Fl. Afric. 5: 458 (1894).
Senegal to Uganda and Angola. 22 BEN GAM GHA GNB GUI IVO LBR NGA NGR SEN SIE 23 CAF CMN ZAI 24 SUD 25 UGA 26 ANG.
    Calamus barteri Drude, Bot. Jahrb. Syst. 21: 134 (1895).
    Calamus heudelotii Becc. & Drude, Bot. Jahrb. Syst. 21: 134 (1895).
    Calamus leprieurii Becc., Rec. Bot. Surv. India 2: 200 (1902).
    Calamus perrottetii Becc., Rec. Bot. Surv. India 2: 200 (1902).
    Calamus laurentii De Wild., Ann. Mus. Congo Belge, Bot., V, 1: 97 (1904).
    Calamus akimensis Becc., Ann. Roy. Bot. Gard. (Calcutta) 11(1): 162 (1908).
    Calamus falabensis Becc., Ann. Roy. Bot. Gard. (Calcutta) 11(1): 157 (1908).
    Calamus schweinfurthii Becc. in G.W.J.Mildbraed (ed.), Wiss. Erg. Deut. Zentr.-Afr. Exped., Bot. 2: 54 (1910).

Calamus delessertianus Becc., Ann. Roy. Bot. Gard. (Calcutta) 11(1): 276 (1908).
SW. India. 40 IND.

Calamus delicatulus Thwaites, Enum. Pl. Zeyl.: 330 (1864). *Palmijuncus delicatulus* (Thwaites) Kuntze, Revis. Gen. Pl. 2: 733 (1891).
SW. Sri Lanka. 40 SRL.

Calamus densiflorus Becc. in J.D.Hooker, Fl. Brit. India 6: 445 (1893).
Pen. Thailand to Pen. Malaysia. 41 THA 42 MLY.
    Calamus neglectus Becc. in J.D.Hooker, Fl. Brit. India 6: 458 (1893).

Calamus depauperatus Ridl., Trans. Linn. Soc. London, Bot. 9: 234 (1916).
New Guinea. 43 NWG.

Calamus depressiusculus Miq. ex H.Wendl. = **Daemonorops depressiuscula** (Miq. ex H.Wendl.) Becc.

Calamus dianbaiensis C.F.Wei, Guihaia 6: 24 (1986).
China (Guangxi, Guangdong). 36 CHS.

Calamus didymocarpus Warb. ex Becc., Ann. Roy. Bot. Gard. (Calcutta) 11(1): 467 (1908).
Sulawesi. 42 SUL.

Calamus didymophyllus (Becc.) Ridl. = **Daemonorops didymophylla** Becc.

Calamus diepenhorstii Miq., Fl. Ned. Ind., Eerste Bijv.: 594 (1861). *Palmijuncus diepenhorstii* (Miq.) Kuntze, Revis. Gen. Pl. 2: 733 (1891).
Pen. Thailand to W. & C. Malesia. 41 THA 42 BOR MLY PHI SUM.

var. **diepenhorstii**
Pen. Thailand to W. Malesia. 41 THA 42 BOR MLY SUM.
    Calamus singaporensis Becc. in J.D.Hooker, Fl. Brit. India 6: 454 (1893). *Calamus diepenhorstii* var. *singaporensis* (Becc.) Becc., Ann. Roy. Bot. Gard. (Calcutta) 11: 325 (1908).
    Calamus pacificus Ridl., J. Fed. Malay States Mus. 6: 59 (1915).

Calamus diepenhorstii var. *kemamanensis* Fernando, Gard. Bull. Singapore 15: 258 (1956).

var. **exulans** Becc., Philipp. J. Sci., Bot. 4: 627 (1909).
Philippines. 42 PHI.

var. **major** J.Dransf., Kew Bull. 36: 806 (1982).
Borneo (Sabah). 42 BOR.

Calamus diepenhorstii var. *kemamanensis* Fernando = **Calamus diepenhorstii** var. **diepenhorstii**

Calamus diepenhorstii var. *singaporensis* (Becc.) Becc. = **Calamus diepenhorstii** var. **diepenhorstii**

Calamus diffusus Becc. = **Calamus oxleyanus** var. **oxleyanus**

Calamus digitatus Becc. in J.D.Hooker, Fl. Brit. India 6: 442 (1892).
SW. Sri Lanka. 40 SRL.

Calamus dilaceratus Becc., Rec. Bot. Surv. India 2: 198 (1902).
Nicobar Is. 41 NCB.

Calamus dimorphacanthus Becc., Rec. Bot. Surv. India 2: 214 (1902).
Philippines. 42 PHI.

var. **benguetensis** Baja-Lapis, Sylvatrop 12: 72 (1987 publ. 1989).
Philippines. 42 PHI.

var. **dimorphacanthus**
Philippines. 42 PHI.

var. **halconensis** (Becc.) Baja-Lapis, Sylvatrop 12: 73 (1987 publ. 1989).
Philippines. 42 PHI.
Calamus halconensis Becc., Philipp. J. Sci., C 4: 633 (1909).

var. **montalbanicus** Becc., Philipp. J. Sci., Bot. 4: 631 (1909).
Philippines (Luzon). 42 PHI.

var. **zambalensis** Becc., Philipp. J. Sci., Bot. 4: 632 (1909).
Philippines (Luzon). 42 PHI.

Calamus dioicus Lour., Fl. Cochinch.: 211 (1790).
Palmijuncus dioicus (Lour.) Kuntze, Revis. Gen. Pl. 2: 732 (1891).
Vietnam. 41 VIE.

Calamus discolor Mart., Hist. Nat. Palm. 3: 341 (1853).
Palmijuncus discolor (Mart.) Kuntze, Revis. Gen. Pl. 2: 733 (1891).
Philippines. 42 PHI.

var. **discolor**
Philippines. 42 PHI.
Calamus lindenii Rodigas, Ill. Hort. 30(1): 499 (1883). *Palmijuncus lindenii* (Rodigas) Kuntze, Revis. Gen. Pl. 2: 733 (1891).

var. **negrosensis** Becc., Philipp. J. Sci., Bot. 4: 635 (1909).
Philippines. 42 PHI.

Calamus distans Ridl. = **Calamus luridus** Becc.

Calamus distentus Burret, J. Arnold Arbor. 20: 194 (1939).
New Guinea. 43 NWG.

Calamus distichoideus Furtado = **Calamus viridispinus** var. **viridispinus**

Calamus distichus Ridl. = **Calamus viridispinus** var. **viridispinus**

Calamus distichus var. *shangsiensis* S.J.Pei & S.Y.Chen = **Calamus viridispinus** var. **viridispinus**

Calamus divaricatus Becc., Ann. Roy. Bot. Gard. (Calcutta) 11(App.): 10 (1913).
Borneo. 42 BOR.

var. **contrarius** J.Dransf., Kew Bull. 45: 89 (1990).
Borneo (Sarawak). 42 BOR.

var. **divaricatus**
Borneo. 42 BOR.

Calamus docilis (Becc.) Becc. = **Calamus interruptus** Becc.

Calamus dongnaiensis Pierre ex Becc., Rec. Bot. Surv. India 2: 198 (1902).
S. Vietnam. 41 VIE.

Calamus doriaei Becc. in J.D.Hooker, Fl. Brit. India 6: 456 (1892).
Myanmar (Mt. Korin). 41 MYA.

Calamus draco Willd. = **Daemonorops draco** (Willd.) Blume

Calamus draconis Oken = **Daemonorops draco** (Willd.) Blume

Calamus dransfieldii Renuka, Kew Bull. 42: 433 (1987).
SW. India. 40 IND.

Calamus dumetorum Ridl. = **Calamus palustris** var. **malaccensis** Becc.

Calamus egregius Burret, Notizbl. Bot. Gart. Berlin-Dahlem 13: 599 (1937).
Hainan. 36 CHH.

Calamus elegans H.Wendl. = **?** [41 VIE]

Calamus elegans Becc. ex Ridl. = **Calamus viridispinus** var. **viridispinus**

Calamus elmerianus Becc. ex Elmer, Leafl. Philipp. Bot. 2: 647 (1909).
Philippines. 42 PHI.

Calamus elongatus (Blume) Miq. = **Daemonorops elongata** Blume

Calamus elopurensis J.Dransf., Kew Bull. 36: 787 (1982).
Borneo (Sabah). 42 BOR.

Calamus endauensis J.Dransf., Malaysian Forester 41: 330 (1978).
Pen. Malaysia (Johore). 42 MLY.

Calamus epetiolaris Mart., Hist. Nat. Palm. 3: 336 (1853). *Palmijuncus epetiolaris* (Mart.) Kuntze, Revis. Gen. Pl. 2: 733 (1891).
Jawa. 42 JAW. — Provisionally accepted.

Calamus equestris Blume = **Calamus javensis** Blume

Calamus equestris Willd., Sp. Pl. 2: 204 (1799). *Palmijuncus equestris* (Willd.) Kuntze, Revis. Gen. Pl. 2: 732 (1891). *Rotang equestris* (Willd.) Baill., Hist. Pl. 13: 299 (1895).
Maluku. 42 MOL.

Calamus erectus Roxb., Fl. Ind. ed. 1832, 3: 774 (1832). *Palmijuncus erectus* (Roxb.) Kuntze, Revis. Gen. Pl. 2: 733 (1891).
Sikkim to China (Yunnan) and Indo-China. 36 CHC 40 ASS BAN EHM 41 LAO MYA THA.
Calamus collinus Griff., Calcutta J. Nat. Hist. 5: 31 (1845). *Palmijuncus collinus* (Griff.) Kuntze, Revis. Gen. Pl. 2: 733 (1891). *Calamus erectus* var. *collinus* (Griff.) Becc. in J.D.Hooker, Fl. Brit. India 6: 439 (1892).

Calamus schizospathus Griff., Calcutta J. Nat. Hist. 5: 32 (1845). *Palmijuncus schizospathus* (Griff.) Kuntze, Revis. Gen. Pl. 2: 733 (1891). *Calamus erectus* var. *schizospathus* (Griff.) Becc., Ann. Roy. Bot. Gard. (Calcutta) 11: 125 (1908).

Calamus macrocarpus Griff. ex Mart., Hist. Nat. Palm. 3: 333 (1853). *Palmijuncus macrocarpus* (Griff. ex Mart.) Kuntze, Revis. Gen. Pl. 2: 733 (1891). *Calamus erectus* var. *macrocarpus* (Griff. ex Mart.) Becc. in J.D.Hooker, Fl. Brit. India 6: 439 (1892).

Calamus erectus var. *birmanicus* Becc., Rec. Bot. Surv. India 2: 197 (1902).

Calamus erectus var. *birmanicus* Becc. = **Calamus erectus** Roxb.

Calamus erectus var. *collinus* (Griff.) Becc. = **Calamus erectus** Roxb.

Calamus erectus var. *macrocarpus* (Griff. ex Mart.) Becc. = **Calamus erectus** Roxb.

Calamus erectus var. *schizospathus* (Griff.) Becc. = **Calamus erectus** Roxb.

Calamus erinaceus (Becc.) J.Dransf., Kew Bull. 32: 484 (1978).
Pen. Thailand to W. Malesia and Philippines (Palawan). 41 CBD THA 42 BOR MLY PHI SUM.
Daemonorops erinacea Becc., Rec. Bot. Surv. India 2: 225 (1902).
Calamus aquatilis Ridl., J. Straits Branch Roy. Asiat. Soc. 41: 43 (1903).

Calamus erioacanthus Becc., For. Borneo: 610 (1902).
Borneo (Sarawak). 42 BOR.

Calamus essigii W.J.Baker, Kew Bull. 57: 720 (2002).
SE. New Guinea. 43 NWG.

Calamus exilis Griff., Palms Brit. E. Ind.: 51 (1850). *Palmijuncus exilis* (Griff.) Kuntze, Revis. Gen. Pl. 2: 733 (1891).
Pen. Thailand to Sumatera. 41 THA 42 MLY SUM.
Calamus curtisii Ridl., Mat. Fl. Malay. Penins. 2: 204 (1907).
Calamus ciliaris var. *peninsularis* Furtado, Gard. Bull. Singapore 15: 60 (1956).

Calamus eximius Burret, J. Arnold Arbor. 20: 193 (1939).
New Guinea. 43 NWG.

Calamus extensus Roxb. = ? **[40 BAN]**

Calamus extensus Mart. = **Calamus viminalis** Willd.

Calamus faberi Becc., Ann. Roy. Bot. Gard. (Calcutta) 11(1): 274 (1908).
SE. China to Hainan. 36 CHH CHS.

Calamus faberi var. *brevispicatus* (C.F.Wei) S.J.Pei & S.Y.Chen = **Calamus walkeri** Hance

Calamus falabensis Becc. = **Calamus deerratus** G.Mann & H.Wendl.

Calamus farinosus Linden, Ill. Hort. 19: t. 109 (1872). *Palmijuncus farinosus* (Linden) Kuntze, Revis. Gen. Pl. 2: 733 (1891).
Sumatera. 42 SUM. — Provisionally accepted.
Calamus fasciculatus Roxb. = **Calamus viminalis** Willd.

Calamus feanus Becc. = **Calamus acanthospathus** Griff.

Calamus feanus var. *medogensis* S.J.Pei & S.Y.Chen = **Calamus acanthospathus** Griff.

Calamus fernandezii H.Wendl. = **Calamus oxleyanus** var. **oxleyanus**

Calamus ferrugineus Becc. = **Calamus mattanensis** Becc.

Calamus fertilis Becc., Ann. Roy. Bot. Gard. (Calcutta) 11(1): 492 (1908).
New Guinea. 43 NWG.

Calamus filiformis Becc. = **Calamus javensis** Blume

Calamus filipendulus Becc. in J.D.Hooker, Fl. Brit. India 6: 443 (1892).
Pen. Malaysia. 42 MLY.
Calamus pauciflorus Ridl., Fl. Malay Penins. 5: 57 (1925).

Calamus filispadix Becc., Philipp. J. Sci., C 6: 230 (1911).
Philippines. 42 PHI.

Calamus fimbriatus Van Valk., Blumea 40: 461 (1995).
Borneo (Kalimantan). 42 BOR.

Calamus fissijugatus Burret, Notizbl. Bot. Gart. Berlin-Dahlem 15: 804 (1943).
Sumatera. 42 SUM.

Calamus fissus (Blume) Miq. = **Daemonorops fissa** Blume

Calamus flabellatus Becc., Malesia 3: 62 (1886). *Palmijuncus flabellatus* (Becc.) Kuntze, Revis. Gen. Pl. 2: 733 (1891).
W. Malesia. 42 BOR MLY SUM.
Calamus flabelloides Furtado, Gard. Bull. Singapore 15: 173 (1956).

Calamus flabelloides Furtado = **Calamus flabellatus** Becc.

Calamus flagellum Griff. ex Mart., Hist. Nat. Palm. 3: 333 (1853). *Palmijuncus flagellum* (Griff. ex Mart.) Kuntze, Revis. Gen. Pl. 2: 733 (1891).
Sikkim to China (Yunnan) and Indo-China. 36 CHC CHT 40 ASS BAN EHM 41 LAO MYA THA VIE.

var. **flagellum**
Sikkim to China (Yunnan) and Indo-China. 36 CHC CHT 40 ASS BAN EHM 41 LAO MYA THA VIE.
Calamus polygamus Roxb., Fl. Ind. ed. 1832, 3: 780 (1832). Provisional synonym. *Palmijuncus polygamus* (Roxb.) Kuntze, Revis. Gen. Pl. 2: 733 (1891).
Calamus jenkinsianus Griff., Palms Brit. E. Ind.: 40 (1850), nom. illeg. *Palmijuncus jenkinsianus* Kuntze, Revis. Gen. Pl. 2: 733 (1891).
Calamus flagellum var. *karinensis* Becc., Ann. Roy. Bot. Gard. (Calcutta) 11(1): 129 (1908). *Calamus karinensis* (Becc.) S.J.Pei & S.Y.Chen, Acta Phytotax. Sin. 27: 133 (1989).

var. **furvifurfuraceus** S.J.Pei & S.Y.Chen, Acta Phytotax. Sin. 27: 133 (1989).
China (S. Yunnan) to Vietnam. 36 CHC 41 VIE.

Calamus flagellum var. *karinensis* Becc. = **Calamus flagellum** var. **flagellum**

Calamus floribundus Griff., Calcutta J. Nat. Hist. 5: 56 (1845). *Palmijuncus floribundus* (Griff.) Kuntze, Revis. Gen. Pl. 2: 733 (1891).
E. Himalaya to Myanmar. 40 ASS BAN EHM 41 MYA.
Calamus mishmeensis Griff., Calcutta J. Nat. Hist. 5: 55 (1845). *Palmijuncus mishmeensis* (Griff.) Kuntze, Revis. Gen. Pl. 2: 733 (1891).
Calamus floribundus var. *depauperatus* Becc., Ann. Roy. Bot. Gard. (Calcutta) 11: 79 (1908).

Calamus floribundus var. *depauperatus* Becc. = **Calamus floribundus** Griff.

Calamus formosanus Becc., Rec. Bot. Surv. India 2: 211 (1902).
Taiwan. 38 TAI.
Calamus orientalis C.E.Chang, Quart. J. Chin. Forest. 21: 108 (1988).

Calamus foxworthyi Becc., Ann. Roy. Bot. Gard. (Calcutta) 11(App.): 81 (1913).
Philippines (Palawan). 42 PHI.

Calamus fuscus Becc., Bot. Jahrb. Syst. 58: 461 (1923).
New Guinea. 43 NWG.

Calamus gamblei Becc. in J.D.Hooker, Fl. Brit. India 6: 453 (1893).
India (Nilgiri Hills). 40 IND.
Calamus gamblei var. *sphaerocarpus* Becc. in J.D. Hooker, Fl. Brit. India 6: 453 (1893).

Calamus gamblei var. *sphaerocarpus* Becc. = **Calamus gamblei** Becc.

Calamus gaudichaudii (Mart.) H.Wendl. = **Daemonorops mollis** (Blanco) Merr.

Calamus geminiflorus Griff. = **Plectocomiopsis geminiflora** (Griff.) Becc.

Calamus geniculatus Griff. = **Daemonorops geniculata** (Griff.) Mart.

Calamus gibbsianus Becc., Ann. Roy. Bot. Gard. (Calcutta) 11(App.): 58 (1913).
Borneo (Sabah, Sarawak). 42 BOR.
Calamus dachangensis Furtado, Gard. Bull. Straits Settlem. 8: 247 (1935).

Calamus giganteus Becc. = **Calamus manan** Miq.

Calamus giganteus var. *robustus* S.J.Pei & S.Y.Chen = **Calamus platyacanthoides** Merr.

Calamus glaucescens Blume = **Calamus caesius** Blume

Calamus glaucescens D.Dietr. = **Ceratolobus glaucescens** Blume

Calamus godefroyi Becc., Ann. Roy. Bot. Gard. (Calcutta) 11(1): 267 (1908).
Indo-China. 41 CBD LAO THA.

Calamus gogolensis Becc., Ann. Roy. Bot. Gard. (Calcutta) 11(1): 261 (1908).
New Guinea. 43 NWG.

Calamus gonospermus Becc., Rec. Bot. Surv. India 2: 202 (1902).
Borneo. 42 BOR. — Fruit edible.

Calamus gracilipes Miq. = **Daemonorops gracilipes** (Miq.) Becc.

Calamus gracilis Blanco = **Calamus usitatus** Blanco

Calamus gracilis Roxb., Fl. Ind. ed. 1832, 3: 781 (1832). *Palmijuncus gracilis* (Roxb.) Kuntze, Revis. Gen. Pl. 2: 733 (1891).
E. India (Andrah Pradesh) to Hainan. 36 CHC CHH 40 ASS BAN IND 41 LAO.
Calamus hainanensis C.C.Chang & L.G.Xu ex R.H.Miau, Acta Sci. Nat. Univ. Sunyatseni 1981(3): 116 (1981), without type.

Calamus gracilis Thwaites = **Calamus pachystemonus** Thwaites

Calamus graminosus Blume, Rumphia 3: 31 (1847). *Palmijuncus graminosus* (Blume) Kuntze, Revis. Gen. Pl. 2: 732 (1891). *Rotang graminosus* (Blume) Baill., Hist. Pl. 13: 299 (1895).
Maluku. 42 MOL. — Provisionally accepted.

Calamus grandifolius Becc., Philipp. J. Sci., C 4: 629 (1909).
Philippines (Luzon). 42 PHI.

Calamus grandis Griff. = **Daemonorops grandis** (Griff.) Mart.

Calamus gregisectus Burret, Notizbl. Bot. Gart. Berlin-Dahlem 15: 811 (1943).
Myanmar. 41 MYA.

Calamus griffithianus Mart. = **Calamus castaneus** Griff.

Calamus griseus J.Dransf., Thai Forest Bull., Bot. 28: 157 (2000).
Thailand to Sumatera. 41 THA 42 MLY SUM.

Calamus guangxiensis C.F.Wei, Guihaia 6: 28 (1986).
China (Guangxi). 36 CHS.

Calamus guruba Buch.-Ham. in C.F.P.von Martius, Hist. Nat. Palm. 3: 211 (1838). *Daemonorops guruba* (Buch.-Ham.) Mart., Hist. Nat. Palm. 3: 330 (1853). *Palmijuncus guruba* (Buch.-Ham.) Kuntze, Revis. Gen. Pl. 2: 732 (1891).
Darjiling to Pen. Malaysia. 36 CHC 40 ASS EHM BAN IND 41 CBD LAO MYA THA 42 MLY.
Calamus mastersianus Griff., Calcutta J. Nat. Hist. 5: 76 (1845).
Calamus nitidus Mart., Hist. Nat. Palm. 3: 334 (1853). *Palmijuncus nitidus* (Mart.) Kuntze, Revis. Gen. Pl. 2: 733 (1891).
Calamus multirameus Ridl., Mat. Fl. Malay. Penins. 2: 202 (1907).
Calamus guruba var. *ellipsoideus* San Y.Chen & K.L.Wang, Acta Bot. Yunnan. 24: 202 (2002).

Calamus guruba var. *ellipsoideus* San Y.Chen & K.L.Wang = **Calamus guruba** Buch.-Ham.

Calamus haenkeanus Mart. = **Calamus usitatus** Blanco

Calamus hainanensis C.C.Chang & L.G.Xu ex R.H.Miau = **Calamus gracilis** Roxb.

Calamus halconensis Becc. = **Calamus dimorphacanthus** var. **halconensis** (Becc.) Baja-Lapis

Calamus halmaherensis Burret, Notizbl. Bot. Gart. Berlin-Dahlem 15: 813 (1943).
Maluku. 42 MOL.

Calamus harmandii Pierre ex Becc., Rec. Bot. Surv. India 2: 216 (1902). *Zalaccella harmandii* (Pierre ex Becc.) Becc., Ann. Roy. Bot. Gard. (Calcutta) 11(1): 496 (1908).
S. Laos. 41 LAO.

Calamus hartmannii Becc., Ann. Roy. Bot. Gard. (Calcutta) 11(1): 494 (1908).
New Guinea. 43 NWG.

Calamus helferianus Kurz, J. Asiat. Soc. Bengal, Pt. 2, Nat. Hist. 43(2): 213 (1874). *Palmijuncus helferianus* (Kurz) Kuntze, Revis. Gen. Pl. 2: 733 (1891).
Myanmar. 41 MYA.

Calamus heliotropium Buch.-Ham. ex Kunth = **Calamus tenuis** Roxb.

Calamus hendersonii Furtado = **Calamus axillaris** Becc.

Calamus henryanus Becc., Rec. Bot. Surv. India 2: 199 (1902).
S. China to Indo-China. 36 CHC CHS 41 LAO THA VIE.
Calamus balansaeanus Becc., Webbia 3: 230 (1910).
Calamus henryanus var. *castaneolepis* C.F.Wei, Guihaia 6: 32 (1986). *Calamus balansaeanus* var. *castaneolepis* (C.F.Wei) S.J.Pei & S.Y.Chen, Acta Phytotax. Sin. 27: 134 (1989).

Calamus henryanus var. *castaneolepis* C.F.Wei = **Calamus henryanus** Becc.

Calamus hepburnii J.Dransf., Kew Bull. 36: 795 (1982). Borneo (Sabah). 42 BOR.

Calamus heteracanthus Zipp. ex Blume, Rumphia 3: 56 (1847). *Palmijuncus heteracanthus* (Zipp. ex Blume) Kuntze, Revis. Gen. Pl. 2: 733 (1891). Maluku to New Guinea. 42 MOL 43 NWG.
> *Daemonorops heteracantha* Blume, Rumphia 3: t. 139 (1847).

Calamus heteroideus Blume, Rumphia 3: 46 (1847). *Palmijuncus heteroideus* (Blume) Kuntze, Revis. Gen. Pl. 2: 733 (1891). *Rotang heteroideus* (Blume) Baill., Hist. Pl. 13: 299 (1895). Sumatera to Jawa. 42 JAW SUM.
> *Calamus pallens* Blume, Rumphia 3: 51 (1847). *Palmijuncus pallens* (Blume) Kuntze, Revis. Gen. Pl. 2: 732 (1891).

Calamus heudelotii Becc. & Drude = **Calamus deerratus** G.Mann & H.Wendl.

Calamus hewittianus Becc. = **Calamus myriacanthus** Becc.

Calamus hirsutus (Blume) Miq. = **Daemonorops hirsuta** Blume

Calamus hispidulus Becc., Rec. Bot. Surv. India 2: 209 (1902). Borneo. 42 BOR.

Calamus hollrungii Becc. = **Calamus aruensis** Becc.

Calamus holttumii Furtado, Gard. Bull. Singapore 15: 228 (1956). Pen. Malaysia (Trengganu). 42 MLY.

Calamus hookeri G.Mann & H.Wendl. = **Eremospatha hookeri** (G.Mann & H.Wendl.) H.Wendl.

Calamus hookerianus Becc., Ann. Roy. Bot. Gard. (Calcutta) 11(1): 226 (1908). S. India. 40 IND.

Calamus hoplites Dunn, J. Linn. Soc., Bot. 38: 369 (1908). SE. China to Hainan. 36 CHH CHS.

Calamus horrens Blume = **Calamus tenuis** Roxb.

Calamus hostilis Wall. ex Voigt = **Calamus arborescens** Griff.

Calamus huegelianus Mart. = **Calamus wightii** Griff.

Calamus humboldtianus Becc. in L.S.Gibbs, Fl. Arfak Mts.: 93 (1917). New Guinea. 43 NWG.

Calamus humilis Roxb. = **Calamus latifolius** Roxb.

Calamus hygrophilus Griff. = **Daemonorops angustifolia** (Griff.) Mart.

Calamus hypertrichosus Becc., Ann. Roy. Bot. Gard. (Calcutta) 11(App.): 17 (1913). Borneo. 42 BOR.

Calamus hypoleucus Kurz, J. Asiat. Soc. Bengal, Pt. 2, Nat. Hist. 43(2): 208 (1874). *Palmijuncus hypoleucus* (Kurz) Kuntze, Revis. Gen. Pl. 2: 733 (1891). Indo-China. 41 LAO MYA.
> *Daemonorops hypoleuca* Kurz, J. Asiat. Soc. Bengal, Pt. 2, Nat. Hist. 43(2): 208 (1874).

Calamus hystrix Griff. = **Daemonorops hirsuta** Blume

Calamus impar Becc., Ann. Roy. Bot. Gard. (Calcutta) 11(App.): 19 (1913). Borneo. 42 BOR.

Calamus inermis T.Anderson = **Calamus latifolius** Roxb.

Calamus inermis var. *menghaiensis* San Y.Chen, S.J.Pei & K.L.Wang = **Calamus latifolius** Roxb.

Calamus inflatus Warb. = **Calamus siphonospathus** var. **siphonospathus**

Calamus inopinatus Furtado, Gard. Bull. Straits Settlem. 9: 184 (1937). Trop. Asia (?). 4+.

Calamus inops Becc. ex K.Heyne, Nutt. Pl. Ned.-Ind., ed. 2, 1: 372 (1922). Sulawesi. 42 SUL.

Calamus insignis Griff., Calcutta J. Nat. Hist. 5: 59 (1845). *Palmijuncus insignis* (Griff.) Kuntze, Revis. Gen. Pl. 2: 733 (1891). Thailand to N. Sumatera. 41 THA 42 MLY SUM.

var. **insignis**
> Pen. Thailand to Pen. Malaysia. 41 THA 42 MLY.
> > *Calamus spathulatus* Becc. in J.D.Hooker, Fl. Brit. India 6: 459 (1893).
> > *Calamus subspathulatus* Ridl., Mat. Fl. Malay. Penins. 2: 194 (1907).

var. **longispinosus** J.Dransf., Malaysian Forester 41: 342 (1978).
> Thailand to N. Sumatera. 41 THA 42 MLY SUM.

var. **robustus** (Becc.) J.Dransf., Malaysian Forester 41: 342 (1978).
> Thailand to Pen. Malaysia (Perak). 41 THA 42 MLY.
> > *Calamus spathulatus* var. *robustus* Becc. in J.D.Hooker, Fl. Brit. India 6: 459 (1893).

Calamus intermedius Griff. = **Daemonorops grandis** (Griff.) Mart.

Calamus interruptus Becc., Malesia 3: 60 (1886). New Guinea. 43 NWG.

Calamus interruptus var. *docilis* Becc., Malesia 2: 60 (1884). *Calamus docilis* (Becc.) Becc., Rec. Bot. Surv. India 2: 204 (1902).

Calamus interruptus var. *docilis* Becc. = **Calamus interruptus** Becc.

Calamus intumescens (Becc.) Ridl. = **Calamus paspalanthus** Becc.

Calamus jaboolum F.M.Bailey = **Calamus australis** Mart.

Calamus jaherianus Becc. = **Calamus myriacanthus** Becc.

Calamus javensis Blume, Rumphia 3: 62 (1847). *Palmijuncus javensis* (Blume) Kuntze, Revis. Gen. Pl. 2: 733 (1891). Pen. Thailand to W. Malesia, Philippines (Palawan). 41 THA 42 BOR JAW MLY PHI SUM.
> *Calamus equestris* Blume in J.J.Roemer & J.A.Schultes, Syst. Veg. 7: 1330 (1830), nom. illeg.
> *Calamus tetrastichus* Blume, Rumphia 3: 62 (1847). *Palmijuncus tetrastichus* (Blume) Kuntze, Revis. Gen. Pl. 2: 733 (1891).
> *Calamus borneensis* Miq., Anal. Bot. Ind. 1: 4 (1850). *Palmijuncus borneensis* (Miq.) Kuntze, Revis. Gen. Pl. 2: 733 (1891).
> *Calamus amplectens* Becc., Malesia 2: 78 (1884). *Palmijuncus amplectens* (Becc.) Kuntze, Revis. Gen. Pl. 2: 733 (1891).
> *Calamus javensis* subvar. *intermedius* Becc. in J.D.Hooker, Fl. Brit. India 6: 443 (1892).
> *Calamus javensis* subvar. *penangianus* Becc. in J.D.Hooker, Fl. Brit. India 6: 443 (1892).

Calamus javensis subvar. *polyphyllus* Becc. in J.D.Hooker, Fl. Brit. India 6: 443 (1892).
Calamus javensis subvar. *purpurascens* Becc. in J.D.Hooker, Fl. Brit. India 6: 443 (1892).
Calamus javensis subvar. *tenuissimus* Becc. in J.D.Hooker, Fl. Brit. India 6: 443 (1892).
Calamus borneensis Becc., Rec. Bot. Surv. India 2: 205 (1902), nom. illeg.
Calamus filiformis Becc., For. Borneo: 609 (1902).
Calamus javensis var. *acicularis* Becc., Ann. Roy. Bot. Gard. (Calcutta) 11(1): 185 (1908).
Calamus kemamanensis Furtado, Gard. Bull. Singapore 15: 170 (1956).

Calamus javensis var. *acicularis* Becc. = **Calamus javensis** Blume

Calamus javensis subvar. *intermedius* Becc. = **Calamus javensis** Blume

Calamus javensis subvar. *penangianus* Becc. = **Calamus javensis** Blume

Calamus javensis subvar. *polyphyllus* Becc. = **Calamus javensis** Blume

Calamus javensis subvar. *purpurascens* Becc. = **Calamus javensis** Blume

Calamus javensis subvar. *tenuissimus* Becc. = **Calamus javensis** Blume

Calamus jenkinsianus Griff. = **Calamus flagellum** var. **flagellum**

Calamus jenkinsianus Griff. = **Daemonorops jenkinsiana** (Griff.) Mart.

Calamus jenningsianus Becc., Philipp. J. Sci., C 4: 623 (1909).
Philippines (Mindoro). 42 PHI.

Calamus kandariensis Becc., Rec. Bot. Surv. India 2: 210 (1902).
Sulawesi. 42 SUL.

Calamus karinensis (Becc.) S.J.Pei & S.Y.Chen = **Calamus flagellum** var. **flagellum**

Calamus karnatakensis Renuka & Lakshmana, R. I. C. Bull. 9: 10 (1990).
SW. India. 40 IND.

Calamus karuensis Ridl., J. Malayan Branch Roy. Asiat. Soc. 1: 104 (1923).
Sumatera. 42 SUM.

Calamus kemamanensis Furtado = **Calamus javensis** Blume

Calamus kerrianus Becc. = **Calamus palustris** var. **palustris**

Calamus keyensis Becc., Ann. Roy. Bot. Gard. (Calcutta) 11(App.): 137 (1913).
Maluku (Kep. Kai). 42 MOL.

Calamus khasianus Becc., Ann. Roy. Bot. Gard. (Calcutta) 11(1): 431 (1908).
Assam to China (Yunnan). 36 CHC 40 ASS EHM?

Calamus kiahii Furtado, Gard. Bull. Straits Settlem. 8: 251 (1935).
Borneo (Sabah, Sarawak). 42 BOR.

Calamus kingianus Becc., Ann. Roy. Bot. Gard. (Calcutta) 11(1): 197 (1908).
Assam, Laos. 40 ASS 41 LAO.

Calamus kjellbergii Furtado, Gard. Bull. Straits Settlem. 8: 252 (1935).
Sulawesi. 42 SUL.

Calamus klossii Ridl., Trans. Linn. Soc. London, Bot. 9: 235 (1916).
New Guinea. 43 NWG.

Calamus koordersianus Becc., Ann. Roy. Bot. Gard. (Calcutta) 11(App.): 65 (1913).
N. Sulawesi. 42 SUL.

Calamus koribanus Furtado = **Calamus viridispinus** var. **viridispinus**

Calamus korthalsii (Blume) Miq. = **Daemonorops korthalsii** Blume

Calamus kunzeanus Becc. = **Pigafetta filaris** (Giseke) Becc.

Calamus lacciferus Lakshmana & Renuka, J. Econ. Taxon. Bot. 14: 707 (1990).
SW. India. 40 IND.

Calamus laceratus Burret = **Calamus zebrinus** Becc.

Calamus laevigatus Mart., Hist. Nat. Palm. 3: 339 (1853). *Palmijuncus laevigatus* (Mart.) Kuntze, Revis. Gen. Pl. 2: 733 (1891). *Ceratolobus laevigatus* (Mart.) Becc. & Hook.f. in J.D.Hooker, Fl. Brit. India 6(2): 477 (1893).
Pen. Thailand to W. Malesia. 41 THA 42 BOR MLY SUM.

var. **laevigatus**
Pen. Thailand to W. Malesia. 41 THA 42 BOR MLY SUM.
Calamus pallidulus Becc. in J.D.Hooker, Fl. Brit. India 6: 457 (1893).
Calamus retrophyllus Becc., Ann. Roy. Bot. Gard. (Calcutta) 11(App.): 123 (1913).

var. **mucronatus** (Becc.) J.Dransf., Bot. J. Linn. Soc. 81: 8 (1980).
Borneo. 42 BOR.
Calamus mucronatus Becc., Rec. Bot. Surv. India 2: 213 (1902).

var. **serpentinus** J.Dransf., Kew Bull. 36: 806 (1982).
Borneo (Sabah). 42 BOR.

Calamus laevis G.Mann & H.Wendl. = **Laccosperma laeve** (G.Mann & H.Wendl.) Kuntze

Calamus lakshmanae Renuka, J. Econ. Taxon. Bot. 14: 703 (1990).
SW. India. 40 IND.

Calamus lambirensis J.Dransf., Kew Bull. 45: 89 (1990).
Borneo (Sarawak, Brunei). 42 BOR.

Calamus lanatus Ridl. = **Calamus perakensis** var. **perakensis**

Calamus laoensis T.Evans & al., Kew Bull. 55: 929 (2000).
C. Laos. 41 LAO.

Calamus latifolius Roxb., Fl. Ind. ed. 1832, 3: 775 (1832). *Palmijuncus latifolius* (Roxb.) Kuntze, Revis. Gen. Pl. 2: 732 (1891).
Nepal to China (Yunnan). 36 CHC 40 ASS BAN EHM NEP 41 MYA.
Calamus humilis Roxb., Fl. Ind. ed. 1832, 3: 773 (1832). *Palmijuncus humilis* (Roxb.) Kuntze, Revis. Gen. Pl. 2: 733 (1891).
Calamus inermis T.Anderson, J. Linn. Soc., Bot. 11: 11 (1871). *Palmijuncus inermis* (T.Anderson) Kuntze, Revis. Gen. Pl. 2: 733 (1891).
Calamus macracanthus T.Anderson, J. Linn. Soc., Bot. 11: 10 (1871). *Palmijuncus macracanthus* (T. Anderson) Kuntze, Revis. Gen. Pl. 2: 733 (1891).

Calamus latifolius var. *marmoratus* Becc., Ann. Roy. Bot. Gard. (Calcutta) 12: 107 (1918).

Calamus inermis var. *menghaiensis* San Y.Chen, S.J.Pei & K.L.Wang, Acta Bot. Yunnan. 24: 202 (2002).

Calamus latifolius Kurz = **Calamus palustris** var. **palustris**

Calamus latifolius var. *marmoratus* Becc. = **Calamus latifolius** Roxb.

Calamus latisectus Burret = **Calamus myriocladus** Burret

Calamus latisectus Burret = **Calamus aruensis** Becc.

Calamus latispinus Miq., Verh. Kon. Ned. Akad. Wetensch., Afd. Natuurk. 11: 29 (1868). *Rotang latispinus* (Miq.) Baill., Hist. Pl. 13: 299 (1895). Sumatera. 42 SUM. — Provisionally accepted.
> *Daemonorops latispina* Teijsm. & Binn., Cat. Hort. Bot. Bogor.: 74 (1866), nom. nud.

Calamus laurentii De Wild. = **Calamus deerratus** G.Mann & H.Wendl.

Calamus lauterbachii Becc., Ann. Roy. Bot. Gard. (Calcutta) 11(1): 491 (1908). New Guinea. 43 NWG.

Calamus laxiflorus Becc. = **Calamus luridus** Becc.

Calamus laxissimus Ridl., Mat. Fl. Malay. Penins. 2: 210 (1907). Pen. Malaysia. 42 MLY.

Calamus ledermannianus Becc., Bot. Jahrb. Syst. 58: 454 (1923). New Guinea. 43 NWG.

Calamus leiocaulis Becc. ex K.Heyne, Nutt. Pl. Ned.-Ind., ed. 2, 1: 375 (1922). Sulawesi. 42 SUL.

Calamus leiospathus Bartlett = **Calamus oxleyanus** var. **oxleyanus**

Calamus leloi J.Dransf., Bot. J. Linn. Soc. 81: 11 (1980). W. Borneo. 42 BOR.

Calamus leprieurii Becc. = **Calamus deerratus** G.Mann & H.Wendl.

Calamus leptopus Griff. = **Daemonorops leptopus** (Griff.) Mart.

Calamus leptospadix Griff., Calcutta J. Nat. Hist. 5: 49 (1845). *Palmijuncus leptospadix* (Griff.) Kuntze, Revis. Gen. Pl. 2: 733 (1891). E. Nepal to Assam. 40 ASS BAN EHM NEP.

Calamus leptostachys Becc. ex K.Heyne, Nutt. Pl. Ned.-Ind., ed. 2, 1: 375 (1922). C. Sulawesi (Buton). 42 SUL.

Calamus leucotes Becc. = **Calamus myrianthus** Becc.

Calamus lewisianus Griff. = **Daemonorops lewisiana** (Griff.) Mart.

Calamus lindenii Rodigas = **Calamus discolor** var. **discolor**

Calamus litoralis Blume = **Calamus viminalis** Willd.

Calamus lobbianus Becc. in J.D.Hooker, Fl. Brit. India 6: 451 (1893). *Cornera lobbiana* (Becc.) Furtado, Gard. Bull. Singapore 14: 521 (1955). Pen. Malaysia, Borneo (Sarawak). 42 BOR MLY.
> *Calamus melanocarpus* Ridl., Trans. Linn. Soc. London, Bot. 3: 392 (1893).

Calamus loeiensis Hodel = **Calamus palustris** var. **palustris**

Calamus longipes Griff. = **Daemonorops longipes** (Griff.) Mart.

Calamus longipinna K.Schum. & Lauterb., Fl. Schutzgeb. Südsee: 203 (1900). Papuasia. 43 BIS NWG SOL.
> *Calamus ralumensis* Warb. ex K.Schum., Notizbl. Königl. Bot. Gart. Berlin 2: 98 (1898), nom. nud.

Calamus longisetus Griff., Calcutta J. Nat. Hist. 5: 36 (1845). *Palmijuncus longisetus* (Griff.) Kuntze, Revis. Gen. Pl. 2: 733 (1891). Indo-China to Pen. Malaysia. 40 BAN41 AND MYA THA 42 MLY.
> *Calamus tigrinus* Kurz, J. Asiat. Soc. Bengal, Pt. 2, Nat. Hist. 43(2): 211 (1874). *Palmijuncus tigrinus* (Kurz) Kuntze, Revis. Gen. Pl. 2: 733 (1891).

Calamus longispathus Ridl., Mat. Fl. Malay. Penins. 2: 209 (1907). Pen. Malaysia. 42 MLY.

Calamus luridus Becc. in J.D.Hooker, Fl. Brit. India 6: 445 (1892). Pen. Thailand to Pen. Malaysia. 41 THA 42 MLY.
> *Calamus laxiflorus* Becc., Ann. Roy. Bot. Gard. (Calcutta) 11(App.): 13 (1913).
> *Calamus distans* Ridl., Fl. Malay Penins. 5: 56 (1925).
> *Calamus belumutensis* Furtado, Gard. Bull. Singapore 15: 223 (1956).

Calamus macgregorii Becc., Ann. Roy. Bot. Gard. (Calcutta) 11(1): 493 (1908). New Guinea. 43 NWG.

Calamus macracanthus T.Anderson = **Calamus latifolius** Roxb.

Calamus macrocarpus Griff. ex Mart. = **Calamus erectus** Roxb.

Calamus macrocarpus G.Mann & H.Wendl. = **Eremospatha macrocarpa** H.Wendl.

Calamus macrochlamys Becc., Ann. Roy. Bot. Gard. (Calcutta) 11(1): 259 (1908). New Guinea. 43 NWG.
> *Calamus macrospadix* Burret, Notizbl. Bot. Gart. Berlin-Dahlem 12: 317 (1935).

Calamus macropterus Miq. = **Daemonorops macroptera** (Miq.) Becc.

Calamus macrorhynchus Burret, Notizbl. Bot. Gart. Berlin-Dahlem 13: 590 (1937). China (Guizhou, Guangxi, Guangdong). 36 CHS.

Calamus macrospadix Burret = **Calamus macrochlamys** Becc.

Calamus macrosphaerion Becc., Ann. Roy. Bot. Gard. (Calcutta) 11(1): 448 (1908). Sulawesi. 42 SUL.

Calamus maiadum J.Dransf., Rattans Brunei: 193 (1997 publ. 1998). Borneo (Brunei). 42 BOR.

Calamus malawaliensis J.Dransf., Kew Bull. 36: 805 (1982). Borneo (Sabah) to Philippines (Palawan). 42 BOR PHI.

Calamus manan Miq., Fl. Ned. Ind., Eerste Bijv.: 595 (1861). *Palmijuncus manan* (Miq.) Kuntze, Revis. Gen. Pl. 2: 733 (1891). *Rotang manan* (Miq.) Baill., Hist. Pl. 13: 299 (1895). Pen. Thailand to W. Malesia. 41 THA 42 BOR MLY SUM.
> *Calamus giganteus* Becc. in J.D.Hooker, Fl. Brit. India 6: 460 (1893).

Calamus manicatus Teijsm. & Binn. ex Miq. = **Daemonorops crinita** Blume

Calamus manillensis (Mart.) H.Wendl. in O.C.E.de Kerchove de Denterghem, Palmiers: 237 (1878).
Philippines. 42 PHI.
*Daemonorops manillensis Mart., Hist. Nat. Palm. 3: 330 (1853). *Palmijuncus manillensis* (Mart.) Kuntze, Revis. Gen. Pl. 2: 733 (1891).

Calamus mannii H.Wendl. = **Oncocalamus mannii** (H.Wendl.) H.Wendl.

Calamus margaritae Hance = **Daemonorops margaritae** (Hance) Becc.

Calamus marginatus (Blume) Mart., Hist. Nat. Palm. 3: 342 (1853).
Sumatera to Philippines (Palawan). 42 BOR PHI SUM.
*Daemonorops marginata Blume, Rumphia 3: 24 (1847). *Palmijuncus marginatus* (Blume) Kuntze, Revis. Gen. Pl. 2: 733 (1891).
Calamus rostratus Furtado, Gard. Bull. Straits Settlem. 8: 257 (1935).
Calamus regularis Burret, Notizbl. Bot. Gart. Berlin-Dahlem 15: 816 (1943).

Calamus maritimus Blume, Rumphia 3: 31 (1847).
Jawa. 42 JAW. — Provisionally accepted.

Calamus martianus Becc. = **Calamus penicillatus** Roxb.

Calamus mastersianus Griff. = **Calamus guruba** Buch.-Ham.

Calamus mattanensis Becc., Rec. Bot. Surv. India 2: 216 (1902).
Borneo (Sarawak, W. Kalimantan). 42 BOR.
Calamus ferrugineus Becc., Rec. Bot. Surv. India 2: 216 (1902).
Calamus mattanensis var. *sabut* Becc., Ann. Roy. Bot. Gard. (Calcutta) 11(App.): 110 (1913).

Calamus mattanensis var. *sabut* Becc. = **Calamus mattanensis** Becc.

Calamus maturbongsii W.J.Baker & J.Dransf., Kew Bull. 57: 725 (2002).
W. New Guinea. 43 NWG.

Calamus mawaiensis Furtado = **Calamus blumei** Becc.

Calamus maximus Reinw. ex Schult.f. = **Plectocomia elongata** var. **elongata**

Calamus maximus Becc. = **Calamus merrillii** var. **merrillii**

Calamus maximus Blanco = **?** [42 PHI]

Calamus maximus Reinw. ex de Vriese = **?**

Calamus mayrii Burret, Notizbl. Bot. Gart. Berlin-Dahlem 11: 704 (1933).
New Guinea. 43 NWG.

Calamus megaphyllus Becc., Ann. Roy. Bot. Gard. (Calcutta) 11(App.): 66 (1913).
Philippines (Mindanao, Leyte). 42 PHI.

Calamus melanacanthus Mart., Hist. Nat. Palm. 3: 333 (1853). *Palmijuncus melanacanthus* (Mart.) Kuntze, Revis. Gen. Pl. 2: 733 (1891).
Myanmar. 41 MYA.

Calamus melanocarpus Ridl. = **Calamus lobbianus** Becc.

Calamus melanochaetes (Blume) Miq. = **Daemonorops melanochaetes** Blume

Calamus melanochrous Burret, Notizbl. Bot. Gart. Berlin-Dahlem 11: 208 (1931).
China (Guangxi). 36 CHS.

Calamus melanolepis (Mart.) H.Wendl. = **Calamus wightii** Griff.

Calamus melanoloma Mart., Hist. Nat. Palm. 3: 209 (1845). *Palmijuncus melanoloma* (Mart.) Kuntze, Revis. Gen. Pl. 2: 733 (1891). *Rotang melanoloma* (Mart.) Baill., Hist. Pl. 13: 300 (1895).
Jawa. 42 JAW.
Calamus anceps Blume, Rumphia 3: 65 (1847).
Calamus calolepis Miq., Pl. Jungh.: 159 (1852). *Palmijuncus calolepis* (Miq.) Kuntze, Revis. Gen. Pl. 2: 733 (1891).

Calamus melanorhynchus Becc., Ann. Roy. Bot. Gard. (Calcutta) 11(App.): 30 (1913).
Philippines (Mindanao). 42 PHI.

Calamus merrillii Becc., Webbia 1: 347 (1905).
Philippines. 42 PHI.
Calamus maximus Becc. in G.H.Perkins & al., Fragm. Fl. Philipp. 1: 45 (1904), nom. illeg.

var. **merrillii**
Philippines. 42 PHI.

var. **merrittianus** (Becc.) Becc., Philipp. J. Sci., C 14: 351 (1919).
Philippines. 42 PHI. Cl.
*Calamus merrittianus Becc., Philipp. J. Sci., C 2: 233 (1907).

var. **nanga** Becc., Philipp. J. Sci., C 14: 351 (1919).
Philippines. 42 PHI. Cl.

Calamus merrittianus Becc. = **Calamus merrillii** var. **merrittianus** (Becc.) Becc.

Calamus mesilauensis J.Dransf., Kew Bull. 36: 797 (1982).
Borneo (Sabah). 42 BOR.

Calamus metzianus Schltdl., Linnaea 26: 727 (1855).
S. India. 40 IND.

Calamus meyenianus Schauer = **Calamus usitatus** Blanco

Calamus micracanthus Griff. = **Daemonorops micracantha** (Griff.) Becc.

Calamus micranthus Blume, Rumphia 3: 53 (1847). *Palmijuncus micranthus* (Blume) Kuntze, Revis. Gen. Pl. 2: 733 (1891).
Sumatera. 42 SUM.

Calamus microcarpus Becc., Rec. Bot. Surv. India 2: 213 (1902).
Philippines. 42 PHI.

var. **diminutus** Becc., Philipp. J. Sci., C 14: 356 (1919).
Philippines. 42 PHI.

var. **longiocrea** Baja-Lapis, Sylvatrop 12: 68 (1987 publ. 1989).
Philippines. 42 PHI.

var. **microcarpus**
Philippines. 42 PHI.

Calamus microsphaerion Becc. in G.H.Perkins & al., Fragm. Fl. Philipp. 1: 45 (1904).
Borneo to Philippines. 42 BOR PHI.

var. **microsphaerion**
Borneo to Philippines. 42 BOR PHI.

var. **spinosior** Becc., Philipp. J. Sci., C 14: 354 (1919).
Philippines (Palawan). 42 PHI.

Calamus minahassae Warb. ex Becc., Ann. Roy. Bot. Gard. (Calcutta) 11(1): 356 (1908).
Sulawesi. 42 SUL.

Calamus mindorensis Becc., Philipp. J. Sci., C 2: 235 (1907).
Philippines (Luzon, Mindoro). 42 PHI.

Calamus minutus J.Dransf., Malaysian Forester 41: 339 (1978).
Pen. Malaysia (Trengganu). 42 MLY.

Calamus mirabilis Mart. = **Daemonorops mirabilis** (Mart.) Mart.

Calamus mishmeensis Griff. = **Calamus floribundus** Griff.

Calamus mitis Becc., Philipp. J. Sci., C 3: 341 (1909).
N. Philippines. 42 PHI.

Calamus modestus T.Evans & T.P.Anh, Kew Bull. 56: 731 (2001).
Vietnam. 41 VIE.

Calamus mogeae J.Dransf., Kew Bull. 55: 717 (2000).
Sumatera. 42 SUM.

Calamus mollis Blanco = **Daemonorops mollis** (Blanco) Merr.

Calamus monoecus Roxb. = **Calamus rotang** L.

Calamus montanus T.Anderson = **Calamus acanthospathus** Griff.

Calamus monticolus Griff. = **Daemonorops monticola** (Griff.) Mart.

Calamus moorhousei Furtado, Gard. Bull. Singapore 15: 207 (1956).
Pen. Malaysia (Negri Sembilan). 42 MLY.

Calamus moseleyanus Becc., Rec. Bot. Surv. India 2: 211 (1902).
Philippines. 42 PHI.

Calamus moszkowskianus Becc., Ann. Roy. Bot. Gard. (Calcutta) 11(App.): 139 (1913).
New Guinea. 43 NWG.

Calamus moti F.M.Bailey, Bull. Dept. Agric. Queensland 13: 13 (1899).
NE. Queensland. 50 QLD.

Calamus mucronatus Becc. = **Calamus laevigatus** var. **mucronatus** (Becc.) J.Dransf.

Calamus muelleri H.Wendl. & Drude, Linnaea 39: 193 (1875). *Palmijuncus muelleri* (H.Wendl. & Drude) Kuntze, Revis. Gen. Pl. 2: 733 (1891).
E. Australia. 50 NSW QLD.
Calamus muelleri var. *macrospermus* H.Wendl., Linnaea 39: 194 (1875).

Calamus muelleri var. *macrospermus* H.Wendl. = **Calamus muelleri** H.Wendl. & Drude

Calamus multinervis Becc., Ann. Roy. Bot. Gard. (Calcutta) 11(App.): 88 (1913).
China (Yunnan), Philippines (Mindanao). 36 CHC 42 PHI.

var. **menglaensis** San Y.Chen, S.J.Pei & K.L.Wang, Acta Bot. Yunnan. 24: 202 (2002).
China (Yunnan). 36 CHC.

var. **multinervis**
Philippines (Mindanao). 42 PHI.

Calamus multirameus Ridl. = **Calamus guruba** Buch.-Ham.

Calamus multisetosus Burret, Notizbl. Bot. Gart. Berlin-Dahlem 15: 807 (1943).
New Guinea. 43 NWG.

Calamus multispicatus Burret, Notizbl. Bot. Gart. Berlin-Dahlem 13: 592 (1937).

China (Guangdong, Hainan). 36 CHH CHS.

Calamus muricatus Becc., For. Borneo: 609 (1902).
Borneo. 42 BOR.
Calamus sphaeruliferus Becc., Ann. Roy. Bot. Gard. (Calcutta) 11(App.): 11 (1913).

Calamus myriacanthus Becc., Rec. Bot. Surv. India 2: 214 (1902).
Borneo. 42 BOR.
Calamus hewittianus Becc., Ann. Roy. Bot. Gard. (Calcutta) 11(App.): 45 (1913).
Calamus jaherianus Becc., Ann. Roy. Bot. Gard. (Calcutta) 11(App.): 46 (1913).

Calamus myrianthus Becc. in J.D.Hooker, Fl. Brit. India 6: 451 (1893).
Indo-China. 41 LAO MYA THA.
Calamus leucotes Becc., Ann. Roy. Bot. Gard. (Calcutta) 11(1): 309 (1908).

Calamus myriocarpus Burret, Notizbl. Bot. Gart. Berlin-Dahlem 15: 806 (1943).
New Guinea. 43 NWG.

Calamus myriocladus Burret, Notizbl. Bot. Gart. Berlin-Dahlem 15: 804 (1943).
Sumatera. 42 SUM.
Calamus latisectus Burret, Notizbl. Bot. Gart. Berlin-Dahlem 15: 190 (1940), nom. illeg.

Calamus nagbettai R.R.Fernald & Dey, Indian Forester 96: 223 (1970).
SW. India. 40 IND.

Calamus nambariensis Becc., Ann. Roy. Bot. Gard. (Calcutta) 11(1): 433 (1908).
Assam to Hainan. 36 CHC CHH 40 ASS 41 LAO THA.
Calamus nambariensis var. *alpinus* S.J.Pei & S.Y.Chen, Acta Phytotax. Sin. 27: 141 (1989).
Calamus nambariensis var. *furfuraceus* S.J.Pei & S.Y.Chen, Acta Phytotax. Sin. 27: 142 (1989).
Calamus nambariensis var. *menglongensis* S.J.Pei & S.Y.Chen, Acta Phytotax. Sin. 27: 141 (1989).
Calamus nambariensis var. *xishuangbannaensis* S.J.Pei & S.Y.Chen, Acta Phytotax. Sin. 27: 141 (1989).
Calamus nambariensis var. *yingjiangensis* S.J.Pei & S.Y.Chen, Acta Phytotax. Sin. 27: 140 (1989).

Calamus nambariensis var. *alpinus* S.J.Pei & S.Y.Chen = **Calamus nambariensis** Becc.

Calamus nambariensis var. *furfuraceus* S.J.Pei & S.Y.Chen = **Calamus nambariensis** Becc.

Calamus nambariensis var. *menglongensis* S.J.Pei & S.Y.Chen = **Calamus nambariensis** Becc.

Calamus nambariensis var. *xishuangbannaensis* S.J.Pei & S.Y.Chen = **Calamus nambariensis** Becc.

Calamus nambariensis var. *yingjiangensis* S.J.Pei & S.Y.Chen = **Calamus nambariensis** Becc.

Calamus nannostachys Burret, J. Arnold Arbor. 12: 264 (1931).
New Guinea. 43 NWG.

Calamus nanodendron J.Dransf., Kew Bull. 45: 90 (1990).
Borneo (Sarawak). 42 BOR.

Calamus nanus Burret = **Calamus tenompokensis** Furtado

Calamus neelagiricus Renuka, Rheedea 7: 69 (1997).
SW. India. 40 IND.

Calamus neglectus Becc. = **Calamus densiflorus** Becc.

Calamus nematospadix Becc., Rec. Bot. Surv. India 2: 204 (1902).
Borneo. 42 BOR.

Calamus nicobaricus Becc. in J.D.Hooker, Fl. Brit. India 6: 446 (1892).
Nicobar Is. 41 NCB.

Calamus nielsenii J.Dransf., Kew Bull. 35: 843 (1981).
Borneo (Sarawak). 42 BOR.

Calamus niger Willd. = **Daemonorops nigra** (Willd.) Blume

Calamus niger J.Braun & K.Schum. = **Oncocalamus mannii** (H.Wendl.) H.Wendl.

Calamus nigricans Van Valk., Blumea 40: 463 (1995).
Borneo (Kalimantan). 42 BOR.

Calamus nitidus Mart. = **Calamus guruba** Buch.-Ham.

Calamus nutantiflorus Griff. = **Daemonorops jenkinsiana** (Griff.) Mart.

Calamus oblongus Reinw. ex Blume = **Daemonorops oblonga** (Reinw. ex Blume) Blume

Calamus obovoideus S.J.Pei & S.Y.Chen, Acta Phytotax. Sin. 27: 142 (1989).
China (Yunnan). 36 CHC.

Calamus obstruens F.Muell. = **Calamus australis** Mart.

Calamus occidentalis Witono & J.Dransf., Kew Bull. 53: 747 (1998).
W. Jawa. 42 JAW.

Calamus ochreatus Miq. = **Korthalsia echinometra** Becc.

Calamus oligostachys T.Evans & al., Kew Bull. 56: 242 (2001).
Laos. 41 LAO.
  Calamus pauciflorus T.Evans & al., Kew Bull. 55: 935 (2000), nom. illeg.

Calamus opacus Blume, Rumphia 3: 59 (1847).
Sumatera. 42 SUM.

Calamus opacus G.Mann & H.Wendl. = **Laccosperma opacum** Drude

Calamus optimus Becc., For. Borneo: 610 (1902).
Borneo. 42 BOR.
  Calamus stramineus Furtado, Gard. Bull. Straits Settlem. 8: 258 (1935).
  Calamus stramineus var. *megalocarpus* Furtado, Gard. Bull. Straits Settlem. 8: 259 (1935).

Calamus oreophilus Furtado = **Calamus viridispinus** var. **viridispinus**

Calamus orientalis C.E.Chang = **Calamus formosanus** Becc.

Calamus ornatus Blume in J.J.Roemer & J.A.Schultes, Syst. Veg. 7: 1326 (1830). *Palmijuncus ornatus* (Blume) Kuntze, Revis. Gen. Pl. 2: 733 (1891). *Rotang ornatus* (Blume) Baill., Hist. Pl. 13: 299 (1895).
Pen. Thailand to W. & C. Malesia. 41 THA 42 BOR JAW MLY PHI SUL SUM.

var. **ornatus**
  Pen. Thailand to W. & C. Malesia. 41 THA 42 BOR JAW MLY PHI SUL SUM.
  Calamus ovatus Reinw. ex Kunth, Enum. Pl. 3: 205 (1841).
  Calamus aureus Reinw. ex Mart., Hist. Nat. Palm. 3: 341 (1853). *Palmijuncus aureus* (Reinw. ex Mart.) Kuntze, Revis. Gen. Pl. 2: 733 (1891).

  Calamus ornatus var. *horridus* Becc. in J.D.Hooker, Fl. Brit. India 6: 460 (1893).
  Calamus ornatus var. *philippinensis* Becc., Webbia 1: 346 (1905).
  Calamus ornatus var. *celebicus* Becc., Ann. Roy. Bot. Gard. (Calcutta) 11(App.): 74 (1913).

var. **pulverulentus** Fernando, Gard. Bull. Singapore 41: 53 (1988 publ. 1989).
  Philippines. 42 PHI.

Calamus ornatus var. *celebicus* Becc. = **Calamus ornatus** var. **ornatus**

Calamus ornatus var. *horridus* Becc. = **Calamus ornatus** var. **ornatus**

Calamus ornatus var. *philippinensis* Becc. = **Calamus ornatus** var. **ornatus**

Calamus orthostachyus Furtado, Gard. Bull. Straits Settlem. 8: 244 (1935).
Sulawesi. 42 SUL.

Calamus ovatus Reinw. ex Kunth = **Calamus ornatus** var. **ornatus**

Calamus ovoideus Thwaites ex Trimen, J. Bot. 23: 269 (1885). *Palmijuncus ovoideus* (Thwaites ex Trimen) Kuntze, Revis. Gen. Pl. 2: 732 (1891).
SW. Sri Lanka. 40 SRL.

Calamus oxleyanus Teijsm. & Binn. ex Miq., Palm. Archip. Ind.: 17 (1868). *Palmijuncus oxleyanus* (Teijsm. & Binn. ex Miq.) Kuntze, Revis. Gen. Pl. 2: 733 (1891).
Pen. Thailand to W. Malesia. 41 THA 42 BOR MLY SUM.

var. **montanus** Furtado, Gard. Bull. Singapore 15: 86 (1956).
  Pen. Malaysia (Trengganu). 42 MLY.

var. **oxleyanus**
  Pen. Thailand to W. Malesia. 41 THA 42 BOR MLY SUM.
  Daemonorops fasciculata Mart., Hist. Nat. Palm. 3: 330 (1853).
  Calamus fernandezii H.Wendl. in O.C.E.de Kerchove de Denterghem, Palmiers: 236 (1878). *Palmijuncus fernandezii* (H.Wendl.) Kuntze, Revis. Gen. Pl. 2: 733 (1891).
  Calamus diffusus Becc. in J.D.Hooker, Fl. Brit. India 6: 447 (1892).
  Calamus oxleyanus var. *obovatus* Becc., Ann. Roy. Bot. Gard. (Calcutta) 11(App.): 112 (1913).
  Calamus leiospathus Bartlett, Pap. Michigan Acad. Sci. 25: 8 (1939 publ. 1940).

Calamus oxleyanus var. *obovatus* Becc. = **Calamus oxleyanus** var. **oxleyanus**

Calamus oxycarpus Becc., Ann. Roy. Bot. Gard. (Calcutta) 11(App.): 138 (1913).
China (Yunnan, Guizhou, Guangxi). 36 CHC CHS.
  Calamus oxycarpus var. *angustifolius* San Y.Chen & K.L.Wang, Acta Bot. Yunnan. 24: 201 (2002).

Calamus oxycarpus var. *angustifolius* San Y.Chen & K.L.Wang = **Calamus oxycarpus** Becc.

Calamus pachypus W.J.Baker & al., Kew Bull. 58: 361 (2003).
New Guinea to Bismarck Arch. 43 BIS NWG.

Calamus pachystachys Warb. ex Becc., Ann. Roy. Bot. Gard. (Calcutta) 11(1): 465 (1908).
Sulawesi. 42 SUL.

Calamus pachystemonus Thwaites, Enum. Pl. Zeyl.: 431 (1864). *Palmijuncus pachystemonus* (Thwaites) Kuntze, Revis. Gen. Pl. 2: 733 (1891).
SW. Sri Lanka. 40 SRL.
Calamus gracilis Thwaites, Enum. Pl. Zeyl.: 330 (1864), nom. illeg.

Calamus pacificus Ridl. = **Calamus diepenhorstii** var. **diepenhorstii**

Calamus padangensis Furtado, Gard. Bull. Singapore 15: 62 (1956).
Pen. Malaysia (Trengganu). 42 MLY.

Calamus palembanicus Becc. = **Calamus ridleyanus** Becc.

Calamus palembanicus (Blume) Miq. = **Daemonorops palembanica** Blume

Calamus pallens Blume = **Calamus heteroideus** Blume

Calamus pallidulus Becc. = **Calamus laevigatus** var. **laevigatus**

Calamus palustris Griff., Calcutta J. Nat. Hist. 5: 60 (1845). *Palmijuncus palustris* (Griff.) Kuntze, Revis. Gen. Pl. 2: 733 (1891).
S. China to Nicobar Is. and Pen. Malaysia. 36 CHC CHS 41 AND CBD LAO MYA NCB THA 42 MLY.

var. **malaccensis** Becc., Ann. Roy. Bot. Gard. (Calcutta) 11(1): 405 (1908).
Myanmar to Pen. Malaysia (Perak, Penang). 41 MYA THA 42 MLY.
Calamus dumetorum Ridl., Mat. Fl. Malay. Penins. 2: 211 (1907).

var. **palustris**
S. China to Nicobar Is. 36 CHC CHS 41 AND CBD LAO MYA NCB THA.
Calamus latifolius Kurz, J. Asiat. Soc. Bengal, Pt. 2, Nat. Hist. 43(2): 20 (1874), nom. illeg.
Calamus palustris var. *amplissimus* Becc., Ann. Roy. Bot. Gard. (Calcutta) 11(1): 405 (1908).
Calamus palustris var. *cochinchinensis* Becc., Ann. Roy. Bot. Gard. (Calcutta) 11(1): 405 (1908).
Calamus kerrianus Becc., Ann. Roy. Bot. Gard. (Calcutta) 11(App.): 140 (1913).
Calamus palustris var. *longistachys* S.J.Pei & S.Y.Chen, Acta Phytotax. Sin. 27: 138 (1989).
Calamus loeiensis Hodel, Palm J. 139: 54 (1998).

Calamus palustris var. *amplissimus* Becc. = **Calamus palustris** var. **palustris**

Calamus palustris var. *cochinchinensis* Becc. = **Calamus palustris** var. **palustris**

Calamus palustris var. *longistachys* S.J.Pei & S.Y.Chen = **Calamus palustris** var. **palustris**

Calamus pandanosmus Furtado, Gard. Bull. Singapore 15: 217 (1956).
Pen. Thailand to W. Malesia. 41 THA 42 BOR MLY SUM.

Calamus papuanus Becc., Malesia 3: 60 (1886). *Palmijuncus papuanus* (Becc.) Kuntze, Revis. Gen. Pl. 2: 733 (1891).
New Guinea. 43 NWG.

Calamus paradoxus Kurz = **Myrialepis paradoxa** (Kurz) J.Dransf.

Calamus paspalanthus Becc. in J.D.Hooker, Fl. Brit. India 6: 450 (1893).
Pen. Malaysia, Borneo. 42 BOR MLY.
Daemonorops intumescens Becc., Rec. Bot. Surv. India 2: 222 (1902). *Calamus intumescens* (Becc.) Ridl., Mat. Fl. Malay. Penins. 2: 200 (1907).

Calamus pauciflorus T.Evans & al. = **Calamus oligostachys** T.Evans & al.

Calamus pauciflorus Ridl. = **Calamus filipendulus** Becc.

Calamus paucijugus Becc. ex K.Heyne, Nutt. Pl. Ned.-Ind., ed. 2, 1: 381 (1922).
Sulawesi. 42 SUL.

Calamus paulii J.Dransf., Kew Bull. 45: 81 (1990).
Borneo (Sarawak). 42 BOR.

Calamus pedicellatus Becc. ex K.Heyne, Nutt. Pl. Ned.-Ind., ed. 2, 1: 381 (1922).
Sulawesi. 42 SUL.

Calamus penangensis Ridl. = **Calamus penicillatus** Roxb.

Calamus penibukanensis Furtado = **Calamus blumei** Becc.

Calamus penicillatus Mart. = **Calamus penicillatus** Roxb.

Calamus penicillatus Roxb., Fl. Ind. ed. 1832, 3: 781 (1832). *Palmijuncus penicillatus* (Roxb.) Kuntze, Revis. Gen. Pl. 2: 732 (1891).
Pen. Malaysia (Penang). 42 MLY.
Calamus penicillatus Mart., Hist. Nat. Palm. 3: 334 (1853), nom. illeg. *Calamus martianus* Becc. in J.D.Hooker, Fl. Brit. India 6: 459 (1893).
Calamus penangensis Ridl., Mat. Fl. Malay. Penins. 2: 192 (1907).

Calamus perakensis Becc. in J.D.Hooker, Fl. Brit. India 6: 451 (1893).
Pen. Malaysia to W. Sumatera. 42 MLY SUM.

var. **crassus** J.Dransf., Malaysian Forester 41: 336 (1978).
Pen. Malaysia (Trengganu). 42 MLY.

var. **niger** J.Dransf., Malaysian Forester 41: 335 (1978).
Pen. Malaysia (Johore). 42 MLY.

var. **perakensis**
Pen. Malaysia to W. Sumatera. 42 MLY SUM.
Calamus lanatus Ridl., Mat. Fl. Malay. Penins. 2: 202 (1907).
Calamus perakensis var. *gracilis* Fernando, Gard. Bull. Singapore 15: 155 (1956).

Calamus perakensis var. *gracilis* Fernando = **Calamus perakensis** var. **perakensis**

Calamus peregrinus Furtado, Gard. Bull. Singapore 15: 66 (1956).
Pen. Thailand to Pen. Malaysia. 41 THA 42 MLY.

Calamus periacanthus (Miq.) Miq. = **Daemonorops periacantha** Miq.

Calamus perrottetii Becc. = **Calamus deerratus** G.Mann & H.Wendl.

Calamus petiolaris Griff. = **Daemonorops calicarpa** (Griff.) Mart.

Calamus petraeus Lour. = **? [41 VIE] Plectocomia ?**

Calamus pholidostachys J.Dransf. & W.J.Baker, Kew Bull. 58: 381 (2003).
Papua New Guinea. 43 NWG.

Calamus pilosellus Becc., Rec. Bot. Surv. India 2: 208 (1902).
Borneo. 42 BOR.

Calamus pilossisimus Becc., Nova Guinea 8: 219 (1909).
New Guinea. 43 NWG.

Calamus pisicarpus Blume, Rumphia 3: 39 (1847). *Palmijuncus pisicarpus* (Blume) Kuntze, Revis. Gen. Pl. 2: 732 (1891). *Rotang pisicarpus* (Blume) Baill., Hist. Pl. 13: 299 (1895).
Maluku. 42 MOL. — Provisionally accepted.

Calamus platyacanthoides Merr., Lingnan Sci. J. 13: 54 (1934).
China (Yunnan) to Indo-China. 36 CHC 41 LAO VIE.
*Calamus platyacanthus Warb. ex Becc., Ann. Roy. Bot. Gard. (Calcutta) 11(1): 442 (1908), nom. illeg.
Calamus giganteus var. *robustus* S.J.Pei & S.Y.Chen, Acta Phytotax. Sin. 27: 143 (1989).
Calamus platyacanthus var. *mediostachys* S.J.Pei & S.Y.Chen, Acta Phytotax. Sin. 27: 143 (1989).
Calamus platyacanthus var. *longicarpus* San Y.Chen & K.L.Wang, Acta Bot. Yunnan. 24: 203 (2002).

Calamus platyacanthus Warb. ex Becc. = **Calamus platyacanthoides** Merr.

Calamus platyacanthus Mart. = **Daemonorops oblonga** (Reinw. ex Blume) Blume

Calamus platyacanthus var. *longicarpus* San Y.Chen & K.L.Wang = **Calamus platyacanthoides** Merr.

Calamus platyacanthus var. *mediostachys* S.J.Pei & S.Y.Chen = **Calamus platyacanthoides** Merr.

Calamus platyspathus Mart. ex Kunth, Enum. Pl. 3: 209 (1841). *Daemonorops platyspatha* (Mart. ex Kunth) Mart., Hist. Nat. Palm. 3: 329 (1853). *Palmijuncus platyspathus* (Mart. ex Kunth) Kuntze, Revis. Gen. Pl. 2: 733 (1891).
Myanmar. 41 MYA.

Calamus plicatus Blume, Rumphia 3: 67 (1847). *Palmijuncus plicatus* (Blume) Kuntze, Revis. Gen. Pl. 2: 733 (1891).
Sulawesi. 42 SUL.

Calamus poensis Becc., Ann. Roy. Bot. Gard. (Calcutta) 11(App.): 43 (1913).
Borneo (Sarawak: Mt. Pueh). 42 BOR.

Calamus pogonacanthus Becc. ex H.J.P.Winkl., Bot. Jahrb. Syst. 48: 91 (1912).
Borneo. 42 BOR.

Calamus poilanei Conrard, Notul. Syst. (Paris) 7: 28 (1938).
Indo-China. 41 LAO THA VIE.

Calamus polycladus Burret, Notizbl. Bot. Gart. Berlin-Dahlem 15: 802 (1943).
New Guinea. 43 NWG.

Calamus polydesmus Becc., Ann. Roy. Bot. Gard. (Calcutta) 11(1): 430 (1908).
Myanmar. 41 MYA.

Calamus polygamus Roxb. = **Calamus flagellum** var. **flagellum**

Calamus polystachys Becc., Ann. Roy. Bot. Gard. (Calcutta) 11(1): 383 (1908).
W. Malesia. 42 JAW MLY SUM.

Calamus praetermissus J.Dransf., Kew Bull. 36: 802 (1982).
Borneo (Sabah, Kalimantan). 42 BOR.

Calamus prasinus Lakshmana & Renuka, J. Econ. Taxon. Bot. 14: 705 (1991).
SW. India. 40 IND.

Calamus prattianus Becc. in L.S.Gibbs, Fl. Arfak Mts.: 97 (1917).
New Guinea. 43 NWG.

Calamus pseudofeanus S.K.Basu, J. Econ. Taxon. Bot. 13: 133 (1989).
S. India. 40 IND.

Calamus pseudomollis Becc., Ann. Roy. Bot. Gard. (Calcutta) 11(App.): 23 (1913).
N. Sulawesi. 42 SUL.

Calamus pseudorivalis Becc., Ann. Roy. Bot. Gard. (Calcutta) 11(1): 222 (1908).
Andaman Is., Nicobar Is. 41 AND NCB.

Calamus pseudorotang Mart. ex Kunth = **Calamus viminalis** Willd.

Calamus pseudoscutellaris Conrard = **Calamus rhabdocladus** Burret

Calamus pseudotenuis Becc. in J.D.Hooker, Fl. Brit. India 6: 445 (1892).
SW. India, Sri Lanka. 40 IND SRL.

Calamus pseudoulur Becc., Ann. Roy. Bot. Gard. (Calcutta) 11(App.): 133 (1913).
Borneo (Sarawak). 42 BOR.

Calamus pseudozebrinus Burret, Notizbl. Bot. Gart. Berlin-Dahlem 12: 319 (1935).
New Guinea. 43 NWG.

Calamus psilocladus J.Dransf., Kew Bull. 45: 96 (1990).
Borneo (Sarawak). 42 BOR.

Calamus pulaiensis Becc., Ann. Roy. Bot. Gard. (Calcutta) 11(App.): 34 (1913).
Pen. Malaysia. 42 MLY. — Provisionally accepted.

Calamus pulchellus Burret, Notizbl. Bot. Gart. Berlin-Dahlem 13: 597 (1937).
China (Guangdong, Hainan). 36 CHH CHS.

Calamus pulcher Miq., Anal. Bot. Ind. 1: 3 (1850). *Palmijuncus pulcher* (Miq.) Kuntze, Revis. Gen. Pl. 2: 733 (1891).
Borneo (Kalimantan). 42 BOR. — Provisionally accepted.

Calamus pycnocarpus (Furtado) J.Dransf., Malaysian Forester 40: 202 (1977).
Pen. Malaysia (Trengganu). 42 MLY.
Cornera pycnocarpa Furtado, Gard. Bull. Singapore 14: 523 (1955).

Calamus pygmaeus Becc., Malesia 3: 8 (1886). *Palmijuncus pygmaeus* (Becc.) Kuntze, Revis. Gen. Pl. 2: 733 (1891).
Borneo (Sarawak). 42 BOR.

Calamus quinquenervius Roxb., Fl. Ind. ed. 1832, 3: 777 (1832). *Palmijuncus quinquenervius* (Roxb.) Kuntze, Revis. Gen. Pl. 2: 733 (1891).
Bangladesh. 40 ASS? BAN. — Provisionally accepted.

Calamus quinquesetinervius Burret, Notizbl. Bot. Gart. Berlin-Dahlem 15: 810 (1943).
Taiwan. 38 TAI.

Calamus radiatus Thwaites, Enum. Pl. Zeyl.: 431 (1864). *Palmijuncus radiatus* (Thwaites) Kuntze, Revis. Gen. Pl. 2: 733 (1891).
SW. Sri Lanka. 40 SRL.

Calamus radicalis H.Wendl. & Drude, Linnaea 39: 195 (1875). *Palmijuncus radicalis* (H.Wendl. & Drude) Kuntze, Revis. Gen. Pl. 2: 732 (1891).
NE. Queensland. 50 QLD.

Calamus radulosus Becc. in J.D.Hooker, Fl. Brit. India 6: 443 (1892).
Thailand to Pen. Malaysia (Perak). 41 THA 42 MLY.

Calamus ralumensis Warb. ex K.Schum. = **Calamus longipinna** K.Schum. & Lauterb.

Calamus ramosissimus Griff. = **Daemonorops longipes** (Griff.) Mart.

Calamus ramulosus Becc. in G.H.Perkins & al., Fragm. Fl. Philipp. 1: 46 (1904).
Philippines. 42 PHI.

Calamus reinwardtii Mart., Hist. Nat. Palm. 3(ed. 2): 208 (1845).
Sumatera to Jawa. 42 JAW SUM.

Calamus regularis Burret = **Calamus marginatus** (Blume) Mart.

Calamus reticulatus Burret, J. Arnold Arbor. 20: 195 (1939).
EC. New Guinea. 43 NWG.

Calamus retrophyllus Becc. = **Calamus laevigatus** var. **laevigatus**

Calamus reyesianus Becc., Philipp. J. Sci., C 2: 237 (1907).
Philippines. 42 PHI.

Calamus rhabdocladus Burret, Notizbl. Bot. Gart. Berlin-Dahlem 10: 884 (1930).
S. China to Indo-China. 36 CHC CHH CHS 41 LAO VIE.
   Calamus pseudoscutellaris Conrard, Notul. Syst. (Paris) 7: 25 (1938).
   Calamus rhabdocladus var. *globulosus* S.J.Pei & S.Y.Chen, Acta Phytotax. Sin. 27: 137 (1989).

Calamus rhabdocladus var. *globulosus* S.J.Pei & S.Y.Chen = **Calamus rhabdocladus** Burret

Calamus rheedei Griff., Calcutta J. Nat. Hist. 5: 73 (1845). *Daemonorops rheedei* (Griff.) Mart., Hist. Nat. Palm. 3: 330 (1853). *Palmijuncus rheedei* (Griff.) Kuntze, Revis. Gen. Pl. 2: 732 (1891).
SW. India. 40 IND.

Calamus rhomboideus Blume in J.J.Roemer & J.A.Schultes, Syst. Veg. 7: 133 (1829). *Palmijuncus rhomboideus* (Blume) Kuntze, Revis. Gen. Pl. 2: 732 (1891). *Rotang rhomboideus* (Blume) Baill., Hist. Pl. 13: 299 (1895).
Sumatera to Jawa. 42 JAW SUM.

Calamus rhytidomus Becc., Ann. Roy. Bot. Gard. (Calcutta) 11(App.): 7 (1913).
Borneo (Kalimantan). 42 BOR.

Calamus ridleyanus Becc., Rec. Bot. Surv. India 2: 205 (1902).
Pen. Malaysia (incl. Singapore) to Sumatera. 42 MLY SUM.
   Calamus palembanicus Becc., Ann. Roy. Bot. Gard. (Calcutta) 11(App.): 75 (1913), nom. illeg.

Calamus riedelianus Miq. = **Daemonorops riedeliana** (Miq.) Becc.

Calamus riparius Furtado = **Calamus axillaris** Becc.

Calamus rivalis Thwaites ex Trimen, J. Bot. 23: 386 (1885). *Palmijuncus rivalis* (Thwaites ex Trimen) Kuntze, Revis. Gen. Pl. 2: 733 (1891).
Sri Lanka. 40 SRL.
   Calamus rudentum Mart., Hist. Nat. Palm. 3: 340 (1853), nom. illeg.

Calamus robinsonianus Becc., Philipp. J. Sci., C 12: 81 (1917).
S. Sulawesi to Maluku. 42 MOL SUL.

Calamus robustus L.Linden & Rodigas = **? [42 BOR]**

Calamus rostratus Furtado = **Calamus marginatus** (Blume) Mart.

Calamus rotang L., Sp. Pl.: 325 (1753). *Rotanga calamus* Crantz, Inst. Rei Herb. 1: 127 (1766). *Rotang linnaei* Baill., Hist. Pl. 13: 299 (1895).
S. India, Sri Lanka. 40 IND SRL 41 MYA.

Calamus roxburghii Griff. = **Calamus rotang** L.

Calamus royleanus Griff. = **Calamus tenuis** Roxb.

Calamus ruber Reinw. ex Mart. = **Daemonorops rubra** (Reinw. ex Mart.) Blume

Calamus rubiginosus Ridl. = **? [42 BOR]**

Calamus rudentum Mart. = **Calamus rivalis** Thwaites ex Trimen

Calamus rudentum Lour., Fl. Cochinch.: 209 (1790). *Palmijuncus rudentum* (Lour.) Kuntze, Revis. Gen. Pl. 2: 732 (1891). *Rotang rudentum* (Lour.) Baill., Hist. Pl. 13: 299 (1895).
Indo-China. 41 CBD LAO THA VIE.

Calamus rugosus Becc. in J.D.Hooker, Fl. Brit. India 6: 443 (1892).
China (Yunnan), W. Malesia. 36 CHC 42 BOR MLY SUM.

Calamus rumphii Blume, Rumphia 3: 38 (1847). *Daemonorops rumphii* (Blume) Mart., Hist. Nat. Palm. 3: 331 (1853).
Maluku. 42 MOL. — Provisionally accepted.

Calamus ruvidus Becc., Rec. Bot. Surv. India 2: 202 (1902).
Borneo (Sarawak). 42 BOR.

Calamus sabalensis J.Dransf., Kew Bull. 45: 83 (1990).
Borneo (Sarawak). 42 BOR.

Calamus sabensis Becc., Ann. Roy. Bot. Gard. (Calcutta) 11(1): 245 (1908).
Borneo (Sabah). 42 BOR.

Calamus salakka Willd. ex Steud. = **Salacca zalacca** (Gaertn.) Voss

Calamus salicifolius Becc., Rec. Bot. Surv. India 2: 206 (1902).
Indo-China. 41 CBD VIE.
   Calamus salicifolius var. *leiophyllus* Becc., Ann. Roy. Bot. Gard. (Calcutta) 11(1): 281 (1908).

Calamus salicifolius var. *leiophyllus* Becc. = **Calamus salicifolius** Becc.

Calamus samian Becc., Ann. Roy. Bot. Gard. (Calcutta) 11(App.): 92 (1913).
Philippines. 42 PHI.

Calamus sarawakensis Becc., Rec. Bot. Surv. India 2: 208 (1902).
Borneo (Sabah, Sarawak). 42 BOR.
   Calamus scabrifolius Becc., Ann. Roy. Bot. Gard. (Calcutta) 11(App.): 56 (1913).

Calamus scabridulus Becc., Rec. Bot. Surv. India 2: 203 (1902).
Pen. Malaysia, Sumatera (Billiton). 42 MLY SUM.

Calamus scabrifolius Becc. = **Calamus sarawakensis** Becc.

Calamus scabrispathus Becc., Bot. Jahrb. Syst. 58: 459 (1923).
New Guinea. 43 NWG.

Calamus schaeferianus Burret, Notizbl. Bot. Gart. Berlin-Dahlem 15: 189 (1940).
Sumatera. 42 SUM.

Calamus schistoacanthus Blume, Rumphia 3: 49 (1847). *Palmijuncus schistoacanthus* (Blume) Kuntze, Revis. Gen. Pl. 2: 733 (1891).
Borneo. 42 BOR.

Calamus schizospathus Griff. = **Calamus erectus** Roxb.

Calamus schlechterianus Becc., Ann. Roy. Bot. Gard. (Calcutta) 11(App.): 119 (1913).
New Guinea. 43 NWG.

Calamus schweinfurthii Becc. = **Calamus deerratus** G.Mann & H.Wendl.

Calamus scipionum Lam. = **?**

Calamus scipionum Lour., Fl. Cochinch.: 210 (1790). *Palmijuncus scipionum* (Lour.) Kuntze, Revis. Gen. Pl. 2: 733 (1891). *Rotang scipionum* (Lour.) Baill., Hist. Pl. 13: 299 (1895).
Pen. Thailand to W. Malesia, Philippines (Palawan). 41 THA 42 BOR MLY PHI SUM.

Calamus scleracanthus Becc. ex K.Heyne, Nutt. Pl. Ned.-Ind., ed. 2, 1: 387 (1922).
N. Sulawesi. 42 SUL.

Calamus scutellaris Becc. = **Calamus thysanolepis** var. **thysanolepis**

Calamus secundiflorus P.Beauv. = **Laccosperma secundiflorum** (P.Beauv.) Kuntze

Calamus sedens J.Dransf., Kew Bull. 33: 528 (1979).
Pen. Thailand to Pen. Malaysia. 41 THA 42 MLY.

Calamus semierectus Renuka & Vij.Kumar, Rheedea 4: 122 (1994).
Nicobar Is. 41 NCB.

Calamus semoi Becc., Ann. Roy. Bot. Gard. (Calcutta) 11(App.): 129 (1913).
Borneo (Sarawak). 42 BOR.

Calamus senalingensis J.Dransf., Malaysian Forester 41: 342 (1978).
Pen. Malaysia (Negri Sembilan). 42 MLY.

Calamus sepikensis Becc., Bot. Jahrb. Syst. 58: 457 (1923).
Papua New Guinea. 43 NWG.

Calamus serrulatus Becc., Malesia 3: 61 (1886). *Palmijuncus serrulatus* (Becc.) Kuntze, Revis. Gen. Pl. 2: 733 (1891).
New Guinea. 43 NWG.

Calamus sessilifolius Burret, Notizbl. Bot. Gart. Berlin-Dahlem 15: 7 (1940).
Papua New Guinea. 43 NWG.

Calamus setiger Burret = **Calamus anomalus** Burret

Calamus setulosus J.Dransf., Malaysian Forester 41: 343 (1978).
Thailand to Pen. Malaysia (Perak). 41 THA 42 MLY.

Calamus shendurunii Anto, Renuka & Sreek., Rheedea 11: 37 (2001).
India (Kerala). 40 IND.

Calamus siamensis Becc., Rec. Bot. Surv. India 2: 203 (1902).
Indo-China to Pen. Malaysia. 41 LAO MYA? THA 42 MLY.
Calamus siamensis var. *malaianus* Furtado, Gard. Bull. Singapore 15: 215 (1956).

Calamus siamensis var. *malaianus* Furtado = **Calamus siamensis** Becc.

Calamus simplex Becc. in J.D.Hooker, Fl. Brit. India 6: 456 (1893).
Pen. Malaysia (Perak, Pahang). 42 MLY.

Calamus simplicifolius C.F.Wei, Guihaia 6: 36 (1986).
China (Guangxi, Guangdong) to Hainan. 36 CHH CHS.

Calamus singaporensis Becc. = **Calamus diepenhorstii** var. **diepenhorstii**

Calamus siphonospathus Mart., Hist. Nat. Palm. 3: 342 (1853). *Palmijuncus siphonospathus* (Mart.) Kuntze, Revis. Gen. Pl. 2: 734 (1891).
Taiwan to C. Malesia. 38 TAI 42 PHI SUL.

var. **dransfieldii** Baja-Lapis, Sylvatrop 12: 80 (1987 publ. 1989).
Philippines to N. Sulawesi. 42 PHI SUL.

var. **farinosus** Becc., Ann. Roy. Bot. Gard. (Calcutta) 11: 474 (1908).
Philippines. 42 PHI.

var. **oligolepis** Becc., Webbia 1: 353 (1905).
Philippines. 42 PHI.

var. **polylepis** Becc., Webbia 1: 354 (1905).
Philippines. 42 PHI.

var. **siphonospathus**
Taiwan to Philippines. 38 TAI 42 PHI.
Calamus inflatus Warb. in G.H.Perkins & al., Fragm. Fl. Philipp. 1: 45 (1904).

var. **sublaevis** Becc., Webbia 1: 354 (1905).
Philippines. 42 PHI.

Calamus siphonospathus var. *batanensis* Becc. = **Calamus batanensis** (Becc.) Baja-Lapis

Calamus slootenii Furtado = **Calamus blumei** Becc.

Calamus solitarius T.Evans & al., Kew Bull. 55: 932 (2000).
Indo-China. 41 LAO THA.

Calamus sordidus J.Dransf., Bot. J. Linn. Soc. 81: 13 (1980).
Borneo. 42 BOR.

Calamus spathulatus Becc. = **Calamus insignis** var. **insignis**

Calamus spathulatus var. *robustus* Becc. = **Calamus insignis** var. **robustus** (Becc.) J.Dransf.

Calamus sphaeruliferus Becc. = **Calamus muricatus** Becc.

Calamus speciosissimus Furtado, Gard. Bull. Singapore 15: 198 (1956).
Pen. Thailand to Sumatera. 41 THA 42 MLY SUM.

Calamus spectabilis Blume, Rumphia 3: 55 (1847). *Palmijuncus spectabilis* (Blume) Kuntze, Revis. Gen. Pl. 2: 734 (1891). *Rotang spectabilis* (Blume) Baill., Hist. Pl. 13: 299 (1895).
Sumatera to Jawa. 42 JAW SUM.

Calamus spectatissimus Furtado, Gard. Bull. Singapore 15: 64 (1956).
Pen. Thailand to W. Malesia. 41 THA 42 BOR MLY SUM.

Calamus spinifolius Becc., Rec. Bot. Surv. India 2: 202 (1902).
Philippines. 42 PHI.

Calamus spinulinervis Becc., Ann. Roy. Bot. Gard. (Calcutta) 11(App.): 12 (1913).
Borneo. 42 BOR.

Calamus steenisii Furtado = **Calamus zebrinus** Becc.

Calamus stipitatus Burret = **Calamus vitiensis** Warb. ex Becc.

Calamus stoloniferus Teijsm. & Binn. = **Calamus tenuis** Roxb.

Calamus stoloniferus Renuka, J. Econ. Taxon. Bot. 14: 701 (1990)
SW. India. 40 IND.

Calamus stramineus Furtado = **Calamus optimus** Becc.

Calamus stramineus var. *megalocarpus* Furtado = **Calamus optimus** Becc.

Calamus strictus (Blume) Miq. = **Daemonorops longipes** (Griff.) Mart.

Calamus suaveolens W.J.Baker & J.Dransf., Kew Bull. 59: 69 (2004).
N. & C. Sulawesi. 42 SUL.

Calamus subangulatus Miq. = **Ceratolobus subangulatus** (Miq.) Becc.

Calamus subinermis H.Wendl. ex Becc., Rec. Bot. Surv. India 2: 212 (1902).
N. Borneo to N. Sulawesi. 42 BOR PHI SUL.

Calamus subspathulatus Ridl. = **Calamus insignis** var. **insignis**

Calamus sumbawensis Burret, Notizbl. Bot. Gart. Berlin-Dahlem 15: 802 (1943).
Lesser Sunda Is. (Sumbawa). 42 LSI.

Calamus symphysipus Mart., Hist. Nat. Palm. 3: 336 (1853). *Palmijuncus symphysipus* (Mart.) Kuntze, Revis. Gen. Pl. 2: 734 (1891).
Philippines to Sulawesi. 42 PHI SUL.

Calamus tanakadatei Furtado, Gard. Bull. Singapore 15: 225 (1956).
Pen. Malaysia. 42 MLY.

Calamus tapa Becc., Ann. Roy. Bot. Gard. (Calcutta) 11(App.): 128 (1913).
Borneo (Kalimantan). 42 BOR.

Calamus temburongii J.Dransf., Rattans Brunei: 194 (1997 publ. 1998).
Borneo (Brunei). 42 BOR.

Calamus temii T.Evans, Kew Bull. 57: 85 (2002).
Thailand. 41 THA.

Calamus tenompokensis Furtado, Gard. Bull. Straits Settlem. 8: 260 (1935).
Borneo (Sabah, Sarawak). 42 BOR.
Calamus nanus Burret, Notizbl. Bot. Gart. Berlin-Dahlem 15: 818 (1943).

Calamus tenuis Roxb., Fl. Ind. ed. 1832, 3: 780 (1832). *Palmijuncus tenuis* (Roxb.) Kuntze, Revis. Gen. Pl. 2: 734 (1891).
Uttaranchal and Jawa to to Indo-China. 40 ASS BAN IND WHM 41 CBD LAO MYA THA VIE 42 JAW SUM.
Calamus amarus Lour., Fl. Cochinch.: 210 (1790). Provisional synonym. *Palmijuncus amarus* (Lour.) Kuntze, Revis. Gen. Pl. 2: 733 (1891).
Calamus heliotropium Buch.-Ham. ex Kunth, Enum. Pl. 3: 210 (1841). *Palmijuncus heliotropium* (Buch.-Ham. ex Kunth) Kuntze, Revis. Gen. Pl. 2: 733 (1891).
Calamus royleanus Griff., Calcutta J. Nat. Hist. 5: 40 (1845). *Rotang royleanus* (Griff.) Baill., Hist. Pl. 13: 299 (1895).
Calamus horrens Blume, Rumphia 3: 43 (1847). *Palmijuncus horrens* (Blume) Kuntze, Revis. Gen. Pl. 2: 733 (1891).
Calamus stoloniferus Teijsm. & Binn., Cat. Hort. Bot. Bogor.: 75 (1866), nom. inval.
Palmijuncus royleanus (Griff.) Kuntze, Revis. Gen. Pl. 2: 732 (1891).

Calamus tetradactyloides Burret, Notizbl. Bot. Gart. Berlin-Dahlem 13: 596 (1937).
S. China to Vietnam. 36 CHC CHH CHS 41 VIE.

Calamus tetradactylus Hance, J. Bot. 13: 289 (1875). *Palmijuncus tetradactylus* (Hance) Kuntze, Revis. Gen. Pl. 2: 732 (1891).
S. China to Indo-China. 36 CHC CHH 41 CBD LAO THA VIE.
Calamus bonianus Becc., Webbia 3: 231 (1910).
Calamus cambojensis Becc., Webbia 3: 232 (1910).

Calamus tetrastichus Blume = **Calamus javensis** Blume

Calamus thwaitesii Becc. in J.D.Hooker, Fl. Brit. India 6: 441 (1892).
SW. India, Sri Lanka. 40 IND SRL.
Calamus thwaitesii var. *canaranus* Becc., Ann. Roy. Bot. Gard. (Calcutta) 11: 71 (1908).

Calamus thwaitesii var. *canaranus* Becc. = **Calamus thwaitesii** Becc.

Calamus thysanolepis Hance, J. Bot. 12: 265 (1874). *Palmijuncus thysanolepis* (Hance) Kuntze, Revis. Gen. Pl. 2: 734 (1891).
SE. China to N. Vietnam. 36 CHS 41 VIE.

var. **polylepis** C.F.Wei, Guihaia 6: 24 (1986).
China (Guangdong). 36 CHS.

var. **thysanolepis**
SE. China to N. Vietnam. 36 CHS 41 VIE.
Calamus scutellaris Becc., Webbia 3: 234 (1910).

Calamus tigrinus Kurz = **Calamus longisetus** Griff.

Calamus timorensis Becc., Ann. Roy. Bot. Gard. (Calcutta) 40(App.): 136 (1913).
Lesser Sunda Is. (Timor). 42 LSI.

Calamus toli-toliensis Becc. ex K.Heyne, Nutt. Pl. Ned.-Ind., ed. 2, 1: 389 (1922).
Sulawesi. 42 SUL.

Calamus tomentosus Becc. in J.D.Hooker, Fl. Brit. India 6: 455 (1893).
Pen. Malaysia, Borneo. 42 BOR MLY.

Calamus tonkinensis Becc. = **Calamus walkeri** Hance

Calamus tonkinensis var. *brevispicatus* C.F.Wei = **Calamus walkeri** Hance

Calamus trachycoleus Becc., Ann. Roy. Bot. Gard. (Calcutta) 11(App.): 108 (1913).
Borneo (Kalimantan). 42 BOR.

Calamus travancoricus Bedd. ex Becc. in J.D.Hooker, Fl. Brit. India 6: 452 (1893).
S. India. 40 IND.

Calamus trichrous (Miq.) Miq. = **Daemonorops trichroa** Miq.

Calamus trinervis W.Watson = **?**

Calamus triqueter Becc. = **Plectocomiopsis triquetra** (Becc.) J.Dransf.

Calamus trispermus Becc. in G.H.Perkins & al., Fragm. Fl. Philipp. 1: 46 (1904).
Philippines (Luzon). 42 PHI.

Calamus tumidus Furtado, Gard. Bull. Singapore 15: 105 (1956).
Pen. Malaysia to Sumatera. 42 MLY SUM.

Calamus turbinatus Ridl. = **Plectocomiopsis geminiflora** (Griff.) Becc.

Calamus ulur Becc., Ann. Roy. Bot. Gard. (Calcutta) 11(App.): 131 (1913).
Pen. Malaysia to Sumatera. 42 MLY SUM.

Calamus unifarius H.Wendl., Bot. Zeitung (Berlin) 17: 158 (1859). *Palmijuncus unifarius* (H.Wendl.) Kuntze, Revis. Gen. Pl. 2: 734 (1891).
Nicobar Is., Sumatera to Jawa. 41 NCB 42 JAW SUM.

var. **pentong** Becc. in J.D.Hooker, Fl. Brit. India 6: 458 (1893).
Nicobar Is. 41 NCB.

var. **unifarius**
Sumatera to Jawa. 42 JAW SUM.

Calamus usitatus Blanco, Fl. Filip.: 265 (1837). *Palmijuncus usitatus* (Blanco) Kuntze, Revis. Gen. Pl. 2: 733 (1891).
Borneo to Philippines. 42 BOR PHI.
   Calamus gracilis Blanco, Fl. Filip.: 267 (1837), nom. illeg. *Calamus blancoi* Kunth, Enum. Pl. 3: 595 (1841). *Palmijuncus blancoi* (Kunth) Kuntze, Revis. Gen. Pl. 2: 733 (1891). *Rotang blancoi* (Kunth) Baill., Hist. Pl. 13: 299 (1895).
   Calamus meyenianus Schauer, Nov. Actorum Acad. Caes. Leop.-Carol. Nat. Cur. 19(Suppl. 1): 425 (1843). *Palmijuncus meyenianus* (Schauer) Kuntze, Revis. Gen. Pl. 2: 733 (1891).
   Calamus brevifrons Mart., Hist. Nat. Palm. 3: 338 (1853). *Palmijuncus brevifrons* (Mart.) Kuntze, Revis. Gen. Pl. 2: 733 (1891).
   Calamus haenkeanus Mart., Hist. Nat. Palm. 3: 337 (1853). *Palmijuncus haenkeanus* (Mart.) Kuntze, Revis. Gen. Pl. 2: 733 (1891).

Calamus vanuatuensis Dowe = **Calamus vitiensis** Warb. ex Becc.

Calamus vattayila Renuka, Curr. Sci. 56: 1012 (1987).
India. 40 IND.

Calamus verschaffeltii H.Wendl. = **Acanthophoenix rubra** (Bory) H.Wendl.

Calamus verticillaris Griff. = **Daemonorops verticillaris** (Griff.) Mart.

Calamus verus Lour. = **?** [41 VIE] **Daemonorops** sp.

Calamus vestitus Becc., Malesia 3: 59 (1886). *Palmijuncus vestitus* (Becc.) Kuntze, Revis. Gen. Pl. 2: 734 (1891).
NW. & NC. New Guinea. 43 NWG.

Calamus vidalianus Becc., Rec. Bot. Surv. India 2: 212 (1902).
Philippines (Luzon). 42 PHI.

Calamus viminalis Willd., Sp. Pl. 2: 203 (1799). *Palmijuncus viminalis* (Willd.) Kuntze, Revis. Gen. Pl. 2: 732 (1891). *Rotang viminalis* (Willd.) Baill., Hist. Pl. 13: 299 (1895).
NE. India to SC. China and Lesser Sunda Is. (Bali). 36 CHC 40 BAN IND 41 AND CBD LAO MYA NCB THA VIE 42 JAW LSI MLY.
   Calamus fasciculatus Roxb., Fl. Ind. ed. 1832, 3: 779 (1832). *Palmijuncus fasciculatus* (Roxb.) Kuntze, Revis. Gen. Pl. 2: 733 (1891). *Calamus viminalis* var. *fasciculatus* (Roxb.) Becc. in J.D.Hooker, Fl. Brit. India 6: 444 (1892).
   Calamus extensus Mart., Hist. Nat. Palm. 3: 210 (1838), nom. illeg.
   Calamus pseudorotang Mart. ex Kunth, Enum. Pl. 3: 207 (1841). *Palmijuncus pseudorotang* (Mart. ex Kunth) Kuntze, Revis. Gen. Pl. 2: 732 (1891).
   Calamus litoralis Blume, Rumphia 3: 43 (1847). *Palmijuncus litoralis* (Blume) Kuntze, Revis. Gen. Pl. 2: 733 (1891).
   Calamus viminalis subvar. *pinangianus* Becc., Ann. Roy. Bot. Gard. (Calcutta) 11(1): 207 (1908).

Calamus viminalis var. *andamanicus* Becc., Ann. Roy. Bot. Gard. (Calcutta) 11(1): 207 (1908).
Calamus viminalis var. *bengalensis* Becc., Ann. Roy. Bot. Gard. (Calcutta) 11(1): 206 (1908).
Calamus viminalis var. *cochinchinensis* Becc., Ann. Roy. Bot. Gard. (Calcutta) 11(1): 207 (1908).

Calamus viminalis Reinw. ex Mart. = **?**

Calamus viminalis var. *andamanicus* Becc. = **Calamus viminalis** Willd.

Calamus viminalis var. *bengalensis* Becc. = **Calamus viminalis** Willd.

Calamus viminalis var. *cochinchinensis* Becc. = **Calamus viminalis** Willd.

Calamus viminalis var. *fasciculatus* (Roxb.) Becc. = **Calamus viminalis** Willd.

Calamus viminalis subvar. *pinangianus* Becc. = **Calamus viminalis** Willd.

Calamus viminalis subsp. *prostratus* Blume = **Calamus buroensis** Mart.

Calamus vinosus Becc., Leafl. Philipp. Bot. 8: 3061 (1919).
Philippines (Mindanao). 42 PHI.

Calamus viridispinus Becc. in J.D.Hooker, Fl. Brit. India 6: 458 (1893).
China (Guangxi), Pen. Thailand to Sumatera. 36 CHS 41 THA 42 MLY SUM.

var. **sumatranus** Becc., Ann. Roy. Bot. Gard. (Calcutta) 11(1): 109 (1908).
Sumatera. 42 SUM.

var. **viridispinus**
China (Guangxi), Pen. Thailand to Pen. Malaysia. 36 CHS 41 THA 42 MLY.
   Calamus brevispadix Ridl., Mat. Fl. Malay. Penins. 2: 207 (1907).
   Calamus distichus Ridl., Mat. Fl. Malay. Penins. 2: 206 (1907).
   Calamus elegans Becc. ex Ridl., Mat. Fl. Malay. Penins. 2: 207 (1907).
   Calamus bubuensis Becc., Ann. Roy. Bot. Gard. (Calcutta) 11(1): 417 (1908).
   Calamus benomensis Furtado, Gard. Bull. Singapore 15: 132 (1956).
   Calamus distichoideus Furtado, Gard. Bull. Singapore 15: 122 (1956).
   Calamus koribanus Furtado, Gard. Bull. Singapore 15: 128 (1956).
   Calamus oreophilus Furtado, Gard. Bull. Singapore 15: 124 (1956).
   Calamus distichus var. *shangsiensis* S.J.Pei & S.Y.Chen, Acta Phytotax. Sin. 27: 140 (1989).

Calamus viridissimus Becc., Ann. Roy. Bot. Gard. (Calcutta) 11(App.): 84 (1913).
Philippines (Mindanao). 42 PHI.

Calamus vitiensis Warb. ex Becc., Ann. Roy. Bot. Gard. (Calcutta) 11(1): 350 (1908).
Solomon Is. to Fiji (Viti Levu, Taveuni). 43 SOL 60 FIJ VAN.
   Calamus stipitatus Burret, Notizbl. Bot. Gart. Berlin-Dahlem 15: 814 (1943).
   Calamus vanuatuensis Dowe, Principes 37: 206 (1993).

Calamus wailong S.J.Pei & S.Y.Chen, Acta Phytotax. Sin. 27: 138 (1989).
China (Yunnan) to Indo-China. 36 CHC 41 LAO THA.

Calamus walkeri Hance, J. Bot. 12: 266 (1874). *Palmijuncus walkeri* (Hance) Kuntze, Revis. Gen. Pl. 2: 734 (1891).
China (Guangdong, Hong Kong) to Vietnam. 36 CHS 41 VIE.
  Calamus tonkinensis Becc., Ann. Roy. Bot. Gard. (Calcutta) 11(1): 275 (1908).
  Calamus tonkinensis var. *brevispicatus* C.F.Wei, Guihaia 6: 31 (1986). *Calamus faberi* var. *brevispicatus* (C.F.Wei) S.J.Pei & S.Y.Chen, Acta Phytotax. Sin. 27: 133 (1989).

Calamus wanggaii W.J.Baker & J.Dransf., Kew Bull. 57: 863 (2002).
W. New Guinea. 43 NWG.

Calamus warburgii K.Schum. in K.M.Schumann & C.A.G.Lauterbach, Fl. Schutzgeb. Südsee: 203 (1900).
Papua New Guinea to N. Queensland. 43 NWG 50 QLD.

Calamus wari-wariensis Becc., Ann. Roy. Bot. Gard. (Calcutta) 11(App.): 71 (1913).
New Guinea. 43 NWG.

Calamus whitmorei J.Dransf., Malaysian Forester 41: 337 (1978).
Pen. Malaysia (Trengganu). 42 MLY.

Calamus wightii Griff., Palms Brit. E. Ind.: t. 216C (1850). *Palmijuncus wightii* (Griff.) Kuntze, Revis. Gen. Pl. 2: 733 (1891).
India (Nilgiri Hills). 40 IND.
  Calamus huegelianus Mart., Hist. Nat. Palm. 3: 338 (1853). *Palmijuncus huegelianus* (Mart.) Kuntze, Revis. Gen. Pl. 2: 733 (1891).
  Daemonorops melanolepis Mart., Hist. Nat. Palm. 3: 331 (1853). *Calamus melanolepis* (Mart.) H.Wendl. in O.C.E.de Kerchove de Denterghem, Palmiers: 237 (1878). *Palmijuncus melanolepis* (Mart.) Kuntze, Revis. Gen. Pl. 2: 733 (1891).

Calamus winklerianus Becc., Bot. Jahrb. Syst. 48: 91 (1912).
Borneo (Kalimantan). 42 BOR.

Calamus wuliangshanensis San Y.Chen, K.L.Wang & S.J.Pei, Acta Bot. Yunnan. 24: 199 (2002).
China (Yunnan). 36 CHC.

  var. **sphaerocarpus** San Y.Chen & K.L.Wang, Acta Bot. Yunnan. 24: 201 (2002).
  China (Yunnan). 36 CHC.

  var. **wuliangshanensis**
  China (Yunnan). 36 CHC.

Calamus yuangchunensis C.F.Wei, Guihaia 6: 26 (1986).
China (Guangdong). 36 CHS.

Calamus yunnanensis Govaerts = **Calamus acanthospathus** Griff.

Calamus yunnanensis var. *densiflorus* S.J.Pei & S.Y.Chen = **Calamus acanthospathus** Griff.

Calamus yunnanensis var. *intermedius* S.J.Pei & S.Y.Chen = **Calamus acanthospathus** Griff.

Calamus zalacca Roxb. = **Salacca wallichiana** Mart.

Calamus zalacca Gaertn. = **Salacca zalacca** (Gaertn.) Voss

Calamus zebrinus Becc., Malesia 3: 59 (1886). *Palmijuncus zebrinus* (Becc.) Kuntze, Revis. Gen. Pl. 2: 733 (1891).
New Guinea. 43 NWG.
  Calamus laceratus Burret, Notizbl. Bot. Gart. Berlin-Dahlem 13: 318 (1936).

Calamus steenisii Furtado, Gard. Bull. Straits Settlem. 9: 182 (1937).

Calamus zeylanicus Becc. in J.D.Hooker, Fl. Brit. India 5: 466 (1893).
SW. Sri Lanka. 40 SRL.

Calamus zollingeri Becc., Ann. Roy. Bot. Gard. (Calcutta) 11(1): 104 (1908).
Sulawesi. 42 SUL.

Calamus zonatus Becc., For. Borneo: 609 (1902).
Borneo. 42 BOR.

Calappa

Calappa Steck = **Cocos** L.

Calappa acaulis (Drude) Kuntze = **Syagrus comosa** (Mart.) Mart.

Calappa acrocomioides (Drude) Kuntze = **Syagrus romanzoffiana** (Cham.) Glassman

Calappa amara (Jacq.) Kuntze = **Syagrus amara** (Jacq.) Mart.

Calappa australis (Mart.) Kuntze = **Syagrus romanzoffiana** (Cham.) Glassman

Calappa botryophora (Mart.) Kuntze = **Syagrus botryophora** (Mart.) Mart.

Calappa campestris (Mart.) Kuntze = **Syagrus flexuosa** (Mart.) Becc.

Calappa capitata (Mart.) Kuntze = **Butia capitata** (Mart.) Becc.

Calappa cocoides Kuntze = **Syagrus cocoides** Mart.

Calappa comosa (Mart.) Kuntze = **Syagrus comosa** (Mart.) Mart.

Calappa coronata (Mart.) Kuntze = **Syagrus coronata** (Mart.) Becc.

Calappa datil (Drude & Griseb.) Kuntze = **Syagrus romanzoffiana** (Cham.) Glassman

Calappa elegantina Kuntze = **Lytocaryum weddellianum** (H.Wendl.) Toledo

Calappa eriospatha (Mart. ex Drude) Kuntze = **Butia eriospatha** (Mart. ex Drude) Becc.

Calappa flexuosa (Mart.) Kuntze = **Syagrus flexuosa** (Mart.) Becc.

Calappa graminifolia (Drude) Kuntze = **Syagrus graminifolia** (Drude) Becc.

Calappa insignis (Drude) Kuntze = **Lytocaryum weddellianum** (H.Wendl.) Toledo

Calappa leiospatha (Barb.Rodr.) Kuntze = **Cocos leiospatha**

Calappa martiana (Drude & Glaz.) Kuntze = **Syagrus romanzoffiana** (Cham.) Glassman

Calappa mikaniana (Mart.) Kuntze = **Syagrus pseudococos** (Raddi) Glassman

Calappa nucifera (L.) Kuntze = **Cocos nucifera** L.

Calappa oleracea (Mart.) Kuntze = **Syagrus oleracea** (Mart.) Becc.

Calappa orinocensis (Spruce) Kuntze = **Syagrus orinocensis** (Spruce) Burret

Calappa petraea (Mart.) Kuntze = **Syagrus petraea** (Mart.) Becc.

Calappa pityrophylla (Mart.) Kuntze = **Cocos pityrophylla**

Calappa plumosa (Hook.f.) Kuntze = **Syagrus romanzoffiana** (Cham.) Glassman

Calappa procopiana (Glaz. ex Drude) Kuntze = **Syagrus macrocarpa** Barb.Rodr.

Calappa romanzoffiana (Cham.) Kuntze = **Syagrus romanzoffiana** (Cham.) Glassman

Calappa sancona (Kunth) Kuntze = **Syagrus sancona** (Kunth) H.Karst.

Calappa schizophylla (Mart.) Kuntze = **Syagrus schizophylla** (Mart.) Glassman

Calappa speciosa (Barb.Rodr.) Kuntze = **Syagrus inajai** (Spruce) Becc.

Calappa weddellii (Drude) Kuntze = **Syagrus cocoides** Mart.

Calappa yatay (Mart.) Kuntze = **Butia yatay** (Mart.) Becc.

Calospatha

Calospatha Becc., Ann. Roy. Bot. Gard. (Calcutta) 12(1): 232 (1911).
1 species, Pen. Malaysia. 42.

Calospatha confusa Furtado = **Calospatha scortechinii** Becc.

Calospatha scortechinii Becc., Ann. Roy. Bot. Gard. (Calcutta) 12(1): 232 (1911).
Pen. Malaysia. 42 MLY.
Daemonorops calospatha Ridl., Mat. Fl. Malay. Penins. 2: 179 (1907).
Calospatha confusa Furtado, Gard. Bull. Singapore 13: 361 (1951).

Calyptrocalyx

Calyptrocalyx Blume, Rumphia 2: 103 (1843).
26 species, Maluku to New Guinea. 42 43.
Linospadix Becc. ex Hook.f. in G.Bentham & J.D.Hooker, Gen. Pl. 3: 503 (1833), nom. illeg.
Paralinospadix Burret, Notizbl. Bot. Gart. Berlin-Dahlem 12: 331 (1935).

Calyptrocalyx albertisianus Becc., Webbia 1: 305 (1905).
New Guinea to Bismarck Arch. 43 BIS NWG.
Ptychosperma normanbyi Becc. in D'Albertis, Nova Guinea 5(2): 399 (1905).
Calyptrocalyx albertisianus var. *minor* Burret, Notizbl. Bot. Gart. Berlin-Dahlem 13: 71 (1936).
Calyptrocalyx minor Burret, Notizbl. Bot. Gart. Berlin-Dahlem 13: 73 (1936), nom. inval.
Calyptrocalyx clemensiae Burret, Notizbl. Bot. Gart. Berlin-Dahlem 15: 9 (1940).

Calyptrocalyx albertisianus var. *minor* Burret = **Calyptrocalyx albertisianus** Becc.

Calyptrocalyx amoenus Dowe & M.D.Ferrero, Blumea 46: 215 (2001).
E. New Guinea (W. Sepik Prov.). 43 NWG.

Calyptrocalyx angustifrons Becc. = **Calyptrocalyx pauciflorus** Becc.

Calyptrocalyx archboldianus Burret = **Calyptrocalyx lauterbachianus** Warb. ex Becc.

Calyptrocalyx arfakianus (Becc.) Dowe & M.D.Ferrero, Blumea 46: 217 (2001).
W. New Guinea. 43 NWG.
Linospadix arfakiana Becc., Malesia 1: 62 (1877).
Bacularia arfakiana (Becc.) F.Muell., Fragm. 11: 58 (1878). *Paralinospadix arfakianus* (Becc.) Burret, Notizbl. Bot. Gart. Berlin-Dahlem 12: 333 (1935).

Linospadix pachystachys Burret, Notizbl. Bot. Gart. Berlin-Dahlem 11: 711 (1933). *Paralinospadix pachystachys* (Burret) Burret, Notizbl. Bot. Gart. Berlin-Dahlem 12: 335 (1935).

Calyptrocalyx australasicus (H.Wendl. & Drude) Scheff. ex B.D.Jacks. = **Laccospadix australasicus** H.Wendl. & Drude

Calyptrocalyx awa Dowe & M.D.Ferrero, Blumea 46: 218 (2001).
E. New Guinea (W. Sepik Prov.). 43 NWG.

Calyptrocalyx bifurcatus Becc. = **Calyptrocalyx elegans** Becc.

Calyptrocalyx clemensiae Burret = **Calyptrocalyx albertisianus** Becc.

Calyptrocalyx caudiculatus (Becc.) Dowe & M.D.Ferrero, Blumea 46: 220 (2001).
W. New Guinea. 43 NWG.
Linospadix caudiculata Becc., Nova Guinea 8: 213 (1909). *Paralinospadix caudiculatus* (Becc.) Burret, Notizbl. Bot. Gart. Berlin-Dahlem 12: 335 (1935).

Calyptrocalyx doxanthus Dowe & M.D.Ferrero, Wodyetia 4(3): 9 (1999).
W. New Guinea (Jayapura Div.). 43 NWG.

Calyptrocalyx elegans Becc. in K.M.Schumann & U.M.Hollrung, Fl. Kais. Wilh. Land: 16 (1889).
New Guinea. 43 NWG.
Calyptrocalyx moszkowskianus Becc., Bot. Jahrb. Syst. 52: 33 (1914).
Calyptrocalyx schultzianus Becc., Bot. Jahrb. Syst. 52: 32 (1914).
Calyptrocalyx bifurcatus Becc., Bot. Jahrb. Syst. 58: 450 (1923).

Calyptrocalyx flabellatus (Becc.) Dowe & M.D.Ferrero, Blumea 46: 224 (2001).
W. New Guinea. 43 NWG.
Linospadix flabellata Becc., Malesia 1: 64 (1877). *Paralinospadix flabellatus* (Becc.) Burret, Notizbl. Bot. Gart. Berlin-Dahlem 12: 334 (1935). *Bacularia flabellata* (Becc.) F.Muell., Fragm. 11: 58 (1978).

Calyptrocalyx forbesii (Ridl.) Dowe & M.D.Ferrero, Wodyetia 4(3): 10 (1999).
EC. & E. New Guinea. 43 NWG.
Linospadix forbesii Ridl., J. Bot. 24: 358 (1886). *Paralinospadix forbesii* (Ridl.) Burret, Notizbl. Bot. Gart. Berlin-Dahlem 12: 334 (1935).
Linospadix petrickiana Sander, Gard. Chron. 1898(2): 298 (1898). *Paralinospadix petrickianus* (Sander) Burret, Notizbl. Bot. Gart. Berlin-Dahlem 12: 334 (1935).
Paralinospadix stenoschistus Burret, Notizbl. Bot. Gart. Berlin-Dahlem 13: 323 (1936).

Calyptrocalyx geonomiformis (Becc.) Dowe & M.D.Ferrero, Blumea 46: 226 (2001).
W. New Guinea (Paniai Div.). 43 NWG.
Linospadix geonomiformis Becc., Nova Guinea 8: 211 (1909). *Paralinospadix geonomiformis* (Becc.) Burret, Notizbl. Bot. Gart. Berlin-Dahlem 12: 335 (1935).

Calyptrocalyx hollrungii (Becc.) Dowe & M.D.Ferrero, Blumea 46: 226 (2001).
E. New Guinea. 43 NWG.
Linospadix hollrungii Becc. in K.M.Schumann & U.M.Hollrung, Fl. Kais. Wilh. Land: 16 (1889). *Paralinospadix hollrungii* (Becc.) Burret, Notizbl. Bot. Gart. Berlin-Dahlem 12: 334 (1935).

Linospadix hellwigiana Warb. ex Becc., Webbia 1: 293 (1905).

Linospadix schlechteri Becc., Webbia 1: 296 (1905).
Paralinospadix schlechteri (Becc.) Burret, Notizbl. Bot. Gart. Berlin-Dahlem 12: 335 (1935).
Paralinospadix clemensiae Burret, Notizbl. Bot. Gart. Berlin-Dahlem 13: 322 (1936).

Calyptrocalyx julianettii (Becc.) Dowe & M.D.Ferrero, Blumea 46: 228 (2001).
New Guinea. 43 NWG.
  **Linospadix julianettii* Becc., Webbia 1: 295 (1905).
  Paralinospadix julianettii (Becc.) Burret, Notizbl. Bot. Gart. Berlin-Dahlem 12: 334 (1935).
  Paralinospadix amischus Burret, Notizbl. Bot. Gart. Berlin-Dahlem 12: 335 (1935).

Calyptrocalyx lauterbachianus Warb. ex Becc., Webbia 4: 158 (1913). *Linospadix lauterbachiana* (Warb. ex Becc.) Becc., Webbia 4: 158 (1913). *Laccospadix lauterbachianus* (Warb. ex Becc.) Burret, Repert. Spec. Nov. Regni Veg. 24: 290 (1928).
E. New Guinea. 43 NWG.
  Calyptrocalyx stenophyllus Becc., Bot. Jahrb. Syst. 52: 32 (1914).
  Calyptrocalyx archboldianus Burret, Notizbl. Bot. Gart. Berlin-Dahlem 12: 323 (1935).

Calyptrocalyx laxiflorus Becc., Webbia 1: 311 (1905).
Papua New Guinea (Torricelli Mts.). 43 NWG.

Calyptrocalyx lepidotus (Burret) Dowe & M.D.Ferrero, Blumea 46: 231 (2001).
C. New Guinea. 43 NWG.
  **Paralinospadix lepidotus* Burret, J. Arnold Arbor. 20: 199 (1939).

Calyptrocalyx leptostachys Becc., Webbia 1: 306 (1905).
Papua New Guinea (Mt. Yule). 43 NWG.

Calyptrocalyx merrillianus (Burret) Dowe & M.D.Ferrero, Blumea 46: 232 (2001).
E. New Guinea. 43 NWG.
  **Paralinospadix merrillianus* Burret, J. Arnold Arbor. 20: 201 (1939).

Calyptrocalyx micholitzii (Ridl.) Dowe & M.D.Ferrero, Blumea 46: 233 (2001).
W. New Guinea. 43 NWG.
  **Linospadix micholitzii* Ridl., Gard. Chron., III, 18: 262 (1895). *Paralinospadix micholitzii* (Ridl.) Burret, Notizbl. Bot. Gart. Berlin-Dahlem 12: 334 (1935).
  Linospadix pauciflora Ridl., Trans. Linn. Soc. London, Bot. 9: 233 (1916). *Paralinospadix pauciflorus* (Ridl.) Burret, Notizbl. Bot. Gart. Berlin-Dahlem 12: 335 (1935).

Calyptrocalyx minor Burret = **Calyptrocalyx albertisianus** Becc.

Calyptrocalyx moszkowskianus Becc. = **Calyptrocalyx elegans** Becc.

Calyptrocalyx multifidus (Becc.) Dowe & M.D.Ferrero, Blumea 46: 234 (2001).
W. New Guinea. 43 NWG.
  **Linospadix multifida* Becc., Malesia 1: 64 (1877). *Paralinospadix multifidus* (Becc.) Burret, Notizbl. Bot. Gart. Berlin-Dahlem 12: 334 (1935).

Calyptrocalyx pachystachys Becc., Webbia 1: 308 (1905).
New Guinea. 43 NWG.
  Calyptrocalyx schlechterianus Becc., Bot. Jahrb. Syst. 52: 33 (1914).

Calyptrocalyx pauciflorus Becc., Bot. Jahrb. Syst. 58: 449 (1923).
NE. New Guinea. 43 NWG.
  Calyptrocalyx angustifrons Becc., Bot. Jahrb. Syst. 58: 449 (1923).

Calyptrocalyx polyphyllus Becc., Bot. Jahrb. Syst. 58: 449 (1923).
NE. New Guinea. 43 NWG.

Calyptrocalyx pusillus (Becc.) Dowe & M.D.Ferrero, Blumea 46: 238 (2001).
E. New Guinea (Milne Bay Prov.). 43 NWG.
  Linospadix parvula Becc., Webbia 1: 293 (1905), orth. var.
  **Linospadix pusilla* Becc., Webbia 1: 295 (1905). *Paralinospadix pusillus* (Becc.) Burret, Notizbl. Bot. Gart. Berlin-Dahlem 12: 334 (1935).

Calyptrocalyx schlechterianus Becc. = **Calyptrocalyx pachystachys** Becc.

Calyptrocalyx schultzianus Becc. = **Calyptrocalyx elegans** Becc.

Calyptrocalyx sessiliflorus Dowe & M.D.Ferrero, Wodyetia 4: 11 (1999).
E. New Guinea. 43 NWG.
  **Linospadix leptostachys* Burret, Notizbl. Bot. Gart. Berlin-Dahlem 11: 711 (1933). *Paralinospadix leptostachys* (Burret) Burret, Notizbl. Bot. Gart. Berlin-Dahlem 12: 335 (1935).

Calyptrocalyx spicatus (Lam.) Blume, Rumphia 2: 103 (1843).
Maluku. 42 MOL.
  *Areca spicata Lam., Encycl. 1: 241 (1783).
  Pinanga globosa G.Nicholson, Ill. Dict. Gard. 3: 130 (1886).

Calyptrocalyx stenophyllus Becc. = **Calyptrocalyx lauterbachianus** Warb. ex Becc.

Calyptrocalyx yamutumene Dowe & M.D.Ferrero, Blumea 46: 241 (2001).
E. New Guinea (W. Sepik Prov.). 43 NWG.

Calyptrogyne

Calyptrogyne H.Wendl., Bot. Zeitung (Berlin) 17: 72 (1859).
9 species, SE. Mexico to C. America. 79 80.

Calyptrogyne allenii (L.H.Bailey) Nevers, Proc. Calif. Acad. Sci., IV, 48: 336 (1995).
W. Panama. 80 PAN.
  *Geonoma allenii L.H.Bailey, Gentes Herb. 6: 204 (1943).

Calyptrogyne anomala Nevers & A.J.Hend., Syst. Bot. 13: 428 (1988).
Panama. 80 PAN.

Calyptrogyne brachystachys H.Wendl. ex Burret = **Calyptrogyne ghiesbreghtiana** (Linden & H.Wendl.) H.Wendl.

Calyptrogyne clementis Léon = **Calyptronoma plumeriana** (Mart.) Lourteig

Calyptrogyne condensata (L.H.Bailey) Wess.Boer, Verh. Kon. Ned. Akad. Wetensch., Afd. Natuurk., Tweede Sect. 58(1): 69 (1968).
Costa Rica to W. Panama. 80 COS PAN.
  **Geonoma condensata* L.H.Bailey, Gentes Herb. 6: 209 (1943).

Calyptrogyne costatifrons (L.H.Bailey) Nevers, Proc. Calif. Acad. Sci., IV, 48: 336 (1995).

E. Panama. 80 PAN.
Geonoma costatifrons L.H.Bailey, Gentes Herb. 6: 206 (1943).

Calyptrogyne dactyloides (H.E.Moore) de Boer = **Pholidostachys dactyloides** H.E.Moore

Calyptrogyne donnell-smithii (Dammer) Burret = **Calyptrogyne ghiesbreghtiana** (Linden & H.Wendl.) H.Wendl.

Calyptrogyne dulcis (C.Wright ex Griseb.) M.Gómez = **Calyptronoma plumeriana** (Mart.) Lourteig

Calyptrogyne elata H.Wendl. = **?**

Calyptrogyne ghiesbreghtiana (Linden & H.Wendl.) H.Wendl., Bot. Zeitung (Berlin) 17: 72 (1859).
SE. Mexico to C. America. 79 MXT 80 BLZ COS GUA HON NIC PAN.
Geonoma ghiesbreghtiana Lindl. & H.Wendl., Linnaea 28: 343 (1856).
Geonoma glauca Oerst., Vidensk. Meddel. Dansk Naturhist. Foren. Kjøbenhavn 1858: 35 (1858).
Calyptrogyne glauca (Oerst.) H.Wendl. in O.C.E.de Kerchove de Denterghem, Palmiers: 238 (1878).
Geonoma spicigera K.Koch, Wochenschr. Gärtnerei Pflanzenk. 1: 244 (1858). *Calyptrogyne spicigera* (K.Koch) H.Wendl., Bot. Zeitung (Berlin) 17: 72 (1859).
Calyptrogyne ghiesbreghtii H.Wendl. in O.C.E.de Kerchove de Denterghem, Palmiers: 238 (1878).
Calyptrogyne sarapiquensis H.Wendl. in O.C.E.de Kerchove de Denterghem, Palmiers: 238 (1878).
Geonoma donnell-smithii Dammer, Bot. Jahrb. Syst. 36(80): 32 (1905). *Calyptrogyne donnell-smithii* (Dammer) Burret, Bot. Jahrb. Syst. 63: 133 (1930).
Calyptrogyne brachystachys H.Wendl. ex Burret, Bot. Jahrb. Syst. 63: 132 (1930).

Calyptrogyne ghiesbreghtii H.Wendl. = **Calyptrogyne ghiesbreghtiana** (Linden & H.Wendl.) H.Wendl.

Calyptrogyne glauca (Oerst.) H.Wendl. = **Calyptrogyne ghiesbreghtiana** (Linden & H.Wendl.) H.Wendl.

Calyptrogyne herrerae Grayum, Phytologia 84: 309 (1998 publ. 1999).
Costa Rica. 80 COS.

Calyptrogyne intermedia M.Gómez ex Léon = **Calyptronoma plumeriana** (Mart.) Lourteig

Calyptrogyne kalbreyeri Burret = **Pholidostachys synanthera** (Mart.) H.E.Moore

Calyptrogyne kunorum Nevers, Proc. Calif. Acad. Sci., IV, 48: 338 (1995).
Panama (San Blas). 80 PAN.

Calyptrogyne microcarpa Léon = **Calyptronoma plumeriana** (Mart.) Lourteig

Calyptrogyne occidentalis (Sw.) M.Gómez = **Calyptronoma occidentalis** (Sw.) H.E.Moore

Calyptrogyne pubescens Nevers, Proc. Calif. Acad. Sci., IV, 48: 336 (1995).
W. Panama. 80 PAN.

Calyptrogyne pulchra Burret = **Pholidostachys pulchra** H.Wendl. ex Burret

Calyptrogyne quisqueyana (L.H.Bailey) Léon = **Calyptronoma rivalis** (O.F.Cook) L.H.Bailey

Calyptrogyne rivalis (O.F.Cook) Léon = **Calyptronoma rivalis** (O.F.Cook) L.H.Bailey

Calyptrogyne robusta (Trail) Burret = **Pholidostachys synanthera** (Mart.) H.E.Moore

Calyptrogyne sarapiquensis H.Wendl. = **Calyptrogyne ghiesbreghtiana** (Linden & H.Wendl.) H.Wendl.

Calyptrogyne spicigera (K.Koch) H.Wendl. = **Calyptrogyne ghiesbreghtiana** (Linden & H.Wendl.) H.Wendl.

Calyptrogyne swartzii Hook.f. = **Calyptronoma occidentalis** (Sw.) H.E.Moore

Calyptrogyne synanthera (Mart.) Burret = **Pholidostachys synanthera** (Mart.) H.E.Moore

Calyptrogyne trichostachys Burret, Bot. Jahrb. Syst. 63: 129, 135 (1930).
Costa Rica to Panama. 80 COS PAN.

Calyptrogyne victorinii Léon = **Calyptronoma occidentalis** (Sw.) H.E.Moore

Calyptrogyne weberbaueri Burret = **Pholidostachys synanthera** (Mart.) H.E.Moore

Calyptronoma

Calyptronoma Griseb., Fl. Brit. W. I.: 518 (1864).
3 species, Caribbean. 81.
Cocops O.F.Cook, Bull. Torrey Bot. Club 28: 568 (1901).

Calyptronoma clementis (Léon) A.D.Hawkes = **Calyptronoma plumeriana** (Mart.) Lourteig

Calyptronoma clementis subsp. *orientensis* O.Muñiz & Borhidi = **Calyptronoma plumeriana** (Mart.) Lourteig

Calyptronoma dulcis (C.Wright ex Griseb.) H.Wendl. = **Calyptronoma plumeriana** (Mart.) Lourteig

Calyptronoma intermedia H.Wendl. = **Calyptronoma plumeriana** (Mart.) Lourteig

Calyptronoma kalbreyeri (Burret) L.H.Bailey = **Pholidostachys synanthera** (Mart.) H.E.Moore

Calyptronoma microcarpa (Léon) A.D.Hawkes = **Calyptronoma plumeriana** (Mart.) Lourteig

Calyptronoma occidentalis (Sw.) H.E.Moore, Gentes Herb. 9: 252 (1963).
Jamaica. 81 JAM.
Elaeis occidentalis Sw., Fl. Ind. Occid. 1: 619 (1797).
Calyptronoma swartzii Griseb., Fl. Brit. W. I.: 518 (1864), nom. illeg. *Geonoma swartzii* Griseb., Cat. Pl. Cub.: 222 (1866), nom. illeg. *Calyptrogyne swartzii* Hook.f., Rep. Progr. Condition Roy. Bot. Gard. Kew 1882: 61 (1884). *Calyptrogyne occidentalis* (Sw.) M.Gómez, Nov. Bot. Sist.: 50 (1893).
Calyptrogyne victorinii Léon, Contr. Ocas. Mus. Hist. Nat. Colegio "De Le Salle" 3: 4 (1944).

Calyptronoma plumeriana (Mart.) Lourteig, Phytologia 65: 484 (1989).
W. & E. Cuba to Hispaniola. 81 CUB DOM HAI.
Geonoma plumeriana Mart. in A.D.d'Orbigny, Voy. Amér. Mér. 7(3): 34 (1843).
Geonoma dulcis C.Wright ex Griseb., Cat. Pl. Cub.: 222 (1866). *Calyptronoma dulcis* (C.Wright ex Griseb.) H.Wendl. in O.C.E.de Kerchove de Denterghem, Palmiers: 238 (1878). *Calyptrogyne dulcis* (C.Wright ex Griseb.) M.Gómez, Dicc. Bot. Nom. Vulg. Cub. Puerto-Riquenos: 72 (1889).
Calyptronoma intermedia H.Wendl. in O.C.E.de Kerchove de Denterghem, Palmiers: 238 (1878). *Geonoma intermedia* (H.Wendl.) B.S.Williams, Cat. 1882: 27 (1882).
Calyptrogyne clementis Léon, Contr. Ocas. Mus. Hist. Nat. Colegio "De Le Salle" 3: 11 (1944).

Calyptronoma clementis (Léon) A.D.Hawkes, Phytologia 3: 145 (1949).

Calyptrogyne intermedia M.Gómez ex Léon, Contr. Ocas. Mus. Hist. Nat. Colegio "De Le Salle" 3: 8 (1944).

Calyptrogyne microcarpa Léon, Contr. Ocas. Mus. Hist. Nat. Colegio "De Le Salle" 3: 10 (1944). *Calyptronoma microcarpa* (Léon) A.D.Hawkes, Phytologia 3: 145 (1949).

Calyptronoma clementis subsp. *orientensis* O.Muñiz & Borhidi, Acta Bot. Acad. Sci. Hung. 28: 342 (1982).

Calyptronoma quisqueyana L.H.Bailey = **Calyptronoma rivalis** (O.F.Cook) L.H.Bailey

Calyptronoma rivalis (O.F.Cook) L.H.Bailey, Gentes Herb. 4: 171 (1938).
Hispaniola to Puerto Rico. 81 DOM HAI PUE.
  *Cocops rivalis O.F.Cook, Bull. Torrey Bot. Club 28: 568 (1901). *Calyptrogyne rivalis* (O.F.Cook) Léon, Contr. Ocas. Mus. Hist. Nat. Colegio "De Le Salle" 3: 12 (1944).
  Calyptronoma quisqueyana L.H.Bailey, Gentes Herb. 4: 169 (1938). *Calyptrogyne quisqueyana* (L.H.Bailey) Léon, Contr. Ocas. Mus. Hist. Nat. Colegio "De Le Salle" 3: 12 (1944).

Calyptronoma robusta Trail = **Pholidostachys synanthera** (Mart.) H.E.Moore

Calyptronoma swartzii Griseb. = **Calyptronoma occidentalis** (Sw.) H.E.Moore

Calyptronoma synanthera (Mart.) L.H.Bailey = **Pholidostachys synanthera** (Mart.) H.E.Moore

Calyptronoma weberbaueri (Burret) L.H.Bailey = **Pholidostachys synanthera** (Mart.) H.E.Moore

Campecarpus

Campecarpus H.Wendl. ex Becc., Palme Nuova Caledonia: 28 (1920).
1 species, New Caledonia. 60.

Campecarpus fulcitus (Brongn.) H.Wendl. ex Becc., Palme Nuova Caledonia: 29 (1920).
S. New Caledonia. 60 NWC.
  *Kentia fulcita Brongn., Compt. Rend. Hebd. Séances Acad. Sci. 77: 399 (1873). *Cyphophoenix fulcita* (Brongn.) Hook.f. ex Salomon, Palmen: 86 (1887).

Carpentaria

Carpentaria Becc., Ann. Jard. Bot. Buitenzorg 2: 128 (1885).
1 species, N. Australia. 50.

Carpentaria acuminata (H.Wendl. & Drude) Becc., Ann. Jard. Bot. Buitenzorg 2: 128 (1885).
Northern Territory. 50 NTA.
  *Kentia acuminata H.Wendl. & Drude, Linnaea 39: 207 (1875).

Carpentaria bleeseri (Burret) Burret = **Ptychosperma bleeseri** Burret

Carpoxylon

Carpoxylon H.Wendl. & Drude, Linnaea 39: 177 (1875).
1 species, Vanuatu. 60.

Carpoxylon macrospermum H.Wendl. & Drude, Linnaea 39: 177 (1875).
Vanuatu (Anatom). 60 VAN.

Caryota

Caryota L., Sp. Pl.: 1189 (1753).
13 species, Trop. & Subtrop. Asia to Vanuatu. 36 (38) 40 41 42 43 50 60.
  Schunda-Pana Adans., Fam. Pl. 2: 24 (1763).
  Thuessinkia Korth. ex Miq., Fl. Ned. Ind. 3: 41 (1855), nom. illeg.

Caryota aequatorialis (Becc.) Ridl. = **Caryota maxima** Blume

Caryota albertii F.Muell. ex H.Wendl. = **Caryota rumphiana** Mart.

Caryota arenga Mezieres ex Desjardins = **?**

Caryota bacsonensis Magalon, Contr. Étud. Palmiers Indoch.: 128 (1930).
Indo-China. 41 LAO THA VIE.

Caryota blancoi Hook.f. = **Caryota cumingii** Lodd. ex Mart.

Caryota cumingii Lodd. ex Mart., Hist. Nat. Palm. 3: 315 (1853).
Philippines. 42 PHI.
  Caryota blancoi Hook.f., Rep. Progr. Condition Roy. Bot. Gard. Kew 1882: 61 (1884).
  Caryota merrillii Becc., Webbia 1: 333 (1905).

Caryota furfuracea Blume = **Caryota mitis** Lour.

Caryota furfuracea var. *caudata* Blume = **Caryota maxima** Blume

Caryota furfuracea var. *furcata* Blume = **Caryota maxima** Blume

Caryota gigas Hahn ex Hodel = **Caryota obtusa** Griff.

Caryota griffithii Becc. = **Caryota mitis** Lour.

Caryota griffithii var. *selebica* Becc. = **Caryota mitis** Lour.

Caryota horrida Jacq. = **Aiphanes horrida** (Jacq.) Burret

Caryota humilis Reinw. ex Kunth = **Arenga porphyrocarpa** (Blume) H.E.Moore

Caryota javanica Zipp. ex Miq. = **Caryota mitis** Lour.

Caryota javanica Osbeck = **Korthalsia sp.**

Caryota kiriwongensis Hodel = **? [41 THA]**

Caryota macrantha Burret = **Caryota maxima** Blume

Caryota majestica Linden = **? [42 PHI]**

Caryota maxima Blume in C.F.P.von Martius, Hist. Nat. Palm. 3: 195 (1838). *Caryota rumphiana* var. *javanica* Becc., Malesia 1: 74 (1877).
S. China to Thailand and W. Malesia. 36 CHC CHH CHS 41 LAO MYA THA VIE 42 JAW MLY SUM.
  Caryota furfuracea var. *caudata* Blume in C.F.P.von Martius, Hist. Nat. Palm. 3(ed. 2): 195 (1845).
  Caryota furfuracea var. *furcata* Blume in C.F.P.von Martius, Hist. Nat. Palm. 3(ed. 2): 195 (1845).
  Caryota obtusa var. *aequatorialis* Becc. in J.D.Hooker, Fl. Brit. India 6: 423 (1892). *Caryota aequatorialis* (Becc.) Ridl., Fl. Malay Penins. 5: 20 (1925).
  Caryota rumphiana var. *oxyodonta* Becc., Philipp. J. Sci. 14: 337 (1919).
  Caryota rumphiana var. *philippinensis* Becc., Philipp. J. Sci. 14: 337 (1919).
  Caryota macrantha Burret, Notizbl. Bot. Gart. Berlin-Dahlem 15: 197 (1940).

Caryota merrillii Becc. = **Caryota cumingii** Lodd. ex Mart.

Caryota mitis Lour., Fl. Cochinch.: 697 (1790).
SE. China to Indo-China and Malesia. 36 CHH CHS
41 AND CBD LAO MYA NCB THA VIE 42 BOR
JAW MLY PHI SUL SUM.
Caryota furfuracea Blume in C.F.P.von Martius, Hist.
Nat. Palm. 3: 195 (1838).
Caryota propinqua Blume in C.F.P.von Martius, Hist.
Nat. Palm. 3: 195 (1838).
Caryota sobolifera Wall. in C.F.P.von Martius, Hist.
Nat. Palm. 3: 194 (1838).
Drymophloeus zippellii Hassk., Tijdschr. Natuurl.
Gesch. Physiol. 9: 170 (1842).
Thuessinkia speciosa Korth., Fl. Ned. Ind. 3: 41 (1855).
Caryota javanica Zipp. ex Miq., Fl. Ned. Ind. 2: 41
(1856), nom. illeg.
Caryota griffithii Becc., Nuovo Giorn. Bot. Ital. 3: 15
(1871).
Caryota griffithii var. *selebica* Becc., Malesia 1: 75
(1877).
Caryota nana Linden, Ill. Hort. 28: 16 (1881).
Caryota speciosa Linden, Ill. Hort. 28: 16 (1881).

Caryota monostachya Becc., Webbia 3: 196 (1910).
S. China to Vietnam. 36 CHC CHS 41 VIE.
Caryota nana Linden = **Caryota mitis** Lour.

Caryota no Becc., Nuovo Giorn. Bot. Ital. 3: 12 (1871).
Caryota rumphiana var. *borneensis* Becc., Malesia 1:
74 (1877).
Borneo. 42 BOR.

Caryota obtusa Griff., Calcutta J. Nat. Hist. 5: 480
(1845). *Caryota rumphiana* var. *indica* Becc., Malesia
1: 75 (1877).
Assam to N. Thailand. 40 ASS 41 THA.
Caryota obtusidentata Griff., Palms Brit. E. Ind.: t.
236A, B (1850).
Caryota gigas Hahn ex Hodel, Palm J. 139: 51 (1998),
without diagnostic latin descr.

Caryota obtusa var. *aequatorialis* Becc. = **Caryota**
maxima Blume

Caryota obtusidentata Griff. = **Caryota obtusa** Griff.

Caryota ochlandra Hance, J. Bot. 17: 174 (1879).
SE. China. 36 CHS.

Caryota onusta Blanco = **Arenga pinnata** (Wurmb)
Merr.

Caryota ophiopellis Dowe, Austral. Syst. Bot. 9: 20
(1996).
Vanuatu. 60 VAN.

Caryota palindan Blanco = **Orania palindan** (Blanco)
Merr.

Caryota princeps Voigt = **?**

Caryota propinqua Blume = **Caryota mitis** Lour.

Caryota rumphiana Mart., Hist. Nat. Palm. 3: 195
(1838).
Philippines to Solomon Is. 42 MOL PHI SUL 43
NWG SOL 50 QLD.
Caryota albertii F.Muell. ex H.Wendl., Linnaea 39:
221 (1875). *Caryota rumphiana* var. *australiensis*
Becc., Malesia 1: 74 (1877). *Caryota rumphiana*
var. *albertii* (F.Muell. ex H.Wendl.) F.M.Bailey,
Queensland Agric. J. 1(3): 233 (1897).
Caryota rumphiana var. *moluccana* Becc., Malesia 1:
70 (1877).
Caryota rumphiana var. *papuana* Becc., Malesia 1: 70
(1877).

Caryota rumphiana var. *albertii* (F.Muell. ex H.Wendl.)
F.M.Bailey = **Caryota rumphiana** Mart.

Caryota rumphiana var. *australiensis* Becc. = **Caryota**
rumphiana Mart.

Caryota rumphiana var. *borneensis* Becc. = **Caryota no**
Becc.

Caryota rumphiana var. *indica* Becc. = **Caryota obtusa**
Griff.

Caryota rumphiana var. *javanica* Becc. = **Caryota**
maxima Blume

Caryota rumphiana var. *moluccana* Becc. = **Caryota**
rumphiana Mart.

Caryota rumphiana var. *oxyodonta* Becc. = **Caryota**
maxima Blume

Caryota rumphiana var. *papuana* Becc. = **Caryota**
rumphiana Mart.

Caryota rumphiana var. *philippinensis* Becc. = **Caryota**
maxima Blume

Caryota sobolifera Wall. = **Caryota mitis** Lour.

Caryota speciosa Linden = **Caryota mitis** Lour.

Caryota sympetala Gagnep., Notul. Syst. (Paris) 6: 151
(1937).
Laos, Vietnam. 41 LAO VIE.

Caryota tremula Blanco = **Arenga tremula** (Blanco)
Becc.

Caryota urens L., Sp. Pl.: 1189 (1753).
India to Pen. Malaysia. 36 CHC CHS (38) oga 40 ASS
BAN IND NEP SRL 41 MYA THA 42 MLY.

Caryota zebrina Hambali & al., Palms 44: 171 (2000).
New Guinea. 43 NWG.

Catis

Catis O.F.Cook = **Euterpe** Mart.

Catis martiana O.F.Cook = **Euterpe oleracea** Mart.

Catoblastus

Catoblastus H.Wendl. = **Wettinia** Poepp. ex Endl.

Catoblastus aequalis (O.F.Cook & Doyle) Burret =
Wettinia aequalis (O.F.Cook & Doyle) R.Bernal

Catoblastus andinus Dugand = **Wettinia praemorsa**
(Willd.) Wess.Boer

Catoblastus anomalus (Burret) Burret = **Wettinia**
anomala (Burret) R.Bernal

Catoblastus cuatrecasasii Dugand = **Wettinia praemorsa**
(Willd.) Wess.Boer

Catoblastus distichus R.Bernal = **Wettinia disticha**
(R.Bernal) R.Bernal

Catoblastus drudei O.F.Cook & Doyle = **Wettinia**
drudei (O.F.Cook & Doyle) A.J.Hend.

Catoblastus dryanderae Burret = **Wettinia radiata**
(O.F.Cook & Doyle) R.Bernal

Catoblastus engelii H.Wendl. ex Burret = **Wettinia**
praemorsa (Willd.) Wess.Boer

Catoblastus inconstans (Dugand) Glassman = **Wettinia**
kalbreyeri (Burret) R.Bernal

Catoblastus kalbreyeri Burret = **Wettinia kalbreyeri**
(Burret) R.Bernal

Catoblastus maynensis (Spruce) Drude = **Wettinia**
maynensis Spruce

Catoblastus megalocarpus Burret = **Wettinia kalbreyeri**
(Burret) R.Bernal

Catoblastus mesocarpus Burret = **Wettinia praemorsa** (Willd.) Wess.Boer

Catoblastus microcarpus Burret = **Wettinia microcarpa** (Burret) R.Bernal

Catoblastus microcaryus Burret = **Wettinia kalbreyeri** (Burret) R.Bernal

Catoblastus praemorsus (Willd.) H.Wendl. = **Wettinia praemorsa** (Willd.) Wess.Boer

Catoblastus pubescens (H.Karst.) H.Wendl. = **Wettinia praemorsa** (Willd.) Wess.Boer

Catoblastus radiatus (O.F.Cook & Doyle) Burret = **Wettinia radiata** (O.F.Cook & Doyle) R.Bernal

Catoblastus sphaerocarpus Burret = **Wettinia kalbreyeri** (Burret) R.Bernal

Catoblastus velutinus Burret = **Wettinia aequalis** (O.F.Cook & Doyle) R.Bernal

Catostigma

Catostigma O.F.Cook & Doyle = **Wettinia** Poepp. ex Endl.

Catostigma aequale (O.F.Cook & Doyle) Burret = **Wettinia aequalis** (O.F.Cook & Doyle) R.Bernal

Catostigma anomalum Burret = **Wettinia anomala** (Burret) R.Bernal

Catostigma drudei (O.F.Cook & Doyle) Burret = **Wettinia drudei** (O.F.Cook & Doyle) A.J.Hend.

Catostigma dryanderae Burret = **Wettinia radiata** (O.F.Cook & Doyle) R.Bernal

Catostigma inconstans Dugand = **Wettinia kalbreyeri** (Burret) R.Bernal

Catostigma kalbreyeri Burret = **Wettinia kalbreyeri** (Burret) R.Bernal

Catostigma megalocarpum Burret = **Wettinia kalbreyeri** (Burret) R.Bernal

Catostigma microcaryum Burret = **Wettinia kalbreyeri** (Burret) R.Bernal

Catostigma radiatum O.F.Cook & Doyle = **Wettinia radiata** (O.F.Cook & Doyle) R.Bernal

Catostigma sphaerocarpum Burret = **Wettinia kalbreyeri** (Burret) R.Bernal

Catostigma sphaerocarpum var. *microcaryum* Burret = **Wettinia kalbreyeri** (Burret) R.Bernal

Ceratolobus

Ceratolobus Blume ex Schult. & Schult.f., Syst. Veg. 7: lxxx (1830).
6 species, Pen. Thailand to W. Malesia. 41 42.

Ceratolobus concolor Blume, Rumphia 2: 165 (1843).
Sumatera, Borneo. 42 BOR SUM.

Ceratolobus discolor Becc., Malesia 3: 63 (1886).
W. Sumatera, Borneo. 42 BOR SUM.
    Ceratolobus hallierianus Becc. ex K.Heyne, Nutt. Pl. Ned.-Ind.: 93 (1913).

Ceratolobus forgetiana auct. = **?**

Ceratolobus hallierianus Becc. ex K.Heyne = **Ceratolobus discolor** Becc.

Ceratolobus glaucescens Blume in J.J.Roemer & J.A.Schultes, Syst. Veg. 7: 1334 (1830).
Pen. Thailand, W. Jawa. 41 THA 42 JAW.
    Calamus glaucescens D.Dietr., Syn. Pl. 2: 1064 (1840).

Ceratolobus javanicus (Osbeck) Merr. = **Caryota javanica**

Ceratolobus kingianus Becc. & Hook.f. in J.D.Hooker, Fl. Brit. India 6(2): 477 (1893).
Pen. Malaysia. 42 MLY.

Ceratolobus laevigatus (Mart.) Becc. & Hook.f. = **Calamus laevigatus** Mart.

Ceratolobus laevigatus var. *angustifolius* Becc. = **Ceratolobus subangulatus** (Miq.) Becc.

Ceratolobus laevigatus var. *borneensis* Becc. = **Ceratolobus subangulatus** (Miq.) Becc.

Ceratolobus laevigatus var. *divaricatus* Becc. = **Ceratolobus subangulatus** (Miq.) Becc.

Ceratolobus laevigatus var. *major* Becc. = **Ceratolobus subangulatus** (Miq.) Becc.

Ceratolobus laevigatus var. *regularis* Becc. = **Ceratolobus subangulatus** (Miq.) Becc.

Ceratolobus laevigatus var. *subangulatus* (Miq.) Becc. = **Ceratolobus subangulatus** (Miq.) Becc.

Ceratolobus micholtziana auct. = **Plectocomia sp. ?**

Ceratolobus plicatus Zipp. ex Blume = **Korthalsia zippelii** Blume

Ceratolobus pseudoconcolor J.Dransf., Kew Bull. 34: 19 (1979).
S. Sumatera to W. Jawa. 42 JAW SUM.

Ceratolobus rostratus (Blume) Becc. = **Korthalsia rostrata** Blume

Ceratolobus subangulatus (Miq.) Becc., Ann. Roy. Bot. Gard. (Calcutta) 11(App.): iii (1913).
Pen. Thailand to W. Malesia. 41 THA 42 BOR MLY SUM.
    Calamus subangulatus Miq., Fl. Ned. Ind., Eerste Bijv.: 594 (1861). *Palmijuncus subangulatus* (Miq.) Kuntze, Revis. Gen. Pl. 2: 734 (1891). *Ceratolobus laevigatus* var. *subangulatus* (Miq.) Becc., Ann. Roy. Bot. Gard. (Calcutta) 12(2): 16 (1918).
    Ceratolobus laevigatus var. *angustifolius* Becc. in J.D.Hooker, Fl. Brit. India 6: 477 (1893).
    Ceratolobus laevigatus var. *borneensis* Becc., Ann. Roy. Bot. Gard. (Calcutta) 12(2): 16 (1918).
    Ceratolobus laevigatus var. *divaricatus* Becc., Ann. Roy. Bot. Gard. (Calcutta) 12(2): 16 (1918).
    Ceratolobus laevigatus var. *major* Becc., Ann. Roy. Bot. Gard. (Calcutta) 12(2): 16 (1918).
    Ceratolobus laevigatus var. *regularis* Becc., Ann. Roy. Bot. Gard. (Calcutta) 12(2): 16 (1918).

Ceratolobus zippelii Blume = **Korthalsia zippelii** Blume

Ceroxylon

Ceroxylon Bonpl. ex DC., Bull. Sci. Soc. Philom. Paris 3: 239 (1804).
11 species, Venezuela to W. South America. 82 83.
    Klopstockia H.Karst., Linnaea 28: 251 (1856).
    Beethovenia Engel, Linnaea 33: 677 (1865).

Ceroxylon alpinum Bonpl. ex DC., Bull. Sci. Soc. Philom. Paris 3: 239 (1804).
Venezuela to Ecuador. 82 VEN 83 CLM ECU.

subsp. **alpinum**
Colombia to Venezuela. 82 VEN 83 CLM.
    Ceroxylon andicolum Humb. & Bonpl., Pl. Aequinoct. 1: 1 (1805). *Iriartea andicola* (Humb. & Bonpl.) Spreng., Syst. Veg. 2: 623 (1825).
    Ceroxylon ferrugineum Regel, Gartenflora 28: 163 (1879).

subsp. **ecuadorense** Galeano, Caldasia 17: 395 (1995). W. Ecuador. 83 ECU.

Ceroxylon amazonicum Galeano, Caldasia 17: 398 (1995).
SE. Ecuador. 83 ECU.

Ceroxylon andicolum Humb. & Bonpl. = **Ceroxylon alpinum** subsp. **alpinum**

Ceroxylon australe Mart. = **Juania australis** (Mart.) Drude ex Hook.f.

Ceroxylon beethovenia Burret = **Ceroxylon ceriferum** (H.Karst.) Pittier

Ceroxylon ceriferum (H.Karst.) Pittier, Bol. Ci. Technol. 1: 10 (1926).
Colombia to Venezuela. 82 VEN 83 CLM.
Ceroxylon klopstockia Mart., Hist. Nat. Palm. 3: 314 (1849), nom. rejic. prop. *Iriartea klopstockia* (Mart.) W.Watson, Gard. Chron., n.s., 23: 338 (1885).
*Klopstockia cerifera H.Karst., Linnaea 28: 251 (1856).
Beethovenia cerifera Engl., Linnaea 33: 677 (1865).
Iriartea nivea W.Watson, Gard. Chron., n.s., 24: 750 (1885).
Ceroxylon beethovenia Burret, Notizbl. Bot. Gart. Berlin-Dahlem 10: 845 (1929).
Ceroxylon schultzei Burret, Notizbl. Bot. Gart. Berlin-Dahlem 10: 846 (1929).

Ceroxylon coarctatum (Engel) H.Wendl. = **Ceroxylon vogelianum** (Engel) H.Wendl.

Ceroxylon crispum Burret = **Ceroxylon vogelianum** (Engel) H.Wendl.

Ceroxylon echinulatum Galeano, Caldasia 17: 399 (1995).
Ecuador. 83 ECU.

Ceroxylon ferrugineum Regel = **Ceroxylon alpinum** subsp. **alpinum**

Ceroxylon flexuosum Galeano & R.Bernal = **Ceroxylon vogelianum** (Engel) H.Wendl.

Ceroxylon floccosum Burret = **Ceroxylon quindiuense** (H.Karst.) H.Wendl.

Ceroxylon hexandrum Dugand = **Ceroxylon vogelianum** (Engel) H.Wendl.

Ceroxylon interruptum (H.Karst.) H.Wendl. = **Kloptockia interrupta**

Ceroxylon klopstockia Mart. = **Ceroxylon ceriferum** (H.Karst.) Pittier

Ceroxylon latisectum Burret = **Ceroxylon parvifrons** (Engel) H.Wendl.

Ceroxylon mooreanum Galeano & R.Bernal = **Ceroxylon parvifrons** (Engel) H.Wendl.

Ceroxylon niveum H.Wendl. = **Polyandrococos caudescens** (Mart.) Barb.Rodr.

Ceroxylon parvifrons (Engel) H.Wendl. in O.C.E.de Kerchove de Denterghem, Palmiers: 239 (1878).
Venezuela to W. South America. 82 VEN 83 BOL CLM ECU PER.
*Klopstockia parvifrons Engel, Linnaea 33: 674 (1865).
Ceroxylon latisectum Burret, Notizbl. Bot. Gart. Berlin-Dahlem 10: 844 (1929).
Ceroxylon sclerophyllum Dugand, Mutisia 14: 4 (1953).
Ceroxylon mooreanum Galeano & R.Bernal, Principes 26: 180 (1982).

Ceroxylon parvum Galeano, Caldasia 17: 403 (1995).
Ecuador to Bolivia. 83 BOL ECU PER.

Ceroxylon pityrophyllum (Mart.) Mart. ex H.Wendl. = **Cocos pityrophylla**

Ceroxylon quindiuense (H.Karst.) H.Wendl., Bonplandia 8: 70 (1860).
Colombia. 83 CLM.
*Klopstockia quindiuensis H.Karst., Fl. Columb. 1: 1 (1859).
Ceroxylon floccosum Burret, Notizbl. Bot. Gart. Berlin-Dahlem 10: 851 (1929).

Ceroxylon sasaimae Galeano, Caldasia 17: 404 (1995).
Colombia (Cundinamerca). 83 CLM.

Ceroxylon schultzei Burret = **Ceroxylon ceriferum** (H.Karst.) Pittier

Ceroxylon sclerophyllum Dugand = **Ceroxylon parvifrons** (Engel) H.Wendl.

Ceroxylon utile (H.Karst.) H.Wendl. = **Klopstockia utilis**

Ceroxylon ventricosum Burret, Notizbl. Bot. Gart. Berlin-Dahlem 10: 847 (1929).
SW. Colombia to S. Ecuador. 83 CLM ECU.

Ceroxylon verruculosum Burret = **Ceroxylon vogelianum** (Engel) H.Wendl.

Ceroxylon vogelianum (Engel) H.Wendl. in O.C.E.de Kerchove de Denterghem, Palmiers: 239 (1878).
Venezuela to W. South America. 82 VEN 83 BOL CLM ECU PER.
Klopstockia coarctata Engel, Linnaea 33: 676 (1865).
Ceroxylon coarctatum (Engel) H.Wendl. in O.C.E.de Kerchove de Denterghem, Palmiers: 238 (1878).
*Klopstockia vogeliana Engel, Linnaea 33: 673 (1865).
Ceroxylon crispum Burret, Notizbl. Bot. Gart. Berlin-Dahlem 10: 849 (1929).
Ceroxylon verruculosum Burret, Notizbl. Bot. Gart. Berlin-Dahlem 10: 850 (1929).
Ceroxylon hexandrum Dugand, Mutisia 14: 1 (1953).
Ceroxylon flexuosum Galeano & R.Bernal, Principes 26: 178 (1982).

Ceroxylon weberbaueri Burret, Notizbl. Bot. Gart. Berlin-Dahlem 10: 848 (1929).
Peru (Cusco, Pasco). 83 PER.

Chamaedorea

Chamaedorea Willd., Sp. Pl. 4: 638 (1806).
111 species, Mexico to C. & S. Trop. America. 79 80 82 83 84.
Morenia Ruiz & Pav., Fl. Peruv. Prodr.: 150 (1794).
Nunnezharia Ruiz & Pav., Fl. Peruv. Prodr.: 147 (1794).
Nunnezia Willd., Sp. Pl. 4: 1154 (1806).
Kunthia Humb. & Bonpl., Pl. Aequinoct. 2: 128 (1813).
Collinia (Liebm.) Liebm. ex Oerst., Vidensk. Meddel. Dansk Naturhist. Foren. Kjøbenhavn 1845: 8 (1846).
Stachyophorbe (Liebm.) Liebm. ex Klotzsch, Overs. Kongel. Danske Vidensk. Selsk. Forh. Medlemmers Arbeider 1846: 8 (1846).
Dasystachys Oerst., Vidensk. Meddel. Dansk Naturhist. Foren. Kjøbenhavn 1858: 25 (1858).
Spathoscaphe Oerst., Vidensk. Meddel. Dansk Naturhist. Foren. Kjøbenhavn 1858: 29 (1858).
Stephanostachys Klotzsch ex Oerst., Vidensk. Meddel. Dansk Naturhist. Foren. Kjøbenhavn 1858: 26 (1858).
Eleutheropetalum (H.Wendl.) H.Wendl. ex Oerst., Vidensk. Meddel. Dansk Naturhist. Foren. Kjøbenhavn 1858: 6 (1859).
Kinetostigma Dammer, Notizbl. Königl. Bot. Gart. Berlin 4: 171 (1905).

Tuerckheimia Dammer in J.D.Smith, Enum. Pl. Guatem. 7: 53 (1905), nom. nud.

Neanthe O.F.Cook, Science, n.s., 86: 120 (1937).

Edanthe O.F.Cook & Doyle, Natl. Hort. Mag. 18: 172 (1939), nom. inval.

Omanthe O.F.Cook, Science, n.s., 90: 298 (1939), no latin descr.

Anothea O.F.Cook, Natl. Hort. Mag. 22: 135 (1943), no latin descr.

Cladandra O.F.Cook, Natl. Hort. Mag. 22: 148, 150 (1943), no latin descr.

Discoma O.F.Cook, Natl. Hort. Mag. 22: 150 (1943), nom. inval.

Docanthe O.F.Cook, Natl. Hort. Mag. 22: 150 (1943), nom. inval.

Legnea O.F.Cook, Natl. Hort. Mag. 22: 150 (1943), no latin descr.

Lobia O.F.Cook, Natl. Hort. Mag. 22: 152 (1943), no latin descr.

Lophothele O.F.Cook, Natl. Hort. Mag. 22: 152 (1943), no latin descr.

Mauranthe O.F.Cook, Natl. Hort. Mag. 22: 152 (1943), no latin descr.

Meiota O.F.Cook, Natl. Hort. Mag. 22: 152 (1943), no latin descr.

Migandra O.F.Cook, Natl. Hort. Mag. 22: 152 (1943).

Paranthe O.F.Cook, Natl. Hort. Mag. 22: 139 (1943), no latin descr.

Platythea O.F.Cook, Natl. Hort. Mag. 26: 228 (1947), no latin descr.

Vadia O.F.Cook, Natl. Hort. Mag. 26: 12 (1947), no latin descr.

Chamaedorea adscendens (Dammer) Burret, Notizbl. Bot. Gart. Berlin-Dahlem 11: 737 (1933).
Belize to Guatemala. 80 BLZ GUA.
*Kinetostigma adscendens Dammer, Notizbl. Königl. Bot. Gart. Berlin 4: 172 (1905).
Tuerckheimia ascendens Dammer in J.D.Smith, Enum. Pl. Guatem. 7: 53 (1905), nom. nud.

Chamaedorea aequalis Standl. & Steyerm. = **Chamaedorea liebmannii** Mart.

Chamaedorea affinis Liebm. = **Chamaedorea elatior** Mart.

Chamaedorea aguilariana Standl. & Steyerm. = **Chamaedorea pinnatifrons** (Jacq.) Oerst.

Chamaedorea allenii L.H.Bailey, Gentes Herb. 6: 241 (1943).
Panama to NW. Colombia. 80 PAN 83 CLM.

Chamaedorea alternans H.Wendl. in E.von Regel, Gartenflora 29: 104 (1880). *Nunnezharia alternans* (H.Wendl.) Kuntze, Revis. Gen. Pl. 2: 730 (1891).
Mexico (Veracruz). 79 MXG.

Chamaedorea amabilis H.Wendl. ex Dammer, Gard. Chron., III, 1904(2): 245 (1904). *Nunnezharia amabilis* (H.Wendl. ex Dammer) Kuntze, Revis. Gen. Pl. 2: 731 (1891).
Costa Rica to Panama. 80 COS PAN.
Chamaedorea coclensis L.H.Bailey, Gentes Herb. 6: 236 (1943).

Chamaedorea amazonica (Kuntze) Dammer = **Chamaedorea pauciflora** Mart.

Chamaedorea andreana Linden = **?** [83 CLM]

Chamaedorea anemophila Hodel, Principes 39: 14 (1995).
Panama. 80 PAN.

Chamaedorea angustisecta Burret, Notizbl. Bot. Gart. Berlin-Dahlem 11: 318 (1932).
Peru to Bolivia. 83 BOL PER.
Chamaedorea leonis H.E.Moore, Gentes Herb. 12: 30 (1980).

Chamaedorea anomospadix Burret = **Chamaedorea tepejilote** Liebm.

Chamaedorea arenbergiana H.Wendl., Index Palm.: 66 (1854). *Spathoscaphe arenbergiana* (H.Wendl.) Oerst., Vidensk. Meddel. Dansk Naturhist. Foren. Kjøbenhavn 1858: 30 (1858). *Nunnezharia arenbergiana* (H.Wendl.) Kuntze, Revis. Gen. Pl. 2: 730 (1891).
Mexico (Veracruz, Oaxaca, Chiapas) to C. America. 79 MXG MXS MXT 80 BLZ COS? ELS? GUA HON NIC? PAN 83 CLM?
Chamaedorea latifrons H.Wendl., Index Palm.: 11 (1854). *Nunnezharia latifrons* (H.Wendl.) Kuntze, Revis. Gen. Pl. 2: 731 (1891).
Chamaedorea latifolia W.Watson, Gard. Chron., n.s., 1885(1): 410 (1885).
Chamaedorea densiflora Guillaumin, J. Soc. Natl. Hort. France, IV, 22: 226 (1923).

Chamaedorea atrovirens Mart., Flora 35: 721 (1852). *Nunnezharia atrovirens* (Mart.) Kuntze, Revis. Gen. Pl. 2: 730 (1891). *Vadia atrovirens* (Mart.) O.F.Cook, Natl. Hort. Mag. 26: 26 (1947).
Mexico (Oaxaca). 79 MXS. — Provisionally accepted.

Chamaedorea atrovirens H.Wendl. = **Chamaedorea cataractarum** Mart.

Chamaedorea aurantiaca Brongn. ex Neumann = **Chamaedorea sartorii** Liebm.

Chamaedorea bambusoides (H.Wendl. ex Dammer) Gérôme = **Chamaedorea elatior** Mart.

Chamaedorea bambusoides var. *graminifolia* Gérôme = **Chamaedorea elatior** Mart.

Chamaedorea bambusoides var. *juncea* Gérôme = **Chamaedorea elatior** Mart.

Chamaedorea bartlingiana H.Wendl. = **Chamaedorea pinnatifrons** (Jacq.) Oerst.

Chamaedorea benziei Hodel, Principes 36: 188 (1992).
Mexico (Chiapas). 79 MXT.

Chamaedorea bifurcata Oerst. = **Chamaedorea pinnatifrons** (Jacq.) Oerst.

Chamaedorea biloba H.Wendl. = **Chamaedorea oblongata** Mart.

Chamaedorea binderi Hodel, Principes 40: 215 (1996).
Costa Rica. 80 COS.

Chamaedorea biolleyi Guillaumin = **Chamaedorea costaricana** Oerst.

Chamaedorea boliviensis Dammer = **Chamaedorea pinnatifrons** (Jacq.) Oerst.

Chamaedorea brachyclada H.Wendl., Gartenflora 29: 101 (1880). *Nunnezharia brachyclada* (H.Wendl.) Kuntze, Revis. Gen. Pl. 2: 730 (1891).
Costa Rica to Panama. 80 COS PAN.

Chamaedorea brachypoda Standl. & Steyerm., Publ. Field Mus. Nat. Hist., Bot. Ser. 23: 198 (1947).
Guatemala to Honduras. 80 GUA HON.

Chamaedorea bracteata H.Wendl. = **Chamaedorea pinnatifrons** (Jacq.) Oerst.

Chamaedorea brevifrons H.Wendl. = **Chamaedorea pinnatifrons** (Jacq.) Oerst.

Chamaedorea carchensis Standl. & Steyerm., Publ. Field Mus. Nat. Hist., Bot. Ser. 23: 199 (1947).
Guatemala. 80 GUA.

Chamaedorea casperiana Klotzsch = **Chamaedorea tepejilote** Liebm.

Chamaedorea castillo-montii Hodel, Phytologia 68: 397 (1990).
Guatemala. 80 GUA.

Chamaedorea cataractarum Mart., Hist. Nat. Palm. 3: 309 (1849). *Stachyophorbe cataractarum* (Mart.) Liebm. ex Klotzsch, Overs. Kongel. Danske Vidensk. Selsk. Forh. Medlemmers Arbeider 1846: 8 (1846). *Nunnezharia cataractarum* (Mart.) Kuntze, Revis. Gen. Pl. 2: 730 (1891).
S. Mexico. 79 MXS MXT.
  Chamaedorea lindeniana H.Wendl., Allg. Gartenzeitung 21: 139 (1853).
  Chamaedorea martiana H.Wendl., Allg. Gartenzeitung 21: 137 (1853). *Stephanostachys martiana* (H.Wendl.) Oerst., Vidensk. Meddel. Dansk Naturhist. Foren. Kjøbenhavn 1858: 29 (1858). *Nunnezharia martiana* (H.Wendl.) Kuntze, Revis. Gen. Pl. 2: 730 (1891).
  Chamaedorea flexuosa H.Wendl., Index Palm.: 14 (1854). *Nunnezharia flexuosa* (H.Wendl.) Kuntze, Revis. Gen. Pl. 2: 731 (1891).
  Chamaedorea atrovirens H.Wendl. in O.C.E.de Kerchove de Denterghem, Palmiers: 239 (1878).
  Vadia jotolana O.F.Cook, Natl. Hort. Mag. 26: 34 (1947), no latin descr.

Chamaedorea chazdoniae Hodel, Principes 35: 73 (1991).
Costa Rica. 80 COS.

Chamaedorea christinae Hodel, Novon 7: 36 (1997).
Colombia. 83 CLM.

Chamaedorea coclensis L.H.Bailey = **Chamaedorea amabilis** H.Wendl. ex Dammer

Chamaedorea columbica Burret = **Chamaedorea tepejilote** Liebm.

Chamaedorea concinna Burret = **Chamaedorea pinnatifrons** (Jacq.) Oerst.

Chamaedorea concolor Mart. = **Chamaedorea pinnatifrons** (Jacq.) Oerst.

Chamaedorea conocarpa Mart. = **Chamaedorea pinnatifrons** (Jacq.) Oerst.

Chamaedorea coralliformis Hodel = **Chamaedorea crucensis** Hodel

Chamaedorea corallina Hook.f. = **Chamaedorea oblongata** Mart.

Chamaedorea correae Hodel & N.W.Uhl, Principes 34: 125 (1990).
Panama. 80 PAN.

Chamaedorea costaricana Oerst., Vidensk. Meddel. Dansk Naturhist. Foren. Kjøbenhavn 1858: 19 (1858). *Nunnezharia costaricana* (Oerst.) Kuntze, Revis. Gen. Pl. 2: 730 (1891). *Omanthe costaricana* (Oerst.) O.F.Cook, Science, n.s., 90: 298 (1939).
SE. Mexico to C. America. 79 MXT 80 COS ELS GUA HON NIC PAN.
  Chamaedorea biolleyi Guillaumin, Bull. Mus. Natl. Hist. Nat. 28: 543 (1922).
  Chamaedorea linearia L.H.Bailey, Gentes Herb. 6: 249 (1943).
  Chamaedorea seibertii L.H.Bailey, Gentes Herb. 6: 238 (1943).

  Legnea laciniata O.F.Cook, Natl. Hort. Mag. 22: 134 (1943), nom. inval.

Chamaedorea crucensis Hodel, Principes 34: 166 (1990).
Costa Rica. 80 COS.
  Chamaedorea coralliformis Hodel, Principes 40: 212 (1996).

Chamaedorea crucifolia Hook.f. = **Chamaedorea glaucifolia** H.Wendl.

Chamaedorea dammeriana Burret, Notizbl. Bot. Gart. Berlin-Dahlem 11: 737 (1933).
Costa Rica to Panama. 80 COS PAN.
  Chamaedorea variabilis H.Wendl. ex Burret, Notizbl. Bot. Gart. Berlin-Dahlem 11: 726 (1933).
  Chamaedorea wedeliana L.H.Bailey, Gentes Herb. 6: 247 (1943).

Chamaedorea deckeriana (Klotzsch) Hemsl., Biol. Cent.-Amer., Bot. 3: 404 (1885).
Costa Rica to Panama. 80 COS PAN.
  Stachyophorbe deckeriana Klotzsch, Allg. Gartenzeitung 20: 364 (1852). *Dasystachys deckeriana* (Klotzsch) Oerst., Vidensk. Meddel. Dansk Naturhist. Foren. Kjøbenhavn 1858: 26 (1858). *Nunnezharia deckeriana* (Klotzsch) Kuntze, Revis. Gen. Pl. 2: 730 (1891).

Chamaedorea deneversiana Grayum & Hodel, Principes 35: 133 (1991).
Panama, Ecuador. 80 PAN 83 ECU.

Chamaedorea densiflora Guillaumin = **Chamaedorea arenbergiana** H.Wendl.

Chamaedorea depauperata Dammer = **Chamaedorea pinnatifrons** (Jacq.) Oerst.

Chamaedorea deppeana Klotzsch = **Chamaedorea elegans** Mart.

Chamaedorea desmoncoides H.Wendl. = **Chamaedorea elatior** Mart.

Chamaedorea digitata Standl. & Steyerm. = **Chamaedorea parvisecta** Burret

Chamaedorea donnell-smithii Dammer = **Chamaedorea seifrizii** Burret

Chamaedorea dryanderae Burret = **Chamaedorea pinnatifrons** (Jacq.) Oerst.

Chamaedorea elatior Mart., Linnaea 5: 205 (1830). *Nunnezharia elatior* (Mart.) Kuntze, Revis. Gen. Pl. 2: 730 (1891).
Mexico to C. America. 79 MXC MXG MXS MXT 80 GUA HON.
  Chamaedorea affinis Liebm. in C.F.P.von Martius, Hist. Nat. Palm. 3: 308 (1849). *Nunnezharia affinis* (Liebm.) Kuntze, Revis. Gen. Pl. 2: 730 (1891).
  Chamaedorea montana Liebm. in C.F.P.von Martius, Hist. Nat. Palm. 3: 308 (1849).
  Chamaedorea scandens Liebm. in C.F.P.von Martius, Hist. Nat. Palm. 3: 308 (1849). *Anothea scandens* (Liebm.) O.F.Cook, Natl. Hort. Mag. 22: 135 (1943).
  Chamaedorea desmoncoides H.Wendl., Allg. Gartenzeitung 21: 177 (1853). *Nunnezharia desmoncoides* (H.Wendl.) Kuntze, Revis. Gen. Pl. 2: 730 (1891).
  Chamaedorea resinifera H.Wendl., Allg. Gartenzeitung 21: 179 (1853). *Nunnezharia resinifera* (H.Wendl.) Kuntze, Revis. Gen. Pl. 2: 731 (1891).
  Chamaedorea regia H.Wendl., Index Palm.: 11 (1854). *Nunnezharia regia* (H.Wendl.) Kuntze, Revis. Gen. Pl. 2: 731 (1891).

Chamaedorea repens H.Wendl., Index Palm.: 11 (1854). *Nunnezharia repens* (H.Wendl.) Kuntze, Revis. Gen. Pl. 2: 731 (1891).

Chamaedorea robusta H.Wendl., Index Palm.: 11 (1854). *Nunnezharia robusta* (H.Wendl.) Kuntze, Revis. Gen. Pl. 2: 731 (1891).

Nunnezharia oaxacensis Kuntze, Revis. Gen. Pl. 2: 730 (1891).

Chamaedorea elatior var. *bambusoides* H.Wendl. ex Dammer, Gard. Chron., III, 38: 42 (1905). *Chamaedorea bambusoides* (H.Wendl. ex Dammer) Gérôme, Rev. Hort. 83: 571 (1911).

Chamaedorea elatior var. *desmoncoides* H.Wendl. ex Dammer, Gard. Chron., III, 38: 42 (1905).

Chamaedorea bambusoides var. *graminifolia* Gérôme, Rev. Hort. 83: 570 (1911).

Chamaedorea bambusoides var. *juncea* Gérôme, Rev. Hort. 83: 571 (1911).

Platythea graminea O.F.Cook, Natl. Hort. Mag. 26: 228 (1947), no latin descr.

Chamaedorea elatior var. *bambusoides* H.Wendl. ex Dammer = **Chamaedorea elatior** Mart.

Chamaedorea elatior var. *desmoncoides* H.Wendl. ex Dammer = **Chamaedorea elatior** Mart.

Chamaedorea elegans Mart., Linnaea 5: 204 (1830). *Collinia elegans* (Mart.) Liebm. ex Oerst., Overs. Kongel. Danske Vidensk. Selsk. Forh. Medlemmers Arbeider 1845: 8 (1846). *Nunnezharia elegans* (Mart.) Kuntze, Revis. Gen. Pl. 2: 730 (1891). *Neanthe elegans* (Mart.) O.F.Cook, Science, n.s., 86: 122 (1937).

Mexico to Guatemala. 79 MXC MXE MXG MXS MXT 80 BLZ GUA.

Kunthia deppii Zucc., Allg. Gartenzeitung 2: 245 (1834).

Chamaedorea elegans var. *angustifolia* M.Martens & Galeotti, Bull. Acad. Roy. Sci. Bruxelles 10(1): 122 (1843).

Collinia humilis Liebm. ex Oerst., Overs. Kongel. Danske Vidensk. Selsk. Forh. Medlemmers Arbeider 1845: 8 (1846). *Chamaedorea humilis* (Liebm. ex Oerst.) Mart., Hist. Nat. Palm. 3: 308 (1849). *Nunnezharia humilis* (Liebm. ex Oerst.) Kuntze, Revis. Gen. Pl. 2: 730 (1891).

Chamaedorea deppeana Klotzsch, Allg. Gartenzeitung 20: 362 (1852).

Chamaedorea helleriana Klotzsch, Allg. Gartenzeitung 20: 362 (1852).

Collinia deppeana Klotzsch, Allg. Gartenzeitung 20: 362 (1852).

Chamaedorea pulchella Linden, Cat. Gén. 117: 4 (1887). *Nunnezharia pulchella* (Linden) Kuntze, Revis. Gen. Pl. 2: 731 (1891).

Neanthe bella O.F.Cook, Science, n.s., 86: 122 (1937).

Neanthe neesiana O.F.Cook, Science, n.s., 86: 122 (1937).

Chamaedorea elegans var. *angustifolia* M.Martens & Galeotti = **Chamaedorea elegans** Mart.

Chamaedorea ernesti-augusti H.Wendl., Allg. Gartenzeitung 20: 73 (1852). *Morenia ernesti-augusti* (H.Wendl.) H.Wendl., Allg. Gartenzeitung 21: 3 (1853). *Eleutheropetalum ernesti-augusti* (H.Wendl.) Oerst., Vidensk. Meddel. Dansk Naturhist. Foren. Kjøbenhavn 1858: 7 (1859). *Nunnezharia ernesti-augusti* (H.Wendl.) Kuntze, Revis. Gen. Pl. 2: 730 (1891).

Mexico to C. America. 79 MXG MXS MXT 80 BLZ GUA HON.

Chamaedorea simplicifrons Heynh., Alph. Aufz. Gew.

2: 135 (1846), nom. nud. *Nunnezharia simplicifrons* (Heynh.) Kuntze, Revis. Gen. Pl. 2: 731 (1891), nom. inval.

Geonoma corallifera C.Morren, Belgique Hort. 5: 168 (1855).

Chamaedorea glazioviana Drude ex Guillaumin, J. Soc. Natl. Hort. France, IV, 24: 231 (1925).

Geonoma latifrons Burret, Notizbl. Bot. Gart. Berlin-Dahlem 11: 728 (1933), pro syn.

Chamaedorea erumpens H.E.Moore = **Chamaedorea seifrizii** Burret

Chamaedorea exorrhiza Dammer ex Guillaumin = **Chamaedorea tepejilote** Liebm.

Chamaedorea falcaria L.H.Bailey = **Hyospathe elegans** Mart.

Chamaedorea falcifera H.E.Moore, Principes 2: 68 (1958).

Guatemala. 80 GUA.

Chamaedorea fenestrata H.Wendl. = **Chamaedorea geonomiformis** H.Wendl.

Chamaedorea ferruginea H.E.Moore = **Chamaedorea liebmannii** Mart.

Chamaedorea fibrosa H.Wendl. = **Synechanthus fibrosus** (H.Wendl.) H.Wendl.

Chamaedorea flavovirens H.Wendl. = **Chamaedorea pinnatifrons** (Jacq.) Oerst.

Chamaedorea flexuosa H.Wendl. = **Chamaedorea cataractarum** Mart.

Chamaedorea formosa W.Bull = **Chamaedorea linearis** (Ruiz & Pav.) Mart.

Chamaedorea foveata Hodel, Phytologia 68: 403 (1990). Mexico (Oaxaca, Chiapas). 79 MXS MXT.

Chamaedorea fractiflexa Hodel & Cast.Mont, Principes 35: 6 (1991).

Mexico (Chiapas) to Guatemala. 79 MXT 80 GUA.

Chamaedorea fragrans Mart., Hist. Nat. Palm. 2: 4 (1823).

Peru. 83 PER.

Chamaedorea gratissima Linden, Cat.

Nunnezharia fragrans Ruiz & Pav., Syst. Veg. Fl. Peruv. Chil.: 294 (1798). *Nunnezia fragrans* (Ruiz & Pav.) Willd., Sp. Pl. 4: 1154 (1806).

Chamaedorea verschaffeltii Kerch., Palmiers(1878). *Nunnezharia verschaffeltii* (Kerch.) Kuntze, Revis. Gen. Pl. 2: 731 (1891).

Chamaedorea pavoniana H.Wendl., Gard. Chron., III, 1904(2): 246 (1904).

Chamaedorea ruizii H.Wendl., Gard. Chron., III, 1904(2): 246 (1904).

Chamaedorea frondosa Hodel, Cast.Mont & Zúñiga, Principes 39: 186 (1995).

Honduras. 80 HON.

Chamaedorea fusca Standl. & Steyerm. = **Chamaedorea oblongata** Mart.

Chamaedorea geonomoides (Spruce) Drude = **Chamaedorea pinnatifrons** (Jacq.) Oerst.

Chamaedorea geonomiformis H.Wendl., Allg. Gartenzeitung 20: 1 (1852). *Nunnezharia geonomiformis* (H.Wendl.) Hook.f., Bot. Mag. 100: t. 6088 (1874).

S. Mexico to C. America. 79 MXS MXT 80 BLZ GUA HON.

Chamaedorea humilis H.Wendl., nom. illeg. *Nunnezharia humilis* Kuntze, Revis. Gen. Pl. 2: 731 (1891), nom. illeg.

Chamaedorea fenestrata H.Wendl., Index Palm.: 28 (1854). *Geonoma fenestrata* (H.Wendl.) H.Wendl. in O.C.E.de Kerchove de Denterghem, Palmiers: 245 (1878). *Nunnezharia fenestrata* (H.Wendl.) Kuntze, Revis. Gen. Pl. 2: 731 (1891). *Geonoma humilis* auct., Gard. Chron., III, 36: 202 (1904).

Chamaedorea glauca Linden = ? [8]

Chamaedorea glaucifolia H.Wendl., Index Palm.: 64 (1854). *Nunnezharia glaucifolia* (H.Wendl.) Kuntze, Revis. Gen. Pl. 2: 730 (1891). *Discoma glaucifolia* (H.Wendl.) O.F.Cook, Natl. Hort. Mag. 22: 137 (1943). Mexico (Chiapas). 79 MXT.
> *Chamaedorea crucifolia* Hook.f., Rep. Progr. Condition Roy. Bot. Gard. Kew 1882: 59 (1884).

Chamaedorea glazioviana Drude ex Guillaumin = **Chamaedorea ernesti-augusti** H.Wendl.

Chamaedorea gracilis Willd. = **Chamaedorea pinnatifrons** (Jacq.) Oerst.

Chamaedorea graminifolia H.Wendl., Index Palm.: 62 (1854). *Nunnezharia graminifolia* (H.Wendl.) Kuntze, Revis. Gen. Pl. 2: 730 (1891). N. Costa Rica.80 COS NIC?

Chamaedorea gratissima Linden = **Chamaedorea fragrans** Mart.

Chamaedorea guntheriana Hodel & N.W.Uhl, Principes 34: 126 (1990). Panama. 80 PAN.

Chamaedorea hageniorum L.H.Bailey = **Chamaedorea pittieri** L.H.Bailey

Chamaedorea hartwegii W.Watson = **Chamaedorea sartorii** Liebm.

Chamaedorea heilbornii Burret = **Chamaedorea pinnatifrons** (Jacq.) Oerst.

Chamaedorea helleriana Klotzsch = **Chamaedorea elegans** Mart.

Chamaedorea herrerae Burret = **Chamaedorea pinnatifrons** (Jacq.) Oerst.

Chamaedorea hodelii Grayum, Phytologia 84: 312 (1998 publ. 1999). Costa Rica. 80 COS.

Chamaedorea holmgrenii Burret = **Chamaedorea pinnatifrons** (Jacq.) Oerst.

Chamaedorea hooperiana Hodel, Principes 35: 188 (1991). Mexico (Veracruz). 79 MXG.

Chamaedorea hoppii Burret = **Chamaedorea pinnatifrons** (Jacq.) Oerst.

Chamaedorea humilis (Liebm. ex Oerst.) Mart. = **Chamaedorea elegans** Mart.

Chamaedorea humilis H.Wendl. = **Chamaedorea geonomiformis** H.Wendl.

Chamaedorea ibarrae Hodel, Principes 36: 191 (1992). Mexico (Chiapas) to Guatemala. 79 MXT 80 GUA.

Chamaedorea incrustata Hodel, G.Herrera & Casc., Palm J. 137: 40 (1997). Costa Rica. 80 COS.

Chamaedorea integrifolia (Trail) Dammer = **Chamaedorea pauciflora** Mart.

Chamaedorea kalbreyeriana H.Wendl. ex Burret = **Chamaedorea pinnatifrons** (Jacq.) Oerst.

Chamaedorea karwinskyana H.Wendl. = **Chamaedorea pochutlensis** Liebm.

Chamaedorea × *katzeri* Loebner = **C. ernesti-augusti** × **C. sartorii**

Chamaedorea keelerorum Hodel & Cast.Mont, Principes 36: 194 (1992). Mexico (Chiapas) to Guatemala. 79 MXT 80 GUA.

Chamaedorea klotzschiana H.Wendl., Index Palm.: 63 (1854). *Nunnezharia klotzschiana* (H.Wendl.) Kuntze, Revis. Gen. Pl. 2: 730 (1891). Mexico (Veracruz). 79 MXG.

Chamaedorea lanceolata (Ruiz & Pav.) Kunth = **Chamaedorea pinnatifrons** (Jacq.) Oerst.

Chamaedorea latifolia W.Watson = **Chamaedorea arenbergiana** H.Wendl.

Chamaedorea latifrons H.Wendl. = **Chamaedorea arenbergiana** H.Wendl.

Chamaedorea latipinna L.H.Bailey = **Chamaedorea warscewiczii** H.Wendl.

Chamaedorea latisecta (H.E.Moore) A.H.Gentry, Ann. Missouri Bot. Gard. 73: 163 (1986). Colombia. 83 CLM.
> *Morenia latisecta* H.E.Moore, Gentes Herb. 8: 203 (1949).

Chamaedorea lechleriana H.Wendl. = **Chamaedorea pauciflora** Mart.

Chamaedorea lehmannii Burret, Notizbl. Bot. Gart. Berlin-Dahlem 11: 857 (1933). Guatemala. 80 GUA.

Chamaedorea leonis H.E.Moore = **Chamaedorea angustisecta** Burret

Chamaedorea lepidota H.Wendl. = **Chamaedorea liebmannii** Mart.

Chamaedorea liebmannii Mart., Hist. Nat. Palm. 3: 308 (1849). *Nunnezharia liebmannii* (Mart.) Kuntze, Revis. Gen. Pl. 2: 730 (1891). Mexico to Guatemala. 79 MXC MXG MXS MXT 80 GUA.
> *Collinia elatior* Liebm. ex Oerst., Overs. Kongel. Danske Vidensk. Selsk. Forh. Medlemmers Arbeider 1845: 8 (1846).
> *Chamaedorea lepidota* H.Wendl., Allg. Gartenzeitung 21: 138 (1853). *Nunnezharia lepidota* (H.Wendl.) Kuntze, Revis. Gen. Pl. 2: 730 (1891).
> *Chamaedorea velutina* H.Wendl., Index Palm.: 13 (1854). *Nunnezharia velutina* (H.Wendl.) Kuntze, Revis. Gen. Pl. 2: 731 (1891).
> *Lophothele ramea* O.F.Cook, Natl. Hort. Mag. 22: 142 (1943), no latin descr.
> *Chamaedorea aequalis* Standl. & Steyerm., Publ. Field Mus. Nat. Hist., Bot. Ser. 23: 196 (1947).
> *Chamaedorea ferruginea* H.E.Moore, Gentes Herb. 8: 236 (1951).

Chamaedorea lindeniana H.Wendl. = **Chamaedorea cataractarum** Mart.

Chamaedorea linearia L.H.Bailey = **Chamaedorea costaricana** Oerst.

Chamaedorea linearis (Ruiz & Pav.) Mart., Hist. Nat. Palm. 2: 5 (1823). Venezuela to Bolivia. 82 VEN 83 BOL CLM ECU PER.
> *Martinezia linearis* Ruiz & Pav., Syst. Veg. Fl. Peruv. Chil.: 297 (1798). *Nunnezharia linearis* (Ruiz & Pav.) Kuntze, Revis. Gen. Pl. 2: 730 (1891). *Morenia linearis* (Ruiz & Pav.) Burret, Notizbl. Bot. Gart. Berlin-Dahlem 11: 316 (1932).
> *Morenia fragrans* Ruiz & Pav., Syst. Veg. Fl. Peruv. Chil.: 299 (1798).

Kunthia montana Humb. & Bonpl., Pl. Aequinoct. 2: 128 (1813). *Nunnezharia montana* (Humb. & Bonpl.) Kuntze, Revis. Gen. Pl. 2: 730 (1891). *Morenia montana* (Humb. & Bonpl.) Burret, Notizbl. Bot. Gart. Berlin-Dahlem 11: 316 (1932).

Morenia poeppigiana Mart., Hist. Nat. Palm. 3: 161 (1838). *Nunnezharia poeppigiana* (Mart.) Kuntze, Revis. Gen. Pl. 2: 730 (1891). *Chamaedorea poeppigiana* (Mart.) A.H.Gentry, Ann. Missouri Bot. Gard. 73: 162 (1986).

Morenia corallina H.Karst., Linnaea 28: 274 (1856). *Nunnezharia corallina* (H.Karst.) Kuntze, Revis. Gen. Pl. 2: 731 (1891).

Morenia lindeniana H.Wendl., Bot. Zeitung (Berlin) 17: 17 (1859). *Nunnezharia lindeniana* (H.Wendl.) Kuntze, Revis. Gen. Pl. 2: 730 (1891).

Chamaedorea formosa W.Bull, Gard. Chron., n.s., 1876(1): 724 (1876). *Nunnezharia formosa* (W.Bull) Kuntze, Revis. Gen. Pl. 2: 731 (1891).

Nunnezharia morenia Kuntze, Revis. Gen. Pl. 2: 730 (1891).

Chamaedorea polyclada Burret, Notizbl. Bot. Gart. Berlin-Dahlem 13: 341 (1936).

Morenia caudata Burret, Notizbl. Bot. Gart. Berlin-Dahlem 13: 337 (1936).

Morenia macrocarpa Burret, Notizbl. Bot. Gart. Berlin-Dahlem 13: 333 (1936).

Morenia robusta Burret, Notizbl. Bot. Gart. Berlin-Dahlem 13: 335 (1936).

Morenia microspadix Burret, Notizbl. Bot. Gart. Berlin-Dahlem 15: 33 (1940).

Chamaedorea megaphylla A.H.Gentry, Ann. Missouri Bot. Gard. 73: 163 (1986).

Chamaedorea lucidifrons L.H.Bailey, Gentes Herb. 6: 244 (1943).
Nicaragua to Panama. 80 COS NIC PAN.
Chamaedorea selvae Hodel, Principes 35: 79 (1991).

Chamaedorea lunata Liebm. = **Chamaedorea oblongata** Mart.

Chamaedorea macroloba Galeano = **Chamaedorea murriensis** Galeano

Chamaedorea macroloba Burret = **Chamaedorea pinnatifrons** (Jacq.) Oerst.

Chamaedorea macrospadix Oerst., Vidensk. Meddel. Dansk Naturhist. Foren. Kjøbenhavn 1858: 20 (1858). *Nunnezharia macrospadix* (Oerst.) Kuntze, Revis. Gen. Pl. 2: 730 (1891).
Costa Rica. 80 COS.

Chamaedorea martiana H.Wendl. = **Chamaedorea cataractarum** Mart.

Chamaedorea matae Hodel, Principes 35: 75 (1991).
Costa Rica to Panama. 80 COS PAN.

Chamaedorea megaphylla A.H.Gentry = **Chamaedorea linearis** (Ruiz & Pav.) Mart.

Chamaedorea membranacea Oerst. = **Chamaedorea pinnatifrons** (Jacq.) Oerst.

Chamaedorea metallica O.F.Cook ex H.E.Moore, Principes 10: 45 (1966).
Mexico (Veracruz, Oaxaca). 79 MXG MXS.

Chamaedorea microphylla H.Wendl., Bot. Zeitung (Berlin) 17: 102 (1859). *Nunnezharia microphylla* (H.Wendl.) Kuntze, Revis. Gen. Pl. 2: 730 (1891).
Panama. 80 PAN.

Chamaedorea mexicana Heynh. = **Chamaedorea sartorii** Liebm.

Chamaedorea micrantha Burret = **Chamaedorea pinnatifrons** (Jacq.) Oerst.

Chamaedorea microspadix Burret, Notizbl. Bot. Gart. Berlin-Dahlem 11: 734 (1933).
Mexico (Hidalgo, Querétaro, San Luis Potosí, Veracruz). 79 MXE MXG.

Chamaedorea minima Hodel, Principes 35: 72 (1991).
Costa Rica. 80 COS.

Chamaedorea minor Burret = **Chamaedorea pinnatifrons** (Jacq.) Oerst.

Chamaedorea moliniana Hodel, Cast.Mont & Zúñiga, Principes 39: 183 (1995).
Honduras. 80 HON.

Chamaedorea monostachys Burret = **Chamaedorea oreophila** Mart.

Chamaedorea montana Liebm. = **Chamaedorea elatior** Mart.

Chamaedorea murriensis Galeano, Principes 31: 143 (1987).
Panama to Colombia. 80 PAN 83 CLM.
Chamaedorea macroloba Galeano, Brittonia 38: 60 (1986).

Chamaedorea nana N.E.Br. = **Chamaedorea pumila** H.Wendl. ex Dammer

Chamaedorea nationsiana Hodel & Cast.Mont, Principes 35: 4 (1991).
Guatemala. 80 GUA.

Chamaedorea neurochlamys Burret, Notizbl. Bot. Gart. Berlin-Dahlem 11: 744 (1933).
SE. Mexico to Honduras. 79 MXT 80 BLZ GUA HON.

Chamaedorea nubium Standl. & Steyerm., Publ. Field Mus. Nat. Hist., Bot. Ser. 23: 202 (1947).
S. Mexico to C. America. 79 MXS MXT 80 ELS GUA HON.

Chamaedorea oblongata Mart., Hist. Nat. Palm. 3: 160 (1838). *Nunnezharia oblongata* (Mart.) Kuntze, Revis. Gen. Pl. 2: 730 (1891).
Mexico to C. America. 79 MXC MXE MXG MXS MXT 80 BLZ GUA HON NIC.
Chamaedorea lunata Liebm. in C.F.P.von Martius, Hist. Nat. Palm. 3: 307 (1849). *Nunnezharia lunata* (Liebm.) Kuntze, Revis. Gen. Pl. 2: 730 (1891). *Mauranthe lunata* (Liebm.) O.F.Cook, Natl. Hort. Mag. 22: 96 (1943).
Chamaedorea biloba H.Wendl., Index Palm.: 11 (1854). *Nunnezharia biloba* (H.Wendl.) Kuntze, Revis. Gen. Pl. 2: 731 (1891).
Morenia corallocarpa H.Wendl., Index Palm.: 29 (1854). *Nunnezharia corallocarpa* (H.Wendl.) Kuntze, Revis. Gen. Pl. 2: 731 (1891).
Chamaedorea paradoxa H.Wendl., Bot. Zeitung (Berlin) 17: 29 (1859). *Nunnezharia paradoxa* (H.Wendl.) Kuntze, Revis. Gen. Pl. 2: 730 (1891).
Chamaedorea corallina Hook.f., Rep. Progr. Condition Roy. Bot. Gard. Kew 1882: 58 (1884).
Chamaedorea fusca Standl. & Steyerm., Publ. Field Mus. Nat. Hist., Bot. Ser. 23: 201 (1947).

Chamaedorea oerstedii O.F.Cook & Doyle = **Chamaedorea pinnatifrons** (Jacq.) Oerst.

Chamaedorea oreophila Mart., Hist. Nat. Palm. 3: 309 (1849). *Nunnezharia oreophila* (Mart.) Kuntze, Revis. Gen. Pl. 2: 730 (1891). *Stachyophorbe oreophila* (Mart.) O.F.Cook, Natl. Hort. Mag. 22: 146 (1943).
Mexico (Veracruz, Oaxaca). 79 MXG MXS.

Stachyophorbe montana Liebm. ex Oerst., Overs. Kongel. Danske Vidensk. Selsk. Forh. Medlemmers Arbeider 1846: 8 (1846).

Chamaedorea monostachys Burret, Notizbl. Bot. Gart. Berlin-Dahlem 11: 761 (1933).

Stachyophorbe filipes O.F.Cook, Natl. Hort. Mag. 22: 144 (1943), no latin descr.

Chamaedorea pacaya Oerst. = **Chamaedorea pinnatifrons** (Jacq.) Oerst.

Chamaedorea pachecoana Standl. & Steyerm., Publ. Field Mus. Nat. Hist., Bot. Ser. 23: 203 (1947). Guatemala. 80 GUA.

Chamaedorea palmeriana Hodel & N.W.Uhl, Principes 34: 122 (1990). Costa Rica to Panama. 80 COS PAN.

Chamaedorea paradoxa H.Wendl. = **Chamaedorea oblongata** Mart.

Chamaedorea parvifolia Burret, Notizbl. Bot. Gart. Berlin-Dahlem 11: 746 (1933). Costa Rica. 80 COS.

Chamaedorea parvisecta Burret, Notizbl. Bot. Gart. Berlin-Dahlem 11: 742 (1933). Mexico (Chiapas) to Guatemala. 79 MXT 80 GUA.

   Chamaedorea pulchra Burret, Notizbl. Bot. Gart. Berlin-Dahlem 11: 741 (1933).

   Paranthe violacea O.F.Cook, Natl. Hort. Mag. 22: 140 (1943), no latin descr.

   Chamaedorea digitata Standl. & Steyerm., Publ. Field Mus. Nat. Hist., Bot. Ser. 23: 200 (1947).

Chamaedorea pauciflora Mart., Hist. Nat. Palm. 2: 5 (1823). *Morenia pauciflora* (Mart.) Drude in C.F.P.von Martius & auct. suc. (eds.), Fl. Bras. 3(2): 526 (1882). *Nunnezharia pauciflora* (Mart.) Kuntze, Revis. Gen. Pl. 2: 730 (1891). W. South America to W. Brazil. 83 CLM ECU PER 84 BZC BZN.

   Morenia integrifolia Trail, J. Bot. 14: 331 (1876). *Nunnezharia integrifolia* (Trail) Kuntze, Revis. Gen. Pl. 2: 730 (1891). *Chamaedorea integrifolia* (Trail) Dammer, Verh. Bot. Vereins Prov. Brandenburg 48: 125 (1906 publ. 1907).

   Nunnezharia amazonica Kuntze, Revis. Gen. Pl. 2: 730 (1891). *Chamaedorea amazonica* (Kuntze) Dammer, Notizbl. Königl. Bot. Gart. Berlin 6: 263 (1915).

   Chamaedorea lechleriana H.Wendl., Gard. Chron., III, 1904(2): 246 (1904).

Chamaedorea pavoniana H.Wendl. = **Chamaedorea fragrans** Mart.

Chamaedorea pedunculata Hodel & N.W.Uhl, Principes 34: 131 (1990). Costa Rica. 80 COS.

Chamaedorea pinnatifrons (Jacq.) Oerst., Vidensk. Meddel. Dansk Naturhist. Foren. Kjøbenhavn 1858: 14 (1858). S. Mexico to Bolivia. 79 MXS MXT 80 COS ELS GUA HON NIC PAN 82 VEN 83 BOL CLM ECU PER 84 BZN.

   Borassus pinnatifrons Jacq., Pl. Hort. Schoenbr. 2: 65 (1797). *Nunnezharia pinnatifrons* (Jacq.) Kuntze, Revis. Gen. Pl. 2: 730 (1891).

   Martinezia lanceolata Ruiz & Pav., Syst. Veg. Fl. Peruv. Chil.: 297 (1798). *Chamaedorea lanceolata* (Ruiz & Pav.) Kunth, Enum. Pl. 3: 172 (1841). *Nunnezharia lanceolata* (Ruiz & Pav.) Kuntze, Revis. Gen. Pl. 2: 730 (1891).

Chamaedorea gracilis Willd., Sp. Pl. 4: 800 (1806).

Chamaedorea concolor Mart., Hist. Nat. Palm. 3: 160 (1838). *Nunnezharia concolor* (Mart.) Kuntze, Revis. Gen. Pl. 2: 730 (1891).

Chamaedorea conocarpa Mart. in A.D.d'Orbigny, Voy. Amér. Mér. 7(3): 6 (1842). *Nunnezharia conocarpa* (Mart.) Kuntze, Revis. Gen. Pl. 2: 730 (1891).

Hyospathe montana Mart. in A.D.d'Orbigny, Voy. Amér. Mér. 8: t. 6, f. 1 (1842).

Chamaedorea bartlingiana H.Wendl., Index Palm.: 60 (1854). *Nunnezharia bartlingiana* (H.Wendl.) Kuntze, Revis. Gen. Pl. 2: 730 (1891).

Chamaedorea brevifrons H.Wendl., Index Palm.: 61 (1854). *Nunnezharia brevifrons* (H.Wendl.) Kuntze, Revis. Gen. Pl. 2: 730 (1891).

Chamaedorea flavovirens H.Wendl., Index Palm.: 60 (1854). *Nunnezharia flavovirens* (H.Wendl.) Kuntze, Revis. Gen. Pl. 2: 730 (1891).

Chamaedorea bifurcata Oerst., Vidensk. Meddel. Dansk Naturhist. Foren. Kjøbenhavn 1858: 13 (1858). *Nunnezharia bifurcata* (Oerst.) Kuntze, Revis. Gen. Pl. 2: 730 (1891).

Chamaedorea membranacea Oerst., Vidensk. Meddel. Dansk Naturhist. Foren. Kjøbenhavn 1858: 22 (1858). *Nunnezharia membranacea* (Oerst.) Kuntze, Revis. Gen. Pl. 2: 730 (1891).

Chamaedorea pacaya Oerst., Vidensk. Meddel. Dansk Naturhist. Foren. Kjøbenhavn 1858: 12 (1858). *Nunnezharia pacaya* (Oerst.) Kuntze, Revis. Gen. Pl. 2: 730 (1891).

Chamaedorea bracteata H.Wendl., Bot. Zeitung (Berlin) 17: 29 (1859). *Nunnezharia bracteata* (H.Wendl.) Kuntze, Revis. Gen. Pl. 2: 730 (1891).

Nunnezharia geonomoides Spruce, J. Linn. Soc., Bot. 11: 122 (1871). *Chamaedorea geonomoides* (Spruce) Drude in C.F.P.von Martius & auct. suc. (eds.), Fl. Bras. 3(2): 531 (1882).

Chamaedorea boliviensis Dammer, Notizbl. Königl. Bot. Gart. Berlin 6: 262 (1915).

Chamaedorea depauperata Dammer, Notizbl. Königl. Bot. Gart. Berlin 6: 263 (1915).

Chamaedorea minor Burret, Notizbl. Bot. Gart. Berlin-Dahlem 11: 2 (1930).

Chamaedorea concinna Burret, Notizbl. Bot. Gart. Berlin-Dahlem 11: 747 (1933).

Chamaedorea dryanderae Burret, Notizbl. Bot. Gart. Berlin-Dahlem 11: 754 (1933).

Chamaedorea heilbornii Burret, Notizbl. Bot. Gart. Berlin-Dahlem 11: 750 (1933).

Chamaedorea herrerae Burret, Notizbl. Bot. Gart. Berlin-Dahlem 11: 748 (1933).

Chamaedorea holmgrenii Burret, Notizbl. Bot. Gart. Berlin-Dahlem 11: 749 (1933).

Chamaedorea hoppii Burret, Notizbl. Bot. Gart. Berlin-Dahlem 11: 755 (1933).

Chamaedorea kalbreyeriana H.Wendl. ex Burret, Notizbl. Bot. Gart. Berlin-Dahlem 11: 753 (1933).

Chamaedorea macroloba Burret, Notizbl. Bot. Gart. Berlin-Dahlem 11: 757 (1933).

Chamaedorea micrantha Burret, Notizbl. Bot. Gart. Berlin-Dahlem 11: 749 (1933).

Chamaedorea rhombea Burret, Notizbl. Bot. Gart. Berlin-Dahlem 11: 753 (1933).

Chamaedorea oerstedii O.F.Cook & Doyle, Natl. Hort. Mag. 18: 168 (1939).

Docanthe alba O.F.Cook, Natl. Hort. Mag. 22: 96 (1943), no latin descr.

Chamaedorea aguilariana Standl. & Steyerm., Publ. Field Mus. Nat. Hist., Bot. Ser. 23: 197 (1947).

Chamaedorea piscifolia Hodel, G.Herrera & Casc., Palm J. 137: 32 (1997).
Costa Rica. 80 COS.

Chamaedorea pittieri L.H.Bailey, Gentes Herb. 6: 252 (1943).
Costa Rica to Panama. 80 COS PAN.
Chamaedorea hageniorum L.H.Bailey, Gentes Herb. 6: 247 (1943).

Chamaedorea plumosa Hodel, Principes 36: 197 (1992).
Mexico (Chiapas). 79 MXT.

Chamaedorea pochutlensis Liebm. in C.F.P.von Martius, Hist. Nat. Palm. 3: 308 (1849). *Nunnezharia pochutlensis* (Liebm.) Kuntze, Revis. Gen. Pl. 2: 730 (1891).
Mexico (Sinaloa to Oaxaca). 79 MXN MXS.
Chamaedorea karwinskyana H.Wendl., Allg. Gartenzeitung 21: 179 (1853). *Nunnezharia karwinskyana* (H.Wendl.) Kuntze, Revis. Gen. Pl. 2: 730 (1891).

Chamaedorea poeppigiana (Mart.) A.H.Gentry = **Chamaedorea linearis** (Ruiz & Pav.) Mart.

Chamaedorea polyclada Burret = **Chamaedorea linearis** (Ruiz & Pav.) Mart.

Chamaedorea ponderosa Hodel, Novon 7: 35 (1997).
Panama. 80 PAN.

Chamaedorea pringlei S.Watson = **Chamaedorea radicalis** Mart.

Chamaedorea pulchella Linden = **Chamaedorea elegans** Mart.

Chamaedorea pulchra Burret = **Chamaedorea parvisecta** Burret

Chamaedorea pumila H.Wendl. ex Dammer, Gard. Chron., III, 1904(2): 246 (1904). *Nunnezharia pumila* (H.Wendl. ex Dammer) Kuntze, Revis. Gen. Pl. 2: 731 (1891).
Costa Rica. 80 COS.
Chamaedorea nana N.E.Br., Bull. Misc. Inform. Kew 1914: 156 (1914). *Kinetostigma nanum* (N.E.Br.) Burret, Notizbl. Bot. Gart. Berlin-Dahlem 11: 318 (1932).

Chamaedorea pygmaea H.Wendl., Allg. Gartenzeitung 20: 217 (1852). *Stachyophorbe pygmaea* (H.Wendl.) Oerst., Vidensk. Meddel. Dansk Naturhist. Foren. Kjøbenhavn 1858: 10 (1858). *Nunnezharia pygmaea* (H.Wendl.) Kuntze, Revis. Gen. Pl. 2: 731 (1891). *Cladandra pygmaea* (H.Wendl.) O.F.Cook, Natl. Hort. Mag. 22: 148 (1943).
Costa Rica to NW. Colombia. 80 COS PAN 83 CLM.
Chamaedorea terryorum Standl., Publ. Field Mus. Nat. Hist., Bot. Ser. 22: 326 (1940).

Chamaedorea queroana Hodel, Phytologia 68: 406 (1990).
Mexico (Oaxaca). 79 MXS.

Chamaedorea quezalteca Standl. & Steyerm., Publ. Field Mus. Nat. Hist., Bot. Ser. 23: 204 (1947).
Guatemala. 80 GUA.

Chamaedorea radicalis Mart., Hist. Nat. Palm. 3: 308 (1849). *Nunnezharia radicalis* (Mart.) Kuntze, Revis. Gen. Pl. 2: 731 (1891).
NE. Mexico. 79 MXE.
Chamaedorea pringlei S.Watson, Proc. Amer. Acad. Arts 26: 157 (1891).

Chamaedorea recurvata Hodel, Principes 39: 16 (1995).
Panama. 80 PAN.

Chamaedorea regia H.Wendl. = **Chamaedorea elatior** Mart.

Chamaedorea repens H.Wendl. = **Chamaedorea elatior** Mart.

Chamaedorea resinifera H.Wendl. = **Chamaedorea elatior** Mart.

Chamaedorea rhizomatosa Hodel, Phytologia 68: 401 (1990).
Mexico (Oaxaca). 79 MXS.

Chamaedorea rhombea Burret = **Chamaedorea pinnatifrons** (Jacq.) Oerst.

Chamaedorea ricardoi Bernal et al., Palms 48: 27 (2004).
Colombia. 83 CLM.

Chamaedorea rigida H.Wendl. ex Dammer, Gard. Chron., III, 1904(2): 246 (1904). *Nunnezharia rigida* (H.Wendl. ex Dammer) Kuntze, Revis. Gen. Pl. 2: 731 (1891).
Mexico (Oaxaca). 79 MXS.

Chamaedorea robertii Hodel & N.W.Uhl, Principes 34: 120 (1990).
Costa Rica to Panama. 80 COS PAN.

Chamaedorea robusta H.Wendl. = **Chamaedorea elatior** Mart.

Chamaedorea rojasiana Standl. & Steyerm., Publ. Field Mus. Nat. Hist., Bot. Ser. 23: 205 (1947).
Mexico (Chiapas) to Guatemala. 79 MXT 80 GUA.

Chamaedorea × romana Guillaumin = **Chamaedorea × katzeri**

Chamaedorea rosibeliae Hodel, G.Herrera & Casc., Palm J. 137: 43 (1997).
Costa Rica (Limón). 80 COS.

Chamaedorea rossteniorum Hodel, G.Herrera & Casc., Palm J. 137: 34 (1997).
Costa Rica to Panama. 80 COS PAN.

Chamaedorea ruizii H.Wendl. = **Chamaedorea fragrans** Mart.

Chamaedorea sartorii Liebm. in C.F.P.von Martius, Hist. Nat. Palm. 3: 308 (1849). *Eleutheropetalum sartorii* (Liebm.) Oerst., Amér. Centr.: 13 (1863). *Nunnezharia sartorii* (Liebm.) Kuntze, Revis. Gen. Pl. 2: 731 (1891).
Mexico (Veracruz, Puebla, Oaxaca), Honduras. 79 MXC MXG MXS 80 HON.
Chamaedorea mexicana Heynh., Alph. Aufz. Gew. 2: 135 (1846), nom. nud. *Nunnezharia mexicana* Kuntze, Revis. Gen. Pl. 2: 731 (1891), nom. inval.
Chamaedorea aurantiaca Brongn. ex Neumann, Rev. Hort., III, 1: 86 (1847), nom. subnud.
Morenia oblongata H.Wendl., Allg. Gartenzeitung 21: 3 (1853).
Chamaedorea hartwegii W.Watson, Gard. Chron., n.s., 1885(1): 410 (1885).
Chamaedorea wobstiana Linden, Cat. Gén. 117: 4 (1887). *Nunnezharia wobstiana* (Linden) Kuntze, Revis. Gen. Pl. 2: 731 (1891).
Nunnezharia aurantiaca (Brongn.) Kuntze, Revis. Gen. Pl. 2: 731 (1891).

Chamaedorea scheryi L.H.Bailey, Gentes Herb. 6: 252 (1943).
Costa Rica to Panama. 80 COS PAN.

Chamaedorea scandens Liebm. = **Chamaedorea elatior** Mart.

Chamaedorea schiedeana Mart., Linnaea 5: 204 (1830). *Nunnezharia schiedeana* (Mart.) Kuntze,

Revis. Gen. Pl. 2: 731 (1891).
Mexico (Veracruz, Puebla, Oaxaca). 79 MXC MXG MXS.
Chamaedorea speciosa H.Wendl., Index Palm.: 15 (1854). *Nunnezharia speciosa* (H.Wendl.) Kuntze, Revis. Gen. Pl. 2: 731 (1891).
Kunthia xalapensis Otto & A.Dietr. ex H.Wendl. in O.C.E.de Kerchove de Denterghem, Palmiers: 249 (1878).

Chamaedorea schippii Burret, Notizbl. Bot. Gart. Berlin-Dahlem 11: 1038 (1934).
Mexico (Chiapas) to Guetamala. 79 MXT 80 BLZ GUA.

Chamaedorea seibertii L.H.Bailey = **Chamaedorea costaricana** Oerst.

Chamaedorea seifrizii Burret, Notizbl. Bot. Gart. Berlin-Dahlem 14: 268 (1938).
SE. Mexico to Honduras. 79 MXT 80 BLZ GUA HON.
Chamaedorea donnell-smithii Dammer, Gard. Chron., III, 38: 43 (1905), nom. rejec.
Meiota campechana O.F.Cook, Natl. Hort. Mag. 22: 138 (1943), no latin descr.
Chamaedorea erumpens H.E.Moore, Gentes Herb. 8: 232 (1951).

Chamaedorea selvae Hodel = **Chamaedorea lucidifrons** L.H.Bailey

Chamaedorea serpens Hodel, Principes 35: 77 (1991).
Panama. 80 PAN.

Chamaedorea simplex Burret, Notizbl. Bot. Gart. Berlin-Dahlem 11: 758 (1933).
Mexico (Chiapas) to Guatemala. 79 MXT 80 GUA.

Chamaedorea simplicifrons Heynh. = **Chamaedorea ernesti-augusti** H.Wendl.

Chamaedorea skutchii Standl. & Steyerm., Publ. Field Mus. Nat. Hist., Bot. Ser. 23: 206 (1947).
Guatemala. 80 GUA.

Chamaedorea smithii A.H.Gentry, Ann. Missouri Bot. Gard. 73: 164 (1986).
Peru (Junín). 83 PER.

Chamaedorea speciosa H.Wendl. = **Chamaedorea schiedeana** Mart.

Chamaedorea sphaerocarpa Burret = **Chamaedorea tepejilote** Liebm.

Chamaedorea stenocarpa Standl. & Steyerm., Publ. Field Mus. Nat. Hist., Bot. Ser. 23: 206 (1947).
C. America. 80 COS GUA PAN.

Chamaedorea stolonifera H.Wendl. ex Hook.f., Bot. Mag. 188: t. 7265 (1892).
Mexico (Chiapas). 79 MXT.

Chamaedorea stricta Standl. & Steyerm., Publ. Field Mus. Nat. Hist., Bot. Ser. 23: 207 (1947).
Mexico (Chiapas) to C. America. 79 MXT 80 COS GUA PAN.

Chamaedorea subjectifolia Hodel, Principes 39: 18 (1995).
Panama. 80 PAN.

Chamaedorea sullivaniorum Hodel & N.W.Uhl, Principes 34: 128 (1990).
Costa Rica to W. Colombia. 80 COS PAN 83 CLM.

Chamaedorea tenella H.Wendl., Gartenflora 29: 102 (1880). *Nunnezharia tenella* (H.Wendl.) Hook.f., Bot. Mag. 107: t. 6584 (1881).
Mexico to C. America. 79 MXG MXT 80 COS.

Chamaedorea tenerrima Burret, Notizbl. Bot. Gart. Berlin-Dahlem 11: 858 (1933).
Guatemala. 80 GUA.
Lobia erosa O.F.Cook, Natl. Hort. Mag. 22: 148 (1943), no latin descr.

Chamaedorea tepejilote Liebm. in C.F.P.von Martius, Hist. Nat. Palm. 3: 308 (1849). *Stephanostachys tepejilote* (Liebm.) Oerst., Vidensk. Meddel. Dansk Naturhist. Foren. Kjøbenhavn 1858: 28 (1858). *Nunnezharia tepejilote* (Liebm.) Kuntze, Revis. Gen. Pl. 2: 731 (1891).
Mexico (Veracruz, Oaxaca, Chiapas) to W. Colombia. 79 MXG MXS MXG 80 BLZ COS GUA HON NIC PAN 83 CLM.
Chamaedorea casperiana Klotzsch, Allg. Gartenzeitung 20: 363 (1852). *Nunnezharia casperiana* (Klotzsch) Kuntze, Revis. Gen. Pl. 2: 730 (1891).
Stephanostachys wendlandiana Oerst., Vidensk. Meddel. Dansk Naturhist. Foren. Kjøbenhavn 1858: 28 (1858). *Chamaedorea wendlandiana* (Oerst.) Hemsl., Biol. Cent.-Amer., Bot. 3: 407 (1885). *Nunnezharia wendlandiana* (Oerst.) Kuntze, Revis. Gen. Pl. 2: 730 (1891).
Chamaedorea exorrhiza Dammer ex Guillaumin, Bull. Mus. Natl. Hist. Nat. 28: 542 (1922).
Chamaedorea anomospadix Burret, Notizbl. Bot. Gart. Berlin-Dahlem 11: 763 (1933).
Chamaedorea sphaerocarpa Burret, Notizbl. Bot. Gart. Berlin-Dahlem 11: 762 (1933).
Chamaedorea columbica Burret, Notizbl. Bot. Gart. Berlin-Dahlem 12: 42 (1934).
Edanthe veraepacis O.F.Cook, Natl. Hort. Mag. 18: 172 (1939), nom. inval.

Chamaedorea terryorum Standl. = **Chamaedorea pygmaea** H.Wendl.

Chamaedorea tuerckheimii (Dammer) Burret, Notizbl. Bot. Gart. Berlin-Dahlem 11: 766 (1933).
Mexico (Veracruz, Oaxaca) to Honduras. 79 MXG MXS 80 GUA HON.
*Malortiea tuerckheimii Dammer, Gard. Chron. 1905(1): 19 (1905). *Kinetostigma tuerckbeimii* (Dammer) Burret, Notizbl. Bot. Gart. Berlin-Dahlem 11: 317 (1932).

Chamaedorea undulatifolia Hodel & N.W.Uhl, Principes 34: 116 (1990).
Costa Rica. 80 COS.

Chamaedorea variabilis H.Wendl. ex Burret = **Chamaedorea dammeriana** Burret

Chamaedorea velutina H.Wendl. = **Chamaedorea liebmannii** Mart.

Chamaedorea ventricosa Hook.f. = **Gaussia princeps** H.Wendl.

Chamaedorea verapazensis Hodel & Cast.Mont, Phytologia 68: 390 (1990).
Guatemala. 80 GUA.

Chamaedorea verecunda Grayum & Hodel, Principes 35: 135 (1991).
Panama. 80 PAN.

Chamaedorea verschaffeltii Kerch. = **Chamaedorea fragrans** Mart.

Chamaedorea vistae Hodel & N.W.Uhl = **Chamaedorea woodsoniana** L.H.Bailey

Chamaedorea volcanensis Hodel & Cast.Mont, Phytologia 68: 393 (1990).
Guatemala. 80 GUA.

Chamaedorea vulgata Standl. & Steyerm., Publ. Field Mus. Nat. Hist., Bot. Ser. 23: 208 (1947).
Mexico (Chiapas) to Guatemala. 79 MXT 80 GUA.

Chamaedorea warscewiczii H.Wendl., Bonplandia 10: 37 (1862). *Nunnezharia warscewiczii* (H.Wendl.) Kuntze, Revis. Gen. Pl. 2: 731 (1891).
Costa Rica to Panama. 80 COS PAN.
    Chamaedorea latipinna L.H.Bailey, Gentes Herb. 6: 244 (1943).

Chamaedorea wallisii Linden = **?** [8]

Chamaedorea wedeliana L.H.Bailey = **Chamaedorea dammeriana** Burret

Chamaedorea wendlandiana (Oerst.) Hemsl. = **Chamaedorea tepejilote** Liebm.

Chamaedorea wendlandii H.Wendl. = **?**

Chamaedorea whitelockiana Hodel & N.W.Uhl, Principes 34: 61 (1990).
Mexico (Oaxaca, Chiapas). 79 MXS MXT.

Chamaedorea wobstiana Linden = **Chamaedorea sartorii** Liebm.

Chamaedorea woodsoniana L.H.Bailey, Gentes Herb. 6: 238 (1943).
Mexico to NW. Colombia. 79 MXG MXS MXT 80 BLZ COS GUA HON NIC PAN 83 CLM.
    Chamaedorea vistae Hodel & N.W.Uhl, Principes 34: 58 (1990).

Chamaedorea zamorae Hodel, Principes 34: 173 (1990).
Costa Rica. 80 COS.

Chamaephoenix

Chamaephoenix H.Wendl. ex Curtiss = **Pseudophoenix** H.Wendl. ex Sarg.

Chamaephoenix sargentii (H.Wendl. ex Sarg.) Curtiss = **Pseudophoenix sargentii** H.Wendl. ex Sarg.

Chamaeriphe

Chamaeriphe Steck = **Chamaerops** L.

Chamaeriphes

Chamaeriphes Dill. ex Kuntze = **Hyphaene** Gaertn.

Chamaeriphes Ponted. ex Gaertn. = **Chamaerops** L.

Chamaeriphes benguelensis (Welw. ex H.Wendl.) Kuntze = **Hyphaene petersiana** Klotzsch ex Mart.

Chamaeriphes compressa (H.Wendl.) Kuntze = **Hyphaene compressa** H.Wendl.

Chamaeriphes coriacea (Gaertn.) Kuntze = **Hyphaene coriacea** Gaertn.

Chamaeriphes crinita (Gaertn.) Kuntze = **Hyphaene thebaica** (L.) Mart.

Chamaeriphes guineensis (Schumach. & Thonn.) Kuntze = **Hyphaene guineensis** Schumach. & Thonn.

Chamaeriphes macrosperma (H.Wendl.) Kuntze = **Hyphaene macrosperma** H.Wendl.

Chamaeriphes shatan (Bojer ex Dammer) Kuntze = **Hyphaene coriacea** Gaertn.

Chamaeriphes thebaica (L.) Kuntze = **Hyphaene thebaica** (L.) Mart.

Chamaeriphes turbinata (H.Wendl.) Kuntze = **Hyphaene coriacea** Gaertn.

Chamaeriphes ventricosa (Kirk) Kuntze = **Hyphaene petersiana** Klotzsch ex Mart.

Chamaerops

Chamaerops L., Sp. Pl.: 1187 (1753).
1 species, W. & C. Medit. 12 13 20.
    Chamaeriphe Steck, Sagu: 20 (1757).
    Chamaeriphes Ponted. ex Gaertn., Fruct. Sem. Pl. 1: 25 (1788).

Chamaerops acaulis Michx. = **Sabal minor** (Jacq.) Pers.

Chamaerops antillarum Desc. = **Zombia antillarum** (Desc.) L.H.Bailey

Chamaerops arborescens (Poir.) Steud. = **Chamaerops humilis** var. **humilis**

Chamaerops arundinacea (Aiton) Sm. = **Sabal minor** (Jacq.) Pers.

Chamaerops bilaminata Gentil = **Chamaerops humilis** var. **humilis**

Chamaerops biroo Siebold ex Mart. = **Livistona rotundifolia** (Lam.) Mart.

Chamaerops cochinchinensis Lour. = **Livistona saribus** (Lour.) Merr. ex A.Chev.

Chamaerops conduplicata J.Kickx f. = **Chamaerops humilis** var. **humilis**

Chamaerops depressa Chabaud = **Chamaerops humilis** var. **humilis**

Chamaerops elegans Hook.f. = **Chamaerops humilis** var. **humilis**

Chamaerops excelsa Thunb. = **Rhapis excelsa** (Thunb.) Henry

Chamaerops excelsa var. *humilior* Thunb. = **Rhapis humilis** Blume

Chamaerops excelsior Bojer = **Latania loddigesii** Mart.

Chamaerops fortunei Hook. = **Trachycarpus fortunei** (Hook.) H.Wendl.

Chamaerops ghiesbreghtii H.Wendl. = **Sabal ?**

Chamaerops glabra Mill. = **Sabal minor** (Jacq.) Pers.

Chamaerops griffithii Lodd. ex Verl. = **Trachycarpus martianus** (Wall. ex Mart.) H.Wendl.

Chamaerops humilis L., Sp. Pl.: 1187 (1753). *Phoenix humilis* (L.) Cav., Icon. 2: 12 (1793).
W. & C. Medit. 12 BAL FRA POR SAR SPA 13 ITA SIC 20 ALG LBY MOR TUN.

  var. **argentea** André, Rev. Hort. 1885: 231 (1885).
  Morocco. 20 MOR.
    Chamaerops humilis var. *cerifera* Becc., Webbia 5(1): 65 (1920).

  var. **humilis**
  W. & C. Medit. 12 BAL FRA POR SAR SPA 13 ITA SIC 20 ALG LBY MOR TUN.
    Chamaerops humilis var. *arborescens* Pers. in J.B.A.M.de Lamarck, Encycl. 1: 400 (1785).
    Chamaerops arborescens (Poir.) Steud., Nomencl. Bot. 1: 183 (1821).
    Chamaerops conduplicata J.Kickx f., Bull. Acad. Roy. Sci. Bruxelles 5: 61 (1838).
    Chamaerops humilis var. *depressa* Mart., Hist. Nat. Palm. 3: 248 (1838).
    Chamaerops humilis var. *elata* Mart., Hist. Nat. Palm. 3: 248 (1838).
    Chamaerops macrocarpa Tineo in G.Gussone, Fl. Sicul. Syn. 2: 883 (1844). *Chamaerops humilis* var. *macrocarpa* (Tineo) Becc., Webbia 5(1): 64 (1920).

Chamaerops elegans Hook.f., Rep. Progr. Condition Roy. Bot. Gard. Kew 1882: 64 (1884).

Chamaerops humilis var. *dactylocarpa* Becc. ex Martelli, Nuovo Giorn. Bot. Ital. 21: 412 (1889).

Chamaerops bilaminata Gentil, Pl. Cult. Serres Jard. Bot. Brux.: 52 (1907), nom. inval.

Chamaerops depressa Chabaud, Palmiers: 54 (1915).

Chamaerops humilis var. *decipiens* Becc., Webbia 5(1): 63 (1920).

Chamaerops humilis var. *hystrix* Becc., Webbia 5(1): 64 (1920).

Chamaerops humilis var. *lusitanica* Becc., Webbia 5(1): 63 (1920).

Chamaerops humilis var. *sardoa* Becc., Webbia 5(1): 63 (1920).

Chamaerops humilis var. *sicula* Becc., Webbia 5(1): 64 (1920).

Chamaerops humilis f. *inermis* Regel ex Becc., Webbia 5: 63 (1920).

Chamaerops humilis f. *mitis* Maire & Weiller, in Fl. Afr. Nord 4: 198 (1957).

Chamaerops humilis var. *arborescens* Pers. = **Chamaerops humilis** var. **humilis**

Chamaerops humilis var. *cerifera* Becc. = **Chamaerops humilis** var. **argentea** André

Chamaerops humilis var. *dactylocarpa* Becc. ex Martelli = **Chamaerops humilis** var. **humilis**

Chamaerops humilis var. *decipiens* Becc. = **Chamaerops humilis** var. **humilis**

Chamaerops humilis var. *depressa* Mart. = **Chamaerops humilis** var. **humilis**

Chamaerops humilis var. *elata* Mart. = **Chamaerops humilis** var. **humilis**

Chamaerops humilis var. *hystrix* Becc. = **Chamaerops humilis** var. **humilis**

Chamaerops humilis f. *inermis* Regel ex Becc. = **Chamaerops humilis** L.

Chamaerops humilis var. *lusitanica* Becc. = **Chamaerops humilis** var. **humilis**

Chamaerops humilis var. *macrocarpa* (Tineo) Becc. = **Chamaerops humilis** var. **humilis**

Chamaerops humilis f. *mitis* Maire & Weiller = **Chamaerops humilis** L.

Chamaerops humilis var. *sardoa* Becc. = **Chamaerops humilis** var. **humilis**

Chamaerops humilis var. *sicula* Becc. = **Chamaerops humilis** var. **humilis**

Chamaerops hystrix (Frazer ex Thouin) Pursh = **Rhapidophyllum hystrix** (Frazer ex Thouin) H.Wendl. & Drude

Chamaerops khasyana Griff. = **Trachycarpus martianus** (Wall. ex Mart.) H.Wendl.

Chamaerops kwanwortsik Siebold ex H.Wendl. = **Rhapis excelsa** (Thunb.) Henry

Chamaerops louisiana Darby = **Sabal minor** (Jacq.) Pers.

Chamaerops macrocarpa Tineo = **Chamaerops humilis** var. **humilis**

Chamaerops macrocarpa Linden = **?**

Chamaerops martiana Wall. ex Mart. = **Trachycarpus martianus** (Wall. ex Mart.) H.Wendl.

Chamaerops mitis J.Mey. = **?**

Chamaerops mocinoi Kunth = **Cryosophila nana** (Kunth) Blume

Chamaerops nana (Becc.) Chabaud = **Trachycarpus nanus** Becc.

Chamaerops nepalensis Lodd. ex Schult. & Schult.f. = **Trachycarpus martianus** (Wall. ex Mart.) H.Wendl.

Chamaerops palmetto (Walter) Michx. = **Sabal palmetto** (Walter) Lodd. ex Schult. & Schult.f.

Chamaerops ritchiana Griff. = **Nannorrhops ritchiana** (Griff.) Aitch.

Chamaerops sabaloides Baldwin ex Darl. = **Sabal minor** (Jacq.) Pers.

Chamaerops serrulata Michx. = **Serenoa repens** (W.Bartram) Small

Chamaerops sirotsik H.Wendl. = **Rhapis humilis** Blume

Chamaerops stauracantha Heynh. = **Cryosophila stauracantha** (Heynh.) R.J.Evans

Chamaerops tomentosa C.Morren = **Trachycarpus martianus** (Wall. ex Mart.) H.Wendl.

Chamaethrinax

Chamaethrinax H.Wendl. ex R.Pfister = **Trithrinax** Mart.

Chamaethrinax hookeriana H.Wendl. ex R.Pfister = **Trithrinax campestris** (Burmeist.) Drude & Griseb.

Chambeyronia

Chambeyronia Vieill., Bull. Soc. Linn. Normandie, II, 6: 229 (1872 publ. 1873).
2 species, New Caledonia. 60.

Chambeyronia hookeri Becc. = **Chambeyronia macrocarpa** (Brongn.) Vieill. ex Becc.

Chambeyronia lepidota H.E.Moore, Gentes Herb. 11: 291 (1978).
NE. New Caledonia. 60 NWC.

Chambeyronia macrocarpa (Brongn.) Vieill. ex Becc., Palme Nuova Caledonia: 13 (1920).
New Caledonia. 60 NWC.
  Kentia macrocarpa Vieill. ex Brongn., Compt. Rend. Hebd. Séances Acad. Sci. 77: 398 (1873).
  *Kentiopsis macrocarpa Brongn., Compt. Rend. Hebd. Séances Acad. Sci. 77: 398 (1873). *Cyphokentia macrocarpa* (Brongn.) auct., Gard. Chron. 1878(1): 440 (1878).
  Kentia rubricaulis Linden ex Salomon, Gard. Chron. 1876(1): 603 (1876).
  Kentia lindenii Linden ex André, Ill. Hort. 24: t. 276 (1877).
  Kentia lucianii Linden ex Rodigas, Gard. Chron., n.s., 9: 440 (1878). *Kentiopsis lucianii* (Linden ex Rodigas) Rodigas, Ill. Hort. 29: t. 451 (1882).
  Chambeyronia hookeri Becc., Webbia 5: 81, 85 (1921).

Chelyocarpus

Chelyocarpus Dammer, Notizbl. Bot. Gart. Berlin-Dahlem 7: 395 (1920).
4 species, W. South America to N. Brazil. 83 84.
  Tessmanniophoenix Burret, Notizbl. Bot. Gart. Berlin-Dahlem 10: 397 (1928).
  Tessmanniodoxa Burret, Notizbl. Bot. Gart. Berlin-Dahlem 15: 336 (1941).

Chelyocarpus chuco (Mart.) H.E.Moore, Principes 16: 73 (1972).
Brazil (Acre, Rondônia), Bolivia. 83 BOL 84 BZN.
Thrinax chuco Mart. in A.D.d'Orbigny, Voy. Amér. Mér. 7(3): 45 (1844). *Trithrinax chuco* (Mart.) Walp., Ann. Bot. Syst. 1: 1005 (1849). *Acanthorrhiza chuco* (Mart.) Drude in C.F.P.von Martius & auct. suc. (eds.), Fl. Bras. 3(2): 554 (1882). *Tessmanniophoenix chuco* (Mart.) Burret, Notizbl. Bot. Gart. Berlin-Dahlem 10: 400 (1928). *Tessmanniodoxa chuco* (Mart.) Burret, Notizbl. Bot. Gart. Berlin-Dahlem 15: 337 (1941).

Chelyocarpus dianeurus (Burret) H.E.Moore, Principes 16: 74 (1972).
Colombia. 83 CLM.
Tessmanniophoenix dianeura Burret, Notizbl. Bot. Gart. Berlin-Dahlem 11: 499 (1932).

Chelyocarpus repens F.Kahn & K.Mejia, Principes 32: 69 (1988).
Peru (Loreto). 83 PER.

Chelyocarpus ulei Dammer, Notizbl. Bot. Gart. Berlin-Dahlem 7: 395 (1920).
W. South America to Brazil (Acre). 83 CLM ECU PER 84 BZN.
Tessmanniophoenix longibracteata Burret, Notizbl. Bot. Gart. Berlin-Dahlem 10: 398 (1928).

Chelyocarpus wallisii (H.Wendl.) Burret = **Acanthorrhiza wallisii**

Chrysalidocarpus

Chrysalidocarpus H.Wendl. = **Dypsis** Noronha ex Mart.

Chrysalidocarpus acuminum Jum. = **Dypsis acuminum** (Jum.) Beentje & J.Dransf.

Chrysalidocarpus ambolo Jum. = **Dypsis linearis** Jum.

Chrysalidocarpus ankaizinensis Jum. = **Dypsis ankaizinensis** (Jum.) Beentje & J.Dransf.

Chrysalidocarpus arenarum Jum. = **Dypsis arenarum** (Jum.) Beentje & J.Dransf.

Chrysalidocarpus auriculatus Jum. = **Dypsis perrieri** (Jum.) Beentje & J.Dransf.

Chrysalidocarpus baronii Becc. = **Dypsis baronii** (Becc.) Beentje & J.Dransf.

Chrysalidocarpus baronii var. *littoralis* Jum. & H.Perrier = **Dypsis lutescens** (H.Wendl.) Beentje & J.Dransf.

Chrysalidocarpus brevinodis H.Perrier = **Dypsis onilahensis** (Jum. & H.Perrier) Beentje & J.Dransf.

Chrysalidocarpus cabadae H.E.Moore = **Dypsis cabadae** (H.E.Moore) Beentje & J.Dransf.

Chrysalidocarpus canescens Jum. & H.Perrier = **Dypsis canescens** (Jum. & H.Perrier) Beentje & J.Dransf.

Chrysalidocarpus decipiens Becc. = **Dypsis decipiens** (Becc.) Beentje & J.Dransf.

Chrysalidocarpus fibrosus Jum. = **Dypsis mananjarensis** (Jum. & H.Perrier) Beentje & J.Dransf.

Chrysalidocarpus glaucescens Waby = **Dypsis lutescens** (H.Wendl.) Beentje & J.Dransf.

Chrysalidocarpus humblotiana (Baill.) Becc. = **Dypsis humblotiana** (Baill.) Beentje & J.Dransf.

Chrysalidocarpus lanceolatus Becc. = **Dypsis lanceolata** (Becc.) Beentje & J.Dransf.

Chrysalidocarpus lucubensis Becc. = **Dypsis madagascariensis**

Chrysalidocarpus lutescens H.Wendl. = **Dypsis lutescens** (H.Wendl.) Beentje & J.Dransf.

Chrysalidocarpus madagascariensis Becc. = **Dypsis madagascariensis**

Chrysalidocarpus madagascariensis var. *lucumbensis* (Becc.) Jum. = **Dypsis madagascariensis**

Chrysalidocarpus madagascariensis f. *oleraceus* (Jum. & H.Perrier) Jum. = **Dypsis madagascariensis**

Chrysalidocarpus madagascariensis var. *oleraceus* (Jum. & H.Perrier) Jum. = **Dypsis madagascariensis**

Chrysalidocarpus mananjarensis Jum. & H.Perrier = **Dypsis mananjarensis** (Jum. & H.Perrier) Beentje & J.Dransf.

Chrysalidocarpus midongensis Jum. = **Dypsis onilahensis** (Jum. & H.Perrier) Beentje & J.Dransf.

Chrysalidocarpus nossibensis Becc. = **Dypsis nossibensis** (Becc.) Beentje & J.Dransf.

Chrysalidocarpus oleraceus Jum. & H.Perrier = **Dypsis madagascariensis**

Chrysalidocarpus oligostachyus Becc. = **Dypsis boiviniana** Baill.

Chrysalidocarpus onilahensis Jum. & H.Perrier = **Dypsis onilahensis** (Jum. & H.Perrier) Beentje & J.Dransf.

Chrysalidocarpus paucifolius Jum. = **Dypsis piluliferus**

Chrysalidocarpus pembanus H.E.Moore = **Dypsis pembana** (H.E.Moore) Beentje & J.Dransf.

Chrysalidocarpus piluliferus Becc. = **Dypsis pilulifera** (Becc.) Beentje & J.Dransf.

Chrysalidocarpus propinquus Jum. = **Dypsis baronii** (Becc.) Beentje & J.Dransf.

Chrysalidocarpus rivularis Jum. & H.Perrier = **Dypsis rivularis** (Jum. & H.Perrier) Beentje & J.Dransf.

Chrysalidocarpus ruber Jum. = **Dypsis perrieri** (Jum.) Beentje & J.Dransf.

Chrysalidocarpus sahanofensis (Jum. & H.Perrier) Jum. = **Dypsis sahanofensis** (Jum. & H.Perrier) Beentje & J.Dransf.

Chrysalidocarpus sambiranensis (Jum. & H.Perrier) Jum. = **Dypsis pinnatifrons** Mart.

Chrysallidosperma

Chrysallidosperma H.E.Moore = **Syagrus** Mart.

Chrysallidosperma smithii H.E.Moore = **Syagrus smithii** (H.E.Moore) Glassman

Chuniophoenix

Chuniophoenix Burret, Notizbl. Bot. Gart. Berlin-Dahlem 13: 580 (1937).
2 species, Indo-China, Hainan, S. China. 36 41.

Chuniophoenix hainanensis Burret, Notizbl. Bot. Gart. Berlin-Dahlem 13: 583 (1937).
SE. China to Hainan. 36 CHH CHS.

Chuniophoenix humilis C.Z.Tang & T.L.Wu = **Chuniophoenix nana** Burret

Chuniophoenix nana Burret, Notizbl. Bot. Gart. Berlin-Dahlem 15: 97 (1940).
Hainan to Vietnam. 36 CHH 41 VIE.
Chuniophoenix humilis C.Z.Tang & T.L.Wu, Acta Phytotax. Sin. 15: 111 (1977).

Cladandra

Cladandra O.F.Cook = **Chamaedorea** Willd.

Cladandra pygmaea (H.Wendl.) O.F.Cook = **Chamaedorea pygmaea** H.Wendl.

Cladosperma

Cladosperma Griff. = **Pinanga** Blume

Cleophora

Cleophora Gaertn. = **Latania** Comm. ex Juss.

Cleophora commersonii (J.F.Gmel.) O.F.Cook = **Latania lontaroides** (Gaertn.) H.E.Moore

Cleophora dendriformis Lodd. ex Baker = **Latania loddigesii** Mart.

Cleophora loddigesii (Mart.) O.F.Cook = **Latania loddigesii** Mart.

Cleophora lontaroides Gaertn. = **Latania lontaroides** (Gaertn.) H.E.Moore

Cleophora verschaffieltii (Lem.) O.F.Cook = **Latania verschaffieltii**

Clinosperma

Clinosperma Becc., Palme Nuova Caledonia: 52 (1920).
1 species, New Caledonia. 60.

Clinosperma bracteale (Brongn.) Becc., Palme Nuova Caledonia: 52 (1920).
C. & S. New Caledonia. 60 NWC.
Cyphokentia bractealis Brongn., Compt. Rend. Hebd. Séances Acad. Sci. 77: 400 (1873). *Clinostigma bracteale* (Brongn.) Becc., Malesia 1: 40 (1877).

Clinostigma

Clinostigma H.Wendl., Bonplandia 10: 196 (1862).
11 species, Ogasawara-shoto, Bismarck Arch. to W. Pacific. 38 43 60 62.
Exorrhiza Becc., Ann. Jard. Bot. Buitenzorg 2: 128 (1885).
Bentinickiopsis Becc., Webbia 5: 113 (1921).
Clinostigmopsis Becc., Atti Soc. Tosc. Sci. Nat. Pisa Processi Verbali 44: 161 (1934).

Clinostigma billardieri Becc. = **Basselinia gracilis** (Brongn. & Gris) Vieill.

Clinostigma bracteale (Brongn.) Becc. = **Clinosperma bracteale** (Brongn.) Becc.

Clinostigma carolinense (Becc.) H.E.Moore & Fosberg, Gentes Herb. 8: 462 (1956).
Caroline Is. (Chuuk). 62 CRL.
Cyphokentia carolinensis Becc., Bot. Jahrb. Syst. 52: 4 (1914). *Bentinickiopsis carolinensis* (Becc.) Becc., Webbia 5: 113 (1921). *Exorrhiza carolinensis* (Becc.) Burret, Repert. Spec. Nov. Regni Veg. 24: 296 (1928). *Cyphophoenix carolinensis* (Becc.) Kaneh. & Hatus., J. Dept. Agric. Kyushu Imp. Univ. 4: 432 (1935).

Clinostigma collegarum J.Dransf., Principes 26: 73 (1982).
Bismarck Arch. 43 BIS.

Clinostigma deplanchei (Brongn. & Gris) Becc. = **Basselinia deplanchei** (Brongn. & Gris) Vieill.

Clinostigma eriostachys Becc. = **Basselinia gracilis** (Brongn. & Gris) Vieill.

Clinostigma exorrhizum (H.Wendl.) Becc., Nuovo

Giorn. Bot. Ital., n.s., 42: 53 (1935).
Fiji. 60 FIJ.
Kentia exorrhiza H.Wendl., Bonplandia 10: 191 (1862). *Exorrhiza wendlandiana* Becc., Ann. Jard. Bot. Buitenzorg 2: 128 (1885).
Clinostigma thurstonii Becc., Webbia 3: 145 (1910). *Clinostigmopsis thurstonii* (Becc.) Becc., Atti Soc. Tosc. Sci. Nat. Pisa Processi Verbali 44: 162 (1934). *Exorrhiza thurstonii* (Becc.) Burret, Notizbl. Bot. Gart. Berlin-Dahlem 12: 593 (1935).
Clinostigma seemannii Becc., Nuovo Giorn. Bot. Ital., n.s., 42: 53 (1935).
Exorrhiza smithii Burret, Occas. Pap. Bernice Pauahi Bishop Mus. 11(4): 3 (1935). *Clinostigma smithii* (Burret) H.E.Moore & Fosberg, Gentes Herb. 8: 462 (1956).

Clinostigma gracile (Brongn. & Gris) Becc. = **Basselinia gracilis** (Brongn. & Gris) Vieill.

Clinostigma gronophyllum H.E.Moore, Principes 13: 71 (1969).
Solomon Is. 43 SOL.

Clinostigma haerestigma H.E.Moore, Principes 13: 73 (1969).
Solomon Is. 43 SOL.

Clinostigma harlandii Becc., Webbia 3: 150 (1910). *Clinostigmopsis harlandii* (Becc.) Becc., Atti Soc. Tosc. Sci. Nat. Pisa Processi Verbali 44: 163 (1934). *Exorrhiza harlandii* (Becc.) Burret, Notizbl. Bot. Gart. Berlin-Dahlem 12: 593 (1935).
Vanuatu. 60 VAN.

Clinostigma humboldtianum (Brongn.) Becc. = **Basselinia humboldtiana** (Brongn.) H.E.Moore

Clinostigma macrostachyum (Brongn.) Becc. = **Cyphokentia macrostachya** Brongn.

Clinostigma mooreanum (F.Muell.) H.Wendl. & Drude = **Lepidorrhachis mooreana** (F.Muell.) O.F.Cook

Clinostigma onchorhynchum Becc., Webbia 4: 284 (1914). *Exorrhiza onchorhyncha* (Becc.) Burret, Repert. Spec. Nov. Regni Veg. 24: 293 (1928).
Samoa. 60 SAM.
Lepidorrhachis onchorhyncha Becc. ex Martelli, Nuovo Giorn. Bot. Ital., n.s., 42: 56 (1935), pro syn.

Clinostigma pancheri (Brongn. & Gris) Becc. = **Basselinia pancheri** (Brongn. & Gris) Vieill.

Clinostigma ponapense (Becc.) H.E.Moore & Fosberg, Gentes Herb. 8: 463 (1956).
Caroline Is. (Pohnpei). 62 CRL.
Bentnickiopsis ponapensis Becc., Webbia 5: 113 (1921). *Exorrhiza ponapensis* (Becc.) Burret, Repert. Spec. Nov. Regni Veg. 24: 296 (1928).

Clinostigma powellianum Becc. = **Clinostigma samoense** H.Wendl.

Clinostigma robustum (Brongn.) Becc. = **Cyphokentia macrostachya** Brongn.

Clinostigma samoense H.Wendl., Bonplandia 10: 196 (1862). *Cyphokentia samoensis* (H.Wendl.) Warb., Bot. Jahrb. Syst. 25: 588 (1898).
Samoa. 60 SAM.
Clinostigma powellianum Becc., Webbia 4: 286 (1914).

Clinostigma savaiiense Christoph., Bernice P. Bishop Mus. Bull. 128: 28 (1935).
Samoa. 60 SAM.
Clinostigma warburgii Becc., Atti Soc. Tosc. Sci. Nat. Pisa Processi Verbali 44: 155 (1934).

Clinostigma vaupelii Burret, Notizbl. Bot. Gart. Berlin-Dahlem 12: 593 (1935).
Exorrhiza vaupelii Burret, Occas. Pap. Bernice Pauahi Bishop Mus. 11(4): 4 (1935).

Clinostigma savoryanum (Rehder & E.H.Wilson) H.E.Moore & Fosberg, Gentes Herb. 8: 465 (1956). Ogasawara-shoto. 38 OGA.
*Cyphokentia savoryana Rehder & E.H.Wilson, J. Arnold Arbor. 1: 115 (1919). *Bentnickiopsis savoryana* (Rehder & E.H.Wilson) Becc., Webbia 5: 113 (1921). *Exorrhiza savoryana* (Rehder & E.H.Wilson) Burret, Repert. Spec. Nov. Regni Veg. 24: 296 (1928).

Clinostigma seemannii Becc. = **Clinostigma exorrhizum** (H.Wendl.) Becc.

Clinostigma smithii (Burret) H.E.Moore & Fosberg = **Clinostigma exorrhizum** (H.Wendl.) Becc.

Clinostigma surculosum (Brongn.) Becc. = **Basselinia deplanchei** (Brongn. & Gris) Vieill.

Clinostigma thurstonii Becc. = **Clinostigma exorrhizum** (H.Wendl.) Becc.

Clinostigma vaginatum (Brongn.) Becc. = **Brongniartikentia vaginata** (Brongn.) Becc.

Clinostigma vaupelii Burret = **Clinostigma savaiiense** Christoph.

Clinostigma warburgii Becc. = **Clinostigma savaiiense** Christoph.

Clinostigmopsis

Clinostigmopsis Becc. = **Clinostigma** H.Wendl.

Clinostigmopsis harlandii (Becc.) Becc. = **Clinostigma harlandii** Becc.

Clinostigmopsis thurstonii (Becc.) Becc. = **Clinostigma exorrhizum** (H.Wendl.) Becc.

Coccos

Coccos Gaertn. = **Cocos** L.

Coccus

Coccus Mill. = **Cocos** L.

Cocops

Cocops O.F.Cook = **Calyptronoma** Griseb.

Cocops rivalis O.F.Cook = **Calyptronoma rivalis** (O.F.Cook) L.H.Bailey

Coccothrinax

Coccothrinax Sarg., Bot. Gaz. 27: 87 (1899).
50 species, Florida, Mexico, Caribbean, Colombia. 78 79 81 83.
Thrincoma O.F.Cook, Bull. Torrey Bot. Club 28: 539 28 (1901).
Thringis O.F.Cook, Bull. Torrey Bot. Club 28: 544 (1901).
Antia O.F.Cook, Natl. Hort. Mag. 20: 34 (1941), no latin descr.
Beata O.F.Cook, Natl. Hort. Mag. 20: 52 (1941), nom. inval.
Pithodes O.F.Cook, Natl. Hort. Mag. 20: 52 (1941), no latin descr.
Haitiella L.H.Bailey, Contr. Gray Herb., n.s., 165: 7 (1947).

Coccothrinax acunana Léon, Mem. Soc. Cub. Hist. Nat. "Felipe Poey" 13: 128 (1939). Cuba (Pico Turquino). 81 CUB.

Coccothrinax alexandri Léon, Mem. Soc. Cub. Hist. Nat. "Felipe Poey" 13: 122 (1939). E. Cuba. 81 CUB.

subsp. **alexandri**
E. Cuba. 81 CUB.

subsp. **nitida** (Léon) Borhidi & O.Muñiz, Bot. Közlem. 58: 175 (1971).
E. Cuba. 81 CUB.
*Coccothrinax alexandri var. *nitida* Léon, Mem. Soc. Cub. Hist. Nat. "Felipe Poey" 13: 123 (1939).

Coccothrinax alexandri var. *nitida* Léon = **Coccothrinax alexandri** subsp. **nitida** (Léon) Borhidi & O.Muñiz

Coccothrinax alta (O.F.Cook) Becc. = **Coccothrinax barbadensis** (Lodd. ex Mart.) Becc.

Coccothrinax anomala Becc. = **Zombia antillarum** (Desc.) L.H.Bailey

Coccothrinax argentata (Jacq.) L.H.Bailey, Gentes Herb. 4: 223 (1939).
S. Florida, SE. Mexico, Caribbean, Colombia. 78 FLA 81 BAH SWC 83 CLM.
*Palma argentata Jacq., Fragm. Bot.: 38 (1804)
Thrinax garberi Chapm., Bot. Gaz. 3: 12 (1878). *Coccothrinax garberi* (Chapm.) Sarg., Bot. Gaz. 27: 90 (1899).
Coccothrinax jucunda Sarg., Bot. Gaz. 27: 89 (1899).
Coccothrinax jucunda var. *macrosperma* Becc., Webbia 2: 312 (1907).
Coccothrinax jucunda var. *marquesensis* Becc., Webbia 2: 313 (1907).
Thrinax altissima N.Taylor in L.H.Bailey, Stand. Cycl. Hort. 6: 3334 (1917).

Coccothrinax argentea (Lodd. ex Schult. & Schult.f.) Sarg. ex Becc., Just's Bot. Jahresber. 27(1): 469 (1901).
*Thrinax argentea Lodd. ex Schult. & Schult.f., Syst. Veg. 7: 1300 (1830). *Acanthorrhiza argentea* (Lodd. ex Schult. & Schult.f.) O.F.Cook, Natl. Hort. Mag. 20: 50 (1941).
Thrinax multiflora Mart., Hist. Nat. Palm. 3: t. 103, f. 1 (1853).
Thrinax graminifolia H.Wendl., Index Palm.: 39 (1854).
Thrinax longistyla Becc. in I.Urban, Symb. Antill. 7: 170 (1912).

Coccothrinax argentea subsp. *guantanamensis* (Léon) Borhidi & O.Muñiz = **Coccothrinax guantanamensis** (Léon) O.Muñiz & Borhidi

Coccothrinax argentea var. *guantanamensis* Léon = **Coccothrinax guantanamensis** (Léon) O.Muñiz & Borhidi

Coccothrinax australis L.H.Bailey = **Coccothrinax barbadensis** (Lodd. ex Mart.) Becc.

Coccothrinax baileyana O.F.Cook = **?**

Coccothrinax baracoensis Borhidi & O.Muñiz, Acta Bot. Acad. Sci. Hung. 27: 440 (1981). SE. Cuba. 81 CUB.

Coccothrinax barbadensis (Lodd. ex Mart.) Becc., Webbia 2: 328 (1908).
Puerto Rico to Trinidad. 81 LEE PUE TRT VNA WIN.

Thrinax parviflora Maycock, Fl. Barbad.: 146 (1830), nom. illeg.

*Thrinax barbadensis Lodd. ex Mart., Hist. Nat. Palm. 3: 320 (1853). *Copernicia barbadensis* (Lodd. ex Mart.) H.Wendl. in O.C.E.de Kerchove de Denterghem, Palmiers: 241 (1878).

Thrincoma alta O.F.Cook, Bull. Torrey Bot. Club 28: 540 (1901). *Coccothrinax alta* (O.F.Cook) Becc., Webbia 2: 331 (1908).

Thringis latifrons O.F.Cook, Bull. Torrey Bot. Club 28: 545 (1901). *Coccothrinax latifrons* (O.F.Cook) Becc., Webbia 2: 326 (1908).

Thringis laxa O.F.Cook, Bull. Torrey Bot. Club 28: 545 (1901). *Coccothrinax laxa* (O.F.Cook) Becc., Webbia 2: 333 (1908).

Coccothrinax eggersiana Becc., Webbia 2: 321 (1908).

Coccothrinax eggersiana var. *sanctaecrucis* Becc., Webbia 2: 323 (1908).

Coccothrinax martinicaensis Becc., Webbia 2: 324 (1908).

Coccothrinax sanctae-thomae Becc., Webbia 2: 303 (1908).

Coccothrinax australis L.H.Bailey, Gentes Herb. 7: 365 (1947).

Coccothrinax boxii L.H.Bailey, Gentes Herb. 8: 113 (1949).

Coccothrinax discreta L.H.Bailey, Gentes Herb. 8: 104 (1949).

Coccothrinax dussiana L.H.Bailey, Gentes Herb. 8: 109 (1949).

Coccothrinax sabana L.H.Bailey, Gentes Herb. 8: 110 (1949).

Coccothrinax bermudezii Léon, Mem. Soc. Cub. Hist. Nat. "Felipe Poey" 13: 124 (1939).
SE. Cuba. 81 CUB.

Coccothrinax borhidiana O.Muñiz, Acta Agron. Acad. Sci. Hung. 27: 437 (1978).
Cuba (Matanzas). 81 CUB.

Coccothrinax boschiana M.M.Mejía & R.García, Moscosoa 9: 1 (1997).
Dominican Rep. 81 DOM.

Coccothrinax boxii L.H.Bailey = **Coccothrinax barbadensis** (Lodd. ex Mart.) Becc.

Coccothrinax camagueyana Borhidi & O.Muñiz, Acta Bot. Acad. Sci. Hung. 27: 441 (1981).
EC. Cuba. 81 CUB.

Coccothrinax clarensis Léon, Mem. Soc. Cub. Hist. Nat. "Felipe Poey" 13: 147 (1939).
E. Cuba. 81 CUB.

subsp. **brevifolia** (Léon) Borhidi & O.Muñiz, Bot. Közlem. 58: 176 (1971).
E. Cuba. 81 CUB.
*Coccothrinax clarensis var. *brevifolia* Léon, Mem. Soc. Cub. Hist. Nat. "Felipe Poey" 13: 148 (1939).

subsp. **clarensis**
E. Cuba. 81 CUB.

subsp. **perrigida** (Léon) Borhidi & O.Muñiz, Bot. Közlem. 58: 176 (1971).
E. Cuba. 81 CUB.
*Coccothrinax clarensis var. *perrigida* Léon, Mem. Soc. Cub. Hist. Nat. "Felipe Poey" 13: 149 (1939).

Coccothrinax clarensis var. *brevifolia* Léon = **Coccothrinax clarensis** subsp. **brevifolia** (Léon) Borhidi & O.Muñiz

Coccothrinax clarensis var. *perrigida* Léon = **Coccothrinax clarensis** subsp. **perrigida** (Léon) Borhidi & O.Muñiz

Coccothrinax concolor Burret, Kongl. Svenska Vetenskapsakad. Handl., III, 6(7): 13 (1929).
Haiti. 81 HAI.

Coccothrinax crinita (Griseb. & H.Wendl. ex C.H.Wright) Becc., Webbia 2: 334 (1908). *Antia crinita* (Becc.) O.F.Cook, Natl. Hort. Mag. 20: 50 (1941).
Cuba. 81 CUB.
*Thrinax crinita Griseb. & H.Wendl. ex C.H.Wright in O.C.E.de Kerchove de Denterghem, Palmiers: 258 (1878).

subsp. **brevicrinis** Borhidi & O.Muñiz, Acta Bot. Acad. Sci. Hung. 27: 448 (1981).
C. Cuba. 81 CUB.

subsp. **crinita**
W. Cuba. 81 CUB.

Coccothrinax cupularis (Léon) O.Muñiz & Borhidi, Acta Bot. Acad. Sci. Hung. 27: 449 (1981).
S. Cuba. 81 CUB.
*Coccothrinax miraguama var. *cupularis* Léon, Mem. Soc. Cub. Hist. Nat. "Felipe Poey" 13: 117 (1939). *Coccothrinax miraguama* subsp. *cupularis* (Léon) Borhidi & O.Muñiz, Bot. Közlem. 58: 175 (1971).

Coccothrinax discreta L.H.Bailey = **Coccothrinax barbadensis** (Lodd. ex Mart.) Becc.

Coccothrinax dussiana L.H.Bailey = **Coccothrinax barbadensis** (Lodd. ex Mart.) Becc.

Coccothrinax eggersiana Becc. = **Coccothrinax barbadensis** (Lodd. ex Mart.) Becc.

Coccothrinax eggersiana var. *sanctaecrucis* Becc. = **Coccothrinax barbadensis** (Lodd. ex Mart.) Becc.

Coccothrinax ekmanii Burret, Kongl. Svenska Vetenskapsakad. Handl., III, 6(7): 11 (1929). *Haitiella ekmanii* (Burret) L.H.Bailey, Contr. Gray Herb., n.s., 165: 7 (1947).
Hispaniola. 81 DOM HAI.

Coccothrinax elegans O.Muñiz & Borhidi, Acta Bot. Acad. Sci. Hung. 27: 442 (1981).
Cuba. 81 CUB.

Coccothrinax fagildei Borhidi & O.Muñiz, Acta Bot. Hung. 31: 227 (1985).
Cuba. 81 CUB.

Coccothrinax fragrans Burret, Kongl. Svenska Vetenskapsakad. Handl., III, 6(7): 15 (1929).
E. Cuba to Hispaniola. 81 CUB HAI.

Coccothrinax garberi (Chapm.) Sarg. = **Coccothrinax argentata** (Jacq.) L.H.Bailey

Coccothrinax garciana Léon, Mem. Soc. Cub. Hist. Nat. "Felipe Poey" 13: 143 (1939).
Cuba (Holguín). 81 CUB.

Coccothrinax gracilis Burret, Kongl. Svenska Vetenskapsakad. Handl., III, 6(7): 14 (1929).
Hispaniola. 81 DOM HAI.

Coccothrinax guantanamensis (Léon) O.Muñiz & Borhidi, Acta Bot. Acad. Sci. Hung. 27: 449 (1981).
E. Cuba. 81 CUB.
*Coccothrinax argentea var. *guantanamensis* Léon, Mem. Soc. Cub. Hist. Nat. "Felipe Poey" 13: 134 (1939). *Coccothrinax argentea* subsp. *guantanamensis* (Léon) Borhidi & O.Muñiz, Bot. Közlem. 58: 176 (1971).

Coccothrinax gundlachii Léon, Mem. Soc. Cub. Hist. Nat. "Felipe Poey" 13: 149 (1939).
C. & E. Cuba. 81 CUB.

Coccothrinax hioramii Léon, Mem. Soc. Cub. Hist. Nat. "Felipe Poey" 13: 135 (1939).
E. Cuba. 81 CUB.

Coccothrinax inaguensis Read, Principes 10: 30 (1966).
Bahamas. 81 BAH.

Coccothrinax jamaicensis Read, Principes 10: 133 (1966).
Jamaica. 81 JAM.

Coccothrinax jucunda Sarg. = **Coccothrinax argentata** (Jacq.) L.H.Bailey

Coccothrinax jucunda var. *macrosperma* Becc. = **Coccothrinax argentata** (Jacq.) L.H.Bailey

Coccothrinax jucunda var. *marquesensis* Becc. = **Coccothrinax argentata** (Jacq.) L.H.Bailey

Coccothrinax latifrons (O.F.Cook) Becc. = **Coccothrinax barbadensis** (Lodd. ex Mart.) Becc.

Coccothrinax laxa (O.F.Cook) Becc. = **Coccothrinax barbadensis** (Lodd. ex Mart.) Becc.

Coccothrinax leonis O.Muñiz & Borhidi, Acta Bot. Acad. Sci. Hung. 27: 443 (1981).
Cuba. 81 CUB.

Coccothrinax littoralis Léon, Mem. Soc. Cub. Hist. Nat. "Felipe Poey" 13: 138 (1939).
Cuba. 81 CUB.

Coccothrinax macroglossa (Léon) O.Muñiz & Borhidi, Acta Bot. Acad. Sci. Hung. 27: 450 (1981).
E. Cuba. 81 CUB.

> *Coccothrinax miraguama* var. *macroglossa* Léon, Mem. Soc. Cub. Hist. Nat. "Felipe Poey" 13: 118 (1939). *Coccothrinax miraguama* subsp. *macroglossa* (Léon) Borhidi & O.Muñiz, Bot. Közlem. 58: 175 (1971).

Coccothrinax martii (Griseb.) Becc. = **Thrinax radiata** Lodd. ex Schult. & Schult.f.

Coccothrinax martinicaensis Becc. = **Coccothrinax barbadensis** (Lodd. ex Mart.) Becc.

Coccothrinax microphylla Borhidi & O.Muñiz, Acta Bot. Acad. Sci. Hung. 27: 444 (1981).
E. Cuba. 81 CUB.

Coccothrinax miraguama (Kunth) Becc., Webbia 2: 295 (1908).
Cuba. 81 CUB.

> *Corypha miraguama* Kunth in F.W.H.von Humboldt, A.J.A.Bonpland & C.S.Kunth, Nov. Gen. Sp. 1: 298 (1816). *Copernicia miraguama* (Kunth) Mart., Hist. Nat. Palm. 3: 243 (1849). *Thrinax miraguama* (Kunth) Mart., Hist. Nat. Palm. 3: 320 (1853).

subsp. **arenicola** (Léon) Borhidi & O.Muñiz, Bot. Közlem. 58: 175 (1971).
W. Cuba. 81 CUB.

> *Coccothrinax miraguama* var. *arenicola* Léon, Mem. Soc. Cub. Hist. Nat. "Felipe Poey" 13: 114 (1939).

subsp. **havanensis** (Léon) Borhidi & O.Muñiz, Bot. Közlem. 58: 175 (1971).
W. Cuba. 81 CUB.

> *Coccothrinax miraguama* var. *havanensis* Léon, Mem. Soc. Cub. Hist. Nat. "Felipe Poey" 13: 116 (1939).

subsp. **miraguama**
Cuba. 81 CUB.

> *Thrinax stellata* Lodd. ex Mart., Hist. Nat. Palm. 3: 320 (1853).
> *Thrinax acuminata* Griseb. & H.Wendl. ex Sarg., Bot. Gaz. 27: 89 (1899). *Coccothrinax acuminata* (Griseb. & H.Wendl. ex Sarg.) Becc., Just's Bot. Jahresber. 27(1): 469 (1901).

subsp. **roseocarpa** (Léon) Borhidi & O.Muñiz, Bot. Közlem. 58: 175 (1971).
EC. Cuba. 81 CUB.

> *Coccothrinax miraguama* var. *roseocarpa* Léon, Mem. Soc. Cub. Hist. Nat. "Felipe Poey" 13: 117 (1939).

Coccothrinax miraguama var. *arenicola* Léon = **Coccothrinax miraguama** subsp. **arenicola** (Léon) Borhidi & O.Muñiz

Coccothrinax miraguama var. *cupularis* Léon = **Coccothrinax cupularis** (Léon) O.Muñiz & Borhidi

Coccothrinax miraguama subsp. *cupularis* (Léon) Borhidi & O.Muñiz = **Coccothrinax cupularis** (Léon) O.Muñiz & Borhidi

Coccothrinax miraguama var. *havanensis* Léon = **Coccothrinax miraguama** subsp. **havanensis** (Léon) Borhidi & O.Muñiz

Coccothrinax miraguama subsp. *macroglossa* (Léon) Borhidi & O.Muñiz = **Coccothrinax macroglossa** (Léon) O.Muñiz & Borhidi

Coccothrinax miraguama var. *macroglossa* Léon = **Coccothrinax macroglossa** (Léon) O.Muñiz & Borhidi

Coccothrinax miraguama var. *roseocarpa* Léon = **Coccothrinax miraguama** subsp. **roseocarpa** (Léon) Borhidi & O.Muñiz

Coccothrinax moaensis (Borhidi & O.Muñiz) O.Muñiz, Acta Bot. Acad. Sci. Hung. 27: 451 (1981).
E. Cuba. 81 CUB.

> *Coccothrinax yuraguana* subsp. *moaensis* Borhidi & O.Muñiz, Acta Bot. Acad. Sci. Hung. 17: 1 (1971 publ. 1972).

Coccothrinax montana Burret, Kongl. Svenska Vetenskapsakad. Handl., III, 6(7): 17 (1929).
Haiti. 81 HAI.

Coccothrinax munizii Borhidi, Acta Bot. Acad. Sci. Hung. 17: 2 (1971 publ. 1972). *Haitiella munizii* (Borhidi) Borhidi, Acta Bot. Acad. Sci. Hung. 25: 2 (1979).
E. Cuba. 81 CUB.

Coccothrinax muricata Léon, Mem. Soc. Cub. Hist. Nat. "Felipe Poey" 13: 129 (1939).
EC. Cuba. 81 CUB.

Coccothrinax muricata subsp. *savannarum* (Léon) Borhidi & O.Muñiz = **Coccothrinax pauciramosa** Burret

Coccothrinax muricata var. *savannarum* Léon = **Coccothrinax pauciramosa** Burret

Coccothrinax nipensis Borhidi & O.Muñiz, Acta Bot. Acad. Sci. Hung. 27: 446 (1981).
E. Cuba. 81 CUB.

Coccothrinax orientalis (Léon) O.Muñiz & Borhidi, Acta Bot. Acad. Sci. Hung. 27: 451 (1981).
E. Cuba. 81 CUB.

> *Coccothrinax yuraguana* var. *orientalis* Léon, Mem. Soc. Cub. Hist. Nat. "Felipe Poey" 13: 119 (1939). *Coccothrinax yuraguana* subsp. *orientalis* (Léon) Borhidi, Acta Bot. Acad. Sci. Hung. 17: 2 (1971 publ. 1972).

Coccothrinax pauciramosa Burret, Kongl. Svenska Vetenskapsakad. Handl., III, 6(7): 12 (1929).
E. Cuba. 81 CUB.
Coccothrinax muricata var. *savannarum* Léon, Mem. Soc. Cub. Hist. Nat. "Felipe Poey" 13: 130 (1939).
Coccothrinax muricata subsp. *savannarum* (Léon) Borhidi & O.Muñiz, Bot. Közlem. 58: 176 (1971).
Coccothrinax savannarum (Léon) Borhidi & O.Muñiz, Acta Bot. Acad. Sci. Hung. 27: 452 (1981).

Coccothrinax proctorii Read, Phytologia 46: 285 (1980).
Cayman Is. 81 CAY.

Coccothrinax pseudorigida Léon, Mem. Soc. Cub. Hist. Nat. "Felipe Poey" 13: 145 (1939).
EC. Cuba. 81 CUB.
Coccothrinax pseudorigida var. *acaulis* Léon, Mem. Soc. Cub. Hist. Nat. "Felipe Poey" 13: 146 (1939).

Coccothrinax pseudorigida var. *acaulis* Léon = **Coccothrinax pseudorigida** Léon

Coccothrinax pumila Borhidi & J.A.Hern., Acta Bot. Hung. 38: 195 (1993-1994 publ. 1995).
Cuba. 81 CUB.

Coccothrinax radiata (Lodd. ex Schult. & Schult.f.) Sarg. = **Thrinax radiata** Lodd. ex Schult. & Schult.f.

Coccothrinax readii H.J.Quero, Principes 24: 118 (1980).
SE. Mexico. 79 MXT.

Coccothrinax rigida (Griseb. & H.Wendl.) Becc., Webbia 2: 299 (1908).
E. Cuba. 81 CUB.
Thrinax rigida Griseb. & H.Wendl., Cat. Pl. Cub.: 221 (1866).

Coccothrinax sabana L.H.Bailey = **Coccothrinax barbadensis** (Lodd. ex Mart.) Becc.

Coccothrinax salvatoris Léon, Mem. Soc. Cub. Hist. Nat. "Felipe Poey" 13: 125 (1939).
EC. & E. Cuba. 81 CUB.

subsp. **loricata** (Léon) Borhidi & O.Muñiz, Bot. Közlem. 58: 175 (1971).
E. Cuba. 81 CUB.
Coccothrinax salvatoris var. *loricata* Léon, Mem. Soc. Cub. Hist. Nat. "Felipe Poey" 13: 127 (1939).

subsp. **salvatoris**
EC. & E. Cuba. 81 CUB.

Coccothrinax salvatoris var. *loricata* Léon = **Coccothrinax salvatoris** subsp. **loricata** (Léon) Borhidi & O.Muñiz

Coccothrinax sanctae-thomae Becc. = **Coccothrinax barbadensis** (Lodd. ex Mart.) Becc.

Coccothrinax savannarum (Léon) Borhidi & O.Muñiz = **Coccothrinax pauciramosa** Burret

Coccothrinax saxicola Léon, Mem. Soc. Cub. Hist. Nat. "Felipe Poey" 13: 141 (1939).
E. Cuba. 81 CUB.

Coccothrinax scoparia Becc., Repert. Spec. Nov. Regni Veg. 6: 95 (1908).
Haiti. 81 HAI.

Coccothrinax spissa L.H.Bailey, Gentes Herb. 4: 253 (1939). *Pithodes spissa* (L.H.Bailey) O.F.Cook, Natl. Hort. Mag. 20: 52 (1941).
Hispaniola. 81 DOM HAI.

Coccothrinax trinitensis Borhidi & O.Muñiz, Acta Bot. Hung. 31: 228 (1985).
EC. Cuba. 81 CUB.

Coccothrinax victorini Léon, Mem. Soc. Cub. Hist. Nat. "Felipe Poey" 13: 139 (1939).
E. Cuba. 81 CUB.

Coccothrinax yunquensis Borhidi & O.Muñiz, Acta Bot. Acad. Sci. Hung. 27: 447 (1981).
S. Cuba. 81 CUB.

Coccothrinax yuraguana (A.Rich.) Léon, Mem. Soc. Cub. Hist. Nat. "Felipe Poey" 13: 119 (1939).
W. Cuba. 81 CUB.
Thrinax yuraguana A.Rich. in R.de la Sagra, Hist. Fis. Cuba, Bot. 11: 278 (1850).

Coccothrinax yuraguana subsp. *moaensis* Borhidi & O.Muñiz = **Coccothrinax moaensis** (Borhidi & O.Muñiz) O.Muñiz

Coccothrinax yuraguana subsp. *orientalis* (Léon) Borhidi = **Coccothrinax orientalis** (Léon) O.Muñiz & Borhidi

Coccothrinax yuraguana var. *orientalis* Léon = **Coccothrinax orientalis** (Léon) O.Muñiz & Borhidi

Cocos

Cocos L., Sp. Pl.: 1188 (1753).
1 species, C. Malesia to SW. Pacific. (29) (38) (40) (41) 42 43 50 60 (61) (62) (63) (80) (81).
Coccus Mill., Gard. Dict. Abr. ed. 4(1754).
Calappa Steck, Sagu: 9 (1757).
Coccos Gaertn., Fruct. Sem. Pl. 1: 15 (1788), orth. var.

Cocos acaulis Drude = **Syagrus comosa** (Mart.) Mart.

Cocos acaulis var. *glabra* Drude = **Syagrus comosa** (Mart.) Mart.

Cocos acicularis Sw. = **Bactris guineensis** (L.) H.E.Moore

Cocos acrocomioides Drude = **Syagrus romanzoffiana** (Cham.) Glassman

Cocos aculeata Jacq. = **Acrocomia aculeata** (Jacq.) Lodd. ex Mart.

Cocos aequatorialis Barb.Rodr. = **Syagrus inajai** (Spruce) Becc.

Cocos amadelpha Barb.Rodr. = **Butia paraguayensis** (Barb.Rodr.) L.H.Bailey

Cocos amara Jacq. = **Syagrus amara** (Jacq.) Mart.

Cocos apaensis Barb.Rodr. = **Syagrus campylospatha** (Barb.Rodr.) Becc.

Cocos arechavaletana Barb.Rodr. = **Syagrus romanzoffiana** (Cham.) Glassman

Cocos arenaria M.Gómez = **Allagoptera arenaria** (M.Gómez) Kuntze

Cocos arenicola Barb.Rodr. = **Butia paraguayensis** (Barb.Rodr.) L.H.Bailey

Cocos argentea Engel = **Syagrus sancona** (Kunth) H.Karst.

Cocos aricui Wied-Neuw. = **Syagrus schizophylla** (Mart.) Glassman

Cocos arikuryroba Barb.Rodr. = **Syagrus schizophylla** (Mart.) Glassman

Cocos australis Mart. = **Syagrus romanzoffiana** (Cham.) Glassman

Cocos australis Drude & Brandt = **Syagrus romanzoffiana** (Cham.) Glassman

Cocos barbosii Barb.Rodr. = **?**

Cocos botryophora Mart. = **Syagrus botryophora** (Mart.) Mart.

Cocos butyracea Mutis ex L.f. = **Attalea butyracea** (Mutis ex L.f.) Wess.Boer

Cocos butyrosa H.Wendl. = **Attalea humilis** Mart. ex Spreng.

Cocos campestris Mart. = **Syagrus flexuosa** (Mart.) Becc.

Cocos campicola Barb.Rodr. = **Butia campicola** (Barb.Rodr.) Noblick

Cocos × campos-portoana Bondar = **Syagrus × campos-portoana** (Bondar) Glassman

Cocos campylospatha Barb.Rodr. = **Syagrus campylospatha** (Barb.Rodr.) Becc.

Cocos capanemae (Barb.Rodr.) Drude = **Syagrus schizophylla** (Mart.) Glassman

Cocos capitata Mart. = **Butia capitata** (Mart.) Becc.

Cocos catechucarpa Barb.Rodr. = **Syagrus picrophylla** Barb.Rodr.

Cocos chavesiana Barb.Rodr. ex Becc. = **Syagrus inajai** (Spruce) Becc.

Cocos chilensis (Molina) Molina = **Jubaea chilensis** (Molina) Baill.

Cocos chiragua (H.Karst.) Becc. = **Syagrus sancona** (Kunth) H.Karst.

Cocos chloroleuca Barb.Rodr. = **?**

Cocos cocoyule Mart. = **Attalea guacuyule** (Liebm. ex Mart.) Zona

Cocos cogniauxiana Barb.Rodr. = **?**

Cocos comosa Mart. = **Syagrus comosa** (Mart.) Mart.

Cocos coronata Mart. = **Syagrus coronata** (Mart.) Becc.

Cocos crispa Kunth = **Gastrococos crispa** (Kunth) H.E.Moore

Cocos datil Drude & Griseb. = **Syagrus romanzoffiana** (Cham.) Glassman

Cocos drudei Becc. = **Syagrus cocoides** Mart.

Cocos dyeriana Barb.Rodr. = **Butia paraguayensis** (Barb.Rodr.) L.H.Bailey

Cocos edulis Barb.Rodr. = **?**

Cocos elegantissima H.Wendl. = **?**

Cocos elegantissima Chabaud = **Butia capitata** (Mart.) Becc.

Cocos equatorialis Barb.Rodr. = **Syagrus inajai** (Spruce) Becc.

Cocos eriospatha Mart. ex Drude = **Butia eriospatha** (Mart. ex Drude) Becc.

Cocos erythrospatha Chabaud = **Butia capitata** (Mart.) Becc.

Cocos flexuosa Mart. = **Syagrus flexuosa** (Mart.) Becc.

Cocos fusiformis Sw. = **Acrocomia aculeata** (Jacq.) Lodd. ex Mart.

Cocos gaertneri W.Watson = **?**

Cocos geriba Barb.Rodr. = **Syagrus romanzoffiana** (Cham.) Glassman

Cocos getuliana Bondar = **Syagrus macrocarpa** Barb.Rodr.

Cocos glazioviana Dammer = **Syagrus petraea** (Mart.) Becc.

Cocos graminifolia Drude = **Syagrus graminifolia** (Drude) Becc.

Cocos guacuyule Liebm. ex Mart. = **Attalea guacuyule** (Liebm. ex Mart.) Zona

Cocos guineensis L. = **Bactris guineensis** (L.) H.E.Moore

Cocos hassleriana Barb.Rodr. = **Syagrus campylospatha** (Barb.Rodr.) Becc.

Cocos iagua Sessé & Moç. = **?**

Cocos inajai (Spruce) Trail = **Syagrus inajai** (Spruce) Becc.

Cocos indica Royle = **Cocos nucifera** L.

Cocos insignis Mart. ex H.Wendl. = **Lytocaryum weddellianum** (H.Wendl.) Toledo

Cocos lapidea Gaertn. = **? [84]**

Cocos leiospatha Barb.Rodr. = **Butia capitata** (Mart.) Becc.

Cocos lilaceiflora Chabaud = **Butia capitata** (Mart.) Becc.

Cocos lilliputiana Barb.Rodr. = **Syagrus graminifolia** (Drude) Becc.

Cocos maldivica J.F.Gmel. = **Lodoicea maldivica** (J.F.Gmel.) Pers. ex H.Wendl.

Cocos mamillaris Blanco = **?**

Cocos maritima Comm. ex H.Wendl. = **Lodoicea maldivica** (J.F.Gmel.) Pers. ex H.Wendl.

Cocos marocarpa (Barb.Rodr.) Barb.Rodr. = **Syagrus macrocarpa** Barb.Rodr.

Cocos martiana Drude & Glaz. = **Syagrus romanzoffiana** (Cham.) Glassman

Cocos × mataforme Bondar = **Syagrus × mataforme** (Bondar) A.D.Hawkes

Cocos mikaniana Mart. = **Syagrus pseudococos** (Raddi) Glassman

Cocos naja Arruda ex Kunth = **?**

Cocos nana Griff. = **Cocos nucifera** L.

Cocos nehrlingiana Abbott ex Nehrl. = **Butia capitata** (Mart.) Becc.

Cocos nolaia-assu Wied-Neuw. = **?**

Cocos normanbyi W.Hill ex F.Muell. = **Normanbya normanbyi** (F.Muell.) L.H.Bailey

Cocos nucifera L., Sp. Pl.: 1188 (1753). *Palma cocos* Mill., Gard. Dict. ed. 8: 2 (1768), nom. illeg. *Calappa nucifera* (L.) Kuntze, Revis. Gen. Pl. 2: 982 (1891).
C. Malesia to SW. Pacific, widely introduced elsewere. (29) mdg (38) oga (40) ind srl (41) tha 42 bor jaw mly MOL PHI 43 NWG 50 QLD 60 VAN (61) mrq sci tua (62) crl mcs mrn mrs (63) haw (80) blz (81). *Cocos indica* Royle, Ill. Bot. Himal. Mts.: 395 (1840).
Cocos nana Griff., Not. Pl. Asiat. 3: 166 (1851).

Cocos nypa Lour. = **Nypa fruticans** Wurmb

Cocos odorata Barb.Rodr. = **Butia capitata** (Mart.) Becc.

Cocos oleracea Mart. = **Syagrus oleracea** (Mart.) Becc.

Cocos oleracea var. *platyphylla* Drude = **Syagrus oleracea** (Mart.) Becc.

Cocos orbignyana Becc. = **?**

Cocos orinocensis Spruce = **Syagrus orinocensis** (Spruce) Burret

Cocos paraguayensis Barb.Rodr. = **Butia paraguayensis** (Barb.Rodr.) L.H.Bailey

Cocos petraea Mart. = **Syagrus petraea** (Mart.) Becc.

Cocos picrophylla Barb.Rodr. ex Becc. = **Syagrus oleracea** (Mart.) Becc.

Cocos pityrophylla Mart. = **?**

Cocos plumosa Hook.f. = **Syagrus romanzoffiana** (Cham.) Glassman

Cocos plumosa Lodd. ex Loudon = **Syagrus comosa** (Mart.) Mart.

Cocos poni Hauman = **Butia yatay** (Mart.) Becc.

Cocos procopiana Glaz. ex Drude = **Syagrus macrocarpa** Barb.Rodr.

Cocos pulposa Barb.Rodr. = **Butia capitata** (Mart.) Becc.

Cocos purusana Huber = **?** [84 BZN]

Cocos pynaertii auct. = **Lytocaryum weddellianum** (H.Wendl.) Toledo

Cocos quinquefaria Barb.Rodr. = **Syagrus coronata** (Mart.) Becc.

Cocos regia Liebm. = **Attalea liebmannii** (Becc.) Zona

Cocos romanzoffanopulposa Barb.Rodr. = **?**

Cocos romanzoffiana Cham. = **Syagrus romanzoffiana** (Cham.) Glassman

Cocos rupestris Barb.Rodr. = **Syagrus petraea** (Mart.) Becc.

Cocos ruschiana Bondar = **Syagrus ruschiana** (Bondar) Glassman

Cocos sancona (Kunth) Hook.f. = **Syagrus sancona** (Kunth) H.Karst.

Cocos sapida Barb.Rodr. = **?**

Cocos schizophylla Mart. = **Syagrus schizophylla** (Mart.) Glassman

Cocos speciosa Barb.Rodr. = **Syagrus inajai** (Spruce) Becc.

Cocos stolonifera Barb.Rodr. = **Butia stolonifera** (Barb.Rodr.) Becc.

Cocos syagrus Drude = **Syagrus cocoides** Mart.

Cocos × tostana Bondar = **Syagrus × tostana** (Bondar) Glassman

Cocos urbaniana Dammer = **Syagrus flexuosa** (Mart.) Becc.

Cocos vagans Bondar = **Syagrus vagans** (Bondar) A.D.Hawkes

Cocos ventricosa Arruda = **?**

Cocos vinifera (Mart.) Mart. = **Pseudophoenix vinifera** (Mart.) Becc.

Cocos virgata A.Usteri = **?**

Cocos weddelliana H.Wendl. = **Lytocaryum weddellianum** (H.Wendl.) Toledo

Cocos weddellii Drude = **Syagrus cocoides** Mart.

Cocos wildemaniana Barb.Rodr. = **Butia paraguayensis** (Barb.Rodr.) L.H.Bailey

Cocos yatay Mart. = **Butia yatay** (Mart.) Becc.

Cocos yurumaguas H.Wendl. = **?**

Codda-pana

Codda-Pana Adans. = **Corypha** L.

Coelococcus

Coelococcus H.Wendl. = **Metroxylon** Rottb.

Coelococcus amicarum (H.Wendl.) W.Wight = **Metroxylon amicarum** (H.Wendl.) Hook.f.

Coelococcus carolinensis Dingler = **Metroxylon amicarum** (H.Wendl.) Hook.f.

Coelococcus salomonensis Warb. = **Metroxylon salomonense** (Warb.) Becc.

Coelococcus vitiensis H.Wendl. = **Metroxylon vitiense** (H.Wendl.) Hook.f.

Coelococcus warburgii Heimerl = **Metroxylon warburgii** (Heim) Becc.

Coleospadix

Coleospadix Becc. = **Drymophloeus** Zipp.

Coleospadix angustifolius (Blume) Burret = **Drymophloeus litigiosus** (Becc.) H.E.Moore

Coleospadix beguinii Burret = **Drymophloeus litigiosus** (Becc.) H.E.Moore

Coleospadix gracilis (Giseke) Burret = **Drymophloeus oliviformis** (Giseke) Mart.

Coleospadix litigiosa (Becc.) Becc. = **Drymophloeus litigiosus** (Becc.) H.E.Moore

Coleospadix oninensis (Becc.) Becc. = **Drymophloeus oninensis** (Becc.) H.E.Moore

Coleospadix porrectus Burret = **Drymophloeus litigiosus** (Becc.) H.E.Moore

Collinia

Collinia (Liebm.) Liebm. ex Oerst. = **Chamaedorea** Willd.

Collinia deppeana Klotzsch = **Chamaedorea elegans** Mart.

Collinia elatior Liebm. ex Oerst. = **Chamaedorea liebmannii** Mart.

Collinia elegans (Mart.) Liebm. ex Oerst. = **Chamaedorea elegans** Mart.

Collinia fibrosa (H.Wendl.) Oerst. = **Synechanthus fibrosus** (H.Wendl.) H.Wendl.

Collinia humilis Liebm. ex Oerst. = **Chamaedorea elegans** Mart.

Colpothrinax

Colpothrinax Griseb. & H.Wendl., Bot. Zeitung (Berlin) 37: 147 (1879).
3 species, C. America, Cuba. 80 81.

Colpothrinax aphanopetala R.Evans, Palms 45: 189 (2001).
SE. Nicaragua to Panama. 80 COS NIC PAN.

Colpothrinax cookii Read, Principes 13: 13 (1969).
Belize to Honduras. 80 BLZ GUA HON.

Colpothrinax wrightii Griseb. & H.Wendl. ex Voss, Vilm. Blumengärtn. ed. 3, 1: 1147 (1895). *Pritchardia wrightii* (Griseb. & H.Wendl. ex Voss) Becc., Webbia 2: 203 (1908).
SW. Cuba (incl. I. de la Juventud). 81 CUB.

Copernicia

Copernicia Mart. ex Endl., Gen. Pl.: 253 (1837).
21 species, Caribbean, S. Trop. America. 81 82 83 84 85.

Arrudaria Macedo, Not. Palm. Carnauba: 5 (1867), nom. inval.

Coryphomia Rojas Acosta, Bull. Acad. Int. Géogr. Bot. 28: 158 (1918).

Copernicia alba Morong, Ann. New York Acad. Sci. 7: 246 (1893).
WC. Brazil to NE. Argentina. 83 BOL 84 BZC 85 AGE PAR.
Copernicia nigra Morong, Ann. Mus. Natl. Hist. Nat. 7: 245 (1893).
Copernicia rubra Morong, Ann. Mus. Natl. Hist. Nat. 7: 247 (1893).
Copernicia australis Becc., Webbia 2: 158 (1908).
Coryphomia tectorum Rojas, Bull. Acad. Int. Géogr. Bot. 28: 158 (1918).
Copernicia ramulosa Burret, Notizbl. Bot. Gart. Berlin-Dahlem 10: 403 (1928).

Copernicia australis Becc. = **Copernicia alba** Morong

Copernicia baileyana Léon, Revista Soc. Geogr. Cuba 4: 22 (1931).
C. & E. Cuba. 81 CUB.
Copernicia baileyana f. *bifida* Léon, Mem. Soc. Cub. Hist. Nat. "Felipe Poey" 10: 255 (1936).
Copernicia baileyana var. *laciniosa* Léon, Mem. Soc. Cub. Hist. Nat. "Felipe Poey" 10: 224 (1936).

Copernicia baileyana f. *bifida* Léon = **Copernicia baileyana** Léon

Copernicia baileyana var. *laciniosa* Léon = **Copernicia baileyana** Léon

Copernicia barbadensis (Lodd. ex Mart.) H.Wendl. = **Coccothrinax barbadensis** (Lodd. ex Mart.) Becc.

Copernicia berteroana Becc., Webbia 2: 150 (1908).
Hispaniola. 81 DOM HAI nla (82) ven.

Copernicia brittonorum Léon, Revista Soc. Geogr. Cuba 4: 19 (1931).
W. & WC. Cuba. 81 CUB.
Copernicia brittonorum var. *acuta* Léon, Mem. Soc. Cub. Hist. Nat. "Felipe Poey" 10: 222 (1936).
Copernicia brittonorum var. *sabaloense* Léon, Mem. Soc. Cub. Hist. Nat. "Felipe Poey" 10: 223 (1936).

Copernicia brittonorum var. *acuta* Léon = **Copernicia brittonorum** Léon

Copernicia brittonorum var. *sabaloense* Léon = **Copernicia brittonorum** Léon

Copernicia × burretiana Léon, Mem. Soc. Cub. Hist. Nat. "Felipe Poey" 10: 208 (1936). C. hospita × C. macroglossa.
Cuba. 81 CUB.

Copernicia campestris Burmeist. = **Trithrinax campestris** (Burmeist.) Drude & Griseb.

Copernicia cerifera (Arruda) Mart. = **Copernicia prunifera** (Mill.) H.E.Moore

Copernicia clarensis Léon = **Copernicia curtissii** Becc.

Copernicia clarkii Léon = **Copernicia longiglossa** Léon

Copernicia cowellii Britton & P.Wilson, Bull. Torrey Bot. Club 41: 17 (1914).
Cuba (Camagüei). 81 CUB.

Copernicia curbeloi Léon, Revista Soc. Geogr. Cuba 4: 23 (1931).
Cuba. 81 CUB.
Copernicia molineti var. *cuneata* Léon, Mem. Soc. Cub. Hist. Nat. "Felipe Poey" 10: 216 (1936).
Copernicia sueroana var. *semiorbicularis* Léon, Mem. Soc. Cub. Hist. Nat. "Felipe Poey" 10: 216 (1936).

Copernicia curtissii Becc., Webbia 2: 176 (1908).
Cuba. 81 CUB.
Copernicia pauciflora Burret, Kongl. Svenska Vetenskapsakad. Handl., III, 6(7): 8 (1929).
Copernicia clarensis Léon, Revista Soc. Geogr. Cuba 4: 15 (1931).

Copernicia depressa Liebm. ex Dalgrem = **Brahea dulcis** (Kunth) Mart.

Copernicia ekmanii Burret, Kongl. Svenska Vetenskapsakad. Handl., III, 6(7): 5 (1929).
N. Haiti. 81 HAI.

Copernicia × escarzana Léon, Revista Soc. Geogr. Cuba 4: 12 (1931). .
Cuba. 81 CUB.

Copernicia excelsa Léon = **Copernicia gigas** Ekman ex Burret

Copernicia fallaensis Léon, Revista Soc. Geogr. Cuba 4: 21 (1931).
Cuba. 81 CUB.

Copernicia gigas Ekman ex Burret, Kongl. Svenska Vetenskapsakad. Handl., III, 6(7): 3 (1929).
E. Cuba. 81 CUB.
Copernicia excelsa Léon, Revista Soc. Geogr. Cuba 4: 26 (1931).

Copernicia glabrescens H.Wendl. ex Becc., Webbia 2: 170 (1908).
W. & WC. Cuba. 81 CUB.

var. glabrescens
W. & WC. Cuba. 81 CUB.
Copernicia glabrescens var. *havanensis* León, Mem. Soc. Cub. Hist. Nat. "Felipe Poey" 10: 217 (1936).

var. **ramosissima** (Burret) O.Muñiz & Borhidi, Acta Bot. Acad. Sci. Hung. 28: 332 (1982).
W. Cuba. 81 CUB.
*Copernicia ramosissima Burret, Kongl. Svenska Vetenskapsakad. Handl., III, 6(7): 8 (1929).

Copernicia glabrescens var. *havanensis* León = **Copernicia glabrescens** var. **glabrescens**

Copernicia holguinensis Léon = **Copernicia yarey** Burret

Copernicia hospita Mart., Hist. Nat. Palm. 3: 243 (1838).
Cuba. 81 CUB.

Copernicia humicola Léon, Mem. Soc. Cub. Hist. Nat. "Felipe Poey" 10: 22 (1936).
Cuba. 81 CUB.

Copernicia leoniana Dahlgren & Glassman = **Copernicia macroglossa** H.Wendl. ex Becc.

Copernicia longiglossa Léon, Mem. Soc. Cub. Hist. Nat. "Felipe Poey" 10: 210 (1936).
E. Cuba. 81 CUB.
Copernicia clarkii Léon, Mem. Soc. Cub. Hist. Nat. "Felipe Poey" 10: 213 (1936).

Copernicia macroglossa H.Wendl. ex Becc., Webbia 2: 177 (1907).
W. & C. Cuba. 81 CUB.
Copernicia torreana Léon, Revista Soc. Geogr. Cuba 4: 10 (1931).
Copernicia leoniana Dahlgren & Glassman, Principes 2: 103 (1958).

Copernicia maritima (Kunth) Mart. = **Sabal maritima** (Kunth) Burret

Copernicia miraguama (Kunth) Mart. = **Coccothrinax miraguama** (Kunth) Becc.

Copernicia molineti Léon, Revista Soc. Geogr. Cuba 4: 25 (1931).
Cuba. 81 CUB.

Copernicia molineti var. *cuneata* Léon = **Copernicia curbeloi** Léon

Copernicia nana (Kunth) Liebm. ex Hemsl. = **Cryosophila nana** (Kunth) Blume

Copernicia nigra Morong = **Copernicia alba** Morong

Copernicia × occidentalis Léon, Mem. Soc. Cub. Hist. Nat. "Felipe Poey" 10: 218 (1936). C. brittonanum × C. hospita.
Cuba. 81 CUB.

Copernicia oxycalyx Burret = **Copernicia rigida** Britton & P.Wilson

Copernicia pauciflora Burret = **Copernicia curtissii** Becc.

Copernicia prunifera (Mill.) H.E.Moore, Gentes Herb. 9: 24a (1963).
C. & NE. Brazil. 84 BZC BZE BZN.
   *Palma prunifera Mill., Gard. Dict. ed. 8: 7 (1768).
   Corypha cerifera Arruda in H.Koster, Trav. Brazil: 494 (1816). *Copernicia cerifera* (Arruda) Mart., Hist. Nat. Palm. 3: 242 (1838). *Arrudaria cerifera* (Arruda) Macedo, Not. Palm. Carnauba: 5 (1867).

Copernicia pumos (Kunth) Mart. = **Sabal pumos** (Kunth) Burret

Copernicia ramosissima Burret = **Copernicia glabrescens** var. **ramosissima** (Burret) O.Muñiz & Borhidi

Copernicia ramulosa Burret = **Copernicia alba** Morong

Copernicia rigida Britton & P.Wilson, Bull. Torrey Bot. Club 41: 17 (1914).
C. & E. Cuba. 81 CUB.
   Copernicia oxycalyx Burret, Kongl. Svenska Vetenskapsakad. Handl., III, 6(7): 6 (1929).
   Copernicia rigida f. *fissilingua* Léon, Mem. Soc. Cub. Hist. Nat. "Felipe Poey" 10: 210 (1936).

Copernicia rigida f. *fissilingua* Léon = **Copernicia rigida** Britton & P.Wilson

Copernicia robusta Linden = **Licuala ?**

Copernicia roigii Léon, Revista Soc. Geogr. Cuba 4: 17 (1931).
Cuba. 81 CUB.

Copernicia rubra Morong = **Copernicia alba** Morong

Copernicia sanctae-martae Becc. = **Copernicia tectorum** (Kunth) Mart.

Copernicia × shaferi Dahlgren & Glassman, Principes 3: 88 (1959). C. cowellii × C. hospita.
Cuba. 81 CUB.

Copernicia × sueroana Léon, Revista Soc. Geogr. Cuba 4: 14 (1931). C. hospita × C. rigida.
Cuba. 81 CUB.

Copernicia × sueroana var. *semiorbicularis* Léon = **Copernicia curbeloi** Léon

Copernicia tectorum (Kunth) Mart., Hist. Nat. Palm. 3: 243 (1838).
Colombia, N. Venezuela. (81) nla 82 VEN 83 CLM.
   *Corypha tectorum Kunth in F.W.H.von Humboldt, A.J.A.Bonpland & C.S.Kunth, Nov. Gen. Sp. 1: 299 (1816).
   Copernicia sanctae-martae Becc., Webbia 2: 154 (1908).

Copernicia × textilis Léon, Revista Soc. Geogr. Cuba 4: 24 (1931). C. baileyana × C. hospita.
Cuba. 81 CUB.

Copernicia torreana Léon = **Copernicia macroglossa** H.Wendl. ex Becc.

Copernicia × vespertilionum Léon, Revista Soc. Geogr. Cuba 4: 27 (1931). C. gigas × C. rigida.
Cuba. 81 CUB.

Copernicia wrightii Griseb. & H.Wendl. = **Acoelorrhaphe wrightii** (Griseb. & H.Wendl.) H.Wendl. ex Becc.

Copernicia yarey Burret, Kongl. Svenska Vetenskapsakad. Handl., III, 6(7): 7 (1929).
Cuba. 81 CUB.
   Copernicia holguinensis Léon, Revista Soc. Geogr. Cuba 4: 18 (1931).
   Copernicia yarey var. *robusta* León, Mem. Soc. Cub. Hist. Nat. "Felipe Poey" 10: 221 (1936).

Copernicia yarey var. *robusta* León = **Copernicia yarey** Burret

Cornera

Cornera Furtado = **Calamus** L.

Cornera conirostris (Becc.) Furtado = **Calamus conirostris** Becc.

Cornera lobbiana (Becc.) Furtado = **Calamus lobbianus** Becc.

Cornera pycnocarpa Furtado = **Calamus pycnocarpus** (Furtado) J.Dransf.

Corozo

Corozo Jacq. ex Giseke = **Elaeis** Jacq.

Corozo oleifera (Kunth) L.H.Bailey = **Elaeis oleifera** (Kunth) Cortés

Corypha

Corypha L., Sp. Pl.: 1187 (1753). *Codda-Pana* Adans., Fam. Pl. 2: 25 (1763), nom. illeg.
6 species, Trop. Asia to N. Australia. 40 41 42 43 50.
   Taliera Mart., Palm. Fam.: 10 (1824).
   Gembanga Blume, Flora 8: 580 (1825).
   Bessia Raf., Sylva Tellur.: 13 (1838).
   Dendrema Raf., Sylva Tellur.: 14 (1838).

Corypha africana Lour. = **Hyphaene coriacea** Gaertn.

Corypha australis R.Br. = **Livistona australis** (R.Br.) Mart.

Corypha careyana Becc. = **Corypha taliera** Roxb.

Corypha cerifera Arruda = **Copernicia prunifera** (Mill.) H.E.Moore

Corypha decora W.Bull = **Livistona decora** (W.Bull) Dowe

Corypha dulcis Kunth = **Brahea dulcis** (Kunth) Mart.

Corypha elata Roxb. = **Corypha utan** Lam.

Corypha frigida Mohl ex Mart. = **Brahea dulcis** (Kunth) Mart.

Corypha gebang Mart. = **Corypha utan** Lam.

Corypha gembanga (Blume) Blume = **Corypha utan** Lam.

Corypha glaucescens Lodd. ex Loudon = **Sabal sp.**

Corypha griffithiana Becc., Webbia 5: 7 (1921).
Myanmar. 41 MYA.

Corypha guineensis L. = **Corypha umbraculifera** L.

Corypha hystrix Frazer ex Thouin = **Rhapidophyllum hystrix** (Frazer ex Thouin) H.Wendl. & Drude

Corypha laevis (Lour.) A.Chev. = **Pandanus laevis**

Corypha lecomtei Becc. ex Lecomte, Bull. Soc. Bot. France 63: 79 (1917).
Indo-China. 41 CBD LAO THA VIE.

Corypha licuala Lam. = **Licuala rumphii** Blume

Corypha macrophylla Roster = **Corypha utan** Lam.

Corypha macropoda Kurz ex Linden = **Corypha utan** Lam.

Corypha macropoda Linden ex Kurz = **Corypha utan** Lam.

Corypha maritima Kunth = **Sabal maritima** (Kunth) Burret

Corypha martiana Becc. ex Hook.f. = **Corypha taliera** Roxb.

Corypha microclada Becc., Webbia 5: 7 (1921).
Philippines. 42 PHI.

Corypha minor Jacq. = **Sabal minor** (Jacq.) Pers.

Corypha miraguama Kunth = **Coccothrinax miraguama** (Kunth) Becc.

Corypha nana Kunth = **Cryosophila nana** (Kunth) Blume

Corypha obliqua W.Bartram = **Serenoa repens** (W.Bartram) Small

Corypha palmacea Steud. = **Thrinax parviflora** subsp. **parviflora**

Corypha palmetto Walter = **Sabal palmetto** (Walter) Lodd. ex Schult. & Schult.f.

Corypha pilearia Lour. = **Licuala spinosa** Wurmb

Corypha pumila Walter = **Sabal minor** (Jacq.) Pers.

Corypha pumos Kunth = **Sabal pumos** (Kunth) Burret

Corypha repens W.Bartram = **Serenoa repens** (W.Bartram) Small

Corypha rotundifolia Lam. = **Livistona rotundifolia** (Lam.) Mart.

Corypha saribus Lour. = **Livistona saribus** (Lour.) Merr. ex A.Chev.

Corypha sylvestris (Blume) Mart. = **Corypha utan** Lam.

Corypha tectorum Kunth = **Copernicia tectorum** (Kunth) Mart.

Corypha taliera Roxb., Pl. Coromandel 3: 51 (1820).
India (West Bengal) to Myanmar. 40 BAN IND 41 MYA.
  Taliera bengalensis Spreng., Syst. Veg. 2: 18 (1825).
  Taliera tali Mart. ex Blume in J.J.Roemer & J.A.Schultes, Syst. Veg. 7: 1306 (1830).
  Corypha martiana Becc. ex Hook.f., Fl. Brit. India 6: 429 (1892).
  Corypha careyana Becc., Webbia 5: 7 (1921).

Corypha thebaica L. = **Hyphaene thebaica** (L.) Mart.

Corypha umbraculifera Jacq. = **Sabal palmetto** (Walter) Lodd. ex Schult. & Schult.f.

Corypha umbraculifera L., Sp. Pl.: 1178 (1753).
SW. India. 40 IND srl (41) mya tha.
  Corypha guineensis L., Mant. Pl. 1: 137 (1767).
  Bessia sanguinolenta Raf., Sylva Tellur.: 13 (1838).

Corypha utan Lam., Encycl. 2: 131 (1786). *Taliera sylvestris* Blume in J.J.Roemer & J.A.Schultes, Syst. Veg. 7: 1307 (1830), nom. illeg. *Corypha sylvestris* (Blume) Mart., Hist. Nat. Palm. 3: 233 (1838).
NE. India to N. Australia. 40 ASS BAN IND 41 AND CBD LAO MYA THA VIE 42 BOR JAW LSI MLY

MOL PHI SUL SUM 43 NWG 50 NTA QLD.
  Corypha elata Roxb., Fl. Ind. 2: 176 (1824). *Taliera elata* (Roxb.) Wall., Rep. Calcutta Bot. Gard. to G.A. Bushby: 29 (1840).
  Gembanga rotundifolia Blume, Flora 8: 580 (1825).
  Taliera gembanga Blume in J.J.Roemer & J.A.Schultes, Syst. Veg. 7: 1307 (1830), nom. illeg. *Corypha gembanga* (Blume) Blume, Rumphia 2: 59 (1839).
  Corypha gebang Mart., Hist. Nat. Palm. 3: 233 (1838).
  Corypha macropoda Kurz ex Linden, Cat. Gén. 1871: 87 (1871).
  Corypha macropoda Linden ex Kurz, J. Asiat. Soc. Bengal, Pt. 2, Nat. Hist. 43(2): 197 (1874).
  Corypha macrophylla Roster, Bull. Soc. Tosc. Ortic. 29: 81 (1904).
  Livistona vidalii Becc., Webbia 1: 343 (1905).

Coryphomia

Coryphomia Rojas Acosta = **Copernicia** Mart. ex Endl.

Coryphomia tectorum Rojas = **Copernicia alba** Morong

Cryosophila

Cryosophila Blume, Rumphia 2: 53 (1838).
10 species, Mexico to N. Colombia. 79 80 83.
  Acanthorrhiza H.Wendl., Gartenflora 18: 241 (1869).

Cryosophila albida Bartlett = **Cryosophila williamsii** P.H.Allen

Cryosophila argentea Bartlett = **Cryosophila stauracantha** (Heynh.) R.J.Evans

Cryosophila bartlettii R.J.Evans, Syst. Bot. Monogr. 46: 38 (1995).
C. Panama. 80 PAN.

Cryosophila bifurcata Lundell = **Cryosophila stauracantha** (Heynh.) R.J.Evans

Cryosophila cookii Bartlett, Publ. Carnegie Inst. Wash. 461: 39 (1935).
Costa Rica (Limón). 80 COS.

Cryosophila grayumii R.J.Evans, Syst. Bot. Monogr. 46: 42 (1995).
Costa Rica (Puntarenas). 80 COS.

Cryosophila guagara P.H.Allen, Ceiba 3: 174 (1953).
Costa Rica to Panama. 80 COS PAN.

Cryosophila kalbreyeri (Dammer ex Burret) Dahlgren, Field Mus. Nat. Hist., Bot. Ser. 14: 134 (1936).
SE. Panama to Colombia. 80 PAN 83 CLM.
  *Acanthorrhiza kalbreyeri Dammer ex Burret, Notizbl. Bot. Gart. Berlin-Dahlem 11: 313 (1932).

subsp. **cogolloi** R.J.Evans, Syst. Bot. Monogr. 46: 50 (1995).
Colombia. 83 CLM.

subsp. **kalbreyeri**
SE. Panama to NW. Colombia. 80 PAN 83 CLM.

Cryosophila macrocarpa R.J.Evans, Novon 2: 58 (1992).
Colombia (Chocó). 83 CLM.

Cryosophila mocinoi (Kunth) R.R.Fernandez = **Cryosophila nana** (Kunth) Blume

Cryosophila nana (Kunth) Blume, Rumphia 2: 53 (1838).
W. & S. Mexico. 79 MXN MXS MXT.
  *Corypha nana Kunth in F.W.H.von Humboldt, A.J.A.Bonpland & C.S.Kunth, Nov. Gen. Sp. 1: 299 (1816). *Copernicia nana* (Kunth) Liebm. ex Hemsl., Biol. Cent.-Amer., Bot. 3: 411 (1885).

Chamaerops mocinoi Kunth in F.W.H.von Humboldt, A.J.A.Bonpland & C.S.Kunth, Nov. Gen. Sp. 1: 300 (1817). *Acanthorrhiza mocinoi* (Kunth) Benth. & Hook.f., Gen. Pl. 3: 925 (1883). *Cryosophila mocinoi* (Kunth) R.R.Fernandez, Trees Mumbai: 269 (1999). *Trithrinax aculeata* Liebm. ex Mart., Hist. Nat. Palm. 3: 320 (1853). *Acanthorrhiza aculeata* (Liebm. ex Mart.) H.Wendl., Gartenflora 18: 241 (1869).

Cryosophila stauracantha (Heynh.) R.J.Evans, Syst. Bot. Monogr. 46: 57 (1995).
SE. Mexico to C. America. 79 MXT 80 BLZ GUA.
*Chamaerops stauracantha Heynh., Alph. Aufz. Gew. 2: 136 (1846). *Acanthorrhiza stauracantha* (Heynh.) H.Wendl. ex Linden, Cat. Gén. 87: (1871).
Cryosophila argentea Bartlett, Publ. Carnegie Inst. Wash. 461: 40 (1935). *Acanthorrhiza collinsii* O.F.Cook, Natl. Hort. Mag. 20: 50 (1941).
Cryosophila bifurcata Lundell, Wrightia 1: 53 (1945).

Cryosophila warscewiczii (H.Wendl.) Bartlett, Publ. Carnegie Inst. Wash. 461: 38 (1935).
C. America. 80 COS NIC PAN.
*Acanthorrhiza warscewiczii H.Wendl., Gartenflora 18: 242 (1869).

Cryosophila williamsii P.H.Allen, Ceiba 3: 174 (1953).
WC. Honduras. 80 HON.
Cryosophila albida Bartlett, Publ. Carnegie Inst. Wash. 461: 40 (1935).

Cuatrecasea

Cuatrecasea Dugand = **Iriatrella**

Cuatrecasea spruceana (Barb.Rodr.) Dugand = **Iriartella setigera** (Mart.) H.Wendl.

Cuatrecasea vaupesana Dugand = **Iriartella setigera** (Mart.) H.Wendl.

Cucifera

Cucifera Delile = **Hyphaene** Gaertn.

Cucifera thebaica (L.) Delile = **Hyphaene thebaica** (L.) Mart.

Curima

Curima O.F.Cook = **Aiphanes** Willd.

Curima colophylla O.F.Cook = **Aiphanes minima** (Gaertn.) Burret

Curima corallina (Mart.) O.F.Cook = **Aiphanes minima** (Gaertn.) Burret

Cyclospathe

Cyclospathe O.F.Cook = **Pseudophoenix** H.Wendl. ex Sarg.

Cyclospathe northropii O.F.Cook = **Pseudophoenix sargentii** H.Wendl. ex Sarg.

Cyphokentia

Cyphokentia Brongn., Compt. Rend. Hebd. Séances Acad. Sci. 77: 397 (1873).
1 species, New Caledonia. 60.
Dolichokentia Becc., Webbia 5: 113 (1921).

Cyphokentia balansae Brongn. = **Cyphosperma balansae** (Brongn.) H.Wendl. ex Salomon

Cyphokentia billardieri Brongn. = **Basselinia gracilis** (Brongn. & Gris) Vieill.

Cyphokentia bractealis Brongn. = **Clinosperma bracteale** (Brongn.) Becc.

Cyphokentia carolinensis Becc. = **Clinostigma carolinense** (Becc.) H.E.Moore & Fosberg

Cyphokentia deplanchei (Brongn. & Gris) Brongn. = **Basselinia deplanchei** (Brongn. & Gris) Vieill.

Cyphokentia eriostachys Brongn. = **Basselinia gracilis** (Brongn. & Gris) Vieill.

Cyphokentia gracilis (Brongn. & Gris) Brongn. = **Basselinia gracilis** (Brongn. & Gris) Vieill.

Cyphokentia heanei auct. = **?**

Cyphokentia humboldtiana Brongn. = **Basselinia humboldtiana** (Brongn.) H.E.Moore

Cyphokentia macrocarpa (Brongn.) auct. = **Chambeyronia macrocarpa** (Brongn.) Vieill. ex Becc.

Cyphokentia macrostachya Brongn., Compt. Rend. Hebd. Séances Acad. Sci. 77: 399 (1873). *Kentia macrostachya* (Brongn.) Pancher ex Brongn., Compt. Rend. Hebd. Séances Acad. Sci. 77: 399 (1873). *Clinostigma macrostachyum* (Brongn.) Becc., Malesia 1: 40 (1877).
C. & S. New Caledonia. 60 NWC.

> *Cyphokentia robusta* Brongn., Compt. Rend. Hebd. Séances Acad. Sci. 77: 400 (1873). *Clinostigma robustum* (Brongn.) Becc., Malesia 1: 40 (1877). *Kentia robusta* (Brongn.) Linden ex H.Wendl. in O.C.E.de Kerchove de Denterghem, Palmiers: 248 (1878). *Dolichokentia robusta* (Brongn.) Becc., Webbia 5: 114 (1921).

Cyphokentia pancheri (Brongn. & Gris) Brongn. = **Basselinia pancheri** (Brongn. & Gris) Vieill.

Cyphokentia robusta Brongn. = **Cyphokentia macrostachya** Brongn.

Cyphokentia samoensis (H.Wendl.) Warb. = **Clinostigma samoense** H.Wendl.

Cyphokentia savoryana Rehder & E.H.Wilson = **Clinostigma savoryanum** (Rehder & E.H.Wilson) H.E.Moore & Fosberg

Cyphokentia surculosa Brongn. = **Basselinia deplanchei** (Brongn. & Gris) Vieill.

Cyphokentia tete (Becc.) Becc. = **Physokentia tete** (Becc.) Becc.

Cyphokentia thurstonii (Becc.) Becc. = **Physokentia thurstonii** (Becc.) Becc.

Cyphokentia vaginata Brongn. = **Brongniartikentia vaginata** (Brongn.) Becc.

Cyphophoenix

Cyphophoenix H.Wendl. ex Hook.f. in G.Bentham & J.D.Hooker, Gen. Pl. 3: 893 (1883).
2 species, New Caledonia. 60.

Cyphophoenix carolinensis (Becc.) Kaneh. & Hatus. = **Clinostigma carolinense** (Becc.) H.E.Moore & Fosberg

Cyphophoenix elegans (Brongn. & Gris) H.Wendl. ex Salomon, Palmen: 86 (1887).
NE. New Caledonia. 60 NWC.
*Kentia elegans Brongn. & Gris, Bull. Soc. Bot. France 11: 312 (1864).

Cyphophoenix fulcita (Brongn.) Hook.f. ex Salomon = **Campecarpus fulcitus** (Brongn.) H.Wendl. ex Becc.

Cyphophoenix nucele H.E.Moore, Gentes Herb. 11: 165 (1976).
New Caledonia (Is. Loyauté). 60 NWC.

Cyphosperma

Cyphosperma H.Wendl. ex Hook.f. in G.Bentham & J.D.Hooker, Gen. Pl. 3: 895 (1883).
4 species, SW. Pacific. 60.
Taveunia Burret, Occas. Pap. Bernice Pauahi Bishop Mus. 11(4): 12 (1935).

Cyphosperma balansae (Brongn.) H.Wendl. ex Salomon, Palmen: 87 (1887).
NE. & C. New Caledonia. 60 NWC.
*Cyphokentia balansae Brongn., Compt. Rend. Hebd. Séances Acad. Sci. 77: 400 (1873).

Cyphosperma tanga (H.E.Moore) H.E.Moore, Principes 21: 88 (1977).
Fiji (Viti Levu). 60 FIJ.
*Taveunia tanga H.E.Moore, Candollea 20: 98 (1965).

Cyphosperma tete Becc. = **Physokentia tete** (Becc.) Becc.

Cyphosperma thurstonii Becc. = **Physokentia thurstonii** (Becc.) Becc.

Cyphosperma trichospadix (Burret) H.E.Moore, Principes 21: 88 (1977).
Fiji (Vanua Levu, Taveuni). 60 FIJ.
*Taveunia trichospadix Burret, Occas. Pap. Bernice Pauahi Bishop Mus. 11(4): 13 (1935).

Cyphosperma vieillardii (Brongn. & Gris) H.Wendl. ex Salomon = **Burretiokentia vieillardii** (Brongn. & Gris) Pic.Serm.

Cyphosperma voutmelense Dowe, Principes 37: 209 (1993).
Vanuatu. 60 VAN.

Cyrtostachys

Cyrtostachys Blume, Bull. Sci. Phys. Nat. Néerl. 1: 66 (1838).
11 species, Pen. Thailand to Papuasia. 41 42 43.

Cyrtostachys brassii Burret, Notizbl. Bot. Gart. Berlin-Dahlem 12: 328 (1935).
New Guinea. 43 NWG.

Cyrtostachys ceramica (Miq.) H.Wendl. = **Rhopaloblaste ceramica** (Miq.) Burret

Cyrtostachys compsoclada Burret, Notizbl. Bot. Gart. Berlin-Dahlem 13: 325 (1936).
New Guinea. 43 NWG.

Cyrtostachys elegans Burret, Notizbl. Bot. Gart. Berlin-Dahlem 13: 472 (1937).
New Guinea. 43 NWG.

Cyrtostachys glauca H.E.Moore, Principes 10: 86 (1966).
New Guinea. 43 NWG.

Cyrtostachys kisu Becc., Webbia 4: 289 (1914).
Solomon Is. 43 SOL.

Cyrtostachys lakka Becc. = **Cyrtostachys renda** Blume

Cyrtostachys ledermanniana Becc., Bot. Jahrb. Syst. 58: 450 (1923).
New Guinea. 43 NWG.

Cyrtostachys loriae Becc., Webbia 1: 303 (1905).
New Guinea. 43 NWG.

Cyrtostachys microcarpa Burret, J. Arnold Arbor. 20: 203 (1939).
New Guinea. 43 NWG.

Cyrtostachys peekeliana Becc., Bot. Jahrb. Syst. 52: 28 (1914).
Bismarck Arch. 43 BIS.

Cyrtostachys phanerolepis Burret, Notizbl. Bot. Gart. Berlin-Dahlem 13: 324 (1936).
New Guinea. 43 NWG.

Cyrtostachys renda Blume, Bull. Sci. Phys. Nat. Néerl. 1: 66 (1838). *Bentinckia renda* (Blume) Mart., Hist. Nat. Palm. 3: 316 (1853).
Pen. Thailand to W. Malesia. 41 THA 42 BOR MLY SUM.
Areca erythropoda Miq., J. Bot. Néerl. 1: 6 (1861).
Pinanga purpurea Miq., Fl. Ned. Ind., Eerste Bijv.: 590 (1861).
Ptychosperma coccinea Teijsm. & Binn., Cat. Hort. Bot. Bogor.: 69 (1866).
Areca erythrocarpa H.Wendl. in O.C.E.de Kerchove de Denterghem, Palmiers: 231 (1878).
Cyrtostachys lakka Becc., Ann. Jard. Bot. Buitenzorg 2: 141 (1885).
Pinanga rubricaulis Linden, Cat. Pl. Hort. Contin. 1885: 61, n. 117 (1885).

Dachel

Dachel Adans. = **Phoenix** L.

Daemonorops

Daemonorops Blume in J.J.Roemer & J.A.Schultes, Syst. Veg. 7: 1333 (1830).
102 species, Trop. & Subtrop. Asia. 36 38 40 41 42 43.

Daemonorops acamptostachys Becc., Ann. Roy. Bot. Gard. (Calcutta) 12(1): 209 (1911).
Borneo (Sarawak). 42 BOR.

Daemonorops acanthobola Becc. = **Daemonorops collarifera** Becc.

Daemonorops accedens Blume = **Daemonorops rubra** (Reinw. ex Mart.) Blume

Daemonorops accedens var. *brevispatha* Blume = **Daemonorops gracilipes** (Miq.) Becc.

Daemonorops acehensis Rustiami, Gard. Bull. Singapore 54: 202 (2002).
Sumatera. 42 SUM.

Daemonorops aciculata Ridl. = **Daemonorops sepal** Becc.

Daemonorops adspersa Blume = **Calamus adspersus** (Blume) Blume

Daemonorops affinis Becc., Leafl. Philipp. Bot. 8: 3042 (1919).
Philippines (Mindanao). 42 PHI.

Daemonorops angustifolia (Griff.) Mart., Hist. Nat. Palm. 3: 327 (1853).
Pen. Thailand to Pen. Malaysia. 41 THA 42 MLY.
*Calamus angustifolius Griff., Calcutta J. Nat. Hist. 5: 89 (1845).
Calamus hygrophilus Griff., Palms Brit. E. Ind.: t. 213 C (1850). *Daemonorops hygrophila* (Griff.) Mart., Hist. Nat. Palm. 3: 328 (1853). *Palmijuncus hygrophilus* (Griff.) Kuntze, Revis. Gen. Pl. 2: 733 (1891).
Daemonorops carcharodon Ridl., Mat. Fl. Malay. Penins. 2: 178 (1907).
Daemonorops angustispatha Furtado, Gard. Bull. Straits Settlem. 9: 161 (1937).

Daemonorops angustispatha Furtado = **Daemonorops angustifolia** (Griff.) Mart.

Daemonorops annulata Becc. = **Daemonorops sabut** Becc.

Daemonorops aruensis Becc., Ann. Roy. Bot. Gard. (Calcutta) 12(1): 50 (1911).
New Guinea (Kep. Aru). 43 NWG.

Daemonorops asteracantha Becc., Ann. Roy. Bot. Gard. (Calcutta) 12(1): 227 (1911).
Borneo. 42 BOR.

Daemonorops atra J.Dransf., Bot. J. Linn. Soc. 81: 15 (1980).
Borneo. 42 BOR.

Daemonorops aurea Renuka & Vijayak., Rheedea 4: 122 (1994).
Andaman Is. 41 AND.

Daemonorops bakauensis Becc. = **Daemonorops lewisiana** (Griff.) Mart.

Daemonorops banggiensis J.Dransf., Kew Bull. 36: 813 (1982).
N. Borneo. 42 BOR.

Daemonorops barbata (Zipp. ex Blume) Mart. = **Calamus barbatus** Zipp. ex Blume

Daemonorops beguinii Burret, Notizbl. Bot. Gart. Berlin-Dahlem 11: 204 (1931).
Maluku. 42 MOL.

Daemonorops binnendijkii Becc., Ann. Roy. Bot. Gard. (Calcutta) 12(1): 67 (1911).
Sumatera. 42 SUM.

Daemonorops brachystachys Furtado, Gard. Bull. Straits Settlem. 8: 344 (1935).
Pen. Malaysia to N. Sumatera. 42 MLY SUM.

Daemonorops calapparia (Mart.) Blume, Rumphia 3: 7 (1847).
Maluku. 42 MOL.
  Calamus calapparius Mart., Hist. Nat. Palm. 3: 209 (1838). *Palmijuncus calapparius* (Mart.) Kuntze, Revis. Gen. Pl. 2: 731 (1891).
  Calamus amboinensis Miq., Verh. Kon. Ned. Akad. Wetensch., Afd. Natuurk. 11: 20 (1868). *Palmijuncus amboinensis* (Miq.) Kuntze, Revis. Gen. Pl. 2: 733 (1891).

Daemonorops calicarpa (Griff.) Mart., Hist. Nat. Palm. 3: 326 (1853).
Pen. Malaysia to N. Sumatera. 42 MLY SUM.
  Calamus calicarpus Griff., Calcutta J. Nat. Hist. 5: 92 (1845). *Palmijuncus calicarpus* (Griff.) Kuntze, Revis. Gen. Pl. 2: 733 (1891).
  Calamus petiolaris Griff., Calcutta J. Nat. Hist. 5: 93 (1845). *Daemonorops petiolaris* (Griff.) Mart., Hist. Nat. Palm. 3: 326 (1853). *Palmijuncus petiolaris* (Griff.) Kuntze, Revis. Gen. Pl. 2: 732 (1891).
  Daemonorops microthamnus Becc., Rec. Bot. Surv. India 2: 221 (1902).

Daemonorops calospatha Ridl. = **Calospatha scortechinii** Becc.

Daemonorops calothyrsa Furtado = **Daemonorops longipes** (Griff.) Mart.

Daemonorops carcharodon Ridl. = **Daemonorops angustifolia** (Griff.) Mart.

Daemonorops clemensiana Becc., Philipp. J. Sci., C 4: 636 (1909).
Philippines (Mindanao). 42 PHI.

Daemonorops cochleata Teijsm. & Binn. ex Miq. = **Daemonorops didymophylla** Becc.

Daemonorops collarifera Becc., Rec. Bot. Surv. India 2: 227 (1902).
NW. Borneo. 42 BOR.

Daemonorops acanthobola Becc., Rec. Bot. Surv. India 2: 228 (1902).

Daemonorops congesta Ridl. = **Daemonorops leptopus** (Griff.) Mart.

Daemonorops confusa Furtado, Gard. Bull. Straits Settlem. 8: 347 (1935).
Sumatera. 42 SUM.

Daemonorops crinita Blume, Rumphia 2: viii (1838).
Calamus crinitus (Blume) Miq., Anal. Bot. Ind. 1: 6 (1850). *Palmijuncus crinitus* (Blume) Kuntze, Revis. Gen. Pl. 2: 733 (1891). *Rotang crinitus* (Blume) Baill., Hist. Pl. 13: 300 (1895).
Sumatera to S. Borneo. 42 BOR SUM.
  Calamus manicatus Teijsm. & Binn. ex Miq., Fl. Ned. Ind. 3: 135 (1855). *Palmijuncus manicatus* (Teijsm. & Binn. ex Miq.) Kuntze, Revis. Gen. Pl. 2: 733 (1891).

Daemonorops cristata Becc., Nelle Forest. Borneo: 608 (1902).
Borneo (Sarawak). 42 BOR.
  Daemonorops diversispina Becc., Rec. Bot. Surv. India 2: 229 (1902).

Daemonorops curranii Becc., Philipp. J. Sci., C 2: 238 (1907).
Philippines (Palawan). 42 PHI.

Daemonorops curtisii Furtado = **Daemonorops lewisiana** (Griff.) Mart.

Daemonorops depressiuscula (Miq. ex H.Wendl.) Becc., Rec. Bot. Surv. India 2: 226 (1902).
Sumatera. 42 SUM.
  Calamus depressiusculus Miq. ex H.Wendl. in O.C.E.de Kerchove de Denterghem, Palmiers: 236 (1878). *Palmijuncus depressiusculus* (Miq. ex H.Wendl.) Kuntze, Revis. Gen. Pl. 2: 733 (1891).

Daemonorops didymophylla Becc. in J.D.Hooker, Fl. Brit. India 6: 468 (1893). *Calamus didymophyllus* (Becc.) Ridl., J. Straits Branch Roy. Asiat. Soc. 30: 221 (1897).
Pen. Thailand to W. Malesia. 41 THA 42 BOR MLY SUM.
  Calamus cochleatus Miq., Verh. Kon. Ned. Akad. Wetensch., Afd. Natuurk. 11: 29 (1868), nom. nud. *Palmijuncus cochleatus* (Miq.) Kuntze, Revis. Gen. Pl. 2: 733 (1891), nom. inval.
  Daemonorops cochleata Teijsm. & Binn. ex Miq., Verh. Kon. Ned. Akad. Wetensch., Afd. Natuurk., Tweede Sect. 11(5): 29 (1868), nom. nud.
  Daemonorops mattanensis Becc., Nelle Forest. Borneo: 608 (1902).
  Daemonorops motleyi Becc., Rec. Bot. Surv. India 2: 224 (1902).

Daemonorops dissitophylla Becc. = **Daemonorops periacantha** Miq.

Daemonorops diversispina Becc. = **Daemonorops cristata** Becc.

Daemonorops draco (Willd.) Blume, Rumphia 2: viii (1838).
S. Sumatera, Borneo (Sarawak). 42 BOR SUM.
  Calamus draco Willd., Sp. Pl. 2: 203 (1799). *Palmijuncus draco* (Willd.) Kuntze, Revis. Gen. Pl. 2: 732 (1891).
  Calamus draconis Oken, Allg. Naturgesch. 3(1): 648 (1841).

Daemonorops draconcella Becc. = **Daemonorops micracantha** (Griff.) Becc.

Daemonorops dracuncula Ridl., Bull. Misc. Inform. Kew 1926: 91 (1926).
Sumatera (Kep. Mentawai). 42 SUM.

Daemonorops dransfieldii Rustiami, Gard. Bull. Singapore 54: 199 (2002).
Sumatera. 42 SUM.

Daemonorops elongata Blume, Rumphia 2: iii (1838). *Calamus elongatus* (Blume) Miq., Anal. Bot. Ind. 1: 6 (1850). *Palmijuncus elongatus* (Blume) Kuntze, Revis. Gen. Pl. 2: 733 (1891).
Borneo. 42 BOR.

Daemonorops erinacea Becc. = **Calamus erinaceus** (Becc.) J.Dransf.

Daemonorops fasciculata Mart. = **Calamus oxleyanus** var. **oxleyanus**

Daemonorops fissa Blume, Rumphia 3: 17 (1847). *Calamus fissus* (Blume) Miq., Anal. Bot. Ind. 1: 6 (1850). *Palmijuncus fissus* (Blume) Kuntze, Revis. Gen. Pl. 2: 733 (1891).
Borneo. 42 BOR.
  Daemonorops hallieriana Becc., Ann. Roy. Bot. Gard. (Calcutta) 12(1): 218 (1911).

Daemonorops florida Becc. = **Daemonorops periacantha** Miq.

Daemonorops forbesii Becc., Rec. Bot. Surv. India 2: 227 (1902).
Sumatera. 42 SUM.

Daemonorops formicaria Becc., Nelle Forest. Borneo: 608 (1902).
NW. Borneo. 42 BOR.

Daemonorops fusca Mart. = **Daemonorops mollis** (Blanco) Merr.

Daemonorops gaudichaudii Mart. = **Daemonorops mollis** (Blanco) Merr.

Daemonorops geniculata (Griff.) Mart., Hist. Nat. Palm. 3: 329 (1853).
Pen. Thailand to Sumatera. 41 THA 42 MLY SUM.
  *Calamus geniculatus Griff., Calcutta J. Nat. Hist. 5: 67 (1845). Palmijuncus geniculatus (Griff.) Kuntze, Revis. Gen. Pl. 2: 733 (1891).

Daemonorops gracilipes (Miq.) Becc., Rec. Bot. Surv. India 2: 225 (1902).
W. Sumatera. 42 SUM.
  Daemonorops accedens var. *brevispatha* Blume, Rumphia 3: 13 (1838).
  Daemonorops longipes Miq., Fl. Ned. Ind., Eerste Bijv.: 592 (1861), nom. illeg.
  *Calamus gracilipes Miq., Verh. Kon. Ned. Akad. Wetensch., Afd. Natuurk. 11: 28 (1868). Palmijuncus gracilipes (Miq.) Kuntze, Revis. Gen. Pl. 2: 733 (1891).

Daemonorops gracilis Becc., Leafl. Philipp. Bot. 8: 3044 (1919).
Philippines (Palawan). 42 PHI.

Daemonorops grandis (Griff.) Mart., Hist. Nat. Palm. 3: 327 (1853).
Pen. Thailand to Pen. Malaysia. 41 THA 42 MLY.
  *Calamus grandis Griff., Calcutta J. Nat. Hist. 5: 84 (1845). Palmijuncus grandis (Griff.) Kuntze, Revis. Gen. Pl. 2: 733 (1891). Palmijuncus grandis (Griff.) Kuntze, Revis. Gen. Pl. 2: 733 (1891).
  Calamus intermedius Griff., Calcutta J. Nat. Hist. 5: 86 (1845). *Daemonorops intermedia* (Griff.) Mart., Hist. Nat. Palm. 3: 327 (1853). *Palmijuncus intermedius* (Griff.) Kuntze, Revis. Gen. Pl. 2: 733 (1891).

Calamus acanthopis Griff., Palms Brit. E. Ind.: t. 216 (1850).
Daemonorops kirtong Griff., Palms Brit. E. Ind.: 102 (1850).
Daemonorops malaccensis Mart., Hist. Nat. Palm. 3: 327 (1853). *Palmijuncus malaccensis* (Mart.) Kuntze, Revis. Gen. Pl. 2: 733 (1891).
Daemonorops grandis var. *megacarpus* Furtado, Gard. Bull. Singapore 14: 67 (1953).
Daemonorops laciniata Furtado, Gard. Bull. Singapore 14: 75 (1953).

Daemonorops grandis var. *megacarpus* Furtado = **Daemonorops grandis** (Griff.) Mart.

Daemonorops guruba (Buch.-Ham.) Mart. = **Calamus guruba** Buch.-Ham.

Daemonorops hallieriana Becc. = **Daemonorops fissa** Blume

Daemonorops heteracantha Blume = **Calamus heteracanthus** Zipp. ex Blume

Daemonorops hirsuta Blume, Rumphia 2: t. 135 (1843). *Calamus hirsutus* (Blume) Miq., Verh. Kon. Ned. Akad. Wetensch., Afd. Natuurk. 11: 28 (1868). *Palmijuncus hirsutus* (Blume) Kuntze, Revis. Gen. Pl. 2: 733 (1891).
W. Malesia. 42 BOR MLY SUM.
  Calamus hystrix Griff., Calcutta J. Nat. Hist. 5: 70 (1845). *Daemonorops hystrix* (Griff.) Mart., Hist. Nat. Palm. 3: 328 (1853). *Palmijuncus hystrix* (Griff.) Kuntze, Revis. Gen. Pl. 2: 733 (1891).
  Daemonorops hystrix var. *exulans* Becc., Rec. Bot. Surv. India 2: 224 (1902).
  Daemonorops hystrix var. *minor* Becc., Ann. Roy. Bot. Gard. (Calcutta) 12(1): 138 (1911).

Daemonorops horrida Burret, Notizbl. Bot. Gart. Berlin-Dahlem 15: 193 (1940).
Sumatera. 42 SUM. — Provisionally accepted.

Daemonorops hygrophila (Griff.) Mart. = **Daemonorops angustifolia** (Griff.) Mart.

Daemonorops hypoleuca Kurz = **Calamus hypoleucus** Kurz

Daemonorops hystrix (Griff.) Mart. = **Daemonorops hirsuta** Blume

Daemonorops hystrix var. *exulans* Becc. = **Daemonorops hirsuta** Blume

Daemonorops hystrix var. *minor* Becc. = **Daemonorops hirsuta** Blume

Daemonorops imbellis Becc. = **Daemonorops sepal** Becc.

Daemonorops ingens J.Dransf., Bot. J. Linn. Soc. 81: 20 (1980).
Borneo. 42 BOR.

Daemonorops intermedia (Griff.) Mart. = **Daemonorops grandis** (Griff.) Mart.

Daemonorops intumescens Becc. = **Calamus paspalanthus** Becc.

Daemonorops javanica Furtado = **Daemonorops melanochaetes** Blume

Daemonorops jenkinsiana (Griff.) Mart., Hist. Nat. Palm. 3: 327 (1853).
E. Himalaya to Indo-China. 36 CHS 40 ASS BAN EHM 41 CBD LAO MYA THA VIE.
  *Calamus jenkinsianus Griff., Calcutta J. Nat. Hist. 5: 81 (1845). Palmijuncus jenkinsianus Kuntze, Revis. Gen. Pl. 2: 733 (1891).

Calamus nutantiflorus Griff., Calcutta J. Nat. Hist. 5: 79 (1845). *Daemonorops nutantiflora* (Griff.) Mart., Hist. Nat. Palm. 3: 326 (1853). *Palmijuncus nutantiflorus* (Griff.) Kuntze, Revis. Gen. Pl. 2: 732 (1891).
Daemonorops pierreana Becc., Rec. Bot. Surv. India 2: 220 (1902).
Daemonorops schmidtiana Becc., Bot. Tidsskr. 29: 98 (1909).
Daemonorops jenkinsiana var. *tenasserimica* Becc., Ann. Roy. Bot. Gard. (Calcutta) 12(1): 44 (1911).

Daemonorops jenkinsiana var. *tenasserimica* Becc. = **Daemonorops jenkinsiana** (Griff.) Mart.

Daemonorops kiahii Furtado = **Daemonorops sepal** Becc.

Daemonorops kirtong Griff. = **Daemonorops grandis** (Griff.) Mart.

Daemonorops korthalsii Blume, Rumphia 3: 23 (1847).
Calamus korthalsii (Blume) Miq., Anal. Bot. Ind. 1: 6 (1850). *Palmijuncus korthalsii* (Blume) Kuntze, Revis. Gen. Pl. 2: 733 (1891).
Borneo. 42 BOR.

Daemonorops kunstleri Becc. in J.D.Hooker, Fl. Brit. India 6: 469 (1893).
Pen. Thailand to Pen. Malaysia. 41 THA 42 MLY SUM?
Daemonorops vagans Becc. in J.D.Hooker, Fl. Brit. India 6: 469 (1893).

Daemonorops kurziana Hook.f. ex Becc. in J.D.Hooker, Fl. Brit. India 6: 463 (1893).
S. Andaman Is., W. Thailand. 41 AND MYA? THA.

Daemonorops laciniata Furtado = **Daemonorops grandis** (Griff.) Mart.

Daemonorops lamprolepis Becc., Rec. Bot. Surv. India 2: 223 (1902).
Sulawesi. 42 SUL.

Daemonorops lasiospatha Furtado = **Daemonorops scapigera** Becc.

Daemonorops latispina Teijsm. & Binn. = **Calamus latispinus** Miq.

Daemonorops leptopus (Griff.) Mart., Hist. Nat. Palm. 3: 329 (1853).
Pen. Thailand to Pen. Malaysia. 41 THA 42 MLY.
Calamus leptopus Griff., Calcutta J. Nat. Hist. 5: 73 (1845). *Palmijuncus leptopus* (Griff.) Kuntze, Revis. Gen. Pl. 2: 733 (1891).
Daemonorops congesta Ridl., Mat. Fl. Malay. Penins. 2: 179 (1901).

Daemonorops lewisiana (Griff.) Mart., Hist. Nat. Palm. 3: 327 (1853).
Pen. Thailand to Sumatera. 41 THA 42 MLY SUM.
Calamus lewisianus Griff., Calcutta J. Nat. Hist. 5: 87 (1845). *Palmijuncus lewisianus* (Griff.) Kuntze, Revis. Gen. Pl. 2: 733 (1891).
Daemonorops pseudosepal Becc. in J.D.Hooker, Fl. Brit. India 6: 465 (1893).
Daemonorops tabacina Becc. in J.D.Hooker, Fl. Brit. India 6: 466 (1893).
Daemonorops bakauensis Becc., Ann. Roy. Bot. Gard. (Calcutta) 12(1): 220 (1911).
Daemonorops curtisii Furtado, Gard. Bull. Straits Settlem. 9: 164 (1937).

Daemonorops loheriana Becc., Philipp. J. Sci., C 4: 637 (1909). *Daemonorops loheriana* (Becc.) Becc., Ann. Roy. Bot. Gard. (Calcutta) 12(1): 104 (1911).
Philippines. 42 PHI.

Daemonorops loheriana (Becc.) Becc. = **Daemonorops loheriana** Becc.

Daemonorops longipedunculata Furtado = **Daemonorops longipes** (Griff.) Mart.

Daemonorops longipes Miq. = **Daemonorops gracilipes** (Miq.) Becc.

Daemonorops longipes (Griff.) Mart., Hist. Nat. Palm. 3: 329 (1853).
W. Malesia to Philippines (Palawan). 42 BOR MLY PHI SUM.
Calamus longipes Griff., Calcutta J. Nat. Hist. 5: 68 (1845). *Rotang longipes* (Griff.) Baill., Hist. Pl. 13: 300 (1895).
Calamus ramosissimus Griff., Calcutta J. Nat. Hist. 5: 78 (1845). *Daemonorops ramosissima* (Griff.) Mart., Hist. Nat. Palm. 3: 330 (1853). *Palmijuncus ramosissimus* (Griff.) Kuntze, Revis. Gen. Pl. 2: 733 (1891).
Daemonorops stricta Blume, Rumphia 3: 19 (1847). *Calamus strictus* (Blume) Miq., Verh. Kon. Ned. Akad. Wetensch., Afd. Natuurk. 11: 28 (1868).
Daemonorops virescens Becc. in G.H.Perkins & al., Fragm. Fl. Philipp. 1: 47 (1904).
Daemonorops sabensis Becc. ex Gibbs, J. Linn. Soc., Bot. 42: 169 (1914).
Daemonorops calothyrsa Furtado, Gard. Bull. Straits Settlem. 8: 345 (1935).
Daemonorops longipedunculata Furtado, Gard. Bull. Straits Settlem. 8: 353 (1935).

Daemonorops longispatha Becc., Rec. Bot. Surv. India 2: 230 (1902).
Borneo. 42 BOR.

Daemonorops longispinosa Burret, Notizbl. Bot. Gart. Berlin-Dahlem 15: 196 (1940).
Sumatera. 42 SUM. — Provisionally accepted.

Daemonorops longistipes Burret, Notizbl. Bot. Gart. Berlin-Dahlem 15: 798 (1943).
Borneo (Sabah, Sarawak). 42 BOR.
Daemonorops pleioclada Burret, Notizbl. Bot. Gart. Berlin-Dahlem 15: 797 (1943).

Daemonorops macrophylla Becc. in J.D.Hooker, Fl. Brit. India 6: 470 (1893).
Pen. Thailand to Pen. Malaysia. 41 THA 42 MLY.

Daemonorops macroptera (Miq.) Becc., Rec. Bot. Surv. India 2: 223 (1902).
Sulawesi. 42 SUL.
Calamus macropterus Miq., Verh. Kon. Ned. Akad. Wetensch., Afd. Natuurk. 11: 19 (1868). *Palmijuncus macropterus* (Miq.) Kuntze, Revis. Gen. Pl. 2: 733 (1891).

Daemonorops maculata J.Dransf., Kew Bull. 45: 76 (1990).
NW. Borneo. 42 BOR.

Daemonorops malaccensis Mart. = **Daemonorops grandis** (Griff.) Mart.

Daemonorops manii Becc. in J.D.Hooker, Fl. Brit. India 6: 463 (1893).
Andaman Is. 41 AND.

Daemonorops manillensis Mart. = **Calamus manillensis** (Mart.) H.Wendl.

Daemonorops margaritae (Hance) Becc., Rec. Bot. Surv. India 2: 220 (1902).
SE. China to Philippines. 36 CHH CHS 38 TAI 42 PHI.
Calamus margaritae Hance, J. Bot. 12: 266 (1874).

Palmijuncus margaritae (Hance) Kuntze, Revis. Gen. Pl. 2: 733 (1891).

var. **margaritae**
SE. China, Hainan, Taiwan. 36 CHH CHS 38 TAI.

var. **palawanica** Becc., Ann. Roy. Bot. Gard. (Calcutta) 12(1): 57 (1911).
Philippines (Palawan). 42 PHI.

Daemonorops marginata Blume = **Calamus marginatus** (Blume) Mart.

Daemonorops mattanensis Becc. = **Daemonorops didymophylla** Becc.

Daemonorops megalocarpa Burret, Notizbl. Bot. Gart. Berlin-Dahlem 15: 194 (1940).
Sumatera. 42 SUM.

Daemonorops melanochaetes Blume in J.J.Roemer & J.A.Schultes, Syst. Veg. 7: 1333 (1830). *Calamus melanochaetes* (Blume) Miq., Verh. Kon. Ned. Akad. Wetensch., Afd. Natuurk. 11: 28 (1868).
Pen. Thailand to W. Malesia. 41 THA 42 JAW MLY SUM.
Daemonorops javanica Furtado, Gard. Bull. Straits Settlem. 9: 170 (1937).

Daemonorops melanolepis Mart. = **Calamus wightii** Griff.

Daemonorops micracantha (Griff.) Becc. in J.D.Hooker, Fl. Brit. India 6: 467 (1893).
Pen. Malaysia, Borneo. 42 BOR MLY.
Calamus micracanthus Griff., Calcutta J. Nat. Hist. 5: 62 (1845). *Palmijuncus micracanthus* (Griff.) Kuntze, Revis. Gen. Pl. 2: 733 (1891). *Rotang micracanthus* (Griff.) Baill., Hist. Pl. 13: 299 (1895). *Daemonorops draconcella* Becc., Nelle Forest. Borneo: 608 (1902).

Daemonorops microcarpa Burret, Notizbl. Bot. Gart. Berlin-Dahlem 15: 195 (1940).
Sumatera. 42 SUM. — Provisionally accepted.

Daemonorops microstachys Becc., Rec. Bot. Surv. India 2: 225 (1902).
Borneo. 42 BOR.

Daemonorops microthamnus Becc. = **Daemonorops calicarpa** (Griff.) Mart.

Daemonorops mirabilis (Mart.) Mart., Hist. Nat. Palm. 3: 329 (1853).
S. Borneo. 42 BOR. — Provisionally accepted.
Calamus mirabilis Mart., Hist. Nat. Palm. 3: 213 (1838). *Palmijuncus mirabilis* (Mart.) Kuntze, Revis. Gen. Pl. 2: 733 (1891).

Daemonorops mollis (Blanco) Merr., Sp. Blancoan.: 86 (1918).
Philippines. 42 PHI.
Calamus mollis Blanco, Fl. Filip.: 264 (1837). *Palmijuncus mollis* (Blanco) Kuntze, Revis. Gen. Pl. 2: 733 (1891). *Daemonorops fusca* Mart., Hist. Nat. Palm. 3: 331 (1853). *Daemonorops gaudichaudii* Mart., Hist. Nat. Palm. 3: 331 (1853). *Calamus gaudichaudii* (Mart.) H.Wendl. in O.C.E.de Kerchove de Denterghem, Palmiers: 236 (1878). *Palmijuncus gaudichaudii* (Mart.) Kuntze, Revis. Gen. Pl. 2: 733 (1891).

Daemonorops mollispina J.Dransf., Kew Bull. 56: 662 (2001).
Vietnam. 41 VIE.

Daemonorops monticola (Griff.) Mart., Hist. Nat. Palm. 3: 328 (1853).
Pen. Thailand to Pen. Malaysia. 41 THA 42 MLY.

Calamus monticolus Griff., Calcutta J. Nat. Hist. 5: 90 (1845). *Palmijuncus monticolus* (Griff.) Kuntze, Revis. Gen. Pl. 2: 733 (1891).

Daemonorops motleyi Becc. = **Daemonorops didymophylla** Becc.

Daemonorops nigra (Willd.) Blume, Rumphia 3: 5 (1847).
Maluku. 42 MOL. — Provisionally accepted.
Calamus niger Willd., Sp. Pl. 2: 203 (1799). *Palmijuncus niger* (Willd.) Kuntze, Revis. Gen. Pl. 2: 732 (1891). *Rotang niger* (Willd.) Baill., Hist. Pl. 13: 299 (1895).

Daemonorops nurii Furtado = **Daemonorops sepal** Becc.

Daemonorops nutantiflora (Griff.) Mart. = **Daemonorops jenkinsiana** (Griff.) Mart.

Daemonorops oblata J.Dransf., Bot. J. Linn. Soc. 81: 18 (1980).
Borneo. 42 BOR.

Daemonorops oblonga (Reinw. ex Blume) Blume, Rumphia 3: 25 (1847).
SE. Sumatera ?, W. Jawa. 42 JAW SUM?
Calamus oblongus Reinw. ex Blume in J.J.Roemer & J.A.Schultes, Syst. Veg. 7: 1323 (1830). *Calamus platyacanthus* Mart., Hist. Nat. Palm. 3: 206 (1838). *Daemonorops platyacantha* (Mart.) Mart., Hist. Nat. Palm. 3(ed. 2): 204 (1845). *Palmijuncus platyacanthus* (Mart.) Kuntze, Revis. Gen. Pl. 2: 732 (1891).

Daemonorops ochreata Teijsm. & Binn. = **Korthalsia echinometra** Becc.

Daemonorops ochrolepis Becc. in G.H.Perkins & al., Fragm. Fl. Philipp. 1: 47 (1904).
Philippines. 42 PHI.

Daemonorops oligolepis Becc., Leafl. Philipp. Bot. 8: 3035 (1919).
Philippines (Mindanao). 42 PHI.

Daemonorops oligophylla Becc. in J.D.Hooker, Fl. Brit. India 6: 467 (1893).
Pen. Malaysia (Perak). 42 MLY.

Daemonorops oxycarpa Becc., Nelle Forest. Borneo: 607 (1902).
Borneo. 42 BOR.

Daemonorops pachyrostris Becc., Ann. Roy. Bot. Gard. (Calcutta) 12(1): 217 (1911).
Borneo (Kalimantan). 42 BOR.

Daemonorops palembanica Blume, Rumphia 3: 20 (1847). *Calamus palembanicus* (Blume) Miq., Verh. Kon. Ned. Akad. Wetensch., Afd. Natuurk. 11: 29 (1868). *Palmijuncus palembanicus* (Blume) Kuntze, Revis. Gen. Pl. 2: 732 (1891). *Rotang palembanicus* (Blume) Baill., Hist. Pl. 13: 300 (1895).
Sumatera. 42 SUM.

Daemonorops pannosa Becc., Leafl. Philipp. Bot. 8: 3033 (1919).
Philippines (Mindanao). 42 PHI.

Daemonorops pedicellaris Becc., Leafl. Philipp. Bot. 8: 3040 (1919).
Philippines (Leyte, Mindanao). 42 PHI.

Daemonorops periacantha Miq., Fl. Ned. Ind., Eerste Bijv.: 593 (1861). *Calamus periacanthus* (Miq.) Miq., Verh. Kon. Ned. Akad. Wetensch., Afd. Natuurk. 11: 28 (1868). *Palmijuncus periacanthus* (Miq.) Kuntze, Revis. Gen. Pl. 2: 732 (1891). *Rotang periacanthus*

(Miq.) Baill., Hist. Pl. 13: 300 (1895).
W. Malesia. 42 BOR MLY SUM.

Daemonorops dissitophylla Becc., Nelle Forest. Borneo: 608 (1902).

Daemonorops florida Becc., Ann. Roy. Bot. Gard. (Calcutta) 12(1): 230 (1911).

Daemonorops petiolaris (Griff.) Mart. = **Daemonorops calicarpa** (Griff.) Mart.

Daemonorops pierreana Becc. = **Daemonorops jenkinsiana** (Griff.) Mart.

Daemonorops plagiocycla Burret, Notizbl. Bot. Gart. Berlin-Dahlem 15: 194 (1940).
Sumatera. 42 SUM. — Provisionally accepted.

Daemonorops platyacantha (Mart.) Mart. = **Daemonorops oblonga** (Reinw. ex Blume) Blume

Daemonorops platyspatha (Mart. ex Kunth) Mart. = **Calamus platyspathus** Mart. ex Kunth

Daemonorops pleioclada Burret = **Daemonorops longistipes** Burret

Daemonorops plumosus W.Bull = **?**

Daemonorops poilanei J.Dransf., Kew Bull. 56: 663 (2001).
Vietnam. 41 VIE.

Daemonorops polita Fernando, Gard. Bull. Singapore 41: 56 (1988 publ. 1989).
Philippines (Mindanao). 42 PHI.

Daemonorops propinqua Becc. in J.D.Hooker, Fl. Brit. India 6: 467 (1893).
Pen. Thailand to Sumatera. 41 THA 42 MLY SUM.

Daemonorops pseudomirabilis Becc. = **Daemonorops sabut** Becc.

Daemonorops pseudosepal Becc. = **Daemonorops lewisiana** (Griff.) Mart.

Daemonorops pumila Van Valk., Blumea 40: 465 (1995).
Borneo (Kalimantan). 42 BOR.

Daemonorops ramosissima (Griff.) Mart. = **Daemonorops longipes** (Griff.) Mart.

Daemonorops rarispinosa Renuka & Vijayak., Rheedea 4: 125 (1994).
Andaman Is. 41 AND.

Daemonorops rheedei (Griff.) Mart. = **Calamus rheedei** Griff.

Daemonorops riedeliana (Miq.) Becc., Rec. Bot. Surv. India 2: 226 (1902).
N. Sulawesi. 42 SUL.

  Calamus riedelianus Miq., Verh. Kon. Ned. Akad. Wetensch., Afd. Natuurk. 11: 18 (1868). *Palmijuncus riedelianus* (Miq.) Kuntze, Revis. Gen. Pl. 2: 733 (1891).

Daemonorops robusta Warb. ex Becc., Ann. Roy. Bot. Gard. (Calcutta) 12(1): 101 (1911).
N. Sulawesi to Maluku. 42 MOL SUL.

Daemonorops rubra (Reinw. ex Mart.) Blume, Rumphia 3: 6 (1847).
W. Jawa. 42 JAW.

  Calamus ruber Reinw. ex Mart., Hist. Nat. Palm. 3: 209 (1838). *Palmijuncus ruber* (Reinw. ex Mart.) Kuntze, Revis. Gen. Pl. 2: 732 (1891).

  Daemonorops accedens Blume, Rumphia 2: viii (1838). *Calamus accedens* (Blume) Miq., Verh. Kon. Ned. Akad. Wetensch., Afd. Natuurk. 11: 28

(1868). *Palmijuncus accedens* (Blume) Kuntze, Revis. Gen. Pl. 2: 734 (1891). *Rotang accedens* (Blume) Baill., Hist. Pl. 13: 300 (1895).

Daemonorops rumphii (Blume) Mart. = **Calamus rumphii** Blume

Daemonorops ruptilis Becc., Rec. Bot. Surv. India 2: 230 (1902).
Borneo. 42 BOR.

  var. **acaulescens** J.Dransf., Bot. J. Linn. Soc. 81: 23 (1980).
  Borneo. 42 BOR.

  var. **ruptilis**
  N. Borneo. 42 BOR.

Daemonorops sabensis Becc. ex Gibbs = **Daemonorops longipes** (Griff.) Mart.

Daemonorops sabut Becc. in J.D.Hooker, Fl. Brit. India 6: 469 (1893).
Pen. Thailand to W. Malesia. 41 THA 42 BOR MLY SUM?

  Daemonorops annulata Becc., Rec. Bot. Surv. India 2: 227 (1902).

  Daemonorops pseudomirabilis Becc., Rec. Bot. Surv. India 2: 226 (1902).

  Daemonorops turbinata Becc., Ann. Roy. Bot. Gard. (Calcutta) 12(1): 225 (1911).

Daemonorops sarasinorum Warb. ex Becc., Ann. Roy. Bot. Gard. (Calcutta) 12(1): 100 (1911).
N. Sulawesi. 42 SUL.

Daemonorops scapigera Becc., Rec. Bot. Surv. India 2: 228 (1902).
Pen. Malaysia (Johor), Borneo (incl. Kep. Natuna), Sumatera. 42 BOR MLY SUM.

  Daemonorops lasiospatha Furtado, Gard. Bull. Straits Settlem. 8: 351 (1935).

  Calamus bifacialis Burret, Notizbl. Bot. Gart. Berlin-Dahlem 15: 809 (1943).

Daemonorops schlechteri Burret, Notizbl. Bot. Gart. Berlin-Dahlem 15: 799 (1943).
Sulawesi. 42 SUL. — Provisionally accepted.

Daemonorops sepal Becc. in J.D.Hooker, Fl. Brit. India 6: 465 (1893).
Pen. Thailand to Pen. Malaysia. 41 THA 42 MLY.

  Daemonorops imbellis Becc., Rec. Bot. Surv. India 2: 220 (1902).

  Daemonorops aciculata Ridl., Mat. Fl. Malay. Penins. 2: 176 (1907).

  Daemonorops scortechinii Becc., Ann. Roy. Bot. Gard. (Calcutta) 12(1): 81 (1911).

  Daemonorops kiahii Furtado, Gard. Bull. Singapore 14: 73 (1953).

  Daemonorops nurii Furtado, Gard. Bull. Singapore 14: 85 (1953).

Daemonorops schmidtiana Becc. = **Daemonorops jenkinsiana** (Griff.) Mart.

Daemonorops scortechinii Becc. = **Daemonorops sepal** Becc.

Daemonorops serpentina J.Dransf., Kew Bull. 36: 810 (1982).
N. Borneo. 42 BOR.

Daemonorops setigera Ridl. = **Daemonorops verticillaris** (Griff.) Mart.

Daemonorops siberutensis Rustiami, Kew Bull. 57: 729 (2002).
Sumatera (Siberut). 42 SUM.

Daemonorops singalana Becc., Rec. Bot. Surv. India 2: 219 (1902).
Sumatera. 42 SUM.

Daemonorops sparsiflora Becc., Rec. Bot. Surv. India 2: 224 (1902).
Borneo. 42 BOR.

Daemonorops spectabilis Becc., Ann. Roy. Bot. Gard. (Calcutta) 12(1): 228 (1911).
Borneo. 42 BOR.

Daemonorops stenophylla Becc., Rec. Bot. Surv. India 2: 220 (1902).
Sumatera. 42 SUM.

Daemonorops stipitata Furtado = **Daemonorops verticillaris** (Griff.) Mart.

Daemonorops stricta Blume = **Daemonorops longipes** (Griff.) Mart.

Daemonorops tabacina Becc. = **Daemonorops lewisiana** (Griff.) Mart.

Daemonorops treubiana Becc., Ann. Roy. Bot. Gard. (Calcutta) 12(1): 75 (1911).
Malesia (?). 42 +.

Daemonorops trichroa Miq., Fl. Ned. Ind., Eerste Bijv.: 952 (1861). *Calamus trichrous* (Miq.) Miq., Verh. Kon. Ned. Akad. Wetensch., Afd. Natuurk. 11: 28 (1868).
Sumatera. 42 SUM.

Daemonorops turbinata Becc. = **Daemonorops sabut** Becc.

Daemonorops unijuga J.Dransf., Bot. J. Linn. Soc. 81: 24 (1980).
Borneo (Sarawak). 42 BOR.

Daemonorops urdanetana Becc., Leafl. Philipp. Bot. 8: 3038 (1919).
Philippines (Mindanao). 42 PHI.

Daemonorops ursina Becc. = **Pogonotium ursinum** (Becc.) J.Dransf.

Daemonorops uschdraweitiana Burret, Notizbl. Bot. Gart. Berlin-Dahlem 15: 191 (1940).
Sumatera. 42 SUM.

Daemonorops vagans Becc. = **Daemonorops kunstleri** Becc.

Daemonorops verticillaris (Griff.) Mart., Hist. Nat. Palm. 3: 329 (1853).
Pen. Thailand to Sumatera. 41 THA 42 MLY SUM.
    Calamus verticillaris Griff., Calcutta J. Nat. Hist. 5: 63 (1845). *Palmijuncus verticillaris* (Griff.) Kuntze, Revis. Gen. Pl. 2: 733 (1891).
    Daemonorops setigera Ridl., Fl. Malay Penins. 5: 45 (1925).
    Daemonorops stipitata Furtado, Gard. Bull. Singapore 14: 142 (1953).

Daemonorops virescens Becc. = **Daemonorops longipes** (Griff.) Mart.

Daemonorops wrightmyoensis Renuka & Vijayak., Rheedea 4: 125 (1994).
Andaman Is. 41 AND.

Dahlgrenia

Dahlgrenia Steyerm. = **Dictyocaryum** H.Wendl.

Dahlgrenia ptariana Steyerm. = **Dictyocaryum ptarianum** (Steyerm.) H.E.Moore & Steyerm.

Dammera

Dammera K.Schum. & Lauterb. = **Licuala** Wurmb

Dammera ramosa K.Schum. & Lauterb. = **Licuala beccariana** (K.Schum. & Lauterb.) Furtado

Dammera simplex K.Schum. & Lauterb. = **Licuala simplex** (K.Schum. & Lauterb.) Becc.

Dasystachys

Dasystachys Oerst. = **Chamaedorea** Willd.

Dasystachys deckeriana (Klotzsch) Oerst. = **Chamaedorea deckeriana**

Deckenia

Deckenia H.Wendl. ex Seem., Gard. Chron. 1870: 561 (1870).
1 species, Seychelles. 29.

Deckenia nobilis H.Wendl. ex Seem., Gard. Chron. 1870: 561 (1870).
Seychelles. 29 SEY.

Deckeria

Deckeria H.Karst. = **Iriartea** Ruiz & Pav.

Deckeria corneto H.Karst. = **Iriartea deltoidea** Ruiz & Pav.

Deckeria elegans Linden = **?**

Deckeria lamarckiana (Mart.) H.Karst. = **Dictyocaryum lamarckianum** (Mart.) H.Wendl.

Deckeria nobilis H.Wendl. ex Seem. = **?**

Deckeria phaeocarpa (Mart.) H.Karst. = **Iriartea deltoidea** Ruiz & Pav.

Deckeria ventricosa (Mart.) H.Karst. = **Iriartea deltoidea** Ruiz & Pav.

Dendrema

Dendrema Raf. = **Corypha** L.

Denea

Denea O.F.Cook = **Howea** Becc.

Denea forsteriana (C.Moore & F.Muell.) O.F.Cook = **Howea forsteriana** (C.Moore & F.Muell.) Becc.

Desmonchus

Desmonchus Desf. = **Desmoncus** Mart.

Desmoncus

Desmoncus Mart., Hist. Nat. Palm. 2: 84 (1824). *Atitara* Barrère ex Kuntze, Revis. Gen. Pl. 2: 727 (1891), nom. illeg.
12 species, Trop. America. 79 80 81 82 83 84.
    Desmonchus Desf., Tabl. École Bot., ed. 3: 30 (1829), orth. var.

Desmoncus aereus Drude = **Desmoncus polyacanthos** var. **polyacanthos**

Desmoncus andicola Pasq. = **?** [83]

Desmoncus angustisectus Burret = **Desmoncus orthacanthos** Mart.

Desmoncus anomalus Bartlett, J. Wash. Acad. Sci. 25: 84 (1935).
Guatemala. 80 GUA.

Desmoncus apureanus L.H.Bailey = **Desmoncus orthacanthos** Mart.

Desmoncus ataxacanthus Barb.Rodr. = **Desmoncus orthacanthos** Mart.

Desmoncus brevisectus Burret = **Desmoncus polyacanthos** var. **polyacanthos**

Desmoncus brittonii L.H.Bailey = **Desmoncus orthacanthos** Mart.

Desmoncus caespitosus Barb.Rodr. = **Desmoncus polyacanthos** var. **polyacanthos**

Desmoncus campylacanthus Burret = **Desmoncus polyacanthos** var. **polyacanthos**

Desmoncus chinantlensis Liebm. ex Mart., Hist. Nat. Palm. 3: 321 (1853). *Atitara chinantlensis* (Liebm. ex Mart.) Kuntze, Revis. Gen. Pl. 2: 727 (1891).
SE. Mexico. 79 MXT.

Desmoncus cirrhiferus A.H.Gentry & Zardini, Ann. Missouri Bot. Gard. 75: 1436 (1988 publ. 1989).
Colombia to Ecuador. 83 CLM ECU.

Desmoncus costaricensis (Kuntze) Burret, Repert. Spec. Nov. Regni Veg. 36: 202 (1934).
Costa Rica to W. Panama. 80 COS PAN.
Atitara costaricensis Kuntze, Revis. Gen. Pl. 2: 726 (1891).

Desmoncus cuyabaensis Barb.Rodr. = **Desmoncus orthacanthos** Mart.

Desmoncus dasyacanthus Burret = **Desmoncus polyacanthos** var. **polyacanthos**

Desmoncus demeraranus L.H.Bailey & H.E.Moon = **Desmoncus orthacanthos** Mart.

Desmoncus duidensis Steyerm. = **Desmoncus polyacanthos** var. **polyacanthos**

Desmoncus ferox Bartlett = **Desmoncus schippii** Burret

Desmoncus giganteus A.J.Hend., Palms Amazon: 225 (1995).
W. South America to N. Brazil. 83 CLM ECU PER 84 BZN.

Desmoncus granatensis W.Bull = ? **[83 CLM]**

Desmoncus grandifolius Linden = ? **[83 CLM]**

Desmoncus hartii L.H.Bailey = **Desmoncus orthacanthos** Mart.

Desmoncus horridus Splitg. ex Mart. = **Desmoncus orthacanthos** Mart.

Desmoncus huebneri Burret = **Desmoncus orthacanthos** Mart.

Desmoncus inermis Barb.Rodr. = **Desmoncus polyacanthos** var. **polyacanthos**

Desmoncus intermedius Mart. ex H.Wendl. = **?**

Desmoncus isthmius L.H.Bailey, Gentes Herb. 6: 211 (1943).
E. Panama. 80 PAN.

Desmoncus kaieteurensis L.H.Bailey = **Desmoncus phoenicocarpus** Barb.Rodr.

Desmoncus kuhlmannii Burret = **Desmoncus orthacanthos** Mart.

Desmoncus latifrons W.Bull = **?**

Desmoncus latisectus Burret = **Desmoncus polyacanthos** var. **polyacanthos**

Desmoncus leiorhachis Burret = **Desmoncus schippii** Burret

Desmoncus leptochaete Burret = **Desmoncus orthacanthos** Mart.

Desmoncus leptoclonos Drude = **Desmoncus polyacanthos** var. **polyacanthos**

Desmoncus leptospadix Mart. = **Desmoncus mitis** var. **leptospadix** (Mart.) A.J.Hend.

Desmoncus longifolius Mart. = **Desmoncus orthacanthos** Mart.

Desmoncus longisectus Burret = **Desmoncus polyacanthos** var. **polyacanthos**

Desmoncus lophacanthos Mart. = **Desmoncus orthacanthos** Mart.

Desmoncus luetzelburgii Burret = **Desmoncus orthacanthos** Mart.

Desmoncus lundellii Bartlett = **Desmoncus schippii** Burret

Desmoncus macroacanthos Mart. = **Desmoncus polyacanthos** var. **polyacanthos**

Desmoncus macrocarpus Barb.Rodr. = **Desmoncus orthacanthos** Mart.

Desmoncus macrodon Barb.Rodr. = **Desmoncus phoenicocarpus** Barb.Rodr.

Desmoncus maguirei L.H.Bailey = **Desmoncus polyacanthos** var. **polyacanthos**

Desmoncus major Crueg. ex Griseb. = **Desmoncus orthacanthos** Mart.

Desmoncus melanacanthos Mart. ex Drude = **Desmoncus orthacanthos** Mart.

Desmoncus mirandanus L.H.Bailey = **Desmoncus polyacanthos** var. **polyacanthos**

Desmoncus mitis Mart., Hist. Nat. Palm. 2: 90 (1824). *Atitara mitis* (Mart.) Kuntze, Revis. Gen. Pl. 2: 727 (1891).
S. Trop. America. 82 VEN 83 BOL CLM ECU PER 84 BZN.

var. **leptoclonos** A.J.Hend., Palms Amazon: 227 (1995).
N. Brazil to Bolivia. 83 BOL 84 BZN.

var. **leptospadix** (Mart.) A.J.Hend., Palms Amazon: 228 (1995).
Colombia to Peru. 83 CLM PER.
Desmoncus leptospadix Mart. in A.D.d'Orbigny, Voy. Amér. Mér. 7(3): 52 (1844). *Atitara leptospadix* (Mart.) Kuntze, Revis. Gen. Pl. 2: 727 (1891).

var. **mitis**
S. Trop. America. 82 VEN 83 CLM ECU PER 84 BZN.
Desmoncus pumilus Trail, J. Bot. 14: 353 (1876). *Atitara pumila* (Trail) Kuntze, Revis. Gen. Pl. 2: 727 (1891).
Desmoncus setosus var. *mitescens* Mart. ex Drude in C.F.P.von Martius & auct. suc. (eds.), Fl. Bras. 3(2): 316 (1881).

var. **rurrenabaquensis** A.J.Hend., Palms Amazon: 228 (1995).
Peru to Bolivia. 83 BOL PER.

var. **tenerrimus** (Mart. ex Drude) A.J.Hend., Palms Amazon: 228 (1995).
W. South America to N. Brazil. 83 CLM PER 84 BZN.
Bactris tenerrima Mart. ex Drude in C.F.P.von Martius & auct. suc. (eds.), Fl. Bras. 3(2): 328 (1881). *Desmoncus tenerrimus* (Mart. ex Drude)

Mart. ex Burret, Repert. Spec. Nov. Regni Veg. 34: 236 (1934).

Desmoncus vacivus L.H.Bailey, Gentes Herb. 8: 186 (1949).

Desmoncus multijugus Steyerm. = **Desmoncus orthacanthos** Mart.

Desmoncus myriacanthos Dugand = **Desmoncus orthacanthos** Mart.

Desmoncus nemorosus Barb.Rodr. = **Desmoncus phoenicocarpus** Barb.Rodr.

Desmoncus oligacanthus Barb.Rodr. = **Desmoncus polyacanthos** var. **polyacanthos**

Desmoncus orthacanthos Mart., Hist. Nat. Palm. 2: 87 (1824). *Atitara orthacantha* (Mart.) Kuntze, Revis. Gen. Pl. 2: 727 (1891).
Trinidad, S. Trop. America. 81 TRT 82 FRG GUY SUR VEN 83 BOL CLM ECU PER 84 BZC BZL BZN.
Desmoncus horridus Splitg. ex Mart. in A.D.d'Orbigny, Voy. Amér. Mér. 7(3): 51 (1844). *Atitara horrida* (Splitg. ex Mart.) Kuntze, Revis. Gen. Pl. 2: 727 (1891).
Desmoncus longifolius Mart. in A.D.d'Orbigny, Voy. Amér. Mér. 7(3): 52 (1844).
Desmoncus lophacanthos Mart. in A.D.d'Orbigny, Voy. Amér. Mér. 7(3): 50 (1844). *Atitara lophacantha* (Mart.) Barb.Rodr., Contr. Jard. Bot. Rio de Janeiro 3: 76 (1902).
Desmoncus rudentum Mart. in A.D.d'Orbigny, Voy. Amér. Mér. 7(3): 48 (1844). *Atitara rudenta* (Mart.) Barb.Rodr., Contr. Jard. Bot. Rio de Janeiro 3: 75 (1902).
Desmoncus major Crueg. ex Griseb., Fl. Brit. W. I.: 519 (1864). *Atitara major* (Crueg. ex Griseb.) Kuntze, Revis. Gen. Pl. 2: 727 (1891).
Desmoncus ataxacanthus Barb.Rodr., Enum. Palm. Nov.: 25 (1875). *Atitara ataxacantha* (Barb.Rodr.) Kuntze, Revis. Gen. Pl. 2: 727 (1891).
Desmoncus palustris Trail, J. Bot. 14: 353 (1876). *Atitara palustris* (Trail) Kuntze, Revis. Gen. Pl. 2: 727 (1891).
Desmoncus melanacanthos Mart. ex Drude in C.F.P.von Martius & auct. suc. (eds.), Fl. Bras. 3(2): 305 (1881).
Desmoncus macrocarpus Barb.Rodr., Vellosia 1: 34 (1888). *Atitara macrocarpa* (Barb.Rodr.) Barb. Rodr., Contr. Jard. Bot. Rio de Janeiro 3: 75 (1902).
Atitara chinantlensis (Liebm. ex Mart.) Kuntze, Revis. Gen. Pl. 2: 727 (1891).
Atitara drudeana Kuntze, Revis. Gen. Pl. 2: 727 (1891).
Desmoncus cuyabaensis Barb.Rodr., Palm. Mattogross.: 30 (1898). *Atitara cuyabaensis* (Barb.Rodr.) Barb.Rodr., Contr. Jard. Bot. Rio de Janeiro 2: 75 (1902).
Desmoncus prostratus Lindm., Bih. Kongl. Svenska Vetensk.-Akad. Handl. 26(5): 8 (1901). *Atitara prostrata* (Lindm.) Barb.Rodr., Contr. Jard. Bot. Rio de Janeiro 3: 75 (1902).
Desmoncus angustisectus Burret, Notizbl. Bot. Gart. Berlin-Dahlem 10: 1025 (1930).
Desmoncus luetzelburgii Burret, Notizbl. Bot. Gart. Berlin-Dahlem 10: 1025 (1930).
Desmoncus werdermannii Burret, Repert. Spec. Nov. Regni Veg. 32: 114 (1933).
Desmoncus huebneri Burret, Repert. Spec. Nov. Regni Veg. 36: 200 (1934).
Desmoncus leptochaete Burret, Repert. Spec. Nov. Regni Veg. 36: 204 (1934).

Desmoncus kuhlmannii Burret, Notizbl. Bot. Gart. Berlin-Dahlem 14: 267 (1938).
Desmoncus myriacanthos Dugand, Caldasia 2: 75 (1943).
Desmoncus brittonii L.H.Bailey, Gentes Herb. 7: 371 (1947).
Desmoncus hartii L.H.Bailey, Gentes Herb. 7: 369 (1947).
Desmoncus tobagonis L.H.Bailey, Gentes Herb. 7: 371 (1947).
Desmoncus apureanus L.H.Bailey, Gentes Herb. 8: 183 (1949).
Desmoncus demeraranus L.H.Bailey & H.E.Moon, Gentes Herb. 8: 181 (1949).
Desmoncus velezii L.H.Bailey, Gentes Herb. 8: 186 (1949).
Desmoncus multijugus Steyerm., Fieldiana, Bot. 28(1): 85 (1951).

Desmoncus oxyacanthos Mart. = **Desmoncus polyacanthos** var. **polyacanthos**

Desmoncus palustris Trail = **Desmoncus orthacanthos** Mart.

Desmoncus panamensis Linden = **?**

Desmoncus paraensis (Barb.Rodr.) Barb.Rodr. = **Desmoncus polyacanthos** var. **polyacanthos**

Desmoncus parvulus L.H.Bailey = **Desmoncus phoenicocarpus** Barb.Rodr.

Desmoncus peraltus L.H.Bailey = **Desmoncus polyacanthos** var. **polyacanthos**

Desmoncus phengophyllus Drude = **Desmoncus polyacanthos** var. **polyacanthos**

Desmoncus philippianus Barb.Rodr. = **Desmoncus polyacanthos** var. **polyacanthos**

Desmoncus phoenicocarpus Barb.Rodr., Enum. Palm. Nov.: 24 (1875). *Atitara phoenicocarpa* (Barb.Rodr.) Kuntze, Revis. Gen. Pl. 2: 727 (1891).
N. South America to Brazil. 82 FRG GUY SUR VEN 84 BZE BZN.
Desmoncus macrodon Barb.Rodr., Vellosia 1: 39 (1888). *Atitara macrodon* (Barb.Rodr.) Barb.Rodr., Contr. Jard. Bot. Rio de Janeiro 3: 75 (1902).
Desmoncus nemorosus Barb.Rodr., Vellosia 1: 36 (1888). *Atitara nemorosa* (Barb.Rodr.) Barb.Rodr., Contr. Jard. Bot. Rio de Janeiro 3: 75 (1902).
Desmoncus kaieteurensis L.H.Bailey, Bull. Torrey Bot. Club 75: 115 (1948).
Desmoncus parvulus L.H.Bailey, Bull. Torrey Bot. Club 75: 115 (1948).

Desmoncus polyacanthos Mart., Hist. Nat. Palm. 2: 85 (1824). *Atitara polyacantha* (Mart.) Kuntze, Revis. Gen. Pl. 2: 726 (1891).
S. Caribbean to S. Trop. America. 81 WIN TRT 82 FRG GUY SUR VEN 83 BOL CLM ECU PER 84 BZC BZE BZL BZN.

var. **polyacanthos**
S. Caribbean to S. Trop. America. 81 WIN TRT 82 FRG GUY SUR VEN 83 BOL CLM ECU PER 84 BZC BZE BZL BZN.
Desmoncus macroacanthos Mart., Hist. Nat. Palm. 2: 86 (1824). *Atitara macroacantha* (Mart.) Kuntze, Revis. Gen. Pl. 2: 727 (1891).
Desmoncus oxyacanthos Mart., Hist. Nat. Palm. 2: 88 (1824). *Atitara oxyacantha* (Mart.) Kuntze, Revis. Gen. Pl. 2: 727 (1891).
Desmoncus pycnacanthos Mart., Hist. Nat. Palm. 2: 89 (1824). *Atitara pycnacantha* (Mart.) Kuntze,

Revis. Gen. Pl. 2: 727 (1891).

Desmoncus setosus Mart., Hist. Nat. Palm. 2: 89 (1824). *Atitara setosa* (Mart.) Kuntze, Revis. Gen. Pl. 2: 727 (1891).

Desmoncus riparius Spruce, J. Linn. Soc., Bot. 11: 156 (1871). *Atitara riparia* (Spruce) Kuntze, Revis. Gen. Pl. 2: 727 (1891).

Desmoncus oligacanthus Barb.Rodr., Enum. Palm. Nov.: 24 (1875). *Atitara oligacantha* (Barb.Rodr.) Kuntze, Revis. Gen. Pl. 2: 727 (1891).

Desmoncus aereus Drude in C.F.P.von Martius & auct. suc. (eds.), Fl. Bras. 3(2): 307 (1881). *Atitara aerea* (Drude) Barb.Rodr., Contr. Jard. Bot. Rio de Janeiro 3: 75 (1902).

Desmoncus leptoclonos Drude in C.F.P.von Martius & auct. suc. (eds.), Fl. Bras. 3(2): 315 (1881). *Atitara leptoclona* (Drude) Barb.Rodr., Contr. Jard. Bot. Rio de Janeiro 3: 76 (1902).

Desmoncus phengophyllus Drude in C.F.P.von Martius & auct. suc. (eds.), Fl. Bras. 3(2): 314 (1881). *Atitara phengophylla* (Drude) Kuntze, Revis. Gen. Pl. 2: 727 (1891).

Desmoncus caespitosus Barb.Rodr., Vellosia 1: 37 (1888). *Atitara caespitosa* (Barb.Rodr.) Barb.Rodr., Contr. Jard. Bot. Rio de Janeiro 3: 76 (1902).

Desmoncus philippianus Barb.Rodr., Vellosia 1: 38 (1888). *Atitara philippiana* (Barb.Rodr.) Barb.Rodr., Contr. Jard. Bot. Rio de Janeiro 3: 76 (1902).

Atitara dubia Kuntze, Revis. Gen. Pl. 2: 727 (1891). Provisional synonym.

Desmoncus inermis Barb.Rodr., Contr. Jard. Bot. Rio de Janeiro 1: 17 (1901). *Atitara inermis* (Barb.Rodr.) Barb.Rodr., Contr. Jard. Bot. Rio de Janeiro 3: 76 (1902).

Atitara paraensis Barb.Rodr., Contr. Jard. Bot. Rio de Janeiro 3: 76 (1902). *Desmoncus paraensis* (Barb.Rodr.) Barb.Rodr., Sert. Palm. Brasil. 2: 57 (1903).

Desmoncus ulei Dammer, Verh. Bot. Vereins Prov. Brandenburg 48: 129 (1906 publ. 1907).

Desmoncus brevisectus Burret, Repert. Spec. Nov. Regni Veg. 36: 215 (1934).

Desmoncus campylacanthus Burret, Repert. Spec. Nov. Regni Veg. 36: 210 (1934).

Desmoncus dasyacanthus Burret, Repert. Spec. Nov. Regni Veg. 36: 213 (1934).

Desmoncus latisectus Burret, Repert. Spec. Nov. Regni Veg. 36: 215 (1934).

Desmoncus longisectus Burret, Repert. Spec. Nov. Regni Veg. 36: 212 (1934).

Desmoncus prestoei L.H.Bailey, Gentes Herb. 6: 215 (1943).

Desmoncus peraltus L.H.Bailey, Gentes Herb. 7: 373 (1947).

Desmoncus maguirei L.H.Bailey, Bull. Torrey Bot. Club 75: 108 (1948).

Desmoncus mirandanus L.H.Bailey, Gentes Herb. 8: 183 (1949).

Desmoncus duidensis Steyerm., Fieldiana, Bot. 28(1): 85 (1951).

var. **prunifer** (Poepp. ex Mart.) A.J.Hend., Palms Amazon: 233 (1995).
W. South America. 83 CLM ECU PER.
    **Desmoncus prunifer* Poepp. ex Mart., Hist. Nat. Palm. 2: 148 (1837). *Atitara prunifera* (Poepp. ex Mart.) Kuntze, Revis. Gen. Pl. 2: 727 (1891).

Desmoncus prestoei L.H.Bailey = **Desmoncus polyacanthos** var. **polyacanthos**

Desmoncus prostratus Lindm. = **Desmoncus orthacanthos** Mart.

Desmoncus prunifer Poepp. ex Mart. = **Desmoncus polyacanthos** var. **prunifer** (Poepp. ex Mart.) A.J.Hend.

Desmoncus pumilus Trail = **Desmoncus mitis** var. **mitis**

Desmoncus pycnacanthos Mart. = **Desmoncus polyacanthos** var. **polyacanthos**

Desmoncus quasillarius Bartlett = **Desmoncus schippii** Burret

Desmoncus riparius Spruce = **Desmoncus polyacanthos** var. **polyacanthos**

Desmoncus rudentum Mart. = **Desmoncus orthacanthos** Mart.

Desmoncus schippii Burret, Repert. Spec. Nov. Regni Veg. 36: 202 (1934).
C. America. 80 BLZ COS ELS GUA HON NIC.
    Desmoncus leiorhachis Burret, Repert. Spec. Nov. Regni Veg. 36: 203 (1934).
    Desmoncus ferox Bartlett, J. Wash. Acad. Sci. 25: 87 (1935).
    Desmoncus lundellii Bartlett, J. Wash. Acad. Sci. 25: 84 (1935).
    Desmoncus quasillarius Bartlett, J. Wash. Acad. Sci. 25: 85 (1935).
    Desmoncus uaxactunensis Bartlett, J. Wash. Acad. Sci. 25: 86 (1935).

Desmoncus setosus Mart. = **Desmoncus polyacanthos** var. **polyacanthos**

Desmoncus setosus var. *mitescens* Mart. ex Drude = **Desmoncus mitis** var. **mitis**

Desmoncus stans Grayum & Nevers, Principes 32: 106 (1988).
S. Costa Rica. 80 COS.

Desmoncus tenerrimus (Mart. ex Drude) Mart. ex Burret = **Desmoncus mitis** var. **tenerrimus** (Mart. ex Drude) A.J.Hend.

Desmoncus tobagonis L.H.Bailey = **Desmoncus orthacanthos** Mart.

Desmoncus uaxactunensis Bartlett = **Desmoncus schippii** Burret

Desmoncus ulei Dammer = **Desmoncus polyacanthos** var. **polyacanthos**

Desmoncus vacivus L.H.Bailey = **Desmoncus mitis** var. **tenerrimus** (Mart. ex Drude) A.J.Hend.

Desmoncus velezii L.H.Bailey = **Desmoncus orthacanthos** Mart.

Desmoncus wallisii Linden = ? [84]

Desmoncus werdermannii Burret = **Desmoncus orthacanthos** Mart.

Dicrosperma

Dicrosperma W.Watson = **Dictyosperma** H.Wendl. & Drude

Dictyocaryum

Dictyocaryum H.Wendl., Bonplandia 8: 106 (1860).
3 species, Panama to S. Trop. America. 80 82 83 84.
    Dahlgrenia Steyerm., Fieldiana, Bot. 28(1): 82 (1951).

Dictyocaryum fuscum (H.Wendl.) H.Wendl., Bot. Zeitung (Berlin) 21: 131 (1863).
Venezuela. 82 VEN.
　Socratea fusca H.Karst., Fl. Columb. 1: 109 (1860).
　Iriartea fusca (H.Karst.) Drude in H.G.A.Engler & K.A.E.Prantl (eds.), Nat. Pflanzenfam. 2(3): 60 (1887).
　Iriartea altissima Klotzsch ex Al.Jahn, Palm. Fl. Venez.: 52 (1908).

Dictyocaryum glaucescens Linden = **?**

Dictyocaryum globiferum Dugand = **Dictyocaryum lamarckianum** (Mart.) H.Wendl.

Dictyocaryum lamarckianum (Mart.) H.Wendl., Bot. Zeitung (Berlin) 21: 131 (1863).
Panama to Bolivia. 80 PAN 82 VEN 83 BOL CLM ECU PER.
　Iriartea lamarckiana Mart., Hist. Nat. Palm. 3: 190 (1838). *Deckeria lamarckiana* (Mart.) H.Karst., Linnaea 28: 259 (1856).
　Dictyocaryum platysepalum Burret, Notizbl. Bot. Gart. Berlin-Dahlem 10: 927 (1930).
　Dictyocaryum schultzei Burret, Notizbl. Bot. Gart. Berlin-Dahlem 10: 925 (1930).
　Dictyocaryum globiferum Dugand, Caldasia 1(1): 13 (1940).
　Dictyocaryum superbum Burret, Notizbl. Bot. Gart. Berlin-Dahlem 15: 29 (1940).

Dictyocaryum platysepalum Burret = **Dictyocaryum lamarckianum** (Mart.) H.Wendl.

Dictyocaryum ptarianum (Steyerm.) H.E.Moore & Steyerm., Acta Bot. Venez. 2: 139 (1967).
S. Trop. America. 82 GUY VEN 83 CLM PER 84 BZN.
　Dahlgrenia ptariana Steyerm., Fieldiana, Bot. 28(1): 82 (1951).

Dictyocaryum schultzei Burret = **Dictyocaryum lamarckianum** (Mart.) H.Wendl.

Dictyocaryum superbum Burret = **Dictyocaryum lamarckianum** (Mart.) H.Wendl.

Dictyocaryum wallisii H.Wendl. = **?**

Dictyosperma

Dictyosperma H.Wendl. & Drude, Linnaea 39: 181 (1875).
1 species, Mascarenes. 29.
　Dicrosperma W.Watson, Gard. Chron. 1885(2): 362 (1885), orth. var.
　Linoma O.F.Cook, J. Wash. Acad. Sci. 7: 123 (1917).

Dictyosperma album (Bory) Scheff., Ann. Jard. Bot. Buitenzorg 1: 157 (1876).
Mascarenes. 29 MAU REU ROD.
　Areca alba Bory, Voy. îles Afrique 1: 306 (1804).
　Linoma alba (Bory) O.F.Cook, J. Wash. Acad. Sci. 7: 123 (1917).

var. **album**
Mauritius, Réunion. 29 MAU REU.
　Sublimia palmicaulis Comm. ex Mart., Hist. Nat. Palm. 3: 175 (1838), nom. inval.
　Areca borbonica Kunth, Enum. Pl. 3: 186 (1841).
　Areca lactea Miq., Fl. Ned. Ind. 3: 10 (1855).
　Areca propria Miq., Fl. Ned. Ind. 3: 10 (1855).
　Areca furfuracea H.Wendl. in O.C.E.de Kerchove de Denterghem, Palmiers: 231 (1878).
　Areca rubra H.Wendl. in O.C.E.de Kerchove de Denterghem, Palmiers: 232 (1878).

Dictyosperma furfuraceum H.Wendl. & Drude in O.C.E.de Kerchove de Denerghem, Palmiers: 243 (1878).
Dictyosperma rubrum H.Wendl. & Drude in O.C.E.de Kerchove de Denerghem, Palmiers: 243 (1878). *Dictyosperma album* var. *rubrum* (H.Wendl. & Drude) L.H.Bailey, Hortus: 215 (1930).
Areca pisifera Lodd. ex Hook.f., Rep. Progr. Condition Roy. Bot. Gard. Kew 1882: 54 (1884).

var. **aureum** Balf.f. in J.G.Baker, Fl. Mauritius: 384 (1877). *Dictyosperma aureum* (Balf.f.) G.Nicholson, Ill. Dict. Gard. 1: 470 (1884).
Rodrigues. 29 ROD.
　Areca aurea Van Houtte, Ann. Gén. Hort. 7: 43 (1867).

var. **conjugatum** H.E.Moore & J.Guého, Gentes Herb. 12: 15 (1980).
Mauritius (Î. Ronde). 29 MAU.

Dictyosperma album var. *rubrum* (H.Wendl. & Drude) L.H.Bailey = **Dictyosperma album** var. **album**

Dictyosperma aureum (Balf.f.) G.Nicholson = **Dictyosperma album** var. **aureum** Balf.f.

Dictyosperma fibrosum C.H.Wright = **Dypsis fibrosa** (C.H.Wright) Beentje & J.Dransf.

Dictyosperma furfuraceum H.Wendl. & Drude = **Dictyosperma album** var. **album**

Dictyosperma rubrum H.Wendl. & Drude = **Dictyosperma album** var. **album**

Didymosperma

Didymosperma H.Wendl. & Drude ex Hook.f. = **Arenga** Labill. ex DC.

Didymosperma borneense Becc. = **Arenga hastata** (Becc.) Whitmore

Didymosperma caudatum (Lour.) H.Wendl. & Drude ex B.D.Jacks. = **Arenga caudata** (Lour.) H.E.Moore

Didymosperma caudatum var. *tonkinense* Becc. = **Arenga caudata** (Lour.) H.E.Moore

Didymosperma distichum (T.Anderson) Hook.f. = **Wallichia disticha** T.Anderson

Didymosperma engleri (Becc.) Warb. = **Arenga engleri** Becc.

Didymosperma gracile Hook.f. = **?** [40 ASS]

Didymosperma hastatum Becc. = **Arenga hastata** (Becc.) Whitmore

Didymosperma hookerianum Becc. = **Arenga hookeriana** (Becc.) Whitmore

Didymosperma horsfieldii (Blume) H.Wendl. & Drude = **Arenga porphyrocarpa** (Blume) H.E.Moore

Didymosperma humile K.Schum. & Lauterb. = **?** [43 NWG]

Didymosperma microcarpum (Becc.) Warb. ex K.Schum. & Lauterb. = **Arenga microcarpa** Becc.

Didymosperma nanum (Griff.) H.Wendl. & Drude = **Arenga nana** (Griff.) H.E.Moore

Didymosperma porphyrocarpum (Blume) H.Wendl. & Drude ex Hook.f. = **Arenga porphyrocarpa** (Blume) H.E.Moore

Didymosperma reinwardtianum (Miq.) H.Wendl. & Drude ex B.D.Jacks. = **Arenga porphyrocarpa** (Blume) H.E.Moore

Didymosperma tonkinense (Becc.) Becc. ex Gagnep. = **Arenga caudata** (Lour.) H.E.Moore

Didymosperma tremulum (Blanco) H.Wendl. & Drude ex B.D.Jacks. = **Arenga tremula** (Blanco) Becc.

Diglossophyllum

Diglossophyllum H.Wendl. ex Salomon = **Serenoa** Hook.f.

Diglossophyllum serrulatum (Michx.) H.Wendl. ex Salomon = **Serenoa repens** (W.Bartram) Small

Diodosperma

Diodosperma H.Wendl. = **Trithrinax** Mart.

Diodosperma burity H.Wendl. = **Trithrinax schizophylla** Drude

Diplothemium

Diplothemium Mart. = **Allagoptera** Nees

Diplothemium anisitsii Barb.Rodr. = **Allagoptera leucocalyx** (Drude) Kuntze

Diplothemium arenarium (M.Gómez) Vasc. & Franco = **Allagoptera arenaria** (M.Gómez) Kuntze

Diplothemium campestre Mart. = **Allagoptera campestris** (Mart.) Kuntze

Diplothemium caudescens Mart. = **Polyandrococos caudescens** (Mart.) Barb.Rodr.

Diplothemium hasslerianum Barb.Rodr. = **Allagoptera leucocalyx** (Drude) Kuntze

Diplothemium henryanum F.Br. = **?** [61 MRQ]

Diplothemium jangadense S.Moore = **Allagoptera leucocalyx** (Drude) Kuntze

Diplothemium leucocalyx Drude = **Allagoptera leucocalyx** (Drude) Kuntze

Diplothemium littorale Mart. = **Allagoptera arenaria** (M.Gómez) Kuntze

Diplothemium maritimum Mart. = **Allagoptera arenaria** (M.Gómez) Kuntze

Diplothemium pectinatum Barb.Rodr. = **Polyandrococos caudescens** (Mart.) Barb.Rodr.

Diplothemium torallyi Mart. = **Parajubaea torallyi** (Mart.) Burret

Diplothenium

Diplothenium Voigt = **Allagoptera** Nees

Discoma

Discoma O.F.Cook = **Chamaedorea** Willd.

Discoma glaucifolia (H.Wendl.) O.F.Cook = **Chamaedorea glaucifolia** H.Wendl.

Docanthe

Docanthe O.F.Cook = **Chamaedorea** Willd.

Docanthe alba O.F.Cook = **Chamaedorea pinnatifrons** (Jacq.) Oerst.

Dolichokentia

Dolichokentia Becc. = **Cyphokentia** Brongn.

Dolichokentia robusta (Brongn.) Becc. = **Cyphokentia macrostachya** Brongn.

Doma

Doma Lam. = **Hyphaene** Gaertn.

Douma

Douma Poir. = **Hyphaene** Gaertn.

Douma thebaica (L.) Poir. = **Hyphaene thebaica** (L.) Mart.

Drymophloeus

Drymophloeus Zipp., Alg. Konst- Lett.-Bode 1: 297 (1829).
   8 species, Maluku to Papuasia. 42 43.
      Coleospadix Becc., Ann. Jard. Bot. Buitenzorg 2: 90 (1885).
      Saguaster Kuntze, Revis. Gen. Pl. 2: 734 (1891).
      Rehderophoenix Burret, Notizbl. Bot. Gart. Berlin-Dahlem 13: 86 (1936).

Drymophloeus ambiguus Becc. = **Ptychosperma ambiguum** (Becc.) Becc. ex Martelli

Drymophloeus angustifolius (Blume) Mart. = **Drymophloeus litigiosus** (Becc.) H.E.Moore

Drymophloeus appendiculatus (Blume) Miq. = **Drymophloeus oliviformis** (Giseke) Mart.

Drymophloeus beguinii (Burret) H.E.Moore = **Drymophloeus litigiosus** (Becc.) H.E.Moore

Drymophloeus bifidus Becc. = **Drymophloeus oliviformis** (Giseke) Mart.

Drymophloeus ceramensis Scheff. = **Drymophloeus oliviformis** (Giseke) Mart.

Drymophloeus ceramensis Miq. = **Drymophloeus oliviformis** (Giseke) Mart.

Drymophloeus communis (Zipp. ex Blume) Miq. = **Drymophloeus litigiosus** (Becc.) H.E.Moore

Drymophloeus divaricatus (Brongn.) Benth. & Hook.f. ex Becc. = **Actinokentia divaricata** (Brongn.) Dammer

Drymophloeus filiferus (H.Wendl.) Scheff. = **Veitchia filifera** (H.Wendl.) H.E.Moore

Drymophloeus hentyi (Essig) Zona, Blumea 44: 13 (1999).
   Bismarck Arch. (New Britain). 43 BIS.
      Ptychosperma hentyi Essig, Principes 31: 113 (1987).

Drymophloeus jaculatorius Mart. = **Drymophloeus oliviformis** (Giseke) Mart.

Drymophloeus kerstenianus Sander ex Burret = **Balaka seemannii** (H.Wendl.) Becc.

Drymophloeus lepidotus H.E.Moore, Principes 13: 75 (1969).
   Solomon Is. 43 SOL.

Drymophloeus leprosus Becc. = **Drymophloeus oliviformis** (Giseke) Mart.

Drymophloeus litigiosus (Becc.) H.E.Moore, Principes 13: 76 (1969).
   Maluku to NW. New Guinea. 42 MOL 43 NWG.
      Areca communis Zipp. ex Blume, Rumphia 2: 73 (1839). Provisional synonym. *Seaforthia communis* (Zipp. ex Blume) Mart., Hist. Nat. Palm. 3: 313 (1849). *Ptychosperma communis* (Zipp. ex Blume) Miq., Fl. Ned. Ind. 3: 31 (1855). *Drymophloeus communis* (Zipp. ex Blume) Miq., Palm. Archip. Ind.: 24 (1868).
      Areca litoralis Blume, Rumphia 2: 123 (1843), nom. inval.

Ptychosperma angustifolium Blume, Rumphia 2: 122 (1843). Provisional synonym. *Drymophloeus angustifolius* (Blume) Mart., Hist. Nat. Palm. 3: 314 (1849). *Saguaster angustifolius* (Blume) Kuntze, Revis. Gen. Pl. 2: 735 (1891). *Coleospadix angustifolius* (Blume) Burret, Repert. Spec. Nov. Regni Veg. 24: 286 (1928). *Actinophloeus angustifolius* (Blume) L.H.Bailey, Gentes Herb. 3: 424 (1935).

*Ptychosperma litigiosum Becc., Malesia 1: 50 (1877). *Coleospadix litigiosa* (Becc.) Becc., Ann. Jard. Bot. Buitenzorg 2: 90 (1885).

Coleospadix beguinii Burret, Repert. Spec. Nov. Regni Veg. 24: 286 (1928). *Drymophloeus beguinii* (Burret) H.E.Moore, Gentes Herb. 8: 304 (1953).

Coleospadix porrectus Burret, Repert. Spec. Nov. Regni Veg. 24: 287 (1928). *Drymophloeus porrectus* (Burret) H.E.Moore, Gentes Herb. 8: 307 (1953).

Drymophloeus mambare F.M.Bailey = **Ptychosperma mambare** (F.M.Bailey) Becc. ex Martelli

Drymophloeus minutus Rech. = **Balaka minuta** Burret

Drymophloeus montanus K.Schum. & Lauterb. = **Ptychosperma caryotoides** Ridl.

Drymophloeus mooreanus auct. = **?**

Drymophloeus normanbyi (F.Muell.) Benth. & Hook.f. ex Becc. = **Normanbya normanbyi** (F.Muell.) L.H.Bailey

Drymophloeus oliviformis (Giseke) Mart., Hist. Nat. Palm. 3: 314 (1849).

Maluku to NW. New Guinea. 42 MOL 43 NWG.

Areca oliviformis var. *gracilis* Giseke, Prael. Ord. Nat. Pl.: 80 (1792). *Ptychosperma appendiculatum* Blume, Rumphia 2: 122 (1843). *Seaforthia appendiculata* (Blume) Juss. ex Kunth, Enum. Pl. 3: 192 (1841). *Drymophloeus appendiculatus* (Blume) Miq., Palm. Archip. Ind.: 24 (1868). *Saguaster appendiculatus* (Blume) Kuntze, Revis. Gen. Pl. 2: 734 (1891). *Coleospadix gracilis* (Giseke) Burret, Repert. Spec. Nov. Regni Veg. 24: 285 (1928), nom. illeg.

Areca vaginata Giseke, Prael. Ord. Nat. Pl.: 78 (1792). Provisional synonym. *Drymophloeus jaculatorius* Mart., Hist. Nat. Palm. 3: 314 (1849), nom. illeg.

*Areca oliviformis Giseke, Ord. Nat. Pl.: 79 (1830). *Seaforthia oliviformis* (Giseke) Mart., Hist. Nat. Palm. 3: 314 (1849). *Drymophloeus rumphii* Blume ex Scheff., Ann. Jard. Bot. Buitenzorg 1: 52 (1876), nom. illeg. *Saguaster oliviformis* (Giseke) Kuntze, Revis. Gen. Pl. 2: 734 (1891).

Iriartea leprosa Zipp., Bijdr. Natuurk. Wetensch. 5: 178 (1830).

Iriartea monogyna Zipp., Bijdr. Natuurk. Wetensch. 5: 178 (1830).

Areca elaeocarpa Reinw. ex Kunth, Enum. Pl. 3: 195 (1841).

Seaforthia blumei Juss. ex Kunth, Enum. Pl. 3: 192 (1841).

Ptychosperma rumphii Blume, Rumphia 2: 119 (1843). *Harina rumphii* (Blume) Mart., Hist. Nat. Palm. 3: 314 (1849).

Seaforthia jaculatoria Mart., Hist. Nat. Palm. 3: 314 (1849).

Drymophloeus ceramensis Miq., Verh. Kon. Ned. Akad. Wetensch., Afd. Natuurk., Tweede Sect. 11(5): 5 (1868).

Drymophloeus ceramensis Scheff., Ann. Jard. Bot. Buitenzorg 1: 121 (1876).

Drymophloeus bifidus Becc., Malesia 1: 44 (1877). *Saguaster bifidus* (Becc.) Kuntze, Revis. Gen. Pl. 2: 735 (1891).

Drymophloeus leprosus Becc., Ann. Jard. Bot. Buitenzorg 2: 119 (1885). *Saguaster leprosus* (Becc.) Kuntze, Revis. Gen. Pl. 2: 735 (1891).

Drymophloeus oninensis (Becc.) H.E.Moore, Principes 13: 76 (1969).

New Guinea. 43 NWG.

*Ptychosperma litigiosum var. *oninense* Becc., Malesia 1: 52 (1877). *Coleospadix oninensis* (Becc.) Becc., Ann. Jard. Bot. Buitenzorg 2: 90 (1885). *Saguaster oninensis* (Becc.) Kuntze, Revis. Gen. Pl. 2: 735 (1891).

Drymophloeus pachycladus (Burret) H.E.Moore, Principes 13: 76 (1969).

Solomon Is. 43 SOL.

*Rehderophoenix pachyclada Burret, Notizbl. Bot. Gart. Berlin-Dahlem 13: 87 (1936).

Drymophloeus paradoxus Scheff. = **Ptychococcus paradoxus** (Scheff.) Becc.

Drymophloeus pauciflorus (H.Wendl.) Becc. = **Balaka pauciflora** (H.Wendl.) H.E.Moore

Drymophloeus porrectus (Burret) H.E.Moore = **Drymophloeus litigiosus** (Becc.) H.E.Moore

Drymophloeus propinquus Becc. = **Ptychosperma propinquum** (Becc.) Becc. ex Martelli

Drymophloeus propinquus var. *keiensis* Becc. = **Ptychosperma propinquum** (Becc.) Becc. ex Martelli

Drymophloeus puniceus (Zipp. ex Blume) Becc. = **Pinanga rumphiana** (Mart.) J.Dransf. & Govaerts

Drymophloeus reineckei Warb. = **Balaka tahitensis** (H.Wendl.) Becc.

Drymophloeus rumphianus Mart. = **Pinanga rumphiana** (Mart.) J.Dransf. & Govaerts

Drymophloeus rumphii Blume ex Scheff. = **Drymophloeus oliviformis** (Giseke) Mart.

Drymophloeus samoensis (Rech.) Becc. ex Martelli = **Solfia samoensis** Rech.

Drymophloeus saxatilis (Burm.f.) Mart. = **Areca oryziformis var. saxatilis**

Drymophloeus schumannii (Becc.) Warb. ex K.Schum. & Lauterb. = **Brassiophoenix schumannii** (Becc.) Essig

Drymophloeus seemannii (H.Wendl.) Becc. ex Martelli = **Balaka seemannii** (H.Wendl.) Becc.

Drymophloeus singaporensis (Becc.) Hook.f. = **Rhopaloblaste singaporensis** (Becc.) Hook.f.

Drymophloeus subdistichus (H.E.Moore) H.E.Moore, Principes 13: 76 (1969).

Solomon Is. 43 SOL.

*Rehderophoenix subdisticha H.E.Moore, Principes 10: 93 (1966).

Drymophloeus vestiarius Miq. = **Areca vestiaria** Giseke

Drymophloeus whitmeeanus Becc., Webbia 4: 261 (1914). *Solfia whitmeeana* (Becc.) Burret, Repert. Spec. Nov. Regni Veg. 24: 281 (1928).

Samoa. 60 SAM.

Drymophloeus zippellii Hassk. = **Caryota mitis** Lour.

Drypsis

Drypsis Duch. = **Dypsis** Noronha ex Mart.

Dypsidium

Dypsidium Baill. = **Dypsis** Noronha ex Mart.

Dypsidium catatianum Baill. = **Dypsis catatiana** (Baill.) Beentje & J.Dransf.

Dypsidium emirnense Baill. = **Dypsis heterophylla** Baker

Dypsidium vilersianum Baill. = **Dypsis heterophylla** Baker

Dypsis

Dypsis Noronha ex Mart., Hist. Nat. Palm. 3: 180 (1838).
140 species, Tanzania (Pemba), Comoros, Madagascar. 25 29.
Drypsis Duch. in C.V.D.d'Orbigny, Dict. Univ. Hist. Nat. 9: 424 (1849).
Chrysalidocarpus H.Wendl., Bot. Zeitung (Berlin) 36: 117 (1878).
Phloga Noronha ex Hook.f. in G.Bentham & J.D.Hooker, Gen. Pl. 3(2): 909 (1883).
Dypsidium Baill., Bull. Mens. Soc. Linn. Paris 2: 1172 (1893).
Haplodypsis Baill., Bull. Mens. Soc. Linn. Paris 2: 1167 (1894).
Haplophloga Baill., Bull. Mens. Soc. Linn. Paris 2: 1168 (1894).
Neodypsis Baill., Bull. Mens. Soc. Linn. Paris 2: 1172 (1894).
Neophloga Baill., Bull. Mens. Soc. Linn. Paris 2: 1173 (1894).
Phlogella Baill., Bull. Mens. Soc. Linn. Paris 2: 1175 (1894).
Trichodypsis Baill., Bull. Mens. Soc. Linn. Paris 2: 1165 (1894).
Adelodypsis Becc., Bot. Jahrb. Syst. 38(87): 16 (1906).
Vonitra Becc., Bot. Jahrb. Syst. 38(87): 18 (1906).
Macrophloga Becc., Palme Madagascar: 47 (1914).
Antongilia Jum., Ann. Inst. Bot.-Géol. Colon. Marseille, IV, 6(2): 17 (1928).

Dypsis acaulis J.Dransf. in J.Dransfield & H.Beentje, Palms Madagascar: 409 (1995).
NE. Madagascar. 29 MDG.

Dypsis acuminum (Jum.) Beentje & J.Dransf., Palms Madagascar: 211 (1995).
N. Madagascar. 29 MDG.
*Chrysalidocarpus acuminum Jum., Ann. Inst. Bot.-Géol. Colon. Marseille, III, 5(3): 16 (1922).

Dypsis albofarinosa Hodel & Marcus, Palms 48: 91 (2004).
Madagascar (?). 29 MDG.

Dypsis ambanjae Beentje in J.Dransfield & H.Beentje, Palms Madagascar: 299 (1995).
NW. Madagascar. 29 MDG.
*Phloga sambiranensis Jum., Ann. Inst. Bot.-Géol. Colon. Marseille, V, 1(3): 18 (1933).

Dypsis ambilaensis J.Dransf. in J.Dransfield & H.Beentje, Palms Madagascar: 382 (1995).
CE. Madagascar. 29 MDG.

Dypsis ambositrae Beentje in J.Dransfield & H.Beentje, Palms Madagascar: 195 (1995).
C. Madagascar. 29 MDG.

Dypsis ampasindavae Beentje in J.Dransfield & H.Beentje, Palms Madagascar: 153 (1995).
NW. Madagascar (incl. Nosy Bé). 29 MDG.
*Neodypsis loucoubensis Jum., Ann. Inst. Bot.-Géol. Colon. Marseille, V, 1(1): 17 (1933).

Dypsis andapae Beentje in J.Dransfield & H.Beentje, Palms Madagascar: 300 (1995).
N. Madagascar. 29 MDG.

Dypsis andrianatonga Beentje in J.Dransfield & H.Beentje, Palms Madagascar: 203 (1995).
N. Madagascar. 29 MDG.

Dypsis angusta Jum., Ann. Inst. Bot.-Géol. Colon. Marseille, III, 6(1): 34 (1918).
ESE. Madagascar. 29 MDG.

Dypsis angustifolia (H.Perrier) Beentje & J.Dransf., Palms Madagascar: 336 (1995).
CE. Madagascar. 29 MDG.
*Dypsis humbertii var. *angustifolia* H.Perrier, Notul. Syst. (Paris) 8: 47 (1939).

Dypsis ankaizinensis (Jum.) Beentje & J.Dransf., Palms Madagascar: 182 (1995).
N. Madagascar. 29 MDG.
*Chrysalidocarpus ankaizinensis Jum., Ann. Inst. Bot.-Géol. Colon. Marseille, II, 10(3): 23 (1922).
Neodypsis lobata Jum., Ann. Inst. Bot.-Géol. Colon. Marseille, IV, 2(2): 13 (1924).

Dypsis anovensis J.Dransf. = **Dypsis linearis** Jum.

Dypsis antanambensis Beentje in J.Dransfield & H.Beentje, Palms Madagascar: 368 (1995).
NE. Madagascar. 29 MDG.

Dypsis aquatilis Beentje in J.Dransfield & H.Beentje, Palms Madagascar: 372 (1995).
SE. Madagascar. 29 MDG.

Dypsis arenarum (Jum.) Beentje & J.Dransf., Palms Madagascar: 215 (1995).
ENE. Madagascar. 29 MDG.
*Chrysalidocarpus arenarum Jum., Ann. Inst. Bot.-Géol. Colon. Marseille, II, 10(3): 17 (1922).

Dypsis baronii (Becc.) Beentje & J.Dransf., Palms Madagascar: 198 (1995).
Madagascar. 29 MDG.
*Chrysalidocarpus baronii Becc., Bot. Jahrb. Syst. 38(87): 33 (1906). *Neodypsis baronii* (Becc.) Jum., Compt. Rend. Hebd. Séances Acad. Sci. 179: 250 (1924).
Chrysalidocarpus propinquus Jum., Ann. Inst. Bot.-Géol. Colon. Marseille, II, 10(3): 19 (1922).
Neodypsis compacta Jum., Ann. Inst. Bot.-Géol. Colon. Marseille, V, 1(1): 13 (1933).

Dypsis basilonga (Jum. & H.Perrier) Beentje & J.Dransf., Palms Madagascar: 193 (1995).
E. Madagascar (Mt. Vatovavy). 29 MDG.
*Neodypsis basilonga Jum. & H.Perrier, Ann. Inst. Bot.-Géol. Colon. Marseille, III, 1: 16 (1913).

Dypsis beentjei J.Dransf. in J.Dransfield & H.Beentje, Palms Madagascar: 401 (1995).
NE. Madagascar. 29 MDG.

Dypsis bejofo Beentje in J.Dransfield & H.Beentje, Palms Madagascar: 146 (1995).
NE. Madagascar. 29 MDG.

Dypsis bernierana (Baill.) Beentje & J.Dransf., Palms Madagascar: 304 (1995).
NE. & CE. Madagascar. 29 MDG.
*Haplophloga bernierana Baill., Bull. Mens. Soc. Linn. Paris 2: 1171 (1894).

Dypsis betamponensis (Jum.) Beentje & J.Dransf., Palms Madagascar: 295 (1995).
ENE. Madagascar (Betampona). 29 MDG.
Neophloga betamponensis Jum., Ann. Inst. Bot.-Géol. Colon. Marseille, IV, 6(3): 33 (1928 publ. 1929).

Dypsis boiviniana Baill., Bull. Mens. Soc. Linn. Paris 2: 1164 (1893). *Adelodypsis boiviniana* (Baill.) Becc., Bot. Jahrb. Syst. 38(87): 17 (1906).
NE. Madagascar. 29 MDG.
Chrysalidocarpus oligostachyus Becc., Bot. Jahrb. Syst. 38(87): 37 (1906). *Neophloga oligostachya* (Becc.) H.Perrier, in Fl. Madag. 30: 81 (1945).

Dypsis bonsai Beentje in J.Dransfield & H.Beentje, Palms Madagascar: 252 (1995).
NE. Madagascar. 29 MDG.

Dypsis bosseri J.Dransf. in J.Dransfield & H.Beentje, Palms Madagascar: 393 (1995).
ENE. Madagascar (W. of Mahavelona). 29 MDG.

Dypsis brevicaulis (Guillaumet) Beentje & J.Dransf., Palms Madagascar: 323 (1995).
SE. Madagascar. 29 MDG.
Neophloga brevicaulis Guillaumet, Adansonia, n.s., 13: 343 (1973).

Dypsis cabadae (H.E.Moore) Beentje & J.Dransf., Palms Madagascar: 219 (1995).
Comoros. 29 COM.
Chrysalidocarpus cabadae H.E.Moore, Principes 6: 108 (1962).

Dypsis canaliculata (Jum.) Beentje & J.Dransf., Palms Madagascar: 149 (1995).
NW. & CE. Madagascar. 29 MDG.
Neodypsis canaliculata Jum., Ann. Inst. Bot.-Géol. Colon. Marseille, IV, 2: 9 (1924).

Dypsis canescens (Jum. & H.Perrier) Beentje & J.Dransf., Palms Madagascar: 410 (1995).
NW. Madagascar. 29 MDG.
Chrysalidocarpus canescens Jum. & H.Perrier, Ann. Inst. Bot.-Géol. Colon. Marseille, III, 1: 38 (1913).

Dypsis carlsmithii J.Dransf. & Marcus, Palms 46: 48 (2002).
Madagascar (?). 29 MDG.

Dypsis catatiana (Baill.) Beentje & J.Dransf., Palms Madagascar: 308 (1995).
N. & E. Madagascar. 29 MDG.
Dypsidium catatianum Baill., Bull. Mens. Soc. Linn. Paris 2: 1173 (1893). *Neophloga catatiana* (Baill.) Becc., Bot. Jahrb. Syst. 38(87): 25 (1906).
Neophloga indivisa Jum. & H.Perrier, Ann. Inst. Bot.-Géol. Colon. Marseille, III, 1: 29 (1913).

Dypsis caudata Beentje in J.Dransfield & H.Beentje, Palms Madagascar: 254 (1995).
NE. Madagascar. 29 MDG.

Dypsis ceracea (Jum.) Beentje & J.Dransf., Palms Madagascar: 151 (1995).
NE. Madagascar. 29 MDG.
Neodypsis ceracea Jum., Ann. Inst. Bot.-Géol. Colon. Marseille, V, 1(1): 18 (1933).

Dypsis commersoniana (Baill.) Beentje & J.Dransf., Palms Madagascar: 236 (1995).
SE. Madagascar. 29 MDG.
Neophloga commersoniana Baill., Bull. Mens. Soc. Linn. Paris 2: 1173 (1894).
Neophloga pygmaea Pic.Serm., Webbia 11: 149 (1955).

Dypsis concinna Baker, J. Linn. Soc., Bot. 22: 526 (1887). *Neophloga concinna* (Baker) Becc., Bot. Jahrb. Syst. 38(87): 27 (1906).
CE. Madagascar. 29 MDG.
Neophloga tenuisecta Jum. & H.Perrier, Ann. Inst. Bot.-Géol. Colon. Marseille, III, 1: 30 (1913).
Neophloga triangularis Jum. & H.Perrier, Ann. Inst. Bot.-Géol. Colon. Marseille, III, 1: 32 (1913). *Neophloga concinna* f. *triangularis* (Jum. & H.Perrier) Jum., Ann. Inst. Bot.-Géol. Colon. Marseille, IV, 6(3): 40 (1928 publ. 1929).
Neophloga microphylla Becc., Palme Madagascar: 36 (1914).

Dypsis confusa Beentje in J.Dransfield & H.Beentje, Palms Madagascar: 288 (1995).
NE. & ENE. Madagascar. 29 MDG.

Dypsis cookei J.Dransf. in J.Dransfield & H.Beentje, Palms Madagascar: 399 (1995).
NE. Madagascar. 29 MDG.

Dypsis coriacea Beentje in J.Dransfield & H.Beentje, Palms Madagascar: 311 (1995).
NE. Madagascar. 29 MDG.

Dypsis corniculata (Becc.) Beentje & J.Dransf., Palms Madagascar: 280 (1995).
ENE. Madagascar. 29 MDG.
Neophloga corniculata Becc., Bot. Jahrb. Syst. 38(87): 24 (1906).

Dypsis coursii Beentje in J.Dransfield & H.Beentje, Palms Madagascar: 230 (1995).
NE. Madagascar. 29 MDG.

Dypsis crinita (Jum. & H.Perrier) Beentje & J.Dransf., Palms Madagascar: 361 (1995).
N. & NE. Madagascar. 29 MDG.
Vonitra crinita Jum. & H.Perrier, Agric. Prat. Pays Chauds 10(1): 293 (1910).

Dypsis curtisii Baker, J. Linn. Soc., Bot. 22: 526 (1887). *Neophloga curtisii* (Baker) Becc., Bot. Jahrb. Syst. 38(87): 30 (1906).
N. Madagascar. 29 MDG.

Dypsis decaryi (Jum.) Beentje & J.Dransf., Palms Madagascar: 187 (1995).
SE. Madagascar. 29 MDG.
Neodypsis decaryi Jum., Ann. Inst. Bot.-Géol. Colon. Marseille, V, 1(1): 15 (1933).

Dypsis decipiens (Becc.) Beentje & J.Dransf., Palms Madagascar: 191 (1995).
C. Madagascar. 29 MDG.
Chrysalidocarpus decipiens Becc., Bot. Jahrb. Syst. 38(87): 36 (1906). *Macrophloga decipiens* Becc., Palme Madagascar: 47 (1914).

Dypsis digitata (Becc.) Beentje & J.Dransf., Palms Madagascar: 320 (1995).
ENE. Madagascar. 29 MDG.
Neophloga digitata Becc., Palme Madagascar: 36 (1914).

Dypsis dransfieldii Beentje in J.Dransfield & H.Beentje, Palms Madagascar: 355 (1995).
NE. Madagascar. 29 MDG.

Dypsis elegans Beentje in J.Dransfield & H.Beentje, Palms Madagascar: 271 (1995).
ESE. Madagascar. 29 MDG.

Dypsis eriostachys J.Dransf. in J.Dransfield & H.Beentje, Palms Madagascar: 291 (1995).
E. Madagascar (Mt. Vatovavy). 29 MDG.

Dypsis faneva Beentje in J.Dransfield & H.Beentje, Palms Madagascar: 257 (1995).
NE. Madagascar. 29 MDG.

Dypsis fanjana Beentje in J.Dransfield & H.Beentje, Palms Madagascar: 259 (1995).
NE. Madagascar. 29 MDG.

Dypsis fasciculata Jum., Ann. Inst. Bot.-Géol. Colon. Marseille, III, 6(1): 37 (1918).
NE. & ENE. Madagascar. 29 MDG.

Dypsis fibrosa (C.H.Wright) Beentje & J.Dransf., Palms Madagascar: 366 (1995).
N. & E. Madagascar. 29 MDG.
  *Dictyosperma fibrosum C.H.Wright, Bull. Misc. Inform. Kew 1894: 358 (1894). *Vonitra fibrosa* (C.H.Wright) Becc., Agric. Colon. 5: 322 (1911).

Dypsis forficifolia Noronha ex Mart., Hist. Nat. Palm. 3: 180 (1838).
NE. Madagascar. 29 MDG.
  Dypsis hirtula Mart., Hist. Nat. Palm. 3: 181 (1838).
  Dypsis littoralis Jum., Ann. Inst. Bot.-Géol. Colon. Marseille, III, 6(1): 34 (1918).
  Dypsis masoalensis Jum., Ann. Inst. Bot.-Géol. Colon. Marseille, III, 6(1): 36 (1918).

Dypsis furcata J.Dransf. in J.Dransfield & H.Beentje, Palms Madagascar: 394 (1995).
CE. Madagascar. 29 MDG.

Dypsis glabrescens (Becc.) Becc., Palme Madagascar: 16 (1912).
ENE. Madagascar. 29 MDG.
  *Trichodypsis glabrescens Becc., Bot. Jahrb. Syst. 38(87): 15 (1906).

Dypsis gracilis Bory ex Mart. = **Dypsis pinnatifrons** Mart.

Dypsis gracilis var. *sambiranensis* Jum. & H.Perrier = **Dypsis pinnatifrons** Mart.

Dypsis heteromorpha (Jum.) Beentje & J.Dransf., Palms Madagascar: 198 (1995).
N. Madagascar. 29 MDG.
  *Neodypsis heteromorpha Jum., Ann. Inst. Bot.-Géol. Colon. Marseille, IV, 2(2): 20 (1924).

Dypsis heterophylla Baker, J. Linn. Soc., Bot. 22: 552 (1887). *Neophloga heterophylla* (Baker) Becc., Bot. Jahrb. Syst. 38(87): 28 (1906).
N. & E. Madagascar. 29 MDG.
  Dypsis rhodotricha Baker, J. Linn. Soc., Bot. 22: 525 (1887). *Neophloga rhodotricha* (Baker) Becc., Bot. Jahrb. Syst. 38(87): 29 (1906).
  Dypsidium emirnense Baill., Bull. Mens. Soc. Linn. Paris 2: 1173 (1893). *Neophloga emirnensis* (Baill.) Becc., Bot. Jahrb. Syst. 38(87): 28 (1906).
  Dypsidium vilersianum Baill., Bull. Mens. Soc. Linn. Paris 2: 1173 (1893).
  Neophloga majorana Becc., Bot. Jahrb. Syst. 38(87): 23 (1906).
  Neophloga linearis var. *disticha* Jum., Ann. Inst. Bot.-Géol. Colon. Marseille, IV, 6(3): 44 (1928 publ. 1929).

Dypsis hiarakae Beentje in J.Dransfield & H.Beentje, Palms Madagascar: 286 (1995).
N. & NE. Madagascar. 29 MDG.

Dypsis hildebrandtii (Baill.) Becc., Palme Madagascar: 14 (1912).
EC. & CE. Madagascar. 29 MDG.
  *Trichodypsis hildebrandtii Baill., Bull. Mens. Soc. Linn. Paris 2: 1165 (1894).

Dypsis hirtula Mart. = **Dypsis forficifolia** Noronha ex Mart.

Dypsis hovomantsina Beentje in J.Dransfield & H.Beentje, Palms Madagascar: 150 (1995).
NE. Madagascar. 29 MDG.

Dypsis humbertii H.Perrier, Notul. Syst. (Paris) 8: 46 (1939). *Dypsis zahamenae* J.Dransf. in J.Dransfield & H.Beentje, Palms Madagascar: 336 (1995), nom. illeg.
ENE. Madagascar. 29 MDG.
  Neophloga humbertii Jum., Ann. Inst. Bot.-Géol. Colon. Marseille, V, 1(1): 20 (1933). *Dypsis humbertii* (Jum.) Beentje & J.Dransf., Palms Madagascar: 239 (1995), nom. illeg.

Dypsis humbertii (Jum.) Beentje & J.Dransf. = **? [29 MDG]**

Dypsis humbertii var. *angustifolia* H.Perrier = **Dypsis angustifolia** (H.Perrier) Beentje & J.Dransf.

Dypsis humblotiana (Baill.) Beentje & J.Dransf., Palms Madagascar: 221 (1995).
Comoros (Njazidja). 29 MDG.
  *Phlogella humblotiana Baill., Bull. Mens. Soc. Linn. Paris 2: 1175 (1894). *Chrysalidocarpus humblotiana* (Baill.) Becc., Bot. Jahrb. Syst. 38(87): 33 (1906).

Dypsis ifanadianae Beentje in J.Dransfield & H.Beentje, Palms Madagascar: 171 (1995).
CE. Madagascar. 29 MDG.

Dypsis integra (Jum.) Beentje & J.Dransf., Palms Madagascar: 319 (1995).
E. Madagascar. 29 MDG.
  *Neophloga integra Jum., Ann. Inst. Bot.-Géol. Colon. Marseille, IV, 6(3): 13 (1928 publ. 1929).

Dypsis intermedia Beentje in J.Dransfield & H.Beentje, Palms Madagascar: 243 (1995).
ESE. Madagascar. 29 MDG.

Dypsis interrupta J.Dransf. in J.Dransfield & H.Beentje, Palms Madagascar: 327 (1995).
E. Madagascar (Ifanadiana area). 29 MDG.

Dypsis jumelleana Beentje & J.Dransf., Palms Madagascar: 247 (1995).
CE. Madagascar. 29 MDG.
  *Neophloga lanceolata Jum., Ann. Inst. Bot.-Géol. Colon. Marseille, IV, 6(3): 45 (1928 publ. 1939).

Dypsis laevis J.Dransf. in J.Dransfield & H.Beentje, Palms Madagascar: 386 (1995).
ESE. Madagascar. 29 MDG.

Dypsis lanceolata (Becc.) Beentje & J.Dransf., Palms Madagascar: 223 (1995).
Comoros. 29 COM.
  *Chrysalidocarpus lanceolatus Becc., Bot. Jahrb. Syst. 38(87): 34 (1906).

Dypsis lantzeana Baill., Bull. Mens. Soc. Linn. Paris 2: 1163 (1893).
NE. Madagascar. 29 MDG.

Dypsis lantzeana var. *simplicifrons* Becc., Bot. Jahrb. Syst. 38(87): 13 (1906).

Dypsis lantzeana var. *simplicifrons* Becc. = **Dypsis lantzeana** Baill.

Dypsis lanuginosa J.Dransf. in J.Dransfield & H.Beentje, Palms Madagascar: 396 (1995).
N. & NE. Madagascar. 29 MDG.

Dypsis lastelliana (Baill.) Beentje & J.Dransf., Palms Madagascar: 175 (1995).
N. Madagascar. 29 MDG.

Neodypsis lastelliana Baill., Bull. Mens. Soc. Linn. Paris 2: 1172 (1894).

Dypsis leptocheilos (Hodel) Beentje & J.Dransf., Palms Madagascar: 176 (1995).
N. Madagascar. 29 MDG.
Neodypsis leptocheilos Hodel, Palm J. 139: 9 (1993).

Dypsis ligulata (Jum.) Beentje & J.Dransf., Palms Madagascar: 178 (1995).
NW. Madagascar. 29 MDG.
Neodypsis ligulata Jum., Ann. Inst. Bot.-Géol. Colon. Marseille, IV, 2(2): 19 (1924).

Dypsis linearis Jum., Ann. Inst. Bot.-Géol. Colon. Marseille, III, 6(1): 35 (1918). *Dypsis anovensis* J.Dransf. in J.Dransfield & H.Beentje, Palms Madagascar: 380 (1995), nom. illeg.
NE. Madagascar. 29 MDG.
Neophloga mananjarensis Jum. & H.Perrier, Ann. Inst. Bot.-Géol. Colon. Marseille, III, 1: 26 (1913). *Chrysalidocarpus ambolo* Jum., Ann. Inst. Bot.-Géol. Colon. Marseille, IV, 6(3): 8 (1928 publ. 1929). *Neophloga procumbens* Jum. & H.Perrier, Ann. Inst. Bot.-Géol. Colon. Marseille, III, 1: 27 (1913).

Dypsis linearis (Becc.) Beentje & J.Dransf. = **Dypsis procumbens**

Dypsis littoralis Jum. = **Dypsis forficifolia** Noronha ex Mart.

Dypsis lokohensis J.Dransf. in J.Dransfield & H.Beentje, Palms Madagascar: 350 (1995).
N. Madagascar. 29 MDG.

Dypsis longipes Jum. = **Dypsis procera** Jum.

Dypsis louvelii Jum. & H.Perrier, Ann. Inst. Bot.-Géol. Colon. Marseille, III, 1: 21 (1913).
CE. Madagascar. 29 MDG.

Dypsis lucens (Jum.) Beentje & J.Dransf., Palms Madagascar: 315 (1995).
NE. Madagascar. 29 MDG.
Neophloga lucens Jum., Ann. Inst. Bot.-Géol. Colon. Marseille, IV, 6(3): 15 (1928 publ. 1929).

Dypsis lutea (Jum.) Beentje & J.Dransf., Palms Madagascar: 293 (1995).
NE. & CE. Madagascar. 29 MDG.
Neophloga lutea Jum., Ann. Inst. Bot.-Géol. Colon. Marseille, IV, 6(3): 32 (1928 publ. 1929). *Neophloga lutea* var. *transiens* Jum. & H.Perrier, in Fl. Madag. 30: 76 (1945).

Dypsis lutescens (H.Wendl.) Beentje & J.Dransf., Palms Madagascar: 212 (1995).
E. Madagascar. 29 MDG.
Chrysalidocarpus lutescens H.Wendl., Bot. Zeitung (Berlin) 36: 171 (1878). *Areca flavescens* Voss, Vilm. Blumengärtn. ed. 3, 1: 1153 (1895). *Chrysalidocarpus baronii* var. *littoralis* Jum. & H.Perrier, Ann. Inst. Bot.-Géol. Colon. Marseille, III, 1(1): 35 (1913). *Chrysalidocarpus glaucescens* Waby, Bull. Misc. Inform. Kew 1923: 376 (1923).

Dypsis madagascariensis (Mart.) G.Nicholson = **Areca madagascariensis**

Dypsis madagascariensis (Becc.) Beentje & J.Dransf. = **? [29 MDG]**

Dypsis mahia Beentje in J.Dransfield & H.Beentje, Palms Madagascar: 296 (1995).
SE. Madagascar. 29 MDG.

Dypsis malcomberi Beentje in J.Dransfield & H.Beentje, Palms Madagascar: 165 (1995).
SE. Madagascar. 29 MDG.

Dypsis mananjarensis (Jum. & H.Perrier) Beentje & J.Dransf., Palms Madagascar: 163 (1995).
E. Madagascar. 29 MDG.
Chrysalidocarpus mananjarensis Jum. & H.Perrier, Ann. Inst. Bot.-Géol. Colon. Marseille, III, 1: 33 (1913). *Chrysalidocarpus fibrosus* Jum., Ann. Inst. Bot.-Géol. Colon. Marseille, II, 10(3): 10 (1922).

Dypsis manaranensis Jum. = **Dypsis mocquerysiana** (Becc.) Becc.

Dypsis mangorensis (Jum.) Beentje & J.Dransf., Palms Madagascar: 264 (1995).
NE. & CE. Madagascar. 29 MDG.
Neophloga littoralis Jum., Ann. Inst. Bot.-Géol. Colon. Marseille, IV, 6(3): 41 (1928 publ. 1929). *Neophloga mangorensis* Jum., Ann. Inst. Bot.-Géol. Colon. Marseille, IV, 6(3): 34 (1928 publ. 1929).

Dypsis marojejyi Beentje in J.Dransfield & H.Beentje, Palms Madagascar: 234 (1995).
NE. Madagascar. 29 MDG.

Dypsis masoalensis Jum. = **Dypsis forficifolia** Noronha ex Mart.

Dypsis mcdonaldiana Beentje in J.Dransfield & H.Beentje, Palms Madagascar: 245 (1995).
SE. Madagascar. 29 MDG.

Dypsis minuta Beentje in J.Dransfield & H.Beentje, Palms Madagascar: 313 (1995).
NE. Madagascar. 29 MDG.

Dypsis mirabilis J.Dransf. in J.Dransfield & H.Beentje, Palms Madagascar: 345 (1995).
NE. Madagascar. 29 MDG.

Dypsis mocquerysiana (Becc.) Becc., Palme Madagascar: 15 (1912).
E. Madagascar. 29 MDG.
Trichodypsis mocquerysiana Becc., Bot. Jahrb. Syst. 38(87): 15 (1906). *Dypsis manaranensis* Jum., Ann. Inst. Bot.-Géol. Colon. Marseille, III, 6(1): 32 (1918).

Dypsis monostachya Jum., Ann. Inst. Bot.-Géol. Colon. Marseille, III, 6(1): 36 (1918).
NE. Madagascar. 29 MDG.

Dypsis montana (Jum.) Beentje & J.Dransf., Palms Madagascar: 303 (1995).
N. Madagascar. 29 MDG.
Neophloga montana Jum., Ann. Inst. Bot.-Géol. Colon. Marseille, IV, 6(3): 29 (1928 publ. 1929).

Dypsis moorei Beentje in J.Dransfield & H.Beentje, Palms Madagascar: 354 (1995).
NE. Madagascar. 29 MDG.

Dypsis nauseosa (Jum. & H.Perrier) Beentje & J.Dransf., Palms Madagascar: 156 (1995).
SE. Madagascar. 29 MDG.
Neodypsis nauseosa Jum. & H.Perrier, Ann. Inst. Bot.-Géol. Colon. Marseille, III, 1: 19 (1913).

Dypsis nodifera Mart., Hist. Nat. Palm. 3: 312 (1849). *Phloga nodifera* (Mart.) Pic.Serm., Webbia 11: 142 (1955).
N. & E. Madagascar. 29 MDG.
Dypsis polystachya Baker, J. Linn. Soc., Bot. 22: 525 (1887). *Phloga polystachya* (Baker) Noronha ex Baill., Bull. Mens. Soc. Linn. Paris 148: 1175 (1895).

Dypsis vilersiana Baill., Bull. Mens. Soc. Linn. Paris 2: 1165 (1893).
Phloga polystachya var. *stenophylla* Becc., Bot. Jahrb. Syst. 38(87): 11 (1906).

Dypsis nossibensis (Becc.) Beentje & J.Dransf., Palms Madagascar: 358 (1995).
N. & EC. Madagascar. 29 MDG.
Haplophloga loucoubensis Baill., Bull. Mens. Soc. Linn. Paris 2: 1171 (1894). Provisional synonym.
Vonitra loucoubensis (Baill.) Jum., Rev. Bot. Appl. Agric. Colon. 2: 160 (1922).
*Chrysalidocarpus nossibensis Becc., Bot. Jahrb. Syst. 38(87): 34 (1906). *Vonitra nossibensis* (Becc.) H.Perrier, in Fl. Madag. 30: 130 (1945).

Dypsis occidentalis (Jum.) Beentje & J.Dransf., Palms Madagascar: 301 (1995).
N. Madagascar. 29 MDG.
*Neophloga occidentalis Jum., Ann. Inst. Bot.-Géol. Colon. Marseille, IV, 6(3): 24 (1928 publ. 1929).

Dypsis onilahensis (Jum. & H.Perrier) Beentje & J.Dransf., Palms Madagascar: 207 (1995).
Madagascar. 29 MDG.
*Chrysalidocarpus onilahensis Jum. & H.Perrier, Ann. Inst. Bot.-Géol. Colon. Marseille, III, 1: 37 (1913).
Chrysalidocarpus midongensis Jum., Ann. Inst. Bot.-Géol. Colon. Marseille, II, 10(3): 17 (1922).
Chrysalidocarpus brevinodis H.Perrier, Notul. Syst. (Paris) 8: 47 (1939).

Dypsis oreophila Beentje in J.Dransfield & H.Beentje, Palms Madagascar: 227 (1995).
N. & NE. Madagascar. 29 MDG.
*Neodypsis gracilis Jum., Ann. Inst. Bot.-Géol. Colon. Marseille, V, 1(1): 19 (1933). *Phloga gracilis* (Jum.) H.Perrier, in Fl. Madag. 30: 126 (1945).

Dypsis oropedionis Beentje in J.Dransfield & H.Beentje, Palms Madagascar: 159 (1995).
C. Madagascar. 29 MDG.

Dypsis ovobontsira Beentje in J.Dransfield & H.Beentje, Palms Madagascar: 180 (1995).
NE. Madagascar. 29 MDG.

Dypsis pachyramea J.Dransf. in J.Dransfield & H.Beentje, Palms Madagascar: 404 (1995).
NE. Madagascar. 29 MDG.

Dypsis paludosa J.Dransf. in J.Dransfield & H.Beentje, Palms Madagascar: 343 (1995).
NE. & CE. Madagascar. 29 MDG.

Dypsis pembana (H.E.Moore) Beentje & J.Dransf., Palms Madagascar: 219 (1995).
Tanzania (Pemba). 25 TAN.
*Chrysalidocarpus pembanus H.E.Moore, Principes 6: 109 (1962).

Dypsis perrieri (Jum.) Beentje & J.Dransf., Palms Madagascar: 351 (1995).
NE. Madagascar. 29 MDG.
*Antongilia perrieri Jum., Ann. Inst. Bot.-Géol. Colon. Marseille, IV, 6(2): 17 (1928).
Chrysalidocarpus auriculatus Jum., Ann. Inst. Bot.-Géol. Colon. Marseille, V, 1(1): 25 (1933).
Chrysalidocarpus ruber Jum., Ann. Inst. Bot.-Géol. Colon. Marseille, V, 1(1): 23 (1933).

Dypsis pervillei (Jum.) Beentje & J.Dransf., Palms Madagascar: 269 (1995).
CE. Madagascar (Betampona). 29 MDG.
*Haplodypsis pervillei Baill., Bull. Mens. Soc. Linn. Paris 2: 1167 (1894). *Neophloga pervillei* (Baill.) Becc., Bot. Jahrb. Syst. 38(87): 26 (1906).

Dypsis pilulifera (Becc.) Beentje & J.Dransf., Palms Madagascar: 161 (1995).
N. & EC. Madagascar. 29 MDG.
*Chrysalidocarpus piluliferus Becc., Bot. Jahrb. Syst. 38(87): 37 (1906).

Dypsis pinnatifrons Mart., Hist. Nat. Palm. 3: 180 (1838).
N. & E. Madagascar. 29 MDG.
Areca gracilis Thouars ex Kunth, Enum. Pl. 3: 188 (1841), pro syn.
Dypsis gracilis Bory ex Mart., Hist. Nat. Palm., ed. 2, 3: 181 (1845). *Adelodypsis gracilis* (Bory ex Mart.) Becc., Bot. Jahrb. Syst. 38(87): 17 (1906).
Dypsis gracilis var. *sambiranensis* Jum. & H.Perrier, Ann. Inst. Bot.-Géol. Colon. Marseille, III, 1: 24 (1913). *Dypsis sambiranensis* (Jum. & H.Perrier) Jum., Ann. Inst. Bot.-Géol. Colon. Marseille, V, 1(3): 15 (1933). *Chrysalidocarpus sambiranensis* (Jum. & H.Perrier) Jum., Ann. Inst. Bot.-Géol. Colon. Marseille, V, 1(1): 21 (1933). *Adelodypsis sambiranensis* (Jum. & H.Perrier) H.P.Guérin, Ann. Sci. Nat., Bot., XI, 12: 28 (1950).

Dypsis plurisecta Jum., Ann. Inst. Bot.-Géol. Colon. Marseille, III, 6(1): 35 (1918).
NE. Madagascar. 29 MDG.

Dypsis poivreana (Baill.) Beentje & J.Dransf., Palms Madagascar: 307 (1995).
ENE. Madagascar. 29 MDG.
*Haplophloga poivreana Baill., Bull. Mens. Soc. Linn. Paris 2: 1168 (1894). *Neophloga poivreana* (Baill.) Becc., Bot. Jahrb. Syst. 38(87): 24 (1906).

Dypsis polystachya Baker = **Dypsis nodifera** Mart.

Dypsis prestoniana Beentje in J.Dransfield & H.Beentje, Palms Madagascar: 167 (1995).
CE. & SE. Madagascar. 29 MDG.

Dypsis procera Jum., Ann. Inst. Bot.-Géol. Colon. Marseille, III, 6(1): 33 (1918).
NE. & CE. Madagascar. 29 MDG.
Dypsis longipes Jum., Ann. Inst. Bot.-Géol. Colon. Marseille, III, 6(1): 37 (1918).

Dypsis psammophila Beentje in J.Dransfield & H.Beentje, Palms Madagascar: 216 (1995).
ENE. Madagascar. 29 MDG.

Dypsis pulchella J.Dransf. in J.Dransfield & H.Beentje, Palms Madagascar: 297 (1995).
CE. Madagascar. 29 MDG.

Dypsis pumila Beentje in J.Dransfield & H.Beentje, Palms Madagascar: 223 (1995).
N. Madagascar. 29 MDG.

Dypsis pusilla Beentje in J.Dransfield & H.Beentje, Palms Madagascar: 370 (1995).
NE. Madagascar. 29 MDG.

Dypsis ramentacea J.Dransf. in J.Dransfield & H.Beentje, Palms Madagascar: 411 (1995).
NE. Madagascar. 29 MDG.

Dypsis remotiflora J.Dransf. in J.Dransfield & H.Beentje, Palms Madagascar: 331 (1995).
SE. Madagascar. 29 MDG.

Dypsis rhodotricha Baker = **Dypsis heterophylla** Baker

Dypsis rivularis (Jum. & H.Perrier) Beentje & J.Dransf., Palms Madagascar: 232 (1995).
NW. & N. Madagascar. 29 MDG.
*Chrysalidocarpus rivularis Jum. & H.Perrier, Ann. Inst. Bot.-Géol. Colon. Marseille, III, 1: 40 (1913).

Dypsis sahanofensis (Jum. & H.Perrier) Beentje & J.Dransf., Palms Madagascar: 290 (1995).
E. Madagascar. 29 MDG.
Neophloga sahanofensis Jum. & H.Perrier, Ann. Inst. Bot.-Géol. Colon. Marseille, III, 1: 32 (1913).
Chrysalidocarpus sahanofensis (Jum. & H.Perrier) Jum., Ann. Inst. Bot.-Géol. Colon. Marseille, II, 10(3): 6 (1922).

Dypsis saintelucei Beentje in J.Dransfield & H.Beentje, Palms Madagascar: 178 (1995).
SE. Madagascar. 29 MDG.

Dypsis sambiranensis (Jum. & H.Perrier) Jum. = **Dypsis pinnatifrons** Mart.

Dypsis sanctaemariae J.Dransf. in J.Dransfield & H.Beentje, Palms Madagascar: 264 (1995).
NE. Madagascar (Î. Sainte Marie). 29 MDG.

Dypsis scandens J.Dransf. in J.Dransfield & H.Beentje, Palms Madagascar: 255 (1995).
E. Madagascar (Ifanadiana area). 29 MDG.

Dypsis schatzii Beentje in J.Dransfield & H.Beentje, Palms Madagascar: 278 (1995).
ENE. Madagascar (Betampona). 29 MDG.

Dypsis scottiana (Becc.) Beentje & J.Dransf., Palms Madagascar: 239 (1995).
SE. Madagascar. 29 MDG.
Phloga scottiana Becc., J. Linn. Soc., Bot. 29: 61 (1891). *Neophloga scottiana* (Becc.) Becc., Bot. Jahrb. Syst. 38(87): 22 (1906).
Neophloga affinis Becc., Bot. Jahrb. Syst. 38(87): 22 (1906).

Dypsis serpentina Beentje in J.Dransfield & H.Beentje, Palms Madagascar: 206 (1995).
NE. Madagascar. 29 MDG.

Dypsis simianensis (Jum.) Beentje & J.Dransf., Palms Madagascar: 317 (1995).
E. Madagascar. 29 MDG.
Neophloga simianensis Jum., Ann. Inst. Bot.-Géol. Colon. Marseille, IV, 6(3): 13 (1928 publ. 1929).

Dypsis singularis Beentje in J.Dransfield & H.Beentje, Palms Madagascar: 242 (1995).
ESE. Madagascar. 29 MDG.

Dypsis soanieranae Beentje in J.Dransfield & H.Beentje, Palms Madagascar: 266 (1995).
ENE. Madagascar. 29 MDG.

Dypsis spicata J.Dransf. in J.Dransfield & H.Beentje, Palms Madagascar: 407 (1995).
NE. Madagascar. 29 MDG.

Dypsis tanalensis (Jum. & H.Perrier) Beentje & J.Dransf., Palms Madagascar: 182 (1995).
SE. Madagascar (S. of Manakara). 29 MDG.
Neodypsis tanalensis Jum. & H.Perrier, Ann. Inst. Bot.-Géol. Colon. Marseille, III, 1: 18 (1913).

Dypsis tenuissima Beentje in J.Dransfield & H.Beentje, Palms Madagascar: 315 (1995).
SE. Madagascar. 29 MDG.

Dypsis thermarum J.Dransf. in J.Dransfield & H.Beentje, Palms Madagascar: 377 (1995).
ESE. Madagascar. 29 MDG.

Dypsis thiryana (Becc.) Beentje & J.Dransf., Palms Madagascar: 282 (1995).
E. Madagascar. 29 MDG.
Neophloga thiryana Becc., Bot. Jahrb. Syst. 38(87): 23 (1906).

Dypsis thouarsiana Baill., Bull. Mens. Soc. Linn. Paris 2: 1163 (1893). *Vonitra thouarsiana* (Baill.) Becc., Bot. Jahrb. Syst. 38(87): 18 (1906).
NE. Madagascar. 29 MDG.

Dypsis tokoravina Beentje in J.Dransfield & H.Beentje, Palms Madagascar: 170 (1995).
NE. Madagascar. 29 MDG.

Dypsis trapezoidea J.Dransf. in J.Dransfield & H.Beentje, Palms Madagascar: 284 (1995).
E. Madagascar (Mt. Vatovavy). 29 MDG.

Dypsis tsaratananensis (Jum.) Beentje & J.Dransf., Palms Madagascar: 226 (1995).
N. Madagascar. 29 MDG.
Neodypsis tsaratananensis Jum., Ann. Inst. Bot.-Géol. Colon. Marseille, IV, 2(2): 15 (1924).

Dypsis tsaravoasira Beentje in J.Dransfield & H.Beentje, Palms Madagascar: 154 (1995).
NE. Madagascar. 29 MDG.

Dypsis turkii J.Dransf., Palms 47: 27 (2003).
Madagascar. 29 MDG.

Dypsis utilis (Jum.) Beentje & J.Dransf., Palms Madagascar: 364 (1995).
CE. Madagascar. 29 MDG.
Vonitra utilis Jum., Compt. Rend. Hebd. Séances Acad. Sci. 164: 921 (1917).

Dypsis vilersiana Baill. = **Dypsis nodifera** Mart.

Dypsis viridis Jum., Ann. Inst. Bot.-Géol. Colon. Marseille, III, 6(1): 35 (1918).
ENE. Madagascar. 29 MDG.

Dypsis zahamenae J.Dransf. = **Dypsis humbertii** H.Perrier

Edanthe

Edanthe O.F.Cook & Doyle = **Chamaedorea** Willd.

Edanthe veraepacis O.F.Cook = **Chamaedorea tepejilote** Liebm.

Elaeis

Elaeis Jacq., Select. Stirp. Amer. Hist.: 280 (1763).
2 species, Trop. Africa, C. & S. Trop. America. 22 23 25 26 80 82 83 84 (29) (40) (42).
Corozo Jacq. ex Giseke, Prael. Ord. Nat. Pl.: 42 (1792).
Alfonsia Kunth in F.W.H.von Humboldt, A.J.A. Bonpland & C.S.Kunth, Nov. Gen. Sp. 1: 306 (1816).

Elaeis dybowskii Hua = **Elaeis guineensis** Jacq.

Elaeis guineensis Jacq., Select. Stirp. Amer. Hist.: 280 (1763). *Elaeis guineensis* subsp. *nigrescens* A.Chev., Veg. Ut. Afr. Trop. Franç. 7: 46 (1910), nom. inval.
Trop. Africa, widely cultivated elsewere. 22 BEN GHA GUI IVO LBR NGA SEN SIE TOG 23 BUR CAF CMN CON GAB RWA ZAI 25 KEN TAN UGA 26 ANG (29) srl (40) (42) mly sum.
Palma oleosa Mill., Gard. Dict. ed. 8: 6 (1768).
Elaeis melanococca Gaertn., Fruct. Sem. Pl. 1: 18 (1788).
Elaeis melanococca var. *semicircularis* Oerst., Vidensk. Meddel. Dansk Naturhist. Foren. Kjobenhavn 1858: 51 (1859).
Elaeis dybowskii Hua, Bull. Mus. Hist. Nat. (Paris) 1: 315 (1895).
Elaeis guineensis f. *androgyna* A.Chev., Veg. Ut. Afr. Trop. Franç. 7: 66 (1910).
Elaeis guineensis f. *dioica* A.Chev., Veg. Ut. Afr. Trop. Franç. 7: 66 (1910).

Elaeis guineensis f. *ramosa* A.Chev., Veg. Ut. Afr. Trop. Franç. 7: 66 (1910).

Elaeis guineensis subsp. *virescens* A.Chev., Veg. Ut. Afr. Trop. Franç. 7: 60 (1910). *Elaeis virescens* (A.Chev.) D.Prain, Index Kew., Suppl. 4: 77 (1913).

Elaeis guineensis var. *ceredia* A.Chev., Veg. Ut. Afr. Trop. Franç. 7: 66 (1910).

Elaeis guineensis var. *gracilinux* A.Chev., Veg. Ut. Afr. Trop. Franç. 7: 66 (1910).

Elaeis guineensis var. *idolatrica* A.Chev., Veg. Ut. Afr. Trop. Franç. 7: 66 (1910).

Elaeis guineensis var. *intermedia* A.Chev., Veg. Ut. Afr. Trop. Franç. 7: 66 (1910).

Elaeis guineensis var. *macrocarpa* A.Chev., Veg. Ut. Afr. Trop. Franç. 7: 66 (1910).

Elaeis guineensis var. *macrophylla* A.Chev., Veg. Ut. Afr. Trop. Franç. 7: 66 (1910).

Elaeis guineensis var. *pisifera* A.Chev., Veg. Ut. Afr. Trop. Franç. 7: 66 (1910).

Elaeis guineensis var. *repanda* A.Chev., Veg. Ut. Afr. Trop. Franç. 7: 66 (1910).

Elaeis guineensis var. *sempernigra* A.Chev., Veg. Ut. Afr. Trop. Franç. 7: 66 (1910).

Elaeis guineensis var. *spectabilis* A.Chev., Veg. Ut. Afr. Trop. Franç. 7: 66 (1910).

Elaeis guineensis var. *madagascariensis* Jum. & H.Perrier, Mat. Gross: 6 (1911). *Elaeis madagascariensis* (Jum. & H.Perrier) Becc., Palme Madagascar: 55 (1914).

Elaeis nigrescens (A.Chev.) D.Prain, Index Kew., Suppl. 4: 77 (1913).

Elaeis guineensis f. *caryolitica* Becc., Contr. Conosc. Palma Olio: 65 (1914).

Elaeis guineensis f. *dura* Becc., Contr. Conosc. Palma Olio: 37 (1914).

Elaeis guineensis f. *fatua* Becc., Contr. Conosc. Palma Olio: 54 (1914).

Elaeis guineensis f. *semidura* Becc., Contr. Conosc. Palma Olio: 56 (1914).

Elaeis guineensis f. *tenera* Becc., Contr. Conosc. Palma Olio: 38 (1914).

Elaeis guineensis var. *albescens* Becc., Contr. Conosc. Palma Olio: 62 (1914).

Elaeis guineensis var. *angulosa* Becc., Contr. Conosc. Palma Olio: 49 (1914).

Elaeis guineensis var. *compressa* Becc., Contr. Conosc. Palma Olio: 71 (1914).

Elaeis guineensis var. *leucocarpa* Becc., Contr. Conosc. Palma Olio: 40 (1914).

Elaeis guineensis var. *macrocarya* Becc., Contr. Conosc. Palma Olio: 71 (1914).

Elaeis guineensis var. *rostrata* Becc., Contr. Conosc. Palma Olio: 50 (1914).

Elaeis macrophylla A.Chev., Explor. Bot. Afrique Occ. Franç. 1: 676 (1920), nom. nud.

Elaeis guineensis var. *albescens* Becc. = **Elaeis guineensis** Jacq.

Elaeis guineensis f. *androgyna* A.Chev. = **Elaeis guineensis** Jacq.

Elaeis guineensis var. *angulosa* Becc. = **Elaeis guineensis** Jacq.

Elaeis guineensis f. *caryolitica* Becc. = **Elaeis guineensis** Jacq.

Elaeis guineensis var. *ceredia* A.Chev. = **Elaeis guineensis** Jacq.

Elaeis guineensis var. *compressa* Becc. = **Elaeis guineensis** Jacq.

Elaeis guineensis f. *dioica* A.Chev. = **Elaeis guineensis** Jacq.

Elaeis guineensis f. *dura* Becc. = **Elaeis guineensis** Jacq.

Elaeis guineensis f. *fatua* Becc. = **Elaeis guineensis** Jacq.

Elaeis guineensis var. *gracilinux* A.Chev. = **Elaeis guineensis** Jacq.

Elaeis guineensis var. *idolatrica* A.Chev. = **Elaeis guineensis** Jacq.

Elaeis guineensis var. *intermedia* A.Chev. = **Elaeis guineensis** Jacq.

Elaeis guineensis var. *leucocarpa* Becc. = **Elaeis guineensis** Jacq.

Elaeis guineensis var. *macrocarpa* A.Chev. = **Elaeis guineensis** Jacq.

Elaeis guineensis var. *macrocarya* Becc. = **Elaeis guineensis** Jacq.

Elaeis guineensis var. *macrophylla* A.Chev. = **Elaeis guineensis** Jacq.

Elaeis guineensis var. *madagascariensis* Jum. & H.Perrier = **Elaeis guineensis** Jacq.

Elaeis guineensis subsp. *nigrescens* A.Chev. = **Elaeis guineensis** Jacq.

Elaeis guineensis var. *pisifera* A.Chev. = **Elaeis guineensis** Jacq.

Elaeis guineensis f. *ramosa* A.Chev. = **Elaeis guineensis** Jacq.

Elaeis guineensis var. *repanda* A.Chev. = **Elaeis guineensis** Jacq.

Elaeis guineensis var. *rostrata* Becc. = **Elaeis guineensis** Jacq.

Elaeis guineensis f. *semidura* Becc. = **Elaeis guineensis** Jacq.

Elaeis guineensis var. *sempernigra* A.Chev. = **Elaeis guineensis** Jacq.

Elaeis guineensis var. *spectabilis* A.Chev. = **Elaeis guineensis** Jacq.

Elaeis guineensis f. *tenera* Becc. = **Elaeis guineensis** Jacq.

Elaeis guineensis subsp. *virescens* A.Chev. = **Elaeis guineensis** Jacq.

Elaeis macrophylla A.Chev. = **Elaeis guineensis** Jacq.

Elaeis madagascariensis (Jum. & H.Perrier) Becc. = **Elaeis guineensis** Jacq.

Elaeis melanococca Gaertn. = **Elaeis guineensis** Jacq.

Elaeis melanococca Mart. = **Elaeis oleifera** (Kunth) Cortés

Elaeis melanococca var. *semicircularis* Oerst. = **Elaeis guineensis** Jacq.

Elaeis nigrescens (A.Chev.) D.Prain = **Elaeis guineensis** Jacq.

Elaeis occidentalis Sw. = **Calyptronoma occidentalis** (Sw.) H.E.Moore

Elaeis odora Trail = **Barcella odora** (Trail) Drude

Elaeis oleifera (Kunth) Cortés, Fl. Colomb. 1: 203 (1897). C. & S. Trop. America. 80 COS HON NIC PAN 82 FRG SUR 83 CLM ECU PER 84 BZN.

Elaeis pernambucana Lodd. ex G.Don = ? [84]

Elaeis spectabilis Lodd. ex Sweet = ? [40]

Elaeis virescens (A.Chev.) D.Prain = **Elaeis guineensis** Jacq.

Elate

Elate L. = **Phoenix** L.

Elate sylvestris L. = **Phoenix sylvestris** (L.) Roxb.

Elate versicolor Salisb. = **Phoenix sylvestris** (L.) Roxb.

Eleiodoxa

Eleiodoxa (Becc.) Burret, Notizbl. Bot. Gart. Berlin-Dahlem 15: 733 (1942).
1 species, Pen. Thailand to W. Malesia. 41 42.

Eleiodoxa conferta (Griff.) Burret, Notizbl. Bot. Gart. Berlin-Dahlem 15: 734 (1942).
Pen. Thailand to W. Malesia. 41 THA 42 BOR MLY SUM.
Salacca conferta Griff., Calcutta J. Nat. Hist. 5: 16 (1845).
Salacca scortechinii Becc., Ann. Roy. Bot. Gard. (Calcutta) 12(3): 97 (1919). *Eleiodoxa scortechinii* (Becc.) Burret, Notizbl. Bot. Gart. Berlin-Dahlem 15: 735 (1942).
Eleiodoxa microcarpa Burret, Notizbl. Bot. Gart. Berlin-Dahlem 15: 735 (1942).
Eleiodoxa orthoschista Burret, Notizbl. Bot. Gart. Berlin-Dahlem 15: 734 (1942).
Eleiodoxa xantholepis Burret, Notizbl. Bot. Gart. Berlin-Dahlem 15: 735 (1942).

Eleiodoxa microcarpa Burret = **Eleiodoxa conferta** (Griff.) Burret

Eleiodoxa orthoschista Burret = **Eleiodoxa conferta** (Griff.) Burret

Eleiodoxa scortechinii (Becc.) Burret = **Eleiodoxa conferta** (Griff.) Burret

Eleiodoxa xantholepis Burret = **Eleiodoxa conferta** (Griff.) Burret

Elephantusia

Elephantusia Willd. = **Phytelephas** Ruiz & Pav.

Elephantusia macrocarpa (Ruiz & Pav.) Willd. = **Phytelephas macrocarpa** Ruiz & Pav.

Elephantusia microcarpa (Ruiz & Pav.) Willd. = **Phytelephas macrocarpa** Ruiz & Pav.

Eleutheropetalum

Eleutheropetalum (H.Wendl.) H.Wendl. ex Oerst. = **Chamaedorea** Willd.

Eleutheropetalum ernesti-augusti (H.Wendl.) Oerst. = **Chamaedorea ernesti-augusti** H.Wendl.

Eleutheropetalum sartorii (Liebm.) Oerst. = **Chamaedorea sartorii** Liebm.

Englerophoenix

Englerophoenix Kuntze = **Attalea** Kunth

Englerophoenix attaleoides (Barb.Rodr.) Barb.Rodr. = **Attalea attaleoides** (Barb.Rodr.) Wess.Boer

Englerophoenix caribaeum (Griseb. & H.Wendl.) Kuntze = **Attalea maripa** (Aubl.) Mart.

Englerophoenix insignis (Mart.) Kuntze = **Attalea insignis** (Mart.) Drude

Englerophoenix longirostrata (Barb.Rodr.) Barb.Rodr. = **Attalea maripa** (Aubl.) Mart.

Englerophoenix maripa (Aubl.) Kuntze = **Attalea maripa** (Aubl.) Mart.

Englerophoenix regia (Mart.) Kuntze = **Attalea maripa** (Aubl.) Mart.

Eora

Eora O.F.Cook = **Rhopalostylis** H.Wendl. & Drude

Eora baueri (Hook.f. ex Lem.) O.F.Cook = **Rhopalostylis baueri** (Hook.f. ex Lem.) H.Wendl. & Drude

Eora cheesemanii (Becc. ex Cheeseman) O.F.Cook = **Rhopalostylis baueri** var. **cheesemanii** (Becc. ex Cheeseman) Sykes

Eora sapida (Sol. ex G.Forst.) O.F.Cook = **Rhopalostylis sapida** (Sol. ex G.Forst.) H.Wendl. & Drude

Eora ultima O.F.Cook = **Rhopalostylis baueri** var. **cheesemanii** (Becc. ex Cheeseman) Sykes

Eremospatha

Eremospatha (G.Mann & H.Wendl.) H.Wendl. in O.C.E. de Kerchove de Denterghem, Palmiers: 244 (1878).
10 species, W. Trop. Africa to Zambia. 22 23 26.

Eremospatha barendii Sunderl., J. Bamboo Rattan 1: 361 (2002).
Cameroon. 23 CMN.

Eremospatha cabrae (De Wild. & T.Durand) De Wild., Ann. Mus. Congo Belge, Bot., V, 1: 95 (1903).
WC. Trop. Africa to Angola. 23 CAB CAF CON GAB ZAI 26 ANG.
Calamus cabrae De Wild. & T.Durand, Bull. Soc. Roy. Bot. Belgique, Compt. Rend. 38: 151 (1899).
Eremospatha rhomboidea Burret, Notizbl. Bot. Gart. Berlin-Dahlem 15: 751 (1942).
Eremospatha suborhicularis Burret, Notizbl. Bot. Gart. Berlin-Dahlem 15: 750 (1942).

Eremospatha cuspidata (G.Mann & H.Wendl.) H.Wendl. in O.C.E.de Kerchove de Denterghem, Palmiers: 244 (1878).
WC. Trop. Africa to NW. Zambia. 23 CAF CMN CON EQG GAB ZAI 26 ANG ZAM.
Calamus cuspidatus G.Mann & H.Wendl., Trans. Linn. Soc. London 24: 434 (1864).

Eremospatha deerrata (G.Mann & H.Wendl.) T.Durand & Schinz = **Calamus deerratus** G.Mann & H.Wendl.

Eremospatha haullevilleana De Wild., Ann. Mus. Congo Belge, Bot., V, 1: 96 (1903).
WC. & E. Trop. Africa. 23 BUR CAF CMN CON GAB ZAI 25 TAN UGA 26 ANG.

Eremospatha hookeri (G.Mann & H.Wendl.) H.Wendl. in O.C.E.de Kerchove de Denterghem, Palmiers: 244 (1878).
W. & WC. Trop. Africa. 22 GHA IVO NGA SIE 23 CAB CMN CON EQG GAB ZAI.
Calamus hookeri G.Mann & H.Wendl., Trans. Linn. Soc. London 24: 434 (1864).

Eremospatha korthalsiifolia Becc. = **Eremospatha wendlandiana** Dammer ex Becc.

Eremospatha laurentii De Wild., Bull. Jard. Bot. État 5: 147 (1916).
W. & WC. Trop. Africa. 22 LBR NGA SIE 23 CAF CMN EQG GAB ZAI.

Eremospatha macrocarpa H.Wendl. in O.C.E.de Kerchove de Denterghem, Palmiers: 244 (1878).
W. & WC. Trop. Africa. 22 BEN GHA GUI IVO LBR NGA SIE 23 CAF CMN EQG ZAI.

Calamus macrocarpus G.Mann & H.Wendl., Trans. Linn. Soc. London 24: 435 (1864), nom. illeg.
Eremospatha sapinii De Wild., Bull. Jard. Bot. État 5: 147 (1916).

Eremospatha quinquecostulata Becc., Webbia 3: 279 (1910).
Nigeria to Cameroon. 22 NGA 23 CMN.

Eremospatha rhomboidea Burret = **Eremospatha cabrae** (De Wild. & T.Durand) De Wild.

Eremospatha sapinii De Wild. = **Eremospatha macrocarpa** H.Wendl.

Eremospatha suborhicularis Burret = **Eremospatha cabrae** (De Wild. & T.Durand) De Wild.

Eremospatha tessmanniana Becc., Webbia 3: 278 (1910).
WC. Trop. Africa. 23 CMN CON EQG.

Eremospatha wendlandiana Dammer ex Becc., Webbia 3: 290 (1910).
S. Nigeria to Angola. 22 NGA 23 CAF CMN EQG GAB ZAI 26 ANG.
Eremospatha korthalsiifolia Becc., Webbia 3: 292 (1910).

Erythea

Erythea S.Watson = **Brahea** Mart.

Erythea aculeata Brandegee = **Brahea aculeata** (Brandegee) H.E.Moore

Erythea armata S.Watson = **Brahea armata** S.Watson

Erythea brandegeei Purpus = **Brahea brandegeei** (Purpus) H.E.Moore

Erythea clara L.H.Bailey = **Brahea armata** S.Watson

Erythea edulis (H.Wendl. ex S.Watson) S.Watson = **Brahea edulis** H.Wendl. ex S.Watson

Erythea elegans Franceschi ex Becc. = **Brahea armata** S.Watson

Erythea loretensis M.E.Jones = **Sabal mexicana** Mart.

Erythea pimo (Becc.) H.E.Moore = **Brahea pimo** Becc.

Erythea roezlii (Linden) Becc. ex Martelli = **Brahea armata** S.Watson

Erythea salvadorensis (H.Wendl. ex Becc.) H.E.Moore = **Brahea dulcis** (Kunth) Mart.

Ethnora

Ethnora O.F.Cook = **Maximiliana**

Ethnora maripa (Mart.) O.F.Cook = **Maximiliana maripa**

Eugeissona

Eugeissona Griff., Calcutta J. Nat. Hist. 5: 101 (1845).
6 species, Pen. Thailand to W. Malesia. 41 42.

Eugeissona ambigua Becc., Ann. Roy. Bot. Gard. (Calcutta) 12(2): 200 (1918).
Borneo. 42 BOR.

Eugeissona brachystachys Ridl., J. Fed. Malay States Mus. 6: 189 (1915).
Pen. Malaysia. 42 MLY.

Eugeissona insignis Becc., Nuovo Giorn. Bot. Ital. 3: 22 (1871).
Borneo (Sarawak). 42 BOR.
Eugeissona major Becc., Malesia 3: 3 (1886).

Eugeissona major Becc. = **Eugeissona insignis** Becc.

Eugeissona minor Becc., Nuovo Giorn. Bot. Ital. 3: 18 (1871).
Borneo. 42 BOR.

Eugeissona pachycarpa Burret = **Eugeissona utilis** Becc.

Eugeissona tristis Griff., Calcutta J. Nat. Hist. 5: 101 (1845).
Pen. Thailand to Pen. Malaysia. 41 THA 42 MLY.

Eugeissona utilis Becc., Nuovo Giorn. Bot. Ital. 3: 26 (1871).
Borneo. 42 BOR.
Eugeissona pachycarpa Burret, Notizbl. Bot. Gart. Berlin-Dahlem 15: 729 (1942).
Eugeissona wendlandiana Burret, Notizbl. Bot. Gart. Berlin-Dahlem 15: 728 (1942).

Eugeissona wendlandiana Burret = **Eugeissona utilis** Becc.

Eupritchardia

Eupritchardia Kuntze = **Pritchardia** Seem. & H.Wendl.

Eupritchardia gaudichaudii (Mart.) Kuntze = **Pritchardia martii** (Gaudich.) H.Wendl.

Eupritchardia hillebrandtii (Becc.) Kuntze = **Pritchardia hildebrandtii**

Eupritchardia kamapuaana (Caum) L.H.Bailey = **Pritchardia kamapuaana**

Eupritchardia lanigera (Becc.) Kuntze = **Pritchardia lanigera** Becc.

Eupritchardia martii (Gaudich.) Kuntze = **Pritchardia martii** (Gaudich.) H.Wendl.

Eupritchardia pacifica (Seem. & H.Wendl.) Kuntze = **Pritchardia pacifica** Seem. & H.Wendl.

Eupritchardia pericularum (H.Wendl.) Kuntze = **Pritchardia pericularum** H.Wendl.

Eupritchardia remota (Kuntze) Kuntze = **Pritchardia remota** (Kuntze) Becc.

Eupritchardia thurstonii (F.Muell. & Drude) Kuntze = **Pritchardia thurstonii** F.Muell. & Drude

Eupritchardia vluylstekeana (H.Wendl.) Kuntze = **Pritchardia vuylstekeana** H.Wendl.

Euterpe

Euterpe Mart., Hist. Nat. Palm. 2: 28 (1823), nom. cons.
7 species, Trop. America. 80 81 82 83 84 85.
Catis O.F.Cook, Bull. Torrey Bot. Club 28: 557 (1901).
Plectis O.F.Cook, Bull. Torrey Bot. Club 31: 352 (1904).
Rooseveltia O.F.Cook, Smithsonian Misc. Collect. 98: 21 (1939).

Euterpe Gaertn. = **Prestoea** Hook.f.

Euterpe aculeata (Willd.) Spreng. = **Aiphanes horrida** (Jacq.) Burret

Euterpe acuminata (Willd.) H.Wendl. = **Prestoea acuminata** (Willd.) H.E.Moore

Euterpe andicola Brongn. ex Mart. = **Prestoea acuminata** var. **acuminata**

Euterpe andina Burret = **Prestoea acuminata** var. **acuminata**

Euterpe antioquensis Linden = **Prestoea acuminata** var. **acuminata**

Euterpe aphanolepis Burret = **Prestoea acuminata** var. **acuminata**

Euterpe aurantiaca H.E.Moore = **Euterpe catinga** var. **catinga**

Euterpe badiocarpa Barb.Rodr. = **Euterpe oleracea** Mart.

Euterpe beardii L.H.Bailey = **Euterpe oleracea** Mart.

Euterpe brachyclada Burret = **Prestoea carderi** (W.Bull) Hook.f.

Euterpe brachyspatha Burret = **Prestoea longipetiolata** var. **longipetiolata**

Euterpe brasiliana Oken = **Euterpe oleracea** Mart.

Euterpe brevicaulis Burret = **Prestoea carderi** (W.Bull) Hook.f.

Euterpe brevivaginata Mart. = **Prestoea acuminata** var. **acuminata**

Euterpe broadwayana Becc. = **Euterpe broadwayi** Becc. ex Broadway

Euterpe broadwayi Becc. ex Broadway, Bull. Dept. Agric. Trinidad Tobago 15: 174 (1916).
Windward Is. to Trinidad and Tobago. 81 TRT WIN.
Euterpe broadwayana Becc., Repert. Spec. Nov. Regni Veg. 16: 436 (1920), orth. var.
Euterpe dominicana L.H.Bailey, Gentes Herb. 4: 375 (1940).

Euterpe caatinga Spruce = **Euterpe catinga** var. **catinga**

Euterpe carderi (Hook.f.) Burret = **Prestoea carderi** (W.Bull) Hook.f.

Euterpe caribaea Spreng. = **Roystonea oleracea** (Jacq.) O.F.Cook

Euterpe catinga Wallace, Palm Trees Amazon: 27 (1853).
S. Trop. America. 82 GUY VEN 83 CLM PER 84 BZN.

var. **catinga**
Venezuela to Peru. 82 VEN 83 CLM PER 84 BZN.
Euterpe caatinga Spruce, J. Linn. Soc., Bot. 11: 137 (1869), orth. var.
Euterpe mollissima Barb.Rodr., Enum. Palm. Nov.: 16 (1875).
Euterpe controversa Barb.Rodr., Palmiers: 34 (1882), nom. illeg.
Euterpe concinna Burret, Bot. Jahrb. Syst. 63: 69 (1929), nom. provis.
Euterpe aurantiaca H.E.Moore, Principes 13: 137 (1969).

var. **roraimae** (Dammer) A.J.Hend. & Galeano, Fl. Neotrop. Monogr. 72: 28 (1996).
S. Trop. America. 82 GUY VEN 83 ECU PER 84 BZN.
Euterpe roraimae Dammer, Notizbl. Königl. Bot. Gart. Berlin 6: 264 (1915).
Euterpe montis-duida Burret, Bull. Torrey Bot. Club 58: 319 (1931).
Euterpe ptariana Steyerm., Fieldiana, Bot. 28(1): 87 (1951).
Euterpe erubescens H.E.Moore, Principes 13: 138 (1969).

Euterpe chaunostachys Burret = **Prestoea acuminata** var. **acuminata**

Euterpe concinna Burret = **Euterpe catinga** var. **catinga**

Euterpe confertiflora L.H.Bailey = **Euterpe precatoria** var. **precatoria**

Euterpe controversa Barb.Rodr. = **Euterpe catinga** var. **catinga**

Euterpe cuatrecasana Dugand = **Euterpe oleracea** Mart.

Euterpe dasystachys Burret = **Prestoea acuminata** var. **dasystachys** (Burret) A.J.Hend. & Galeano

Euterpe decurrens H.Wendl. ex Burret = **Prestoea decurrens** (H.Wendl. ex Burret) H.E.Moore

Euterpe disticha H.Wendl. ex Linden = **? [83 CLM]**

Euterpe dominicana L.H.Bailey = **Euterpe broadwayi** Becc. ex Broadway

Euterpe edulis Mart., Hist. Nat. Palm. 2: 33 (1824).
E. & S. Brazil to Argentina (Misiones). 84 BZE BZL BZS 85 AGE PAR.
Euterpe egusquizae Bertoni ex Hauman, Physis 4: 606 (1919).
Euterpe edulis var. *clausa* Mattos, Loefgrenia 71: 1 (1977).
Euterpe espiritosantensis H.Q.B.Fernald, Acta Bot. Brasil. 3(2): 43 (1989 publ. 1990).

Euterpe edulis var. *clausa* Mattos = **Euterpe edulis** Mart.

Euterpe egusquizae Bertoni ex Hauman = **Euterpe edulis** Mart.

Euterpe elegans Linden = **? [83 CLM]**

Euterpe ensiformis (Ruiz & Pav.) Mart. = **Prestoea ensiformis** (Ruiz & Pav.) H.E.Moore

Euterpe erubescens H.E.Moore = **Euterpe catinga** var. **roraimae** (Dammer) A.J.Hend. & Galeano

Euterpe espiritosantensis H.Q.B.Fernald = **Euterpe edulis** Mart.

Euterpe filamentosa Kunth = **Oncosperma tigillarium** (Jack) Ridl.

Euterpe frigida (Kunth) Burret = **Prestoea acuminata** var. **acuminata**

Euterpe globosa Gaertn. = **Prestoea acuminata** var. **acuminata**

Euterpe gracilis Linden = **? [84]**

Euterpe haenkeana Brongn. ex Mart. = **Prestoea acuminata** var. **acuminata**

Euterpe jatapuensis Barb.Rodr. = **Euterpe precatoria** var. **precatoria**

Euterpe jenmanii C.H.Wright = **Roystonea regia** (Kunth) O.F.Cook

Euterpe kalbreyeri Burret = **Euterpe precatoria** var. **longivaginata** (Mart.) A.J.Hend.

Euterpe karsteniana Engel = **Euterpe precatoria** var. **longivaginata** (Mart.) A.J.Hend.

Euterpe langloisii Burret = **Euterpe precatoria** var. **precatoria**

Euterpe latisecta Burret = **Prestoea carderi** (W.Bull) Hook.f.

Euterpe leucospadix H.Wendl. ex Hemsl. = **Euterpe precatoria** var. **longivaginata** (Mart.) A.J.Hend.

Euterpe longibracteata Barb.Rodr., Enum. Palm. Nov.: 17 (1875).
N. South America to Brazil. 82 GUY VEN 84 BZC BZN.

Euterpe longipetiolata Oerst. = **Prestoea longipetiolata** (Oerst.) H.E.Moore

Euterpe longivaginata Mart. = **Euterpe precatoria** var. **longivaginata** (Mart.) A.J.Hend.

Euterpe luminosa A.J.Hend., Galeano & Meza, Brittonia 43: 178 (1991).
Peru (Pasco). 83 PER.

Euterpe macrospadix Oerst. = **Euterpe precatoria** var. **longivaginata** (Mart.) A.J.Hend.

Euterpe manaele (Mart.) Griseb. & H.Wendl. = **Prestoea acuminata** var. **montana** (Graham) A.J.Hend. & Galeano

Euterpe megalochlamys Burret = **Prestoea acuminata** var. **acuminata**

Euterpe microcarpa Burret = **Euterpe precatoria** var. **longivaginata** (Mart.) A.J.Hend.

Euterpe microspadix Burret = **Prestoea acuminata** var. **acuminata**

Euterpe mollissima Barb.Rodr. = **Euterpe catinga** var. **catinga**

Euterpe mollissima Spruce = **Euterpe precatoria** var. **precatoria**

Euterpe montana Graham = **Prestoea acuminata** var. **montana** (Graham) A.J.Hend. & Galeano

Euterpe montis-duida Burret = **Euterpe catinga** var. **roraimae** (Dammer) A.J.Hend. & Galeano

Euterpe oleracea Engel = **Euterpe precatoria** var. **precatoria**

Euterpe oleracea Mart., Hist. Nat. Palm. 2: 29 (1824).
Trinidad to S. Trop. America. 81 TRT 82 FRG GUY SUR VEN 83 CLM ECU 84 BZE BZN.
Euterpe brasiliana Oken, Allg. Naturgesch. 3(1): 674 (1841).
Catis martiana O.F.Cook, Bull. Torrey Bot. Club 28: 557 (1901).
Euterpe badiocarpa Barb.Rodr., Contr. Jard. Bot. Rio de Janeiro 1: 12 (1901).
Euterpe beardii L.H.Bailey, Gentes Herb. 7: 426 (1947).
Euterpe cuatrecasana Dugand, Revista Acad. Colomb. Ci. Exact. 8: 393 (1951).

Euterpe oocarpa Burret = **Prestoea acuminata** var. **acuminata**

Euterpe panamensis Burret = **Euterpe precatoria** var. **longivaginata** (Mart.) A.J.Hend.

Euterpe parviflora Burret = **Prestoea carderi** (W.Bull) Hook.f.

Euterpe pertenuis L.H.Bailey = **Prestoea acuminata** var. **montana** (Graham) A.J.Hend. & Galeano

Euterpe petiolata Burret = **Euterpe precatoria** var. **precatoria**

Euterpe pisifera Gaertn. = **Heterospathe elata** var. **elata**

Euterpe praga (Kunth) Mart. = **Prestoea acuminata** var. **acuminata**

Euterpe precatoria Mart. in A.D.d'Orbigny, Voy. Amér. Mér. 7(3): 10 (1842).
Trinidad to C. & S. Trop. America. 80 BLZ COS CPI GUA HON NIC PAN 81 TRT 82 FRG GUY SUR VEN 83 BOL CLM ECU PER 84 BZN.
var. **longivaginata** (Mart.) A.J.Hend., Palms Amazon: 111 (1995).
C. & S. Trop. America. 80 BLZ COS CPI GUA HON NIC PAN 82 VEN 83 BOL CLM ECU PER 84 BZN.
*Euterpe longivaginata Mart. in A.D.d'Orbigny, Voy. Amér. Mér. 7(3): 11 (1842).
Euterpe macrospadix Oerst., Vidensk. Meddel. Dansk Naturhist. Foren. Kjøbenhavn 1858: 31 (1858).
Euterpe karsteniana Engel, Linnaea 33: 670 (1865).
Euterpe leucospadix H.Wendl. ex Hemsl., Biol. Cent.-Amer., Bot. 3: 401 (1885).

Plectis oweniana O.F.Cook, Bull. Torrey Bot. Club 31: 353 (1904).
Euterpe kalbreyeri Burret, Bot. Jahrb. Syst. 63: 71 (1929).
Euterpe microcarpa Burret, Bot. Jahrb. Syst. 63: 72 (1929).
Euterpe panamensis Burret, Notizbl. Bot. Gart. Berlin-Dahlem 11: 864 (1933).
Rooseveltia frankliniana O.F.Cook, Smithsonian Misc. Collect. 98: 21 (1939).
Euterpe rhodoxyla Dugand, Revista Acad. Colomb. Ci. Exact. 8: 394 (1951).

var. **precatoria**
Trinidad to S. Trop. America. 81 TRT 82 FRG GUY SUR VEN 83 BOL CLM ECU PER 84 BZN.
Euterpe oleracea Engel, Linnaea 33: 671 (1865), nom. illeg.
Euterpe mollissima Spruce, J. Linn. Soc., Bot. 11: 139 (1869), nom. illeg.
Euterpe jatapuensis Barb.Rodr., Contr. Jard. Bot. Rio de Janeiro 1: 12 (1901).
Euterpe stenophylla Trail ex Burret, Bot. Jahrb. Syst. 63: 64 (1929).
Euterpe langloisii Burret, Notizbl. Bot. Gart. Berlin-Dahlem 13: 346 (1936).
Euterpe petiolata Burret, Notizbl. Bot. Gart. Berlin-Dahlem 15: 101 (1940).
Euterpe subruminata Burret, Notizbl. Bot. Gart. Berlin-Dahlem 15: 3 (1940).
Euterpe confertiflora L.H.Bailey, Gentes Herb. 7: 427 (1947).

Euterpe ptariana Steyerm. = **Euterpe catinga** var. **roraimae** (Dammer) A.J.Hend. & Galeano

Euterpe pubigera (Griseb. & H.Wendl.) Burret = **Prestoea pubigera** (Griseb. & H.Wendl.) Hook.f. ex B.D.Jacks.

Euterpe purpurea Engel = **Prestoea acuminata** var. **acuminata**

Euterpe puruensis Linden = **? [84]**

Euterpe rhodoxyla Dugand = **Euterpe precatoria** var. **longivaginata** (Mart.) A.J.Hend.

Euterpe roraimae Dammer = **Euterpe catinga** var. **roraimae** (Dammer) A.J.Hend. & Galeano

Euterpe roseospadix L.H.Bailey = **Prestoea longipetiolata** var. **roseospadix** (L.H.Bailey) A.J.Hend. & Galeano

Euterpe schultzeana Burret = **Prestoea schultzeana** (Burret) H.E.Moore

Euterpe simiarum (Standl. & L.O.Williams) H.E.Moore = **Prestoea longipetiolata** var. **longipetiolata**

Euterpe simplicifrons Burret = **Prestoea carderi** (W.Bull) Hook.f.

Euterpe stenophylla Trail ex Burret = **Euterpe precatoria** var. **precatoria**

Euterpe subruminata Burret = **Euterpe precatoria** var. **precatoria**

Euterpe tenuiramosa Dammer = **Prestoea tenuiramosa** (Dammer) H.E.Moore

Euterpe tobagonis L.H.Bailey = **Prestoea acuminata** var. **montana** (Graham) A.J.Hend. & Galeano

Euterpe trichoclada Burret = **Prestoea acuminata** var. **acuminata**

Euterpe ventricosa C.H.Wright = **Roystonea regia** (Kunth) O.F.Cook

Euterpe vinifera Mart. = **Pseudophoenix vinifera** (Mart.) Becc.

Euterpe williamsii Glassman = **Prestoea longipetiolata** var. **longipetiolata**

Euterpe zephyria Dugand = **Prestoea acuminata** var. **acuminata**

Exorrhiza

Exorrhiza Becc. = **Clinostigma** H.Wendl.

Exorrhiza carolinensis (Becc.) Burret = **Clinostigma carolinense** (Becc.) H.E.Moore & Fosberg

Exorrhiza harlandii (Becc.) Burret = **Clinostigma harlandii** Becc.

Exorrhiza onchorhyncha (Becc.) Burret = **Clinostigma onchorhynchum** Becc.

Exorrhiza ponapensis (Becc.) Burret = **Clinostigma ponapense** (Becc.) H.E.Moore & Fosberg

Exorrhiza savoryana (Rehder & E.H.Wilson) Burret = **Clinostigma savoryanum** (Rehder & E.H.Wilson) H.E.Moore & Fosberg

Exorrhiza smithii Burret = **Clinostigma exorrhizum** (H.Wendl.) Becc.

Exorrhiza thurstonii (Becc.) Burret = **Clinostigma exorrhizum** (H.Wendl.) Becc.

Exorrhiza vaupelii Burret = **Clinostigma savaiiense** Cnristoph.

Exorrhiza wendlandiana Becc. = **Clinostigma exorrhizum** (H.Wendl.) Becc.

Fulchironia

Fulchironia Lesch. = **Phoenix** L.

Fulchironia senegalensis Lesch. = **Phoenix reclinata** Jacq.

Gastrococos

Gastrococos Morales, Repert. Fis.-Nat. Isla Cuba 1: 57 (1866).
1 species, Cuba. 81.

Gastrococos armentalis Morales = **Gastrococos crispa** (Kunth) H.E.Moore

Gastrococos crispa (Kunth) H.E.Moore, Principes 11: 121 (1968).
Cuba. 81 CUB.
  *Cocos crispa Kunth in F.W.H.von Humboldt, A.J.A.Bonpland & C.S.Kunth, Nov. Gen. Sp. 1: 302 (1817). *Acrocomia crispa* (Kunth) C.F.Baker ex Becc., Pomona Coll. J. Econ. Bot. 2: 364 (1912). *Astrocaryum crispum* (Kunth) M.Gómez & Roig, Fl. Cuba: 12 (1914).
  Gastrococos armentalis Morales, Repert. Fis.-Nat. Isla Cuba 1: 57 (1866). *Acrocomia armentalis* (Morales) L.H.Bailey & E.Z.Bailey, Hort. Sec.: 22 (1941).

Gaussia

Gaussia H.Wendl., Nachr. Königl. Ges. Wiss. Georg-Augusts-Univ. 1865: 327 (1865).
5 species, SE. Mexico to Guatemala, Cuba, Puerto Rico. 79 80 81.
  Aeria O.F.Cook, Bull. Torrey Bot. Club 28: 547 (1901).
  Opsiandra O.F.Cook, J. Wash. Acad. Sci. 13: 181 (1923).

Gaussia attenuata (O.F.Cook) Becc., Pomona Coll. J. Econ. Bot. 2: 275 (1912).
Puerto Rico. 81 PUE.
  Gaussia portoricensis H.Wendl. in O.C.E.de Kerchove de Denterghem, Palmiers: 245 (1878), nom. nud.
  *Aeria attenuata O.F.Cook, Bull. Torrey Bot. Club 28: 548 (1901).

Gaussia ghiesbreghtii H.Wendl. = **?** [81]

Gaussia gomez-pompae (H.J.Quero) H.J.Quero, Syst. Bot. 11: 153 (1986).
Mexico (Oaxaca, Chiapas, Veracruz). 79 MXG MXS MXT.
  *Opsiandra gomez-pompae H.J.Quero, Principes 26: 145 (1982).

Gaussia maya (O.F.Cook) H.J.Quero & Read, Syst. Bot. 11: 152 (1986).
SE. Mexico to Guatemala. 79 MXT 80 BLZ GUA.
  *Opsiandra maya O.F.Cook, J. Wash. Acad. Sci. 13: 182 (1923).

Gaussia princeps H.Wendl., Nachr. Königl. Ges. Wiss. Georg-Augusts-Univ. 1865: 328 (1865).
W. Cuba. 81 CUB.
  Chamaedorea ventricosa Hook.f., Rep. Progr. Condition Roy. Bot. Gard. Kew 1882: 59 (1884).

Gaussia portoricensis H.Wendl. = **Gaussia attenuata** (O.F.Cook) Becc.

Gaussia spirituana Moya & Leiva, Revista Jard. Bot. Nac. Univ. Habana 12: 16 (1991 publ. 1993).
EC. Cuba (Sierra de Jatibonico). 81 CUB.

Gaussia vinifera (Mart.) H.Wendl. = **Pseudophoenix vinifera** (Mart.) Becc.

Gembanga

Gembanga Blume = **Corypha** L.

Gembanga rotundifolia Blume = **Corypha utan** Lam.

Geonoma

Geonoma Willd., Sp. Pl. 4: 593 (1805).
59 species, Mexico to Trop. America. 79 80 81 82 83 84 85.
  Vouay Aubl., Hist. Pl. Guiane 2(App.): 99 (1775).
  Gynestum Poit., Mém. Mus. Hist. Nat. 9: 387 (1822).
  Roebelia Engel, Linnaea 33: 680 (1865).
  Kalbreyera Burret, Bot. Jahrb. Syst. 63: 142 (1930).
  Taenianthera Burret, Bot. Jahrb. Syst. 63: 267 (1930).

Geonoma acaulis (Poit.) Burret = **Geonoma macrostachys** var. **poiteauana** (Kunth) A.J.Hend.

Geonoma acaulis Mart. = **Geonoma macrostachys** var. **acaulis** (Mart.) Andrew Hend.

Geonoma acaulis subsp. *tapajotensis* Trail = **Geonoma macrostachys** var. **acaulis** (Mart.) Andrew Hend.

Geonoma acutangula Burret = **Geonoma jussieuana** Mart.

Geonoma acutiflora Mart. = **Geonoma baculifera** (Poit.) Kunth

Geonoma adscendens Dammer ex Burret = **Geonoma jussieuana** Mart.

Geonoma allenii L.H.Bailey = **Calyptrogyne allenii** (L.H.Bailey) Nevers

Geonoma altissima Barb.Rodr. = **Geonoma brevispatha** var. **brevispatha**

Geonoma amabilis H.Wendl. ex Dahlgren = **Pholidostachys pulchra** H.Wendl. ex Burret

Geonoma amazonica H.Wendl. = ? **[84]**

Geonoma ambigua Spruce = **Geonoma maxima** var. **ambigua** (Spruce) A.J.Hend.

Geonoma amoena Burret = **Geonoma jussieuana** Mart.

Geonoma andicola Dammer ex Burret = **Geonoma weberbaueri** Dammer ex Burret

Geonoma andina Burret = **Geonoma orbignyana** Mart.

Geonoma anomoclada Burret = **Geonoma orbignyana** Mart.

Geonoma appuniana Spruce, J. Linn. Soc., Bot. 11: 106 (1871).
N. South America to Brazil (Amazonas). 82 FRG GUY SUR VEN 84 BZN.
Geonoma roraimae Dammer, Notizbl. Königl. Bot. Gart. Berlin 6: 261 (1915).

Geonoma aricanga Barb.Rodr. = **Geonoma brevispatha** var. **brevispatha**

Geonoma arundinacea Mart., Hist. Nat. Palm. 2: 17 (1823).
W. Amazon Reg. 83 CLM ECU PER 84 BZN.
Geonoma uleana Dammer, Verh. Bot. Vereins Prov. Brandenburg 48: 122 (1906 publ. 1907).

Geonoma aspidiifolia Spruce, J. Linn. Soc., Bot. 11: 112 (1871).
SE. Colombia to Guyana. 82 GUY 83 CLM 84 BAN.
Geonoma fusca Wess.Boer, Mem. New York Bot. Gard. 23: 93 (1972).

Geonoma atrovirens Borchs. & Balslev, Nordic J. Bot. 21: 342 (2001 publ. 2002).
E. Ecuador. 83 ECU.

Geonoma aulacophylla Burret = **Geonoma orbignyana** Mart.

Geonoma baculifera (Poit.) Kunth, Enum. Pl. 3: 233 (1841).
N. South America to N. Brazil. 82 FRG GUY SUR VEN 84 BZN.
*Gynestum baculiferum Poit., Mém. Mus. Hist. Nat. 9: 389 (1822).
Geonoma acutiflora Mart., Hist. Nat. Palm. 2: 10 (1823).
Geonoma macrospatha Spruce, J. Linn. Soc., Bot. 11: 105 (1871).
Geonoma estevaniana Burret, Notizbl. Bot. Gart. Berlin-Dahlem 14: 256 (1938).

Geonoma barbigera Barb.Rodr. = **Geonoma pohliana** Mart.

Geonoma barbosiana Burret = **Geonoma pohliana** Mart.

Geonoma barthia Engel = **Geonoma densa** Linden & H.Wendl.

Geonoma bartletii Dammer ex Burret = **Geonoma deversa** (Poit.) Kunth

Geonoma beccariana Barb.Rodr. = **Geonoma laxiflora** Mart.

Geonoma bella Burret = **Geonoma stricta** var. **trailii** (Burret) A.J.Hend.

Geonoma bifurca Drude & H.Wendl. = **Geonoma pauciflora** Mart.

Geonoma bijugata Barb.Rodr. = **Geonoma maxima** var. **chelidonura** (Spruce) A.J.Hend.

Geonoma binervia Oerst. = **Geonoma interrupta** var. **interrupta**

Geonoma blanchetiana H.Wendl. ex Drude = **Geonoma pohliana** Mart.

Geonoma bluntii auct. = ?

Geonoma brachyfoliata Barb.Rodr. = **Geonoma maxima** var. **chelidonura** (Spruce) A.J.Hend.

Geonoma brachystachys Burret = **Geonoma jussieuana** Mart.

Geonoma brenesii Grayum, Phytologia 84: 322 (1998 publ. 1999).
Costa Rica. 80 COS.

Geonoma brevispatha Barb.Rodr., Enum. Palm. Nov.: 41 (1875).
Brazil to Paraguay. 83 BOL PER 84 BZC BZE BZL BZN 85 PAR.

var. **brevispatha**
Brazil to Paraguay. 83 BOL 84 BZC BZE BZL 85 PAR.
Geonoma aricanga Barb.Rodr., Enum. Palm. Nov.: 40 (1875).
Geonoma calophyta Barb.Rodr., Palmiers: 48 (1882).
Geonoma rupestris Barb.Rodr., Palmiers: 47 (1882).
Geonoma schottiana var. *palustris* Warm. ex Drude in C.F.P.von Martius & auct. suc. (eds.), Fl. Bras. 3(2): 493 (1882). *Geonoma warmingii* A.D.Hawkes, Arq. Bot. Estado São Paulo, n.s., f.m., 2: 189 (1952).
Geonoma weddelliana H.Wendl. ex Drude in C.F.P.von Martius & auct. suc. (eds.), Fl. Bras. 3(2): 494 (1882).
Geonoma caudulata Loes., Bot. Jahrb. Syst. 21: 423 (1896).
Geonoma altissima Barb.Rodr., Palm. Mattogross.: 6 (1898).
Geonoma chapadensis Barb.Rodr., Palm. Mattogross.: 4 (1898).
Geonoma stenoschista Burret, Bot. Jahrb. Syst. 63: 233 (1930).
Geonoma decussata Burret, Repert. Spec. Nov. Regni Veg. 32: 103 (1933).
Geonoma plurinervia Burret, Notizbl. Bot. Gart. Berlin-Dahlem 15: 99 (1940).

var. **occidentalis** A.J.Hend., Palms Amazon: 264 (1995).
Peru to Bolivia and Brazil (Rondônia). 83 BOL PER 84 BZN.

Geonoma brongniartii Mart. in A.D.d'Orbigny, Voy. Amér. Mér. 7(3): 24 (1843).
W. South America to N. Brazil. 83 BOL CLM ECU PER 84 BZN.
Geonoma metensis H.Karst., Linnaea 28: 409 (1856).
Geonoma werdermannii Burret, Bot. Jahrb. Syst. 63: 173 (1930).
Geonoma cuneifolia Burret, Notizbl. Bot. Gart. Berlin-Dahlem 11: 199 (1931).

Geonoma caespitosa H.Wendl. ex Drude = **Geonoma pauciflora** Mart.

Geonoma calophyta Barb.Rodr. = **Geonoma brevispatha** var. **brevispatha**

Geonoma calyptrogynoidea Burret = **Geonoma congesta** H.Wendl. ex Spruce

Geonoma camana Trail, J. Bot. 14: 324 (1876).
Taenianthera camana (Trail) Burret, Bot. Jahrb. Syst. 63: 270 (1930).
W. South America to N. Brazil. 83 CLM ECU PER 84 BZN.
Geonoma lagesiana Dammer, Verh. Bot. Vereins Prov. Brandenburg 48: 121 (1906 publ. 1907).

Taenianthera lagesiana (Dammer) Burret, Bot. Jahrb. Syst. 63: 270 (1930).

Geonoma camptoneura Burret = **Geonoma maxima** var. **maxima**

Geonoma campyloclada Burret = **Geonoma orbignyana** Mart.

Geonoma campylostachys Burret = **Geonoma jussieuana** Mart.

Geonoma capanemae Barb.Rodr. = **Geonoma maxima** var. **maxima**

Geonoma carderi W.Bull = **Prestoea carderi** (W.Bull) Hook.f.

Geonoma caudescens H.Wendl. ex Drude = **?** [84]

Geonoma caudulata Loes. = **Geonoma brevispatha** var. **brevispatha**

Geonoma cernua Burret = **Geonoma jussieuana** Mart.

Geonoma chapadensis Barb.Rodr. = **Geonoma brevispatha** var. **brevispatha**

Geonoma chaunostachys Burret = **Geonoma macrostachys** var. **poiteauana** (Kunth) A.J.Hend.

Geonoma chelidonura Spruce = **Geonoma maxima** var. **chelidonura** (Spruce) A.J.Hend.

Geonoma chiriquensis Linden ex Hook.f. = **?**

Geonoma chlamydostachys Galeano-Garcés, Principes 30: 71 (1986).
Colombia (Magdalena valley). 83 CLM.

Geonoma chococola Wess.Boer, Verh. Kon. Ned. Akad. Wetensch., Afd. Natuurk., Tweede Sect. 58(1): 103 (1968).
Panama to W. Colombia. 80 PAN 83 CLM.

Geonoma concinna Burret, Bot. Jahrb. Syst. 63: 229 (1930).
Panama to Colombia (Antioquia). 80 PAN 83 CLM.

Geonoma condensata L.H.Bailey = **Calyptrogyne condensata** (L.H.Bailey) de Boer

Geonoma congesta H.Wendl. ex Spruce, J. Linn. Soc., Bot. 11: 112 (1871).
C. America to NW. Colombia. 80 COS HON NIC PAN 83 CLM.
Geonoma calyptrogynoidea Burret, Bot. Jahrb. Syst. 63: 223 (1930).

Geonoma congestissima Burret = **?** [83 PER]

Geonoma corallifera C.Morren = **Chamaedorea ernestiaugusti** H.Wendl.

Geonoma costatifrons L.H.Bailey = **Calyptrogyne costatifrons** (L.H.Bailey) Nevers

Geonoma cuneata H.Wendl. ex Spruce, J. Linn. Soc., Bot. 11: 104 (1871).
C. America to Ecuador. 80 COS NIC PAN 82 CLM ECU.

var. **cuneata**
C. America to Ecuador. 80 COS NIC PAN 82 CLM ECU.
Geonoma decurrens H.Wendl. ex Burret, Bot. Jahrb. Syst. 63: 162 (1930).

var. **gracilis** (H.Wendl.) Skov, HyperTax. Rev. Geonoma: 122 (1989).
C. America. 80 COS.
Geonoma obovata H.Wendl. ex Spruce, J. Linn. Soc., Bot. 11: 104 (1871).
*Geonoma gracilis H.Wendl. in O.C.E.de Kerchove de Denterghem, Palmiers: 245 (1878).

var. **procumbens** (H.Wendl. ex Spruce) Skov, HyperTax. Rev. Geonoma: 121 (1989).
C. America. 80 COS.
*Geonoma procumbens H.Wendl. ex Spruce, J. Linn. Soc., Bot. 11: 105 (1871).

var. **sodiroi** (Dammer ex Burret) Skov, HyperTax. Rev. Geonoma: 122 (1989).
Ecuador. 83 ECU.
*Geonoma sodiroi Dammer ex Burret, Bot. Jahrb. Syst. 63: 165 (1930).
Geonoma gibbosa Burret, Notizbl. Bot. Gart. Berlin-Dahlem 13: 342 (1936).

Geonoma cuneatoidea Burret = **Geonoma jussieuana** Mart.

Geonoma cuneifolia Burret = **Geonoma brongniartii** Mart.

Geonoma dammeri Huber = **Geonoma macrostachys** var. **poiteauana** (Kunth) A.J.Hend.

Geonoma dasystachys Burret = **Geonoma maxima** var. **chelidonura** (Spruce) A.J.Hend.

Geonoma decora L.Linden & Rodigas = **?** [84]

Geonoma decurrens H.Wendl. ex Burret = **Geonoma cuneata** var. **cuneata**

Geonoma decussata Burret = **Geonoma brevispatha** var. **brevispatha**

Geonoma demarastei Pritz. = **Geonoma deversa** (Poit.) Kunth

Geonoma densa Linden & H.Wendl., Linnaea 28: 333 (1856).
W. South America to Venezuela. 82 VEN 83 BOL CLM ECU PER.
Geonoma barthia Engel, Linnaea 33: 688 (1865).
Geonoma uncibracteata Burret, Bot. Jahrb. Syst. 63: 215 (1930).
Geonoma pulchra Engel, Linnaea 33: 681 (1965).

Geonoma densiflora Spruce = **Geonoma maxima** var. **chelidonura** (Spruce) A.J.Hend.

Geonoma desmarestii Mart. = **Geonoma deversa** (Poit.) Kunth

Geonoma deversa (Poit.) Kunth, Enum. Pl. 3: 231 (1841).
C. & S. Trop. America. 80 BLZ COS HON NIC PAN 82 FRG GUY SUR VEN 83 BOL CLM ECU PER 84 BZN.
*Gynestum deversum Poit., Mém. Mus. Hist. Nat. 9: 390 (1822).
Geonoma paniculigera Mart., Hist. Nat. Palm. 2: 11 (1823).
Geonoma desmarestii Mart. in A.D.d'Orbigny, Voy. Amér. Mér. 7(3): 23 (1843).
Geonoma rectifolia Wallace, Palm Trees Amazon: 67 (1853).
Geonoma demarastei Pritz., Icon. Bot. Index: 486 (1854), orth. var.
Geonoma longipetiolata Oerst., Vidensk. Meddel. Dansk Naturhist. Foren. Kjøbenhavn 1858: 36 (1858).
Geonoma flaccida H.Wendl. ex Spruce, J. Linn. Soc., Bot. 11: 108 (1871).
Geonoma microspatha Spruce, J. Linn. Soc., Bot. 11: 108 (1871).
Geonoma trijugata Barb.Rodr., Enum. Palm. Nov.: 12 (1875).
Geonoma yauaperyensis Barb.Rodr., Contr. Jard. Bot. Rio de Janeiro 2: 88 (1902).
Geonoma bartletii Dammer ex Burret, Bot. Jahrb. Syst. 63: 183 (1930).

Geonoma leptostachys Burret, Notizbl. Bot. Gart. Berlin-Dahlem 10: 1014 (1930).

Geonoma macropoda Burret, Notizbl. Bot. Gart. Berlin-Dahlem 10: 1015 (1930).

Geonoma major Burret, Notizbl. Bot. Gart. Berlin-Dahlem 10: 1016 (1930).

Geonoma tessmannii Burret, Bot. Jahrb. Syst. 63: 181 (1930).

Geonoma killipii Burret, Notizbl. Bot. Gart. Berlin-Dahlem 11: 320 (1932).

Geonoma dicranospadix Burret = **Geonoma jussieuana** Mart.

Geonoma discolor Spruce = **Geonoma maxima** var. **maxima**

Geonoma divisa H.E.Moore, Gentes Herb. 12: 25 (1980).
Panama to NW. Colombia. 80 PAN 83 CLM.

Geonoma dominicana L.H.Bailey = **Geonoma interrupta** var. **interrupta**

Geonoma donnell-smithii Dammer = **Calyptrogyne ghiesbreghtiana** (Linden & H.Wendl.) H.Wendl.

Geonoma dryanderae Burret = **Geonoma interrupta** var. **interrupta**

Geonoma dulcis C.Wright ex Griseb. = **Calyptronoma plumeriana** (Mart.) Lourteig

Geonoma dussiana Becc. = **Geonoma undata** Klotzsch

Geonoma edulis H.Wendl. ex Spruce = **Geonoma interrupta** var. **interrupta**

Geonoma elegans Mart. = **Geonoma pauciflora** Mart.

Geonoma elegans var. *amazonica* Trail = **Geonoma stricta** var. **trailii** (Burret) A.J.Hend.

Geonoma elegans var. *robusta* Drude = **Geonoma pauciflora** Mart.

Geonoma epetiolata H.E.Moore, Gentes Herb. 12: 28 (1980).
Costa Rica to Panama. 80 COS PAN.

Geonoma erythrospadice Barb.Rodr. = **Geonoma schottiana** Mart.

Geonoma estevaniana Burret = **Geonoma baculifera** (Poit.) Kunth

Geonoma euspatha Burret = **Geonoma interrupta** var. **euspatha** (Burret) A.J.Hend.

Geonoma euterpoidea Burret = **Geonoma orbignyana** Mart.

Geonoma falcata Barb.Rodr. = **Geonoma maxima** var. **chelidonura** (Spruce) A.J.Hend.

Geonoma fendleriana Spruce = **Geonoma simplicifrons** Willd.

Geonoma fenestrata (H.Wendl.) H.Wendl. = **Chamaedorea geonomiformis** H.Wendl.

Geonoma ferruginea H.Wendl. ex Spruce, J. Linn. Soc., Bot. 11: 110 (1871).
C. America. 80 COS GUA HON NIC PAN.
Geonoma microspadix H.Wendl. ex Spruce, J. Linn. Soc., Bot. 11: 110 (1871).
Geonoma versiformis H.Wendl. ex Spruce, J. Linn. Soc., Bot. 11: 109 (1871).
Geonoma microstachys H.Wendl. ex Burret, Bot. Jahrb. Syst. 63: 228 (1930).

Geonoma fiscellaria Mart. ex Drude = **Geonoma pohliana** Mart.

Geonoma flaccida H.Wendl. ex Spruce = **Geonoma deversa** (Poit.) Kunth

Geonoma floccosa Dammer ex Burret = **Geonoma orbignyana** Mart.

Geonoma frigida Linden = **?** [83 CLM]

Geonoma frontinensis Burret = **Geonoma jussieuana** Mart.

Geonoma furcifolia Barb.Rodr. = **Geonoma maxima** var. **chelidonura** (Spruce) A.J.Hend.

Geonoma furcifrons Drude = **Geonoma maxima** var. **chelidonura** (Spruce) A.J.Hend.

Geonoma fusca Wess.Boer = **Geonoma aspidiifolia** Spruce

Geonoma gamiova Barb.Rodr., Contr. Jard. Bot. Rio de Janeiro 6: 13 (1907).
SE. & S. Brazil. 84 BZL BZS.

Geonoma gastoniana Glaz. ex Drude in C.F.P.von Martius & auct. suc. (eds.), Fl. Bras. 3(2): 496 (1882). Brazil (Goiás, Rio de Janeiro). 84 BZC BZL.
Geonoma wittigiana Glaz. ex Drude in C.F.P.von Martius & auct. suc. (eds.), Fl. Bras. 3(2): 499 (1882).

Geonoma ghiesbreghtiana Lindl. & H.Wendl. = **Calyptrogyne ghiesbreghtiana** (Linden & H.Wendl.) H.Wendl.

Geonoma gibbosa Burret = **Geonoma cuneata** var. **sodiroi** (Dammer ex Burret) Skov

Geonoma glauca Oerst. = **Calyptrogyne ghiesbreghtiana** (Linden & H.Wendl.) H.Wendl.

Geonoma goniocarpa Burret = **Geonoma orbignyana** Mart.

Geonoma gracilipes Dammer ex Burret = **?** [83 PER]

Geonoma gracilis H.Wendl. = **Geonoma cuneata** var. **gracilis** (H.Wendl.) Skov

Geonoma gracillima Burret = **Geonoma jussieuana** Mart.

Geonoma grandifrons Burret = **Geonoma jussieuana** Mart.

Geonoma grandisecta Burret = **Geonoma maxima** var. **spixiana** (Mart.) A.J.Hend.

Geonoma granditrijuga Burret = **Geonoma jussieuana** Mart.

Geonoma heinrichsiae Burret = **Geonoma orbignyana** Mart.

Geonoma helminthoclada Burret = **Geonoma undata** Klotzsch

Geonoma helminthostachys Burret = **Geonoma jussieuana** Mart.

Geonoma herbstii auct. = **?**

Geonoma herthae Burret = **Geonoma stricta** var. **piscicauda** (Dammer) A.J.Hend.

Geonoma hexasticha Spruce = **Geonoma maxima** var. **maxima**

Geonoma hodgeorum L.H.Bailey = **Geonoma undata** Klotzsch

Geonoma hoehnei Burret = **Geonoma schottiana** Mart.

Geonoma hoffmanniana H.Wendl. ex Spruce = **Geonoma orbignyana** Mart.

Geonoma hoppii Burret = **?** [83 ECU]

Geonoma huebneri Burret = **Geonoma maxima** var. **chelidonura** (Spruce) A.J.Hend.

Geonoma hugonis Grayum & Nevers, Principes 42: 94 (1998).
Panama. 80 PAN.

Geonoma humilis auct. = **Chamaedorea geonomiformis** H.Wendl.

Geonoma imperialis Linden = ? **[83 CLM]**

Geonoma insignis Burret = ? **[83 ECU]**

Geonoma intermedia (H.Wendl.) B.S.Williams = **Calyptronoma plumeriana** (Mart.) Lourteig

Geonoma interrupta (Ruiz & Pav.) Mart., Hist. Nat. Palm. 2: 8 (1823).
Mexico to Trop. America. 79 MXG MXS MXT 80 BLZ COS GUA HON NIC PAN 81 HAI TRT WIN 82 FRG GUY SUR VEN 83 BOL CLM ECU PER 84 BZN.

var. **euspatha** (Burret) A.J.Hend., Palms Amazon: 270 (1995).
S. Trop. America. 82 FRG GUY SUR VEN 83 BOL CLM ECU 84 BZN.
  *Geonoma euspatha Burret, Notizbl. Bot. Gart. Berlin-Dahlem 11: 10 (1930).
  Geonoma karuaiana Steyerm., Fieldiana, Bot. 28(1): 88 (1951).

var. **interrupta**
Mexico to Trop. America. 79 MXG MXS MXT 80 BLZ COS GUA HON NIC PAN 81 HAI TRT WIN 82 FRG GUY SUR VEN 83 BOL CLM ECU PER 84 BZN.
  Martinezia interrupta Ruiz & Pav., Syst. Veg. Fl. Peruv. Chil.: 296 (1798).
  Geonoma pinnatifrons Willd., Sp. Pl. 4: 593 (1805).
  Geonoma martinicensis Mart. in A.D.d'Orbigny, Voy. Amér. Mér. 7(3): 28 (1843).
  Geonoma oxycarpa Mart. in A.D.d'Orbigny, Voy. Amér. Mér. 7(3): 30 (1843).
  Geonoma pleeana Mart. in A.D.d'Orbigny, Voy. Amér. Mér. 7(3): 33 (1843).
  Geonoma mexicana Liebm. ex Mart., Hist. Nat. Palm. 3: 316 (1853).
  Geonoma magnifica Linden & H.Wendl., Linnaea 28: 335 (1856).
  Geonoma binervia Oerst., Vidensk. Meddel. Dansk Naturhist. Foren. Kjøbenhavn 1858: 33 (1858).
  Geonoma vaga Griseb. & H.Wendl. in A.H.R.Grisebach, Fl. Brit. W. I.: 517 (1864).
  Geonoma edulis H.Wendl. ex Spruce, J. Linn. Soc., Bot. 11: 100 (1871).
  Geonoma membranacea H.Wendl. ex Spruce, J. Linn. Soc., Bot. 11: 106 (1871).
  Geonoma purdieana Spruce, J. Linn. Soc., Bot. 11: 109 (1871).
  Geonoma saga Spruce, J. Linn. Soc., Bot. 11: 109 (1871), orth. var.
  Geonoma megaloptila Burret, Bot. Jahrb. Syst. 63: 247 (1930).
  Geonoma preussii Burret, Bot. Jahrb. Syst. 63: 242 (1930).
  Geonoma ramosissima Burret, Bot. Jahrb. Syst. 63: 249 (1930).
  Geonoma rivalis Kalbreyer & Burret, Bot. Jahrb. Syst. 63: 241 (1930).
  Geonoma platybothros Burret, Notizbl. Bot. Gart. Berlin-Dahlem 11: 200 (1931).
  Geonoma leptoclada Burret, Notizbl. Bot. Gart. Berlin-Dahlem 11: 863 (1933).

Geonoma dryanderae Burret, Notizbl. Bot. Gart. Berlin-Dahlem 12: 615 (1935).
Geonoma dominicana L.H.Bailey, Gentes Herb. 4: 232 (1939).
Geonoma polyclada Burret, Notizbl. Bot. Gart. Berlin-Dahlem 15: 26 (1940).

Geonoma iodolepis Burret = **Geonoma orbignyana** Mart.

Geonoma iodoneura Burret = **Geonoma orbignyana** Mart.

Geonoma iraze Linden = ? **[82 VEN]**

Geonoma irena Brochs., Nordic J. Bot. 16: 605 (1996 publ. 1997).
Ecuador. 83 ECU.

Geonoma juruana Dammer = **Geonoma maxima** var. **chelidonura** (Spruce) A.J.Hend.

Geonoma jussieuana Mart. in A.D.d'Orbigny, Voy. Amér. Mér. 7(3): 24 (1843).
Costa Rica to Bolivia. 80 COS PAN 82 VEN 83 BOL CLM ECU PER.
  Geonoma acutangula Burret, Bot. Jahrb. Syst. 63: 177 (1930).
  Geonoma adscendens Dammer ex Burret, Bot. Jahrb. Syst. 63: 175 (1930).
  Geonoma cuneatoidea Burret, Bot. Jahrb. Syst. 63: 167 (1930).
  Geonoma dicranospadix Burret, Bot. Jahrb. Syst. 63: 169 (1930).
  Geonoma frontinensis Burret, Bot. Jahrb. Syst. 63: 170 (1930).
  Geonoma gracillima Burret, Bot. Jahrb. Syst. 63: 165 (1930).
  Geonoma grandifrons Burret, Bot. Jahrb. Syst. 63: 163 (1930).
  Geonoma granditrijuga Burret, Bot. Jahrb. Syst. 63: 171 (1930).
  Geonoma helminthostachys Burret, Bot. Jahrb. Syst. 63: 176 (1930).
  Geonoma kalbreyeri Burret, Bot. Jahrb. Syst. 63: 168 (1930).
  Geonoma lehmannii Dammer ex Burret, Bot. Jahrb. Syst. 63: 180 (1930).
  Geonoma mucronata Burret, Bot. Jahrb. Syst. 63: 171 (1930).
  Geonoma parvifrons Burret, Bot. Jahrb. Syst. 63: 178 (1930).
  Taenianthera multisecta Burret, Notizbl. Bot. Gart. Berlin-Dahlem 11: 13 (1930). *Geonoma multisecta* (Burret) Burret, Notizbl. Bot. Gart. Berlin-Dahlem 12: 155 (1934).
  Taenianthera weberbaueri Burret, Bot. Jahrb. Syst. 63: 269 (1930).
  Geonoma pleioneura Burret, Notizbl. Bot. Gart. Berlin-Dahlem 11: 234 (1931).
  Geonoma amoena Burret, Notizbl. Bot. Gart. Berlin-Dahlem 11: 862 (1933).
  Geonoma brachystachys Burret, Notizbl. Bot. Gart. Berlin-Dahlem 15: 23 (1940).
  Geonoma campylostachys Burret, Notizbl. Bot. Gart. Berlin-Dahlem 15: 24 (1940).
  Geonoma cernua Burret, Notizbl. Bot. Gart. Berlin-Dahlem 15: 24 (1940).

Geonoma kalbreyeri Burret = **Geonoma jussieuana** Mart.

Geonoma karuaiana Steyerm. = **Geonoma interrupta** var. **euspatha** (Burret) A.J.Hend.

Geonoma killipii Burret = **Geonoma deversa** (Poit.) Kunth

Geonoma kuhlmannii Burret = **Geonoma pohliana** Mart.

Geonoma lacerata auct. = **?** [80]

Geonoma lagesiana Dammer = **Geonoma camana** Trail

Geonoma lakoi Burret = **Geonoma maxima** var. **chelidonura** (Spruce) A.J.Hend.

Geonoma lanceolata Burret = **Geonoma stricta** var. **stricta**

Geonoma latifolia Burret = **Geonoma pohliana** Mart.

Geonoma latifrons Burret = **Chamaedorea ernesti-augusti** H.Wendl.

Geonoma latisecta Burret = **Geonoma maxima** var. **maxima**

Geonoma laxiflora Mart., Hist. Nat. Palm. 2: 12 (1823).
W. South America to N. Brazil. 83 BOL CLM ECU PER 84 BZN.
> *Geonoma beccariana* Barb.Rodr., Vellosia 1: 33 (1888).

Geonoma lehmannii Dammer ex Burret = **Geonoma jussieuana** Mart.

Geonoma lepidota Burret = **Geonoma orbignyana** Mart.

Geonoma leptoclada Burret = **Geonoma interrupta** var. **interrupta**

Geonoma leptospadix Trail, J. Bot. 14: 327 (1876).
S. Trop. America. 82 FRG GUY SUR VEN 83 BOL CLM ECU PER 84 BZE BZN.
> *Geonoma saramaccana* L.H.Bailey, Bull. Torrey Bot. Club 75: 104 (1948).

Geonoma leptostachys Burret = **Geonoma deversa** (Poit.) Kunth

Geonoma leucotricha Burret = **Geonoma orbignyana** Mart.

Geonoma lindeniana H.Wendl. = **Geonoma orbignyana** Mart.

Geonoma linearifolia H.Karst. = **Geonoma orbignyana** Mart.

Geonoma linearis Burret, Notizbl. Bot. Gart. Berlin-Dahlem 11: 861 (1933).
W. Colombia to NW. Ecuador. 83 CLM ECU.

Geonoma longipedunculata Burret, Notizbl. Bot. Gart. Berlin-Dahlem 11: 8 (1930).
C. Colombia to SC. Peru. 83 CLM ECU PER.

Geonoma longipetiolata Oerst. = **Geonoma deversa** (Poit.) Kunth

Geonoma longisecta Burret = **Geonoma maxima** var. **chelidonura** (Spruce) A.J.Hend.

Geonoma longivaginata H.Wendl. ex Spruce, J. Linn. Soc., Bot. 11: 109 (1871).
Costa Rica to Panama. 80 COS PAN.

Geonoma luetzelburgii Burret = **Geonoma pohliana** Mart.

Geonoma macroclada Burret = **Geonoma undata** Klotzsch

Geonoma macroclona Drude = **Geonoma pohliana** Mart.

Geonoma macrophylla Burret = **?** [83 CLM]

Geonoma macropoda Burret = **Geonoma deversa** (Poit.) Kunth

Geonoma macrosiphon Burret = **Geonoma undata** Klotzsch

Geonoma macrospatha Spruce = **Geonoma baculifera** (Poit.) Kunth

Geonoma macrostachys Mart., Hist. Nat. Palm. 2: 19 (1823). *Taenianthera macrostachys* (Mart.) Burret, Bot. Jahrb. Syst. 63: 163 (1930).
S. Trop. America. 82 FRG GUY SUR VEN 83 BOL CLM ECU PER 84 BZN.

var. **acaulis** (Mart.) Andrew Hend., Palms Amazon: 274 (1995).
W. South America to N. Brazil. 82 VEN 83 BOL CLM ECU PER 84 BZN.
> *Geonoma acaulis Mart., Hist. Nat. Palm. 2: 18 (1823). *Taenianthera acaulis* (Mart.) Burret, Bot. Jahrb. Syst. 63: 269 (1930).
> *Geonoma acaulis* subsp. *tapajotensis* Trail, J. Bot. 14: 324 (1876).
> *Geonoma tapajotensis* (Trail) Drude in C.F.P.von Martius & auct. suc. (eds.), Fl. Bras. 3(2): 508 (1882).
> *Taenianthera gracilis* Burret, Notizbl. Bot. Gart. Berlin-Dahlem 11: 14 (1930).
> *Taenianthera tapajotensis* (Trail) Burret, Bot. Jahrb. Syst. 63: 269 (1930).
> *Taenianthera oligosticha* Burret, Notizbl. Bot. Gart. Berlin-Dahlem 11: 201 (1931).
> *Taenianthera minor* Burret, Notizbl. Bot. Gart. Berlin-Dahlem 14: 324 (1939).

var. **macrostachys**
W. South America to N. Brazil. 82 VEN 83 BOL CLM ECU PER 84 BZN.
> *Geonoma tamandua* Trail, J. Bot. 14: 323 (1876). *Taenianthera tamandua* (Trail) Burret, Bot. Jahrb. Syst. 63: 268 (1930).
> *Geonoma woronowii* Burret, Notizbl. Bot. Gart. Berlin-Dahlem 11: 6 (1930).

var. **poiteauana** (Kunth) A.J.Hend., Palms Amazon: 277 (1995).
N. South America to N. Brazil. 82 FRG GUY SUR VEN 84 BZN.
> *Gynestum acaule* Poit., Mém. Mus. Hist. Nat. 9: 391 (1822). *Geonoma acaulis* (Poit.) Burret, Bot. Jahrb. Syst. 63: 162 (1930), nom. illeg.
> *Geonoma poiteauana* Kunth, Enum. Pl. 3: 233 (1841).
> *Geonoma dammeri* Huber, Bol. Mus. Paraense Hist. Nat. Ethnogr. 3: 409 (1902). *Taenianthera dammeri* (Huber) Burret, Notizbl. Bot. Gart. Berlin-Dahlem 11: 13 (1930).
> *Taenianthera lakoi* Burret, Notizbl. Bot. Gart. Berlin-Dahlem 11: 11 (1930).
> *Geonoma chaunostachys* Burret, Bull. Torrey Bot. Club 58: 318 (1931).

Geonoma macroura Burret = **Geonoma orbignyana** Mart.

Geonoma magnifica Linden & H.Wendl. = **Geonoma interrupta** var. **interrupta**

Geonoma maguirei L.H.Bailey = **Geonoma stricta** var. **stricta**

Geonoma major Burret = **Geonoma deversa** (Poit.) Kunth

Geonoma margaritoides Engel = **Geonoma undata** Klotzsch

Geonoma margyraffia Engel = **Geonoma orbignyana** Mart.

Geonoma martiana H.Wendl. = **Asterogyne martiana** (H.Wendl.) H.Wendl. ex Drude

Geonoma martinicensis Mart. = **Geonoma interrupta** var. **interrupta**

Geonoma maxima (Poit.) Kunth, Enum. Pl. 3: 229 (1841).
S. Trop. America. 82 FRG GUY SUR VEN 83 BOL CLM ECU PER 84 BZN.
*Gynestum maximum Poit., Mém. Mus. Hist. Nat. 9: 388 (1822).

var. **ambigua** (Spruce) A.J.Hend., Palms Amazon: 278 (1995).
N. South America to N. Brazil (Roraima). 82 FRG GUY SUR VEN 84 BZN.
*Geonoma ambigua Spruce, J. Linn. Soc., Bot. 11: 111 (1871).
Geonoma schomburgkiana Spruce, J. Linn. Soc., Bot. 11: 111 (1871).
Geonoma robusta Burret, Bot. Jahrb. Syst. 63: 259 (1930).

var. **chelidonura** (Spruce) A.J.Hend., Palms Amazon: 279 (1995).
W. South America to N. Brazil. 82 VEN 83 BOL CLM PER 84 BZN.
*Geonoma chelidonura Spruce, J. Linn. Soc., Bot. 11: 111 (1871).
Geonoma densiflora Spruce, J. Linn. Soc., Bot. 11: 112 (1871).
Geonoma personata Spruce, J. Linn. Soc., Bot. 11: 112 (1871).
Geonoma tuberculata Spruce, J. Linn. Soc., Bot. 11: 112 (1871).
Geonoma bijugata Barb.Rodr., Enum. Palm. Nov.: 10 (1875).
Geonoma brachyfoliata Barb.Rodr., Enum. Palm. Nov.: 10 (1875).
Geonoma falcata Barb.Rodr., Enum. Palm. Nov.: 10 (1875).
Geonoma furcifolia Barb.Rodr., Enum. Palm. Nov.: 11 (1875).
Geonoma palustris Barb.Rodr., Enum. Palm. Nov.: 11 (1875).
Geonoma speciosa Barb.Rodr., Enum. Palm. Nov.: 9 (1875).
Geonoma spruceana Trail, J. Bot. 14: 328 (1876), nom. illeg.
Geonoma spruceana subsp. *tuberculata* (Spruce) Trail, J. Bot. 14: 329 (1876).
Geonoma spruceana var. *heptasticha* Trail, J. Bot. 14: 329 (1876).
Geonoma furcifrons Drude in C.F.P.von Martius & auct. suc. (eds.), Fl. Bras. 3(2): 502 (1882), orth. var.
Geonoma juruana Dammer, Verh. Bot. Vereins Prov. Brandenburg 48: 119 (1906 publ. 1907).
Geonoma dasystachys Burret, Bot. Jahrb. Syst. 63: 251 (1930).
Geonoma huebneri Burret, Bot. Jahrb. Syst. 63: 254 (1930).
Geonoma lakoi Burret, Bot. Jahrb. Syst. 63: 253 (1930).
Geonoma longisecta Burret, Bot. Jahrb. Syst. 63: 257 (1930).
Geonoma parvisecta Burret, Notizbl. Bot. Gart. Berlin-Dahlem 10: 1018 (1930).

var. **maxima**
S. Trop. America. 82 FRG VEN 83 BOL CLM ECU PER 84 BZN.
Geonoma multiflora Mart., Hist. Nat. Palm. 2: 7 (1823).

Geonoma discolor Spruce, J. Linn. Soc., Bot. 11: 110 (1871).
Geonoma hexasticha Spruce, J. Linn. Soc., Bot. 11: 110 (1871).
Geonoma negrensis Spruce, J. Linn. Soc., Bot. 11: 113 (1871).
Geonoma paraensis Spruce, J. Linn. Soc., Bot. 11: 112 (1871).
Geonoma capanemae Barb.Rodr., Enum. Palm. Nov.: 9 (1875).
Geonoma uliginosa Barb.Rodr., Enum. Palm. Nov.: 11 (1875).
Geonoma latisecta Burret, Bot. Jahrb. Syst. 63: 255 (1930).
Geonoma camptoneura Burret, Notizbl. Bot. Gart. Berlin-Dahlem 11: 201 (1931).

var. **spixiana** (Mart.) A.J.Hend., Palms Amazon: 281 (1995).
SE. Colombia to Brazil (Amazonas). 83 CLM 84 BZN.
*Geonoma spixiana Mart., Hist. Nat. Palm. 2: 15 (1823).
Geonoma grandisecta Burret, Bot. Jahrb. Syst. 63: 258 (1930).

Geonoma megaloptila Burret = **Geonoma interrupta** var. **interrupta**

Geonoma megalospatha Burret = **Geonoma weberbaueri** Dammer ex Burret

Geonoma membranacea H.Wendl. ex Spruce = **Geonoma interrupta** var. **interrupta**

Geonoma metensis H.Karst. = **Geonoma brongniartii** Mart.

Geonoma mexicana Liebm. ex Mart. = **Geonoma interrupta** var. **interrupta**

Geonoma microclada Burret = **Geonoma orbignyana** Mart.

Geonoma microspadix H.Wendl. ex Spruce = **Geonoma ferruginea** H.Wendl. ex Spruce

Geonoma microspatha Spruce = **Geonoma deversa** (Poit.) Kunth

Geonoma microstachys H.Wendl. ex Burret = **Geonoma ferruginea** H.Wendl. ex Spruce

Geonoma molinae Glassman = **Geonoma undata** Klotzsch

Geonoma molinillo Burret = **Geonoma orbignyana** Mart.

Geonoma monospatha de Nevers, Principes 42: 98 (1998).
Costa Rica to Panama. 80 COS PAN.

Geonoma mooreana de Nevers & Grayum, Novon 5: 354 (1995).
Panama. 80 PAN.

Geonoma mucronata Burret = **Geonoma jussieuana** Mart.

Geonoma multiflora Mart. = **Geonoma maxima** var. **maxima**

Geonoma multisecta (Burret) Burret = **Geonoma jussieuana** Mart.

Geonoma myriantha Dammer, Verh. Bot. Vereins Prov. Brandenburg 48: 120 (1906 publ. 1907).
Brazil (Acre). 84 BZN.

Geonoma negrensis Spruce = **Geonoma maxima** var. **maxima**

Geonoma obovata H.Wendl. ex Spruce = **Geonoma cuneata** var. **gracilis** (H.Wendl.) Skov

Geonoma oldemanii Granv., Adansonia, n.s., 14: 553 (1974 publ. 1975).
French Guiana to Brazil (Amapa, Pará). 82 FRG 84 BZN.

Geonoma olfersiana Klotzsch ex Drude = **Geonoma pauciflora** Mart.

Geonoma oligoclada Burret = **Geonoma poeppigiana** Mart.

Geonoma oligoclona Trail, J. Bot. 14: 325 (1876).
SE. Colombia, S. Venezuela, N. Brazil. 82 VEN 83 CLM 84 BZN.

Geonoma orbignyana Mart. in A.D.d'Orbigny, Voy. Amér. Mér. 7(3): 22 (1843).
C. America to Bolivia. 80 COS NIC PAN 82 VEN 83 BOL CLM ECU PER.
Geonoma lindeniana H.Wendl., Linnaea 28: 331 (1856).
Geonoma linearifolia H.Karst., Linnaea 28: 411 (1856).
Geonoma pumila Linden & H.Wendl., Linnaea 28: 331 (1856).
Geonoma margyraffia Engel, Linnaea 33: 685 (1865).
Geonoma ramosa Engel, Linnaea 33: 684 (1865).
Geonoma hoffmanniana H.Wendl. ex Spruce, J. Linn. Soc., Bot. 11: 106 (1871).
Geonoma andina Burret, Bot. Jahrb. Syst. 63: 188 (1930).
Geonoma aulacophylla Burret, Bot. Jahrb. Syst. 63: 216 (1930).
Geonoma campyloclada Burret, Bot. Jahrb. Syst. 63: 189 (1930).
Geonoma euterpoidea Burret, Bot. Jahrb. Syst. 63: 196 (1930).
Geonoma floccosa Dammer ex Burret, Bot. Jahrb. Syst. 63: 203 (1930).
Geonoma goniocarpa Burret, Bot. Jahrb. Syst. 63: 185 (1930).
Geonoma iodolepis Burret, Bot. Jahrb. Syst. 63: 198 (1930).
Geonoma iodoneura Burret, Bot. Jahrb. Syst. 63: 210 (1930).
Geonoma lepidota Burret, Bot. Jahrb. Syst. 63: 191 (1930).
Geonoma leucotricha Burret, Bot. Jahrb. Syst. 63: 204 (1930).
Geonoma macroura Burret, Bot. Jahrb. Syst. 63: 202 (1930).
Geonoma microclada Burret, Bot. Jahrb. Syst. 63: 190 (1930).
Geonoma pachydicrana Burret, Bot. Jahrb. Syst. 63: 206 (1930).
Geonoma paleacea Burret, Bot. Jahrb. Syst. 63: 199 (1930).
Geonoma plicata Burret, Bot. Jahrb. Syst. 63: 217 (1930).
Geonoma pulcherrima Burret, Bot. Jahrb. Syst. 63: 195 (1930).
Geonoma rhytidocarpa Burret, Bot. Jahrb. Syst. 63: 189 (1930).
Geonoma wendlandiana Burret, Bot. Jahrb. Syst. 63: 192 (1930).
Geonoma heinrichsiae Burret, Notizbl. Bot. Gart. Berlin-Dahlem 12: 43 (1934).
Geonoma anomoclada Burret, Notizbl. Bot. Gart. Berlin-Dahlem 12: 615 (1935).
Geonoma molinillo Burret, Notizbl. Bot. Gart. Berlin-Dahlem 13: 491 (1937).
Geonoma tenuifolia Burret, Notizbl. Bot. Gart. Berlin-Dahlem 15: 25 (1940).

Geonoma oxycarpa Mart. = **Geonoma interrupta** var. **interrupta**

Geonoma pachyclada Burret = **Geonoma undata** Klotzsch

Geonoma pachydicrana Burret = **Geonoma orbignyana** Mart.

Geonoma paleacea Burret = **Geonoma orbignyana** Mart.

Geonoma palustris Barb.Rodr. = **Geonoma maxima** var. **chelidonura** (Spruce) A.J.Hend.

Geonoma paniculigera Mart. = **Geonoma deversa** (Poit.) Kunth

Geonoma paradoxa Burret, Notizbl. Bot. Gart. Berlin-Dahlem 11: 1040 (1934).
SW. Colombia to N. Peru. 83 CLM ECU PER.

Geonoma paraensis Spruce = **Geonoma maxima** var. **maxima**

Geonoma paraguanensis H.Karst., Linnaea 28: 410 (1856).
NW. Venezuela (Falcón). 82 VEN.

Geonoma parvifrons Burret = **Geonoma jussieuana** Mart.

Geonoma parvisecta Burret = **Geonoma maxima** var. **chelidonura** (Spruce) A.J.Hend.

Geonoma pauciflora Mart., Hist. Nat. Palm. 2: 12 (1823).
E. & S. Brazil. 84 BZE BZL BZS.
Geonoma elegans Mart., Hist. Nat. Palm. 2: 144 (1826).
Geonoma porteana H.Wendl., Linnaea 28: 340 (1856).
Geonoma bifurca Drude & H.Wendl. in C.F.P.von Martius & auct. suc. (eds.), Fl. Bras. 3(2): 504 (1882).
Geonoma caespitosa H.Wendl. ex Drude in C.F.P.von Martius & auct. suc. (eds.), Fl. Bras. 3(2): 500 (1882).
Geonoma elegans var. *robusta* Drude in C.F.P.von Martius & auct. suc. (eds.), Fl. Bras. 3(2): 506 (1882).
Geonoma olfersiana Klotzsch ex Drude in C.F.P.von Martius & auct. suc. (eds.), Fl. Bras. 3(2): 506 (1882).

Geonoma personata Spruce = **Geonoma maxima** var. **chelidonura** (Spruce) A.J.Hend.

Geonoma pilosa Barb.Rodr. = **Geonoma pohliana** Mart.

Geonoma pinnatifrons Willd. = **Geonoma interrupta** var. **interrupta**

Geonoma piscicauda Dammer = **Geonoma stricta var. pisidicauda**

Geonoma platybothros Burret = **Geonoma interrupta** var. **interrupta**

Geonoma platycaula Drude = **Geonoma rubescens** H.Wendl. ex Drude

Geonoma pleeana Mart. = **Geonoma interrupta** var. **interrupta**

Geonoma pleioneura Burret = **Geonoma jussieuana** Mart.

Geonoma plicata Burret = **Geonoma orbignyana** Mart.

Geonoma plumeriana Mart. = **Calyptronoma plumeriana** (Mart.) Lourteig

Geonoma plurinervia Burret = **Geonoma brevispatha** var. **brevispatha**

Geonoma poeppigiana Mart. in A.D.d'Orbigny, Voy. Amér. Mér. 7(3): 35 (1843).
W. Amazon Reg. 83 CLM PER 84 BZN.

Geonoma oligoclada Burret, Notizbl. Bot. Gart. Berlin-Dahlem 11: 9 (1930).

Geonoma pohliana Mart., Hist. Nat. Palm. 2: 142 (1826).
E. Brazil. 84 BZE BZL.
Geonoma barbigera Barb.Rodr., Palmiers: 45 (1882).
Geonoma blanchetiana H.Wendl. ex Drude in C.F.P.von Martius & auct. suc. (eds.), Fl. Bras. 3(2): 494 (1882).
Geonoma fiscellaria Mart. ex Drude in C.F.P.von Martius & auct. suc. (eds.), Fl. Bras. 3(2): 486 (1882).
Geonoma macroclona Drude in C.F.P.von Martius & auct. suc. (eds.), Fl. Bras. 3(2): 486 (1882).
Geonoma pilosa Barb.Rodr., Palmiers: 43 (1882).
Geonoma tomentosa Barb.Rodr., Palmiers: 44 (1882).
Geonoma trigonostyla Barb.Rodr., Palmiers: 46 (1882).
Geonoma luetzelburgii Burret, Bot. Jahrb. Syst. 63: 235 (1930).
Geonoma latifolia Burret, Repert. Spec. Nov. Regni Veg. 32: 102 (1933).
Geonoma barbosiana Burret, Notizbl. Bot. Gart. Berlin-Dahlem 14: 255 (1938).
Geonoma kuhlmannii Burret, Notizbl. Bot. Gart. Berlin-Dahlem 14: 261 (1938).

Geonoma poiteauana Kunth = **Geonoma macrostachys** var. **poiteauana** (Kunth) A.J.Hend.

Geonoma polyandra Skov, Nordic J. Bot. 14: 39 (1994).
S. Colombia to NE. Ecuador. 83 CLM ECU.

Geonoma polyclada Burret = **Geonoma interrupta** var. **interrupta**

Geonoma polyneura Burret = **Geonoma undata** Klotzsch

Geonoma porteana H.Wendl. = **Geonoma pauciflora** Mart.

Geonoma preussii Burret = **Geonoma interrupta** var. **interrupta**

Geonoma princeps Linden = ? [83 PER]

Geonoma procumbens H.Wendl. ex Spruce = **Geonoma cuneata** var. **procumbens** (H.Wendl. ex Spruce) Skov

Geonoma pulchella H.Wendl. ex Linden = ? [83 CLM]

Geonoma pulcherrima Burret = **Geonoma orbignyana** Mart.

Geonoma pulchra Engel = **Geonoma densa** Linden & H.Wendl.

Geonoma pumila Linden & H.Wendl. = **Geonoma orbignyana** Mart.

Geonoma purdieana Spruce = **Geonoma interrupta** var. **interrupta**

Geonoma pycnostachys Mart. = **Geonoma stricta** var. **stricta**

Geonoma pynaertiana Sander = **Iguanura wallichiana** var. **major** Becc. ex Hook.f.

Geonoma raimondii Burret = **Geonoma stricta** var. **trailii** (Burret) A.J.Hend.

Geonoma ramosa Engel = **Geonoma orbignyana** Mart.

Geonoma ramosissima Burret = **Geonoma interrupta** var. **interrupta**

Geonoma rectifolia Wallace = **Geonoma deversa** (Poit.) Kunth

Geonoma rhytidocarpa Burret = **Geonoma orbignyana** Mart.

Geonoma riedeliana H.Wendl. = ? [84]

Geonoma rivalis Kalbreyer & Burret = **Geonoma interrupta** var. **interrupta**

Geonoma robusta Burret = **Geonoma maxima** var. **ambigua** (Spruce) A.J.Hend.

Geonoma rodeiensis Barb.Rodr. = **Geonoma rubescens** H.Wendl. ex Drude

Geonoma roraimae Dammer = **Geonoma appuniana** Spruce

Geonoma rubescens H.Wendl. ex Drude in C.F.P.von Martius & auct. suc. (eds.), Fl. Bras. 3(2): 491 (1882).
E. Brazil. 84 BZE BZL.
Geonoma platycaula Drude in C.F.P.von Martius & auct. suc. (eds.), Fl. Bras. 3(2): 490 (1882).
Geonoma rodeiensis Barb.Rodr., Palmiers: 42 (1882).
Geonoma trinervis Drude & H.Wendl. in C.F.P.von Martius & auct. suc. (eds.), Fl. Bras. 3(2): 492 (1882).

Geonoma rupestris Barb.Rodr. = **Geonoma brevispatha** var. **brevispatha**

Geonoma saga Spruce = **Geonoma interrupta** var. **interrupta**

Geonoma saramaccana L.H.Bailey = **Geonoma leptospadix** Trail

Geonoma schomburgkiana Spruce = **Geonoma maxima** var. **ambigua** (Spruce) A.J.Hend.

Geonoma schottiana Mart., Hist. Nat. Palm. 2: 143 (1826).
EC. & S. Brazil. 84 BZC BZL BZN BZS.
Geonoma erythrospadice Barb.Rodr., Enum. Palm. Nov.: 41 (1875).
Geonoma hoehnei Burret, Bot. Jahrb. Syst. 63: 231 (1930).

Geonoma schottiana var. *palustris* Warm. ex Drude = **Geonoma brevispatha** var. **brevispatha**

Geonoma scoparia Grayum & Nevers, Principes 32: 111 (1988).
Costa Rica (Puntarenas). 80 COS.

Geonoma seemannii auct. = ? [80]

Geonoma seleri Burret = **Geonoma undata** Klotzsch

Geonoma simplicifrons Willd., Sp. Pl. 4: 594 (1805).
Venezuela. 82 VEN.
Geonoma willdenowii Klotzsch, Bot. Zeitung (Berlin) 4: 112 (1846).
Geonoma fendleriana Spruce, J. Linn. Soc., Bot. 11: 108 (1871).

Geonoma sodiroi Dammer ex Burret = **Geonoma cuneata** var. **sodiroi** (Dammer ex Burret) Skov

Geonoma solitaria (Engel) Al.Jahn ex A.W.Hill = **Geonoma weberbaueri** Dammer ex Burret

Geonoma solitaria (Engel) Al.Jahn = **Geonoma weberbaueri** Dammer ex Burret

Geonoma speciosa Barb.Rodr. = **Geonoma maxima** var. **chelidonura** (Spruce) A.J.Hend.

Geonoma spicigera K.Koch = **Calyptrogyne ghiesbreghtiana** (Linden & H.Wendl.) H.Wendl.

Geonoma spinescens H.Wendl., Bot. Jahrb. Syst. 63: 230 (1930).
N. Venezuela. 82 VEN.
Geonoma tenuis Burret, Notizbl. Bot. Gart. Berlin-Dahlem 13: 478 (1937).

Geonoma spinescens var. *braunii* F.W.Stauffer, Acta Bot. Venez. 20: 5 (1997).

Geonoma spinescens var. *braunii* F.W.Stauffer, Acta Bot. Venez. 20(2): 5 (1997).

Geonoma spinescens var. *braunii* F.W.Stauffer = **Geonoma spinescens** H.Wendl.

Geonoma spinescens var. *braunii* F.W.Stauffer = **Geonoma spinescens** H.Wendl.

Geonoma spixiana Mart. = **Geonoma maxima** var. **spixiana** (Mart.) A.J.Hend.

Geonoma spruceana Trail = **Geonoma maxima** var. **chelidonura** (Spruce) A.J.Hend.

Geonoma spruceana var. *heptasticha* Trail = **Geonoma maxima** var. **chelidonura** (Spruce) A.J.Hend.

Geonoma spruceana subsp. *tuberculata* (Spruce) Trail = **Geonoma maxima** var. **chelidonura** (Spruce) A.J.Hend.

Geonoma stenoschista Burret = **Geonoma brevispatha** var. **brevispatha**

Geonoma stenothyrsa Burret = **?** [83 CLM]

Geonoma stricta (Poit.) Kunth, Enum. Pl. 3: 232 (1841). S. Trop. America. 82 FRG GUY SUR VEN 83 BOL CLM ECU PER 84 BZN.
> *Gynestum strictum* Poit., Mém. Mus. Hist. Nat. 9: 391 (1822).

var. **piscicauda** (Dammer) A.J.Hend., Palms Amazon: 287 (1995). W. South America to N. Brazil. 83 CLM ECU PER 84 BZN.
> *Geonoma wittiana* Dammer, Verh. Bot. Vereins Prov. Brandenburg 48: 124 (1906 publ. 1907).
> *Geonoma herthae* Burret, Notizbl. Bot. Gart. Berlin-Dahlem 14: 325 (1939).

var. **stricta** S. Trop. America. 82 FRG GUY SUR VEN 83 CLM ECU PER 84 BZN.
> *Geonoma pycnostachys* Mart., Hist. Nat. Palm. 2: 16 (1823).
> *Geonoma lanceolata* Burret, Notizbl. Bot. Gart. Berlin-Dahlem 11: 7 (1930).
> *Geonoma maguirei* L.H.Bailey, Bull. Torrey Bot. Club 75: 102 (1948).

var. **trailii** (Burret) A.J.Hend., Palms Amazon: 288 (1995). W. South America to N. Brazil. 83 BOL CLM ECU PER 84 BZN.
> *Geonoma elegans* var. *amazonica* Trail, J. Bot. 14: 324 (1876). *Geonoma trailii* Burret, Bot. Jahrb. Syst. 63: 178 (1930).
> *Geonoma trauniana* Dammer, Verh. Bot. Vereins Prov. Brandenburg 48: 124 (1906 publ. 1907).
> *Geonoma raimondii* Burret, Bot. Jahrb. Syst. 63: 182 (1930).
> *Geonoma bella* Burret, Notizbl. Bot. Gart. Berlin-Dahlem 12: 304 (1935).

Geonoma stuebelii Burret = **Geonoma weberbaueri** Dammer ex Burret

Geonoma supracostata Svenning, Nordic J. Bot. 21: 344 (2001 publ. 2002). E. Ecuador. 83 ECU.

Geonoma swartzii Griseb. = **Calyptronoma occidentalis** (Sw.) H.E.Moore

Geonoma synanthera Mart. = **Pholidostachys synanthera** (Mart.) H.E.Moore

Geonoma talamancana Grayum, Phytologia 84: 324 (1998 publ. 1999). Costa Rica. 80 COS.

Geonoma tamandua Trail = **Geonoma macrostachys** var. **macrostachys**

Geonoma tapajotensis (Trail) Drude = **Geonoma macrostachys** var. **acaulis** (Mart.) Andrew Hend.

Geonoma tenuifolia auct. = **?** [83 PER]

Geonoma tenuifolia Burret = **Geonoma orbignyana** Mart.

Geonoma tenuis Burret = **Geonoma spinescens** H.Wendl.

Geonoma tenuissima H.E.Moore, Principes 26: 204 (1982). Colombia (Valle) to W. Ecuador. 83 CLM ECU.

Geonoma tessmannii Burret = **Geonoma deversa** (Poit.) Kunth

Geonoma tomentosa Barb.Rodr. = **Geonoma pohliana** Mart.

Geonoma trailii Burret = **Geonoma stricta** var. **trailii** (Burret) A.J.Hend.

Geonoma trauniana Dammer = **Geonoma stricta** var. **trailii** (Burret) A.J.Hend.

Geonoma triandra (Burret) Wess.Boer, Verh. Kon. Ned. Akad. Wetensch., Afd. Natuurk., Tweede Sect. 58(1): 85 (1968). Panama to NW. Colombia. 80 PAN 83 CLM.
> *Kalbreyera triandra* Burret, Bot. Jahrb. Syst. 63: 143 (1930).

Geonoma trichostachys Burret = **?** [83 CLM]

Geonoma trifurcata Oerst. = **Asterogyne martiana** (H.Wendl.) H.Wendl. ex Drude

Geonoma triglochin Burret, Notizbl. Bot. Gart. Berlin-Dahlem 11: 8 (1930). S. Colombia to Peru. 83 CLM ECU PER.

Geonoma trigona (Ruiz & Pav.) A.H.Gentry, Ann. Missouri Bot. Gard. 73: 161 (1986). C. Peru. 83 PER.

Geonoma trigonostyla Barb.Rodr. = **Geonoma pohliana** Mart.

Geonoma trijugata Barb.Rodr. = **Geonoma deversa** (Poit.) Kunth

Geonoma trinervis Drude & H.Wendl. = **Geonoma rubescens** H.Wendl. ex Drude

Geonoma tuberculata Spruce = **Geonoma maxima** var. **chelidonura** (Spruce) A.J.Hend.

Geonoma uleana Dammer = **Geonoma arundinacea** Mart.

Geonoma uliginosa Barb.Rodr. = **Geonoma maxima** var. **maxima**

Geonoma umbraculiformis Wess.Boer, Indig. Palms Surin.: 35 (1965). Guianas to Brazil (Amapá, Pará). 82 FRG GUY SUR 84 BZN.

Geonoma uncibracteata Burret = **Geonoma densa** Linden & H.Wendl.

Geonoma undata Klotzsch, Linnaea 20: 452 (1847). C. & W. South America, Windward Is. 80 BLZ COS GUA HON NIC 81 WIN 82 VEN 83 BOL CLM ECU PER.
> *Geonoma margaritoides* Engel, Linnaea 33: 682 (1865).
> *Geonoma dussiana* Becc., Repert. Spec. Nov. Regni Veg. 16: 436 (1920).

Geonoma helminthoclada Burret, Bot. Jahrb. Syst. 63: 222 (1930).

Geonoma macroclada Burret, Bot. Jahrb. Syst. 63: 220 (1930).

Geonoma macrosiphon Burret, Bot. Jahrb. Syst. 63: 214 (1930).

Geonoma pachyclada Burret, Bot. Jahrb. Syst. 63: 214 (1930).

Geonoma seleri Burret, Bot. Jahrb. Syst. 63: 211 (1930).

Geonoma polyneura Burret, Notizbl. Bot. Gart. Berlin-Dahlem 11: 500 (1932).

Geonoma hodgeorum L.H.Bailey, Caribbean Forester 3: 108 (1942).

Geonoma molinae Glassman, Fieldiana, Bot. 31: 7 (1964).

Geonoma vaga Griseb. & H.Wendl. = **Geonoma interrupta** var. **interrupta**

Geonoma ventricosa Engel = **? [83 CLM]**

Geonoma verdugo Linden = **? [83 CLM]**

Geonoma versiformis H.Wendl. ex Spruce = **Geonoma ferruginea** H.Wendl. ex Spruce

Geonoma warmingii A.D.Hawkes = **Geonoma brevispatha** var. **brevispatha**

Geonoma weberbaueri Dammer ex Burret, Bot. Jahrb. Syst. 63: 221 (1930).
W. South America to Venezuela. 82 VEN 83 BOL CLM ECU PER.

Roebelia solitaria Engel, Linnaea 33: 680 (1865). Provisional synonym. *Geonoma solitaria* (Engel) Al.Jahn, Palm. Fl. Venez.: 67 (1908). *Geonoma solitaria* (Engel) Al.Jahn ex A.W.Hill, Index Kew., Suppl. 7: 102 (1929).

Geonoma andicola Dammer ex Burret, Bot. Jahrb. Syst. 63: 218 (1930).

Geonoma megalospatha Burret, Bot. Jahrb. Syst. 63: 218 (1930).

Geonoma stuebelii Burret, Bot. Jahrb. Syst. 63: 220 (1930).

Geonoma weddelliana H.Wendl. ex Drude = **Geonoma brevispatha** var. **brevispatha**

Geonoma wendlandiana Burret = **Geonoma orbignyana** Mart.

Geonoma wendlandii auct. = **?**

Geonoma werdermannii Burret = **Geonoma brongniartii** Mart.

Geonoma willdenowii Klotzsch = **Geonoma simplicifrons** Willd.

Geonoma wittiana Dammer = **Geonoma stricta** var. **piscicauda** (Dammer) A.J.Hend.

Geonoma wittigiana Glaz. ex Drude = **Geonoma gastoniana** Glaz. ex Drude

Geonoma woronowii Burret = **Geonoma macrostachys** var. **macrostachys**

Geonoma yauaperyensis Barb.Rodr. = **Geonoma deversa** (Poit.) Kunth

Geonoma zamorensis Linden ex H.Wendl. = **? [83 ECU]**

Gigliolia

Gigliolia Becc. = **Areca** L.

Gigliolia insignis Becc. = **Areca insignis** (Becc.) J.Dransf.

Gigliolia subacaulis Becc. = **Areca subacaulis** (Becc.) J.Dransf.

Glaucothea

Glaucothea O.F.Cook = **Brahea** Mart.

Glaucothea aculeata (Brandegee) I.M.Johnst. = **Brahea aculeata** (Brandegee) H.E.Moore

Glaucothea armata (S.Watson) O.F.Cook = **Brahea armata** S.Watson

Glaucothea brandegeei (Purpus) I.M.Johnst. = **Brahea brandegeei** (Purpus) H.E.Moore

Glaucothea elegans (Franceschi ex Becc.) I.M.Johnst. = **Brahea armata** S.Watson

Glaziova

Glaziova Mart. ex H.Wendl. = **Lytocaryum** Toledo

Glaziova elegantissima H.Wendl. = **Lytocaryum weddellianum** (H.Wendl.) Toledo

Glaziova insignis Drude = **Lytocaryum weddellianum** (H.Wendl.) Toledo

Glaziova martiana Glaz. ex Drude = **Lytocaryum weddellianum** (H.Wendl.) Toledo

Glaziova treubiana Becc. = **Syagrus coronata** (Mart.) Becc.

Gomutus

Gomutus Corrêa = **Arenga** Labill. ex DC.

Gomutus obtusifolius Blume = **Arenga obtusifolia** Mart.

Gomutus rumphii Corrêa = **Arenga pinnata** (Wurmb) Merr.

Gomutus saccharifer (Labill. ex DC.) Spreng. = **Arenga pinnata** (Wurmb) Merr.

Gomutus vulgaris Oken = **Arenga pinnata** (Wurmb) Merr.

Goniocladus

Goniocladus Burret = **Physokentia** Becc.

Goniocladus petiolatus Burret = **Physokentia petiolata** (Burret) D.Fuller

Goniosperma

Goniosperma Burret = **Physokentia** Becc.

Goniosperma thurstonii (Becc.) Burret = **Physokentia thurstonii** (Becc.) Becc.

Goniosperma vitiense Burret = **Physokentia thurstonii** (Becc.) Becc.

Gorgasia

Gorgasia O.F.Cook = **Roystonea** O.F.Cook

Gorgasia oleracea (Jacq.) O.F.Cook = **Roystonea oleracea** (Jacq.) O.F.Cook

Grisebachia

Grisebachia Drude & H.Wendl. = **Howea** Becc.

Grisebachia belmoreana (C.Moore & F.Muell.) H.Wendl. & Drude = **Howea belmoreana** (C.Moore & F.Muell.) Becc.

Grisebachia forsteriana (C.Moore & F.Muell.) H.Wendl. & Drude = **Howea forsteriana** (C.Moore & F.Muell.) Becc.

Gronophyllum

Gronophyllum Scheff. = **Hydriastele** H.Wendl. & Drude

Gronophyllum affine (Becc.) Essig & B.E.Young = **Hydriastele affinis** (Becc.) W.J.Baker & Loo

Gronophyllum apricum B.E.Young = **Hydriastele aprica** (B.E.Young) W.J.Baker & Loo

Gronophyllum brassii Burret = **Hydriastele brassii** (Burret) W.J.Baker & Loo

Gronophyllum cariosum Dowe & M.D.Ferrero = **Hydriastele cariosa** (Dowe & M.D.Ferrero) W.J.Baker & Loo

Gronophyllum chaunostachys (Burret) H.E.Moore = **Hydriastele chaunostachys** (Burret) W.J.Baker & Loo

Gronophyllum cyclopense Essig & B.E.Young = **Hydriastele cyclopensis** (Essig & B.E.Young) W.J.Baker & Loo

Gronophyllum densiflorum Ridl. = **Hydriastele pinangoides** (Becc.) W.J.Baker & Loo

Gronophyllum flabellatum (Becc.) Essig & B.E.Young = **Hydriastele flabellata** (Becc.) W.J.Baker & Loo

Gronophyllum gibbsianum (Becc.) H.E.Moore = **Hydriastele gibbsiana** (Becc.) W.J.Baker & Loo

Gronophyllum gracile (Burret) Essig & B.E.Young = **Hydriastele gracilis** (Burret) W.J.Baker & Loo

Gronophyllum kjellbergii Burret = **Hydriastele kjellbergii** (Burret) W.J.Baker & Loo

Gronophyllum ledermannianum (Becc.) H.E.Moore = **Hydriastele ledermanniana** (Becc.) W.J.Baker & Loo

Gronophyllum leonardii Essig & B.E.Young = **Hydriastele pinangoides** (Becc.) W.J.Baker & Loo

Gronophyllum luridum Becc. = **Hydriastele lurida** (Becc.) W.J.Baker & Loo

Gronophyllum manusii Essig = **Hydriastele manusii** (Essig) W.J.Baker & Loo

Gronophyllum mayrii (Burret) H.E.Moore = **Hydriastele mayrii** (Burret) W.J.Baker & Loo

Gronophyllum micranthum (Burret) Essig & B.E.Young = **Hydriastele micrantha** (Burret) W.J.Baker & Loo

Gronophyllum microcarpum Scheff. = **Hydriastele microcarpa** (Scheff.) W.J.Baker & Loo

Gronophyllum microspadix Burret = **Hydriastele nannostachys** W.J.Baker & Loo

Gronophyllum montanum (Becc.) Essig & B.E.Young = **Hydriastele montana** (Becc.) W.J.Baker & Loo

Gronophyllum oxypetalum Burret = **Hydriastele oxypetala** (Burret) W.J.Baker & Loo

Gronophyllum pinangoides (Becc.) Essig & B.E.Young = **Hydriastele pinangoides** (Becc.) W.J.Baker & Loo

Gronophyllum pleurocarpum (Burret) Essig & B.E.Young = **Hydriastele pleurocarpa** (Burret) W.J.Baker & Loo

Gronophyllum procerum (Blume) H.E.Moore = **Hydriastele procera** (Blume) W.J.Baker & Loo

Gronophyllum ramsayi (Becc.) H.E.Moore = **Hydriastele ramsayi** (Becc.) W.J.Baker & Loo

Gronophyllum rhopalocarpum (Becc.) Essig & B.E.Young = **Hydriastele rhopalocarpa** (Becc.) W.J.Baker & Loo

Gronophyllum sarasinorum Burret = **Hydriastele sarasinorum** (Burret) W.J.Baker & Loo

Gronophyllum selebicum (Becc.) Becc. = **Hydriastele selebica** (Becc.) W.J.Baker & Loo

Guihaia

Guihaia J.Dransf., S.K.Lee & F.N.Wei, Principes 29: 7 (1985).
2 species, SE. China to Indo-China. 36 41.

Guihaia argyrata (S.K.Lee & F.N.Wei) S.K.Lee, F.N.Wei & J.Dransf., Principes 29: 9 (1985).
China (Guangxi, Guangdong). 36 CHS.
*Trachycarpus argyratus S.K.Lee & F.N.Wei, Guihaia 2: 131 (1982).

Guihaia grossifibrosa (Gagnep.) J.Dransf., S.K.Lee & F.N.Wei, Principes 29: 12 (1985).
SE. China to Indo-China. 36 CHS 41 VIE.
Rhapis filiformis Burret, Notizbl. Bot. Gart. Berlin-Dahlem 13: 586 (1937).
*Rhapis grossifibrosa Gagnep., Notul. Syst. (Paris) 6: 159 (1937).

Guilelma

Guilelma Link = **Bactris** Jacq. ex Scop.

Guilelma microcarpa Huber = **Bactris gasipaes** var. **chichagui** (H.Karst.) A.J.Hend.

Guilielma

Guilielma Mart. = **Bactris** Jacq. ex Scop.

Guilielma caribaea (H.Karst.) H.Wendl. = **Bactris gasipaes** var. **chichagui** (H.Karst.) A.J.Hend.

Guilielma chontaduro H.Karst. & Triana = **Bactris gasipaes** var. **gasipaes**

Guilielma ciliata (Ruiz & Pav.) H.Wendl. = **Bactris gasipaes** var. **gasipaes**

Guilielma gasipaes (Kunth) L.H.Bailey = **Bactris gasipaes** Kunth

Guilielma granatensis H.Karst. = **Bactris pilosa** H.Karst.

Guilielma insignis Mart. = **Bactris gasipaes** var. **gasipaes**

Guilielma macana Mart. = **Bactris gasipaes** var. **chichagui** (H.Karst.) A.J.Hend.

Guilielma mattogrossensis Barb.Rodr. = **Bactris gasipaes** var. **chichagui** (H.Karst.) A.J.Hend.

Guilielma microcarpa Huber = **Bactris gasipaes** var. **chichagui** (H.Karst.) A.J.Hend.

Guilielma piritu H.Karst. = **Bactris guineensis** (L.) H.E.Moore

Guilielma speciosa Mart. = **Bactris gasipaes** var. **gasipaes**

Guilielma speciosa var. *coccinea* Barb.Rodr. = **Bactris gasipaes** var. **gasipaes**

Guilielma speciosa var. *flava* Barb.Rodr. = **Bactris gasipaes** var. **gasipaes**

Guilielma speciosa var. *mitis* Drude = **Bactris gasipaes** var. **gasipaes**

Guilielma speciosa var. *ochracea* Barb.Rodr. = **Bactris gasipaes** var. **gasipaes**

Guilielma tenera H.Karst. = **Bactris brongniartii** Mart.

Guilielma utilis Oerst. = **Bactris gasipaes** var. **gasipaes**

Gulubia

Gulubia Becc. = **Hydriastele** H.Wendl. & Drude

Gulubia affinis Becc. = **Hydriastele costata** F.M.Bailey

Gulubia brassii Burret = **Hydriastele longispatha** (Becc.) W.J.Baker & Loo

Gulubia costata (Becc.) Becc. = **Hydriastele costata** F.M.Bailey

Gulubia costata var. *gracilior* Burret = **Hydriastele costata** F.M.Bailey

Gulubia costata var. *minor* Becc. = **Hydriastele costata** F.M.Bailey

Gulubia costata var. *pisiformis* Becc. = **Hydriastele costata** F.M.Bailey

Gulubia crenata Becc. = **Hydriastele longispatha** (Becc.) W.J.Baker & Loo

Gulubia cylindrocarpa Becc. = **Hydriastele cylindrocarpa** (Becc.) W.J.Baker & Loo

Gulubia gracilior (Burret) Burret ex A.W.Hill & E.J.Salisbury = **Hydriastele costata** F.M.Bailey

Gulubia hombronii Becc. = **Hydriastele hombronii** (Becc.) W.J.Baker & Loo

Gulubia liukiuensis Hatus. = **Satakentia liukiuensis** (Hatus.) H.E.Moore

Gulubia longispatha Becc. = **Hydriastele longispatha** (Becc.) W.J.Baker & Loo

Gulubia macrospadix (Burret) H.E.Moore = **Hydriastele macrospadix** (Burret) W.J.Baker & Loo

Gulubia microcarpa Essig = **Hydriastele vitiensis** W.J.Baker & Loo

Gulubia moluccana (Becc.) Becc. = **Hydriastele moluccana** (Becc.) W.J.Baker & Loo

Gulubia obscura Becc. = **Hydriastele longispatha** (Becc.) W.J.Baker & Loo

Gulubia palauensis (Becc.) H.E.Moore & Fosberg = **Hydriastele palauensis** (Becc.) W.J.Baker & Loo

Gulubia ramsayi Becc. = **Hydriastele ramsayi** (Becc.) W.J.Baker & Loo

Gulubia valida Essig = **Hydriastele valida** (Essig) W.J.Baker & Loo

Gulubiopsis

Gulubiopsis Becc. = **Hydriastele** H.Wendl. & Drude

Gulubiopsis palauensis Becc. = **Hydriastele palauensis** (Becc.) W.J.Baker & Loo

Gynestum

Gynestum Poit. = **Geonoma** Willd.

Gynestum acaule Poit. = **Geonoma macrostachys** var. **poiteauana** (Kunth) A.J.Hend.

Gynestum baculiferum Poit. = **Geonoma baculifera** (Poit.) Kunth

Gynestum deversum Poit. = **Geonoma deversa** (Poit.) Kunth

Gynestum maximum Poit. = **Geonoma maxima** (Poit.) Kunth

Gynestum strictum Poit. = **Geonoma stricta** (Poit.) Kunth

Haitiella

Haitiella L.H.Bailey = **Coccothrinax** Sarg.

Haitiella ekmanii (Burret) L.H.Bailey = **Coccothrinax ekmanii** Burret

Haitiella munizii (Borhidi) Borhidi = **Coccothrinax munizii** Borhidi

Halmoorea

Halmoorea J.Dransf. & N.W.Uhl = **Orania** Zipp.

Halmoorea trispatha J.Dransf. & N.W.Uhl = **Orania trispatha** (J.Dransf. & N.W.Uhl) Beentje & J.Dransf.

Haplodypsis

Haplodypsis Baill. = **Dypsis** Noronha ex Mart.

Haplodypsis pervillei Baill. = **Dypsis pervillei** (Jum.) Beentje & J.Dransf.

Haplophloga

Haplophloga Baill. = **Dypsis** Noronha ex Mart.

Haplophloga bernierana Baill. = **Dypsis bernierana** (Baill.) Beentje & J.Dransf.

Haplophloga comorensis Baill. = **?** [29 COM]

Haplophloga loucoubensis Baill. = **Dypsis nossibensis** (Becc.) Beentje & J.Dransf.

Haplophloga poivreana Baill. = **Dypsis poivreana** (Baill.) Beentje & J.Dransf.

Harina

Harina Buch.-Ham. = **Wallichia** Roxb.

Harina caryotoides (Roxb.) Buch.-Ham. = **Wallichia caryotoides** Roxb.

Harina densiflora (Mart.) Walp. = **Wallichia densiflora** Mart.

Harina nana (Griff.) Griff. = **Arenga nana** (Griff.) H.E.Moore

Harina oblongifolia (Griff.) Griff. = **Wallichia densiflora** Mart.

Harina rumphii (Blume) Mart. = **Drymophloeus oliviformis** (Giseke) Mart.

Harina wallichia Steud. ex Saloman = **Wallichia caryotoides** Roxb.

Hedyscepe

Hedyscepe H.Wendl. & Drude, Linnaea 39: 178 (1875).
1 species, Lord Howe I. 50.

Hedyscepe canterburyana (C.Moore & F.Muell.) H.Wendl. & Drude, Linnaea 39: 204 (1875).
Lord Howe I. 50 NFK.
*Kentia canterburyana C.Moore & F.Muell., Fragm. 7: 101 (1870). *Veitchia canterburyana* (C.Moore & F.Muell.) H.Wendl., Gard. Chron. 1872: 372 (1872).

Hemithrinax

Hemithrinax Hook.f. in G.Bentham & J.D.Hooker, Gen. Pl. 3: 930 (1883).
2 species, E. Cuba. 81.

Hemithrinax compacta (Griseb. & H.Wendl.) Hook.f. in G.Bentham & J.D.Hooker, Gen. Pl. 3: 931 (1883).
E. Cuba (Sierra de Nipe). 81 CUB.

Trithrinax compacta Griseb. & H.Wendl. in H.A.R. Grisebach, Cat. Pl. Cub.: 221 (1866). *Thrinax compacta* (Griseb. & H.Wendl.) Borhidi & O.Muñiz, Acta Bot. Hung. 31: 226 (1985).

Hemithrinax ekmaniana Burret = **Thrinax ekmaniana** (Burret) Borhidi & O.Muñiz

Hemithrinax rivularis Léon, Mem. Soc. Cub. Hist. Nat. "Felipe Poey" 15: 380 (1941). *Thrinax rivularis* (Léon) Borhidi & O.Muñiz, Acta Bot. Hung. 31: 226 (1985).
E. Cuba (Sierra de Moa). 81 CUB.

var. **rivularis**
E. Cuba (Sierra de Moa). 81 CUB.

var. **savannarum** (Léon) O.Muñiz, Acta Bot. Acad. Sci. Hung. 28: 312 (1982).
E. Cuba (Sierra de Moa). 81 CUB.
*Hemithrinax savannarum Léon, Mem. Soc. Cub. Hist. Nat. "Felipe Poey" 15: 381 (1941). *Thrinax rivularis* var. *savannarum* (Léon) Borhidi & O.Muñiz, Acta Bot. Hung. 31: 226 (1985).

Hemithrinax savannarum Léon = **Hemithrinax rivularis** var. **savannarum** (Léon) O.Muñiz

Heptantra

Heptantra O.F.Cook = **Attalea** Kunth

Heptantra phalerata (Mart.) O.F.Cook = **Attalea speciosa** Mart.

Heterospathe

Heterospathe Scheff., Ann. Jard. Bot. Buitenzorg 1: 141 (1876).
38 species, Philippines to W. Pacific. 42 43 60 62.
Ptychandra Scheff., Ann. Jard. Bot. Buitenzorg 1: 140 (1876).
Barkerwebbia Becc., Webbia 1: 281 (1905).

Heterospathe annectens H.E.Moore, Principes 13: 100 (1969).
New Guinea. 43 NWG.

Heterospathe arfakiana (Becc.) H.E.Moore, Principes 14: 91 (1970).
New Guinea. 43 NWG.
*Ptychosperma arfakianum Becc., Malesia 1: 57 (1877).

Heterospathe brevicaulis Fernando, Kew Bull. 45: 220 (1990).
Philippines (Luzon). 42 PHI.

Heterospathe cagayanensis Becc., Philipp. J. Sci., C 4: 611 (1909). *Ptychoraphis cagayanensis* (Becc.) Becc., Philipp. J. Sci. 14: 328 (1919).
Philippines (Luzon). 42 PHI.

Heterospathe califrons Fernando, Palms 45: 118 (2001).
Philippines. 42 PHI.

Heterospathe clemensiae (Burret) H.E.Moore, Principes 12: 101 (1969).
New Guinea. 43 NWG.
*Ptychandra clemensiae Burret, Notizbl. Bot. Gart. Berlin-Dahlem 13: 468 (1937).

Heterospathe delicatula H.E.Moore, Principes 12: 101 (1969).
New Guinea. 43 NWG.

Heterospathe dransfieldii Fernando, Kew Bull. 45: 223 (1990).
Philippines (Palawan). 42 PHI.

Heterospathe elata Scheff., Ann. Jard. Bot. Buitenzorg 1: 162 (1876).
C. Malesia to NW. Pacific. 42 MOL PHI 62 CRL MRN.

var. **elata**
C. Malesia to Marianas. 42 MOL PHI 62 MRN.
Euterpe pisifera Gaertn., Fruct. Sem. Pl. 1: 24 (1788).
Provisional synonym. *Heterospathe pisifera* (Gaertn.) Burret, Bot. Jahrb. Syst. 63: 76 (1929).
Metroxylon elatum Scheff., Ann. Jard. Bot. Buitenzorg 1: 162 (1876), nom. inval.
Heterospathe elata var. *guamensis* Becc., Atti Soc. Tosc. Sci. Nat. Pisa Processi Verbali 44: 707 (1934).

var. **palauensis** (Becc.) Becc., Atti Soc. Tosc. Sci. Nat. Pisa Processi Verbali 44: 140 (1934).
Caroline Is. (Palau). 62 CRL.
*Heterospathe palauensis Becc., Bot. Jahrb. Syst. 52: 4 (1914).

Heterospathe elata var. *guamensis* Becc. = **Heterospathe elata** var. **elata**

Heterospathe elegans (Becc.) Becc., Nova Guinea 8: 205 (1907).
New Guinea. 43 NWG.
*Barkerwebbia elegans Becc., Webbia 1: 283 (1905).

Heterospathe elmeri Becc., Leafl. Philip. Bot. 2: 646 (1909). *Ptychoraphis elmeri* (Becc.) Becc., Philipp. J. Sci. 14: 328 (1919). *Rhopaloblaste elmeri* (Becc.) Becc., Atti Soc. Tosc. Sci. Nat. Pisa Processi Verbali 44: 138 (1934).
Philippines. 42 PHI.
Ptychoraphis microcarpa Becc., Philipp. J. Sci. 14: 327 (1919). *Rhopaloblaste microcarpa* (Becc.) Becc., Atti Soc. Tosc. Sci. Nat. Pisa Processi Verbali 44: 136 (1934).

Heterospathe glabra (Burret) H.E.Moore, Principes 12: 102 (1969).
New Guinea. 43 NWG.
*Ptychandra glabra Burret, Notizbl. Bot. Gart. Berlin-Dahlem 11: 713 (1933).

Heterospathe glauca (Scheff.) H.E.Moore, Principes 12: 103 (1969).
Maluku. 42 MOL.
*Ptychandra glauca Scheff., Ann. Jard. Bot. Buitenzorg 1: 160 (1876).
Ptychandra musschenbroekiana Becc., Malesia 1: 100 (1877).
Ptychosperma musschenbroekianum Becc., Malesia 1: 53 (1877).

Heterospathe humilis Becc., Bot. Jahrb. Syst. 52: 35 (1914). *Barkerwebbia humilis* (Becc.) Becc. ex Martelli, Nuovo Giorn. Bot. Ital., n.s., 42: 31 (1935).
New Guinea. 43 NWG.

Heterospathe intermedia (Becc.) Fernando, Kew Bull. 45: 226 (1990).
Philippines. 42 PHI.
*Ptychoraphis intermedia Becc., Leafl. Philip. Bot. 8: 3011 (1919). *Rhopaloblaste intermedia* (Becc.) Becc., Atti Soc. Tosc. Sci. Nat. Pisa Processi Verbali 44: 137 (1934).

Heterospathe kajewskii Burret, Notizbl. Bot. Gart. Berlin-Dahlem 13: 78 (1936).
Solomon Is. 43 SOL.

Heterospathe ledermanniana Becc., Bot. Jahrb. Syst. 58: 451 (1923).
New Guinea. 43 NWG.

Heterospathe lepidota H.E.Moore, Principes 12: 103 (1969).
New Guinea. 43 NWG.

Heterospathe macgregorii (Becc.) H.E.Moore, Principes 14: 91 (1970).
New Guinea. 43 NWG.
*Rhopaloblaste macgregorii Becc., Atti Soc. Tosc. Sci. Nat. Pisa Processi Verbali 44: 134 (1934).

Heterospathe micrantha (Becc.) H.E.Moore = **Ptychosperma micranthum** Becc.

Heterospathe minor Burret, Notizbl. Bot. Gart. Berlin-Dahlem 13: 75 (1936).
Solomon Is. 43 SOL.

Heterospathe muelleriana (Becc.) Becc., Atti Soc. Tosc. Sci. Nat. Pisa Processi Verbali 44: 139 (1934).
New Guinea. 43 NWG.
*Ptychandra muelleriana Becc., Nuovo Giorn. Bot. Ital. 20: 177 (1888).

Heterospathe negrosensis Becc., Philipp. J. Sci., C 4: 611 (1909).
Philippines. 42 PHI.

Heterospathe obriensis (Becc.) H.E.Moore, Principes 13: 104 (1969).
New Guinea. 43 NWG.
*Ptychandra obriensis Becc., Nuovo Giorn. Bot. Ital. 20: 178 (1888).
Ptychandra montana Burret, Notizbl. Bot. Gart. Berlin-Dahlem 12: 324 (1935).

Heterospathe palauensis Becc. = **Heterospathe elata** var. **palauensis** (Becc.) Becc.

Heterospathe parviflora Essig, Principes 36: 4 (1992).
Bismarck Arch. 43 BIS.

Heterospathe philippinensis (Becc.) Becc., Philipp. J. Sci., C 4: 610 (1909).
Philippines. 42 PHI.
*Ptychoraphis philippinensis Becc., Ann. Jard. Bot. Buitenzorg 2: 90 (1885).

Heterospathe phillipsii D.Fuller & Dowe, Principes 41: 66 (1997).
Fiji. 60 FIJ.

Heterospathe pilosa (Burret) Burret, Notizbl. Bot. Gart. Berlin-Dahlem 12: 328 (1935).
New Guinea. 43 NWG.
*Rhynchocarpa pilosa Burret, Notizbl. Bot. Gart. Berlin-Dahlem 11: 712 (1933).

Heterospathe pisifera (Gaertn.) Burret = **Heterospathe elata** var. **elata**

Heterospathe pulchra H.E.Moore, Principes 13: 104 (1969).
Papua New Guinea. 43 NWG.

Heterospathe ramulosa Burret, Notizbl. Bot. Gart. Berlin-Dahlem 13: 80 (1936).
Solomon Is. 43 SOL.

Heterospathe salomonensis Becc., Webbia 3: 153 (1910).
Solomon Is. 43 SOL.

Heterospathe scitula Fernando, Kew Bull. 45: 228 (1990).
Philippines (Luzon). 42 PHI.

Heterospathe sensisi Becc., Webbia 4: 278 (1914).
Solomon Is. 43 SOL.

Heterospathe sibuyanensis Becc., Leafl. Philipp. Bot. 8: 3014 (1919).
Philippines (Sibuyan). 42 PHI.

Heterospathe sphaerocarpa Burret, Notizbl. Bot. Gart. Berlin-Dahlem 12: 326 (1935).
New Guinea. 43 NWG.

Heterospathe trispatha Fernando, Kew Bull. 45: 231 (1990).
Philippines (Luzon). 42 PHI.

Heterospathe uniformis Dowe, Austral. Syst. Bot. 9: 40 (1996).
Vanuatu. 60 VAN.

Heterospathe versteegiana Becc., Nova Guinea 8: 203 (1909).
New Guinea. 43 NWG.

Heterospathe woodfordiana Becc., Webbia 4: 281 (1914).
Solomon Is. 43 SOL.

Hexopetion

Hexopetion Burret = **Astrocaryum** G.Mey.

Hexopetion mexicanum (Liebm. ex Mart.) Burret = **Astrocaryum mexicanum** Liebm. ex Mart.

Howea

Howea Becc., Malesia 1: 41 (1877).
2 species, Lord Howe I. 50.
Grisebachia Drude & H.Wendl., Nachr. Königl. Ges. Wiss. Georg-Augusts-Univ. 1875: 5460 (1875), nom. illeg.
Denea O.F.Cook, J. Wash. Acad. Sci. 16: 395 (1926).

Howea belmoreana (C.Moore & F.Muell.) Becc., Malesia 1: 66 (1877).
Lord Howe I. 50 NFK.
*Kentia belmoreana C.Moore & F.Muell., Fragm. 7: 99 (1870). *Grisebachia belmoreana* (C.Moore & F.Muell.) H.Wendl. & Drude, Linnaea 39: 202 (1875).

Howea forsteriana (C.Moore & F.Muell.) Becc., Malesia 1: 66 (1877).
Lord Howe I. 50 NFK.
*Kentia forsteriana C.Moore & F.Muell., Fragm. 7: 100 (1870). *Grisebachia forsteriana* (C.Moore & F.Muell.) H.Wendl. & Drude, Linnaea 39: 203 (1875). *Denea forsteriana* (C.Moore & F.Muell.) O.F.Cook, J. Wash. Acad. Sci. 16: 397 (1926).

Hydriastele

Hydriastele H.Wendl. & Drude, Linnaea 39: 208 (1875).
48 species, C. Malesia to W. Pacific. 42 43 50 60 62.
Kentia Blume, Bull. Sci. Phys. Nat. Néerl. 1: 64 (1838), nom. illeg.
Gronophyllum Scheff., Ann. Jard. Bot. Buitenzorg 1: 135 (1876).
Nengella Becc., Malesia 1: 32 (1877).
Adelonenga (Becc.) Hook.f. in G.Bentham & J.D.Hooker, Gen. Pl. 3(2): 885 (1883).
Gulubia Becc., Ann. Jard. Bot. Buitenzorg 2: 131 (1885).
Leptophoenix Becc., Ann. Jard. Bot. Buitenzorg 2: 82 (1885).
Gulubiopsis Becc., Bot. Jahrb. Syst. 59: 11 (1924).
Siphokentia Burret, Notizbl. Bot. Gart. Berlin-Dahlem 10: 198 (1927).
Paragulubia Burret, Notizbl. Bot. Gart. Berlin-Dahlem 13: 84 (1936).

Hydriastele affinis (Becc.) W.J.Baker & Loo, Kew Bull. 59: 62 (2004).
New Guinea. 43 NWG.

*Nenga affinis Becc., Malesia 1: 29 (1877). Leptophoenix affinis (Becc.) Becc., Ann. Jard. Bot. Buitenzorg 2: 82 (1885). Nengella affinis (Becc.) Burret, Notizbl. Bot. Gart. Berlin-Dahlem 13: 316 (1936). Gronophyllum affine (Becc.) Essig & B.E.Young, Principes 29: 136 (1985).

Hydriastele aprica (B.E.Young) W.J.Baker & Loo, Kew Bull. 59: 62 (2004).
Papua New Guinea. 43 NWG.
*Gronophyllum apricum B.E.Young, Principes 29: 139 (1985).

Hydriastele beccariana Burret, Repert. Spec. Nov. Regni Veg. 24: 292 (1928).
S. New Guinea. 43 NWG.

Hydriastele beguinii (Burret) W.J.Baker & Loo, Kew Bull. 59: 62 (2004).
Maluku. 42 MOL.
*Siphokentia beguinii Burret, Notizbl. Bot. Gart. Berlin-Dahlem 10: 198 (1927).
Siphokentia pachypus Burret, Notizbl. Bot. Gart. Berlin-Dahlem 10: 199 (1927).

Hydriastele boumae W.J.Baker & D.Watling, Kew Bull. 59: 62 (2004).
Fiji (Taveuni). 60 FIJ.

Hydriastele brassii (Burret) W.J.Baker & Loo, Kew Bull. 59: 63 (2004).
New Guinea. 43 NWG.
*Gronophyllum brassii Burret, J. Arnold Arbor. 20: 205 (1939).

Hydriastele cariosa (Dowe & M.D.Ferrero) W.J.Baker & Loo, Kew Bull. 59: 63 (2004).
Papua New Guinea. 43 NWG.
*Gronophyllum cariosum Dowe & M.D.Ferrero, Palms 44: 161 (2000).

Hydriastele carrii Burret, Notizbl. Bot. Gart. Berlin-Dahlem 13: 326 (1936).
New Guinea. 43 NWG.

Hydriastele chaunostachys (Burret) W.J.Baker & Loo, Kew Bull. 59: 63 (2004).
New Guinea. 43 NWG.
*Kentia chaunostachys Burret, Notizbl. Bot. Gart. Berlin-Dahlem 13: 328 (1936). Gronophyllum chaunostachys (Burret) H.E.Moore, Gentes Herb. 9: 264 (1963).

Hydriastele costata F.M.Bailey, Queensland Agric. J. 2: 129 (1898).
New Guinea to N. Queensland. 43 BIS NWG 50 QLD.
Kentia costata Becc., Malesia 1: 36 (1877). Gulubia costata (Becc.) Becc., Ann. Jard. Bot. Buitenzorg 2: 134 (1885).
Gulubia costata var. minor Becc., Ann. Jard. Bot. Buitenzorg 2: 135 (1885).
Gulubia costata var. pisiformis Becc., Ann. Jard. Bot. Buitenzorg 2: 136 (1885).
Pinanga pisiformis Teijsm. ex Becc., Ann. Jard. Bot. Buitenzorg 2: 136 (1885).
Kentia microcarpa Warb. ex K.Schum. & Lauterb., Fl. Schutzgeb. Südsee: 207 (1900).
Gulubia affinis Becc., Bot. Jahrb. Syst. 58: 444 (1923).
Gulubia costata var. gracilior Burret, Notizbl. Bot. Gart. Berlin-Dahlem 13: 81 (1936). Gulubia gracilior (Burret) Burret ex A.W.Hill & E.J.Salisbury, Index Kew., Suppl. 10: 102 (1947).

Hydriastele cyclopensis (Essig & B.E.Young) W.J.Baker & Loo, Kew Bull. 59: 64 (2004).
New Guinea. 43 NWG.

Leptophoenix mayrii Burret, Notizbl. Bot. Gart. Berlin-Dahlem 11: 709 (1933). Nengella mayrii (Burret) Burret, Notizbl. Bot. Gart. Berlin-Dahlem 13: 314 (1936). Gronophyllum cyclopense Essig & B.E.Young, Principes 29: 136 (1985).

Hydriastele cylindrocarpa (Becc.) W.J.Baker & Loo, Kew Bull. 59: 64 (2004).
Santa Cruz Is. to Vanuatu. 60 SCZ VAN.
*Gulubia cylindrocarpa Becc., Webbia 3: 156 (1910).

Hydriastele douglasiana F.M.Bailey = **Hydriastele wendlandiana** (F.Muell.) H.Wendl. & Drude

Hydriastele dransfieldii (Hambali & al.) W.J.Baker & Loo, Kew Bull. 59: 64 (2004).
W. New Guinea. 43 NWG.
*Siphokentia dransfieldii Hambali & al., Palms 44: 179 (2000).

Hydriastele flabellata (Becc.) W.J.Baker & Loo, Kew Bull. 59: 64 (2004).
New Guinea. 43 NWG.
*Nengella flabellata Becc., Malesia 1: 34 (1877). Gronophyllum flabellatum (Becc.) Essig & B.E. Young, Principes 29: 134 (1985).

Hydriastele geelvinkiana (Becc.) Burret, Notizbl. Bot. Gart. Berlin-Dahlem 13: 484 (1937).
New Guinea. 43 NWG.
*Nenga geelvinkiana Becc., Malesia 1: 28 (1877). Adelonenga geelvinkiana (Becc.) Becc., Ann. Jard. Bot. Buitenzorg 2: 82 (1885).

Hydriastele gibbsiana (Becc.) W.J.Baker & Loo, Kew Bull. 59: 65 (2004).
New Guinea. 43 NWG.
*Kentia gibbsiana Becc. in L.S.Gibbs, Fl. Arfak Mts.: 91 (1917). Gronophyllum gibbsianum (Becc.) H.E. Moore, Gentes Herb. 9: 264 (1963).

Hydriastele gracilis (Burret) W.J.Baker & Loo, Kew Bull. 59: 65 (2004).
New Guinea. 43 NWG.
*Nengella gracilis Burret, J. Arnold Arbor. 20: 206 (1939). Gronophyllum gracile (Burret) Essig & B.E.Young, Principes 29: 134 (1985).

Hydriastele hombronii (Becc.) W.J.Baker & Loo, Kew Bull. 59: 65 (2004).
Solomon Is. 43 SOL.
*Gulubia hombronii Becc., Webbia 3: 161 (1910).

Hydriastele kasesa (Lauterb.) Burret, Notizbl. Bot. Gart. Berlin-Dahlem 13: 484 (1937).
Bismarck Arch. 43 BIS.
*Ptychosperma kasesa Lauterb., Bot. Jahrb. Syst. 45: 357 (1911). Adelonenga kasesa (Lauterb.) Becc., Bot. Jahrb. Syst. 52: 26 (1914).

Hydriastele kjellbergii (Burret) W.J.Baker & Loo, Kew Bull. 59: 65 (2004).
Sulawesi. 42 SUL.
*Gronophyllum kjellbergii Burret, Notizbl. Bot. Gart. Berlin-Dahlem 13: 203 (1936).

Hydriastele ledermanniana (Becc.) W.J.Baker & Loo, Kew Bull. 59: 65 (2004).
New Guinea. 43 NWG.
*Kentia ledermanniana Becc., Bot. Jahrb. Syst. 58: 442 (1923). Gronophyllum ledermannianum (Becc.) H.E.Moore, Gentes Herb. 9: 264 (1963).

Hydriastele lepidota Burret, J. Arnold Arbor. 20: 204 (1939).
New Guinea. 43 NWG.

Hydriastele longispatha (Becc.) W.J.Baker & Loo, Kew Bull. 59: 65 (2004).
Papua New Guinea. 43 NWG.
   *Gulubia longispatha Becc., Bot. Jahrb. Syst. 52: 25 (1914).
   Gulubia crenata Becc., Bot. Jahrb. Syst. 58: 445 (1923).
   Gulubia obscura Becc., Bot. Jahrb. Syst. 58: 447 (1923).
   Gulubia brassii Burret, Notizbl. Bot. Gart. Berlin-Dahlem 12: 336 (1935).

Hydriastele lurida (Becc.) W.J.Baker & Loo, Kew Bull. 59: 65 (2004).
New Guinea. 43 NWG.
   *Gronophyllum luridum Becc., Nova Guinea 8: 207 (1909).

Hydriastele macrospadix (Burret) W.J.Baker & Loo, Kew Bull. 59: 65 (2004).
Solomon Is. 43 SOL.
   *Paragulubia macrospadix Burret, Notizbl. Bot. Gart. Berlin-Dahlem 13: 84 (1936). *Gulubia macrospadix* (Burret) H.E.Moore, Principes 10: 88 (1966).

Hydriastele manusii (Essig) W.J.Baker & Loo, Kew Bull. 59: 65 (2004).
Bismarck Arch. 43 BIS.
   *Gronophyllum manusii Essig, Principes 39: 100 (1995).

Hydriastele mayrii (Burret) W.J.Baker & Loo, Kew Bull. 59: 65 (2004).
New Guinea. 43 NWG.
   *Kentia mayrii Burret, Notizbl. Bot. Gart. Berlin-Dahlem 11: 707 (1933). *Gronophyllum mayrii* (Burret) H.E.Moore, Gentes Herb. 9: 265 (1963).

Hydriastele micrantha (Burret) W.J.Baker & Loo, Kew Bull. 59: 65 (2004).
New Guinea. 43 NWG.
   *Leptophoenix micrantha Burret, Notizbl. Bot. Gart. Berlin-Dahlem 11: 710 (1933). *Nengella micrantha* (Burret) Burret, Notizbl. Bot. Gart. Berlin-Dahlem 13: 314 (1936). *Gronophyllum micranthum* (Burret) Essig & B.E.Young, Principes 29: 136 (1985).

Hydriastele microcarpa (Scheff.) W.J.Baker & Loo, Kew Bull. 59: 65 (2004).
Maluku (Seram). 42 MOL.
   *Gronophyllum microcarpum Scheff., Ann. Jard. Bot. Buitenzorg 1: 153 (1876).

Hydriastele microspadix (Warb. ex K.Schum. & Lauterb.) Burret, Notizbl. Bot. Gart. Berlin-Dahlem 13: 484 (1937).
New Guinea. 43 NWG.
   *Kentia microspadix Warb. ex K.Schum. & Lauterb., Fl. Schutzgeb. Südsee: 206 (1900). *Adelonenga microspadix* (Warb. ex K.Schum. & Lauterb.) Becc., Bot. Jahrb. Syst. 52: 26 (1914).
   Ptychosperma beccarianum Warb. ex Burret, Repert. Spec. Nov. Regni Veg. 24: 269 (1928), pro syn.

Hydriastele moluccana (Becc.) W.J.Baker & Loo, Kew Bull. 59: 66 (2004).
Maluku. 42 MOL.
   *Kentia moluccana Becc., Malesia 1: 35 (1877). *Gulubia moluccana* (Becc.) Becc., Ann. Jard. Bot. Buitenzorg 2: 131 (1885).

Hydriastele montana (Becc.) W.J.Baker & Loo, Kew Bull. 59: 66 (2004).
New Guinea. 43 NWG.
   *Nengella montana Becc., Malesia 1: 33 (1877). *Gronophyllum montanum* (Becc.) Essig & B.E. Young, Principes 29: 134 (1985).

Kentia beccarii F.Muell., Extra-trop. Pl. Indian ed.: 163 (1880).

Hydriastele nannostachys W.J.Baker & Loo, Kew Bull. 59: 66 (2004).
S. Sulawesi. 42 SUL.
   *Gronophyllum microspadix Burret, Notizbl. Bot. Gart. Berlin-Dahlem 12: 44 (1934).

Hydriastele oxypetala (Burret) W.J.Baker & Loo, Kew Bull. 59: 66 (2004).
Maluku (Palau Mangoeli). 42 MOL.
   *Gronophyllum oxypetalum Burret, Notizbl. Bot. Gart. Berlin-Dahlem 13: 474 (1937).

Hydriastele palauensis (Becc.) W.J.Baker & Loo, Kew Bull. 59: 66 (2004).
Caroline Is. (Palau). 62 CRL.
   *Gulubiopsis palauensis Becc., Bot. Jahrb. Syst. 59: 11 (1924). *Gulubia palauensis* (Becc.) H.E.Moore & Fosberg, Gentes Herb. 8: 455 (1956).

Hydriastele pinangoides (Becc.) W.J.Baker & Loo, Kew Bull. 59: 66 (2004).
New Guinea. 43 NWG.
   *Nenga pinangoides Becc., Malesia 1: 28 (1877). *Leptophoenix pinangoides* (Becc.) Becc., Ann. Jard. Bot. Buitenzorg 2: 82 (1885). *Nengella pinangoides* (Becc.) Burret, Notizbl. Bot. Gart. Berlin-Dahlem 13: 315 (1936). *Gronophyllum pinangoides* (Becc.) Essig & B.E.Young, Principes 29: 135 (1985).
   Nenga calophylla K.Schum. & Lauterb., Fl. Schutzgeb. Südsee: 208 (1900). *Nengella calophylla* (K.Schum. & Lauterb.) Becc., Bot. Jahrb. Syst. 52: 27 (1914).
   Leptophoenix minor Becc., Webbia 1: 298 (1905). *Nengella minor* (Becc.) Burret, Notizbl. Bot. Gart. Berlin-Dahlem 13: 315 (1936).
   Gronophyllum densiflorum Ridl., Trans. Linn. Soc. London, Bot. 9: 232 (1916). *Nengella densiflora* (Ridl.) Burret, Notizbl. Bot. Gart. Berlin-Dahlem 13: 316 (1936). *Leptophoenix densiflora* (Ridl.) Burret, Notizbl. Bot. Gart. Berlin-Dahlem 13: 205 (1936).
   Leptophoenix incompta Becc., Bot. Jahrb. Syst. 58: 452 (1923). *Nengella incompta* (Becc.) Burret, Notizbl. Bot. Gart. Berlin-Dahlem 13: 316 (1936).
   Leptophoenix pterophylla Becc., Atti Soc. Tosc. Sci. Nat. Pisa Processi Verbali 44: 131 (1934). *Nengella pterophylla* (Becc.) Burret, Notizbl. Bot. Gart. Berlin-Dahlem 13: 316 (1936).
   Leptophoenix yulensis Becc., Atti Soc. Tosc. Sci. Nat. Pisa Processi Verbali 44: 130 (1934). *Nengella yulensis* (Becc.) Burret, Notizbl. Bot. Gart. Berlin-Dahlem 13: 316 (1936).
   Leptophoenix brassii Burret, Notizbl. Bot. Gart. Berlin-Dahlem 12: 339 (1935). *Nengella brassii* (Burret) Burret, Notizbl. Bot. Gart. Berlin-Dahlem 13: 316 (1936). *Gronophyllum leonardii* Essig & B.E.Young, Principes 29: 134 (1985).
   Leptophoenix macrocarpa Burret, Notizbl. Bot. Gart. Berlin-Dahlem 12: 340 (1935). *Nengella macrocarpa* (Burret) Burret, Notizbl. Bot. Gart. Berlin-Dahlem 13: 316 (1936).
   Leptophoenix microcarpa Burret, Notizbl. Bot. Gart. Berlin-Dahlem 12: 342 (1935). *Nengella microcarpa* (Burret) Burret, Notizbl. Bot. Gart. Berlin-Dahlem 13: 316 (1936).
   Nengella rhomboidea Burret, J. Arnold Arbor. 20: 207 (1939).

Hydriastele pleurocarpa (Burret) W.J.Baker & Loo, Kew Bull. 59: 66 (2004).

Papua New Guinea. 43 NWG.

Nengella calophylla var. *montana* Becc., Bot. Jahrb. Syst. 52: 27 (1914). *Nengella pleurocarpa* Burret, Notizbl. Bot. Gart. Berlin-Dahlem 13: 314 (1936). *Gronophyllum pleurocarpum* (Burret) Essig & B.E.Young, Principes 29: 136 (1985).

Hydriastele procera (Blume) W.J.Baker & Loo, Kew Bull. 59: 66 (2004).
New Guinea. 43 NWG.

Kentia procera Blume, Bull. Sci. Phys. Nat. Néerl. 1: 65 (1838). *Gronophyllum procerum* (Blume) H.E.Moore, Gentes Herb. 9: 265 (1963).
Areca procera Zipp. ex Blume, Rumphia 2: 95 (1843).

Hydriastele ramsayi (Becc.) W.J.Baker & Loo, Kew Bull. 59: 66 (2004).
N. Northern Territory. 50 NTA.

Gulubia ramsayi Becc., Webbia 3: 159 (1910). *Kentia ramsayi* (Becc.) Becc., Webbia 4: 148 (1913). *Gronophyllum ramsayi* (Becc.) H.E.Moore, Gentes Herb. 9: 265 (1963).

Hydriastele rheophytica Dowe & M.D.Ferrero, Palms 44: 195 (2000).
New Guinea. 43 NWG.

Hydriastele rhopalocarpa (Becc.) W.J.Baker & Loo, Kew Bull. 59: 66 (2004).
NE. New Guinea. 43 NWG.

Nengella calophylla var. *rhopalocarpa* Becc., Bot. Jahrb. Syst. 52: 28 (1914). *Nengella rhopalocarpa* (Becc.) Burret, Notizbl. Bot. Gart. Berlin-Dahlem 13: 314 (1936). *Gronophyllum rhopalocarpum* (Becc.) Essig & B.E.Young, Principes 29: 136 (1985).

Hydriastele rostrata Burret, Notizbl. Bot. Gart. Berlin-Dahlem 13: 484 (1937).
New Guinea. 43 NWG.

Hydriastele sarasinorum (Burret) W.J.Baker & Loo, Kew Bull. 59: 66 (2004).
Sulawesi. 42 SUL.

Gronophyllum sarasinorum Burret, Notizbl. Bot. Gart. Berlin-Dahlem 13: 202 (1936).

Hydriastele selebica (Becc.) W.J.Baker & Loo, Kew Bull. 59: 66 (2004).
Sulawesi. 42 SUL.

Nenga selebica Becc., Malesia 1: 30 (1877). *Gronophyllum selebicum* (Becc.) Becc., Ann. Jard. Bot. Buitenzorg 2: 82 (1885).

Hydriastele valida (Essig) W.J.Baker & Loo, Kew Bull. 59: 66 (2004).
Papua New Guinea. 43 NWG.

Gulubia valida Essig, Principes 26: 169 (1982).

Hydriastele variabilis (Becc.) Burret, Notizbl. Bot. Gart. Berlin-Dahlem 13: 483 (1937).
New Guinea. 43 NWG.

Nenga variabilis Becc., Malesia 1: 26 (1877). *Adelonenga variabilis* (Becc.) Becc., Ann. Jard. Bot. Buitenzorg 2: 82 (1885).
Nenga variabilis var. *sphaerocarpa* Becc., Malesia 1: 27 (1877). *Hydriastele variabilis* var. *sphaerocarpa* (Becc.) Burret, Notizbl. Bot. Gart. Berlin-Dahlem 13: 483 (1937).

Hydriastele variabilis var. *sphaerocarpa* (Becc.) Burret = **Hydriastele variabilis** (Becc.) Burret

Hydriastele vitiensis W.J.Baker & Loo, Kew Bull. 59: 66 (2004).
Fiji. 60 FIJ.

Gulubia microcarpa Essig, Principes 26: 173 (1982).

Hydriastele wendlandiana (F.Muell.) H.Wendl. & Drude, Linnaea 39: 209 (1875).
N. Northern Territory to N. Queensland. 50 NTA QLD.

Kentia wendlandiana F.Muell., Fragm. 7: 102 (1870). *Hydriastele wendlandiana* var. *microcarpa* H.Wendl. & Drude, Linnaea 39: 210 (1875). *Hydriastele douglasiana* F.M.Bailey, Queensland Agric. J. 1: 232 (1897).

Hydriastele wendlandiana var. *microcarpa* H.Wendl. & Drude = **Hydriastele wendlandiana** (F.Muell.) H.Wendl. & Drude

Hyophorbe

Hyophorbe Gaertn., Fruct. Sem. Pl. 2: 186 (1791).
5 species, Mascarenes. 29.

Sublimia Comm. ex Mart., Hist. Nat. Palm. 3: 164 (1838), nom. inval.
Mascarena L.H.Bailey, Gentes Herb. 6: 71 (1942).

Hyophorbe amaricaulis Mart., Hist. Nat. Palm. 3: 309 (1849). *Hyospathe amaricaulis* (Mart.) Hook.f., Rep. Progr. Condition Roy. Bot. Gard. Kew 1882: 59 (1884).
Mauritius. 29 MAU. — Now only known from one individual.

Sublimia aevidaps Comm. ex Mart., Hist. Nat. Palm. 3: 176 (1838), nom. inval.
Sublimia amaricaulis Comm. ex Mart., Hist. Nat. Palm. 3: 309 (1849), nom. inval.
Areca speciosa Lem., Ill. Hort. 13: t. 462 (1866).

Hyophorbe commersoniana Mart. = **Hyophorbe indica** Gaertn.

Hyophorbe indica Gaertn., Fruct. Sem. Pl. 2: 186 (1791).
Réunion. 29 REU.

Areca lutescens Bory, Voy. îles Afrique 2: 296 (1804). *Hyophorbe lutescens* (Bory) Jum., Cat. Pl. Madag., Palm.: 9 (1933), pro syn.
Hyophorbe commersoniana Mart., Hist. Nat. Palm. 3: 164 (1838).
Sublimia vilicaulis Comm. ex Mart., Hist. Nat. Palm. 3: 164 (1838), nom. inval.

Hyophorbe lagenicaulis (L.H.Bailey) H.E.Moore, Principes 20: 119 (1976).
Mauritius. 29 MAU.

Mascarena lagenicaulis L.H.Bailey, Gentes Herb. 6: 74 (1942).
Mascarena revaughanii L.H.Bailey, Gentes Herb. 6: 72 (1942).

Hyophorbe lutescens (Bory) Jum. = **Hyophorbe indica** Gaertn.

Hyophorbe vaughanii L.H.Bailey, Gentes Herb. 6: 70 (1942).
Mauritius. 29 MAU.

Hyophorbe verschaffeltii H.Wendl., Ill. Hort. 13(1): t. 462 (1866). *Mascarena verschaffieltii* (H.Wendl.) L.H.Bailey, Gentes Herb. 6: 76 (1942).
Rodrigues. 29 ROD.

Areca verschaffeltii Lem., Ill. Hort. 13: t. 462 (1866).

Hyospathe

Hyospathe Mart., Hist. Nat. Palm. 2: 1 (1823).
2 species, C. & S. Trop. America. 80 82 83 84.

Hyospathe amaricaulis (Mart.) Hook.f. = **Hyophorbe amaricaulis** Mart.

Hyospathe brevipedunculata Dammer = **Hyospathe elegans** Mart. subsp. **elegans**

Hyospathe chiriquensis Linden ex H.Wendl. = **?** [**80 COS**]

Hyospathe concinna H.E.Moore = **Hyospathe elegans** Mart. subsp. **concinna** (H.E.Moore) A.J.Hend.

Hyospathe elata Hook.f. = **?** [**42 SUL**]

Hyospathe elegans Mart., Hist. Nat. Palm. 2: 1 (1823).
C. & S. Trop. America. 80 COS PAN 82 FRG GUY SUR VEN 83 BOL CLM ECU PER 84 BZN.
Hyospathe ulei Dammer, Verh. Bot. Vereins Prov. Brandenburg 48: 127 (1906 publ. 1907).
Hyospathe micropetala Burret, Notizbl. Bot. Gart. Berlin-Dahlem 10: 857 (1929).
Hyospathe tessmannii Burret, Notizbl. Bot. Gart. Berlin-Dahlem 10: 856 (1929).
Hyospathe weberbaueri Dammer ex Burret, Notizbl. Bot. Gart. Berlin-Dahlem 10: 858 (1929).
Hyospathe lehmannii Burret, Notizbl. Bot. Gart. Berlin-Dahlem 11: 859 (1933).
Hyospathe simplex Burret, Notizbl. Bot. Gart. Berlin-Dahlem 11: 858 (1933).

subsp. **concinna** (H.E.Moore) A.J.Hend., Am. J. Bot. 91: 964 (2004).
Panama. 80 PAN.
Hyospathe concinna H.E.Moore, Gentes Herb. 8: 195 (1949).
Chamaedorea falcaria L.H.Bailey, Gentes Herb. 6: 254 (1943).

subsp. **costaricensis** A.J.Hend., Am. J. Bot. 91: 964 (2004).
Costa Rica. 80 COS.

subsp. **elegans**
S. Trop. America. 82 FRG GUY SUR VEN 83 BOL CLM ECU PER 84 BZN.
Hyospathe gracilis H.Wendl. ex Drude in C.F.P.von Martius & auct. suc. (eds.), Fl. Bras. 3(2): 523 (1882).
Hyospathe filiformis H.Wendl. ex Drude in C.F.P.von Martius & auct. suc. (eds.), Fl. Bras. 3(2): 523 (1882).
Hyospathe brevipedunculata Dammer, Verh. Bot. Vereins Prov. Brandenburg 48: 126 (1906 publ. 1907).
Hyospathe schultzeae Burret, Notizbl. Bot. Gart. Berlin-Dahlem 13: 340 (1936).
Hyospathe pallida H.E.Moore, Gentes Herb. 8: 197 (1949).
Hyospathe maculata Steyerm., Fieldiana, Bot. 28(1): 89 (1951).

subsp. **sanblasensis** A.J.Hend., Am. J. Bot. 91: 964 (2004).
Panama. 80 PAN.

subsp. **sodiroi** (Dammer ex Burret) A.J.Hend., Am. J. Bot. 91: 964 (2004).
Ecuador. 83 ECU.
Hyospathe sodiroi Dammer ex Burret, Notizbl. Bot. Gart. Berlin-Dahlem 10: 857 (1929).

subsp. **tacarcunensis** A.J.Hend., Am. J. Bot. 91: 964 (2004).
Panama. 80 PAN.

Hyospathe filiformis H.Wendl. ex Drude = **Hyospathe elegans** Mart. subsp. **elegans**

Hyospathe frontinensis A.J.Hend., Am. J. Bot. 91: 962 (2004).
Colombia. 83 CLM.

Hyospathe gracilis H.Wendl. ex Drude = **Hyospathe elegans** Mart. subsp. **elegans**

Hyospathe lehmannii Burret = **Hyospathe elegans** Mart.

Hyospathe macrorhachis Burret, Notizbl. Bot. Gart. Berlin-Dahlem 15: 34 (1940).
E. Ecuador. 83 ECU.

Hyospathe maculata Steyerm. = **Hyospathe elegans** Mart. subsp. **elegans**

Hyospathe micropetala Burret = **Hyospathe elegans** Mart.

Hyospathe montana Mart. = **Chamaedorea pinnatifrons** (Jacq.) Oerst.

Hyospathe pallida H.E.Moore = **Hyospathe elegans** Mart. subsp. **elegans**

Hyospathe peruviana A.J.Hend., Am. J. Bot. 91: 962 (2004).
Peru. 83 PER.

Hyospathe pittieri Burret, Notizbl. Bot. Gart. Berlin-Dahlem 14: 137 (1938).
Venezuela. 82 VEN.

Hyospathe pubigera Griseb. & H.Wendl. = **Prestoea pubigera** (Griseb. & H.Wendl.) Hook.f. ex B.D.Jacks.

Hyospathe schultzeae Burret = **Hyospathe elegans** Mart. subsp. **elegans**

Hyospathe simplex Burret = **Hyospathe elegans** Mart.

Hyospathe sodiroi Dammer ex Burret = **Hyospathe elegans** Mart. subsp. **sodiroi** (Dammer ex Burret) A.J.Hend.

Hyospathe tessmannii Burret = **Hyospathe elegans** Mart.

Hyospathe ulei Dammer = **Hyospathe elegans** Mart.

Hyospathe weberbaueri Dammer ex Burret = **Hyospathe elegans** Mart.

Hyospathe wendlandiana Dammer ex Burret, Notizbl. Bot. Gart. Berlin-Dahlem 10: 855 (1929).
Colombia. 83 COL.

Hyphaene

Hyphaene Gaertn., Fruct. Sem. Pl. 2: 13 (1790).
8 species, Trop. & S. Africa to Sri Lanka. 20 22 23 24 25 26 27 29 35 40.
Doma Lam., Tabl. Encycl.: t. 900 (1799), orth. var.
Cucifera Delile, Descr. Egypte, Hist. Nat. 1: 53 (1809), nom. illeg.
Douma Poir. in H.L.Duhamel du Monceau, Traité Arbr. Arbust., nouv. ed. 4: 47 (1809), nom. illeg.
Chamaeriphes Dill. ex Kuntze, Revis. Gen. Pl. 2: 728 (1891).

Hyphaene argun Mart. = **Medemia argun** (Mart.) Wurttenb. ex H.Wendl.

Hyphaene aurantiaca Dammer = **Hyphaene petersiana** Klotzsch ex Mart.

Hyphaene baikieana Furtado = **Hyphaene thebaica** (L.) Mart.

Hyphaene baronii Becc. = **Hyphaene coriacea** Gaertn.

Hyphaene beccariana Furtado = **Hyphaene coriacea** Gaertn.

Hyphaene benadirensis Becc. = **Hyphaene compressa** H.Wendl.

Hyphaene benguelensis Welw. ex H.Wendl. = **Hyphaene petersiana** Klotzsch ex Mart.

Hyphaene bussei Dammer = **Hyphaene petersiana** Klotzsch ex Mart.

Hyphaene carinensis Chiov. = **Livistona carinensis** (Chiov.) J.Dransf. & N.W.Uhl

Hyphaene compressa H.Wendl., Bot. Zeitung (Berlin) 36: 116 (1878). *Chamaeriphes compressa* (H.Wendl.) Kuntze, Revis. Gen. Pl. 2: 728 (1891). *Hyphaene multiformis* subsp. *compressa* (H.Wendl.) Becc., Palme Borass.: 35 (1924).
S. Ethiopia to Mozambique. 24 ETH SOM 25 KEN TAN 26 MOZ.
    Hyphaene benadirensis Becc., Agric. Colon. 2: 164 (1908).
    Hyphaene mangoides Becc., Agric. Colon. 2: 166 (1908).
    Hyphaene multiformis Becc., Palme Borass.: 32 (1924).
    Hyphaene multiformis subsp. *ambigua* Becc., Palme Borass.: 35 (1924).
    Hyphaene multiformis subsp. *deformis* Becc., Palme Borass.: 33 (1924).
    Hyphaene multiformis subsp. *diminuta* Becc., Palme Borass.: 34 (1924).
    Hyphaene multiformis subsp. *gibbosa* Becc., Palme Borass.: 35 (1924).
    Hyphaene multiformis subsp. *intermedia* Becc., Palme Borass.: 34 (1924).
    Hyphaene multiformis subsp. *kilvaensis* Becc., Palme Borass.: 36 (1924). *Hyphaene kilvaensis* (Becc.) Furtado, Trab. Centro Bot. Junta Invest. Ultramar 15: 453 (1967).
    Hyphaene multiformis subsp. *macrocarpa* Becc., Palme Borass.: 33 (1924).
    Hyphaene multiformis subsp. *mahengensis* Becc., Palme Borass.: 34 (1924).
    Hyphaene multiformis subsp. *manca* Becc., Palme Borass.: 36 (1924).
    Hyphaene multiformis subsp. *morogorensis* Becc., Palme Borass.: 36 (1924).
    Hyphaene multiformis subsp. *moshiensis* Becc., Palme Borass.: 34 (1924).
    Hyphaene multiformis subsp. *nasuta* Becc., Palme Borass.: 35 (1924).
    Hyphaene multiformis subsp. *obconica* Becc., Palme Borass.: 35 (1924).
    Hyphaene multiformis subsp. *obesa* Becc., Palme Borass.: 35 (1924).
    Hyphaene multiformis subsp. *odorata* Becc., Palme Borass.: 36 (1924).
    Hyphaene multiformis subsp. *panganensis* Becc., Palme Borass.: 36 (1924).
    Hyphaene multiformis subsp. *plagiosperma* Becc., Palme Borass.: 36 (1924).
    Hyphaene multiformis subsp. *rovumensis* Becc., Palme Borass.: 34 (1924).
    Hyphaene multiformis subsp. *semiplaena* Becc., Palme Borass.: 36 (1924). *Hyphaene semiplaena* (Becc.) Furtado, Trab. Centro Bot. Junta Invest. Ultramar 15: 458 (1967).
    Hyphaene multiformis subsp. *stenosperma* Becc., Palme Borass.: 34 (1924).
    Hyphaene multiformis subsp. *subglobosa* Becc., Palme Borass.: 34 (1924).
    Hyphaene multiformis subsp. *tangatensis* Becc., Palme Borass.: 36 (1924).
    Hyphaene multiformis subsp. *trigibba* Becc., Palme Borass.: 35 (1924).
    Hyphaene multiformis subvar. *lindiensis* Becc., Palme Borass.: 33 (1924).
    Hyphaene incoje Furtado, Trab. Centro Bot. Junta Invest. Ultramar 15: 453 (1967).

Hyphaene megacarpa Furtado, Trab. Centro Bot. Junta Invest. Ultramar 15: 454 (1967).

Hyphaene coriacea Gaertn., Fruct. Sem. Pl. 1: 28 (1788). *Chamaeriphes coriacea* (Gaertn.) Kuntze, Revis. Gen. Pl. 2: 728 (1891).
Ethiopia to S. Africa, W. Madagascar. 24 ETH SOM 25 KEN TAN 26 MLW MOZ 27 TVL NAT 29 MDG.
    Corypha africana Lour., Fl. Cochinch.: 213 (1790).
    Hyphaene natalensis Kuntze, Linnaea 20: 15 (1847).
    Hyphaene turbinata H.Wendl., Bot. Zeitung (Berlin) 39: 92 (1881). *Chamaeriphes turbinata* (H.Wendl.) Kuntze, Revis. Gen. Pl. 2: 728 (1891).
    Hyphaene shatan Bojer ex Dammer, Notes Roy. Bot. Gard. Edinburgh 3: 34 (1900). *Chamaeriphes shatan* (Bojer ex Dammer) Kuntze, Deutsche Bot. Monatsschr. 21: 173 (1903).
    Hyphaene wendlandii Dammer, Bot. Jahrb. Syst. 28: 354 (1900).
    Hyphaene baronii Becc., Bot. Jahrb. Syst. 38(87): 7 (1906).
    Hyphaene hildebrandtii Becc., Bot. Jahrb. Syst. 38(87): 7 (1906).
    Hyphaene oblonga Becc., Agric. Colon. 2: 168 (1908).
    Hyphaene parvula Becc., Agric. Colon. 2: 172 (1908).
    Hyphaene pleuropoda Becc., Agric. Colon. 2: 170 (1908).
    Hyphaene pyrifera Becc., Agric. Colon. 2: 167 (1908).
    Hyphaene spaerulifera Becc., Agric. Colon. 2: 169 (1908).
    Hyphaene spaerulifera var. *gosciaensis* Becc., Agric. Colon. 2: 170 (1908).
    Hyphaene pileata Becc., Palme Borass.: 42 (1924).
    Hyphaene pyrifera var. *arenicola* Becc., Palme Borass.: 37 (1924).
    Hyphaene pyrifera var. *margaritensis* Becc., Palme Borass.: 37 (1924).
    Hyphaene beccariana Furtado, Trab. Centro Bot. Junta Invest. Ultramar 15: 444 (1967).
    Hyphaene tetragonoides Furtado, Trab. Centro Bot. Junta Invest. Ultramar 15: 459 (1967).

Hyphaene crinita Gaertn. = **Hyphaene thebaica** (L.) Mart.

Hyphaene cuciphera Pers. = **?** [2]

Hyphaene dahomeensis Becc. = **Hyphaene thebaica** (L.) Mart.

Hyphaene dankaliensis Becc. = **Hyphaene thebaica** (L.) Mart.

Hyphaene depressa Becc. = **Hyphaene guineensis** Schumach. & Thonn.

Hyphaene dichotoma (White) Furtado, Gard. Bull. Singapore 25: 301 (1970).
NW. India, Sri Lanka. 40 IND SRL.
    Borassus dichotomus White in J.Graham, Cat. Pl. Bombay: 226 (1839).
    Hyphaene indica Becc., Agric. Colon. 22: 173 (1908).
    Hyphaene taprobanica Furtado, Gard. Bull. Singapore 25: 302 (1970).

Hyphaene doreyi Furtado = **Hyphaene guineensis** Schumach. & Thonn.

Hyphaene goetzei Dammer = **Hyphaene petersiana** Klotzsch ex Mart.

Hyphaene gossweileri Furtado = **Hyphaene guineensis** Schumach. & Thonn.

Hyphaene guineensis Schumach. & Thonn. in H.C.F.Schumacher, Beskr. Guin. Pl.: 445 (1827). *Chamaeriphes guineensis* (Schumach. & Thonn.) Kuntze, Revis. Gen. Pl. 2: 728 (1891).

W. Trop. Africa to Angola. 22 GNB 23 ZAI 26 ANG.

Hyphaene depressa Becc., Palme Borass.: 48 (1924).

Hyphaene mateba Becc., Palme Borass.: 47 (1924).

Hyphaene nephrocarpa Becc., Palme Borass.: 47 (1924).

Hyphaene luandensis Gossw., Bol. Serv. Agric. Comerc. Coloniz. Florest. 1935: 77 (1935), nom. nud.

Hyphaene doreyi Furtado, Trab. Centro Bot. Junta Invest. Ultramar 15: 451 (1967).

Hyphaene gossweileri Furtado, Trab. Centro Bot. Junta Invest. Ultramar 15: 452 (1967).

Hyphaene welwitschii Furtado, Trab. Centro Bot. Junta Invest. Ultramar 15: 459 (1967).

Hyphaene hildebrandtii Becc. = **Hyphaene coriacea** Gaertn.

Hyphaene incoje Furtado = **Hyphaene compressa** H.Wendl.

Hyphaene indica Becc. = **Hyphaene dichotoma** (White) Furtado

Hyphaene kilvaensis (Becc.) Furtado = **Hyphaene compressa** H.Wendl.

Hyphaene luandensis Gossw. = **Hyphaene guineensis** Schumach. & Thonn.

Hyphaene macrosperma H.Wendl., Bot. Zeitung (Berlin) 39: 92 (1881). *Chamaeriphes macrosperma* (H.Wendl.) Kuntze, Revis. Gen. Pl. 2: 728 (1891).
W. Trop. Africa. 22 BEN.

Hyphaene mangoides Becc. = **Hyphaene compressa** H.Wendl.

Hyphaene mateba Becc. = **Hyphaene guineensis** Schumach. & Thonn.

Hyphaene megacarpa Furtado = **Hyphaene compressa** H.Wendl.

Hyphaene migiurtina Chiov. = **Hyphaene reptans** Becc.

Hyphaene multiformis Becc. = **Hyphaene compressa** H.Wendl.

Hyphaene multiformis subsp. *ambigua* Becc. = **Hyphaene compressa** H.Wendl.

Hyphaene multiformis subsp. *compressa* (H.Wendl.) Becc. = **Hyphaene compressa** H.Wendl.

Hyphaene multiformis subsp. *deformis* Becc. = **Hyphaene compressa** H.Wendl.

Hyphaene multiformis subsp. *diminuta* Becc. = **Hyphaene compressa** H.Wendl.

Hyphaene multiformis subsp. *gibbosa* Becc. = **Hyphaene compressa** H.Wendl.

Hyphaene multiformis subsp. *intermedia* Becc. = **Hyphaene compressa** H.Wendl.

Hyphaene multiformis subsp. *kilvaensis* Becc. = **Hyphaene compressa** H.Wendl.

Hyphaene multiformis subvar. *lindiensis* Becc. = **Hyphaene compressa** H.Wendl.

Hyphaene multiformis subsp. *macrocarpa* Becc. = **Hyphaene compressa** H.Wendl.

Hyphaene multiformis subsp. *mahengensis* Becc. = **Hyphaene compressa** H.Wendl.

Hyphaene multiformis subsp. *manca* Becc. = **Hyphaene compressa** H.Wendl.

Hyphaene multiformis subsp. *morogorensis* Becc. = **Hyphaene compressa** H.Wendl.

Hyphaene multiformis subsp. *moshiensis* Becc. = **Hyphaene compressa** H.Wendl.

Hyphaene multiformis subsp. *nasuta* Becc. = **Hyphaene compressa** H.Wendl.

Hyphaene multiformis subsp. *obconica* Becc. = **Hyphaene compressa** H.Wendl.

Hyphaene multiformis subsp. *obesa* Becc. = **Hyphaene compressa** H.Wendl.

Hyphaene multiformis subsp. *odorata* Becc. = **Hyphaene compressa** H.Wendl.

Hyphaene multiformis subsp. *panganensis* Becc. = **Hyphaene compressa** H.Wendl.

Hyphaene multiformis subsp. *plagiosperma* Dammer ex Becc. = **Hyphaene petersiana** Klotzsch ex Mart.

Hyphaene multiformis subsp. *plagiosperma* Becc. = **Hyphaene compressa** H.Wendl.

Hyphaene multiformis subsp. *rovumensis* Becc. = **Hyphaene compressa** H.Wendl.

Hyphaene multiformis subsp. *semiplaena* Becc. = **Hyphaene compressa** H.Wendl.

Hyphaene multiformis subsp. *stenosperma* Becc. = **Hyphaene compressa** H.Wendl.

Hyphaene multiformis subsp. *subglobosa* Becc. = **Hyphaene compressa** H.Wendl.

Hyphaene multiformis subsp. *tangatensis* Becc. = **Hyphaene compressa** H.Wendl.

Hyphaene multiformis subsp. *trigibba* Becc. = **Hyphaene compressa** H.Wendl.

Hyphaene natalensis Kuntze = **Hyphaene coriacea** Gaertn.

Hyphaene nephrocarpa Becc. = **Hyphaene guineensis** Schumach. & Thonn.

Hyphaene nodularia Becc. = **Hyphaene thebaica** (L.) Mart.

Hyphaene oblonga Becc. = **Hyphaene coriacea** Gaertn.

Hyphaene obovata Furtado = **Hyphaene petersiana** Klotzsch ex Mart.

Hyphaene occidentalis Becc. = **Hyphaene thebaica** (L.) Mart.

Hyphaene ovata Furtado = **Hyphaene petersiana** Klotzsch ex Mart.

Hyphaene parvula Becc. = **Hyphaene coriacea** Gaertn.

Hyphaene petersiana Klotzsch ex Mart., Hist. Nat. Palm. 3(ed. 2): 227 (1845). *Hyphaene ventricosa* subsp. *petersiana* (Klotzsch ex Mart.) Becc., Palme Borass.: 45 (1924).
Tanzania to Namibia. 23 BUR RWA ZAI 25 TAN 26 ANG MOZ ZIM 27 NAM TVL.

Hyphaene ventricosa Kirk, J. Linn. Soc., Bot. 9: 235 (1867). *Chamaeriphes ventricosa* (Kirk) Kuntze, Revis. Gen. Pl. 2: 728 (1891).

Hyphaene benguelensis Welw. ex H.Wendl., Bot. Zeitung (Berlin) 39: 92 (1881). *Chamaeriphes benguelensis* (Welw. ex H.Wendl.) Kuntze, Revis. Gen. Pl. 2: 728 (1891).

Hyphaene goetzei Dammer, Bot. Jahrb. Syst. 28: 354 (1900). *Hyphaene ventricosa* subsp. *goetzei* (Dammer) Becc., Palme Borass.: 46 (1924).

Hyphaene aurantiaca Dammer, Bot. Jahrb. Syst. 30: 267 (1901). *Hyphaene ventricosa* subsp. *aurantiaca* (Dammer) Becc., Palme Borass.: 45 (1924).

Hyphaene bussei Dammer, Vegetationsbilder, V, 7: t. 44 (1907). *Hyphaene ventricosa* subsp. *bussei* (Dammer) Becc., Palme Borass.: 46 (1924).

Hyphaene plagiocarpa Dammer, Vegetationsbilder, V, 7: t. 44 (1907). *Hyphaene ventricosa* subsp. *plagiocarpa* (Dammer) Becc., Palme Borass.: 45 (1924).
·*Hyphaene multiformis* subsp. *plagiosperma* Dammer ex Becc., Palme Borass.: 36 (1924), pro syn.
Hyphaene ventricosa subsp. *anisopleura* Becc., Palme Borass.: 46 (1924).
Hyphaene ventricosa subsp. *benguellensis* (Welw. ex H.Wendl.) Becc., Palme Borass.: 44 (1924).
Hyphaene ventricosa subsp. *russisiensis* Becc., Palme Borass.: 46 (1924).
Hyphaene ventricosa subsp. *useguhensis* Becc., Palme Borass.: 46 (1924).
Hyphaene obovata Furtado, Trab. Centro Bot. Junta Invest. Ultramar 15: 456 (1967).
Hyphaene ovata Furtado, Trab. Centro Bot. Junta Invest. Ultramar 15: 457 (1967).

Hyphaene pileata Becc. = **Hyphaene coriacea** Gaertn.

Hyphaene plagiocarpa Dammer = **Hyphaene petersiana** Klotzsch ex Mart.

Hyphaene pleuropoda Becc. = **Hyphaene coriacea** Gaertn.

Hyphaene pyrifera Becc. = **Hyphaene coriacea** Gaertn.

Hyphaene pyrifera var. *arenicola* Becc. = **Hyphaene coriacea** Gaertn.

Hyphaene pyrifera var. *margaritensis* Becc. = **Hyphaene coriacea** Gaertn.

Hyphaene reptans Becc., Agric. Colon. 2: 151 (1908). Somalia to N. Kenya, Yemen. 24 SOM 25 KEN 35 YEM.
Hyphaene migiurtina Chiov., Fl. Somala 1: 318 (1929).

Hyphaene santoana Furtado = **Hyphaene thebaica** (L.) Mart.

Hyphaene semiplaena (Becc.) Furtado = **Hyphaene compressa** H.Wendl.

Hyphaene shatan Bojer ex Dammer = **Hyphaene coriacea** Gaertn.

Hyphaene sinaitica Furtado = **Hyphaene thebaica** (L.) Mart.

Hyphaene spaerulifera Becc. = **Hyphaene coriacea** Gaertn.

Hyphaene spaerulifera var. *gosciaensis* Becc. = **Hyphaene coriacea** Gaertn.

Hyphaene taprobanica Furtado = **Hyphaene dichotoma** (White) Furtado

Hyphaene tetragonoides Furtado = **Hyphaene coriacea** Gaertn.

Hyphaene thebaica (L.) Mart., Hist. Nat. Palm. 3: 226 (1838).
W. Trop. Africa to Egypt and Arabian Pen. 20 EGY 22 BEN BKN GAM GHA IVO MLI MTN NGA NGR SEN SIE TOG 23 CAF CMN 24 CHA ERI ETH SOM SUD 34 SIN 35 SAU YEM.
Corypha thebaica L., Sp. Pl.: 1187 (1753). *Palma thebaica* (L.) Jacq., Fragm. Bot.: 83 (1809). *Douma thebaica* (L.) Poir. in H.L.Duhamel du Monceau, Traité Arbr. Arbust., nouv. ed. 4: 48 (1809). *Cucifera thebaica* (L.) Delile, Descr. Egypte, Hist. Nat.: 145 (1813). *Chamaeriphes thebaica* (L.) Kuntze, Revis. Gen. Pl. 2: 728 (1891).
Hyphaene crinita Gaertn., Fruct. Sem. Pl. 2: 13 (1790). *Chamaeriphes crinita* (Gaertn.) Kuntze, Revis. Gen. Pl. 2: 728 (1891).

Hyphaene dankaliensis Becc., Bot. Jahrb. Syst. 38(87): 8 (1906).
Hyphaene nodularia Becc., Agric. Colon. 2: 160 (1908).
Hyphaene dahomeensis Becc., Palme Borass.: 48 (1924).
Hyphaene occidentalis Becc., Palme Borass.: 27 (1924).
Hyphaene togoensis Dammer ex Becc., Palme Borass.: 27 (1924).
Hyphaene santoana Furtado, Trab. Centro Bot. Junta Invest. Ultramar 15: 457 (1967).
Hyphaene baikieana Furtado, Gard. Bull. Singapore 25: 326 (1970).
Hyphaene sinaitica Furtado, Gard. Bull. Singapore 25: 306 (1970).
Hyphaene tuleyana Furtado, Gard. Bull. Singapore 25: 333 (1970).

Hyphaene togoensis Dammer ex Becc. = **Hyphaene thebaica** (L.) Mart.

Hyphaene tuleyana Furtado = **Hyphaene thebaica** (L.) Mart.

Hyphaene turbinata H.Wendl. = **Hyphaene coriacea** Gaertn.

Hyphaene ventricosa Kirk = **Hyphaene petersiana** Klotzsch ex Mart.

Hyphaene ventricosa subsp. *anisopleura* Becc. = **Hyphaene petersiana** Klotzsch ex Mart.

Hyphaene ventricosa subsp. *aurantiaca* (Dammer) Becc. = **Hyphaene petersiana** Klotzsch ex Mart.

Hyphaene ventricosa subsp. *benguellensis* (Welw. ex H.Wendl.) Becc. = **Hyphaene petersiana** Klotzsch ex Mart.

Hyphaene ventricosa subsp. *bussei* (Dammer) Becc. = **Hyphaene petersiana** Klotzsch ex Mart.

Hyphaene ventricosa subsp. *goetzei* (Dammer) Becc. = **Hyphaene petersiana** Klotzsch ex Mart.

Hyphaene ventricosa subsp. *petersiana* (Klotzsch ex Mart.) Becc. = **Hyphaene petersiana** Klotzsch ex Mart.

Hyphaene ventricosa subsp. *plagiocarpa* (Dammer) Becc. = **Hyphaene petersiana** Klotzsch ex Mart.

Hyphaene ventricosa subsp. *russisiensis* Becc. = **Hyphaene petersiana** Klotzsch ex Mart.

Hyphaene ventricosa subsp. *useguhensis* Becc. = **Hyphaene petersiana** Klotzsch ex Mart.

Hyphaene violascens Becc. = **?**

Hyphaene welwitschii Furtado = **Hyphaene guineensis** Schumach. & Thonn.

Hyphaene wendlandii Dammer = **Hyphaene coriacea** Gaertn.

Iguanura

Iguanura Blume, Bull. Sci. Phys. Nat. Néerl. 1: 66 (1838).
32 species, Thailand to W. Malesia. 41 42.
Slackia Griff., Not. Pl. Asiat. 3: 162 (1851).

Iguanura ambigua Becc., Malesia 3: 125 (1886).
Borneo (Sarawak). 42 BOR.

Iguanura arakudensis Furtado = **Iguanura polymorpha** Becc.

Iguanura asli C.K.Lim, Gard. Bull. Singapore 48: 28 (1996 publ. 1998).
Pen. Malaysia. 42 MLY.

Iguanura belumensis C.K.Lim, Gard. Bull. Singapore 48: 52 (1996 publ. 1998).
Pen. Malaysia (Perak). 42 MLY.

Iguanura bicornis Becc., Malesia 3: 188 (1886).
Pen. Thailand to Pen. Malaysia. 41 THA 42 MLY.

Iguanura borneensis Scheff., Ann. Jard. Bot. Buitenzorg 1: 161 (1876).
Borneo (Kalimantan). 42 BOR.
    Iguanura borneensis var. *australis* Becc., Atti Soc. Tosc. Sci. Nat. Pisa Processi Verbali 44: 172 (1934).

Iguanura borneensis var. *australis* Becc. = **Iguanura borneensis** Scheff.

Iguanura brevipes Hook.f. = **Iguanura polymorpha** Becc.

Iguanura chaiana Kiew, Kew Bull. 34: 144 (1979).
Borneo (Sarawak). 42 BOR.

Iguanura corniculata Becc., Malesia 3: 187 (1886).
Pen. Malaysia. 42 MLY.

Iguanura curvata Kiew, Kew Bull. 34: 143 (1979).
Borneo (Sarawak). 42 BOR.

Iguanura diffusa Becc., Malesia 3: 123 (1886).
Pen. Malaysia. 42 MLY.

Iguanura divergens Hodel, Palm J. 137: 7 (1997).
Pen. Thailand. 41 THA.

Iguanura elegans Becc., Malesia 33: 123 (1886).
Borneo (Sarawak). 42 BOR.
    Iguanura ridleyana Becc., Atti Soc. Tosc. Sci. Nat. Pisa Processi Verbali 44: 173 (1934).

Iguanura ferruginea Ridl. = **Iguanura polymorpha** Becc.

Iguanura geonomiformis Mart., Hist. Nat. Palm. 3: 229 (1845). *Slackia geonomiformis* (Mart.) Griff., Not. Pl. Asiat. 3: 162 (1851).
Pen. Malaysia. 42 MLY.
    Iguanura malaccensis Becc., Malesia 3: 123 (1886).
    Iguanura geonomiformis var. *malaccensis* (Becc.) Ridl., Mat. Fl. Malay. Penins. 2: 150 (1907).
    Iguanura wallichiana var. *malaccensis* (Becc.) Kiew, Gard. Bull. Singapore 28: 222 (1976).
    Iguanura wallichiana subsp. *malaccensis* (Becc.) Kiew, Gard. Bull. Singapore 28: 222 (1976).
    Iguanura geonomiformis var. *ramosa* Ridl., Mat. Fl. Malay. Penins. 2: 151 (1907).
    Iguanura wallichiana var. *elatior* Kiew, Gard. Bull. Singapore 28: 224 (1976).

Iguanura geonomiformis var. *malaccensis* (Becc.) Ridl. = **Iguanura geonomiformis** Mart.

Iguanura geonomiformis var. *ramosa* Ridl. = **Iguanura geonomiformis** Mart.

Iguanura humilis (Kiew) C.K.Lim, Gard. Bull. Singapore 48: 37 (1996 publ. 1998).
Pen. Malaysia. 42 MLY.
    *Iguanura wallichiana var. *humilis* Kiew, Gard. Bull. Singapore 28: 223 (1976).

Iguanura kelantanensis C.K.Lim, Gard. Bull. Singapore 48: 30 (1996 publ. 1998).
Pen. Malaysia. 42 MLY.

Iguanura leucocarpa Blume, Bull. Sci. Phys. Nat. Néerl. 1: 66 (1838).
Sumatera. 42 SUM.

Iguanura macrostachya Becc., Malesia 3: 101 (1886).
Borneo (Sarawak). 42 BOR.

Iguanura malaccensis Becc. = **Iguanura geonomiformis** Mart.

Iguanura melinauensis Kiew, Gard. Bull. Singapore 28: 208 (1976).
Borneo (Sarawak). 42 BOR.

Iguanura minor Kiew, Gard. Bull. Singapore 28: 209 (1976).
Borneo (Sarawak). 42 BOR.

Iguanura mirabilis C.K.Lim, Gard. Bull. Singapore 48: 57 (1996 publ. 1998).
Pen. Malaysia (Terengganu). 42 MLY.

Iguanura multifida Hodel = **Iguanura wallichiana** var. **wallichiana**

Iguanura myochodoides Kiew, Gard. Bull. Singapore 28: 210 (1976).
Borneo (Sarawak). 42 BOR.

Iguanura palmuncula Becc., Malesia 3: 106 (1886).
Borneo (Sarawak). 42 BOR.

    var. **magna** Kiew, Gard. Bull. Singapore 28: 212 (1976).
    Borneo (Sarawak). 42 BOR.

    var. **palmuncula**
    Borneo (Sarawak). 42 BOR.
        Iguanura palmuncula var. *angustisecta* Becc., Malesia 3: 107 (1886).

Iguanura palmuncula var. *angustisecta* Becc. = **Iguanura palmuncula** var. **palmuncula**

Iguanura parvula Becc. in J.D.Hooker, Fl. Brit. India 6: 417 (1892).
Pen. Malaysia (Perak, Kedah). 42 MLY.

Iguanura perdana C.K.Lim, Gard. Bull. Singapore 48: 55 (1996 publ. 1998).
Pen. Malaysia (Perak). 42 MLY.

Iguanura piahensis C.K.Lim, Gard. Bull. Singapore 48: 31 (1996 publ. 1998).
Pen. Malaysia (Perak). 42 MLY.

Iguanura polymorpha Becc., Malesia 3: 189 (1886).
Pen. Thailand to Pen. Malaysia. 41 THA 42 MLY.
    Iguanura polymorpha var. *canina* Becc., Malesia 3: 189 (1886).
    Iguanura brevipes Hook.f., Fl. Brit. India 6: 416 (1892).
    Iguanura ferruginea Ridl., J. Straits Branch Roy. Asiat. Soc. 41: 40 (1903).
    Iguanura arakudensis Furtado, Repert. Spec. Nov. Regni Veg. 35: 273 (1934).
    Iguanura polymorpha var. *integra* C.K.Lim, Principes 42: 112 (1998).

Iguanura polymorpha var. *canina* Becc. = **Iguanura polymorpha** Becc.

Iguanura polymorpha var. *integra* C.K.Lim = **Iguanura polymorpha** Becc.

Iguanura prolifera Kiew, Gard. Bull. Singapore 28: 214 (1976).
Borneo (Kalimantan). 42 BOR.

Iguanura remotiflora H.Wendl., Bot. Zeitung (Berlin) 17: 63 (1859).
Borneo (Sarawak). 42 BOR.

Iguanura ridleyana Becc. = **Iguanura elegans** Becc.

Iguanura sanderiana Ridl., Gard. Chron. 1904(1): 50 (1904).
Borneo (Sarawak). 42 BOR.

Iguanura speciosa Hodel, Palm J. 134: 29 (1997).
Pen. Thailand. 41 THA.

Iguanura spectabilis Ridl. = **Iguanura wallichiana** var. **major** Becc. ex Hook.f.

Iguanura speranskyana Bois = **Geonoma ? [84]**

Iguanura tenuis Hodel, Palm J. 136: 11 (1997).
Pen. Thailand. 41 THA.

　var. **khaosokensis** C.K.Lim, Principes 42: 112 (1998).
　Pen. Thailand. 41 THA.

　var. **tenuis**
　Pen. Thailand. 41 THA.

Iguanura thalangensis C.K.Lim, Principes 42: 114 (1998).
Thailand. 41 THA.

Iguanura wallichiana (Mart.) Becc., Malesia 3: 100 (1886).
Pen. Thailand to Sumatera. 41 THA 42 MLY SUM.
　*Areca wallichiana Mart., Hist. Nat. Palm. 3: 178 (1838).

　var. **major** Becc. ex Hook.f., Fl. Brit. India 6: 416 (1892).
　Pen. Malaysia. 42 MLY.
　　Geonoma pynaertiana Sander, Gard. Chron. 1898(1): 258 (1898).
　　Iguanura spectabilis Ridl., J. Straits Branch Roy. Asiat. Soc. 41: 40 (1903).

　var. **rosea** C.K.Lim, Gard. Bull. Singapore 48: 23 (1996 publ. 1998).
　Pen. Malaysia. 42 MLY.

　var. **wallichiana**
　Pen. Thailand to Sumatera. 41 THA 42 MLY SUM.
　Slackia insignis Griff., J. Trav. 2: 187 (1847).
　　Iguanura wallichiana var. *minor* Becc. ex Hook.f., Fl. Brit. India 6: 416 (1892).
　　Iguanura multifida Hodel, Palm J. 136: 8 (1997).

Iguanura wallichiana var. *elatior* Kiew = **Iguanura geonomiformis** Mart.

Iguanura wallichiana var. *humilis* Kiew = **Iguanura humilis** (Kiew) C.K.Lim

Iguanura wallichiana subsp. *malaccensis* (Becc.) Kiew = **Iguanura geonomiformis** Mart.

Iguanura wallichiana var. *malaccensis* (Becc.) Kiew = **Iguanura geonomiformis** Mart.

Iguanura wallichiana var. *minor* Becc. ex Hook.f. = **Iguanura wallichiana** var. **wallichiana**

Inodes

Inodes O.F.Cook = **Sabal** Adans.

Inodes blackburniana (Glazebr.) O.F.Cook = **Sabal blackburnianum**

Inodes causiarum O.F.Cook = **Sabal causiarum** (O.F.Cook) Becc.

Inodes exul O.F.Cook = **Sabal mexicana** Mart.

Inodes glauca Dammer = **Sabal causiarum** (O.F.Cook) Becc.

Inodes mexicana (Mart.) Standl. = **Sabal mexicana** Mart.

Inodes palmetto (Walter) O.F.Cook = **Sabal palmetto** (Walter) Lodd. ex Schult. & Schult.f.

Inodes princeps (Becc.) Cif. & Giacom. = **Sabal bermudana** L.H.Bailey

Inodes rosei O.F.Cook = **Sabal rosei** (O.F.Cook) Becc.

Inodes schwarzii O.F.Cook = **Sabal palmetto** (Walter) Lodd. ex Schult. & Schult.f.

Inodes texana O.F.Cook = **Sabal mexicana** Mart.

Inodes uresana (Trel.) O.F.Cook = **Sabal uresana** Trel.

Inodes vestita O.F.Cook = **? [81]**

Inodes yapa (C.Wright ex Becc.) Standl. = **Sabal yapa** C.Wright ex Becc.

Iriartea

Iriartea Ruiz & Pav., Fl. Peruv. Prodr.: 149 (1794).
1 species, C. & S. Trop. America. 80 82 83 84.
　Deckeria H.Karst., Linnaea 28: 258 (1856).

Iriartea affinis H.Karst. ex Linden = **? [83 CLM]**

Iriartea altissima Klotzsch ex Al.Jahn = **Dictyocaryum fuscum** (H.Wendl.) H.Wendl.

Iriartea andicola (Humb. & Bonpl.) Spreng. = **Ceroxylon alpinum** subsp. **alpinum**

Iriartea corneto (H.Karst.) H.Wendl. = **Iriartea deltoidea** Ruiz & Pav.

Iriartea costata Linden = **? [83 CLM]**

Iriartea deltoidea Ruiz & Pav., Syst. Veg. Fl. Peruv. Chil.: 298 (1798).
C. & S. Trop. America. 80 COS NIC PAN 82 VEN 83 BOL CLM ECU PER 84 BZN.

Iriartea durissima Oerst. = **Socratea exorrhiza** (Mart.) H.Wendl.

Iriartea exorrhiza Mart. = **Socratea exorrhiza** (Mart.) H.Wendl.

Iriartea fusca (H.Karst.) Drude = **Dictyocaryum fuscum** (H.Wendl.) H.Wendl.

Iriartea gigantea H.Wendl. = **Iriartea deltoidea** Ruiz & Pav.

Iriartea glaucescens Linden = **? [83 CLM]**

Iriartea klopstockia (Mart.) W.Watson = **Ceroxylon ceriferum** (H.Karst.) Pittier

Iriartea lamarckiana Mart. = **Dictyocaryum lamarckianum** (Mart.) H.Wendl.

Iriartea leprosa Zipp. = **Drymophloeus oliviformis** (Giseke) Mart.

Iriartea megalocarpa Burret = **Iriartea deltoidea** Ruiz & Pav.

Iriartea monogyna Zipp. = **Drymophloeus oliviformis** (Giseke) Mart.

Iriartea nivea W.Watson = **Ceroxylon ceriferum** (H.Karst.) Pittier

Iriartea nobilis (H.Wendl. ex Seem.) N.E.Br. = **Deckeria nobilis**

Iriartea orbignyana Mart. = **Socratea exorrhiza** (Mart.) H.Wendl.

Iriartea phaeocarpa Mart. = **Iriartea deltoidea** Ruiz & Pav.

Iriartea philonotia Barb.Rodr. = **Socratea exorrhiza** (Mart.) H.Wendl.

Iriartea praemorsa (Willd.) Klotzsch = **Wettinia praemorsa** (Willd.) Wess.Boer

Iriartea pruriens Spruce = **Iriartella setigera** (Mart.) H.Wendl.

Iriartea pubescens H.Karst. = **Wettinia praemorsa** (Willd.) Wess.Boer

Iriartea pygmaea Linden = **? [84]**

Iriartea robusta Verschaff. ex H.Wendl. = **Iriartea deltoidea** Ruiz & Pav.

Iriartea setigera Mart. = **Iriartella setigera** (Mart.) H.Wendl.

Iriartea spruceana Barb.Rodr. = **Iriartella setigera** (Mart.) H.Wendl.

Iriartea stenocarpa (Burret) J.F.Macbr. = **Iriartella stenocarpa** Burret

Iriartea ventricosa Mart. = **Iriartea deltoidea** Ruiz & Pav.

Iriartea weberbaueri Burret = **Iriartea deltoidea** Ruiz & Pav.

Iriartea xanthorhiza Klotzsch ex Linden = **?** [**82 VEN**]

Iriartea zamorensis Linden = **?** [**83 ECU**]

Iriartella

Iriartella H.Wendl., Bonplandia 8: 103 (1860).
2 species, S. Trop. America. 82 83 84.

Iriartella ferreyrae H.E.Moore = **Iriartella stenocarpa** Burret

Iriartella setigera (Mart.) H.Wendl., Bonplandia 8: 104 (1860).
SE. Colombia to N. South America. 82 GUY VEN 83 CLM 84 BZN.
*Iriartea setigera Mart., Hist. Nat. Palm. 2: 39 (1824). *Iriartea pruriens* Spruce, J. Linn. Soc., Bot. 11: 136 (1871). *Iriartea spruceana* Barb.Rodr., Enum. Palm. Nov.: 13 (1875). *Cuatrecasea spruceana* (Barb.Rodr.) Dugand, Caldasia 2: 72 (1943). *Cuatrecasea vaupesana* Dugand, Revista Acad. Colomb. Ci. Exact. 3: 392 (1940).

Iriartella stenocarpa Burret, Notizbl. Bot. Gart. Berlin-Dahlem 11: 233 (1931). *Iriartea stenocarpa* (Burret) J.F.Macbr., Field Mus. Nat. Hist., Bot. Ser. 13(1): 357 (1960).
SE. Colombia to NE. Peru, Brazil (Acre). 83 CLM PER 84 BZN.
Iriartella ferreyrae H.E.Moore, Gentes Herb. 9: 278 (1963).

Itaya

Itaya H.E.Moore, Principes 16: 85 (1972).
1 species, W. South America to N. Brazil. 83 84.

Itaya amicorum H.E.Moore, Principes 16: 87 (1972).
SE. Colombia to NE. Peru. 83 CLM ECU PER 84 BZN.

Jessenia

Jessenia H.Karst. = **Oenocarpus** Mart.

Jessenia amazonum Drude = **Archontophoenix cunninghamiana** (H.Wendl.) H.Wendl. & Drude

Jessenia bataua (Mart.) Burret = **Oenocarpus bataua** var. **bataua**

Jessenia bataua subsp. *oligocarpa* (Griseb. & H.Wendl.) Balick = **Oenocarpus bataua** var. **oligocarpus** (Griseb. & H.Wendl.) A.J.Hend.

Jessenia glazioviana Dammer = **Archontophoenix alexandrae** (F.Muell.) H.Wendl. & Drude

Jessenia oligocarpa Griseb. & H.Wendl. = **Oenocarpus bataua** var. **oligocarpus** (Griseb. & H.Wendl.) A.J.Hend.

Jessenia polycarpa H.Karst. = **Oenocarpus bataua** var. **bataua**

Jessenia repanda Engl. = **Oenocarpus bataua** var. **bataua**

Jessenia weberbaueri Burret = **Oenocarpus bataua** var. **bataua**

Johannesteijsmannia

Johannesteijsmannia H.E.Moore, Principes 5: 116 (1961).
4 species, Pen. Thailand to W. Malesia. 41 42.
*Teysmannia Rchb.f. & Zoll., Linnaea 28: 657 (1858).

Johannesteijsmannia altifrons (Rchb.f. & Zoll.) H.E.Moore, Principes 5: 116 (1961).
Pen. Thailand to W. Malesia. 41 THA 42 BOR MLY SUM.
*Teysmannia altifrons Rchb.f. & Zoll., Linnaea 28: 657 (1858).

Johannesteijsmannia lanceolata J.Dransf., Gard. Bull. Singapore 26: 78 (1972).
Pen. Malaysia (Selangor). 42 MLY.

Johannesteijsmannia magnifica J.Dransf., Gard. Bull. Singapore 26: 75 (1972).
Pen. Malaysia (Selangor). 42 MLY.

Johannesteijsmannia perakensis J.Dransf., Gard. Bull. Singapore 26: 72 (1972).
Pen. Malaysia (Perak). 42 MLY.

Juania

Juania Drude, Nachr. Königl. Ges. Wiss. Georg-Augusts-Univ. 1878: 40 (1878).
1 species, Juan Fernández Is. 85.

Juania australis (Mart.) Drude ex Hook.f., Rep. Kew Gard., App.: 57 (1884).
Juan Fernández Is. (I. Robinson Crusoe). 85 JNF.
*Ceroxylon australe Mart., Hist. Nat. Palm. 3: 314 (1849). *Morenia chonta* Phil., Bot. Zeitung (Berlin) 14: 648 (1856). *Nunnezharia chonta* (Phil.) Kuntze, Revis. Gen. Pl. 2: 731 (1891).

Jubaea

Jubaea Kunth in F.W.H.von Humboldt, A.J.A.Bonpland & C.S.Kunth, Nov. Gen. Sp. 1: 308 (1816).
1 species, C. Chile. 85.
Molinaea Bertero, Merc. Chil. 13: 606 (1829), nom. illeg. *Micrococos* Phil., Bot. Zeitung (Berlin) 17: 362 (1859).

Jubaea chilensis (Molina) Baill., Hist. Pl. 13: 397 (1895).
C. Chile. 85 CLC.
*Palma chilensis Molina, Geogr. Nat. Hist. Chile: 124 (1808). *Cocos chilensis* (Molina) Molina, Sag. Stor. Nat. Chili, English(2) 1: 146, 292 (1809). *Molinaea micrococos* Bertero, Merc. Chil. 13: 606 (1829), nom. illeg. *Micrococos chilensis* (Molina) Phil., Bot. Zeitung (Berlin) 17: 362 (1859). *Jubaea spectabilis* Kunth in F.W.H.von Humboldt, A.J.A.Bonpland & C.S.Kunth, Nov. Gen. Sp. 1: 308 (1816).

Jubaea spectabilis Kunth = **Jubaea chilensis** (Molina) Baill.

Jubaea torallyi (Mart.) H.Wendl. = **Parajubaea torallyi** (Mart.) Burret

Jubaeopsis

Jubaeopsis Becc., Webbia 4: 171 (1913).
1 species, S. Africa. 27.

Jubaeopsis caffra Becc., Webbia 4: 173 (1913).
Cape Prov. to KwaZulu-Natal. 27 CPP NAT.

× Jubautia

× *Jubautia* Demoly = **Butia** × **Jubaea**

Kajewskia

Kajewskia Guillaumin = **Veitchia** H.Wendl.

Kajewskia aneityensis Guillaumin = **Veitchia spiralis** H.Wendl.

Kalbreyera

Kalbreyera Burret = **Geonoma** Willd.

Kalbreyera triandra Burret = **Geonoma triandra** (Burret) Wess.Boer

Kentia

Kentia Blume = **Hydriastele** H.Wendl. & Drude

Kentia acuminata H.Wendl. & Drude = **Carpentaria acuminata** (H.Wendl. & Drude) Becc.

Kentia australis auct. = **Howea fosteriana**

Kentia baueri (Hook.f. ex Lem.) Seem. = **Rhopalostylis baueri** (Hook.f. ex Lem.) H.Wendl. & Drude

Kentia beccarii F.Muell. = **Hydriastele montana** (Becc.) W.J.Baker & Loo

Kentia belmoreana C.Moore & F.Muell. = **Howea belmoreana** (C.Moore & F.Muell.) Becc.

Kentia canterburyana C.Moore & F.Muell. = **Hedyscepe canterburyana** (C.Moore & F.Muell.) H.Wendl. & Drude

Kentia chaunostachys Burret = **Hydriastele chaunostachys** (Burret) W.J.Baker & Loo

Kentia concinna T.Moore = **Microkentia gracilis**

Kentia costata Becc. = **Hydriastele costata** F.M.Bailey

Kentia deplanchei Brongn. & Gris = **Basselinia deplanchei** (Brongn. & Gris) Vieill.

Kentia divaricata Planch. ex Brongn. = **Actinokentia divaricata** (Brongn.) Dammer

Kentia elegans Brongn. & Gris = **Cyphophoenix elegans** (Brongn. & Gris) H.Wendl. ex Salomon

Kentia exorrhiza H.Wendl. = **Clinostigma exorrhizum** (H.Wendl.) Becc.

Kentia forsteriana C.Moore & F.Muell. = **Howea forsteriana** (C.Moore & F.Muell.) Becc.

Kentia fulcita Brongn. = **Campecarpus fulcitus** (Brongn.) H.Wendl. ex Becc.

Kentia gibbsiana Becc. = **Hydriastele gibbsiana** (Becc.) W.J.Baker & Loo

Kentia gracilis Brongn. & Gris = **Basselinia gracilis** (Brongn. & Gris) Vieill.

Kentia humboldtiana (Brongn.) Brongn. ex H.Wendl. = **Basselinia humboldtiana** (Brongn.) H.E.Moore

Kentia joannis (H.Wendl.) F.Muell. = **Veitchia joannis** H.Wendl.

Kentia kersteniana Sander = **Balaka seemannii** (H.Wendl.) Becc.

Kentia ledermanniana Becc. = **Hydriastele ledermanniana** (Becc.) W.J.Baker & Loo

Kentia lindenii Linden ex André = **Chambeyronia macrocarpa** (Brongn.) Vieill. ex Becc.

Kentia lucianii Linden ex Rodigas = **Chambeyronia macrocarpa** (Brongn.) Vieill. ex Becc.

Kentia macarthurii H.Wendl. ex H.J.Veitch = **Ptychosperma macarthurii** (H.Wendl. ex H.J.Veitch) H.Wendl. ex Hook.f.

Kentia macrocarpa Vieill. ex Brongn. = **Chambeyronia macrocarpa** (Brongn.) Vieill. ex Becc.

Kentia macrostachya (Brongn.) Pancher ex Brongn. = **Cyphokentia macrostachya** Brongn.

Kentia mayrii Burret = **Hydriastele mayrii** (Burret) W.J.Baker & Loo

Kentia microcarpa Warb. ex K.Schum. & Lauterb. = **Hydriastele costata** F.M.Bailey

Kentia microspadix Warb. ex K.Schum. & Lauterb. = **Hydriastele microspadix** (Warb. ex K.Schum. & Lauterb.) Burret

Kentia minor F.Muell. = **Linospadix minor** (F.Muell.) F.Muell.

Kentia moluccana Becc. = **Hydriastele moluccana** (Becc.) W.J.Baker & Loo

Kentia monostachya (Mart.) F.Muell. = **Linospadix monostachya** (Mart.) H.Wendl. & Drude

Kentia mooreana F.Muell. = **Lepidorrhachis mooreana** (F.Muell.) O.F.Cook

Kentia oleracea (Jacq.) Seem. ex H.Wendl. = **Roystonea oleracea** (Jacq.) O.F.Cook

Kentia oliviformis Brongn. & Gris = **Kentiopsis oliviformis** (Brongn. & Gris) Brongn.

Kentia pancheri Brongn. & Gris = **Basselinia pancheri** (Brongn. & Gris) Vieill.

Kentia paradoxa (Griff.) Mart. = **Pinanga paradoxa** (Griff.) Scheff.

Kentia polystemon Pancher ex H.Wendl. = **Actinokentia divaricata** (Brongn.) Dammer

Kentia procera Blume = **Hydriastele procera** (Blume) W.J.Baker & Loo

Kentia ramsayi (Becc.) Becc. = **Hydriastele ramsayi** (Becc.) W.J.Baker & Loo

Kentia robusta (Brongn.) Linden ex H.Wendl. = **Cyphokentia macrostachya** Brongn.

Kentia rubra A.Usteri = **?**

Kentia rubricaulis Linden ex Salomon = **Chambeyronia macrocarpa** (Brongn.) Vieill. ex Becc.

Kentia rupicola Linden = **?**

Kentia sanderiana (Ridl.) Sander = **Ptychosperma sanderianum** Ridl.

Kentia sapida (Sol. ex G.Forst.) Mart. = **Rhopalostylis sapida** (Sol. ex G.Forst.) H.Wendl. & Drude

Kentia storckii (H.Wendl.) F.Muell. ex H.Wendl. = **Neoveitchia storckii** (H.Wendl.) Becc.

Kentia subglobosa (H.Wendl.) F.Muell. = **Veitchia subglobosa**

Kentia vieillardii Brongn. & Gris = **Burretiokentia vieillardii** (Brongn. & Gris) Pic.Serm.

Kentia wendlandiana F.Muell. = **Hydriastele wendlandiana** (F.Muell.) H.Wendl. & Drude

Kentiopsis

Kentiopsis Brongn., Compt. Rend. Hebd. Séances Acad. Sci. 77: 398 (1873).
4 species, New Caledonia. 60.
> *Mackeea* H.E.Moore, Gentes Herb. 11: 304 (1978).

Kentiopsis divaricata Brongn. = **Actinokentia divaricata** (Brongn.) Dammer

Kentiopsis lucianii (Linden ex Rodigas) Rodigas = **Chambeyronia macrocarpa** (Brongn.) Vieill. ex Becc.

Kentiopsis macrocarpa Brongn. = **Chambeyronia macrocarpa** (Brongn.) Vieill. ex Becc.

Kentiopsis magnifica (H.E.Moore) Pintaud & Hodel, Principes 42: 42 (1998).
NE. New Caledonia. 60 NWC.
> **Mackeea magnifica* H.E.Moore, Gentes Herb. 11: 304 (1978).

Kentiopsis oliviformis (Brongn. & Gris) Brongn., Compt. Rend. Hebd. Séances Acad. Sci. 77: 398 (1873).
C. New Caledonia. 60 NWC.
> **Kentia oliviformis* Brongn. & Gris, Compt. Rend. Hebd. Séances Acad. Sci. 77: 399 (1873).

Kentiopsis piersoniorum Pintaud & Hodel, Principes 42: 45 (1998).
NE. New Caledonia (E. Mt. Panié). 60 NWC.

Kentiopsis pyriformis Pintaud & Hodel, Principes 42: 49 (1998).
SE. New Caledonia. 60 NWC.

Keppleria

Keppleria Meisn. = **Oncosperma** Blume

Keppleria Mart. ex Endl. = **Bentinckia** Berry ex Roxb.

Keppleria tigillaria (Jack) Meisn. = **Oncosperma tigillarium** (Jack) Ridl.

Kerriodoxa

Kerriodoxa J.Dransf., Principes 27: 4 (1983).
1 species, Pen. Thailand. 41.

Kerriodoxa elegans J.Dransf., Principes 27: 4 (1983).
Pen. Thailand. 41 THA.

Kinetostigma

Kinetostigma Dammer = **Chamaedorea** Willd.

Kinetostigma adscendens Dammer = **Chamaedorea adscendens** (Dammer) Burret

Kinetostigma nanum (N.E.Br.) Burret = **Chamaedorea pumila** H.Wendl. ex Dammer

Kinetostigma tuerckbeimii (Dammer) Burret = **Chamaedorea tuerckheimii** (Dammer) Burret

Klopstockia

Klopstockia H.Karst. = **Ceroxylon** Bonpl. ex DC.

Klopstockia cerifera H.Karst. = **Ceroxylon ceriferum** (H.Karst.) Pittier

Klopstockia coarctata Engel = **Ceroxylon vogelianum** (Engel) H.Wendl.

Klopstockia interrupta H.Karst. = **? [83 CLM]**

Klopstockia parvifrons Engel = **Ceroxylon parvifrons** (Engel) H.Wendl.

Klopstockia quindiuensis H.Karst. = **Ceroxylon quindiuense** (H.Karst.) H.Wendl.

Klopstockia utilis H.Karst. = **? [83 CLM]**

Klopstockia vogeliana Engel = **Ceroxylon vogelianum** (Engel) H.Wendl.

Korthalsia

Korthalsia Blume, Rumphia 2: 166 (1843).
27 species, Indo-China to New Guinea. 41 42 43.
> *Calamosagus* Griff., Calcutta J. Nat. Hist. 5: 22 (1845).

Korthalsia andamanensis Becc. = **Korthalsia laciniosa** (Griff.) Mart.

Korthalsia angustifolia Blume, Rumphia 2: 172 (1843).
Borneo. 42 BOR.

Korthalsia angustifolia var. *gracilis* Miq. = **Korthalsia echinometra** Becc.

Korthalsia bejaudii Gagnep. ex Humbert, Notul. Syst. (Paris) 6: 152 (1937).
Cambodia. 41 CBD.

Korthalsia brassii Burret, J. Arnold Arbor. 20: 191 (1939).
New Guinea. 43 NWG.

Korthalsia celebica Becc., Ann. Roy. Bot. Gard. (Calcutta) 12(2): 130 (1918).
Sulawesi to Maluku (Sula Is.). 42 MOL SUL.

Korthalsia cheb Becc., Malesia 2: 67 (1884).
Borneo. 42 BOR.

Korthalsia concolor Burret, Notizbl. Bot. Gart. Berlin-Dahlem 15: 736 (1942).
N. Borneo. 42 BOR.

Korthalsia debilis Blume, Rumphia 2: 169 (1843).
Sumatera to NW. Borneo. 42 BOR SUM.

Korthalsia echinometra Becc., Malesia 2: 66 (1884).
W. Malesia. 42 BOR MLY SUM.
> *Daemonorops ochreata* Teijsm. & Binn., Cat. Hort. Bot. Bogor.: 74 (1866), nom. nud.
> *Calamus ochreatus* Miq., Verh. Kon. Ned. Akad. Wetensch., Afd. Natuurk. 11: 29 (1868), nom. nud.
> *Korthalsia angustifolia* var. *gracilis* Miq., Palm. Archip. Ind.: 16 (1868).
> *Korthalsia horrida* Becc., Malesia 2: 66 (1884).

Korthalsia ferox Becc., Malesia 2: 73 (1884).
Borneo. 42 BOR.

Korthalsia ferox var. *malayana* Becc. = **Korthalsia rigida** Blume

Korthalsia flagellaris Miq., Fl. Ned. Ind., Eerste Bijv.: 591 (1861).
Pen. Thailand to W. Malesia. 41 THA 42 BOR MLY SUM.
> *Korthalsia rubiginosa* Becc., Malesia 2: 72 (1884).

Korthalsia furcata Becc., Ann. Roy. Bot. Gard. (Calcutta) 12(2): 120 (1918).
W. Borneo. 42 BOR.

Korthalsia furtadoana J.Dransf., Kew Bull. 36: 185 (1981).
Borneo (Sabah). 42 BOR.

Korthalsia grandis Ridl. = **Korthalsia laciniosa** (Griff.) Mart.

Korthalsia hallieriana Becc. = **Korthalsia rigida** Blume

Korthalsia hispida Becc., Malesia 2: 71 (1884).
W. Malesia. 42 BOR MLY SUM.

Korthalsia horrida Becc. = **Korthalsia echinometra** Becc.

Korthalsia jala J.Dransf., Kew Bull. 36: 183 (1981).
Borneo (Sabah, Sarawak). 42 BOR.

Korthalsia junghuhnii Miq., Pl. Jungh.: 162 (1852).
W. Jawa. 42 JAW.

Korthalsia laciniosa (Griff.) Mart., Hist. Nat. Palm. 3: 211 (1845).
Indo-China to Philippines. 41 AND CBD LAO MYA NCB THA VIE 42 JAW MLY PHI SUM.
> *Calamosagus laciniosus* Griff., Calcutta J. Nat. Hist. 5: 23 (1845).
> *Calamosagus wallichiifolius* Griff., Calcutta J. Nat. Hist. 5: 24 (1845). *Korthalsia wallichiifolia* (Griff.) H.Wendl. in O.C.E.de Kerchove de Denterghem, Palmiers: 248 (1878).
> *Calamosagus harinifolius* Griff., Palms Brit. E. Ind.: 29 (1850).
> *Korthalsia teysmannii* Miq., Fl. Ned. Ind., Eerste Bijv.: 591 (1861).
> *Korthalsia scaphigera* Kurz, Forest Fl. Burma 2: 513 (1877), nom. illeg.
> *Korthalsia andamanensis* Becc., Malesia 2: 76 (1884).
> *Korthalsia grandis* Ridl., Mat. Fl. Malay. Penins. 2: 217 (1907).

Korthalsia lanceolata J.Dransf., Malaysian Forester 41: 325 (1978).
Pen. Malaysia (Perak). 42 MLY.

Korthalsia lobbiana H.Wendl. = **Korthalsia rostrata** Blume

Korthalsia machadonis Ridl. = **Korthalsia rostrata** Blume

Korthalsia macrocarpa Becc. = **Korthalsia robusta** Blume

Korthalsia merrillii Becc., Ann. Roy. Bot. Gard. (Calcutta) 12(2): 128 (1918).
Philippines (Palawan). 42 PHI.

Korthalsia paludosa Furtado = **Korthalsia rigida** Blume

Korthalsia paucijuga Becc., Ann. Roy. Bot. Gard. (Calcutta) 12(2): 121 (1918).
Sumatera to Borneo. 42 BOR SUM.

Korthalsia polystachya Mart. = **Korthalsia rigida** Blume

Korthalsia rigida Blume, Rumphia 2: 167 (1843).
Pen. Thailand to Philippines (Palawan). 41 THA 42 BOR MLY PHI SUM.
> *Korthalsia polystachya* Mart., Hist. Nat. Palm. 3: 210 (1845). *Calamosagus polystachys* (Mart.) H.Wendl. in O.C.E.de Kerchove de Denterghem, Palmiers: 235 (1878).
> *Calamosagus ochriger* Griff., Palms Brit. E. Ind.: 31 (1850).
> *Korthalsia ferox* var. *malayana* Becc. in J.D.Hooker, Fl. Brit. India 6: 476 (1893).
> *Korthalsia hallieriana* Becc., Ann. Roy. Bot. Gard. (Calcutta) 12(2): 142 (1918).
> *Korthalsia paludosa* Furtado, Gard. Bull. Singapore 13: 313 (1951).

Korthalsia robusta Blume, Rumphia 2: 170 (1843).
Sumatera to Philippines. 42 BOR PHI SUM.
> *Korthalsia squarrosa* Becc., Philipp. J. Sci., C 4: 620 (1909).
> *Korthalsia macrocarpa* Becc., Ann. Roy. Bot. Gard. (Calcutta) 12(2): 149 (1918).

Korthalsia rogersii Becc., Ann. Roy. Bot. Gard. (Calcutta) 12(2): 131 (1918).
Andaman Is. 41 AND.

Korthalsia rostrata Blume, Rumphia 2: 168 (1843).
Ceratolobus rostratus (Blume) Becc., Ann. Roy. Bot. Gard. (Calcutta) 12(2): 11 (1918).
Pen. Thailand to W. Malesia. 41 THA 42 BOR MLY SUM.
> *Korthalsia scaphigera* Mart., Hist. Nat. Palm. 3(ed. 2): 211 (1845). *Calamosagus scaphiger* (Mart.) Griff., Palms Brit. E. Ind.: 30 (1850).
> *Korthalsia lobbiana* H.Wendl., Bot. Zeitung (Berlin) 17: 174 (1859).
> *Korthalsia machadonis* Ridl., Mat. Fl. Malay. Penins. 2: 216 (1907).

Korthalsia rubiginosa Becc. = **Korthalsia flagellaris** Miq.

Korthalsia scaphigera Mart. = **Korthalsia rostrata** Blume

Korthalsia scaphigera Kurz = **Korthalsia laciniosa** (Griff.) Mart.

Korthalsia scaphigeroides Becc., Philipp. J. Sci., C 4: 619 (1909).
Philippines (Basilan, Mindanao). 42 PHI.

Korthalsia scortechinii Becc. in J.D.Hooker, Fl. Brit. India 6: 475 (1893).
Pen. Thailand to Pen. Malaysia. 41 THA 42 MLY.

Korthalsia squarrosa Becc. = **Korthalsia robusta** Blume

Korthalsia tenuissima Becc., Malesia 2: 275 (1886).
Pen. Malaysia. 42 MLY.

Korthalsia teysmannii Miq. = **Korthalsia laciniosa** (Griff.) Mart.

Korthalsia wallichiifolia (Griff.) H.Wendl. = **Korthalsia laciniosa** (Griff.) Mart.

Korthalsia zippelii Blume, Rumphia 2: 171 (1843).
New Guinea. 43 NWG.
> *Ceratolobus plicatus* Zipp. ex Blume, Rumphia 2: t. 171 (1843), nom. inval.
> *Ceratolobus zippelii* Blume, Rumphia 2: t. 171 (1843), nom. inval.
> *Korthalsia zippelii* var. *aruensis* Becc., Ann. Roy. Bot. Gard. (Calcutta) 12(2): 147 (1918).

Korthalsia zippelii var. *aruensis* Becc. = **Korthalsia zippelii** Blume

Kunthia

Kunthia Humb. & Bonpl. = **Chamaedorea** Willd.

Kunthia deppii Zucc. = **Chamaedorea elegans** Mart.

Kunthia montana Humb. & Bonpl. = **Chamaedorea linearis** (Ruiz & Pav.) Mart.

Kunthia xalapensis Otto & A.Dietr. ex H.Wendl. = **Chamaedorea schiedeana** Mart.

Laccospadix

Laccospadix H.Wendl. & Drude, Linnaea 39: 205 (1875).
1 species, NE. Australia. 50.

Laccospadix australasicus H.Wendl. & Drude, Linnaea
39: 206 (1875). *Calyptrocalyx australasicus*
(H.Wendl. & Drude) Scheff. ex B.D.Jacks., Index
Kew. 1: 397 (1893).
NE. Queensland. 50 QLD.
Ptychosperma laccospadix Benth., Fl. Austral. 7: 140
(1878).

Laccospadix lauterbachianus (Warb. ex Becc.) Burret =
Calyptrocalyx lauterbachianus Warb. ex Becc.

Laccosperma

Laccosperma Drude, Bot. Zeitung (Berlin) 35: 632
(1877).
5 species, W. Trop. Africa to Angola. 22 23 26.
Ancistrophyllum (G.Mann & H.Wendl.) H.Wendl. in
O.C.E.de Kerchove de Denterghem, Palmiers: 230
(1878), nom. illeg.
Neoancistrophyllum Rauschert, Taxon 31: 557
(1982).

Laccosperma acutiflorum (Becc.) J.Dransf., Kew Bull.
37: 456 (1982).
W. & WC. Trop. Africa. 22 GHA NGA SIE 23 CMN
EQG ZAI.
Ancistrophyllum acutiflorum Becc., Webbia 3: 255
(1910). *Neoancistrophyllum acutiflorum* (Becc.)
Rauschert, Taxon 31: 557 (1982).

Laccosperma korupense Sunderl., Kew Bull. 58: 989
(2004).
Cameroon. 23 CMN.

Laccosperma laeve (G.Mann & H.Wendl.) Kuntze,
Revis. Gen. Pl. 2: 729 (1891).
W. & WC. Trop. Africa to Angola. 22 GHA IVO LBR
23 CMN CON GAB ZAI 26 ANG.
Calamus laevis G.Mann & H.Wendl., Trans. Linn.
Soc. London 24: 430 (1864). *Ancistrophyllum laeve*
(G.Mann & H.Wendl.) Drude, Bot. Jahrb. Syst. 21:
111 (1895). *Neoancistrophyllum laeve* (G.Mann &
H.Wendl.) Rauschert ex J.Dransf., Kew Bull. 37:
456 (1982).

Laccosperma laurentii (De Wild.) J.Dransf. =
Laccosperma secundiflorum (P.Beauv.) Kuntze

Laccosperma majus (Burret) J.Dransf. = **Laccosperma
secundiflorum** (P.Beauv.) Kuntze

Laccosperma opacum Drude, Bot. Zeitung (Berlin) 35:
635 (1877). *Palmijuncus opacus* (Drude) Kuntze,
Revis. Gen. Pl. 2: 733 (1891). *Ancistrophyllum
opacum* (Drude) Drude, Bot. Jahrb. Syst. 21: 111
(1895). *Neoancistrophyllum opacum* (Drude)
Rauschert ex J.Dransf., Kew Bull. 37: 456 (1982).
W. & WC. Trop. Africa. 22 GHA GUI LBR NGA 23
CAF CMN EQG GAB GGI ZAI.
Calamus opacus G.Mann & H.Wendl., Trans. Linn.
Soc. London 24: 431 (1864), nom. illeg.

Laccosperma robustum (Burret) J.Dransf., Kew Bull.
37: 457 (1982).
Nigeria to Angola. 22 NGA 23 CAF CMN EQG GAB
ZAI 26 ANG.
Ancistrophyllum robustum Burret, Notizbl. Bot. Gart.
Berlin-Dahlem 15: 746 (1942). *Neoancistrophyllum
robustum* (Burret) Rauschert, Taxon 31: 557
(1982).

Laccosperma secundiflorum (P.Beauv.) Kuntze, Revis.
Gen. Pl. 2: 729 (1891).
W. & WC. Trop. Africa. 22 BEN GHA GNB GUI IVO
LBR NGA NGR SEN SIE TOG 23 CAB CAF
CMN GAB ZAI.

Calamus secundiflorus P.Beauv., Fl. Oware 1: 15
(1805). *Ancistrophyllum secundiflorum* (P.Beauv.)
G.Mann & H.Wendl. in O.C.E.de Kerchove de
Denterghem, Palmiers: 230 (1878).
Neoancistrophyllum secundiflorum (P.Beauv.)
Rauschert, Taxon 31: 557 (1982).
Ancistrophyllum laurentii De Wild., Bull. Jard. Bot.
État 5: 148 (1916). *Neoancistrophyllum laurentii*
(De Wild.) Rauschert, Taxon 31: 557 (1982).
Laccosperma laurentii (De Wild.) J.Dransf., Kew
Bull. 37: 456 (1982).
Ancistrophyllum majus Burret, Notizbl. Bot. Gart.
Berlin-Dahlem 15: 747 (1942). *Neoancistrophyllum
majus* (Burret) Rauschert, Taxon 31: 557 (1982).
Laccosperma majus (Burret) J.Dransf., Kew Bull.
37: 456 (1982).

Langsdorffia

Langsdorffia Raddi = **Syagrus** Mart.

Langsdorffia pseudococos Raddi = **Syagrus pseudococos**
(Raddi) Glassman

Latania

Latania Comm. ex Juss., Gen. Pl.: 39 (1789).
3 species, Mascarenes. 29.
Cleophora Gaertn., Fruct. Sem. Pl. 2: 185 (1791).

Latania aurea Duncan = **Latania verschaffeltii** Lem.

Latania borbonica Lam. = **Latania lontaroides**
(Gaertn.) H.E.Moore

Latania chinensis Jacq. = **Livistona chinensis** (Jacq.)
R.Br. ex Mart.

Latania commersonii J.F.Gmel. = **Latania lontaroides**
(Gaertn.) H.E.Moore

Latania glaucophylla Devansaye = **Latania loddigesii**
Mart.

Latania loddigesii Mart., Hist. Nat. Palm. 3: 226 (1838).
Cleophora loddigesii (Mart.) O.F.Cook, Natl. Hort.
Mag. 20: 52 (1941).
Mauritius. 29 MAU.
Chamaerops excelsior Bojer, Hortus Maurit.: 307
(1837), nom. nud.
Latania glaucophylla Devansaye, Rev. Hort. 47: 34
(1875), nom. nud.
Cleophora dendriformis Lodd. ex Baker, Fl. Mauritius:
381 (1877), nom. nud.

Latania lontaroides (Gaertn.) H.E.Moore, Principes 7:
85 (1963).
Réunion. 29 REU.
Cleophora lontaroides Gaertn., Fruct. Sem. Pl. 2: 185
(1791).
Latania borbonica Lam., Encycl. 3: 427 (1792).
Latania commersonii J.F.Gmel., Syst. Nat.: 1035
(1792). *Cleophora commersonii* (J.F.Gmel.)
O.F.Cook, Natl. Hort. Mag. 20: 52 (1941).
Latania rubra Jacq., Fragm. Bot.: 13 (1800).
Latania plagicoma Comm. ex Balf.f. in J.G.Baker, Fl.
Mauritius: 381 (1877), nom. inval.
Latania vera Voss, Vilm. Blumengärtn. ed. 3, 1: 1148
(1895), pro syn.

Latania plagicoma Comm. ex Balf.f. = **Latania
lontaroides** (Gaertn.) H.E.Moore

Latania rubra Jacq. = **Latania lontaroides** (Gaertn.)
H.E.Moore

Latania vera Voss = **Latania lontaroides** (Gaertn.)
H.E.Moore

Latania verschaffeltii Lem., Ill. Hort. 6: t. 229 (1859). Rodrigues. 29 ROD.
Latania aurea Duncan, Cat. Hort. Maur.: 56 , nom. nud.

Lavoixia

Lavoixia H.E.Moore, Gentes Herb. 11: 296 (1978).
1 species, New Caledonia. 60.

Lavoixia macrocarpa H.E.Moore, Gentes Herb. 11: 297 (1978).
NE. New Caledonia (E. Mt. Panié). 60 NWC.

Legnea

Legnea O.F.Cook = **Chamaedorea** Willd.

Legnea laciniata O.F.Cook = **Chamaedorea costaricana** Oerst.

Lemurophoenix

Lemurophoenix J.Dransf., Kew Bull. 46: 61 (1991).
1 species, Madagascar. 29.

Lemurophoenix halleuxii J.Dransf., Kew Bull. 46: 62 (1991).
NE. Madagascar. 29 MDG.

Leopoldinia

Leopoldinia Mart., Hist. Nat. Palm. 2: 58 (1824).
3 species, SE. Colombia to S. Venezuela and N. Brazil. 82 83 84.

Leopoldinia major Wallace, Palm Trees Amazon: 15 (1853).
SE. Colombia to S. Venezuela and N. Brazil. 82 VEN 83 CLM 84 BZN.

Leopoldinia piassaba Wallace ex Archer, Hooker's J. Bot. Kew Gard. Misc. 7: 213 (1855).
SE. Colombia to S. Venezuela and N. Brazil. 82 VEN 83 CLM 84 BZN.

Leopoldinia pulchra Mart., Hist. Nat. Palm. 2: 59 (1824).
SE. Colombia to S. Venezuela and N. Brazil. 82 VEN 83 CLM 84 BZN.
Leopoldinia insignis Mart., Hist. Nat. Palm. 2: 60 (1824).

Leopoldinia insignis Mart. = **Leopoldinia pulchra** Mart.

Lepidocaryum

Lepidocaryum Mart., Hist. Nat. Palm. 2: 49 (1824).
1 species, W. Amazon Reg. 82 83 84.

Lepidocaryum allenii Dugand = **Lepidocaryum tenue** Mart. var. **tenue**

Lepidocaryum casiquiarense (Spruce) Drude = **Lepidocaryum tenue** var. **casiquiarense** (Spruce) A.J.Hend.

Lepidocaryum enneaphyllum Barb.Rodr. = **Lepidocaryum tenue** var. **gracile** (Mart.) A.J.Hend.

Lepidocaryum gracile Mart. = **Lepidocaryum tenue** var. **gracile** (Mart.) A.J.Hend.

Lepidocaryum guainiense (Spruce) Drude = **Lepidocaryum tenue** var. **casiquiarense** (Spruce) A.J.Hend.

Lepidocaryum gujanense Becc. = **Lepidocaryum tenue** var. **casiquiarense** (Spruce) A.J.Hend.

Lepidocaryum macrocarpum (Drude) Becc. = **Lepidocaryum tenue** var. **gracile** (Mart.) A.J.Hend.

Lepidocaryum quadripartitum (Spruce) Drude = **Lepidocaryum tenue** Mart. var. **tenue**

Lepidocaryum sexpartitum Trail & Barb.Rodr. = **Lepidocaryum tenue** var. **gracile** (Mart.) A.J.Hend.

Lepidocaryum sexpartitum var. *macrocarpum* Drude = **Lepidocaryum tenue** var. **gracile** (Mart.) A.J.Hend.

Lepidocaryum tenue Mart., Hist. Nat. Palm. 2: 51 (1824). *Mauritia tenuis* (Mart.) Spruce, J. Linn. Soc., Bot. 11: 169 (1869).
W. Amazon Reg. 82 GUY VEN 83 CLM PER 84 BZC BZN.

var. **casiquiarense** (Spruce) A.J.Hend., Palms Amazon: 78 (1995).
SE. Colombia to S. Venezuela. 82 VEN 83 CLM 84 BZN.
*Mauritia casiquiarensis Spruce, J. Linn. Soc., Bot. 11: 173 (1869).
Mauritia guainiensis Spruce, J. Linn. Soc., Bot. 11: 174 (1869). *Lepidocaryum guainiense* (Spruce) Drude in C.F.P.von Martius & auct. suc. (eds.), Fl. Bras. 3(2): 300 (1881).
Lepidocaryum casiquiarense (Spruce) Drude in C.F.P.von Martius & auct. suc. (eds.), Fl. Bras. 3(2): 300 (1881).
Lepidocaryum gujanense Becc., Ann. Roy. Bot. Gard. (Calcutta) 12(3): 221 (1918).

var. **gracile** (Mart.) A.J.Hend., Palms Amazon: 79 (1995).
Guyana to C. Brazil. 82 GUY 84 BZC BZN.
*Lepidocaryum gracile Mart., Hist. Nat. Palm. 2: 50 (1824). *Mauritia gracilis* (Mart.) Spruce, J. Linn. Soc., Bot. 11: 169 (1869).
Lepidocaryum enneaphyllum Barb.Rodr., Enum. Palm. Nov.: 19 (1875).
Lepidocaryum sexpartitum Trail & Barb.Rodr., Enum. Palm. Nov.: 19 (1875).
Lepidocaryum sexpartitum var. *macrocarpum* Drude in C.F.P.von Martius & auct. suc. (eds.), Fl. Bras. 3(2): 299 (1881). *Lepidocaryum macrocarpum* (Drude) Becc., Ann. Roy. Bot. Gard. (Calcutta) 12(2): 221 (1918).

var. **tenue**
W. Amazon Reg. 83 CLM PER 84 BZN.
Mauritia quadripartita Spruce, J. Linn. Soc., Bot. 11: 172 (1869). *Lepidocaryum quadripartitum* (Spruce) Drude in C.F.P.von Martius & auct. suc. (eds.), Fl. Bras. 3(2): 298 (1881).
Lepidocaryum tessmannii Burret, Notizbl. Bot. Gart. Berlin-Dahlem 10: 771 (1929).
Lepidocaryum allenii Dugand, Caldasia 2: 389 (1944).

Lepidocaryum tessmannii Burret = **Lepidocaryum tenue** Mart. var. **tenue**

Lepidococcus

Lepidococcus H.Wendl. & Drude = **Mauritiella** Burret

Lepidococcus aculeatus (Kunth) H.Wendl. & Drude = **Mauritiella aculeata** (Kunth) Burret

Lepidococcus armatus (Mart.) H.Wendl. & Drude = **Mauritiella armata** (Mart.) Burret

Lepidococcus duckei (Burret) A.D.Hawkes = **Mauritiella armata** (Mart.) Burret

Lepidococcus huebneri (Burret) A.D.Hawkes = **Mauritiella armata** (Mart.) Burret

Lepidococcus intermedius (Burret) A.D.Hawkes = **Mauritiella armata** (Mart.) Burret

Lepidococcus macrocladus (Burret) A.D.Hawkes = **Mauritiella macroclada** (Burret) Burret

Lepidococcus martianus (Spruce) A.D.Hawkes = **Mauritiella armata** (Mart.) Burret

Lepidococcus peruvianus (Becc.) A.D.Hawkes = **Mauritiella armata** (Mart.) Burret

Lepidococcus pumilus (Wallace) H.Wendl. & Drude = **Mauritiella armata** (Mart.) Burret

Lepidococcus subinermis (Spruce) A.D.Hawkes = **Mauritiella armata** (Mart.) Burret

Lepidorrhachis

Lepidorrhachis (H.Wendl. & Drude) O.F.Cook, J. Heredity 18: 408 (1927).
1 species, Lord Howe I. 50.

Lepidorrhachis mooreana (F.Muell.) O.F.Cook, J. Heredity 18: 408 (1927).
Lord Howe I. 50 NFK.
*Kentia mooreana F.Muell., Fragm. 7: 101 (1870).
Clinostigma mooreanum (F.Muell.) H.Wendl. & Drude, Linnaea 39: 218 (1875).

Lepidorrhachis onchorhyncha Becc. ex Martelli = **Clinostigma onchorhynchum** Becc.

Leptophoenix

Leptophoenix Becc. = **Hydriastele** H.Wendl. & Drude

Leptophoenix affinis (Becc.) Becc. = **Hydriastele affinis** (Becc.) W.J.Baker & Loo

Leptophoenix brassii Burret = **Hydriastele pinangoides** (Becc.) W.J.Baker & Loo

Leptophoenix densiflora (Ridl.) Burret = **Hydriastele pinangoides** (Becc.) W.J.Baker & Loo

Leptophoenix incompta Becc. = **Hydriastele pinangoides** (Becc.) W.J.Baker & Loo

Leptophoenix macrocarpa Burret = **Hydriastele pinangoides** (Becc.) W.J.Baker & Loo

Leptophoenix mayrii Burret = **Hydriastele cyclopensis** (Essig & B.E.Young) W.J.Baker & Loo

Leptophoenix micrantha Burret = **Hydriastele micrantha** (Burret) W.J.Baker & Loo

Leptophoenix microcarpa Burret = **Hydriastele pinangoides** (Becc.) W.J.Baker & Loo

Leptophoenix minor Becc. = **Hydriastele pinangoides** (Becc.) W.J.Baker & Loo

Leptophoenix pinangoides (Becc.) Becc. = **Hydriastele pinangoides** (Becc.) W.J.Baker & Loo

Leptophoenix pterophylla Becc. = **Hydriastele pinangoides** (Becc.) W.J.Baker & Loo

Leptophoenix yulensis Becc. = **Hydriastele pinangoides** (Becc.) W.J.Baker & Loo

Liberbaileya

Liberbaileya Furtado = **Maxburretia** Furtado

Liberbaileya gracilis (Burret) Burret & Potztal = **Maxburretia gracilis** (Burret) J.Dransf.

Liberbaileya lankawiensis Furtado = **Maxburretia gracilis** (Burret) J.Dransf.

Licuala

Licuala Wurmb, Verh. Batav. Genootsch. Kunsten 2: 469 (1780).
136 species, Bhutan to Vanuatu. 36 40 41 42 43 50 60.
Pericycla Blume, Rumphia 2: 47 (1838).
Dammera K.Schum. & Lauterb., Fl. Schutzgeb. Südsee: 201 (1900).

Licuala acuminata Burret, Notizbl. Bot. Gart. Berlin-Dahlem 15: 330 (1941).
Borneo. 42 BOR.

Licuala acutifida Mart., Hist. Nat. Palm. 3: 237 (1838).
Pen. Malaysia (Penang). 42 MLY.

Licuala acutifida var. *peninsularis* Becc. = **Licuala spinosa** Wurmb

Licuala ahlidurii Saw, Sandakania 10: 30 (1997).
Pen. Malaysia (Terengganu). 42 BOR.

Licuala amplifrons Miq. = **Licuala paludosa** Griff.

Licuala angustiloba Burret, J. Arnold Arbor. 20: 188 (1939).
New Guinea. 43 NWG.

Licuala anomala Becc., Webbia 5: 40 (1921).
New Guinea. 43 NWG.

Licuala arbuscula Becc., Malesia 3: 79 (1886).
Borneo. 42 BOR.

Licuala aruensis Becc., Malesia 1: 83 (1877).
New Guinea. 43 NWG.

Licuala aurantiaca Hodel = **Licuala paludosa** Griff.

Licuala bacularia Becc., Malesia 1: 82 (1877).
New Guinea. 43 NWG.

Licuala bayana Saw, Sandakania 10: 21 (1997).
Pen. Malaysia. 42 MLY.

Licuala beccariana (K.Schum. & Lauterb.) Furtado, Gard. Bull. Straits Settlem. 11: 37 (1940).
New Guinea. 43 NWG.
*Dammera ramosa K.Schum. & Lauterb., Fl. Schutzgeb. Südsee: 201 (1900). *Licuala ramosa* (K.Schum. & Lauterb.) Becc., Webbia 5: 52 (1921), nom. illeg.

Licuala bellatula Becc., Nova Guinea 8: 215 (1909).
New Guinea. 43 NWG.

Licuala bidentata Becc., Malesia 3: 80 (1886).
Borneo. 42 BOR.

Licuala bintulensis Becc., Malesia 3: 75 (1886).
Borneo. 42 BOR.

Licuala bissula Miq., Fl. Ned. Ind. 3: 57 (1855).
Sulawesi. 42 SUL.

Licuala borneensis Becc., Malesia 3: 85 (1886).
Borneo. 42 BOR.

Licuala bracteata Gagnep., Notul. Syst. (Paris) 6: 153 (1937).
S. Vietnam. 41 VIE.

Licuala brevicalyx Becc., Nova Guinea 8: 218 (1909).
New Guinea. 43 NWG.

Licuala cabalionii Dowe, Principes 37: 205 (1993).
Vanuatu. 60 VAN.

Licuala calciphila Becc., Webbia 3: 216 (1910).
N. Vietnam. 41 VIE.

Licuala cameronensis Saw, Sandakania 10: 76 (1997).
Pen. Malaysia (Pahang). 42 MLY.

Licuala celebica Miq., Verh. Kon. Ned. Akad. Wetensch., Afd. Natuurk. 11(5): 11 (1868).
Sulawesi. 42 SUL.

Licuala concinna Burret, J. Arnold Arbor. 20: 187 (1939).
New Guinea. 43 NWG.

Licuala confusa Furtado = **Licuala ridleyana** Becc.

Licuala cordata Becc., Malesia 3: 84 (1886).
Borneo. 42 BOR.

Licuala corneri Furtado, Gard. Bull. Straits Settlem. 11: 47 (1940).
Pen. Malaysia (Terengganu). 42 MLY.

Licuala crassiflora Barfod, Palms 44: 199 (2000).
E. New Guinea. 43 NWG.

Licuala dasyantha Burret, Notizbl. Bot. Gart. Berlin-Dahlem 15: 334 (1941).
S. China. 36 CHC CHS.

Licuala debilis Becc., Nova Guinea 8: 216 (1909).
New Guinea. 43 NWG.

Licuala delicata Hodel = **Licuala scortechinii** Becc.

Licuala densiflora Becc., Webbia 5: 44 (1921).
Borneo. 42 BOR.

Licuala distans Ridl., J. Fed. Malay States Mus. 10: 123 (1920).
Pen. Thailand. 41 THA.

Licuala dransfieldii Kiew = **Licuala ridleyana** Becc.

Licuala egregia Saw, Sandakania 10: 16 (1997).
Pen. Malaysia (Pahang). 42 MLY.

Licuala elegans Blume, Rumphia 2: 42 (1838).
Sumatera. 42 SUM.

Licuala elegantissima Ridl., Gard. Chron. 1904(1): 50 (1904).
Borneo. 42 BOR.

Licuala fatua Becc., Webbia 3: 218 (1910).
N. Vietnam. 41 VIE.

Licuala ferruginea Becc. in J.D.Hooker, Fl. Brit. India 6: 432 (1892).
Pen. Malaysia. 42 MLY.

Licuala ferruginoides Becc., Webbia 5: 45 (1921).
Sumatera. 42 SUM.

Licuala filiformis Hodel = **Licuala triphylla** Griff.

Licuala flabellum Mart., Hist. Nat. Palm. 3: 237 (1838).
Sulawesi. 42 SUL.

Licuala flavida Ridl., Trans. Linn. Soc. London, Bot. 9: 234 (1916).
New Guinea. 43 NWG.

Licuala flexuosa Burret, Notizbl. Bot. Gart. Berlin-Dahlem 15: 331 (1941).
Maluku. 42 MOL.

Licuala fordiana Becc., Malesia 3: 198 (1889).
SE. China to Hainan. 36 CHH CHS.

Licuala fractiflexa Saw, Sandakania 10: 32 (1997).
Pen. Malaysia (Terengganu). 42 MLY.

Licuala furcata Becc., Malesia 3: 198 (1889).
Borneo. 42 BOR.

Licuala gjellerupii Becc., Webbia 5: 41 (1921).
New Guinea. 43 NWG.

Licuala glaberrima Gagnep., Notul. Syst. (Paris) 6: 154 (1937).
S. Vietnam. 41 VIE.

Licuala glabra Griff., Calcutta J. Nat. Hist. 5: 329 (1845).
Pen. Thailand to Pen. Malaysia. 41 THA 42 MLY.

var. **glabra**
Pen. Thailand to Pen. Malaysia. 41 THA 42 MLY.
Licuala longipedunculata Ridl., J. Straits Branch Roy. Asiat. Soc. 41: 42 (1903).

var. **selangorensis** Becc., Webbia 5: 35 (1921).
Pen. Malaysia. 42 MLY.

Licuala gracilis Blume in J.J.Roemer & J.A.Schultes, Syst. Veg. 7: 1303 (1830).
Jawa. 42 JAW.

Licuala grandiflora Ridl., Trans. Linn. Soc. London, Bot. 9: 233 (1916).
New Guinea. 43 NWG.

Licuala grandis H.Wendl., Ill. Hort. 27: t. 412 (1880).
Pritchardia grandis (H.Wendl.) W.Bull, Gard. Chron., n.s., 1: 415 (1874).
Santa Cruz Is. to Vanuatu. (42) mly 60 SCZ VAN.

Licuala hallieriana Becc., Webbia 5: 51 (1921).
Borneo. 42 BOR.

Licuala hexasepala Gagnep., Notul. Syst. (Paris) 6: 155 (1937).
S. Vietnam. 41 VIE.

Licuala hirta Hodel = **Licuala merguensis** Becc.

Licuala horrida Blume = **Licuala spinosa** Wurmb

Licuala hospita Burret = ? [**42 MLY**]

Licuala insignis Becc., Malesia 1: 80 (1877).
New Guinea. 43 NWG.

Licuala jeanneneyi André = **Licuala ramsayi** (F.Muell.) Domin

Licuala kamarudinii Saw, Sandakania 10: 24 (1997).
Pen. Malaysia (SC. Pahang). 42 MLY.

Licuala kemamanensis Furtado, Gard. Bull. Straits Settlem. 11: 50 (1940).
Pen. Malaysia (Terengganu). 42 MLY.

Licuala kersteniana André = ?

Licuala khoonmengii Saw, Sandakania 10: 43 (1997).
Pen. Malaysia (Terengganu). 42 MLY.

Licuala kiahii Furtado, Gard. Bull. Straits Settlem. 11: 52 (1940).
Pen. Malaysia. 42 MLY.

Licuala kingiana Becc., Malesia 3: 193 (1889).
Pen. Malaysia (C. Perak). 42 MLY.

Licuala kirsteniana André = ?

Licuala klossii Ridl., Trans. Linn. Soc. London, Bot. 9: 234 (1916).
New Guinea. 43 NWG.

Licuala kunstleri Becc. in J.D.Hooker, Fl. Brit. India 6: 433 (1892).
Pen. Thailand to Pen. Malaysia. 41 THA 42 MLY.

Licuala lanata J.Dransf., Bot. J. Linn. Soc. 81: 27 (1980).
Borneo (Sarawak). 42 BOR.

Licuala lanuginosa Ridl., J. Straits Branch Roy. Asiat. Soc. 44: 203 (1905).
Pen. Malaysia (Johor). 42 MLY.

Licuala lauterbachii Dammer & K.Schum. in K.M.Schumann & C.A.G.Lauterbach, Fl. Schutzgeb. Südsee: 199 (1900).
New Guinea. 43 NWG.

Licuala leprosa Dammer ex Becc., Webbia 5: 42 (1921).
New Guinea. 43 NWG.

Licuala leptocalyx Burret, Notizbl. Bot. Gart. Berlin-Dahlem 11: 706 (1933).
New Guinea. 43 NWG.

Licuala linearis Burret, Notizbl. Bot. Gart. Berlin-Dahlem 12: 314 (1935).
New Guinea. 43 NWG.

Licuala longicalycata Furtado, Gard. Bull. Straits Settlem. 11: 54 (1940).
S. Pen. Malaysia. 42 MLY.

Licuala longipedunculata Ridl. = **Licuala glabra** var. **glabra**

Licuala longipes Griff., Calcutta J. Nat. Hist. 5: 330 (1845).
SW. Pen. Malaysia , SE. Sumatera. 42 MLY SUM.

Licuala longispadix Banka & Bradford, Kew Bull. 59: 73 (2004).
Papua New Guinea. 43 NWG.

Licuala macrantha Burret, Notizbl. Bot. Gart. Berlin-Dahlem 15: 327 (1941).
New Guinea. 43 NWG.

Licuala magna Burret, J. Arnold Arbor. 20: 188 (1929).
New Guinea. 43 NWG.

Licuala malajana Becc., Malesia 3: 197 (1889).
Pen. Thailand to Pen. Malaysia. 41 THA 42 MLY.

var. **humilis** Saw, Sandakania 10: 90 (1997).
Pen. Malaysia (Terengganu). 42 MLY.

var. **malajana**
Pen. Thailand to Pen. Malaysia. 41 THA 42 MLY.

Licuala mattanensis Becc., Malesia 3: 86 (1889).
Borneo. 42 BOR.

Licuala merguensis Becc., Webbia 5: 47 (1921).
S. Myanmar to Pen. Thailand. 41 MYA THA.
Licuala hirta Hodel, Palm J. 136: 12 (1997).

Licuala micholitzii Ridl., Gard. Chron. 1904(1): 50 (1904).
Borneo. 42 BOR.

Licuala micrantha Becc., Webbia 1: 289 (1905).
New Guinea. 43 NWG.

Licuala mirabilis Furtado, Gard. Bull. Straits Settlem. 11: 58 (1940).
NW. Pen. Malaysia. 42 MLY.

Licuala modesta Becc., Malesia 3: 195 (1889).
Pen. Thailand to Pen. Malaysia. 41 THA 42 MLY.
Licuala wrayi Becc. ex Ridl., J. Straits Branch Roy. Asiat. Soc. 82: 201 (1920).

Licuala montana Dammer & K.Schum. in K.M.Schumann & C.A.G.Lauterbach, Fl. Schutzgeb. Südsee: 200 (1900).
New Guinea. 43 NWG.

Licuala moszkowskiana Becc., Bot. Jahrb. Syst. 52: 38 (1914).
New Guinea. 43 NWG.

Licuala moyseyi Furtado, Gard. Bull. Straits Settlem. 11: 61 (1940).
Pen. Malaysia (Terengganu). 42 MLY.

Licuala muelleri H.Wendl. & Drude = **Licuala ramsayi** (F.Muell.) Domin

Licuala mustapana Saw, Sandakania 10: 55 (1997).
Pen. Malaysia (Terengganu). 42 MLY.

Licuala nana Blume, Rumphia 2: 46 (1838).
Sumatera. 42 SUM.

Licuala naumoniensis Becc., Bot. Jahrb. Syst. 52: 39 (1914).
New Guinea. 43 NWG.

Licuala nauroannii Burret, Notizbl. Bot. Gart. Berlin-Dahlem 12: 312 (1935).
Solomon Is. 43 SOL.

Licuala olivifera Becc., Malesia 3: 78 (1889).
Borneo. 42 BOR.

Licuala oliviformis Becc., Nelle Forest. Borneo: 385 (1902).
Borneo. 42 BOR.

Licuala oninensis Becc., Webbia 5: 41 (1921).
New Guinea. 43 NWG.

Licuala orbicularis Becc., Malesia 3: 83 (1889).
Borneo. 42 BOR.
Pritchardia grandis H.J.Veitch, Cat. 1885: 54 (1885), nom. illeg.
Licuala veitchii W.Watson ex Hook.f., Bot. Mag. 115: t. 7053 (1889).

Licuala oxleyi H.Wendl. = **Licuala paludosa** Griff.

Licuala pachycalyx Burret, Notizbl. Bot. Gart. Berlin-Dahlem 11: 705 (1933).
New Guinea. 43 NWG.

Licuala pahangensis Furtado, Gard. Bull. Straits Settlem. 11: 63 (1940).
Pen. Malaysia (SW. Pahang). 42 MLY.

Licuala palas Saw, Sandakania 10: 59 (1997).
Pen. Malaysia (Terengganu). 42 MLY.

Licuala paludosa Griff., Calcutta J. Nat. Hist. 5: 323 (1844).
Indo-China to W. Malesia. 41 CBD THA VIE 42 BOR MLY SUM.
Licuala amplifrons Miq., Fl. Ned. Ind., Eerste Bijv.: 591 (1861).
Licuala oxleyi H.Wendl. in O.C.E.de Kerchove de Denterghem, Palmiers: 249 (1878).
Licuala paniculata Ridl., J. Straits Branch Roy. Asiat. Soc. 41: 42 (1903).
Licuala aurantiaca Hodel, Palm J. 134: 30 (1997).

Licuala paniculata Ridl. = **Licuala paludosa** Griff.

Licuala parviflora Dammer ex Becc., Webbia 5: 50 (1921).
New Guinea. 43 NWG.

Licuala patens Ridl., J. Straits Branch Roy. Asiat. Soc. 82: 202 (1920).
Pen. Malaysia (Perak). 42 MLY.

Licuala paucisecta Burret, Notizbl. Bot. Gart. Berlin-Dahlem 12: 311 (1935).
New Guinea. 43 NWG.

Licuala peekelii Lauterb., Bot. Jahrb. Syst. 45: 356 (1911).
Bismarck Arch. 43 BIS.

Licuala peltata Roxb. ex Buch.-Ham., Mem. Wern. Nat. Hist. Soc. 5: 313 (1826).
Bhutan to Pen. Malaysia. 40 ASS BAN EHM 41 AND MYA NCB THA 42 MLY.

var. **peltata**
Bhutan to Pen. Malaysia. 40 ASS BAN EHM 41 AND MYA NCB THA 42 MLY.

var. **sumawongii** Saw, Sandakania 10: 10 (1997).
Pen. Thailand to Pen. Malaysia. 41 THA 42 MLY.

Licuala penduliflora (Blume) Zipp. ex Blume, Rumphia 2: 47 (1838).
New Guinea. 43 NWG.
  Pericycla penduliflora Blume, Rumphia 2: 47 (1838).

Licuala petiolulata Becc., Malesia 3: 77 (1889).
Borneo. 42 BOR.

Licuala pilearia (Lour.) Blume = **Licuala spinosa** Wurmb

Licuala platydactyla Becc., Webbia 5: 42 (1921).
New Guinea. 43 NWG.

Licuala polyschista K.Schum. & Lauterb., Fl. Schutzgeb. Südsee: 199 (1900).
New Guinea. 43 NWG.

Licuala poonsakii Hodel, Palm J. 134: 32 (1997).
E. & SE. Thailand. 41 THA.

Licuala pulchella Burret, Notizbl. Bot. Gart. Berlin-Dahlem 12: 315 (1935).
New Guinea. 43 NWG.

Licuala pumila Blume in J.J.Roemer & J.A.Schultes, Syst. Veg. 7: 1302 (1830).
Jawa. 42 JAW.

Licuala punctulata Burret, Notizbl. Bot. Gart. Berlin-Dahlem 15: 329 (1941).
Borneo. 42 BOR.

Licuala pusilla Becc., Malesia 3: 194 (1889).
Pen. Thailand to WC. Pen. Malaysia. 41 THA 42 MLY.
  Licuala tansachana Hodel, Palm J. 134: 34 (1997).

Licuala pygmaea Merr. = **Licuala triphylla** Griff.

Licuala radula Gagnep., Notul. Syst. (Paris) 6: 155 (1937).
S. Vietnam. 41 VIE.

Licuala ramosa Blume = **Licuala spinosa** Wurmb

Licuala ramosa (K.Schum. & Lauterb.) Becc. = **Licuala beccariana** (K.Schum. & Lauterb.) Furtado

Licuala ramsayi (F.Muell.) Domin, Biblioth. Bot. 85: 500 (1915).
N. & NE. Queensland. 50 QLD.
  Livistona ramsayi F.Muell., Fragm. 8: 221 (1874).
  Licuala muelleri H.Wendl. & Drude, Linnaea 39: 223 (1875).
  Licuala jeanneneyi André, Rev. Hort. 70: 263 (1898).

Licuala reptans Becc., Webbia 5: 45 (1921).
Borneo. 42 BOR.

Licuala ridleyana Becc., Webbia 5: 44 (1921).
S. Pen. Malaysia. 42 MLY.
  Licuala confusa Furtado, Gard. Bull. Straits Settlem. 11: 44 (1940).
  Licuala dransfieldii Kiew, Malayan Nat. J. 42: 263 (1989).

Licuala robinsoniana Becc., Webbia 5: 48 (1921).
S. Vietnam. 41 VIE.

Licuala robusta Warb. ex K.Schum. & Lauterb., Fl. Schutzgeb. Südsee: 199 (1900).
New Guinea. 43 NWG.

Licuala rotundifolia (Lam.) Blume = **Livistona rotundifolia** (Lam.) Mart.

Licuala rumphii Blume, Rumphia 2: 41 (1838).
Maluku. 42 MOL.
  Corypha licuala Lam., Encycl. 2: 131 (1786).

Licuala ruthiae Saw, Sandakania 10: 41 (1997).
Pen. Malaysia (S. Terengganu). 42 MLY.

Licuala sallehana Saw, Sandakania 10: 11 (1997).
Pen. Malaysia (Terengganu). 42 MLY.

var. **incisifolia** Saw, Sandakania 10: 13 (1997).
Pen. Malaysia (Terengganu). 42 MLY.

var. **sallehana**
Pen. Malaysia (Terengganu). 42 MLY.

Licuala sarawakensis Becc., Malesia 3: 81 (1889).
Borneo. 42 BOR.

Licuala scortechinii Becc., Malesia 3: 192 (1889).
Pen. Thailand to NW. Pen. Malaysia. 41 THA 42 MLY.
  Licuala delicata Hodel, Palm J. 136: 11 (1997).

Licuala simplex (K.Schum. & Lauterb.) Becc., Webbia 5: 52 (1921).
New Guinea. 43 NWG.
  Dammera simplex K.Schum. & Lauterb., Fl. Schutzgeb. Südsee: 201 (1900).

Licuala spathellifera Becc., Malesia 3: 76 (1889).
Borneo. 42 BOR.

Licuala spectabilis Miq., Pl. Jungh.: 163 (1852).
Jawa. 42 JAW.

Licuala spicata Becc., Malesia 3: 88 (1889).
Borneo. 42 BOR.

Licuala spinosa Wurmb, Verh. Batav. Genootsch. Kunsten 2: 474 (1780).
Hainan, Indo-China to Philippines. 36 CHH 41 AND CBD MYA NCB THA VIE 42 BOR JAW MLY PHI SUM.
  Corypha pilearia Lour., Fl. Cochinch.: 213 (1790).
  Licuala pilearia (Lour.) Blume, Rumphia 2: 42 (1838).
  Licuala ramosa Blume in J.J.Roemer & J.A.Schultes, Syst. Veg. 7: 1303 (1830).
  Licuala horrida Blume, Rumphia 2: 39 (1838).
  Licuala spinosa var. *brevidens* Becc., Malesia 3: 74 (1886).
  Licuala spinosa var. *cochinchinensis* Becc., Malesia 3: 74 (1886).
  Licuala acutifida var. *peninsularis* Becc., Webbia 5: 44 (1921).

Licuala spinosa var. *brevidens* Becc. = **Licuala spinosa** Wurmb

Licuala spinosa var. *cochinchinensis* Becc. = **Licuala spinosa** Wurmb

Licuala steinii Burret, Notizbl. Bot. Gart. Berlin-Dahlem 11: 707 (1933).
New Guinea. 43 NWG.

Licuala stenophylla Hodel = **Licuala triphylla** Griff.

Licuala stipitata Burret, Notizbl. Bot. Gart. Berlin-Dahlem 15: 333 (1941).
Trop. Asia (?). 4 +.

Licuala stongensis Saw, Sandakania 10: 90 (1997).
Pen. Malaysia (Kelantan). 42 MLY.

Licuala tansachana Hodel = **Licuala pusilla** Becc.

Licuala tanycola H.E.Moore, Principes 13: 105 (1969).
New Guinea. 43 NWG.

Licuala taynguyensis Barfod & Borchs., Brittonia 52: 354 (2000 publ. 2001).
Vietnam. 41 VIE.

Licuala telifera Becc., Malesia 1: 81 (1877).

Licuala tenuissima Saw, Sandakania 10: 37 (1997).
Pen. Malaysia. 42 MLY.

Licuala terengganuensis Saw, Sandakania 10: 61 (1997).
Pen. Malaysia (Terengganu). 42 MLY.

Licuala ternata Griff. ex Mart. = **Licuala triphylla** Griff.

Licuala thoana Saw & J.Dransf., Gard. Bull. Singapore 42: 71 (1989 publ. 1990).
Pen. Malaysia. 42 MLY.

Licuala tiomanensis Furtado, Gard. Bull. Straits Settlem. 11: 68 (1940).
Pen. Malaysia (Tioman I.). 42 MLY.

Licuala tomentosa Burret, Notizbl. Bot. Gart. Berlin-Dahlem 15: 98 (1940).
N. Vietnam. 41 VIE.

Licuala tonkinensis Becc., Webbia 3: 214 (1910).
N. Vietnam. 41 VIE.

Licuala triphylla Griff., Calcutta J. Nat. Hist. 5: 332 (1844).
Pen. Thailand to Pen. Malaysia, Borneo. 41 THA 42 BOR MLY.
    Licuala ternata Griff. ex Mart., Hist. Nat. Palm. 3(ed. 2): 238 (1849).
    Licuala triphylla var. *integrifolia* Ridl., Mat. Fl. Malay. Penins. 2: 164 (1907).
    Licuala pygmaea Merr., Univ. Calif. Publ. Bot. 15: 20 (1929).
    Licuala filiformis Hodel, Palm J. 134: 32 (1997).
    Licuala stenophylla Hodel, Palm J. 137: 16 (1997).

Licuala triphylla var. *integrifolia* Ridl. = **Licuala triphylla** Griff.

Licuala valida Becc., Bot. Jahrb. Syst. 48: 90 (1912).
Borneo. 42 BOR.

Licuala veitchii W.Watson ex Hook.f. = **Licuala orbicularis** Becc.

Licuala waraguh Blume = **Rhapis humilis** Blume

Licuala whitmorei Saw, Sandakania 10: 78 (1997).
Pen. Malaysia (W. Johor). 42 MLY.

Licuala wixu Blume = **Rhapis humilis** Blume

Licuala wrayi Becc. ex Ridl. = **Licuala modesta** Becc.

Linoma

Linoma O.F.Cook = **Dictyosperma** H.Wendl. & Drude

Linoma alba (Bory) O.F.Cook = **Dictyosperma album** (Bory) Scheff.

Linospadix

Linospadix H.Wendl., Linnaea 39: 177 (1875).
9 species, New Guinea to E. Australia. 43 50.
    Bacularia F.Muell. ex Hook.f., Bot. Mag. 108: t. 6644 (1882).

Linospadix Becc. ex Hook.f. = **Calyptrocalyx** Blume

Linospadix aequisegmentosa (Domin) Burret, Notizbl. Bot. Gart. Berlin-Dahlem 12: 331 (1935).
NE. Queensland. 50 QLD.
    Bacularia aequisegmentosa Domin, Biblioth. Bot. 85: 500 (1915).

Linospadix albertisiana (Becc.) Burret, Notizbl. Bot. Gart. Berlin-Dahlem 12: 331 (1935).
New Guinea. 43 NWG.
    Bacularia albertisiana Becc., Malesia 3: 108 (1886).
    Bacularia angustisecta Becc., Webbia 1: 294 (1905).
    Linospadix angustisecta (Becc.) Burret, Notizbl. Bot. Gart. Berlin-Dahlem 12: 331 (1935).
    Bacularia longicruris Becc., Bot. Jahrb. Syst. 52: 35 (1914). *Linospadix longicruris* (Becc.) Burret, Notizbl. Bot. Gart. Berlin-Dahlem 12: 331 (1935).

Linospadix angustisecta (Becc.) Burret = **Linospadix albertisiana** (Becc.) Burret

Linospadix apetiolata Dowe & A.K.Irvine, Principes 41: 215 (1997).
Queensland. 50 QLD.

Linospadix arfakiana Becc. = **Calyptrocalyx arfakianus** (Becc.) Dowe & M.D.Ferrero

Linospadix canina (Becc.) Burret, Notizbl. Bot. Gart. Berlin-Dahlem 12: 331 (1935).
W. New Guinea. 43 NWG.
    Bacularia canina Becc., Nova Guinea 8: 209 (1909).
    Linospadix elegans Ridl., Trans. Linn. Soc. London, Bot. 9: 233 (1916).

Linospadix caudiculata Becc. = **Calyptrocalyx caudiculatus** (Becc.) Dowe & M.D.Ferrero

Linospadix elegans Ridl. = **Linospadix canina** (Becc.) Burret

Linospadix flabellata Becc. = **Calyptrocalyx flabellatus** (Becc.) Dowe & M.D.Ferrero

Linospadix forbesii Ridl. = **Calyptrocalyx forbesii** (Ridl.) Dowe & M.D.Ferrero

Linospadix geonomiformis Becc. = **Calyptrocalyx geonomiformis** (Becc.) Dowe & M.D.Ferrero

Linospadix hellwigiana Warb. ex Becc. = **Calyptrocalyx hollrungii** (Becc.) Dowe & M.D.Ferrero

Linospadix hollrungii Becc. = **Calyptrocalyx hollrungii** (Becc.) Dowe & M.D.Ferrero

Linospadix julianettii Becc. = **Calyptrocalyx julianettii** (Becc.) Dowe & M.D.Ferrero

Linospadix lauterbachiana (Warb. ex Becc.) Becc. = **Calyptrocalyx lauterbachianus** Warb. ex Becc.

Linospadix leopoldii Sander = **?**

Linospadix leptostachys Burret = **Calyptrocalyx sessiliflorus** Dowe & M.D.Ferrero

Linospadix longicruris (Becc.) Burret = **Linospadix albertisiana** (Becc.) Burret

Linospadix micholitzii Ridl. = **Calyptrocalyx micholitzii** (Ridl.) Dowe & M.D.Ferrero

Linospadix microcarya (Domin) Burret, Notizbl. Bot. Gart. Berlin-Dahlem 12: 331 (1935).
NE. Queensland. 50 QLD.
    Bacularia microcarya Domin, Biblioth. Bot. 85: 499 (1915).
    Bacularia sessilifolia Becc., Atti Soc. Tosc. Sci. Nat. Pisa Processi Verbali 44: 133 (1934).

Linospadix microspadix Becc., Bot. Jahrb. Syst. 52: 34 (1914). *Paralinospadix microspadix* (Becc.) Burret, Notizbl. Bot. Gart. Berlin-Dahlem 12: 335 (1935).
New Guinea. 43 NWG. — Provisionally accepted.

Linospadix minor (F.Muell.) F.Muell., Fragm. 11: 58 (1878).
NE. Queensland. 50 QLD.
    Areca minor W.Hill, Rep. Brisbane Bot. Gard.: 6 (1874).
    Kentia minor F.Muell., Fragm. 8: 235 (1874). *Bacularia minor* (F.Muell.) F.Muell., Fragm. 11: 58 (1880).
    Bacularia intermedia C.T.White, Proc. Roy. Soc. Queensland 47: 83 (1935 publ. 1936).

Linospadix monostachya (Mart.) H.Wendl. & Drude, Linnaea 39: 199 (1875).
SE. Queensland to New South Wales. 50 NSW QLD.
    Areca monostachya Mart., Hist. Nat. Palm. 3: 178 (1838). *Kentia monostachya* (Mart.) F.Muell.,

Fragm. 7: 82 (1870). *Bacularia monostachya* (Mart.) F.Muell., Fragm. 7: 103 (1870).

Linospadix multifida Becc. = **Calyptrocalyx multifidus** (Becc.) Dowe & M.D.Ferrero

Linospadix pachystachys Burret = **Calyptrocalyx arfakianus** (Becc.) Dowe & M.D.Ferrero

Linospadix palmeriana (F.M.Bailey) Burret, Notizbl. Bot. Gart. Berlin-Dahlem 12: 331 (1935).
NE. Queensland. 50 QLD.
*Bacularia palmeriana F.M.Bailey, Rep. Exped. Bellenden-Ker: 61 (1889).

Linospadix parvula Becc. = **Calyptrocalyx pusillus** (Becc.) Dowe & M.D.Ferrero

Linospadix pauciflora Ridl. = **Calyptrocalyx micholitzii** (Ridl.) Dowe & M.D.Ferrero

Linospadix petrickiana Sander = **Calyptrocalyx forbesii** (Ridl.) Dowe & M.D.Ferrero

Linospadix pusilla Becc. = **Calyptrocalyx pusillus** (Becc.) Dowe & M.D.Ferrero

Linospadix schlechteri Becc. = **Calyptrocalyx hollrungii** (Becc.) Dowe & M.D.Ferrero

Lithocarpos

Lithocarpos O.Targ.Tozz. ex Steud. = **Attalea** Kunth

Lithocarpos cocciformis O.Targ.Tozz. ex Steud. = **Attalea funifera** Mart.

Livistona

Livistona R.Br., Prodr.: 267 (1810).
33 species, NE. Trop. Africa, S. Yemen, Bangladesh to Japan and Australia. 24 35 36 38 40 41 42 43 50.
Saribus Blume, Rumphia 2: 48 (1838).
Wissmannia Burret, Bot. Jahrb. Syst. 73: 182 (1943).

Livistona alfredii F.Muell., Victorian Naturalist 9: 112 (1892).
NW. Western Australia. 50 WAU.

Livistona altissima Zoll. = **Livistona rotundifolia** (Lam.) Mart.

Livistona australis (R.Br.) Mart., Hist. Nat. Palm. 3: 242 (1838).
E. & SE. Australia. 50 NSW QLD VIC.
*Corypha australis R.Br., Prodr.: 267 (1810).

Livistona beccariana Burret = **Livistona woodfordii** Ridl.

Livistona benthamii F.M.Bailey, Queensl. Fl. 5: 1683 (1902).
New Guinea to N. Australia. 43 NWG 50 NTA QLD.
Livistona holtzei Becc., Webbia 5: 18 (1921).
Livistona melanocarpa Burret, J. Arnold Arbor. 20: 190 (1939).

Livistona bissula Mart. = ? [42 SUL] **Licuala** sp.

Livistona blancoi Merr. = **Livistona rotundifolia** (Lam.) Mart.

Livistona boninensis (Becc.) Nakai = **Livistona chinensis** var. **boninensis** Becc.

Livistona brassii Burret = **Livistona muelleri** F.M.Bailey

Livistona brevifolia Dowe & Mogea, Palms 48: 201 (2004).
W. New Guinea. 43 NWG.

Livistona carinensis (Chiov.) J.Dransf. & N.W.Uhl, Kew Bull. 38: 200 (1983).
NE. Trop. Africa, S. Yemen. 24 DJI SOM 35 YEM.
*Hyphaene carinensis Chiov., Fl. Somala 1: 318 (1929). *Wissmannia carinensis* (Chiov.) Burret, Bot. Jahrb. Syst. 73: 184 (1943).

Livistona chinensis (Jacq.) R.Br. ex Mart., Hist. Nat. Palm. 3: 240 (1838).
S. China to Vietnam, Temp. E. Asia. 36 CHS 41 VIE 38 JAP kzn NNS OGA TAI (42) jaw (62) mrn.
*Latania chinensis Jacq., Fragm. Bot.: 16 (1801).
Saribus chinensis (Jacq.) Blume, Rumphia 2: 48 (1838).

var. **boninensis** Becc., Webbia 5: 12 (1912). *Livistona boninensis* (Becc.) Nakai, Bull. Biogeogr. Soc. Japan 1(3): 255 (1930).
Ogasawara-shoto. 38 kzn OGA.

var. **chinensis**
S. China to Vietnam. 36 CHS 41 VIE.
Livistona mauritiana Wall. ex Mart., Hist. Nat. Palm. 3: 240 (1838), pro syn.
Saribus oliviformis Hassk., Tijdschr. Natuurl. Gesch. Physiol. 9: 176 (1842). *Livistona oliviformis* (Hassk.) Mart., Hist. Nat. Palm. 3: 319 (1853).
Livistona sinensis Griff., Palms Brit. E. Ind.: 131 (1850), orth. var.

var. **subglobosa** (Hassk.) Becc., Webbia 5: 16 (1920).
S. Japan to Taiwan (Chishan I.). 38 JAP NNS TAI (42) jaw (62) mrn.
*Saribus subglobosus Hassk., Tijdschr. Natuurl. Gesch. Physiol. 9: 177 (1842). *Livistona subglobosa* (Hassk.) Mart., Hist. Nat. Palm. 3: 319 (1853).
Livistona japonica Nakai ex Masam., Prelim. Rep. Veg. Yakus.: 50 (1929).

Livistona chocolatina Dowe & Mogea, Palms 48: 199 (2004).
Papua New Guinea. 43 NWG.

Livistona cochinchinensis (Lour.) Mart. = **Livistona saribus** (Lour.) Merr. ex A.Chev.

Livistona concinna Dowe & Barfod, Austrobaileya 6: 166 (2001).
Queensland. 50 QLD.

Livistona crustacea Burret = **Livistona muelleri** F.M.Bailey

Livistona decipiens Becc. = **Livistona decora** (W. Bull) Dowe

Livistona decora (W. Bull) Dowe, Austrobaileya 6: 979 (2004). E. Queensland. 50 QLD.
*Corypha decora W.Bull, Cat. 1887.
Livistona decipiens Becc., Webbia 3: 301 (1910)
Livistona decipiens var. *polyantha* Becc., Webbia 5: 14 (1920).

Livistona decipiens var. *polyantha* Becc. = **Livistona decora** (W. Bull) Dowe

Livistona diepenhorstii Hassk. = **Livistona saribus** (Lour.) Merr. ex A.Chev.

Livistona drudei F.Muell. ex Drude, Bot. Jahrb. Syst. 16(39): 11 (1893).
NE. Queensland. 50 QLD.

Livistona eastonii C.A.Gardner, For. Dept. Bull., W. Austral. 32: 36 (1923).
N. Western Australia. 50 WAU.

Livistona endauensis J.Dransf. & K.M.Wong, Malayan Nat. J. 41: 121 (1987).
Pen. Malaysia. 42 MLY.

Livistona enervis auct. = **?**

Livistona exigua J.Dransf., Kew Bull. 31: 760 (1977).

Livistona fengkaiensis X.W.Wei & M.Y.Xiao = **Livistona saribus** (Lour.) Merr. ex A.Chev.

Livistona filamentosa (H.Wendl. ex Franceschi) Pfister = **Washingtonia filifera** (Linden ex André) H.Wendl. ex de Bary

Livistona fulva Rodd, Telopea 8: 103 (1998).
Queensland (Blackdown Tableland). 50 QLD.

Livistona gaudichaudii Mart. = **Pritchardia martii** (Gaudich.) H.Wendl.

Livistona halongensis T.H.Nguyen & Kiew, Gard. Bull. Singapore 52: 198 (2000).
Vietnam. 41 VIE.

Livistona hasseltii (Hassk.) Hassk. ex Miq. = **Livistona saribus** (Lour.) Merr. ex A.Chev.

Livistona holtzei Becc. = **Livistona benthamii** F.M. Bailey

Livistona hoogendorpii Teijsm. & Binn. ex Miq. = **Livistona saribus** (Lour.) Merr. ex A.Chev.

Livistona humilis R.Br., Prodr.: 268 (1810). *Saribus humilis* (R.Br.) Kuntze, Revis. Gen. Pl. 2: 736 (1891).
N. Northern Territory. 50 NTA.

Livistona leichhardtii F.Muell., Fragm. 5: 49 (1865).

Livistona humilis var. *novoguineensis* Becc. = **Livistona muelleri** F.M.Bailey

Livistona humilis var. *sclirophylla* Becc. = **Livistona muelleri** F.M.Bailey

Livistona inaequisecta Becc. = **Livistona saribus** (Lour.) Merr. ex A.Chev.

Livistona inermis R.Br., Prodr.: 268 (1810). *Saribus inermis* (R.Br.) Kuntze, Revis. Gen. Pl. 2: 736 (1891).
N. & NE. Northern Territory to NW. Queensland. 50 NTA QLD.

Livistona japonica Nakai ex Masam. = **Livistona chinensis** var. **subglobosa** (Hassk.) Becc.

Livistona jenkinsiana Griff., Calcutta J. Nat. Hist. 5: 334 (1845). *Saribus jenkensii* (Griff.) Kuntze, Revis. Gen. Pl. 2: 736 (1891).
Bhutan to N. Thailand. 36 CHC 40 ASS EHM BAN 41 MYA THA 42 MLY.
  Livistona jenkinsii Griff. ex Mart., Hist. Nat. Palm. 3(ed. 2): 242 (1849), orth. var.
  Livistona speciosa Kurz, J. Asiat. Soc. Bengal, Pt. 2, Nat. Hist. 43(2): 204 (1874).
  Saribus speciosus (Kurz) Kuntze, Revis. Gen. Pl. 2: 736 (1891).

Livistona jenkinsii Griff. ex Mart. = **Livistona jenkinsiana** Griff.

Livistona kimberleyana Rodd, Telopea 8: 121 (1998).
N. Western Australia. 50 WAU.

Livistona kingiana Becc. = **Pholidocarpus kingianus** (Becc.) Ridl.

Livistona lanuginosa Rodd, Telopea 8: 82 (1998).
NE. Queensland. 50 QLD.

Livistona leichhardtii F.Muell. = **Livistona humilis** R.Br.

Livistona loriphylla Becc., Webbia 5: 18 (1921).
NW. Western Australia to N. Northern Territory. 50 NTA WAU.

Livistona macrophylla Roster = **?**

Livistona mariae F.Muell., Fragm. 8: 283 (1874). *Saribus mariae* (F.Muell.) Kuntze, Revis. Gen. Pl. 2: 736 (1891).
Western Australia to W. Queensland. 50 NTA QLD WAU.

subsp. **mariae**
S. Northern Territory to W. Queensland. 50 NTA QLD.

subsp. **rigida** (Becc.) Rodd, Telopea 8: 80 (1998).
Northern Territory to NW. Queensland. 50 NTA QLD.
  *Livistona rigida Becc., Webbia 5: 19 (1921).

Livistona mariae subsp. *occidentalis* Rodd = **Livistona nasmophila** Dowe & D.L.Jones

Livistona martiana Mart. = **Pritchardia martii** (Gaudich.) H.Wendl.

Livistona martii Gaudich. = **Pritchardia martii** (Gaudich.) H.Wendl.

Livistona mauritiana Wall. ex Mart. = **Livistona chinensis** var. **chinensis**

Livistona melanocarpa Burret = **Livistona benthamii** F.M.Bailey

Livistona merrillii Becc. in J.R.Perkins & al., Fragm. Fl. Philipp.: 45 (1904).
Philippines. 42 PHI.
  Livistona whitfordii Becc., Webbia 1: 341 (1905).

Livistona microcarpa Becc. = **Livistona rotundifolia** (Lam.) Mart.

Livistona mindorensis Becc. = **Livistona rotundifolia** (Lam.) Mart.

Livistona moluccana H.Wendl. = **?** **[42]**

Livistona muelleri F.M.Bailey, Queensl. Fl. 5: 1683 (1902).
New Guinea to N. Queensland. 43 NWG 50 QLD.
  Livistona humilis var. *novoguineensis* Becc., Webbia 5: 16 (1920).
  Livistona humilis var. *sclirophylla* Becc., Webbia 5: 16 (1920).
  Livistona brassii Burret, Notizbl. Bot. Gart. Berlin-Dahlem 12: 309 (1935).
  Livistona crustacea Burret, J. Arnold Arbor. 20: 189 (1939).

Livistona nasmophila Dowe & D.L.Jones, Austrobaileya 6: 980 (2004).
Western Australia. 50 WAU.
  Livistona mariae subsp. *occidentalis* Rodd, Telopea 8: 81 (1998).

Livistona nitida Rodd, Telopea 8: 96 (1998).
SE. Queensland. 50 QLD.

Livistona occidentalis Hook.f. = **Brahea dulcis** (Kunth) Mart.

Livistona oliviformis (Hassk.) Mart. = **Livistona chinensis** var. **chinensis**

Livistona papuana Becc., Malesia 1: 84 (1877). *Saribus papuanus* (Becc.) Kuntze, Revis. Gen. Pl. 2: 736 (1891).
New Guinea. 43 NWG.

Livistona ramsayi F.Muell. = **Licuala ramsayi** (F.Muell.) Domin

Livistona rigida Becc. = **Livistona mariae** subsp. **rigida** (Becc.) Rodd

Livistona robinsoniana Becc., Philipp. J. Sci., C 6: 230 (1911).
Philippines (Polillo). 42 PHI.

Livistona rotundifolia (Lam.) Mart., Hist. Nat. Palm. 3: 241 (1838).
S. & C. Malesia. 42 JAW LSI mly MOL PHI SUL.
Corypha rotundifolia Lam., Encycl. 2: 131 (1786).
Licuala rotundifolia (Lam.) Blume in J.J.Roemer & J.A.Schultes, Syst. Veg. 7: 1305 (1830). *Saribus rotundifolius* (Lam.) Blume, Rumphia 2: 49 (1838).
Chamaerops biroo Siebold ex Mart., Hist. Nat. Palm. 3: 252 (1838).
Livistona altissima Zoll., Tijdschr. Ned.-Indië 14: 150 (1857).
Livistona microcarpa Becc., Philipp. J. Sci., C 2: 231 (1907). *Livistona rotundifolia* var. *microcarpa* (Becc.) Becc., Philipp. J. Sci. 14: 339 (1919).
Livistona mindorensis Becc., Philipp. J. Sci., C 4: 615 (1909). *Livistona rotundifolia* var. *mindorensis* (Becc.) Becc., Philipp. J. Sci. 14: 339 (1919).
Livistona blancoi Merr., Sp. Blancoan.: 84 (1918).
Livistona rotundifolia var. *luzonensis* Becc., Philipp. J. Sci. 14: 339 (1919).

Livistona rotundifolia var. *luzonensis* Becc. = **Livistona rotundifolia** (Lam.) Mart.

Livistona rotundifolia var. *microcarpa* (Becc.) Becc. = **Livistona rotundifolia** (Lam.) Mart.

Livistona rotundifolia var. *mindorensis* (Becc.) Becc. = **Livistona rotundifolia** (Lam.) Mart.

Livistona rupicola Ridl. = **Maxburretia rupicola** (Ridl.) Furtado

Livistona saribus (Lour.) Merr. ex A.Chev., Bull. Écon. Indochine, n.s., 21: 501 (1919).
SE. China to Indo-China and Philippines. 36 CHS 41 THA VIE 42 BOR JAW MLY PHI.
Chamaerops cochinchinensis Lour., Fl. Cochinch. 2: 657 (1790). *Saribus cochinchinensis* (Lour.) Blume, Rumphia 2: 48 (1838). *Rhapis cochinchinensis* (Lour.) Mart., Hist. Nat. Palm. 3: 254 (1838). *Livistona cochinchinensis* (Lour.) Mart., Hist. Nat. Palm. 3(ed. 2): 242 (1849).
Corypha saribus Lour., Fl. Cochinch.: 212 (1790).
Saribus hasseltii Hassk., Flora 25(Beibl. 2): 16 (1842). *Livistona hasseltii* (Hassk.) Hassk. ex Miq., Palm. Archip. Ind.: 14 (1868).
Livistona spectabilis Griff., Calcutta J. Nat. Hist. 5: 336 (1845).
Livistona diepenhorstii Hassk., Bonplandia 6: 180 (1858). *Pholidocarpus diepenhorstii* (Hassk.) Burret, Notizbl. Bot. Gart. Berlin-Dahlem 15: 327 (1941).
Livistona hoogendorpii Teijsm. & Binn. ex Miq., Palm. Archip. Ind.: 14 (1868). *Saribus hoogendorpii* (Teijsm. & Binn. ex Miq.) Kuntze, Revis. Gen. Pl. 2: 736 (1891). *Sabal hoogendorpii* (Teijsm. & Binn. ex Miq.) L.H.Bailey, Stand. Cycl. Hort. 6: 3045 (1917).
Livistona inaequisecta Becc., Philipp. J. Sci., C 4: 616 (1909).
Livistona vogamii Becc., Webbia 5: 22 (1921).
Livistona tonkinensis Magalon, Contr. Étud. Palmiers Indoch.: 54 (1930).
Livistona fengkaiensis X.W.Wei & M.Y.Xiao, J. S. China Agric. Coll. 8(1): 22 (1982).

Livistona sinensis Griff. = **Livistona chinensis** var. **chinensis**

Livistona speciosa Kurz = **Livistona jenkinsiana** Griff.

Livistona spectabilis Griff. = **Livistona saribus** (Lour.) Merr. ex A.Chev.

Livistona subglobosa (Hassk.) Mart. = **Livistona chinensis** var. **subglobosa** (Hassk.) Becc.

Livistona surru Dowe & Barfod, Austrobaileya 6: 169 (2001).
Papua New Guinea. 43 NWG.

Livistona tahanensis Becc., Webbia 5: 17 (1921).
Pen. Malaysia (Pahang). 42 MLY.

Livistona tonkinensis Magalon = **Livistona saribus** (Lour.) Merr. ex A.Chev.

Livistona tothur Dowe & Barfod, Austrobaileya 6: 171 (2001).
Papua New Guinea. 43 NWG.

Livistona victoriae Rodd, Telopea 8: 123 (1998).
NE. Western Australia to NW. Northern Territory. 50 NTA WAU.

Livistona vidalii Becc. = **Corypha utan** Lam.

Livistona vogamii Becc. = **Livistona saribus** (Lour.) Merr. ex A.Chev.

Livistona whitfordii Becc. = **Livistona merrillii** Becc.

Livistona woodfordii Ridl., Gard. Chron. 1898(1): 177 (1898).
New Guinea to Solomon Is. 43 NWG SOL.
Livistona beccariana Burret, Notizbl. Bot. Gart. Berlin-Dahlem 15: 326 (1941).

Lobia

Lobia O.F.Cook = **Chamaedorea** Willd.

Lobia erosa O.F.Cook = **Chamaedorea tenerrima** Burret

Lodoicea

Lodoicea Comm. ex DC., Bull. Sci. Soc. Philom. Paris 2: 171 (1800).
1 species, Seychelles. 29.

Lodoicea callypige Comm. ex J.St.Hil. = **Lodoicea maldivica** (J.F.Gmel.) Pers. ex H.Wendl.

Lodoicea maldivica (J.F.Gmel.) Pers. ex H.Wendl. in O.C.E.de Kerchove de Denterghem, Palmiers: 250 (1878).
Seychelles (Curieuse, Praslin). 29 SEY.
Cocos maldivica J.F.Gmel., Syst. Nat.: 569 (1791).
Borassus sonneratii Giseke, Prael. Ord. Nat. Pl.: 86 (1792). *Lodoicea sonneratii* (Giseke) Baill., Hist. Pl. 13: 323 (1895).
Lodoicea callypige Comm. ex J.St.Hil., Expos. Fam. Nat. 1: 96 (1805).
Lodoicea sechellarum Labill., Ann. Mus. Natl. Hist. Nat. 9: 140 (1807).
Cocos maritima Comm. ex H.Wendl. in O.C.E.de Kerchove de Denterghem, Palmiers: 241 (1878).

Lodoicea sechellarum Labill. = **Lodoicea maldivica** (J.F.Gmel.) Pers. ex H.Wendl.

Lodoicea sonneratii (Giseke) Baill. = **Lodoicea maldivica** (J.F.Gmel.) Pers. ex H.Wendl.

Lontarus

Lontarus Adans. = **Borassus** L.

Lontarus domestica Gaertn. = **Borassus flabellifer** L.

Lophospatha

Lophospatha Burret = **Salacca** Reinw.

Lophospatha borneensis Burret = **Salacca lophospatha** J.Dransf. & Mogea

Lophothele

Lophothele O.F.Cook = **Chamaedorea** Willd.

Lophothele ramea O.F.Cook = **Chamaedorea liebmannii** Mart.

Loroma

Loroma O.F.Cook = **Archontophoenix** H.Wendl. & Drude

Loroma amethystina O.F.Cook = **Archontophoenix cunninghamiana** (H.Wendl.) H.Wendl. & Drude

Loroma cunninghamiana (H.Wendl.) O.F.Cook = **Archontophoenix cunninghamiana** (H.Wendl.) H.Wendl. & Drude

Louvelia

Louvelia Jum. & H.Perrier = **Ravenea** H.Wendl. ex C.D.Bouché

Louvelia albicans Jum. = **Ravenea albicans** (Jum.) Beentje

Louvelia lakatra Jum. = **Ravenea lakatra** (Jum.) Beentje

Louvelia madagascariensis Jum. & H.Perrier = **Ravenea louvelii** Beentje

Loxococcus

Loxococcus H.Wendl. & Drude, Linnaea 39: 185 (1875).
1 species, Sri Lanka. 40.

Loxococcus rupicola (Thwaites) H.Wendl. & Drude, Linnaea 39: 185 (1875).
SW. Sri Lanka. 40 SRL.
  Ptychosperma rupicola Thwaites, Enum. Pl. Zeyl.: 328 (1864).

Lytocaryum

Lytocaryum Toledo, Arq. Bot. Estado São Paulo, n.s., f.m., 2: 6 (1944).
2 species, SE. Brazil. 84.
  Glaziova Mart. ex H.Wendl., Florist & Pomol. 1871: 116 (1871).
  Microcoelum Burret & Potztal, Willdenowia 1: 378 (1956).

Lytocaryum hoehnei (Burret) Toledo, Arq. Bot. Estado São Paulo, n.s., f.m., 2: 7 (1944).
Brazil (São Paulo). 84 BZL.
  Syagrus hoehnei Burret, Notizbl. Bot. Gart. Berlin-Dahlem 13: 678 (1937).

Lytocaryum insigne (Drude) Toledo = **Lytocaryum weddellianum** (H.Wendl.) Toledo

Lytocaryum weddellianum (H.Wendl.) Toledo, Arq. Bot. Estado São Paulo, n.s., f.m., 2: 8 (1944).
Brazil (Espírito Santo, Rio de Janeiro). 84 BZL.
  Cocos weddelliana H.Wendl., Florist & Pomol. 1871: 114 (1871). *Syagrus weddelliana* (H.Wendl.) Becc., Agric. Colon. 10: 468 (1916). *Microcoelum weddellianum* (H.Wendl.) H.E.Moore, Gentes Herb. 9: 267 (1963).
  Glaziova elegantissima H.Wendl., Florist & Pomol. 1871: 116 (1871).
  Cocos insignis (Drude) Mart. ex H.Wendl. in O.C.E.de Kerchove de Denterghem, Palmiers: 241 (1878). *Glaziova insignis* Drude in C.F.P.von Martius & auct. suc. (eds.), Fl. Bras. 3(2): 398 (1881). *Cocos insignis* (Drude) Mart. ex H.Wendl. in O.C.E.de Kerchove de Denterghem, Palmiers: 241 (1878). *Calappa insignis* (Drude) Kuntze, Revis. Gen. Pl. 2: 982 (1891). *Syagrus insignis* (Drude) Becc., Agric. Colon. 10: 467 (1916). *Lytocaryum insigne* (Drude) Toledo, Arq. Bot. Estado São Paulo, n.s., f.m., 2: 8 (1944). *Microcoelum insigne* (Drude) Burret & Potztal, Willdenowia 1: 388 (1956).
  Glaziova martiana Glaz. ex Drude in C.F.P.von Martius & auct. suc. (eds.), Fl. Bras. 3(2): 397 (1881). *Microcoelum martianum* (Glaz. ex Drude) Burret & Potztal, Willdenowia 1: 388 (1956).
  Calappa elegantina Kuntze, Revis. Gen. Pl. 2: 982 (1891).
  Cocos pynaertii auct., Gard. Chron., III, 1891(1): 683 (1891).

Mackeea

Mackeea H.E.Moore = **Kentiopsis** Brongn.

Mackeea magnifica H.E.Moore = **Kentiopsis magnifica** (H.E.Moore) Pintaud & Hodel

Macrocladus

Macrocladus Griff. = **Orania** Zipp.

Macrocladus sylvicola Griff. = **Orania sylvicola** (Griff.) H.E.Moore

Macrophloga

Macrophloga Becc. = **Dypsis** Noronha ex Mart.

Macrophloga decipiens (Becc.) Becc. = **Dypsis decipiens** (Becc.) Beentje & J.Dransf.

Malortiea

Malortiea H.Wendl. = **Reinhardtia** Liebm.

Malortiea gracilis H.Wendl. = **Reinhardtia gracilis** (H.Wendl.) Burret

Malortiea koschnyana H.Wendl. & Dammer = **Reinhardtia koschnyana** (H.Wendl. & Dammer) Burret

Malortiea lacerata H.Wendl. = **?**

Malortiea latisecta H.Wendl. = **Reinhardtia latisecta** (H.Wendl.) Burret

Malortiea pumila Dugand = **Reinhardtia koschnyana** (H.Wendl. & Dammer) Burret

Malortiea rostrata (Burret) L.H.Bailey = **Reinhardtia gracilis** var. **rostrata** (Burret) H.E.Moore

Malortiea simiarum Standl. & L.O.Williams = **Prestoea longipetiolata** var. **longipetiolata**

Malortiea simplex H.Wendl. = **Reinhardtia simplex** (H.Wendl.) Burret

Malortiea tuerckheimii Dammer = **Chamaedorea tuerckheimii** (Dammer) Burret

Manicaria

Manicaria Gaertn., Fruct. Sem. Pl. 2: 468 (1791).
1 species, Trinidad, C. & S. Trop. America. 80 81 82 83 84.
  Pilophora Jacq., Fragm. Bot.: 32 (1800).

Manicaria atricha Burret = **Manicaria saccifera** Gaertn.

Manicaria martiana Burret = **Manicaria saccifera** Gaertn.

Manicaria plukenetii Griseb. & H.Wendl. = **Manicaria saccifera** Gaertn.

Manicaria saccifera Gaertn., Fruct. Sem. Pl. 2: 468 (1791). *Pilophora saccifera* (Gaertn.) H.Wendl. in O.C.E.de Kerchove de Denterghem, Palmiers: 253 (1878).
  Trinidad, C. & S. Trop. America. 80 BLZ COS GUA HON NIC PAN 81 TRT 82 FRG GUY SUR VEN 83 CLM ECU PER 84 BZN.
  Pilophora testicularis Jacq., Fragm. Bot.: 32 (1800).
  Manicaria plukenetii Griseb. & H.Wendl. in A.H.R.Grisebach, Fl. Brit. W. I.: 518 (1864).
  Manicaria martiana Burret, Notizbl. Bot. Gart. Berlin-Dahlem 10: 392 (1928).
  Manicaria atricha Burret, Notizbl. Bot. Gart. Berlin-Dahlem 10: 1013 (1930).

Marara

Marara H.Karst. = **Aiphanes** Willd.

Marara aculeata (Willd.) H.Karst. ex H.Wendl. = **Aiphanes horrida** (Jacq.) Burret

Marara bicuspidata H.Karst. = **Aiphanes horrida** (Jacq.) Burret

Marara caryotifolia (Kunth) H.Karst. ex H.Wendl. = **Aiphanes horrida** (Jacq.) Burret

Marara erinacea H.Karst. = **Aiphanes erinacea** (H.Karst.) H.Wendl.

Markleya

Markleya Bondar = **Attalea** Kunth

Markleya dahlgreniana Bondar = **Attalea dahlgreniana** (Bondar) Wess.Boer

Marojejya

Marojejya Humbert, Mém. Inst. Sci. Madagascar, Sér. B, Biol. Vég. 6: 92 (1955).
  2 species, Madagascar. 29.

Marojejya darianii J.Dransf. & N.W.Uhl, Principes 28: 151 (1984).
  NE. Madagascar. 29 MDG.

Marojejya insignis Humbert, Mém. Inst. Sci. Madagascar, Sér. B, Biol. Vég. 6: 94 (1955).
  E. Madagascar. 29 MDG.

Martinezia

Martinezia Ruiz & Pav. = **Prestoea** Hook.f.

Martinezia abrupta Ruiz & Pav. = **?** [83 PER]

Martinezia acanthophylla (Mart.) Becc. = **Aiphanes minima** (Gaertn.) Burret

Martinezia aculeata (Willd.) Klotzsch = **Aiphanes horrida** (Jacq.) Burret

Martinezia aiphanes Mart. = **Aiphanes horrida** (Jacq.) Burret

Martinezia caryotifolia Kunth = **Aiphanes horrida** (Jacq.) Burret

Martinezia ciliata Ruiz & Pav. = **Bactris gasipaes** var. **gasipaes**

Martinezia corallina Mart. = **Aiphanes minima** (Gaertn.) Burret

Martinezia disticha Linden = **?** [83 CLM]

Martinezia elegans Linden & H.Wendl. = **Aiphanes horrida** (Jacq.) Burret

Martinezia ensiformis Ruiz & Pav. = **Prestoea ensiformis** (Ruiz & Pav.) H.E.Moore

Martinezia ernestii Burret = **Aiphanes horrida** (Jacq.) Burret

Martinezia erosa (Mart.) Linden = **Aiphanes minima** (Gaertn.) Burret

Martinezia interrupta Ruiz & Pav. = **Geonoma interrupta** var. **interrupta**

Martinezia killipii Burret = **Aiphanes horrida** (Jacq.) Burret

Martinezia lanceolata Ruiz & Pav. = **Chamaedorea pinnatifrons** (Jacq.) Oerst.

Martinezia leucophoeus auct. = **?** [83 CLM]

Martinezia lindeniana H.Wendl. = **Aiphanes lindeniana** (H.Wendl.) H.Wendl.

Martinezia linearis Ruiz & Pav. = **Chamaedorea linearis** (Ruiz & Pav.) Mart.

Martinezia roezlii auct. = **?** [83 CLM]

Martinezia truncata Brongn. ex Mart. = **Aiphanes horrida** (Jacq.) Burret

Martinezia ulei Dammer = **Aiphanes ulei** (Dammer) Burret

Martinezia ulei Dammer = **Aiphanes horrida** (Jacq.) Burret

Mascarena

Mascarena L.H.Bailey = **Hyophorbe** Gaertn.

Mascarena lagenicaulis L.H.Bailey = **Hyophorbe lagenicaulis** (L.H.Bailey) H.E.Moore

Mascarena revaughanii L.H.Bailey = **Hyophorbe lagenicaulis** (L.H.Bailey) H.E.Moore

Mascarena verschaffieltii (H.Wendl.) L.H.Bailey = **Hyophorbe verschaffeltii** H.Wendl.

Masoala

Masoala Jum., Ann. Inst. Bot.-Géol. Colon. Marseille, V, 1(1): 8 (1933).
  2 species, Madagascar. 29.

Masoala kona Beentje in J.Dransfield & H.Beentje, Palms Madagascar: 425 (1995).
  SC. Madagascar (Ifanadiana reg.). 29 MDG.

Masoala madagascariensis Jum., Ann. Inst. Bot.-Géol. Colon. Marseille, V, 1(1): 8 (1933).
  NE. Madagascar. 29 MDG.

Mauranthe

Mauranthe O.F.Cook = **Chamaedorea** Willd.

Mauranthe lunata (Liebm.) O.F.Cook = **Chamaedorea oblongata** Mart.

Mauritia

Mauritia L.f., Suppl. Pl.: 70 (1782).
  2 species, Trinidad to S. Trop. America. 81 82 83 84. *Orophoma* Drude in C.F.P.von Martius & auct. suc. (eds.), Fl. Bras. 3(2): 294 (1881).

Mauritia aculeata Kunth = **Mauritiella aculeata** (Kunth) Burret

Mauritia aculeata Mart. = **Mauritiella armata** (Mart.) Burret

Mauritia amazonica Barb.Rodr. = **Mauritiella aculeata** (Kunth) Burret

Mauritia armata Mart. = **Mauritiella armata** (Mart.) Burret

Mauritia campylostachys (Burret) Balick = **Mauritiella armata** (Mart.) Burret

Mauritia carana Wallace ex Archer, Hooker's J. Bot. Kew Gard. Misc. 7: 213 (1855). *Orophoma carana* (Wallace ex Archer) Spruce ex Drude in C.F.P.von Martius & auct. suc. (eds.), Fl. Bras. 3(2): 294 (1881). Amazon Basin. 82 VEN 83 CLM PER 84 BZN.

Mauritia casiquiarensis Spruce = **Lepidocaryum tenue** var. **casiquiarense** (Spruce) A.J.Hend.

Mauritia cataractarum (Dugand) Balick = **Mauritiella aculeata** (Kunth) Burret

Mauritia duckei (Burret) Balick = **Mauritiella armata** (Mart.) Burret

Mauritia flexuosa L.f., Suppl. Pl.: 454 (1782). Trinidad to S. Trop. America. 81 TRT 82 FRG GUY SUR VEN 83 BOL CLM ECU PER 84 BZN.
  Mauritia vinifera Mart., Hist. Nat. Palm. 2: 42 (1824).
  Mauritia sagus Schult. & Schult.f., Syst. Veg. 7: 1321 (1830). Provisional synonym.
  Mauritia setigera Griseb. & H.Wendl. in A.H.R. Grisebach, Fl. Brit. W. I.: 515 (1864).
  Saguerus americanus H.Wendl. in O.C.E.de Kerchove de Denterghem, Palmiers: 256 (1878).
  Mauritia sphaerocarpa Burret, Notizbl. Bot. Gart. Berlin-Dahlem 10: 569 (1929).
  Mauritia minor Burret, Notizbl. Bot. Gart. Berlin-Dahlem 11: 1 (1930).

Mauritia gracilis (Mart.) Spruce = **Lepidocaryum tenue** var. **gracile** (Mart.) A.J.Hend.

Mauritia gracilis Wallace = **Mauritiella aculeata** (Kunth) Burret

Mauritia guainiensis Spruce = **Lepidocaryum tenue** var. **casiquiarense** (Spruce) A.J.Hend.

Mauritia huebneri Burret = **Mauritiella armata** (Mart.) Burret

Mauritia intermedia Burret = **Mauritiella armata** (Mart.) Burret

Mauritia limnophylla Barb.Rodr. = **Mauritiella aculeata** (Kunth) Burret

Mauritia macroclada Burret = **Mauritiella macroclada** (Burret) Burret

Mauritia macrospadix (Burret) Balick = **Mauritiella armata** (Mart.) Burret

Mauritia martiana Spruce = **Mauritiella armata** (Mart.) Burret

Mauritia minor Burret = **Mauritia flexuosa** L.f.

Mauritia nannostachys (Burret) Balick = **Mauritiella armata** (Mart.) Burret

Mauritia pacifica (Dugand) Balick = **Mauritiella macroclada** (Burret) Burret

Mauritia peruviana Becc. = **Mauritiella armata** (Mart.) Burret

Mauritia piritu Linden = **? [84]**

Mauritia pumila Wallace = **Mauritiella armata** (Mart.) Burret

Mauritia quadripartita Spruce = **Lepidocaryum tenue** Mart. var. **tenue**

Mauritia sagus Schult. & Schult.f. = **Mauritia flexuosa** L.f.

Mauritia setigera Griseb. & H.Wendl. = **Mauritia flexuosa** L.f.

Mauritia sphaerocarpa Burret = **Mauritia flexuosa** L.f.

Mauritia subinermis Spruce = **Mauritiella armata** (Mart.) Burret

Mauritia tenuis (Mart.) Spruce = **Lepidocaryum tenue** Mart.

Mauritia vinifera Mart. = **Mauritia flexuosa** L.f.

Mauritiella

Mauritiella Burret, Notizbl. Bot. Gart. Berlin-Dahlem 12: 609 (1935).
  3 species, S. Trop. America. 82 83 84.
  Lepidococcus H.Wendl. & Drude in O.C.E.de Kerchove de Denterghem, Palmiers: 249 (1878).

Mauritiella aculeata (Kunth) Burret, Notizbl. Bot. Gart. Berlin-Dahlem 12: 609 (1935).
  SE. Colombia to N. Brazil. 82 VEN 83 CLM 84 BZN.
  *Mauritia aculeata Kunth in F.W.H.von Humboldt, A.J.A.Bonpland & C.S.Kunth, Nov. Gen. Sp. 1: 311 (1816). *Lepidococcus aculeatus* (Kunth) H.Wendl. & Drude in O.C.E.de Kerchove de Denterghem, Palmiers: 249 (1878).
  Mauritia gracilis Wallace, Palm Trees Amazon: 57 (1853).
  Mauritia amazonica Barb.Rodr., Enum. Palm. Nov.: 43 (1875).
  Mauritia limnophylla Barb.Rodr., Enum. Palm. Nov.: 18 (1875).
  Mauritiella cataractarum Dugand, Revista Acad. Colomb. Ci. Exact. 8: 385 (1951). *Mauritia cataractarum* (Dugand) Balick, Brittonia 33: 460 (1981).

Mauritiella armata (Mart.) Burret, Notizbl. Bot. Gart. Berlin-Dahlem 12: 611 (1935).
  S. Trop. America. 82 GUY SUR VEN 83 BOL CLM ECU PER 84 BZC BZE BZL BZN.
  Mauritia aculeata Mart., Hist. Nat. Palm. 2: 47 (1824), nom. illeg. *Mauritia martiana* Spruce, J. Linn. Soc., Bot. 11: 171 (1869). *Mauritiella martiana* (Spruce) Burret, Notizbl. Bot. Gart. Berlin-Dahlem 12: 611 (1935). *Lepidococcus martianus* (Spruce) A.D. Hawkes, Arq. Bot. Estado São Paulo, n.s., f.m., 2: 174 (1952).
  *Mauritia armata Mart., Hist. Nat. Palm. 2: 45 (1824). *Lepidococcus armatus* (Mart.) H.Wendl. & Drude in O.C.E.de Kerchove de Denterghem, Palmiers: 249 (1878).
  Mauritia pumila Wallace, Palm Trees Amazon: 59 (1853). *Lepidococcus pumilus* (Wallace) H.Wendl. & Drude in O.C.E.de Kerchove de Denterghem, Palmiers: 249 (1878). *Mauritiella pumila* (Wallace) Burret, Notizbl. Bot. Gart. Berlin-Dahlem 12: 611 (1935).
  Mauritia subinermis Spruce, J. Linn. Soc., Bot. 11: 171 (1869). *Orophoma subinermis* (Spruce) Drude in C.F.P.von Martius & auct. suc. (eds.), Fl. Bras. 3(2): 294 (1881). *Mauritiella subinermis* (Spruce) Burret, Notizbl. Bot. Gart. Berlin-Dahlem 12: 611 (1935). *Lepidococcus subinermis* (Spruce) A.D.Hawkes, Arq. Bot. Estado São Paulo, n.s., f.m., 2: 174 (1952).
  Oenocarpus dealbatus H.Wendl. in O.C.E.de Kerchove de Denterghem, Palmiers: 252 (1878). Provisional synonym.
  Mauritia peruviana Becc., Ann. Roy. Bot. Gard. (Calcutta) 12(2): 225 (1918). *Mauritiella peruviana* (Becc.) Burret, Notizbl. Bot. Gart. Berlin-Dahlem

12: 611 (1935). *Lepidococcus peruvianus* (Becc.)
A.D.Hawkes, Arq. Bot. Estado São Paulo, n.s., f.m.,
2: 174 (1952).

Mauritia huebneri Burret, Notizbl. Bot. Gart. Berlin-
Dahlem 10: 570 (1929). *Mauritiella huebneri* (Burret)
Burret, Notizbl. Bot. Gart. Berlin-Dahlem 12: 611
(1935). *Lepidococcus huebneri* (Burret) A.D.Hawkes,
Arq. Bot. Estado São Paulo, n.s., f.m., 2: 174 (1952).

Mauritia intermedia Burret, Notizbl. Bot. Gart. Berlin-
Dahlem 10: 572 (1929). *Mauritiella intermedia*
(Burret) Burret, Notizbl. Bot. Gart. Berlin-Dahlem
12: 611 (1935). *Lepidococcus intermedius* (Burret)
A.D.Hawkes, Arq. Bot. Estado São Paulo, n.s., f.m.,
2: 174 (1952).

Mauritiella duckei Burret, Notizbl. Bot. Gart. Berlin-
Dahlem 12: 609 (1935). *Lepidococcus duckei*
(Burret) A.D.Hawkes, Arq. Bot. Estado São Paulo,
n.s., f.m., 2: 173 (1952). *Mauritia duckei* (Burret)
Balick, Brittonia 33: 460 (1981).

Mauritiella campylostachys Burret, Notizbl. Bot. Gart.
Berlin-Dahlem 15: 755 (1942). *Mauritia
campylostachys* (Burret) Balick, Brittonia 33: 460
(1981).

Mauritiella macrospadix Burret, Notizbl. Bot. Gart.
Berlin-Dahlem 15: 753 (1942). *Mauritia
macrospadix* (Burret) Balick, Brittonia 33: 460
(1981).

Mauritiella nannostachys Burret, Notizbl. Bot. Gart.
Berlin-Dahlem 15: 754 (1942). *Mauritia
nannostachys* (Burret) Balick, Brittonia 33: 460
(1981).

Mauritiella campylostachys Burret = **Mauritiella
armata** (Mart.) Burret

Mauritiella cataractarum Dugand = **Mauritiella aculeata**
(Kunth) Burret

Mauritiella duckei Burret = **Mauritiella armata** (Mart.)
Burret

Mauritiella huebneri (Burret) Burret = **Mauritiella
armata** (Mart.) Burret

Mauritiella intermedia (Burret) Burret = **Mauritiella
armata** (Mart.) Burret

Mauritiella macroclada (Burret) Burret, Notizbl. Bot.
Gart. Berlin-Dahlem 12: 611 (1935).
W. Colombia to W. Ecuador. 83 CLM ECU.
*Mauritia macroclada Burret, Notizbl. Bot. Gart.
Berlin-Dahlem 10: 574 (1929). *Lepidococcus
macrocladus* (Burret) A.D.Hawkes, Arq. Bot.
Estado São Paulo, n.s., f.m., 2: 174 (1952).
Mauritia pacifica Dugand, Caldasia 2: 387 (1944).
Mauritia pacifica (Dugand) Balick, Brittonia 33:
460 (1981).

Mauritiella macrospadix Burret = **Mauritiella armata**
(Mart.) Burret

Mauritiella martiana (Spruce) Burret = **Mauritiella
armata** (Mart.) Burret

Mauritiella nannostachys Burret = **Mauritiella armata**
(Mart.) Burret

Mauritiella pacifica Dugand = **Mauritiella macroclada**
(Burret) Burret

Mauritiella peruviana (Becc.) Burret = **Mauritiella
armata** (Mart.) Burret

Mauritiella pumila (Wallace) Burret = **Mauritiella
armata** (Mart.) Burret

Mauritiella subinermis (Spruce) Burret = **Mauritiella
armata** (Mart.) Burret

Maxburretia

Maxburretia Furtado, Gard. Bull. Straits Settlem. 11:
240 (1941).
3 species, Pen. Thailand to Pen. Malaysia. 41 42.
Liberbaileya Furtado, Gard. Bull. Straits Settlem. 11:
238 (1941).
Symphyogyne Burret, Notizbl. Bot. Gart. Berlin-
Dahlem 15: 316 (1941), nom. illeg.

Maxburretia furtadoana J.Dransf., Gentes Herb. 11:
195 (1978).
Pen. Thailand. 41 THA.

Maxburretia gracilis (Burret) J.Dransf., Gentes Herb.
11: 194 (1978).
Pen. Malaysia (Langkawi Is.). 42 MLY.
Liberbaileya lankawiensis Furtado, Gard. Bull. Straits
Settlem. 11: 238 (1941).
*Symphyogyne gracilis Burret, Notizbl. Bot. Gart.
Berlin-Dahlem 15: 317 (1941). *Liberbaileya
gracilis* (Burret) Burret & Potztal, Willdenowia 1:
530 (1956).

Maxburretia rupicola (Ridl.) Furtado, Gard. Bull.
Straits Settlem. 11: 240 (1941).
Pen. Malaysia (Selangor). 42 MLY.
*Livistona rupicola Ridl., J. Straits Branch Roy. Asiat.
Soc. 41: 41 (1903). *Symphyogyne rupicola* (Ridl.)
Burret, Notizbl. Bot. Gart. Berlin-Dahlem 15: 318
(1941).

Maximbignya

Maximbignya Glassman = **Attalea** Kunth

Maximbignya dahlgreniana (Bondar) Glassman =
Attalea dahlgreniana (Bondar) Wess.Boer

Maximiliana

Maximiliana Mart. = **Attalea** Kunth

Maximiliana argentinensis Speg. = **?** [85 AG]

Maximiliana attaleoides Barb.Rodr. = **Attalea
attaleoides** (Barb.Rodr.) Wess.Boer

Maximiliana caribaea Griseb. & H.Wendl. = **Attalea
maripa** (Aubl.) Mart.

Maximiliana crassispatha Mart. = **Attalea crassispatha**
(Mart.) Burret

Maximiliana elegans H.Karst. = **Attalea maripa** (Aubl.)
Mart.

Maximiliana inajai Spruce = **Syagrus inajai** (Spruce)
Becc.

Maximiliana insignis Mart. = **Attalea insignis** (Mart.)
Drude

Maximiliana longirostrata Barb.Rodr. = **Attalea maripa**
(Aubl.) Mart.

Maximiliana macrogyne Burret = **Attalea maripa**
(Aubl.) Mart.

Maximiliana macropetala Burret = **Attalea maripa**
(Aubl.) Mart.

Maximiliana maripa (Aubl.) Drude = **Attalea maripa**
(Aubl.) Mart.

Maximiliana martiana H.Karst. = **Attalea maripa**
(Aubl.) Mart.

Maximiliana orinocensis (Spruce) Speg. = **Syagrus
orinocensis** (Spruce) Burret

Maximiliana princeps Mart. = **Attalea phalerata** Mart.
ex Spreng.

Maximiliana regia Mart. = **Attalea maripa** (Aubl.) Mart.

Maximiliana stenocarpa Burret = **Attalea maripa** (Aubl.) Mart.

Maximiliana tetrasticha Drude· = **Attalea maripa** (Aubl.) Mart.

Maximiliana venatorum H.Wendl. = **?** [83 PER]

Medemia

Medemia Wurttenb. ex H.Wendl., Bot. Zeitung (Berlin) 39: 89 (1881).
1 species, S. Egypt to N. Sudan. 20 24.

Medemia abiadensis H.Wendl. = **Medemia argun** (Mart.) Wurttenb. ex H.Wendl.

Medemia argun (Mart.) Wurttenb. ex H.Wendl., Bot. Zeitung (Berlin) 39: 93 (1881).
S. Egypt to N. Sudan. 20 EGY 24 SUD.
Areca passalacquae Kunth, Ann. Sci. Nat. (Paris) 8: 420 (1826), fossil name.
*Hyphaene argun Mart., Hist. Nat. Palm. 3(ed. 2): 227 (1845).
Medemia abiadensis H.Wendl., Bot. Zeitung (Berlin) 39: 93 (1881).

Medemia nobilis (Hildebr. & H.Wendl.) Gall. = **Bismarckia nobilis** Hildebr. & H.Wendl.

Meiota

Meiota O.F.Cook = **Chamaedorea** Willd.

Meiota campechana O.F.Cook = **Chamaedorea seifrizii** Burret

Metasocratea

Metasocratea Dugand = **Socratea** H.Karst.

Metasocratea hecatonandra Dugand = **Socratea hecatonandra** (Dugand) R.Bernal

Metroxylon

Metroxylon Rottb., Nye Saml. Kongel. Dansk. Vidensk. Selsk. Skr. 2: 525 (1783).
7 species, Maluku to W. Pacific. (41) 42 43 69 62.
Sagus Steck, Sagu: 21 (1757).
Coelococcus H.Wendl., Bonplandia 10: 199 (1862).

Metroxylon amicarum (H.Wendl.) Hook.f., Rep. Progr. Condition Roy. Bot. Gard. Kew 1882: 68 (1884).
Caroline Is. (Pohnpei, Chuuk). 62 CRL mrn.
*Sagus amicarum H.Wendl., Bot. Zeitung (Berlin) 36: 115 (1878). *Coelococcus amicarum* (H.Wendl.) W.Wight, Contr. U. S. Natl. Herb. 9: 244 (1905).
Coelococcus carolinensis Dingler, Bot. Centralbl. 32: 349 (1887). *Metroxylon carolinense* (Dingler) Becc., Denkschr. Kaiserl. Akad. Wiss., Wien. Math.-Naturwiss. Kl. 89: 502 (1913 publ. 1914).
Metroxylon amicarum var. *majus* Becc., Ann. Roy. Bot. Gard. (Calcutta) 12(2): 188 (1918).

Metroxylon amicarum var. *majus* Becc. = **Metroxylon amicarum** (H.Wendl.) Hook.f.

Metroxylon bougainvillense Becc. = **Metroxylon salomonense** (Warb.) Becc.

Metroxylon carolinense (Dingler) Becc. = **Metroxylon amicarum** (H.Wendl.) Hook.f.

Metroxylon elatum Scheff. = **Heterospathe elata** var. **elata**

Metroxylon elatum Mart. = **Pigafetta elata** (Mart.) H.Wendl.

Metroxylon filare (Giseke) Mart. = **Pigafetta filaris** (Giseke) Becc.

Metroxylon hermaphroditum Hassk. = **Metroxylon sagu** Rottb.

Metroxylon inerme (Roxb.) Mart. = **Metroxylon sagu** Rottb.

Metroxylon laeve (Giseke) Mart. = **Metroxylon sagu** Rottb.

Metroxylon longispinum (Giseke) Mart. = **Metroxylon sagu** Rottb.

Metroxylon micracanthum Mart. = **Metroxylon sagu** Rottb.

Metroxylon microcarpum Kunth = **Pigafetta filaris** (Giseke) Becc.

Metroxylon microspermum Kunth = **Pigafetta filaris** (Giseke) Becc.

Metroxylon oxybracteatum Warb. ex K.Schum. & Lauterb. = **Metroxylon sagu** Rottb.

Metroxylon paulcoxii McClatchey, Novon 8: 255 (1998).
Western Samoa. 60 SAM.

Metroxylon ruffia (Jacq.) Spreng. = **Raphia farinifera** (Gaertn.) Hyl.

Metroxylon rumphii (Willd.) Mart. = **Metroxylon sagu** Rottb.

Metroxylon sago K.D.Koenig = **Metroxylon sagu** Rottb.

Metroxylon sagu Rottb., Nye Saml. Kongel. Dansk. Vidensk. Selsk. Skr. 2: 527 (1783). *Sagus sagu* (Rottb.) H.Karst., Ill. Repet. Pharm.-Med. Bot.: 412 (1886).
Maluku to New Guinea. (41) tha 42 bor jaw mly MOL sum 43 NWG sol.
Sagus genuina Giseke, Prael. Ord. Nat. Pl.: 94 (1792).
Sagus genuina var. *laevis* Giseke, Prael. Ord. Nat. Pl.: 94 (1792). *Metroxylon laeve* (Giseke) Mart., Hist. Nat. Palm. 3: 215 (1838).
Sagus genuina var. *longispina* Giseke, Prael. Ord. Nat. Pl.: 94 (1792). *Metroxylon longispinum* (Giseke) Mart., Hist. Nat. Palm. 3: 216 (1838). *Sagus longispina* (Giseke) Blume, Rumphia 2: 154 (1843). *Metroxylon sagu* f. *longispinum* (Giseke) Rauwerd., Principes 30: 175 (1986).
Sagus genuina var. *sylvestris* Giseke, Prael. Ord. Nat. Pl.: 94 (1792). *Sagus sylvestris* (Giseke) Blume, Rumphia 2: 153 (1843). *Metroxylon sylvestre* (Giseke) Mart., Hist. Nat. Palm. 3(ed. 2): 215 (1845).
Metroxylon sago K.D.Koenig, Ann. Bot. (König & Sims) 1: 193 (1805), orth. var.
Sagus americana Poir. in J.B.A.M.de Lamarck, Encycl. 6: 395 (1805).
Sagus rumphii Willd., Sp. Pl. 4: 404 (1805). *Metroxylon rumphii* (Willd.) Mart., Hist. Nat. Palm. 3: 214 (1838).
Sagus spinosa Roxb., Hort. Bengal.: 68 (1814), nom. inval.
Sagus inermis Roxb., Fl. Ind. ed. 1832, 3: 623 (1832). *Metroxylon inerme* (Roxb.) Mart., Hist. Nat. Palm. 3(ed. 2): 215 (1845).
Metroxylon hermaphroditum Hassk., Tijdschr. Natuurl. Gesch. Physiol. 9: 175 (1842).
Sagus micracantha (Mart.) Blume, Rumphia 2: 153 (1843).
Metroxylon micracanthum Mart., Hist. Nat. Palm. 3: 216 (1838). *Sagus micracantha* (Mart.) Blume, Rumphia 2: 153 (1843). *Metroxylon sagu* f. *micracanthum* (Mart.) Rauwerd., Principes 30: 175 (1986).

Sagus koenigii Griff., Calcutta J. Nat. Hist. 5: 19 (1845).

Metroxylon oxybracteatum Warb. ex K.Schum. & Lauterb., Fl. Schutzgeb. Südsee: 202 (1900).

Metroxylon squarrosum Becc., Ann. Roy. Bot. Gard. (Calcutta) 12(2): 182 (1918).

Sagus laevis Jack, Malayan Misc. 3: 9 (1923).

Metroxylon sagu f. *tuberatum* Rauwerd., Principes 30: 175 (1986).

Metroxylon sagu f. *longispinum* (Giseke) Rauwerd. = **Metroxylon sagu** Rottb.

Metroxylon sagu f. *micracanthum* (Mart.) Rauwerd. = **Metroxylon sagu** Rottb.

Metroxylon sagu f. *tuberatum* Rauwerd. = **Metroxylon sagu** Rottb.

Metroxylon salomonense (Warb.) Becc., Denkschr. Kaiserl. Akad. Wiss., Wien. Math.-Naturwiss. Kl. 89: 502 (1913 publ. 1914).
E. New Guinea to Vanuatu. 43 BIS NWG SOL 60 SCZ VAN.
*Coelococcus salomonensis Warb., Ber. Deutsch. Bot. Ges. 14: 141 (1896).
Metroxylon bougainvillense Becc., Denkschr. Kaiserl. Akad. Wiss., Wien. Math.-Naturwiss. Kl. 89: 504 (1913 publ. 1914).

Metroxylon squarrosum Becc. = **Metroxylon sagu** Rottb.

Metroxylon sylvestre (Giseke) Mart. = **Metroxylon sagu** Rottb.

Metroxylon taedigerum (Mart.) Spreng. = **Raphia taedigera** (Mart.) Mart.

Metroxylon textile Welw. = **Raphia textilis** Welw.

Metroxylon upoluense Becc., Ann. Roy. Bot. Gard. (Calcutta) 12(2): 184 (1918).
Samoa. 60 SAM. — Provisionally accepted.

Metroxylon viniferum (P.Beauv.) Spreng. = **Raphia vinifera** P.Beauv.

Metroxylon vitiense (H.Wendl.) Hook.f., Rep. Progr. Condition Roy. Bot. Gard. Kew 1882: 68 (1884).
Fiji. 60 FIJ.
*Coelococcus vitiensis H.Wendl., Bonplandia 10: 199 (1862). *Sagus vitiensis* (H.Wendl.) H.Wendl. in B.Seemann, Viti, App.: 444 (1862).

Metroxylon warburgii (Heim) Becc., Ann. Roy. Bot. Gard. (Calcutta) 12(2): 182 (1918).
SW. Pacific. 60 SAM SCZ VAN.
*Coelococcus warburgii Heimerl, Bull. Soc. Bot. France 50: 572 (1903).

Micrococos

Micrococos Phil. = **Jubaea** Kunth

Micrococos chilensis (Molina) Phil. = **Jubaea cilensis**

Microcoelum

Microcoelum Burret & Potztal = **Lytocaryum** Toledo

Microcoelum insigne (Drude) Burret & Potztal = **Lytocaryum weddellianum** (H.Wendl.) Toledo

Microcoelum martianum (Glaz. ex Drude) Burret & Potztal = **Lytocaryum weddellianum** (H.Wendl.) Toledo

Microcoelum weddellianum (H.Wendl.) H.E.Moore = **Lytocaryum weddellianum** (H.Wendl.) Toledo

Microkentia

Microkentia H.Wendl. ex Hook.f. = **Basselinia** Vieill.

Microkentia billardieri (Brongn.) Hook.f. ex Salomon = **Basselinia gracilis** (Brongn. & Gris) Vieill.

Microkentia deplanchei (Brongn. & Gris) Hook.f. ex Salomon = **Basselinia deplanchei** (Brongn. & Gris) Vieill.

Microkentia eriostachys (Brongn.) Hook.f. ex Salomon = **Basselinia gracilis** (Brongn. & Gris) Vieill.

Microkentia gracilis (Brongn. & Gris) Hook.f. ex Salomon = **Basselinia gracilis** (Brongn. & Gris) Vieill.

Microkentia heterophylla Becc. ex Daniker = **Basselinia gracilis** (Brongn. & Gris) Vieill.

Microkentia pancheri (Brongn. & Gris) Hook.f. ex Salomon = **Basselinia pancheri** (Brongn. & Gris) Vieill.

Microkentia schlechteri Dammer = **Basselinia deplanchei** (Brongn. & Gris) Vieill.

Microkentia surculosa (Brongn.) Hook.f. ex Salomon = **Basselinia deplanchei** (Brongn. & Gris) Vieill.

Micronoma

Micronoma H.Wendl. ex Benth. & Hook.f. = **?**

× Microphoenix

× *Microphoenix* Naudin ex Carr. = **Chamaerops** × **Phoenix**

× *Microphoenix decipiens* Naudin ex Carr. = **Chamaerops humilis** × **Phoenix dactylifera**

× *Microphoenix sahuti* Carrière = **Chamaerops humilis** × **Phoenix dactylifera** × **Trachycarpus fortunei**

Migandra

Migandra O.F.Cook = **Chamaedorea** Willd.

Mischophloeus

Mischophloeus Scheff. = **Areca** L.

Mischophloeus paniculatus (Miq.) Scheff. = **Areca vestiaria** Giseke

Mischophloeus vestiarius (Giseke) Merr. = **Areca vestiaria** Giseke

Molinaea

Molinaea Bertero = **Jubaea** Kunth

Molinaea micrococos Bertero = **Jubaea chilensis** (Molina) Baill.

Moratia

Moratia H.E.Moore, Gentes Herb. 12: 18 (1980).
1 species, New Caledonia. 60.

Moratia cerifera H.E.Moore, Gentes Herb. 12: 22 (1980).
NE. & C. New Caledonia. 60 NWC.

Morenia

Morenia Ruiz & Pav. = **Chamaedorea** Willd.

Morenia caudata Burret = **Chamaedorea linearis** (Ruiz & Pav.) Mart.

Morenia chonta Phil. = **Juania australis** (Mart.) Drude ex Hook.f.

Morenia corallina H.Karst. = **Chamaedorea linearis** (Ruiz & Pav.) Mart.

Morenia corallocarpa H.Wendl. = **Chamaedorea oblongata** Mart.

Morenia ernesti-augusti (H.Wendl.) H.Wendl. = **Chamaedorea ernesti-augusti** H.Wendl.

Morenia fragrans Ruiz & Pav. = **Chamaedorea linearis** (Ruiz & Pav.) Mart.

Morenia integrifolia Trail = **Chamaedorea pauciflora** Mart.

Morenia latisecta H.E.Moore = **Chamaedorea latisecta** (H.E.Moore) A.H.Gentry

Morenia lindeniana H.Wendl. = **Chamaedorea linearis** (Ruiz & Pav.) Mart.

Morenia linearis (Ruiz & Pav.) Burret = **Chamaedorea linearis** (Ruiz & Pav.) Mart.

Morenia macrocarpa Burret = **Chamaedorea linearis** (Ruiz & Pav.) Mart.

Morenia microspadix Burret = **Chamaedorea linearis** (Ruiz & Pav.) Mart.

Morenia montana (Humb. & Bonpl.) Burret = **Chamaedorea linearis** (Ruiz & Pav.) Mart.

Morenia oblongata H.Wendl. = **Chamaedorea sartorii** Liebm.

Morenia pauciflora (Mart.) Drude = **Chamaedorea pauciflora** Mart.

Morenia poeppigiana Mart. = **Chamaedorea linearis** (Ruiz & Pav.) Mart.

Morenia robusta Burret = **Chamaedorea linearis** (Ruiz & Pav.) Mart.

Myrialepis

Myrialepis Becc. in J.D.Hooker, Fl. Brit. India 6: 480 (1893).
1 species, Indo-China to Sumatera. 41 42.
Bejaudia Gagnep., Notul. Syst. (Paris) 6: 149 (1937).

Myrialepis floribunda (Becc.) Gagnep. = **Myrialepis paradoxa** (Kurz) J.Dransf.

Myrialepis paradoxa (Kurz) J.Dransf., Kew Bull. 37: 242 (1982).
Indo-China to Sumatera. 41 CBD LAO MYA THA VIE 42 MLY SUM.
Calamus paradoxus Kurz, J. Asiat. Soc. Bengal, Pt. 2, Nat. Hist. 43(2): 213 (1874). *Palmijuncus paradoxus* (Kurz) Kuntze, Revis. Gen. Pl. 2: 733 (1891). *Plectocomiopsis paradoxa* (Kurz) Becc. in J.D.Hooker, Fl. Brit. India 6: 488 (1893).
Myrialepis scortechinii Becc. in J.D.Hooker, Fl. Brit. India 6: 480 (1893). *Plectocomiopsis scortechinii* (Becc.) Ridl., Mat. Fl. Malay. Penins. 2: 213 (1907).
Plectocomiopsis annulata Ridl., Mat. Fl. Malay. Penins. 2: 213 (1907).
Plectocomiopsis floribunda Becc., Webbia 3: 235 (1910). *Myrialepis floribunda* (Becc.) Gagnep. in H.Lecomte, Fl. Indo-Chine 6: 1003 (1937).
Bejaudia cambodiensis Gagnep., Notul. Syst. (Paris) 6: 149 (1937).

Myrialepis scortechinii Becc. = **Myrialepis paradoxa** (Kurz) J.Dransf.

Myrialepis triquetra (Becc.) Becc. = **Plectocomiopsis triquetra** (Becc.) J.Dransf.

Nannorrhops

Nannorrhops H.Wendl., Bot. Zeitung (Berlin) 37: 147 (1879).
1 species, S. Arabian Pen., Iran to Pakistan. 34 35.

Nannorrhops arabica Burret = **Nannorrhops ritchiana** (Griff.) Aitch.

Nannorrhops naudiniana Becc. = **Nannorrhops ritchiana** (Griff.) Aitch.

Nannorrhops ritchiana (Griff.) Aitch., J. Linn. Soc., Bot. 19: 187 (1882).
S. Arabian Pen., Iran to Pakistan. 34 AFG IRN PAK 35 GST OMA YEM.
Chamaerops ritchiana Griff., Calcutta J. Nat. Hist. 5: 342 (1845).
Nannorrhops naudiniana Becc., Webbia 5: 10 (1921).
Nannorrhops stocksiana Becc., Webbia 5: 10 (1921).
Nannorrhops arabica Burret, Bot. Jahrb. Syst. 73: 185 (1943).

Nannorrhops stocksiana Becc. = **Nannorrhops ritchiana** (Griff.) Aitch.

Neanthe

Neanthe O.F.Cook = **Chamaedorea** Willd.

Neanthe bella O.F.Cook = **Chamaedorea elegans** Mart.

Neanthe elegans (Mart.) O.F.Cook = **Chamaedorea elegans** Mart.

Neanthe neesiana O.F.Cook = **Chamaedorea elegans** Mart.

Nenga

Nenga H.Wendl. & Drude, Linnaea 39: 182 (1875).
5 species, S. Indo-China to W. Malesia. 41 42.

Nenga affinis Becc. = **Hydriastele affinis** (Becc.) W.J.Baker & Loo

Nenga banaensis (Magalon) Burret, Notizbl. Bot. Gart. Berlin-Dahlem 13: 347 (1936).
Vietnam. 41 VIE.
Pinanga banaensis Magalon, Contr. Étud. Palmiers Indoch.: 149 (1930). *Areca banaensis* (Magalon) Burret, Notizbl. Bot. Gart. Berlin-Dahlem 13: 198 (1936).
Pinanga nannospadix Burret, Repert. Spec. Nov. Regni Veg. 32: 116 (1933). *Nenga nannospadix* (Burret) Burret, Notizbl. Bot. Gart. Berlin-Dahlem 13: 347 (1936). *Areca microspadix* Burret, Notizbl. Bot. Gart. Berlin-Dahlem 13: 198 (1936).

Nenga calophylla K.Schum. & Lauterb. = **Hydriastele pinangoides** (Becc.) W.J.Baker & Loo

Nenga gajah J.Dransf., Principes 19: 27 (1975).
Sumatera, Borneo (Sabah). 42 BOR SUM.

Nenga geelvinkiana Becc. = **Hydriastele geelvinkiana** (Becc.) Burret

Nenga gracilis (Blume) Becc. = **Pinanga gracilis** Blume

Nenga grandiflora Fernando, Principes 27: 66 (1983).
Pen. Malaysia (Johor). 42 MLY.

Nenga intermedia Becc. = **Nenga pumila** var. **pachystachya** (Blume) Fernando

Nenga latisecta (Blume) Scheff. = **Pinanga latisecta** Blume

Nenga macrocarpa Scort. ex Becc., Malesia 3: 180 (1889).
Pen. Thailand to Pen. Malaysia. 41 THA 42 MLY.

Nenga nagensis (Griff.) Scheff. = **Areca triandra** Roxb. ex Buch.-Ham.

Nenga nannospadix (Burret) Burret = **Nenga banaensis** (Magalon) Burret

Nenga novohibernica Lauterb. = **Areca novohibernica** (Lauterb.) Becc.

Nenga pinangoides Becc. = **Hydriastele pinangoides** (Becc.) W.J.Baker & Loo

Nenga pumila (Blume) H.Wendl. in O.C.E.de Kerchove de Denterghem, Palmiers: 251 (1878).
Pen. Thailand to W. Malesia. 41 THA 42 BOR JAW MLY SUM.
*Areca pumila Blume, Rumphia 2: 71 (1839). *Pinanga pumila* (Blume) Blume, Rumphia 2: 77 (1839).

var. **pachystachya** (Blume) Fernando, Principes 27: 61 (1983).
Pen. Thailand to W. Malesia. 41 THA 42 BOR MLY SUM.
*Pinanga nenga var. *pachystachya* Blume, Rumphia 2: 78 (1839).
Nenga intermedia Becc., Ann. Jard. Bot. Buitenzorg 2: 81 (1885).
Nenga schefferiana Becc., Ann. Jard. Bot. Buitenzorg 2: 81 (1885).
Nenga wendlandiana f. *hexapetala* Becc., Malesia 3: 183 (1889).
Nenga wendlandiana var. *malaccensis* Becc., Malesia 3: 182 (1889).

var. **pumila**
W. Jawa. 42 JAW.
Areca nenga Blume ex Mart., Hist. Nat. Palm. 3: 179 (1838). *Pinanga nenga* (Blume ex Mart.) Blume, Rumphia 2: 77 (1839). *Ptychosperma nenga* (Blume ex Mart.) Teijsm. & Binn., Cat. Hort. Bot. Bogor.: 69 (1866).
Nenga wendlandiana Scheff., Ann. Jard. Bot. Buitenzorg 1: 153 (1876). *Areca wendlandiana* (Scheff.) H.Wendl. in O.C.E.de Kerchove de Denterghem, Palmiers: 332 (1878).

Nenga schefferiana Becc. = **Nenga pumila** var. **pachystachya** (Blume) Fernando

Nenga selebica Becc. = **Hydriastele selebica** (Becc.) W.J.Baker & Loo

Nenga variabilis Becc. = **Hydriastele variabilis** (Becc.) Burret

Nenga variabilis var. *sphaerocarpa* Becc. = **Hydriastele variabilis** (Becc.) Burret

Nenga wendlandiana Scheff. = **Nenga pumila** var. **pumila**

Nenga wendlandiana f. *hexapetala* Becc. = **Nenga pumila** var. **pachystachya** (Blume) Fernando

Nenga wendlandiana var. *malaccensis* Becc. = **Nenga pumila** var. **pachystachya** (Blume) Fernando

Nengella

Nengella Becc. = **Hydriastele** H.Wendl. & Drude

Nengella affinis (Becc.) Burret = **Hydriastele affinis** (Becc.) W.J.Baker & Loo

Nengella brassii (Burret) Burret = **Hydriastele pinangoides** (Becc.) W.J.Baker & Loo

Nengella calophylla (K.Schum. & Lauterb.) Becc. = **Hydriastele pinangoides** (Becc.) W.J.Baker & Loo

Nengella calophylla var. *montana* Becc. = **Hydriastele pleurocarpa** (Burret) W.J.Baker & Loo

Nengella calophylla var. *rhopalocarpa* Becc. = **Hydriastele rhopalocarpa** (Becc.) W.J.Baker & Loo

Nengella densiflora (Ridl.) Burret = **Hydriastele pinangoides** (Becc.) W.J.Baker & Loo

Nengella flabellata Becc. = **Hydriastele flabellata** (Becc.) W.J.Baker & Loo

Nengella gracilis Burret = **Hydriastele gracilis** (Burret) W.J.Baker & Loo

Nengella incompta (Becc.) Burret = **Hydriastele pinangoides** (Becc.) W.J.Baker & Loo

Nengella macrocarpa (Burret) Burret = **Hydriastele pinangoides** (Becc.) W.J.Baker & Loo

Nengella mayrii (Burret) Burret = **Hydriastele cyclopensis** (Essig & B.E.Young) W.J.Baker & Loo

Nengella micrantha (Burret) Burret = **Hydriastele micrantha** (Burret) W.J.Baker & Loo

Nengella microcarpa (Burret) Burret = **Hydriastele pinangoides** (Becc.) W.J.Baker & Loo

Nengella minor (Becc.) Burret = **Hydriastele pinangoides** (Becc.) W.J.Baker & Loo

Nengella montana Becc. = **Hydriastele montana** (Becc.) W.J.Baker & Loo

Nengella paradoxa (Griff.) Becc. = **Pinanga paradoxa** (Griff.) Scheff.

Nengella pinangoides (Becc.) Burret = **Hydriastele pinangoides** (Becc.) W.J.Baker & Loo

Nengella pleurocarpa Burret = **Hydriastele pleurocarpa** (Burret) W.J.Baker & Loo

Nengella pterophylla (Becc.) Burret = **Hydriastele pinangoides** (Becc.) W.J.Baker & Loo

Nengella rhomboidea Burret = **Hydriastele pinangoides** (Becc.) W.J.Baker & Loo

Nengella rhopalocarpa (Becc.) Burret = **Hydriastele rhopalocarpa** (Becc.) W.J.Baker & Loo

Nengella yulensis (Becc.) Burret = **Hydriastele pinangoides** (Becc.) W.J.Baker & Loo

Neoancistrophyllum

Neoancistrophyllum Rauschert = **Laccosperma** Drude

Neoancistrophyllum acutiflorum (Becc.) Rauschert = **Laccosperma acutiflorum** (Becc.) J.Dransf.

Neoancistrophyllum laeve (G.Mann & H.Wendl.) Rauschert ex J.Dransf. = **Laccosperma laeve** (G.Mann & H.Wendl.) Kuntze

Neoancistrophyllum laurentii (De Wild.) Rauschert = **Laccosperma secundiflorum** (P.Beauv.) Kuntze

Neoancistrophyllum majus (Burret) Rauschert = **Laccosperma secundiflorum** (P.Beauv.) Kuntze

Neoancistrophyllum opacum (Drude) Rauschert ex J.Dransf. = **Laccosperma opacum** Drude

Neoancistrophyllum robustum (Burret) Rauschert = **Laccosperma robustum** (Burret) J.Dransf.

Neoancistrophyllum secundiflorum (P.Beauv.) Rauschert = **Laccosperma secundiflorum** (P.Beauv.) Kuntze

Neodypsis

Neodypsis Baill. = **Dypsis** Noronha ex Mart.

Neodypsis baronii (Becc.) Jum. = **Dypsis baronii** (Becc.) Beentje & J.Dransf.

Neodypsis basilonga Jum. & H.Perrier = **Dypsis basilonga** (Jum. & H.Perrier) Beentje & J.Dransf.

Neodypsis canaliculata Jum. = **Dypsis canaliculata** (Jum.) Beentje & J.Dransf.

Neodypsis ceracea Jum. = **Dypsis ceracea** (Jum.) Beentje & J.Dransf.

Neodypsis compacta Jum. = **Dypsis baronii** (Becc.) Beentje & J.Dransf.

Neodypsis decaryi Jum. = **Dypsis decaryi** (Jum.) Beentje & J.Dransf.

Neodypsis gracilis Jum. = **Dypsis oreophila** Beentje

Neodypsis heteromorpha Jum. = **Dypsis heteromorpha** (Jum.) Beentje & J.Dransf.

Neodypsis lastelliana Baill. = **Dypsis lastelliana** (Baill.) Beentje & J.Dransf.

Neodypsis leptocheilos Hodel = **Dypsis leptocheilos** (Hodel) Beentje & J.Dransf.

Neodypsis ligulata Jum. = **Dypsis ligulata** (Jum.) Beentje & J.Dransf.

Neodypsis lobata Jum. = **Dypsis ankaizinensis** (Jum.) Beentje & J.Dransf.

Neodypsis loucoubensis Jum. = **Dypsis ampasindavae** Beentje

Neodypsis nauseosa Jum. & H.Perrier = **Dypsis nauseosa** (Jum. & H.Perrier) Beentje & J.Dransf.

Neodypsis tanalensis Jum. & H.Perrier = **Dypsis tanalensis** (Jum. & H.Perrier) Beentje & J.Dransf.

Neodypsis tsaratananensis Jum. = **Dypsis tsaratananensis** (Jum.) Beentje & J.Dransf.

Neonicholsonia

Neonicholsonia Dammer, Gard. Chron. 1901(2): 178 (1901). *Bisnicholsonia* Kuntze in T.E.von Post, Lex. Gen. Phan.: 621 (1903), nom. illeg.
1 species, C. America. 80.
Woodsonia L.H.Bailey, Gentes Herb. 6: 262 (1943).

Neonicholsonia georgei Dammer = **Neonicholsonia watsonii** Dammer

Neonicholsonia watsonii Dammer, Gard. Chron. 1901(2): 179 (1901).
C. America. 80 COS HON NIC PAN.
Neonicholsonia georgei Dammer, Gard. Chron. 1901(2): 178 (1901).
Woodsonia scheryi L.H.Bailey, Gentes Herb. 6: 262 (1943).

Neophloga

Neophloga Baill. = **Dypsis** Noronha ex Mart.

Neophloga affinis Becc. = **Dypsis scottiana** (Becc.) Beentje & J.Dransf.

Neophloga bernieriana (Baill.) Becc. = **Dypsis bernieriana**

Neophloga betamponensis Jum. = **Dypsis betamponensis** (Jum.) Beentje & J.Dransf.

Neophloga brevicaulis Guillaumet = **Dypsis brevicaulis** (Guillaumet) Beentje & J.Dransf.

Neophloga catatiana (Baill.) Becc. = **Dypsis catatiana** (Baill.) Beentje & J.Dransf.

Neophloga commersoniana Baill. = **Dypsis commersoniana** (Baill.) Beentje & J.Dransf.

Neophloga concinna (Baker) Becc. = **Dypsis concinna** Baker

Neophloga concinna f. *triangularis* (Jum. & H.Perrier) Jum. = **Dypsis concinna** Baker

Neophloga corniculata Becc. = **Dypsis corniculata** (Becc.) Beentje & J.Dransf.

Neophloga curtisii (Baker) Becc. = **Dypsis curtisii** Baker

Neophloga digitata Becc. = **Dypsis digitata** (Becc.) Beentje & J.Dransf.

Neophloga emirnensis (Baill.) Becc. = **Dypsis heterophylla** Baker

Neophloga heterophylla (Baker) Becc. = **Dypsis heterophylla** Baker

Neophloga humbertii Jum. = **Dypsis humbertii** H.Perrier

Neophloga indivisa Jum. & H.Perrier = **Dypsis catatiana** (Baill.) Beentje & J.Dransf.

Neophloga integra Jum. = **Dypsis integra** (Jum.) Beentje & J.Dransf.

Neophloga lanceolata Jum. = **Dypsis jumelleana** Beentje & J.Dransf.

Neophloga linearis Becc. = **Dypsis procumbens**

Neophloga linearis var. *disticha* Jum. = **Dypsis heterophylla** Baker

Neophloga littoralis Jum. = **Dypsis mangorensis** (Jum.) Beentje & J.Dransf.

Neophloga lucens Jum. = **Dypsis lucens** (Jum.) Beentje & J.Dransf.

Neophloga lutea Jum. = **Dypsis lutea** (Jum.) Beentje & J.Dransf.

Neophloga lutea var. *transiens* Jum. & H.Perrier = **Dypsis lutea** (Jum.) Beentje & J.Dransf.

Neophloga majorana Becc. = **Dypsis heterophylla** Baker

Neophloga mananjarensis Jum. & H.Perrier = **Dypsis linearis** Jum.

Neophloga mangorensis Jum. = **Dypsis mangorensis** (Jum.) Beentje & J.Dransf.

Neophloga microphylla Becc. = **Dypsis concinna** Baker

Neophloga montana Jum. = **Dypsis montana** (Jum.) Beentje & J.Dransf.

Neophloga occidentalis Jum. = **Dypsis occidentalis** (Jum.) Beentje & J.Dransf.

Neophloga oligostachya (Becc.) H.Perrier = **Dypsis boiviniana** Baill.

Neophloga pervillei (Baill.) Becc. = **Dypsis pervillei** (Jum.) Beentje & J.Dransf.

Neophloga poivreana (Baill.) Becc. = **Dypsis poivreana** (Baill.) Beentje & J.Dransf.

Neophloga procumbens Jum. & H.Perrier = **Dypsis linearis** Jum.

Neophloga pygmaea Pic.Serm. = **Dypsis commersoniana** (Baill.) Beentje & J.Dransf.

Neophloga rhodotricha (Baker) Becc. = **Dypsis heterophylla** Baker

Neophloga sahanofensis Jum. & H.Perrier = **Dypsis sahanofensis** (Jum. & H.Perrier) Beentje & J.Dransf.

Neophloga scottiana (Becc.) Becc. = **Dypsis scottiana** (Becc.) Beentje & J.Dransf.

Neophloga simianensis Jum. = **Dypsis simianensis** (Jum.) Beentje & J.Dransf.

Neophloga tenuisecta Jum. & H.Perrier = **Dypsis concinna** Baker

Neophloga thiryana Becc. = **Dypsis thiryana** (Becc.) Beentje & J.Dransf.

Neophloga triangularis Jum. & H.Perrier = **Dypsis concinna** Baker

Neoveitchia

Neoveitchia Becc., Palme Nuova Caledonia: 9 (1920).
2 species, SW. Pacific. 60.

Neoveitchia brunnea Dowe, Austral. Syst. Bot. 9: 36 (1996).
Vanuatu. 60 VAN.

Neoveitchia storckii (H.Wendl.) Becc., Palme Nuova Caledonia: 10 (1920).
Fiji (Viti Levu). 60 FIJ.
  **Veitchia storckii* H.Wendl. in B.Seemann, Fl. Vit.: 270 (1868). *Kentia storckii* (H.Wendl.) F.Muell. ex H.Wendl. in O.C.E.de Kerchove de Denterghem, Palmiers: 248 (1878).

Neowashingtonia

Neowashingtonia Sudw. = **Washingtonia** H.Wendl.

Neowashingtonia filamentosa (H.Wendl. ex Franceschi) Sudw. = **Washingtonia filifera** (Linden ex André) H.Wendl. ex de Bary

Neowashingtonia filifera (Linden ex André) Sudw. = **Washingtonia filifera** (Linden ex André) H.Wendl. ex de Bary

Neowashingtonia robusta (H.Wendl.) A.Heller = **Washingtonia robusta** H.Wendl.

Neowashingtonia sonorae (S.Watson) Rose = **Washingtonia robusta** H.Wendl.

Nephrocarpus

Nephrocarpus Dammer = **Basselinia** Vieill.

Nephrocarpus schlechteri Dammer = **Basselinia pancheri** (Brongn. & Gris) Vieill.

Nephrosperma

Nephrosperma Balf.f. in J.G.Baker, Fl. Mauritius: 386 (1877).
1 species, Seychelles. 29.

Nephrosperma van-houtteanum (H.Wendl. ex Van Houtt.) Balf.f. in J.G.Baker, Fl. Mauritius: 386 (1877).
Seychelles. 29 SEY.
  Areca nobilis auct., Gard. Chron. 1868: 349 (1868), pro syn.
  **Oncosperma van-houtteanum* H.Wendl. ex Van Houtte, Ann. Gén. Hort. 17: t. 1798 (1868).

Nipa

Nipa Thunb. = **Nypa** Steck

Nipa arborescens Wurmb ex H.Wendl. = **Nypa fruticans** Wurmb

Nipa fruticans (Wurmb) Thunb. = **Nypa fruticans** Wurmb

Nipa litoralis Blanco = **Nypa fruticans** Wurmb

Normanbya

Normanbya F.Muell. ex Becc., Ann. Jard. Bot. Buitenzorg 2: 91 (1885).
1 species, N. Queensland. 50.

Normanbya australasicus (H.Wendl. & Drude) Baill. = **Arenga australasica** (H.Wendl. & Drude) T.S.Blake ex H.E.Moore

Normanbya merrillii Becc. = **Adonidia merrillii** (Becc.) Becc.

Normanbya muelleri Becc. = **Normanbya normanbyi** (F.Muell.) L.H.Bailey

Normanbya normanbyi (F.Muell.) L.H.Bailey, Gentes Herb. 2: 188 (1930).
N. Queensland. 50 QLD.
  *Areca normanbyi F.Muell., Fragm. 8: 235 (1874). *Ptychosperma normanbyi* (F.Muell.) F.Muell., Fragm. 11: 56 (1880). *Drymophloeus normanbyi* (F.Muell.) Benth. & Hook.f. ex Becc., Ann. Jard. Bot. Buitenzorg 2: 166 (1885). *Saguaster normanbyi* (F.Muell.) Kuntze, Revis. Gen. Pl. 2: 735 (1891). *Normanbya muelleri* Becc., Agric. Colon. 10: 615 (1916).
  Cocos normanbyi W.Hill ex F.Muell., Fragm. 8: 235 (1874), pro syn.

Nunnezharia

Nunnezharia Ruiz & Pav. = **Chamaedorea** Willd.

Nunnezharia affinis (Liebm.) Kuntze = **Chamaedorea elatior** Mart.

Nunnezharia alternans (H.Wendl.) Kuntze = **Chamaedorea alternans** H.Wendl.

Nunnezharia amabilis (H.Wendl. ex Dammer) Kuntze = **Chamaedorea amabilis** H.Wendl. ex Dammer

Nunnezharia amazonica Kuntze = **Chamaedorea pauciflora** Mart.

Nunnezharia andreana (Linden) Kuntze = **Chamaedorea andreana**

Nunnezharia arenbergiana (H.Wendl.) Kuntze = **Chamaedorea arenbergiana** H.Wendl.

Nunnezharia atrovirens (Mart.) Kuntze = **Chamaedorea atrovirens** Mart.

Nunnezharia aurantiaca (Brongn.) Kuntze = **Chamaedorea sartorii** Liebm.

Nunnezharia bartlingiana (H.Wendl.) Kuntze = **Chamaedorea pinnatifrons** (Jacq.) Oerst.

Nunnezharia bifurcata (Oerst.) Kuntze = **Chamaedorea pinnatifrons** (Jacq.) Oerst.

Nunnezharia biloba (H.Wendl.) Kuntze = **Chamaedorea oblongata** Mart.

Nunnezharia brachyclada (H.Wendl.) Kuntze = **Chamaedorea brachyclada** H.Wendl.

Nunnezharia bracteata (H.Wendl.) Kuntze = **Chamaedorea pinnatifrons** (Jacq.) Oerst.

Nunnezharia brevifrons (H.Wendl.) Kuntze = **Chamaedorea pinnatifrons** (Jacq.) Oerst.

Nunnezharia casperiana (Klotzsch) Kuntze = **Chamaedorea tepejilote** Liebm.

Nunnezharia cataractarum (Mart.) Kuntze = **Chamaedorea cataractarum** Mart.

Nunnezharia chonta (Phil.) Kuntze = **Juania australis** (Mart.) Drude ex Hook.f.

Nunnezharia concolor (Mart.) Kuntze = **Chamaedorea pinnatifrons** (Jacq.) Oerst.

Nunnezharia conocarpa (Mart.) Kuntze = **Chamaedorea pinnatifrons** (Jacq.) Oerst.

Nunnezharia corallina (H.Karst.) Kuntze = **Chamaedorea linearis** (Ruiz & Pav.) Mart.

Nunnezharia corallocarpa (H.Wendl.) Kuntze = **Chamaedorea oblongata** Mart.

Nunnezharia costaricana (Oerst.) Kuntze = **Chamaedorea costaricana** Oerst.

Nunnezharia deckeriana (Klotzsch) Kuntze = **Chamaedorea deckeriana** (Klotzsch) Hemsl.

Nunnezharia demaniana Kuntze = **?**

Nunnezharia desmoncoides (H.Wendl.) Kuntze = **Chamaedorea elatior** Mart.

Nunnezharia eburnea Kuntze = **?**

Nunnezharia elatior (Mart.) Kuntze = **Chamaedorea elatior** Mart.

Nunnezharia elegans (Mart.) Kuntze = **Chamaedorea elegans** Mart.

Nunnezharia ernesti-augusti (H.Wendl.) Kuntze = **Chamaedorea ernesti-augusti** H.Wendl.

Nunnezharia fenestrata (H.Wendl.) Kuntze = **Chamaedorea geonomiformis** H.Wendl.

Nunnezharia flavovirens (H.Wendl.) Kuntze = **Chamaedorea pinnatifrons** (Jacq.) Oerst.

Nunnezharia flexuosa (H.Wendl.) Kuntze = **Chamaedorea cataractarum** Mart.

Nunnezharia formosa (W.Bull) Kuntze = **Chamaedorea linearis** (Ruiz & Pav.) Mart.

Nunnezharia fragrans Ruiz & Pav. = **Chamaedorea fragrans** Mart.

Nunnezharia geonomiformis (H.Wendl.) Hook.f. = **Chamaedorea geonomiformis** H.Wendl.

Nunnezharia geonomoides Spruce = **Chamaedorea pinnatifrons** (Jacq.) Oerst.

Nunnezharia glauca (Linden) Kuntze = **Chamaedorea glauca**

Nunnezharia glaucifolia (H.Wendl.) Kuntze = **Chamaedorea glaucifolia** H.Wendl.

Nunnezharia graminifolia (H.Wendl.) Kuntze = **Chamaedorea graminifolia** H.Wendl.

Nunnezharia humilis Kuntze = **Chamaedorea geonomiformis** H.Wendl.

Nunnezharia humilis (Liebm. ex Oerst.) Kuntze = **Chamaedorea elegans** Mart.

Nunnezharia integrifolia (Trail) Kuntze = **Chamaedorea pauciflora** Mart.

Nunnezharia karwinskyana (H.Wendl.) Kuntze = **Chamaedorea pochutlensis** Liebm.

Nunnezharia klotzschiana (H.Wendl.) Kuntze = **Chamaedorea klotzschiana** H.Wendl.

Nunnezharia lanceolata (Ruiz & Pav.) Kuntze = **Chamaedorea pinnatifrons** (Jacq.) Oerst.

Nunnezharia latifrons (H.Wendl.) Kuntze = **Chamaedorea arenbergiana** H.Wendl.

Nunnezharia lepidota (H.Wendl.) Kuntze = **Chamaedorea liebmannii** Mart.

Nunnezharia liboniana Kuntze = **?**

Nunnezharia liebmannii (Mart.) Kuntze = **Chamaedorea liebmannii** Mart.

Nunnezharia lindeniana (H.Wendl.) Kuntze = **Chamaedorea linearis** (Ruiz & Pav.) Mart.

Nunnezharia linearis (Ruiz & Pav.) Kuntze = **Chamaedorea linearis** (Ruiz & Pav.) Mart.

Nunnezharia lunata (Liebm.) Kuntze = **Chamaedorea oblongata** Mart.

Nunnezharia macrospadix (Oerst.) Kuntze = **Chamaedorea macrospadix** Oerst.

Nunnezharia martiana (H.Wendl.) Kuntze = **Chamaedorea cataractarum** Mart.

Nunnezharia membranacea (Oerst.) Kuntze = **Chamaedorea pinnatifrons** (Jacq.) Oerst.

Nunnezharia mexicana Kuntze = **Chamaedorea sartorii** Liebm.

Nunnezharia microphylla (H.Wendl.) Kuntze = **Chamaedorea microphylla** H.Wendl.

Nunnezharia montana (Humb. & Bonpl.) Kuntze = **Chamaedorea linearis** (Ruiz & Pav.) Mart.

Nunnezharia morenia Kuntze = **Chamaedorea linearis** (Ruiz & Pav.) Mart.

Nunnezharia oaxacensis Kuntze = **Chamaedorea elatior** Mart.

Nunnezharia oblongata (Mart.) Kuntze = **Chamaedorea oblongata** Mart.

Nunnezharia oreophila (Mart.) Kuntze = **Chamaedorea oreophila** Mart.

Nunnezharia pacaya (Oerst.) Kuntze = **Chamaedorea pinnatifrons** (Jacq.) Oerst.

Nunnezharia paradoxa (H.Wendl.) Kuntze = **Chamaedorea oblongata** Mart.

Nunnezharia pauciflora (Mart.) Kuntze = **Chamaedorea pauciflora** Mart.

Nunnezharia pinnatifrons (Jacq.) Kuntze = **Chamaedorea pinnatifrons** (Jacq.) Oerst.

Nunnezharia pochutlensis (Liebm.) Kuntze = **Chamaedorea pochutlensis** Liebm.

Nunnezharia poeppigiana (Mart.) Kuntze = **Chamaedorea linearis** (Ruiz & Pav.) Mart.

Nunnezharia polita Kuntze = **?**

Nunnezharia pulchella (Linden) Kuntze = **Chamaedorea elegans** Mart.

Nunnezharia pumila (H.Wendl. ex Dammer) Kuntze = **Chamaedorea pumila** H.Wendl. ex Dammer

Nunnezharia pygmaea (H.Wendl.) Kuntze = **Chamaedorea pygmaea** H.Wendl.

Nunnezharia radicalis (Mart.) Kuntze = **Chamaedorea radicalis** Mart.

Nunnezharia regia (H.Wendl.) Kuntze = **Chamaedorea elatior** Mart.

Nunnezharia repens (H.Wendl.) Kuntze = **Chamaedorea elatior** Mart.

Nunnezharia resinifera (H.Wendl.) Kuntze = **Chamaedorea elatior** Mart.

Nunnezharia rigida (H.Wendl. ex Dammer) Kuntze = **Chamaedorea rigida** H.Wendl. ex Dammer

Nunnezharia robusta (H.Wendl.) Kuntze = **Chamaedorea elatior** Mart.

Nunnezharia sartorii (Liebm.) Kuntze = **Chamaedorea sartorii** Liebm.

Nunnezharia schiedeana (Mart.) Kuntze = **Chamaedorea schiedeana** Mart.

Nunnezharia simplicifrons (Heynh.) Kuntze = **Chamaedorea ernesti-augusti** H.Wendl.

Nunnezharia speciosa (H.Wendl.) Kuntze = **Chamaedorea schiedeana** Mart.

Nunnezharia tenella (H.Wendl.) Hook.f. = **Chamaedorea tenella** H.Wendl.

Nunnezharia tepejilote (Liebm.) Kuntze = **Chamaedorea tepejilote** Liebm.

Nunnezharia velutina (H.Wendl.) Kuntze = **Chamaedorea liebmannii** Mart.

Nunnezharia verschaffeltii (Kerch.) Kuntze = **Chamaedorea fragrans** Mart.

Nunnezharia wallisii (Linden) Kuntze = **Chamaedorea wallisii**

Nunnezharia warscewiczii (H.Wendl.) Kuntze = **Chamaedorea warscewiczii** H.Wendl.

Nunnezharia wendlandiana (Oerst.) Kuntze = **Chamaedorea tepejilote** Liebm.

Nunnezharia wobstiana (Linden) Kuntze = **Chamaedorea sartorii** Liebm.

Nunnezia

Nunnezia Willd. = **Chamaedorea** Willd.

Nunnezia fragrans (Ruiz & Pav.) Willd. = **Chamaedorea fragrans** Mart.

Nypa

Nypa Steck, Sagu: 15 (1757).
1 species, Sri Lanka to Nansei-shoto and Caroline Is. (22) 36 38 40 41 42 43 50 62 (80) (81).
Nipa Thunb., Kongl. Vetensk. Acad. Nya Handl. 3: 231 (1782).

Nypa fruticans Wurmb, Verh. Batav. Genootsch. Kunsten 1: 349 (1779). *Nipa fruticans* (Wurmb) Thunb., Kongl. Vetensk. Acad. Nya Handl. 3: 231 (1782).
Sri Lanka to Nansei-shoto and Caroline Is. (22) nga 36 CHH 38 NNS 40 BAN SRL 41 CBD MYA THA VIE 42 BOR JAW MLY MOL PHI SUL SUM 43 BIS NWG SOL 50 NTA QLD 62 CRL mrn (80) pan (81) trt.
Cocos nypa Lour., Fl. Cochinch.: 567 (1790).
Nipa litoralis Blanco, Fl. Filip.: 662 (1837).
Nipa arborescens Wurmb ex H.Wendl. in O.C.E.de Kerchove de Denterghem, Palmiers: 252 (1878).

Oenocarpus

Oenocarpus Mart., Hist. Nat. Palm. 2: 21 (1823).
9 species, C. & S. Trop. America to Trinidad. 80 81 82 83 84.
Jessenia H.Karst., Linnaea 28: 387 (1856).

Oenocarpus altissimus Klotzsch ex H.Wendl. = ? [82 VEN]

Oenocarpus × andersonii Balick, Bol. Mus. Paraense Emilio Gouldi, Bot. 7: 506 (1991). **O. bacaba × O. minor.**
N. Brazil. 84 BZN.

Oenocarpus bacaba Mart., Hist. Nat. Palm. 2: 24 (1823).
S. Trop. America. 82 FRG GUY SUR VEN 83 CLM 84 BZN.
Areca bacaba Arruda in H.Koster, Trav. Brazil: 490 (1816), nom. nud.
Oenocarpus hoppii Burret, Notizbl. Bot. Gart. Berlin-Dahlem 11: 1041 (1934).

Oenocarpus grandis Burret, Notizbl. Bot. Gart. Berlin-Dahlem 12: 612 (1935). *Oenocarpus bacaba* var. *grandis* (Burret) Wess.Boer, Pittieria 17: 131 (1988).

Oenocarpus bacaba var. *grandis* (Burret) Wess.Boer = **Oenocarpus bacaba** Mart.

Oenocarpus bacaba var. *parvus* Wess.Boer = **Oenocarpus balickii** F.Kahn

Oenocarpus balickii F.Kahn, Candollea 45: 351 (1990).
W. Amazon Reg. 82 VEN 83 BOL CLM PER 84 BZN.
Oenocarpus bacaba var. *parvus* Wess.Boer, Pittieria 17: 130 (1988).

Oenocarpus bataua Mart., Hist. Nat. Palm. 2: 23 (1823). *Jessenia bataua* (Mart.) Burret, Notizbl. Bot. Gart. Berlin-Dahlem 10: 300 (1928).
Trinidad, E. Panama to S. Trop. America. 80 PAN 81 TRT 82 FRG GUY SUR VEN 83 BOL CLM ECU PER 84 BZN.

var. **bataua**
E. Panama to S. Trop. America. 80 PAN 82 FRG GUY SUR VEN 83 BOL CLM ECU PER 84 BZN.
Oenocarpus batawa Wallace, Palm Trees Amazon: 31 (1853), orth. var.
Jessenia polycarpa H.Karst., Linnaea 28: 388 (1856).
Jessenia repanda Engl., Linnaea 33: 691 (1865).
Jessenia bataua (Mart.) Burret, Notizbl. Bot. Gart. Berlin-Dahlem 10: 300 (1928).
Jessenia weberbaueri Burret, Notizbl. Bot. Gart. Berlin-Dahlem 10: 840 (1929).

var. **oligocarpus** (Griseb. & H.Wendl.) A.J.Hend., Palms Amazon: 120 (1995).
Trinidad to N. South America. 81 TRT 82 FRG GUY SUR VEN.
Jessenia oligocarpa Griseb. & H.Wendl. in A.H.R.Grisebach, Fl. Brit. W. I.: 516 (1864).
Oenocarpus oligocarpus (Griseb. & H.Wendl.) Wess.Boer, in Fl. Suriname 5: 58 (1965).
Jessenia bataua subsp. *oligocarpa* (Griseb. & H.Wendl.) Balick, Advances Econ. Bot. 3: 126 (1986).

Oenocarpus batawa Wallace = **Oenocarpus bataua** var. **bataua**

Oenocarpus bolivianus H.Wendl. = ? [83 BOL]

Oenocarpus calaber H.Wendl. = **Prestoea acuminata** var. **acuminata**

Oenocarpus caracasanus H.Wendl. = **Prestoea acuminata** var. **acuminata**

Oenocarpus chiragua H.Wendl. = ? [82 VEN]

Oenocarpus circumtextus Mart., Hist. Nat. Palm. 2: 26 (1823).
SE. Colombia. 83 CLM 84 BZN?

Oenocarpus cubarro H.Wendl. = ?

Oenocarpus dealbatus H.Wendl. = **Mauritiella armata** (Mart.) Burret

Oenocarpus discolor Barb.Rodr. = **Oenocarpus distichus** Mart.

Oenocarpus distichus Mart., Hist. Nat. Palm. 2: 22 (1823).
Brazil to Bolivia. 83 BOL 84 BZC BZE BZN.
Oenocarpus tarampabo Mart. in A.D.d'Orbigny, Voy. Amér. Mér. 7(3): 12 (1842).
Oenocarpus discolor Barb.Rodr., Palm. Mattogross.: 8 (1898).

Oenocarpus dryanderae Burret = **Oenocarpus mapora** H.Karst.

Oenocarpus edulis W.Watson = **?**

Oenocarpus frigidus (Kunth) Spreng. = **Prestoea acuminata** var. **acuminata**

Oenocarpus glaucus Lodd. ex H.Wendl. = **Prestoea acuminata** var. **acuminata**

Oenocarpus grandis Burret = **Oenocarpus bacaba** Mart.

Oenocarpus hoppii Burret = **Oenocarpus bacaba** Mart.

Oenocarpus huebneri Burret = **Oenocarpus minor** Mart.

Oenocarpus intermedius Burret = **Oenocarpus minor** Mart.

Oenocarpus iriartoides H.Karst. & Triana = **?** [83 CLM]

Oenocarpus macrocalyx Burret = **Oenocarpus mapora** H.Karst.

Oenocarpus makeru R.Bernal, Galeano & A.J.Hend., Brittonia 43: 158 (1991).
SE. Colombia. 83 CLM 84 BZN?

Oenocarpus mapora H.Karst., Linnaea 28: 274 (1856).
C. & S. Trop. America. 80 COS PAN 82 VEN 83 BOL CLM ECU PER 84 BZN.
Oenocarpus multicaulis Spruce, J. Linn. Soc., Bot. 11: 42 (1871).
Oenocarpus dryanderae Burret, Notizbl. Bot. Gart. Berlin-Dahlem 11: 865 (1933). *Oenocarpus mapora* subsp. *dryanderae* (Burret) Balick, Syst. Econ. Bot. Oenocarpus-Jessenia: 110 (1986).
Oenocarpus panamanus L.H.Bailey, Gentes Herb. 3: 71 (1933).
Oenocarpus macrocalyx Burret, Notizbl. Bot. Gart. Berlin-Dahlem 11: 1043 (1934).

Oenocarpus mapora subsp. *dryanderae* (Burret) Balick = **Oenocarpus mapora** H.Karst.

Oenocarpus microspadix Burret = **Oenocarpus minor** Mart.

Oenocarpus minor Mart., Hist. Nat. Palm. 2: 25 (1823).
SE. Colombia to N. Brazil. 83 CLM 84 BZN.
Oenocarpus huebneri Burret, Notizbl. Bot. Gart. Berlin-Dahlem 10: 297 (1928).
Oenocarpus intermedius Burret, Notizbl. Bot. Gart. Berlin-Dahlem 10: 298 (1928). *Oenocarpus minor* subsp. *intermedius* (Burret) Balick, Syst. Econ. Bot. Oenocarpus-Jessenia: 113 (1986).
Oenocarpus microspadix Burret, Notizbl. Bot. Gart. Berlin-Dahlem 10: 297 (1928).

Oenocarpus minor subsp. *intermedius* (Burret) Balick = **Oenocarpus minor** Mart.

Oenocarpus multicaulis Spruce = **Oenocarpus mapora** H.Karst.

Oenocarpus oligocarpus (Griseb. & H.Wendl.) Wess.Boer = **Oenocarpus bataua** var. **oligocarpus** (Griseb. & H.Wendl.) A.J.Hend.

Oenocarpus panamanus L.H.Bailey = **Oenocarpus mapora** H.Karst.

Oenocarpus pulchellus Linden = **?** [83 CLM]

Oenocarpus regius (Kunth) Spreng. = **Roystonea regia** (Kunth) O.F.Cook

Oenocarpus sancona (Kunth) Spreng. = **Syagrus sancona** (Kunth) H.Karst.

Oenocarpus simplex R.Bernal, Galeano & A.J.Hend., Brittonia 43: 154 (1991).
SE. Colombia. 83 CLM 84 BZN?

Oenocarpus tarampabo Mart. = **Oenocarpus distichus** Mart.

Oenocarpus utilis Klotzsch = **Prestoea acuminata** var. **acuminata**

Omanthe

Omanthe O.F.Cook = **Chamaedorea** Willd.

Omanthe costaricana (Oerst.) O.F.Cook = **Chamaedorea costaricana** Oerst.

Oncocalamus

Oncocalamus (G.Mann & H.Wendl.) H.Wendl. in O.C.E. de Kerchove de Denterghem, Palmiers: 252 (1878).
5 species, S. Nigeria to Angola. 22 23 26.

Oncocalamus acanthocnemis Drude = **Oncocalamus mannii** (H.Wendl.) H.Wendl.

Oncocalamus djodu De Wild., Bull. Jard. Bot. État 5: 146 (1916).
Zaïre. 23 ZAI. — Provisionally accepted.

Oncocalamus macrospathus Burret, Notizbl. Bot. Gart. Berlin-Dahlem 15: 749 (1942).
WC. Trop. Africa to Angola. 23 CAB CMN EQG GAB ZAI 26 ANG.

Oncocalamus mannii (H.Wendl.) H.Wendl. in O.C.E.de Kerchove de Denterghem, Palmiers: 252 (1878).
S. Nigeria to WC. Trop. Africa. 22 NGA 23 CMN CON EQG GAB ZAI.
Calamus mannii H.Wendl., Trans. Linn. Soc. London 24: 436 (1864).
Calamus niger J.Braun & K.Schum., Mitt. Deutsch. Schutzgeb. 2: 147 (1889), nom. illeg.
Oncocalamus acanthocnemis Drude, Bot. Jahrb. Syst. 21: 133 (1895).
Oncocalamus phaeobalanus Burret, Notizbl. Bot. Gart. Berlin-Dahlem 15: 748 (1942).

Oncocalamus phaeobalanus Burret = **Oncocalamus mannii** (H.Wendl.) H.Wendl.

Oncocalamus tuleyi Sunderl., J. Bamboo Rattan 1: 365 (2002).
Nigeria to Cameroon. 22 NGA 23 CMN.

Oncocalamus wrightianus Hutch., Kew Bull. 17: 181 (1963).
S. Nigeria. 22 NGA.

Oncosperma

Oncosperma Blume, Rumphia 2: 96 (1843).
5 species, Sri Lanka to Philippines. 40 41 42.
Keppleria Meisn., Pl. Vasc. Gen.: 355 (1842), nom. illeg.

Oncosperma cambodianum Hance = **Oncosperma tigillarium** (Jack) Ridl.

Oncosperma fasciculatum Thwaites, Enum. Pl. Zeyl.: 328 (1864).
Sri Lanka. 40 SRL.

Oncosperma filamentosum (Kunth) Blume = **Oncosperma tigillarium** (Jack) Ridl.

Oncosperma gracilipes Becc., Philipp. J. Sci., C 2: 228 (1907).
Philippines. 42 PHI.

Oncosperma horridum (Griff.) Scheff., Tijdschr. Ned.-Indië 32: 191 (1871).
Pen. Thailand to W. & C. Malesia. 41 THA 42 BOR MLY PHI SUL SUM.
Areca horrida Griff., Calcutta J. Nat. Hist. 5: 465 (1845).

Oncosperma platyphyllum Becc., Philipp. J. Sci., C 4: 609 (1909).
Philippines. 42 PHI.

Oncosperma tigillarium (Jack) Ridl., J. Straits Branch Roy. Asiat. Soc. 33: 173 (1864).
Pen. Thailand to W. Malesia. 41 cbd THA 42 BOR JAW MLY SUM.
Areca tigillaria Jack, Malayan Misc. 2(7): 88 (1820).
Keppleria tigillaria (Jack) Meisn., Pl. Vasc. Gen.: 355 (1842).
Areca nibung Mart., Hist. Nat. Palm. 3: 173 (1838).
Areca spinosa Hasselt & Kunth, Enum. Pl. 3: 185 (1841).
Euterpe filamentosa Kunth, Enum. Pl. 3: 185 (1841).
Oncosperma filamentosum (Kunth) Blume, Rumphia 2: 97 (1843).
Oncosperma cambodianum Hance, J. Bot. 14: 261 (1876).

Oncosperma van-houtteanum H.Wendl. ex Van Houtte = **Nephrosperma van-houtteanum** (H.Wendl. ex Van Houtt.) Balf.f.

Oothrinax

Oothrinax (Becc.) O.F.Cook = **Zombia** L.H.Bailey

Oothrinax anomala (Becc.) O.F.Cook = **Zombia antillarum** (Desc.) L.H.Bailey

Ophiria

Ophiria Becc. = **Pinanga** Blume

Ophiria paradoxa (Griff.) Becc. = **Pinanga paradoxa** (Griff.) Scheff.

Opsiandra

Opsiandra O.F.Cook = **Gaussia** H.Wendl.

Opsiandra gomez-pompae H.J.Quero = **Gaussia gomez-pompae** (H.J.Quero) H.J.Quero

Opsiandra maya O.F.Cook = **Gaussia maya** (O.F.Cook) H.J.Quero & Read

Orania

Orania Zipp., Alg. Konst- Lett.-Bode 1: 297 (1829).
18 species, Madagascar, Pen. Thailand to New Guinea. 29 41 42 43.
Macrocladus Griff., Calcutta J. Nat. Hist. 5: 489 (1845).
Sindroa Jum., Ann. Inst. Bot.-Géol. Colon. Marseille, V, 1(1): 11 (1933).
Halmoorea J.Dransf. & N.W.Uhl, Principes 28: 164 (1984).

Orania appendiculata (F.M.Bailey) Domin = **Oraniopsis appendiculata** (F.M.Bailey) J.Dransf., A.K.Irvine & N.W.Uhl

Orania archboldiana Burret, J. Arnold Arbor. 20: 198 (1939).
C. New Guinea. 43 NWG.

Orania aruensis Becc. = **Orania regalis** Zipp. ex Blume

Orania beccarii F.M.Bailey = **Oraniopsis appendiculata** (F.M.Bailey) J.Dransf., A.K.Irvine & N.W.Uhl

Orania brassii Burret = **Orania lauterbachiana** Becc.

Orania clemensiae Burret = **Orania lauterbachiana** Becc.

Orania decipiens Becc., Philipp. J. Sci., C 4: 614 (1909).
Philippines. 42 PHI.

Orania disticha Burret, Notizbl. Bot. Gart. Berlin-Dahlem 12: 32 (1935).
C. New Guinea. 43 NWG.

Orania gagavu Essig, Lyonia 1: 227 (1980).
Papua New Guinea. 43 NWG.

Orania glauca Essig, Lyonia 1: 229 (1980).
Papua New Guinea. 43 NWG.

Orania lauterbachiana Becc., Bot. Jahrb. Syst. 52: 36 (1914).
New Guinea. 43 NWG.
Orania micrantha Becc., Bot. Jahrb. Syst. 52: 36 (1914).
Orania brassii Burret, Notizbl. Bot. Gart. Berlin-Dahlem 13: 68 (1936).
Orania clemensiae Burret, Notizbl. Bot. Gart. Berlin-Dahlem 15: 8 (1940).

Orania longisquama (Jum.) J.Dransf. & N.W.Uhl, Principes 28: 164 (1984).
NW. & E. Madagascar. 29 MDG.
Sindroa longisquama Jum., Ann. Inst. Bot.-Géol. Colon. Marseille, V, 1(1): 11 (1933).

Orania macrocladus Mart. = **Orania sylvicola** (Griff.) H.E.Moore

Orania macropetala K.Schum. & Lauterb., Fl. Schutzgeb. Südsee: 205 (1900).
Papua New Guinea. 43 NWG.

Orania micrantha Becc. = **Orania lauterbachiana** Becc.

Orania moluccana Becc., Ann. Jard. Bot. Buitenzorg 2: 163 (1885).
Maluku. 42 MOL.

Orania nicobarica Kurz = **Bentinckia nicobarica** (Kurz) Becc.

Orania nivea Linden ex W.Watson = **Polyandrococos caudescens** (Mart.) Barb.Rodr.

Orania oreophila Essig, Lyonia 1: 225 (1980).
Papua New Guinea. 43 NWG.

Orania palindan (Blanco) Merr., Publ. Bur. Sci. Gov. Lab. 27: 88 (1905).
Philippines. 42 PHI.
Caryota palindan Blanco, Fl. Filip.: 714 (1837).

Orania philippinensis Scheff. ex Becc., Ann. Jard. Bot. Buitenzorg 2: 156 (1885).

Orania paraguanensis Becc., Webbia 1: 335 (1905).
Borneo to Philippines (Palawan). 42 BOR PHI.

Orania parva Essig, Lyonia 1: 226 (1980).
Papua New Guinea. 43 NWG.

Orania philippinensis Scheff. ex Becc. = **Orania palindan** (Blanco) Merr.

Orania porphyrocarpa Blume = **Arenga porphyrocarpa** (Blume) H.E.Moore

Orania ravaka Beentje in J.Dransfield & H.Beentje, Palms Madagascar: 119 (1995).
NE. Madagascar. 29 MDG.

Orania regalis Zipp. ex Blume, Rumphia 2: 116 (1843).
SW. New Guinea. 43 NWG.
Orania aruensis Becc., Malesia 1: 76 (1877).

Orania rubiginosa Becc., Philipp. J. Sci. 14: 333 (1919).
Philippines. 42 PHI.

Orania sylvicola (Griff.) H.E.Moore, Principes 6: 44 (1962).
Pen. Thailand to Sumatera. 41 THA 42 MLY SUM.
Macrocladus sylvicola Griff., Calcutta J. Nat. Hist. 5: 490 (1845).
Orania macrocladus Mart., Hist. Nat. Palm. 3(ed. 2): 186 (1845).

Orania trispatha (J.Dransf. & N.W.Uhl) Beentje & J.Dransf., Palms Madagascar: 118 (1995).
NE. & SE. Madagascar. 29 MDG.
Halmoorea trispatha J.Dransf. & N.W.Uhl, Principes 28: 166 (1984).

Oraniopsis

Oraniopsis (Becc.) J.Dransf., A.K.Irvine & N.W.Uhl, Principes 29: 57 (1985).
1 species, N. Queensland. 50.

Oraniopsis appendiculata (F.M.Bailey) J.Dransf., A.K.Irvine & N.W.Uhl, Principes 29: 61 (1985).
N. Queensland. 50 QLD.
Areca appendiculata F.M.Bailey, Dept. Agric. Bot. Div. Bull. 4: 18 (1891). *Orania appendiculata* (F.M.Bailey) Domin, Biblioth. Bot. 85: 498 (1915).
Orania beccarii F.M.Bailey, Queensland Agric. J. 23: 35 (1909).

Orbignya

Orbignya Mart. ex Endl. = **Attalea** Kunth

Orbignya agrestis (Barb.Rodr.) Burret = **Attalea microcarpa** Mart.

Orbignya barbosiana Burret = **Attalea speciosa** Mart.

Orbignya brejinhoensis Glassman = **Attalea brejinhoensis** (Glassman) Zona

Orbignya campestris Barb.Rodr. = **Attalea eichleri** (Drude) A.J.Hend.

Orbignya cohune (Mart.) Dahlgren ex Standl. = **Attalea cohune** Mart.

Orbignya crassispatha (Mart.) Glassman = **Attalea crassispatha** (Mart.) Burret

Orbignya cuatrecasana Dugand = **Attalea cuatrecasana** (Dugand) A.J.Hend., Galeano & R.Bernal

Orbignya cuci Kunth ex H.Wendl. = **Attalea speciosa** Mart.

Orbignya dammeriana Barb.Rodr. = **Attalea cohune** Mart.

Orbignya dubia Mart. = **Attalea dubia** (Mart.) Burret

Orbignya eichleri Drude = **Attalea eichleri** (Drude) A.J.Hend.

Orbignya excelsa Barb.Rodr. = **?** [84]

Orbignya guacuyule (Liebm. ex Mart.) Hern.-Xol. = **Attalea guacuyule** (Liebm. ex Mart.) Zona

Orbignya huebneri Burret = **Attalea speciosa** Mart.

Orbignya humilis Mart. = **Attalea eichleri** (Drude) A.J.Hend.

Orbignya longibracteata Barb.Rodr. = **Attalea eichleri** (Drude) A.J.Hend.

Orbignya luetzelburgii Burret = **Attalea luetzelburgii** (Burret) Wess.Boer

Orbignya lydiae Drude = **Attalea speciosa** Mart.

Orbignya macrocarpa Barb.Rodr. = **Attalea eichleri** (Drude) A.J.Hend.

Orbignya macropetala Burret = **Attalea speciosa** Mart.

Orbignya martiana Barb.Rodr. = **Attalea speciosa** Mart.

Orbignya microcarpa (Mart.) Burret = **Attalea microcarpa** Mart.

Orbignya oleifera Burret = **Attalea vitrivir** Zona

Orbignya phalerata Mart. = **Attalea speciosa** Mart.

Orbignya pixuna (Barb.Rodr.) Barb.Rodr. = **Attalea spectabilis** Mart.

Orbignya polysticha Burret = **Attalea microcarpa** Mart.

Orbignya racemosa (Spruce) Drude = **Attalea racemosa** Spruce

Orbignya sabulosa Barb.Rodr. = **Attalea microcarpa** Mart.

Orbignya sagotii Trail ex Thurn = **Attalea microcarpa** Mart.

Orbignya speciosa (Mart.) Barb.Rodr. = **Attalea speciosa** Mart.

Orbignya spectabilis (Mart.) Burret = **Attalea spectabilis** Mart.

Orbignya × teixeirana Bondar = **Attalea × teixeirana** (Bondar) Zona

Orbignya urbaniana Dammer = **Attalea eichleri** (Drude) A.J.Hend.

Oreodoxa

Oreodoxa Willd. = **Prestoea** Hook.f.

Oreodoxa acuminata Willd. = **Prestoea acuminata** (Willd.) H.E.Moore

Oreodoxa borinquena (O.F.Cook) Reasoner ex L.H.Bailey = **Roystonea borinquena** O.F.Cook

Oreodoxa caribaea (Spreng.) Dammer & Urb. = **Roystonea oleracea** (Jacq.) O.F.Cook

Oreodoxa frigida Kunth = **Prestoea acuminata** var. **acuminata**

Oreodoxa ghiesbreghtii (H.Wendl.) auct. = **Gaussia ghiesbrechtii**

Oreodoxa granatensis W.Watson = **?** [83 CLM]

Oreodoxa manaele Mart. = **Prestoea acuminata** var. **montana** (Graham) A.J.Hend. & Galeano

Oreodoxa oleracea (Jacq.) Mart. = **Roystonea oleracea** (Jacq.) O.F.Cook

Oreodoxa praemorsa Willd. = **Wettinia praemorsa** (Willd.) Wess.Boer

Oreodoxa princeps Becc. = **Roystonea princeps** (Becc.) Burret

Oreodoxa regia Kunth = **Roystonea regia** (Kunth) O.F.Cook

Oreodoxa regia var. *jenmanii* Waby = **Roystonea oleracea** (Jacq.) O.F.Cook

Oreodoxa sancona Kunth = **Syagrus sancona** (Kunth) H.Karst.

Oreodoxa ventricosa H.Wendl. = **Gaussia ghiesbreghtii** H.Wendl.

Orophoma

Orophoma Drude = **Mauritia** L.f.

Orophoma carana (Wallace ex Archer) Spruce ex Drude = **Mauritia carana** Wallace ex Archer

Orophoma subinermis (Spruce) Drude = **Mauritiella armata** (Mart.) Burret

Palandra

Palandra O.F.Cook = **Phytelephas** Ruiz & Pav.

Palandra aequatorialis (Spruce) O.F.Cook = **Phytelephas aequatorialis** Spruce

Palma

Palma Mill. = **Phoenix** L.

Palma altissima Mill. = **Roystonea altissima** (Mill.) H.E.Moore

Palma amboinensis Garsault = **?** [42 MOL] **Daemonorops ?**

Palma argentata Jacq. = **Coccothrinax argentata** (Jacq.) L.H.Bailey

Palma avoira Aubl. = **?** [82 FRG]

Palma bache Aubl. = **?** [82 FRG]

Palma chilensis Molina = **Jubaea chilensis** (Molina) Baill.

Palma cocos Mill. = **Cocos nucifera** L.

Palma coman Aubl. = **?** [82 FRG]

Palma dactylifera (L.) Mill. = **Phoenix dactylifera** L.

Palma draco Mill. = **Dracaena draco** L. (Asparagaceae)

Palma elata W.Bartram = **Roystonea regia** (Kunth) O.F.Cook

Palma gracilis Mill. = **Bactris plumeriana** Mart.

Palma major Garsault = **Phoenix dactylifera** L.

Palma maripa Aubl. = **Attalea maripa** (Aubl.) Mart.

Palma mocaia Aubl. = **?** [82 FRG]

Palma oleosa Mill. = **Elaeis guineensis** Jacq.

Palma paripou Aubl. = **?** [82 FRG]

Palma patavoua Aubl. = **?** [82 FRG]

Palma pinao Aubl. = **?** [82 FRG] **Calyptrogyne sp.**

Palma polypodiifolia Mill. = **Cycas circinalis** L. (Cycadaceae)

Palma prunifera Mill. = **Copernicia prunifera** (Mill.) H.E.Moore

Palma pumila Mill. = **Zamia furfuracea** Ait. (Zamiaceae)

Palma sancona (Kunth) Kunth = **Syagrus sancona** (Kunth) H.Karst.

Palma spinosa Mill. = **Acrocomia aculeata** (Jacq.) Lodd. ex Mart.

Palma thebaica (L.) Jacq. = **Hyphaene thebaica** (L.) Mart.

Palma zagueneti Aubl. = **?** [82 FRG]

Palmijuncus

Palmijuncus Rumph. ex Kuntze = **Calamus** L.

Palmijuncus acanthospathus (Griff.) Kuntze = **Calamus acanthospathus** Griff.

Palmijuncus accedens (Blume) Kuntze = **Daemonorops rubra** (Reinw. ex Mart.) Blume

Palmijuncus adspersus (Blume) Kuntze = **Calamus adspersus** (Blume) Blume

Palmijuncus amarus (Lour.) Kuntze = **Calamus tenuis** Roxb.

Palmijuncus amboinensis (Miq.) Kuntze = **Daemonorops calapparia** (Mart.) Blume

Palmijuncus amplectens (Becc.) Kuntze = **Calamus javensis** Blume

Palmijuncus andamanicus (Kurz) Kuntze = **Calamus andamanicus** Kurz

Palmijuncus arborescens (Griff.) Kuntze = **Calamus arborescens** Griff.

Palmijuncus aruensis (Becc.) Kuntze = **Calamus aruensis** Becc.

Palmijuncus asperrimus (Blume) Kuntze = **Calamus asperrimus** Blume

Palmijuncus aureus (Reinw. ex Mart.) Kuntze = **Calamus ornatus** var. **ornatus**

Palmijuncus australis (Mart.) Kuntze = **Calamus australis** Mart.

Palmijuncus barbatus (Zipp. ex Blume) Kuntze = **Calamus barbatus** Zipp. ex Blume

Palmijuncus blancoi (Kunth) Kuntze = **Calamus usitatus** Blanco

Palmijuncus borneensis (Miq.) Kuntze = **Calamus javensis** Blume

Palmijuncus brevifrons (Mart.) Kuntze = **Calamus usitatus** Blanco

Palmijuncus caesius (Blume) Kuntze = **Calamus caesius** Blume

Palmijuncus calapparius (Mart.) Kuntze = **Daemonorops calapparia** (Mart.) Blume

Palmijuncus calicarpus (Griff.) Kuntze = **Daemonorops calicarpa** (Griff.) Mart.

Palmijuncus calolepis (Miq.) Kuntze = **Calamus melanoloma** Mart.

Palmijuncus caryotoides (A.Cunn. ex Mart.) Kuntze = **Calamus caryotoides** A.Cunn. ex Mart.

Palmijuncus ciliaris (Blume) Kuntze = **Calamus ciliaris** Blume

Palmijuncus cochleatus (Miq.) Kuntze = **Daemonorops didymophylla** Becc.

Palmijuncus collinus (Griff.) Kuntze = **Calamus erectus** Roxb.

Palmijuncus concinnus (Mart.) Kuntze = **Calamus concinnus** Mart.

Palmijuncus crinitus (Blume) Kuntze = **Daemonorops crinita** Blume

Palmijuncus deerratus (G.Mann & H.Wendl.) Kuntze = **Calamus deerratus** G.Mann & H.Wendl.

Palmijuncus delicatulus (Thwaites) Kuntze = **Calamus delicatulus** Thwaites

Palmijuncus depressiusculus (Miq. ex H.Wendl.) Kuntze = **Daemonorops depressiuscula** (Miq. ex H.Wendl.) Becc.

Palmijuncus diepenhorstii (Miq.) Kuntze = **Calamus diepenhorstii** Miq.

Palmijuncus dioicus (Lour.) Kuntze = **Calamus dioicus** Lour.

Palmijuncus discolor (Mart.) Kuntze = **Calamus discolor** Mart.

Palmijuncus draco (Willd.) Kuntze = **Daemonorops draco** (Willd.) Blume

Palmijuncus elegans (H.Wendl.) Kuntze = **Calamus elegans**

Palmijuncus elongatus (Blume) Kuntze = **Daemonorops elongata** Blume

Palmijuncus epetiolaris (Mart.) Kuntze = **Calamus epetiolaris** Mart.

Palmijuncus equestris (Willd.) Kuntze = **Calamus equestris** Willd.

Palmijuncus erectus (Roxb.) Kuntze = **Calamus erectus** Roxb.

Palmijuncus exilis (Griff.) Kuntze = **Calamus exilis** Griff.

Palmijuncus extensus (Roxb.) Kuntze = **Calamus extensus** Roxb.

Palmijuncus farinosus (Linden) Kuntze = **Calamus farinosus** Linden

Palmijuncus fasciculatus (Roxb.) Kuntze = **Calamus viminalis** Willd.

Palmijuncus fernandezii (H.Wendl.) Kuntze = **Calamus oxleyanus** var. **oxleyanus**

Palmijuncus fissus (Blume) Kuntze = **Daemonorops fissa** Blume

Palmijuncus flabellatus (Becc.) Kuntze = **Calamus flabellatus** Becc.

Palmijuncus flagellum (Griff. ex Mart.) Kuntze = **Calamus flagellum** Griff. ex Mart.

Palmijuncus floribundus (Griff.) Kuntze = **Calamus floribundus** Griff.

Palmijuncus gaudichaudii (Mart.) Kuntze = **Daemonorops mollis** (Blanco) Merr.

Palmijuncus geniculatus (Griff.) Kuntze = **Daemonorops geniculata** (Griff.) Mart.

Palmijuncus glaucescens Kuntze = **Calamus caesius** Blume

Palmijuncus gracilipes (Miq.) Kuntze = **Daemonorops gracilipes** (Miq.) Becc.

Palmijuncus gracilis (Roxb.) Kuntze = **Calamus gracilis** Roxb.

Palmijuncus graminosus (Blume) Kuntze = **Calamus graminosus** Blume

Palmijuncus grandis (Griff.) Kuntze = **Daemonorops grandis** (Griff.) Mart.

Palmijuncus grandis (Griff.) Kuntze = **Daemonorops grandis** (Griff.) Mart.

Palmijuncus griffithianus (Mart.) Kuntze = **Calamus castaneus** Griff.

Palmijuncus guruba (Buch.-Ham.) Kuntze = **Calamus guruba** Buch.-Ham.

Palmijuncus haenkeanus (Mart.) Kuntze = **Calamus usitatus** Blanco

Palmijuncus helferianus (Kurz) Kuntze = **Calamus helferianus** Kurz

Palmijuncus heliotropium (Buch.-Ham. ex Kunth) Kuntze = **Calamus tenuis** Roxb.

Palmijuncus heteracanthus (Zipp. ex Blume) Kuntze = **Calamus heteracanthus** Zipp. ex Blume

Palmijuncus heteroideus (Blume) Kuntze = **Calamus heteroideus** Blume

Palmijuncus hirsutus (Blume) Kuntze = **Daemonorops hirsuta** Blume

Palmijuncus horrens (Blume) Kuntze = **Calamus tenuis** Roxb.

Palmijuncus huegelianus (Mart.) Kuntze = **Calamus wightii** Griff.

Palmijuncus humilis (Roxb.) Kuntze = **Calamus latifolius** Roxb.

Palmijuncus hygrophilus (Griff.) Kuntze = **Daemonorops angustifolia** (Griff.) Mart.

Palmijuncus hypoleucus (Kurz) Kuntze = **Calamus hypoleucus** Kurz

Palmijuncus hystrix (Griff.) Kuntze = **Daemonorops hirsuta** Blume

Palmijuncus inermis (T.Anderson) Kuntze = **Calamus latifolius** Roxb.

Palmijuncus insignis (Griff.) Kuntze = **Calamus insignis** Griff.

Palmijuncus intermedius (Griff.) Kuntze = **Daemonorops grandis** (Griff.) Mart.

Palmijuncus interuptus (Becc.) Kuntze = **Calamus interuptus** Becc.

Palmijuncus javensis (Blume) Kuntze = **Calamus javensis** Blume

Palmijuncus jenkinsianus Kuntze = **Daemonorops jenkinsiana** (Griff.) Mart.

Palmijuncus korthalsii (Blume) Kuntze = **Daemonorops korthalsii** Blume

Palmijuncus laevigatus (Mart.) Kuntze = **Calamus laevigatus** Mart.

Palmijuncus latifolius (Roxb.) Kuntze = **Calamus latifolius** Roxb.

Palmijuncus leptopus (Griff.) Kuntze = **Daemonorops leptopus** (Griff.) Mart.

Palmijuncus leptospadix (Griff.) Kuntze = **Calamus leptospadix** Griff.

Palmijuncus lewisianus (Griff.) Kuntze = **Daemonorops lewisiana** (Griff.) Mart.

Palmijuncus lindenii (Rodigas) Kuntze = **Calamus discolor** var. **discolor**

Palmijuncus litoralis (Blume) Kuntze = **Calamus viminalis** Willd.

Palmijuncus longisetus (Griff.) Kuntze = **Calamus longisetus** Griff.

Palmijuncus macracanthus (T.Anderson) Kuntze = **Calamus latifolius** Roxb.

Palmijuncus macrocarpus (Griff. ex Mart.) Kuntze = **Calamus erectus** Roxb.

Palmijuncus macropterus (Miq.) Kuntze = **Daemonorops macroptera** (Miq.) Becc.

Palmijuncus malaccensis (Mart.) Kuntze = **Daemonorops grandis** (Griff.) Mart.

Palmijuncus manan (Miq.) Kuntze = **Calamus manan** Miq.

Palmijuncus manicatus (Teijsm. & Binn. ex Miq.) Kuntze = **Daemonorops crinita** Blume

Palmijuncus manillensis (Mart.) Kuntze = **Calamus manillensis** (Mart.) H.Wendl.

Palmijuncus margaritae (Hance) Kuntze = **Daemonorops margaritae** (Hance) Becc.

Palmijuncus marginatus (Blume) Kuntze = **Calamus marginatus** (Blume) Mart.

Palmijuncus maximus Kuntze = **?**

Palmijuncus melanacanthus (Mart.) Kuntze = **Calamus melanacanthus** Mart.

Palmijuncus melanolepis (Mart.) Kuntze = **Calamus wightii** Griff.

Palmijuncus melanoloma (Mart.) Kuntze = **Calamus melanoloma** Mart.

Palmijuncus meyenianus (Schauer) Kuntze = **Calamus usitatus** Blanco

Palmijuncus micracanthus (Griff.) Kuntze = **Daemonorops micracantha** (Griff.) Becc.

Palmijuncus micranthus (Blume) Kuntze = **Calamus micranthus** Blume

Palmijuncus mirabilis (Mart.) Kuntze = **Daemonorops mirabilis** (Mart.) Mart.

Palmijuncus mishmeensis (Griff.) Kuntze = **Calamus floribundus** Griff.

Palmijuncus mollis (Blanco) Kuntze = **Daemonorops mollis Blanco Merr.**

Palmijuncus monoecus (Roxb.) Kuntze = **Calamus rotang** L.

Palmijuncus montanus (T.Anderson) Kuntze = **Calamus acanthospathus** Griff.

Palmijuncus monticolus (Griff.) Kuntze = **Daemonorops monticola** (Griff.) Mart.

Palmijuncus muelleri (H.Wendl. & Drude) Kuntze = **Calamus muelleri** H.Wendl. & Drude

Palmijuncus niger (Willd.) Kuntze = **Daemonorops nigra** (Willd.) Blume

Palmijuncus nitidus (Mart.) Kuntze = **Calamus guruba** Buch.-Ham.

Palmijuncus nivalis (Thwaites ex Trimen) Kuntze = **Calamus nivalis**

Palmijuncus nutantiflorus (Griff.) Kuntze = **Daemonorops jenkinsiana** (Griff.) Mart.

Palmijuncus opacus (Drude) Kuntze = **Laccosperma opacum** Drude

Palmijuncus ornatus (Blume) Kuntze = **Calamus ornatus** Blume

Palmijuncus ovoideus (Thwaites ex Trimen) Kuntze = **Calamus ovoideus** Thwaites ex Trimen

Palmijuncus oxleyanus (Teijsm. & Binn. ex Miq.) Kuntze = **Calamus oxleyanus** Teijsm. & Binn. ex Miq.

Palmijuncus pachystemonus (Thwaites) Kuntze = **Calamus pachystemonus** Thwaites

Palmijuncus palembanicus (Blume) Kuntze = **Daemonorops palembanica** Blume

Palmijuncus pallens (Blume) Kuntze = **Calamus heteroideus** Blume

Palmijuncus palustris (Griff.) Kuntze = **Calamus palustris** Griff.

Palmijuncus papuanus (Becc.) Kuntze = **Calamus papuanus** Becc.

Palmijuncus paradoxus (Kurz) Kuntze = **Myrialepis paradoxa** (Kurz) J.Dransf.

Palmijuncus penicillatus (Roxb.) Kuntze = **Calamus penicillatus** Roxb.

Palmijuncus periacanthus (Miq.) Kuntze = **Daemonorops periacantha** Miq.

Palmijuncus petiolaris (Griff.) Kuntze = **Daemonorops calicarpa** (Griff.) Mart.

Palmijuncus petraeus (Lour.) Kuntze = **Calamus petraeus** Lour.

Palmijuncus pisicarpus (Blume) Kuntze = **Calamus pisicarpus** Blume

Palmijuncus platyacanthus (Mart.) Kuntze = **Daemonorops oblonga** (Reinw. ex Blume) Blume

Palmijuncus platyspathus (Mart. ex Kunth) Kuntze = **Calamus platyspathus** Mart. ex Kunth

Palmijuncus plicatus (Blume) Kuntze = **Calamus plicatus** Blume

Palmijuncus polygamus (Roxb.) Kuntze = **Calamus flagellum** var. **flagellum**

Palmijuncus pseudorotang (Mart. ex Kunth) Kuntze = **Calamus viminalis** Willd.

Palmijuncus pulcher (Miq.) Kuntze = **Calamus pulcher** Miq.

Palmijuncus pygmaeus (Becc.) Kuntze = **Calamus pygmaeus** Becc.

Palmijuncus quinquenervius (Roxb.) Kuntze = **Calamus quinquenervius** Roxb.

Palmijuncus radiatus (Thwaites) Kuntze = **Calamus radiatus** Thwaites

Palmijuncus radicalis (H.Wendl. & Drude) Kuntze = **Calamus radicalis** H.Wendl. & Drude

Palmijuncus ramosissimus (Griff.) Kuntze = **Daemonorops longipes** (Griff.) Mart.

Palmijuncus rheedei (Griff.) Kuntze = **Calamus rheedei** Griff.

Palmijuncus rhomboideus (Blume) Kuntze = **Calamus rhomboideus** Blume

Palmijuncus riedelianus (Miq.) Kuntze = **Daemonorops riedeliana** (Miq.) Becc.

Palmijuncus rivalis (Thwaites ex Trimen) Kuntze = **Calamus rivalis** Thwaites ex Trimen

Palmijuncus royleanus (Griff.) Kuntze = **Calamus tenuis** Roxb.

Palmijuncus ruber (Reinw. ex Mart.) Kuntze = **Daemonorops rubra** (Reinw. ex Mart.) Blume

Palmijuncus rudentum (Lour.) Kuntze = **Calamus rudentum** Lour.

Palmijuncus schistoacanthus (Blume) Kuntze = **Calamus schistoacanthus** Blume

Palmijuncus schizospathus (Griff.) Kuntze = **Calamus erectus** Roxb.

Palmijuncus scipionum (Lour.) Kuntze = **Calamus scipionum** Lour.

Palmijuncus serrulatus (Becc.) Kuntze = **Calamus serrulatus** Becc.

Palmijuncus siphonospathus (Mart.) Kuntze = **Calamus siphonospathus** Mart.

Palmijuncus spectabilis (Blume) Kuntze = **Calamus spectabilis** Blume

Palmijuncus subangulatus (Miq.) Kuntze = **Ceratolobus subangulatus** (Miq.) Becc.

Palmijuncus symphysipus (Mart.) Kuntze = **Calamus symphysipus** Mart.

Palmijuncus tenuis (Roxb.) Kuntze = **Calamus tenuis** Roxb.

Palmijuncus tetradactylus (Hance) Kuntze = **Calamus tetradactylus** Hance

Palmijuncus tetrastichus (Blume) Kuntze = **Calamus javensis** Blume

Palmijuncus thysanolepis (Hance) Kuntze = **Calamus thysanolepis** Hance

Palmijuncus tigrinus (Kurz) Kuntze = **Calamus longisetus** Griff.

Palmijuncus trichrous (Miq.) Kuntze = **Daemonorops trichroa** Miq.

Palmijuncus triqueter (Becc.) Kuntze = **Plectocomiopsis triquetra** (Becc.) J.Dransf.

Palmijuncus unifarius (H.Wendl.) Kuntze = **Calamus unifarius** H.Wendl.

Palmijuncus usitatus (Blanco) Kuntze = **Calamus usitatus** Blanco

Palmijuncus verticillaris (Griff.) Kuntze = **Daemonorops verticillaris** (Griff.) Mart.

Palmijuncus vestitus (Becc.) Kuntze = **Calamus vestitus** Becc.

Palmijuncus viminalis (Willd.) Kuntze = **Calamus viminalis** Willd.

Palmijuncus walkeri (Hance) Kuntze = **Calamus walkeri** Hance

Palmijuncus wightii (Griff.) Kuntze = **Calamus wightii** Griff.

Palmijuncus zebrinus (Becc.) Kuntze = **Calamus zebrinus** Becc.

Paragulubia

Paragulubia Burret = **Hydriastele** H.Wendl. & Drude

Paragulubia macrospadix Burret = **Hydriastele macrospadix** (Burret) W.J.Baker & Loo

Parajubaea

Parajubaea Burret, Notizbl. Bot. Gart. Berlin-Dahlem 11: 48 (1930).
3 species, W. South America. 83.

Parajubaea cocoides Burret, Notizbl. Bot. Gart. Berlin-Dahlem 11: 48 (1930).
S. Colombia to Ecuador. 83 CLM ECU.

Parajubaea sunkha M.Moraes, Novon 6: 85 (1996).
Bolivia. 83 BOL.

Parajubaea torallyi (Mart.) Burret, Notizbl. Bot. Gart. Berlin-Dahlem 11: 50 (1930).
Bolivia. 83 BOL.
    Diplothemium torallyi Mart. in A.D.d'Orbigny, Voy. Amér. Mér. 7(3): 105 (1844). *Jubaea torallyi* (Mart.) H.Wendl. in O.C.E.de Kerchove de Denterghem, Palmiers: 247 (1878). *Allagoptera*

torallyi (Mart.) Kuntze, Revis. Gen. Pl. 2: 726 (1891). *Polyandrococos torallyi* (Mart.) Barb.Rodr., Contr. Jard. Bot. Rio de Janeiro 1: 8 (1901).

var. **microcarpa** M.Moraes, Novon 6: 87 (1996).
Bolivia. 83 BOL.

var. **torallyi**
Bolivia. 83 BOL.

Paralinospadix

Paralinospadix Burret = **Calyptrocalyx** Blume

Paralinospadix amischus Burret = **Calyptrocalyx julianettii** (Becc.) Dowe & M.D.Ferrero

Paralinospadix arfakianus (Becc.) Burret = **Calyptrocalyx arfakianus** (Becc.) Dowe & M.D.Ferrero

Paralinospadix caudiculatus (Becc.) Burret = **Calyptrocalyx caudiculatus** (Becc.) Dowe & M.D.Ferrero

Paralinospadix clemensiae Burret = **Calyptrocalyx hollrungii** (Becc.) Dowe & M.D.Ferrero

Paralinospadix flabellatus (Becc.) Burret = **Calyptrocalyx flabellatus** (Becc.) Dowe & M.D.Ferrero

Paralinospadix forbesii (Ridl.) Burret = **Calyptrocalyx forbesii** (Ridl.) Dowe & M.D.Ferrero

Paralinospadix geonomiformis (Becc.) Burret = **Calyptrocalyx genomiformis**

Paralinospadix hollrungii (Becc.) Burret = **Calyptrocalyx hollrungii** (Becc.) Dowe & M.D.Ferrero

Paralinospadix julianettii (Becc.) Burret = **Calyptrocalyx julianettii** (Becc.) Dowe & M.D.Ferrero

Paralinospadix lepidotus Burret = **Calyptrocalyx lepidotus** (Burret) Dowe & M.D.Ferrero

Paralinospadix leptostachys (Burret) Burret = **Calyptrocalyx sessiliflorus** Dowe & M.D.Ferrero

Paralinospadix merrillianus Burret = **Calyptrocalyx merrillianus** (Burret) Dowe & M.D.Ferrero

Paralinospadix micholitzii (Ridl.) Burret = **Calyptrocalyx micholitzii** (Ridl.) Dowe & M.D.Ferrero

Paralinospadix microspadix (Becc.) Burret = **Linospadix microspadix** Becc.

Paralinospadix multifidus (Becc.) Burret = **Calyptrocalyx multifidus** (Becc.) Dowe & M.D.Ferrero

Paralinospadix pachystachys (Burret) Burret = **Calyptrocalyx arfakianus** (Becc.) Dowe & M.D.Ferrero

Paralinospadix pauciflorus (Ridl.) Burret = **Calyptrocalyx micholitzii** (Ridl.) Dowe & M.D.Ferrero

Paralinospadix petrickianus (Sander) Burret = **Calyptrocalyx forbesii** (Ridl.) Dowe & M.D.Ferrero

Paralinospadix pusillus (Becc.) Burret = **Calyptrocalyx pusillus** (Becc.) Dowe & M.D.Ferrero

Paralinospadix schlechteri (Becc.) Burret = **Calyptrocalyx hollrungii** (Becc.) Dowe & M.D.Ferrero

Paralinospadix stenoschistus Burret = **Calyptrocalyx forbesii** (Ridl.) Dowe & M.D.Ferrero

Paranthe

Paranthe O.F.Cook = **Chamaedorea** Willd.

Paranthe violacea O.F.Cook = **Chamaedorea parvisecta** Burret

Parascheelea

Parascheelea Dugand = **Attalea** Kunth

Parascheelea anchistropetala Dugand = **Attalea luetzelburgii** (Burret) Wess.Boer

Parascheelea luetzelburgii (Burret) Dugand = **Attalea luetzelburgii** (Burret) Wess.Boer

Paripon

Paripon Voigt = **?**

Paripon palmiri Voigt = **?**

Paurotis

Paurotis O.F.Cook = **Acoelorrhaphe** H.Wendl.

Paurotis androsana O.F.Cook = **Acoelorrhaphe wrightii** (Griseb. & H.Wendl.) H.Wendl. ex Becc.

Paurotis arborescens (Sarg.) O.F.Cook = **Acoelorrhaphe wrightii** (Griseb. & H.Wendl.) H.Wendl. ex Becc.

Paurotis psilocalyx (Burret) Lundell = **Acoelorrhaphe wrightii** (Griseb. & H.Wendl.) H.Wendl. ex Becc.

Paurotis schippii Burret = **Acoelorrhaphe wrightii** (Griseb. & H.Wendl.) H.Wendl. ex Becc.

Paurotis wrightii (Griseb. & H.Wendl.) Britton = **Acoelorrhaphe wrightii** (Griseb. & H.Wendl.) H.Wendl. ex Becc.

Pelagodoxa

Pelagodoxa Becc., Rev. Hort., n.s., 15: 302 (1917).
1 species, Marquesas. (43) (60) 61.

Pelagodoxa henryana Becc., Rev. Hort., n.s., 15: 302 (1917).
Marquesas (Nuku Hiva). (43) sol (60) van 61 MRQ.
Pelagodoxa mesocarpa Burret, Notizbl. Bot. Gart. Berlin-Dahlem 10: 288 (1928).

Pelagodoxa mesocarpa Burret = **Pelagodoxa henryana** Becc.

Pericycla

Pericycla Blume = **Licuala** Wurmb

Pericycla penduliflora Blume = **Licuala penduliflora** (Blume) Zipp. ex Blume

Phloga

Phloga Noronha ex Hook.f. = **Dypsis** Noronha ex Mart.

Phloga gracilis (Jum.) H.Perrier = **Dypsis oreophila** Beentje

Phloga microphoenix Baill. = **?** [29 MDG]

Phloga nodifera (Mart.) Pic.Serm. = **Dypsis nodifera** Mart.

Phloga polystachya (Baker) Noronha ex Baill. = **Dypsis nodifera** Mart.

Phloga polystachya var. *stenophylla* Becc. = **Dypsis nodifera** Mart.

Phloga sambiranensis Jum. = **Dypsis ambanjae** Beentje

Phloga scottiana Becc. = **Dypsis scottiana** (Becc.) Beentje & J.Dransf.

Phlogella

Phlogella Baill. = **Dypsis** Noronha ex Mart.

Phlogella humblotiana Baill. = **Dypsis humblotiana** (Baill.) Beentje & J.Dransf.

Phoenicophorium

Phoenicophorium H.Wendl., Ill. Hort. 12(Misc.): 5 (1865).
1 species, Seychelles. 29.
Stevensonia Duncan ex Balf.f. in J.G.Baker, Fl. Mauritius: 388 (1877).

Phoenicophorium borsigianum (K.Koch) Stuntz, U.S.D.A. Bur. Pl. Industr. Invent. Seeds 31: 88 (1914).
Seychelles. 29 SEY.
*Astrocaryum borsigianum K.Koch, Wochenschr. Gärtnerei Pflanzenk. 2: 401 (1859). *Stevensonia borsigiana* (K.Koch) L.H.Bailey, Gentes Herb. 2: 192 (1930).
Phoenicophorium sechellarum H.Wendl., Ill. Hort. 12 (Misc.): 5 (1865). *Astrocaryum sechellarum* (H. Wendl.) Baill., Hist. Pl. 13: 348 (1895). *Areca sechellarum* (H.Wendl.) Baill., Hist. Pl. 13: 348 (1895).
Astrocaryum pictum Balf.f. in J.G.Baker, Fl. Mauritius: 388 (1877).
Stevensonia grandifolia Duncan ex Balf.f. in J.G.Baker, Fl. Mauritius: 388 (1877).

Phoenicophorium sechellarum H.Wendl. = **Phoenicophorium borsigianum** (K.Koch) Stuntz

Phoenicophorium viridifolium H.Wendl. = **Roscheria melanochaetes** (H.Wendl.) H.Wendl. ex Balf.f.

Phoenix

Phoenix L., Sp. Pl.: 1188 (1753).
14 species, Africa, Kriti to W. & C. Malesia. 13 (20) 21 22 23 24 25 26 27 29 34 35 36 38 40 41 42 (76).
Elate L., Sp. Pl.: 1189 (1753).
Palma Mill., Gard. Dict. Abr. ed. 4(1754).
Dachel Adans., Fam. Pl. 2: 25 (1763).
Phoniphora Neck., Elem. Bot. 3: 302 (1790).
Fulchironia Lesch. in R.L.Desfontaines, Tabl. École Bot., ed. 3: 29 (1829).
Zelonops Raf., Fl. Tellur. 2: 102 (1837).

Phoenix abyssinica Drude = **Phoenix reclinata** Jacq.

Phoenix acaulis Roxb., Pl. Coromandel 3: 69 (1820).
Himalaya to S. China. 36 CHS 40 ASS IND NEP WHM.
Phoenix acaulis var. *melanocarpa* Griff., Calcutta J. Nat. Hist. 5: 346 (1845).

Phoenix acaulis var. *melanocarpa* Griff. = **Phoenix acaulis** Roxb.

Phoenix andamanensis S.Barrow, Kew Bull. 53: 538 (1998).
Andaman Is. 41 AND.

Phoenix andamanensis W.Mill., J.G.Sm. & N.Taylor = **Phoenix paludosa** Roxb.

Phoenix arabica Burret = **Phoenix caespitosa** Chiov.

Phoenix atlantica A.Chev., Bull. Mus. Natl. Hist. Nat., II, 7: 137 (1935).

Cape Verde. 21 CVI. Probably identical with P. dactylifera.

Phoenix atlantidis A.Chev., Compt. Rend. Hebd. Séances Acad. Sci. 199: 1153 (1934), nom. nud.

Phoenix atlantica var. *maroccana* A.Chev. = **Phoenix dactylifera** L.

Phoenix atlantidis A.Chev. = **Phoenix atlantica** A.Chev.

Phoenix baoulensis A.Chev. = **Phoenix reclinata** Jacq.

Phoenix caespitosa Chiov., Fl. Somala 1: 317 (1929). Djibouti to N. Somalia, SW. Arabian Pen. 24 DJI SOM 35 SAU YEM.

Phoenix arabica Burret, Bot. Jahrb. Syst. 73: 189 (1943).

Phoenix canariensis Chabaud, Prov. Agric. Hort. Ill. 19: 293 (1882). Canary Is. 21 CNY.

Phoenix dactylifera var. *jubae* Webb & Berthel., Hist. Nat. Iles Canaries 3: 289 (1847). *Phoenix jubae* (Webb & Berthel.) Webb ex H.Christ, Bot. Jahrb. Syst. 6: 469 (1885).

Phoenix tenuis Verschaff., Cat. 84: 13 (1869), nom. nud.

Phoenix cycadifolia Regel, Gartenflora 28: 131 (1879). Provisional synonym.

Phoenix vigieri Naudin, Rev. Hort. 57: 541 (1885).

Phoenix erecta Sauv., Rev. Hort. 66: 495 (1894), nom. inval.

Phoenix macrocarpa Sauv., Rev. Hort. 66: 495 (1894), nom. inval.

Phoenix canariensis var. *porphyrococca* Vasc. & Franco, Portugaliae Act. Biol., Sér. B, Sist. 2: 313 (1948).

Phoenix canariensis var. *porphyrococca* Vasc. & Franco = **Phoenix canariensis** Chabaud

Phoenix chevalieri D.Rivera, S.Ríos & Obón = **Phoenix dactylifera** L.

Phoenix comorensis Becc. = **Phoenix reclinata** Jacq.

Phoenix cycadifolia Regel = **Phoenix canariensis** Chabaud

Phoenix dactylifera L., Sp. Pl.: 1188 (1753). *Palma dactylifera* (L.) Mill., Gard. Dict. ed. 8: 1 (1768). *Phoenix excelsior* Cav., Icon. 2: 13 (1793), nom. illeg. Arabian Pen. to S. Pakistan, widely introduced elsewhere. (20) alg egy lby mor (21) cny cvi mdr (24) soc som (29) mau 34 IRN IRQ SIN tur 35 GST OMA SAU 40 PAK (76) cal.

Palma major Garsault, Fig. Pl. Méd.: t. 47 (1764).

Phoenix dactylifera var. *cylindrocarpa* Mart., Hist. Nat. Palm. 3: 258 (1838).

Phoenix dactylifera var. *gonocarpa* Mart., Hist. Nat. Palm. 3: 258 (1838).

Phoenix dactylifera var. *oocarpa* Mart., Hist. Nat. Palm. 3: 258 (1838).

Phoenix dactylifera var. *oxysperma* Mart., Hist. Nat. Palm. 3: 258 (1838).

Phoenix dactylifera var. *sphaerocarpa* Mart., Hist. Nat. Palm. 3: 258 (1838).

Phoenix dactylifera var. *sphaerosperma* Mart., Hist. Nat. Palm. 3: 258 (1838).

Phoenix dactylifera var. *sylvestris* Mart., Hist. Nat. Palm. 3: 258 (1838).

Phoenix dactylifera var. *adunca* D.H.Christ ex Becc., Malesia 3: 357 (1890).

Phoenix dactylifera var. *costata* Becc., Malesia 3: 357 (1890).

Phoenix atlantica var. *maroccana* A.Chev., Compt. Rend. Hebd. Séances Acad. Sci. 2: 172 (1952).

Phoenix chevalieri D.Rivera, S.Ríos & Obón, Varied. Trad. Frut. Cuenca Río Segura Cat. Etnobot. 1: 79 (1997).

Phoenix iberica D.Rivera, S.Ríos & Obón, Varied. Trad. Frut. Cuenca Río Segura Cat. Etnobot. 1: 73 (1997).

Phoenix dactylifera var. *adunca* D.H.Christ ex Becc. = **Phoenix dactylifera** L.

Phoenix dactylifera var. *costata* Becc. = **Phoenix dactylifera** L.

Phoenix dactylifera var. *cylindrocarpa* Mart. = **Phoenix dactylifera** L.

Phoenix dactylifera var. *gonocarpa* Mart. = **Phoenix dactylifera** L.

Phoenix dactylifera var. *jubae* Webb & Berthel. = **Phoenix canariensis** Chabaud

Phoenix dactylifera var. *oocarpa* Mart. = **Phoenix dactylifera** L.

Phoenix dactylifera var. *oxysperma* Mart. = **Phoenix dactylifera** L.

Phoenix dactylifera var. *sphaerocarpa* Mart. = **Phoenix dactylifera** L.

Phoenix dactylifera var. *sphaerosperma* Mart. = **Phoenix dactylifera** L.

Phoenix dactylifera var. *sylvestris* Mart. = **Phoenix dactylifera** L.

Phoenix djalonensis A.Chev. = **Phoenix reclinata** Jacq.

Phoenix dybowskii A.Chev. = **Phoenix reclinata** Jacq.

Phoenix equinoxialis Bojer = **Phoenix reclinata** Jacq.

Phoenix erecta Sauv. = **Phoenix canariensis** Chabaud

Phoenix excelsior Cav. = **Phoenix dactylifera** L.

Phoenix farinifera Roxb. = **Phoenix pusilla** Gaertn.

Phoenix hanceana Naudin = **Phoenix loureiroi** var. **loureiroi**

Phoenix hanceana var. *formosana* Becc. = **Phoenix loureiroi** var. **loureiroi**

Phoenix hanceana var. *philippinensis* Becc. = **Phoenix loureiroi** var. **loureiroi**

Phoenix humilis (L.) Cav. = **Chamaerops humilis** L.

Phoenix humilis Royle ex Becc. = **Phoenix loureiroi** var. **loureiroi**

Phoenix humilis var. *hanceana* (Naudin) Becc. = **Phoenix loureiroi** var. **loureiroi**

Phoenix humilis var. *loureiroi* (Kunth) Becc. = **Phoenix loureiroi** var. **loureiroi**

Phoenix humilis var. *pedunculata* (Griff.) Becc. = **Phoenix loureiroi** var. **pedunculata** (Griff.) Govaerts

Phoenix humilis var. *robusta* Becc. = **Phoenix loureiroi** var. **pedunculata** (Griff.) Govaerts

Phoenix humilis var. *typica* Becc. = **Phoenix loureiroi** var. **pedunculata** (Griff.) Govaerts

Phoenix iberica D.Rivera, S.Ríos & Obón = **Phoenix dactylifera** L.

Phoenix jubae (Webb & Berthel.) Webb ex H.Christ = **Phoenix canariensis** Chabaud

Phoenix leonensis Lodd. ex Kunth = **Phoenix reclinata** Jacq.

Phoenix loureiroi Kunth, Enum. Pl. 3: 257 (1841).
Phoenix humilis var. *loureiroi* (Kunth) Becc., Malesia 3: 382 (1890).

Indian Subcontinent to S. China and Philippines. 36 CHH CHS 38 TAI 40 ASS BAN EHM IND NEP WHM 41 CBD LAO MYA THA VIE 42 PHI.

Phoenix pusilla Lour., Fl. Cochinch.: 614 (1790), nom. illeg.

var. **loureiroi**

Indo-China to S. China and Philippines. 36 CHH CHS 38 TAI 41 CBD LAO MYA THA VIE 42 PHI.

Phoenix pygmaea Raeusch., Nomencl. Bot., ed. 3: 375 (1797), nom. inval.

Phoenix hanceana Naudin, J. Bot. 17: 174 (1879). *Phoenix humilis* var. *hanceana* (Naudin) Becc., Malesia 3: 392 (1890).

Phoenix humilis Royle ex Becc., Malesia 3: 373 (1890), nom. illeg.

Phoenix humilis var. *loureiroi* (Kunth) Becc., Malesia 3: 382 (1890).

Phoenix hanceana var. *formosana* Becc., Philipp. J. Sci., C 3: 339 (1908).

Phoenix hanceana var. *philippinensis* Becc., Philipp. J. Sci., C 3: 339 (1908).

var. **pedunculata** (Griff.) Govaerts in R.H.A.Govaerts & J.Dransfield, World Checklist of Palms: (2004)

Himalaya to India. 40 ASS BAN EHM IND NEP WHM.

Phoenix ouseleyana Griff., Calcutta J. Nat. Hist. 5: 347 (1845).

Phoenix pedunculata Griff., Palms Brit. E. Ind.: 139 (1850). *Phoenix humilis* var. *pedunculata* (Griff.) Becc., Malesia 3: 384 (1890).

Phoenix humilis var. *robusta* Becc., Malesia 3: 384 (1890). *Phoenix robusta* (Becc.) Hook.f., Fl. Brit. India 6: 427 (1892).

Phoenix humilis var. *typica* Becc., Malesia 3: 380 (1890).

Phoenix loureiroi var. *humilis* S.Barrow, Kew Bull. 53: 563 (1998).

Phoenix loureiroi var. *humilis* S.Barrow = **Phoenix loureiroi** var. **pedunculata** (Griff.) Govaerts

Phoenix macrocarpa Sauv. = **Phoenix canariensis** Chabaud

Phoenix ouseleyana Griff. = **Phoenix loureiroi** var. **pedunculata** (Griff.) Govaerts

Phoenix paludosa Roxb., Fl. Ind. ed. 1832, 3: 789 (1832).

Assam to Sumatera. 40 ASS BAN IND 41 AND CBD MYA NCB THA VIE 42 MLY SUM.

Phoenix siamensis Miq., Verh. Kon. Ned. Akad. Wetensch., Afd. Natuurk. 11(5): 14 (1868).

Phoenix andamanensis W.Mill., J.G.Sm. & N.Taylor in L.H.Bailey, Stand. Cycl. Hort. 5: 2594 (1916), nom. inval.

Phoenix pedunculata Griff. = **Phoenix loureiroi** var. **pedunculata** (Griff.) Govaerts

Phoenix pusilla Gaertn., Fruct. Sem. Pl. 1: 24 (1788). *Zelonops pusilla* (Gaertn.) Raf., Fl. Tellur. 2: 102 (1837).

S. India, Sri Lanka. 40 IND SRL.

Phoenix farinifera Roxb., Pl. Coromandel 1: 53 (1796).

Phoenix zeylanica Trimen, J. Bot. 23: 267 (1885).

Phoenix pusilla Lour. = **Phoenix loureiroi** Kunth

Phoenix pygmaea Raeusch. = **Phoenix loureiroi** var. **loureiroi**

Phoenix reclinata Jacq., Fragm. 1: 27 (1801).

Trop. & S. Africa, Comoros, Madagascar, SW. Arabian Pen. 22 BEN GAM GHA GNB GUI IVO LBR NGA SEN SIE 23 BUR CAF CMN GAB RWA ZAI 24 ERI ETH SOM 25 KEN TAN UGA 26 ANG MLW MOZ ZAM ZIM 27 BOT CPP NAM NAT SWZ TVL 29 COM MDG 35 SAU YEM.

Phoenix spinosa Schumach. & Thonn. in H.C.F.Schumacher, Beskr. Guin. Pl.: 437 (1827).

Fulchironia senegalensis Lesch. in R.L.Desfontaines, Tabl. École Bot., ed. 3: 29 (1829).

Phoenix equinoxialis Bojer, Hortus Maurit.: 306 (1837).

Phoenix leonensis Lodd. ex Kunth, Enum. Pl. 3: 257 (1841).

Phoenix abyssinica Drude, Bot. Jahrb. Syst. 21: 119 (1895).

Phoenix comorensis Becc., Bot. Jahrb. Syst. 38(87): 5 (1906). *Phoenix reclinata* var. *comorensis* (Becc.) Jum. & H.Perrier, in Fl. Madag. 30: 20 (1945).

Phoenix reclinata var. *madagascariensis* Becc., Bot. Jahrb. Syst. 38(87): 4 (1906).

Phoenix reclinata var. *somalensis* Becc. in E.Chiovenda, Result. Sci. Miss. Stefan.-Paoli Somal. Ital: 176 (1916).

Phoenix baoulensis A.Chev., Rev. Int. Bot. Appl. Agric. Colon. 32: 224 (1952).

Phoenix djalonensis A.Chev., Rev. Int. Bot. Appl. Agric. Colon. 32: 223 (1952).

Phoenix dybowskii A.Chev., Rev. Int. Bot. Appl. Agric. Colon. 32: 224 (1952).

Phoenix reclinata var. *comorensis* (Becc.) Jum. & H.Perrier = **Phoenix reclinata** Jacq.

Phoenix reclinata var. *madagascariensis* Becc. = **Phoenix reclinata** Jacq.

Phoenix reclinata var. *somalensis* Becc. = **Phoenix reclinata** Jacq.

Phoenix robusta (Becc.) Hook.f. = **Phoenix loureiroi** var. **pedunculata** (Griff.) Govaerts

Phoenix roebelenii O'Brien, Gard. Chron., III, 6: 475 (1889).

China (Yunnan) to Indo-China. 36 CHC 41 LAO VIE.

Phoenix rupicola T.Anderson, J. Linn. Soc., Bot. 11: 13 (1871).

Bhutan to Darjiling. 40 EHM.

Phoenix siamensis Miq. = **Phoenix paludosa** Roxb.

Phoenix spinosa Schumach. & Thonn. = **Phoenix reclinata** Jacq.

Phoenix sylvestris (L.) Roxb., Hort. Bengal.: 73 (1814).

Pakistan to C. Himalaya, E. India to Bangladesh. (29) mau (36) chs 40 BAN IND NEP PAK srl WHM.

Elate sylvestris L., Sp. Pl.: 1189 (1753).

Elate versicolor Salisb., Prodr. Stirp. Chap. Allerton: 264 (1796).

Phoenix theophrasti Greuter, Bauhinia 3: 243 (1967). Kriti to SW. Turkey. 13 KRI 34 EAI TUR.

Phoenix tenuis Verschaff. = **Phoenix canariensis** Chabaud

Phoenix vigieri Naudin = **Phoenix canariensis** Chabaud

Phoenix zeylanica Trimen = **Phoenix pusilla** Gaertn.

Pholidocarpus

Pholidocarpus Blume in J.J.Roemer & J.A.Schultes, Syst. Veg. 7: 1308 (1830).

6 species, Pen. Thailand to Malesia. 41 42.

Pholidocarpus diepenhorstii (Hassk.) Burret = **Livistona saribus** (Lour.) Merr. ex A.Chev.

Pholidocarpus ihur (Giseke) Blume, Rumphia 3: 90 (1847).
N. Sulawesi to Maluku. 42 MOL SUL.
Borassus ihur Giseke, Prael. Ord. Nat. Pl.: 87 (1792).
Pholidocarpus rumphii Meisn., Pl. Vasc. Gen. 2: 265 (1840), nom. illeg.

Pholidocarpus kingianus (Becc.) Ridl., Mat. Fl. Malay. Penins. 2: 167 (1907).
Pen. Malaysia. 42 MLY.
Livistona kingiana Becc., Malesia 3: 199 (1889).
Saribus kingianus (Becc.) Kuntze, Revis. Gen. Pl. 2: 736 (1891).

Pholidocarpus macrocarpus Becc., Malesia 3: 92 (1886).
Pen. Thailand to Pen. Malaysia. 41 THA 42 MLY.

Pholidocarpus majadum Becc., Malesia 1: 80 (1877).
Borneo. 42 BOR.

Pholidocarpus mucronatus Becc., Malesia 3: 91 (1886).
W. Sumatera. 42 SUM.

Pholidocarpus rumphii Meisn. = **Pholidocarpus ihur** (Giseke) Blume

Pholidocarpus sumatranus Becc., Malesia 3: 92 (1886).
W. Sumatera. 42 SUM.

Pholidocarpus tunicatus (Lour.) H.Wendl. = **Borassus flabellifer** L.

Pholidostachys

Pholidostachys H.Wendl. ex Hook.f., Gen. Pl. 3: 915 (1883).
4 species, C. & W. South America to NW. Brazil. 80 83 84.

Pholidostachys dactyloides H.E.Moore, J. Arnold Arbor. 48: 148 (1967). *Calyptrogyne dactyloides* (H.E.Moore) de Boer, Verh. Kon. Ned. Akad. Wetensch., Afd. Natuurk., Tweede Sect. 58(1): 74 (1968).
E. Panama to Ecuador. 80 PAN 83 CLM ECU.

Pholidostachys kalbreyeri H.Wendl. ex Burret, Bot. Jahrb. Syst. 63: 131 (1930).
Panama to NW. Colombia. 80 PAN 83 CLM.

Pholidostachys pulchra H.Wendl. ex Burret, Bot. Jahrb. Syst. 63: 130 (1930).
C. America to NW. Colombia. 80 COS NIC PAN 83 CLM.
Calyptrogyne pulchra Burret, Bot. Jahrb. Syst. 63: 129 (1930).
Geonoma amabilis H.Wendl. ex Dahlgren, Field Mus. Nat. Hist., Bot. Ser. 14: 158 (1936), nom. nud.

Pholidostachys synanthera (Mart.) H.E.Moore, Taxon 18: 231 (1969).
W. South America to NW. Brazil. 83 CLM ECU PER 84 BZN.
Geonoma synanthera Mart., Hist. Nat. Palm. 2: 13 (1823). *Calyptrogyne synanthera* (Mart.) Burret, Bot. Jahrb. Syst. 63: 129, 137 (1930). *Calyptronoma synanthera* (Mart.) L.H.Bailey, Gentes Herb. 4: 166 (1938).
Calyptronoma robusta Trail, J. Bot. 14: 330 (1876). *Calyptrogyne robusta* (Trail) Burret, Bot. Jahrb. Syst. 63: 129, 137 (1930).

Calyptrogyne kalbreyeri Burret, Bot. Jahrb. Syst. 63: 129, 137 (1930). *Calyptronoma kalbreyeri* (Burret) L.H.Bailey, Gentes Herb. 4: 166 (1938).
Calyptrogyne weberbaueri Burret, Bot. Jahrb. Syst. 63: 129, 139 (1930). *Calyptronoma weberbaueri* (Burret) L.H.Bailey, Gentes Herb. 4: 166 (1938).

Phoniphora

Phoniphora Neck. = **Phoenix** L.

Physokentia

Physokentia Becc., Atti Soc. Tosc. Sci. Nat. Pisa Processi Verbali 44: 152 (1934).
7 species, Bismarck Arch. to SW. Pacific. 43 60.
Goniosperma Burret, Occas. Pap. Bernice Pauahi Bishop Mus. 11(4): 10 (1935).
Goniocladus Burret, Notizbl. Bot. Gart. Berlin-Dahlem 15: 86 (1940).

Physokentia avia H.E.Moore, Principes 21: 86 (1977).
Bismarck Arch. 43 BIS.

Physokentia dennisii H.E.Moore, Principes 13: 131 (1969).
Solomon Is. 43 SOL.

Physokentia insolita H.E.Moore, Principes 13: 133 (1969).
Solomon Is. 43 SOL.

Physokentia petiolata (Burret) D.Fuller, Mem. New York Bot. Gard. 83: 208 (1999).
Fiji (Viti Levu). 60 FIJ.
Goniocladus petiolatus Burret, Notizbl. Bot. Gart. Berlin-Dahlem 15: 87 (1940).
Physokentia rosea H.E.Moore, Principes 10: 90 (1966).

Physokentia rosea H.E.Moore = **Physokentia petiolata** (Burret) D.Fuller

Physokentia tete (Becc.) Becc., Atti Soc. Tosc. Sci. Nat. Pisa Processi Verbali 44: 153 (1934).
Vanuatu. 60 VAN.
Cyphosperma tete Becc., Webbia 3: 137 (1910). *Cyphokentia tete* (Becc.) Becc., Atti Soc. Tosc. Sci. Nat. Pisa Processi Verbali 44: 152 (1934).

Physokentia thurstonii (Becc.) Becc., Atti Soc. Tosc. Sci. Nat. Pisa Processi Verbali 44: 154 (1934).
Fiji (Vanua Levu, Taveuni). 60 FIJ.
Cyphosperma thurstonii Becc., Webbia 4: 272 (1914). *Cyphokentia thurstonii* (Becc.) Becc., Atti Soc. Tosc. Sci. Nat. Pisa Processi Verbali 44: 152 (1934). *Goniosperma thurstonii* (Becc.) Burret, Occas. Pap. - *Goniosperma vitiense* Burret, Occas. Pap. Bernice Pauahi Bishop Mus. 11(4): 11 (1935).

Physokentia whitmorei H.E.Moore, Principes 13: 129 (1969).
Solomon Is. 43 SOL.

Phytelephas

Phytelephas Ruiz & Pav., Syst. Veg. Fl. Peruv. Chil.: 299 (1798).
6 species, Panama to S. Trop. America. 80 82 83.
Elephantusia Willd., Sp. Pl. 4: 1156 (1806).
Palandra O.F.Cook, J. Wash. Acad. Sci. 17: 228 (1927).
Yarina O.F.Cook, J. Wash. Acad. Sci. 17: 223 (1927).

Phytelephas aequatorialis Spruce, J. Linn. Soc., Bot. 11: 179 (1871). *Palandra aequatorialis* (Spruce) O.F.Cook, J. Wash. Acad. Sci. 17: 229 (1927).
W. Ecuador. 83 ECU.

Phytelephas bonplandiana Gaudich. = **?**

Phytelephas brachelus O.F.Cook = **Phytelephas seemannii** O.F.Cook

Phytelephas brachinus O.F.Cook = **Phytelephas seemannii** O.F.Cook

Phytelephas brevipes O.F.Cook = **Phytelephas seemannii** O.F.Cook

Phytelephas cornutus O.F.Cook = **Phytelephas seemannii** O.F.Cook

Phytelephas dasyneura Burret = **Ammandra decasperma** O.F.Cook

Phytelephas decasperma (O.F.Cook) Dahlgren = **Ammandra decasperma** O.F.Cook

Phytelephas endlicheriana Gaudich. = **?**

Phytelephas humboldtiana Gaudich. = **?**

Phytelephas karstenii O.F.Cook = **Phytelephas macrocarpa** Ruiz & Pav.

Phytelephas kunthiana Gaudich. = **?**

Phytelephas longiflora O.F.Cook = **Phytelephas seemannii** O.F.Cook

Phytelephas macrocarpa Ruiz & Pav., Syst. Veg. Fl. Peruv. Chil.: 301 (1798). *Elephantusia macrocarpa* (Ruiz & Pav.) Willd., Sp. Pl. 4: 1156 (1806).
Peru to Bolivia and NW. Brazil. 83 BOL PER 84 BZN.
Phytelephas microcarpa Ruiz & Pav., Syst. Veg. Fl. Peruv. Chil.: 302 (1798). *Elephantusia microcarpa* (Ruiz & Pav.) Willd., Sp. Pl. 4: 1157 (1806). *Yarina microcarpa* (Ruiz & Pav.) O.F.Cook, J. Wash. Acad. Sci. 17: 223 (1927).
Phytelephas karstenii O.F.Cook, J. Wash. Acad. Sci. 17: 227 (1927).

Phytelephas macrocarpa subsp. *schottii* (H.Wendl.) Barfod = **Phytelephas schottii** H.Wendl.

Phytelephas macrocarpa subsp. *tenuicaulis* Barfod = **Phytelephas tenuicaulis** (Barfod) A.J.Hend.

Phytelephas microcarpa Ruiz & Pav. = **Phytelephas macrocarpa** Ruiz & Pav.

Phytelephas orbignyana Gaudich. = **?**

Phytelephas pavonii Gaudich. = **?**

Phytelephas persooniana Gaudich. = **?**

Phytelephas pittieri O.F.Cook = **Phytelephas seemannii** O.F.Cook

Phytelephas poeppigii Gaudich. = **?**

Phytelephas ruizii Gaudich. = **?**

Phytelephas schottii H.Wendl., Bonplandia 8: 119 (1860). *Phytelephas macrocarpa* subsp. *schottii* (H.Wendl.) Barfod, Opera Bot. 109: 60 (1991).
Colombia. 83 CLM.

Phytelephas seemannii O.F.Cook, U.S.D.A. Bur. Pl. Industr. Bull. 242: 68 (1912).
Panama to Colombia. 80 PAN 83 CLM.
Phytelephas brachelus O.F.Cook, J. Wash. Acad. Sci. 3: 142 (1913).
Phytelephas brachinus O.F.Cook, J. Wash. Acad. Sci. 3: 142 (1913).
Phytelephas brevipes O.F.Cook, J. Wash. Acad. Sci. 3: 142 (1913). *Phytelephas seemannii* subsp. *brevipes* (O.F.Cook) Barfod, Opera Bot. 105: 64 (1991), without exact basionym page.
Phytelephas cornutus O.F.Cook, J. Wash. Acad. Sci. 3: 142 (1913).

Phytelephas pittieri O.F.Cook, J. Wash. Acad. Sci. 3: 142 (1913).
Phytelephas longiflora O.F.Cook, J. Wash. Acad. Sci. 17: 227 (1927).

Phytelephas seemannii subsp. *brevipes* (O.F.Cook) Barfod = **Phytelephas seemannii** O.F.Cook

Phytelephas tenuicaulis (Barfod) A.J.Hend., Palms Amazon: 291 (1995).
S. Colombia to N. Peru. 83 CLM ECU PER.
*Phytelephas macrocarpa subsp. *tenuicaulis* Barfod, Opera Bot. 105: 62 (1991).

Phytelephas tumacana O.F.Cook, J. Wash. Acad. Sci. 17: 224 (1927).
Colombia (Nariño). 83 CLM.

Phytelephas willdenowiana Gaudich. = **?**

Pichisermollia

Pichisermollia H.C.Monteiro = **Areca** L.

Pichisermollia insignis (Becc.) H.C.Monteiro = **Areca insignis** (Becc.) J.Dransf.

Pichisermollia insignis var. *moorei* J.Dransf. = **Areca insignis** var. **moorei** (J.Dransf.) J.Dransf.

Pichisermollia subacaulis (Becc.) H.C.Monteiro = **Areca subacaulis** (Becc.) J.Dransf.

Pigafetta

Pigafetta (Blume) Becc., Malesia 1: 89 (1877), nom. cons.
2 species, Sulawesi to New Guinea. 42 43.

Pigafetta elata (Mart.) H.Wendl. in O.C.E.de Kerchove de Denterghem, Palmiers: 253 (1878).
N. Sulawesi. 42 SUL.
*Metroxylon elatum Mart., Hist. Nat. Palm. 3: 216 (1838). *Sagus elata* (Mart.) Reinw. ex Blume, Rumphia 2: 156 (1843).

Pigafetta filifera Merr. = **Pigafetta filaris** (Giseke) Becc.

Pigafetta filaris (Giseke) Becc., Malesia 1: 90 (1877).
Maluku to New Guinea. 42 MOL 43 NWG.
*Sagus filaris Giseke, Prael. Ord. Nat. Pl.: 94 (1792).
Metroxylon filare (Giseke) Mart., Hist. Nat. Palm. 3: 216 (1838).
Sagus microcarpa Zipp. ex Hall, Bijdr. Natuurk. Wetensch. 5: 178 (1830), nom. inval.
Sagus microsperma Zipp. ex Hall, Bijdr. Natuurk. Wetensch. 5: 178 (1830).
Metroxylon microcarpum Kunth, Enum. Pl. 3: 215 (1841).
Metroxylon microspermum Kunth, Enum. Pl. 3: 215 (1841).
Pigafetta papuana Becc., Malesia 1: 89 (1877).
Calamus kunzeanus Becc., Ann. Roy. Bot. Gard. (Calcutta) 11(1): 490 (1908).
Pigafetta filifera Merr., Interpr. Herb. Amboin.: 114 (1917).

Pigafetta papuana Becc. = **Pigafetta filaris** (Giseke) Becc.

Pilophora

Pilophora Jacq. = **Manicaria** Gaertn.

Pilophora saccifera (Gaertn.) H.Wendl. = **Manicaria saccifera** Gaertn.

Pilophora testicularis Jacq. = **Manicaria saccifera** Gaertn.

Pinanga

Pinanga Blume, Rumphia 2: 76 (1839).
132 species, Trop. & Subtrop. Asia to NW. Pacific. 36 38 40 41 42 43 62.
  Cladosperma Griff., Not. Pl. Asiat. 3: 165 (1851).
  Ophiria Becc., Ann. Jard. Bot. Buitenzorg 2: 128 (1885).
  Pseudopinanga Burret, Notizbl. Bot. Gart. Berlin-Dahlem 13: 188 (1936).

Pinanga acaulis Ridl., J. Straits Branch Roy. Asiat. Soc. 44: 202 (1905).
Pen. Malaysia (Perak). 42 MLY.

Pinanga adangensis Ridl., J. Straits Branch Roy. Asiat. Soc. 61: 62 (1912).
Pen. Thailand to Pen. Malaysia. 41 THA 42 MLY.

Pinanga albescens Becc., Bot. Jahrb. Syst. 48: 89 (1912). *Pseudopinanga albescens* (Becc.) Burret, Notizbl. Bot. Gart. Berlin-Dahlem 13: 193 (1936). Borneo. 42 BOR.

Pinanga albescens var. *sarawakensis* Becc. = **Pinanga sessilifolia** Furtado

Pinanga andamanensis Becc., Atti Soc. Tosc. Sci. Nat. Pisa Processi Verbali 44: 121 (1934).
Andaman Is. 41 AND.

Pinanga angustisecta Becc., Malesia 3: 119 (1886).
Borneo. 42 BOR.

Pinanga annamensis Magalon, Contr. Étud. Palmiers Indoch.: 152 (1930).
S. Vietnam. 41 VIE.

Pinanga arinasae Witono, Palms 46: 194 (2002).
Lesser Sunda Is. (Bali). 42 LSI.

Pinanga aristata (Burret) J.Dransf., Kew Bull. 34: 775 (1980).
Borneo. 42 BOR.
  *Pseudopinanga aristata Burret, Notizbl. Bot. Gart. Berlin-Dahlem 13: 191 (1936).

Pinanga arundinacea Ridl., J. Straits Branch Roy. Asiat. Soc. 54: 60 (1910).
Borneo. 42 BOR.

Pinanga auriculata Becc., Malesia 3: 134 (1886).
Myanmar to Borneo. 41 MYA THA 42 BOR MLY.

  var. **auriculata**
  Borneo. 42 BOR.

  var. **leucocarpa** C.K.Lim, Gard. Bull. Singapore 50: 93 (1998).
  Pen. Thailand to Pen. Malaysia. 41 THA 42 MLY.
    Pinanga bowiana Hodel, Palm J. 134: 35 (1997).

  var. **merguensis** (Becc.) C.K.Lim, Gard. Bull. Singapore 50: 89 (1998).
  Myanmar to N. Pen. Malaysia. 41 MYA THA 42 MLY. Nanophan.
    *Pinanga patula var. *merguensis* Becc., Atti Soc. Tosc. Sci. Nat. Pisa Processi Verbali 44: 125 (1934).

Pinanga badia Hodel, Palm J. 136: 16 (1997).
Pen. Thailand to Pen. Malaysia (Perlis). 41 THA 42 MLY.

Pinanga banaensis Magalon = **Nenga banaensis** (Magalon) Burret

inanga baramensis Furtado = **Pinanga lepidota** Rendle

inanga barnesii Becc. = **Pinanga maculata** Porte ex Lem.

inanga barnesii var. *macrocarpa* Becc. = **Pinanga maculata** Porte ex Lem.

inanga barramensis Becc. = **Pinanga lepidota** Rendle

Pinanga basilanensis Becc., Philipp. J. Sci. 14: 322 (1919).
Philippines (Basilan, Mindanao). 42 PHI.

Pinanga batanensis Becc., Philipp. J. Sci., C 3: 340 (1909).
Philippines (Batan). 42 PHI.

Pinanga baviensis Becc., Webbia 3: 193 (1910).
N. Vietnam. 41 VIE.

Pinanga beccariana Furtado = **Pinanga subintegra** var. **beccariana** (Furtado) C.K.Lim

Pinanga bicolana Fernando, Principes 32: 172 (1988).
Philippines. 42 PHI.

Pinanga bifida Blume = **Pinanga disticha** (Roxb.) H.Wendl.

Pinanga borneensis Scheff., Ann. Jard. Bot. Buitenzorg 1: 151 (1876).
Borneo. 42 BOR.

Pinanga bowiana Hodel = **Pinanga auriculata** var. **leucocarpa** C.K.Lim

Pinanga brevipes Becc., Malesia 3: 121 (1886).
Borneo. 42 BOR.

Pinanga brewsteriana Ridl. = **Pinanga polymorpha** Becc.

Pinanga caesia Blume, Rumphia 2: 84 (1839). *Seaforthia caesia* (Blume) Mart., Hist. Nat. Palm. 3: 313 (1849). *Ptychosperma caesia* (Blume) Miq., Fl. Ned. Ind. 3: 20 (1855). *Pseudopinanga caesia* (Blume) Burret, Notizbl. Bot. Gart. Berlin-Dahlem 13: 195 (1936).
N. Sulawesi. 42 SUL.

Pinanga calamifrons Becc. = **Pinanga tenella** var. **tenella**

Pinanga calamifrons var. *tenuissima* Becc. = **Pinanga tenella** var. **tenuissima** (Becc.) J.Dransf.

Pinanga calapparia (Blume) H.Wendl. = **Actinorhytis calapparia** (Blume) H.Wendl. & Drude ex Scheff.

Pinanga canina Becc. = **Pinanga salicifolia** Blume

Pinanga canina f. *intermedia* Becc. = **Pinanga salicifolia** Blume

Pinanga canina f. *major* Becc. = **Pinanga salicifolia** Blume

Pinanga canina f. *minor* Becc. = **Pinanga salicifolia** Blume

Pinanga capitata Becc., J. Linn. Soc., Bot. 42: 168 (1914).
Borneo. 42 BOR.

  var. **capitata**
  Borneo. 42 BOR.
    Pinanga gibbsiana Becc., J. Linn. Soc., Bot. 42: 168 (1914).
    Pinanga clemensii Furtado, Repert. Spec. Nov. Regni Veg. 35: 279 (1934).
    Pinanga dallasensis Furtado, Repert. Spec. Nov. Regni Veg. 35: 275 (1934).
    Pinanga lumuensis Furtado, Repert. Spec. Nov. Regni Veg. 35: 280 (1934).

  var. **divaricata** J.Dransf., Kew Bull. 34: 777 (1980).
  NW. Borneo. 42 BOR.

Pinanga caudata Becc. = **Pinanga rumphiana** (Mart.) J.Dransf. & Govaerts

Pinanga celebica Scheff., Tijdschr. Ned.-Indië 32: 175 (1871).
Sulawesi. 42 SUL.

Pinanga chaiana J.Dransf., Kew Bull. 34: 779 (1980).
NW. Borneo. 42 BOR.

Pinanga chinensis Becc., Webbia 1: 326 (1905).
China (Yunnan). 36 CHC.

Pinanga cleistantha J.Dransf., Principes 26: 127 (1982).
Pen. Malaysia (Terengganu). 42 MLY.

Pinanga clemensiae Furtado = **Pinanga sessilifolia** Furtado

Pinanga clemensii Furtado = **Pinanga capitata** var. **capitata**

Pinanga cochinchinensis Blume = **Pinanga sylvestris** (Lour.) Hodel

Pinanga copelandii Becc., Webbia 1: 317 (1905).
Philippines. 42 PHI.

Pinanga coronata (Blume ex Mart.) Blume, Rumphia 2: 83 (1839).
S. Andaman Is. to Lesser Sunda Is. 41 AND 42 JAW LSI SUM.
  Areca oriziformis var. *gracilis* Giseke, Prael. Ord. Nat. Pl.: 80 (1792).
  *Areca coronata Blume ex Mart., Hist. Nat. Palm. 3: 179 (1838). *Seaforthia coronata* (Blume ex Mart.) Mart., Hist. Nat. Palm. 3(ed. 2): 185 (1845). *Ptychosperma coronatum* (Blume ex Mart.) Miq., Fl. Ned. Ind. 3: 24 (1855).
  Seaforthia montana Blume ex Mart., Hist. Nat. Palm. 3: 185 (1838).
  Seaforthia reinwardtiana Mart., Hist. Nat. Palm. 3: 183 (1838).
  Pinanga costata Blume, Rumphia 2: 80 (1839). *Seaforthia costata* (Blume) Mart., Hist. Nat. Palm. 3: 313 (1849). *Ptychosperma costatum* (Blume) Miq., Fl. Ned. Ind. 3: 25 (1855). *Areca costata* (Blume) Kurz, J. Asiat. Soc. Bengal, Pt. 2, Nat. Hist. 43(2): 200 (1874).
  Pinanga kuhlii Blume, Rumphia 2: 82 (1839). *Seaforthia kuhlii* (Blume) Mart., Hist. Nat. Palm. 3(ed. 2): 185 (1845). *Ptychosperma kuhlii* (Blume) Miq., Fl. Ned. Ind. 3: 21 (1855).
  Pinanga noxa Blume, Rumphia 2: 81 (1839). *Ptychosperma noxa* (Blume) Miq., Fl. Ned. Ind. 3: 23 (1855).
  Ptychosperma album Scheff., Tijdschr. Ned.-Indië 27: 26 (1864).
  Pinanga coronata var. *teijsmannii* Scheff., Natuurk. Tijdschr. Ned.-Indië 32: 181 (1873).
  Pinanga kuhlii var. *alba* Scheff., Natuurk. Tijdschr. Ned.-Indië 32: 183 (1873).
  Pinanga kuhlii var. *sumatrana* Scheff., Natuurk. Tijdschr. Ned.-Indië 32: 183 (1873). *Pinanga sumatrana* (Scheff.) H.Wendl., Ann. Hort. Belge Étrangère 25: 60 (1875).

Pinanga coronata var. *teijsmannii* Scheff. = **Pinanga coronata** (Blume ex Mart.) Blume

Pinanga costata Blume = **Pinanga coronata** (Blume ex Mart.) Blume

Pinanga crassipes Becc., Malesia 3: 120 (1886).
Borneo. 42 BOR.

Pinanga cucullata J.Dransf., Kew Bull. 46: 691 (1991).
Borneo (Sarawak). 42 BOR.

Pinanga curranii Becc., Philipp. J. Sci., C 2: 226 (1907).
Philippines (Palawan). 42 PHI.

Pinanga curvata (Griff.) Becc. = **Pinanga paradoxa** var. **paradoxa**

Pinanga dallasensis Furtado = **Pinanga capitata** var. **capitata**

Pinanga decora L.Linden & Rodigas, Ill. Hort. 34: 171 (1886).
Borneo. 42 BOR.

Pinanga densiflora Becc., Malesia 3: 116 (1886).
Sumatera. 42 SUM.

Pinanga densifolia Ridl. = **Pinanga perakensis** Becc.

Pinanga dicksonii (Roxb.) Blume, Rumphia 2: 77 (1839).
S. India. 40 IND.
  *Areca dicksonii Roxb., Fl. Ind. ed. 1832, 3: 616 (1832). *Seaforthia dicksonii* (Roxb.) Mart., Hist. Nat. Palm. 3: 184 (1838). *Ptychosperma dicksonii* (Roxb.) Miq., Fl. Ned. Ind. 3: 23 (1855).

Pinanga discolor Burret, Notizbl. Bot. Gart. Berlin-Dahlem 13: 187 (1936).
S. China to Hainan. 36 CHC CHH CHS.

Pinanga disticha (Roxb.) H.Wendl. in O.C.E.de Kerchove de Denterghem, Palmiers: 253 (1878).
Pen. Thailand to Sumatera. 41 THA 42 MLY SUM.
  *Areca disticha Roxb., Fl. Ind. ed. 1832, 3: 620 (1832). *Seaforthia disticha* (Roxb.) Mart., Hist. Nat. Palm. 3: 184 (1838). *Ptychosperma distichum* (Roxb.) Miq., Fl. Ned. Ind. 3: 28 (1855).
  Pinanga bifida Blume, Rumphia 2: 92 (1839).
  Areca curvata Griff., Ic. Pl. Asiat. 3: t. 248 (1851). *Pinanga curvata* (Griff.) Becc., Malesia 3: 128 (1886).
  Areca humilis Roxb. ex H.Wendl. in O.C.E.de Kerchove de Denterghem, Palmiers: 231 (1878).

Pinanga dumetosa J.Dransf., Kew Bull. 34: 781 (1980).
Borneo (Sarawak). 42 BOR.

Pinanga duperreana Pierre ex Becc., Malesia 3: 144 (1886).
Cambodia. 41 CBD.

Pinanga egregia Fernando, Kew Bull. 49: 775 (1994).
Philippines. 42 PHI.

Pinanga elmeri Becc. = **Pinanga philippinensis** Becc.

Pinanga forbesii Ridl., J. Bot. 63(Suppl.): 124 (1925).
Sumatera. 42 SUM.

Pinanga fractiflexa Hodel, Palm J. 136: 19 (1997).
Pen. Thailand. 41 THA.

Pinanga fruticans Ridl. = **Pinanga scortechinii** Becc.

Pinanga furfuracea Blume, Rumphia 2: 88 (1839).
  Seaforthia furfuracea (Blume) Mart., Hist. Nat. Palm. 3: 313 (1849). *Ptychosperma furfuraceum* (Blume) Miq., Fl. Ned. Ind. 3: 26 (1855).
Sulawesi. 42 SUL.

Pinanga geonomiformis Becc., Philipp. J. Sci., C 4: 602 (1909).
Philippines (Luzon). 42 PHI.

Pinanga gibbsiana Becc. = **Pinanga capitata** var. **capitata**

Pinanga glauca Ridl., J. Fed. Malay States Mus. 8(4): 120 (1917).
Sumatera. 42 SUM.

Pinanga glaucescens Ridl., J. Straits Branch Roy. Asiat. Soc. 86: 311 (1922).
Pen. Thailand to Pen. Malaysia. 41 THA 42 MLY.

Pinanga glaucifolia Fernando, Kew Bull. 49: 776 (1994).
Philippines. 42 PHI.

Pinanga globosa G.Nicholson = **Calyptrocalyx spicatus** (Lam.) Blume

Pinanga globulifera (Lam.) Blume, Bull. Sci. Phys. Nat. Néerl. 1: 65 (1838).
Maluku. 42 MOL.
  *Areca globulifera Lam., Encycl. 1: 895 (1783).
  Areca oriziformis Gaertn., Fruct. Sem. Pl. 1: 20 (1788).
    Seaforthia oriziformis (Gaertn.) Mart., Hist. Nat. Palm. 3: 185 (1838).

Pinanga gracilis Blume, Rumphia 2: 77 (1839).
  Seaforthia gracilis (Blume) Mart., Hist. Nat. Palm. 3: 185 (1838). *Nenga gracilis* (Blume) Becc., Malesia 1: 25 (1877).
Himalaya to Tibet. 36 CHT 40 ASS BAN EHM 41 MYA.
  *Areca gracilis Buch.-Ham., Mem. Wern. Nat. Hist. Soc. 5(2): 310 (1826).
  *Areca gracilis Roxb., Fl. Ind. ed. 1832, 3: 619 (1832), nom. illeg.

Pinanga gracillima Merr., Sarawak Mus. J. 3: 518 (1928).
Borneo. 42 BOR.

Pinanga grandijuga Burret, Notizbl. Bot. Gart. Berlin-Dahlem 15: 208 (1940).
Sumatera. 42 SUM.

Pinanga grandis Burret, Notizbl. Bot. Gart. Berlin-Dahlem 15: 207 (1940).
Sumatera. 42 SUM.

Pinanga griffithii Becc., Malesia 3: 117 (1886).
Assam. 40 ASS.

Pinanga heterophylla Becc., Philipp. J. Sci. 14: 319 (1919).
Philippines (Luzon, Negros). 42 PHI.

Pinanga hexasticha (Kurz) Scheff., Ann. Jard. Bot. Buitenzorg 1: 148 (1876).
China (Yunnan) to Myanmar. 36 CHC 41 MYA.
  *Areca hexasticha Kurz, J. Asiat. Soc. Bengal, Pt. 2, Nat. Hist. 43(2): 201 (1874).

Pinanga hookeriana Becc., Malesia 3: 141 (1886).
Assam. 40 ASS.

Pinanga hymenospatha Hook.f., Fl. Brit. India 6: 411 (1892).
Myanmar. 41 MYA.

Pinanga inaequalis Blume, Rumphia 2: 91 (1839).
  Seaforthia inaequalis (Blume) Mart., Hist. Nat. Palm. 3: 313 (1849). *Ptychosperma inaequalis* (Blume) Miq., Fl. Ned. Ind. 3: 27 (1855).
Sulawesi. 42 SUL.

Pinanga insignis Becc., Philipp. J. Sci., C 2: 223 (1907).
  Pseudopinanga insignis (Becc.) Burret, Notizbl. Bot. Gart. Berlin-Dahlem 13: 194 (1936).
Philippines to Caroline Is. 42 PHI SUL 62 CRL.
  Pinanga insignis subsp. *loheriana* Becc., Philipp. J. Sci. 14: 322 (1919).
  Pinanga insignis var. *gasterocarpa* Becc., Philipp. J. Sci. 14: 322 (1919).
  Pinanga insignis var. *leptocarpa* Becc., Philipp. J. Sci. 14: 322 (1919).
  Pinanga micronesica Kaneh., J. Jap. Bot. 12: 635 (1936).

Pinanga insignis var. *gasterocarpa* Becc. = **Pinanga insignis** Becc.

Pinanga insignis var. *leptocarpa* Becc. = **Pinanga insignis** Becc.

Pinanga insignis subsp. *loheriana* Becc. = **Pinanga insignis** Becc.

Pinanga isabelensis Becc., Philipp. J. Sci. 14: 318 (1919).
Philippines (Luzon). 42 PHI.

Pinanga jamariensis C.K.Lim, Gard. Bull. Singapore 50: 100 (1998).
Pen. Malaysia (Johor). 42 MLY.

Pinanga javana Blume, Rumphia 2: 85 (1839).
Jawa. 42 JAW.

Pinanga johorensis C.K.Lim & Saw, Gard. Bull. Singapore 50: 104 (1998).
Pen. Malaysia (incl. Singapore). 42 MLY.

Pinanga junghuhnii Miq. = **Pinanga patula** Blume

Pinanga keahii Furtado, Repert. Spec. Nov. Regni Veg. 35: 280 (1934).
Borneo. 42 BOR.

Pinanga kuhlii Blume = **Pinanga coronata** (Blume ex Mart.) Blume

Pinanga kuhlii var. *alba* Scheff. = **Pinanga coronata** (Blume ex Mart.) Blume

Pinanga kuhlii var. *sumatrana* Scheff. = **Pinanga coronata** (Blume ex Mart.) Blume

Pinanga latisecta Blume, Rumphia 2: 79 (1839).
  Seaforthia latisecta (Blume) Mart., Hist. Nat. Palm. 3: 313 (1849). *Ptychosperma latisectum* (Blume) Miq., Fl. Ned. Ind. 3: 20 (1855). *Areca latisecta* (Blume) Scheff., Tijdschr. Ned.-Indië 32: 168 (1871). *Nenga latisecta* (Blume) Scheff., Ann. Jard. Bot. Buitenzorg 1: 120 (1876).
Sumatera. 42 SUM.

Pinanga lepida W.Bull = **?** [40 IND]

Pinanga lepidota Rendle, J. Bot. 39: 177 (1901).
Borneo. 42 BOR.
  Pinanga baramensis Furtado, Repert. Spec. Nov. Regni Veg. 35: 277 (1934).
  Pinanga barramensis Becc., Atti Soc. Tosc. Sci. Nat. Mem., B 44: 124 (1934).

Pinanga ligulata Becc., Malesia 2: 129 (1886).
Borneo. 42 BOR.

Pinanga limosa Ridl., J. Straits Branch Roy. Asiat. Soc. 44: 201 (1905).
Pen. Malaysia. 42 MLY.

var. **limosa**
  Pen. Malaysia (Terengganu). 42 MLY.

var. **montana** C.K.Lim, Folia Malaysiana 2: 253 (2001).
  Pen. Malaysia (Selangor). 42 MLY.

Pinanga lumuensis Furtado = **Pinanga capitata** var. **capitata**

Pinanga macroclada Burret, Notizbl. Bot. Gart. Berlin-Dahlem 13: 188 (1936).
China (Yunnan). 36 CHC.

Pinanga macrospadix Burret, Notizbl. Bot. Gart. Berlin-Dahlem 15: 205 (1940).
Sumatera. 42 SUM.

Pinanga maculata Porte ex Lem., Ill. Hort. 10: 92, t. 361 (1863). *Ptychosperma maculatum* (Porte ex Lem.)

Seem., Gard. Chron. 1870: 697 (1870). *Pseudopinanga maculata* (Porte ex Lem.) Burret, Notizbl. Bot. Gart. Berlin-Dahlem 13: 194 (1936).
Philippines. 42 PHI.
> *Pinanga barnesii* Becc., Webbia 1: 320 (1905). *Pseudopinanga barnesii* (Becc.) Burret, Notizbl. Bot. Gart. Berlin-Dahlem 13: 193 (1936). *Pinanga barnesii* var. *macrocarpa* Becc., Philipp. J. Sci., C 3: 340 (1908).

Pinanga malaiana (Mart.) Scheff., Tijdschr. Ned.-Indië 32: 175 (1871).
Pen. Thailand to Sumatera. 41 THA 42 MLY SUM.
> **Seaforthia malaiana* Mart., Hist. Nat. Palm. 3: 184 (1838). *Areca malaiana* (Mart.) Griff., Calcutta J. Nat. Hist. 5: 457 (1845). *Ptychosperma malaianum* (Mart.) Miq., Fl. Ned. Ind. 3: 25 (1855).
> *Areca haematocarpon* Griff., Not. Pl. Asiat. 3: 165 (1851).

Pinanga manii Becc., Malesia 2: 178 (1889).
S. Andaman Is. to Nicobar Is. 41 AND NCB.

Pinanga megalocarpa Burret, Notizbl. Bot. Gart. Berlin-Dahlem 15: 209 (1940).
Sumatera. 42 SUM.

Pinanga micholitzii auct., Gard. Chron., III, 43: 257 (1908).
Sumatera. 42 SUM.

Pinanga micronesica Kaneh. = **Pinanga insignis** Becc.

Pinanga minor Blume, Rumphia 2: 86 (1839).
> *Seaforthia minor* (Blume) Mart., Hist. Nat. Palm. 3: 313 (1849). *Ptychosperma minor* (Blume) Miq., Fl. Ned. Ind. 3: 23 (1855).
Sulawesi. 42 SUL.

Pinanga minuta Furtado, Repert. Spec. Nov. Regni Veg. 35: 281 (1934).
Borneo. 42 BOR.

Pinanga mirabilis Becc., Malesia 2: 126 (1886).
Borneo. 42 BOR.

Pinanga modesta Becc., Philipp. J. Sci., C 2: 225 (1907).
Philippines. 42 PHI.

Pinanga mooreana J.Dransf., Kew Bull. 34: 783 (1980).
Borneo (Sarawak). 42 BOR.

Pinanga nannospadix Burret = **Nenga banaensis** (Magalon) Burret

Pinanga neglecta Burret = **Nenga pumila** (Blume) H.Wendl. var. **pumila**

Pinanga negrosensis Becc., Leafl. Philipp. Bot. 2: 642 (1909).
Philippines (Dinagat, Negros). 42 PHI.

Pinanga nenga (Blume ex Mart.) Blume = **Nenga pumila** (Blume) H.Wendl. var. **pumila**

Pinanga nenga var. *pachystachya* Blume = **Nenga pumila** var. **pachystachya** (Blume) Fernando

Pinanga noxa Blume = **Pinanga coronata** (Blume ex Mart.) Blume

Pinanga pachycarpa Burret, Notizbl. Bot. Gart. Berlin-Dahlem 13: 185 (1936).
Sumatera. 42 SUM.

Pinanga pachyphylla J.Dransf., Kew Bull. 46: 695 (1991).
Borneo (Sarawak). 42 BOR.

Pinanga palustris Kiew, Gard. Bull. Singapore 50: 106 (1998).
Pen. Malaysia. 42 MLY.

Pinanga pantiensis J.Dransf., Gard. Bull. Singapore 50: 110 (1998).
Pen. Malaysia. 42 MLY.

Pinanga paradoxa (Griff.) Scheff., Tijdschr. Ned.-Indië 32: 179 (1871).
Pen. Thailand to Pen. Malaysia. 41 THA 42 MLY.
> **Areca paradoxa* Griff., Calcutta J. Nat. Hist. 5: 463 (1845). *Kentia paradoxa* (Griff.) Mart., Hist. Nat. Palm. 3: 312 (1849). *Nengella paradoxa* (Griff.) Becc., Malesia 1: 32 (1877). *Ophiria paradoxa* (Griff.) Becc., Ann. Jard. Bot. Buitenzorg 2: 128 (1885).

var. paradoxa
Pen. Thailand to Pen. Malaysia. 41 THA 42 MLY.
> *Areca curvata* Griff., Not. Pl. Asiat. 3: 164 (1851). *Pinanga curvata* (Griff.) Becc., Malesia 3: 128 (1886).

var. unicostata C.K.Lim, Folia Malaysiana 2: 235 (2001).
Pen. Malaysia. 42 MLY.

Pinanga paradoxa var. *multifida* Becc. = **Pinanga subintegra** var. **multifida** (Becc.) C.K.Lim

Pinanga paradoxa var. *subintegra* (Ridl.) Becc. = **Pinanga subintegra** Ridl.

Pinanga parvula Ridl., J. Malayan Branch Roy. Asiat. Soc. 1: 103 (1923).
Sumatera. 42 SUM.

Pinanga patula Blume, Rumphia 2: 87 (1839).
> *Seaforthia patula* (Blume) Mart., Hist. Nat. Palm. 3: 313 (1849). *Ptychosperma patulum* (Blume) Miq., Fl. Ned. Ind. 3: 26 (1855).
Sumatera. 42 SUM.
> *Pinanga junghuhnii* Miq., Pl. Jungh.: 157 (1852). *Ptychosperma junghuhnii* (Miq.) Miq., Fl. Ned. Ind. 3: 28 (1855).
> *Pinanga patula* var. *junghuhnii* (Miq.) Scheff., Natuurk. Tijdschr. Ned.-Indië 32: 178 (1873).

Pinanga patula var. *junghuhnii* (Miq.) Scheff. = **Pinanga patula** Blume

Pinanga patula var. *merguensis* Becc. = **Pinanga auriculata** var. **merguensis** (Becc.) C.K.Lim

Pinanga patula var. *riparia* (Ridl.) Becc. = **Pinanga riparia** Ridl.

Pinanga pectinata Becc. in J.D.Hooker, Fl. Brit. India 6: 410 (1892).
Pen. Malaysia. 42 MLY.

Pinanga perakensis Becc., Malesia 3: 175 (1889).
Pen. Thailand to Pen. Malaysia. 41 THA 42 MLY.
> *Pinanga densifolia* Ridl., J. Fed. Malay States Mus. 4: 85 (1909).

Pinanga philippinensis Becc., Malesia 3: 180 (1889).
Philippines. 42 PHI.
> *Pinanga elmeri* Becc., Webbia 1: 322 (1905).

Pinanga pilosa (Burret) J.Dransf., Kew Bull. 34: 775 (1980).
Borneo. 42 BOR.
> **Pseudopinanga pilosa* Burret, Notizbl. Bot. Gart. Berlin-Dahlem 13: 189 (1936).

Pinanga pisiformis Teijsm. ex Becc. = **Hydriastele costata** F.M.Bailey

Pinanga polymorpha Becc., Malesia 3: 172 (1889).
Pen. Malaysia. 41 THA? 42 MLY.
> *Pinanga robusta* Becc. in J.D.Hooker, Fl. Brit. India 6: 408 (1892).

Pinanga brewsteriana Ridl., J. Fed. Malay States Mus. 6: 188 (1915).

Pinanga wrayi Furtado, Repert. Spec. Nov. Regni Veg. 35: 276 (1934).

Pinanga porrecta Burret, Notizbl. Bot. Gart. Berlin-Dahlem 15: 203 (1940). Sumatera. 42 SUM.

Pinanga pulchella Burret, Notizbl. Bot. Gart. Berlin-Dahlem 15: 200 (1940). Sumatera. 42 SUM.

Pinanga pumila (Blume) Blume = **Nenga pumila** (Blume) H.Wendl.

Pinanga punicea (Zipp. ex Blume) Merr. = **Pinanga rumphiana** (Mart.) J.Dransf. & Govaerts

Pinanga purpurea Miq. = **Cyrtostachys renda** Blume

Pinanga purpurea Hendra, Floribunda 2(2): 31 (2002). Sumatera. 42 SUM.

Pinanga quadrijuga Gagnep., Notul. Syst. (Paris) 6: 156 (1937). S. Vietnam. 41 VIE.

Pinanga ridleyana Becc. ex Furtado, Repert. Spec. Nov. Regni Veg. 35: 282 (1934). Borneo. 42 BOR.

Pinanga rigida Becc., Leafl. Philipp. Bot. 2: 644 (1909). Philippines. 42 PHI.

Pinanga riparia Ridl., J. Straits Branch Roy. Asiat. Soc. 44: 201 (1905). *Pinanga patula* var. *riparia* (Ridl.) Becc., Nuovo Giorn. Bot. Ital., n.s., 42: 71 (1935). Pen. Thailand to Sumatera. 41 THA 42 BOR? MLY SUM.

Pinanga rivularis Becc., Malesia 3: 130 (1886). Borneo (Sarawak). 42 BOR.

Pinanga robusta Becc. = **Pinanga polymorpha** Becc.

Pinanga rubricaulis Linden = **Cyrtostachys renda** Blume

Pinanga rumphiana (Mart.) J.Dransf. & Govaerts World Checklist Palms: (2004) Maluku to New Guinea. 42 MOL 43 NWG.
Seaforthia rumphiana Mart., Hist. Nat. Palm. 3: 186 (1838). *Areca punicea* Zipp. ex Blume, Rumphia 2: 72 (1839), nom. illeg. *Ptychosperma puniceum* (Zipp. ex Blume) Miq., Fl. Ned. Ind. 3: 31 (1855). *Drymophloeus puniceus* (Zipp. ex Blume) Becc., Malesia 1: 47 (1877). *Saguaster puniceus* (Zipp. ex Blume) Kuntze, Revis. Gen. Pl. 2: 735 (1891). *Pinanga punicea* (Zipp. ex Blume) Merr., Interpr. Herb. Amboin.: 122 (1917).
Areca sanguinea Zipp. ex Blume, Rumphia 2: 72 (1839).
Drymophloeus rumphianus Mart., Hist. Nat. Palm. 3: 314 (1849).
Pinanga ternatensis Scheff., Ann. Jard. Bot. Buitenzorg 1: 149 (1876).
Pinanga caudata Becc., Malesia 1: 101 (1877).
Ptychosperma caudatum Becc., Malesia 1: 55 (1877).
Areca gigantea H.Wendl. in O.C.E.de Kerchove de Denterghem, Palmiers: 254 (1878).
Pinanga ternatensis var. *papuana* Becc., Malesia 3: 116 (1886).

Pinanga rupestris J.Dransf., Kew Bull. 46: 693 (1991). Borneo (Sarawak). 42 BOR.

Pinanga salicifolia Blume, Rumphia 2: 93 (1843). *Seaforthia salicifolia* (Blume) Mart., Hist. Nat. Palm. 3: 313 (1849). *Ptychosperma salicifolium* (Blume)

Miq., Fl. Ned. Ind. 3: 28 (1855). Borneo. 42 BOR.
Pinanga canina Becc., Malesia 3: 135 (1886).
Pinanga canina f. *intermedia* Becc., Malesia 3: 136 (1886).
Pinanga canina f. *major* Becc., Malesia 3: 136 (1886).
Pinanga canina f. *minor* Becc., Malesia 3: 136 (1886).

Pinanga samarana Becc., Philipp. J. Sci. 14: 321 (1919). Philippines (Samar). 42 PHI.

Pinanga sanderiana W.Bull = ? **[42 MLY]**

Pinanga sarmentosa Saw, Palms 47: 141 (2003). Pen. Malaysia. 42 MLY.

Pinanga sclerophylla Becc., Philipp. J. Sci., C 4: 603 (1909). Philippines (Mindoro). 42 PHI.

Pinanga scortechinii Becc., Malesia 3: 170 (1889). Pen. Thailand to Pen. Malaysia. 41 THA 42 MLY. *Pinanga fruticans* Ridl., Fl. Malay Penins. 5: 9 (1925).

Pinanga sessilifolia Furtado, Repert. Spec. Nov. Regni Veg. 35: 283 (1934). Borneo. 42 BOR.
Pinanga albescens var. *sarawakensis* Becc., Atti Soc. Tosc. Sci. Nat. Mem., B 44: 126 (1934).
Pinanga clemensiae Furtado, Repert. Spec. Nov. Regni Veg. 35: 278 (1934).

Pinanga sibuyanensis Becc., Philipp. J. Sci. 14: 324 (1919). Philippines (Sibuyan). 42 BOR.

Pinanga sierramadreana Fernando, Kew Bull. 49: 780 (1994). Philippines. 42 PHI.

Pinanga simplicifrons (Miq.) Becc., Malesia 3: 124 (1885). Pen. Thailand to W. Malesia. 41 THA 42 BOR MLY SUM.
Ptychosperma simplicifrons Miq., Fl. Ned. Ind., Eerste Bijv.: 590 (1861).

var. **pinnata** C.K.Lim, Folia Malaysiana 2: 237 (2001). Pen. Malaysia. 42 MLY.

var. **simplicifrons**
Pen. Thailand to W. Malesia. 41 THA 42 BOR MLY SUM.

Pinanga singaporensis Ridl., J. Straits Branch Roy. Asiat. Soc. 41: 38 (1903). Pen. Malaysia. 42 MLY.

Pinanga sinii Burret, Notizbl. Bot. Gart. Berlin-Dahlem 10: 882 (1930). China (Guangxi, Guangdong). 36 CHC.

Pinanga smithii (R.Br.) Scheff. = **Ptychosperma elegans** (R.Br.) Blume

Pinanga sobolifera Fernando, Kew Bull. 49: 782 (1994). Philippines. 42 PHI.

Pinanga speciosa Becc., Webbia 1: 316 (1905). Philippines (Mindanao). 42 PHI.

Pinanga spectabilis W.Bull ex Becc. = ?

Pinanga stricta Becc., Malesia 3: 133 (1886). Borneo. 42 BOR.

Pinanga stylosa Becc., Malesia 3: 177 (1889). *Pseudopinanga stylosa* (Becc.) Burret, Notizbl. Bot. Gart. Berlin-Dahlem 13: 198 (1936). Sumatera. 42 SUM.

Pinanga subintegra Ridl., Mat. Fl. Malay. Penins. 2: 141 (1907). *Pinanga paradoxa* var. *subintegra* (Ridl.) Becc., Atti Soc. Tosc. Sci. Nat. Pisa Processi Verbali 44: 129 (1934).
Pen. Malaysia to Sumatera. 42 MLY SUM.

var. **beccariana** (Furtado) C.K.Lim, Folia Malaysiana 2: 259 (2001).
Pen. Malaysia to Sumatera. 42 MLY SUM.
*Pinanga beccariana Furtado, Repert. Spec. Nov. Regni Veg. 35: 278 (1934).

var. **intermedia** Furtado ex C.K.Lim, Folia Malaysiana 2: 258 (2001).
Pen. Thailand to Pen. Malaysia. 41 THA 42 MLY.

var. **multifida** (Becc.) C.K.Lim, Folia Malaysiana 2: 257 (2001).
Pen. Malaysia (Perak: G. Hijau). 42 MLY.
*Pinanga paradoxa var. *multifida* Becc., Atti Soc. Tosc. Sci. Nat. Pisa Processi Verbali 44: 129 (1934).

Pinanga subruminata Becc., Malesia 3: 174 (1889).
Pen. Malaysia. 42 MLY.

Pinanga sumatrana (Scheff.) H.Wendl. = **Pinanga coronata** (Blume ex Mart.) Blume

Pinanga sylvestris (Lour.) Hodel, Palm J. 139: 55 (1998).
Indo-China. 36? 41 CBD LAO THA VIE.
*Areca sylvestris Lour., Fl. Cochinch. 2: 568 (1790).
Seaforthia sylvestris (Lour.) Blume ex Mart., Hist. Nat. Palm. 3: 185 (1838). *Ptychosperma sylvestris* (Lour.) Miq., Fl. Ned. Ind. 3: 22 (1855).
Pinanga cochinchinensis Blume, Rumphia 2: 85 (1839). *Ptychosperma cochinchinense* (Blume) Miq., Fl. Ned. Ind. 3: 23 (1855).

Pinanga tashiroi Hayata, Icon. Pl. Formosan. 3: 196 (1913). *Pseudopinanga tashiroi* (Hayata) Burret, Notizbl. Bot. Gart. Berlin-Dahlem 13: 194 (1936).
Taiwan (Lan Yü). 38 TAI.

Pinanga tenacinervis J.Dransf., Kew Bull. 34: 785 (1980).
Borneo (Sarawak). 42 BOR.

Pinanga tenella (H.Wendl.) Scheff., Tijdschr. Ned.-Indië 32: 179 (1871).
Borneo. 42 BOR.
*Ptychosperma tenella H.Wendl., Bot. Zeitung (Berlin) 17: 63 (1859).

var. **tenella**
Borneo. 42 BOR.
Pinanga calamifrons Becc., Malesia 3: 132 (1886).

var. **tenuissima** (Becc.) J.Dransf., Kew Bull. 34: 773 (1980).
Borneo. 42 BOR.
*Pinanga calamifrons var. *tenuissima* Becc., Malesia 3: 133 (1886).

Pinanga ternatensis Scheff. = **Pinanga rumphiana** (Mart.) J.Dransf. & Govaerts

Pinanga ternatensis var. *papuana* Becc. = **Pinanga rumphiana** (Mart.) J.Dransf. & Govaerts

Pinanga tomentella Becc., Malesia 3: 126 (1886).
Borneo. 42 BOR.

Pinanga trichoneura Becc., Atti Soc. Tosc. Sci. Nat. Pisa Processi Verbali 44: 120 (1934).
Borneo. 42 BOR.

Pinanga uncinata Burret, Notizbl. Bot. Gart. Berlin-Dahlem 15: 201 (1940).
Sumatera. 42 SUM.

Pinanga urdanetensis Becc., Leafl. Philipp. Bot. 8: 3008 (1919).
Philippines (Mindanao). 42 PHI.

Pinanga urosperma Becc., Philipp. J. Sci., C 3: 341 (1909). *Pseudopinanga urosperma* (Becc.) Burret, Notizbl. Bot. Gart. Berlin-Dahlem 13: 194 (1936).
Philippines. 42 PHI.

Pinanga variegata Becc., Malesia 3: 127 (1886).
Borneo. 42 BOR.

var. **hallieriana** Becc., Atti Soc. Tosc. Nat. Pisa Mem. 44 (1934.
Borneo. 42 BOR.

var. **variegata**
Borneo. 42 BOR.

Pinanga veitchii H.Wendl. ex H.J.Veitch, Cat. 1880: 23 (1880).
Borneo (Sarawak). 42 BOR.

Pinanga vestiaria (Giseke) Blume = **Areca vestiaria** Giseke

Pinanga viridis Burret, Notizbl. Bot. Gart. Berlin-Dahlem 13: 186 (1936).
S. China. 36 CHC CHS.

Pinanga vonmohlii Linden ex W.Watson = **?**

Pinanga watanaiana C.K.Lim, Principes 42: 116 (1998).
Thailand. 41 THA.

Pinanga woodiana Becc., Philipp. J. Sci., C 4: 604 (1909).
Philippines. 42 PHI.

Pinanga wrayi Furtado = **Pinanga polymorpha** Becc.

Pinanga yassinii J.Dransf., Forest. Dept. Occas. Pap., Brunei 2: 2 (1990).
Borneo (Brunei). 42 BOR.

Pindarea

Pindarea Barb.Rodr. = **Attalea** Kunth

Pindarea concinna Barb.Rodr. = **Attalea dubia** (Mart.) Burret

Pindarea dubia (Mart.) A.D.Hawkes = **Attalea dubia** (Mart.) Burret

Pindarea fastuosa Barb.Rodr. = **Attalea dubia** (Mart.) Burret

Pithodes

Pithodes O.F.Cook = **Coccothrinax** Sarg.

Pithodes spissa (L.H.Bailey) O.F.Cook = **Coccothrinax spissa** L.H.Bailey

Platenia

Platenia H.Karst. = **Syagrus** Mart.

Platenia chiragua H.Karst. = **Syagrus sancona** (Kunth) H.Karst.

Platythea

Platythea O.F.Cook = **Chamaedorea** Willd.

Platythea graminea O.F.Cook = **Chamaedorea elatior** Mart.

Plectis

Plectis O.F.Cook = **Euterpe** Mart.

Plectis oweniana O.F.Cook = **Euterpe precatoria** var. **longivaginata** (Mart.) A.J.Hend.

Plectocomia

Plectocomia Mart. & Blume in J.J.Roemer & J.A.Schultes, Syst. Veg. 7(2): 1333 (1830).
16 species, E. Himalaya to Hainan and Malesia. 36 40 41 42.

Plectocomia assamica Griff., Calcutta J. Nat. Hist. 5: 97 (1845).
Arunachal Pradesh to China (S. Yunnan). 36 CHC 40 ASS EHM.

Plectocomia barthiana Hodel = **Plectocomia pierreana** Becc.

Plectocomia billitonensis Becc., Ann. Roy. Bot. Gard. (Calcutta) 12(2): 32 (1918).
Sumatera (Belitung). 42 SUM.

Plectocomia bractealis Becc., Ann. Roy. Bot. Gard. (Calcutta) 12(2): 40 (1918).
Assam. 40 ASS.

Plectocomia cambodiana Gagnep. ex Humbert = **Plectocomia pierreana** Becc.

Plectocomia crinita Gentil ex Chitt. = **Plectocomia elongata** var. **elongata**

Plectocomia dransfieldiana Madulid, Kalikasan 10: 77 (1981).
Pen. Malaysia (Perak). 42 MLY.

Plectocomia elmeri Becc., Ann. Roy. Bot. Gard. (Calcutta) 12(2): 34 (1918).
Philippines (Mindanao: Mt. Apo). 42 PHI.

Plectocomia elongata Mart. & Blume in J.J.Roemer & J.A.Schultes, Syst. Veg. 7: 1333 (1830). *Rotang maximus* Baill., Hist. Pl. 13: 300 (1895).
Thailand to Malesia. 41 THA VIE 42 BOR JAW MLY PHI SUM.

var. **elongata**
Thailand to W. Malesia. 41 THA VIE 42 BOR JAW MLY SUM.
Calamus maximus Reinw. ex Schult.f. in J.J.Roemer & J.A.Schultes, Syst. Veg. 7: 1333 (1830).
Plectocomia sumatrana Miq., Fl. Ned. Ind., Eerste Bijv.: 592 (1861).
Plectocomia icthyospinus auct., Gard. Chron. 5: 735 (1876).
Plectocomia hystrix Linden, Ill. Hort. 28: 32 (1881). Provisional synonym.
Plectocomia maxima Kuntze, Revis. Gen. Pl. 2: 734 (1891), nom. nud.
Plectocomia griffithii Becc. in J.D.Hooker, Fl. Brit. India 6: 478 (1893).
Plectocomia elongata var. *bangkana* Becc., Ann. Roy. Bot. Gard. (Calcutta) 12(2): 26 (1918).
Plectocomia crinita Gentil ex Chitt., Dict. Gard.: 1605 (1956).

var. **philippinensis** Madulid, Kalikasan 10: 52 (1981).
Philippines. 42 PHI.

Plectocomia elongata var. *bangkana* Becc. = **Plectocomia elongata** var. **elongata**

Plectocomia geminiflora (Griff.) H.Wendl. = **Plectocomiopsis geminiflora** (Griff.) Becc.

Plectocomia griffithii Becc. = **Plectocomia elongata** var. **elongata**

Plectocomia himalayana Griff., Calcutta J. Nat. Hist. 5: 100 (1845).
E. Himalaya to N. Laos. 36 CHC 40 EHM 41 LAO THA.
Plectocomia montana Griff. ex T.Anderson, J. Linn. Soc., Bot. 11: 12 (1871).
Plectocomia montana Hook.f. & Thomson, Fl. Brit. India 6: 478 (1893), nom. inval.

Plectocomia hystrix Linden = **Plectocomia elongata** var. **elongata**

Plectocomia icthyospinus auct. = **Plectocomia elongata** var. **elongata**

Plectocomia kerriana Becc., Ann. Roy. Bot. Gard. (Calcutta) 12(2): 41 (1918).
SC. China (S. Yunnan) to N. Thailand. 36 CHC 41 THA.

Plectocomia khasyana Griff., Calcutta J. Nat. Hist. 5: 98 (1845).
Assam. 40 ASS.

Plectocomia longistigma Madulid, Kalikasan 10: 64 (1981).
E. Jawa. 42 JAW.

Plectocomia lorzingii Madulid, Kalikasan 10: 62 (1981).
Sumatera. 42 SUM.

Plectocomia macrostachya Kurz, J. Asiat. Soc. Bengal, Pt. 2, Nat. Hist. 43(2): 207 (1874).
Myanmar. 41 MYA.

Plectocomia maxima Kuntze = **Plectocomia elongata** var. **elongata**

Plectocomia microstachys Burret, Notizbl. Bot. Gart. Berlin-Dahlem 15: 731 (1942).
Hainan. 36 CHH.

Plectocomia minor Ridl. = **Plectocomia mulleri** Blume

Plectocomia montana Hook.f. & Thomson = **Plectocomia himalayana** Griff.

Plectocomia montana Griff. ex T.Anderson = **Plectocomia himalayana** Griff.

Plectocomia mulleri Blume, Rumphia 3: 71 (1847).
Pen. Malaysia, Borneo. 42 BOR MLY.
Plectocomia rigida H.Wendl., Bot. Zeitung (Berlin) 17: 165 (1859).
Plectocomia minor Ridl., J. Straits Branch Roy. Asiat. Soc. 50: 151 (1908).

Plectocomia pierreana Becc., Webbia 3: 236 (1910).
Indo-China. 41 CBD LAO THA.
Plectocomia cambodiana Gagnep. ex Humbert, Notul. Syst. (Paris) 6: 157 (1937).
Plectocomia barthiana Hodel, Palm J. 139: 54 (1998).

Plectocomia pygmaea Madulid, Kalikasan 10: 85 (1981).
Borneo (Kalimantan). 42 BOR.

Plectocomia rigida H.Wendl. = **Plectocomia mulleri** Blume

Plectocomia sumatrana Miq. = **Plectocomia elongata** var. **elongata**

Plectocomiopsis

Plectocomiopsis Becc. in J.D.Hooker, Fl. Brit. India 6: 479 (1893).
5 species, Indo-China to W. Malesia. 41 42.

Plectocomiopsis annulata Ridl. = **Myrialepis paradoxa** (Kurz) J.Dransf.

Plectocomiopsis corneri Furtado, Gard. Bull. Singapore 13: 333 (1951).
Pen. Malaysia to Sumatera. 42 MLY SUM.

Plectocomiopsis dubia Becc. = **Plectocomiopsis wrayi** Becc.

Plectocomiopsis ferox Ridl. = **Calamus concinnus** Mart.

Plectocomiopsis floribunda Becc. = **Myrialepis paradoxa** (Kurz) J.Dransf.

Plectocomiopsis geminiflora (Griff.) Becc. in J.D.Hooker, Fl. Brit. India 6: 479 (1893).
Indo-China to W. Malesia. 41 CBD LAO MYA THA 42 BOR MLY SUM.
Calamus geminiflorus Griff., Palms Brit. E. Ind.: 70 (1850). *Plectocomia geminiflora* (Griff.) H.Wendl. in O.C.E.de Kerchove de Denterghem, Palmiers: 254 (1878).
Calamus turbinatus Ridl., Mat. Fl. Malay. Penins. 2: 212 (1907).
Plectocomiopsis geminiflora var. *billitonensis* Becc., Ann. Roy. Bot. Gard. (Calcutta) 12(2): 51 (1918).
Plectocomiopsis geminiflora var. *borneensis* Becc., Ann. Roy. Bot. Gard. (Calcutta) 12(2): 53 (1918).

Plectocomiopsis geminiflora var. *billitonensis* Becc. = **Plectocomiopsis geminiflora** (Griff.) Becc.

Plectocomiopsis geminiflora var. *borneensis* Becc. = **Plectocomiopsis geminiflora** (Griff.) Becc.

Plectocomiopsis mira J.Dransf., Kew Bull. 37: 247 (1982).
W. Malesia. 42 BOR MLY SUM.

Plectocomiopsis paradoxa (Kurz) Becc. = **Myrialepis paradoxa** (Kurz) J.Dransf.

Plectocomiopsis scortechinii (Becc.) Ridl. = **Myrialepis paradoxa** (Kurz) J.Dransf.

Plectocomiopsis triquetra (Becc.) J.Dransf., Bot. J. Linn. Soc. 81: 6 (1980).
Borneo. 42 BOR.
Calamus triqueter Becc., Malesia 3: 62 (1886). *Palmijuncus triqueter* (Becc.) Kuntze, Revis. Gen. Pl. 2: 734 (1891). *Myrialepis triquetra* (Becc.) Becc. in J.D.Hooker, Fl. Brit. India 6: 480 (1893).

Plectocomiopsis wrayi Becc. in J.D.Hooker, Fl. Brit. India 6: 488 (1893).
Pen. Thailand to Pen. Malaysia. 41 THA 42 MLY.
Plectocomiopsis dubia Becc., Ann. Roy. Bot. Gard. (Calcutta) 12(2): 56 (1918).

Podococcus

Podococcus G.Mann & H.Wendl., Trans. Linn. Soc. London 24: 426 (1864).
1 species, S. Nigeria to WC. Trop. Africa. 22 23.

Podococcus acaulis Hua = **Podococcus barteri** G.Mann & H.Wendl.

Podococcus barteri G.Mann & H.Wendl., Trans. Linn. Soc. London 24: 426 (1864).
S. Nigeria to WC. Trop. Africa. 22 NGA 23 CMN CON EQG GAB ZAI.
Podococcus acaulis Hua, Bull. Mus. Hist. Nat. (Paris) 1: 315 (1895).

Pogonotium

Pogonotium J.Dransf., Kew Bull. 34: 763 (1980).
3 species, Pen. Malaysia to Borneo. 42.

Pogonotium divaricatum J.Dransf., Kew Bull. 34: 766 (1980).

Borneo (Sarawak). 42 BOR.

Pogonotium moorei J.Dransf., Principes 26: 174 (1982).
Borneo (Sarawak). 42 BOR.

Pogonotium ursinum (Becc.) J.Dransf., Kew Bull. 34: 763 (1980).
Pen. Malaysia to Borneo. 42 BOR MLY.
Daemonorops ursina Becc., Rec. Bot. Surv. India 2: 222 (1902).

Polyandrococos

Polyandrococos Barb.Rodr., Contr. Jard. Bot. Rio de Janeiro 1: 7 (1901).
1 species, E. Brazil. 84.

Polyandrococos caudescens (Mart.) Barb.Rodr., Contr. Jard. Bot. Rio de Janeiro 1: 8 (1901).
E. Brazil. 84 BZE BZL.
Diplothemium caudescens Mart., Hist. Nat. Palm. 3: t. 70 (1853). *Allagoptera caudescens* (Mart.) Kuntze, Revis. Gen. Pl. 2: 726 (1891).
Ceroxylon niveum H.Wendl. in O.C.E.de Kerchove de Denterghem, Palmiers: 239 (1878).
Orania nivea Linden ex W.Watson, Gard. Chron. 1887(2): 157 (1887).
Diplothemium pectinatum Barb.Rodr., Palm. Mattogross.: 81 (1898). *Polyandrococos pectinata* (Barb.Rodr.) Barb.Rodr., Contr. Jard. Bot. Rio de Janeiro 1: 8 (1901).

Polyandrococos pectinata (Barb.Rodr.) Barb.Rodr. = **Polyandrococos caudescens** (Mart.) Barb.Rodr.

Polyandrococos torallyi (Mart.) Barb.Rodr. = **Parajubaea torallyi** (Mart.) Burret

Ponapea

Ponapea Becc., Bot. Jahrb. Syst. 59: 13 (1924).
3 species, NW Pacific. 62.

Ponapea hosinoi Kaneh., J. Jap. Bot. 12: 731 (1936).
Caroline Is. (Pohnpei). 62 CRL.
Ptychosperma hosinoi (Kaneh.) H.E.Moore & Fosberg, Gentes Herb. 8: 468 (1956).

Ponapea kusaiensis Burret = **Ponapea ledermanniana** Becc.

Ponapea ledermanniana Becc., Bot. Jahrb. Syst. 59: 14 (1924).
Caroline Is. 62 CRL.
Ptychosperma ledermannianum (Becc.) H.E.Moore & Fosberg, Gentes Herb. 8: 469 (1956).
Ponapea kusaiensis Burret, Notizbl. Bot. Gart. Berlin-Dahlem 15: 92 (1940).

Ponapea palauensis Kaneh., J. Jap. Bot. 12: 732 (1936).
Caroline Is. (Palau). 62 CRL.
Ptychosperma palauense (Kaneh.) H.E.Moore & Fosberg, Gentes Herb. 8: 470 (1956).

Porothrinax

Porothrinax H.Wendl. ex Griseb. = **Thrinax** L.f. ex Sw.

Porothrinax pumilio H.Wendl. ex Griseb. = **Thrinax radiata** Lodd. ex Schult. & Schult.f.

Prestoea

Prestoea Hook.f. in G.Bentham & J.D.Hooker, Gen. Pl. 3: 899 (1883).
10 species, Trop. America. 80 81 82 83 84.
Euterpe Gaertn., Fruct. Sem. Pl. 1: 24 (1788).
Martinezia Ruiz & Pav., Fl. Peruv. Prodr.: 148 (1794).

Oreodoxa Willd., Mém. Acad. Roy. Sci. Hist. (Berlin) 1804: 34 (1807).

Acrista O.F.Cook, Bull. Torrey Bot. Club 28: 555 (1901).

Prestoea acuminata (Willd.) H.E.Moore, Gentes Herb. 9: 286 (1963).

Trop. America. 80 COS NIC PAN 81 CUB DOM HAI LEE PUE TRT WIN 82 VEN 83 BOL CLM ECU PER.

*Oreodoxa acuminata Willd., Mém. Acad. Roy. Sci. Hist. (Berlin) 1804: 35 (1807). *Euterpe acuminata* (Willd.) H.Wendl. in O.C.E.de Kerchove de Denterghem, Palmiers: 244 (1878).

var. **acuminata**

C. & W. South America. 80 COS NIC PAN 82 VEN 83 BOL CLM ECU PER.

Euterpe globosa Gaertn., Fruct. Sem. Pl. 1: 24 (1788). Provisional synonym.

Aiphanes praga Kunth in F.W.H.von Humboldt, A.J.A.Bonpland & C.S.Kunth, Nov. Gen. Sp. 1: 303 (1816). *Euterpe praga* (Kunth) Mart., Hist. Nat. Palm. 2: 24 (1824).

Oreodoxa frigida Kunth in F.W.H.von Humboldt, A.J.A.Bonpland & C.S.Kunth, Nov. Gen. Sp. 1: 304 (1816). *Oenocarpus frigidus* (Kunth) Spreng., Syst. Veg. 2: 140 (1825). *Euterpe frigida* (Kunth) Burret, Notizbl. Bot. Gart. Berlin-Dahlem 10: 300 (1928).

Euterpe andicola Brongn. ex Mart. in A.D.d'Orbigny, Voy. Amér. Mér. 7(3): 8 (1842).

Euterpe haenkeana Brongn. ex Mart. in A.D.d'Orbigny, Voy. Amér. Mér. 7(3): 9 (1842).

Oenocarpus utilis Klotzsch, Linnaea 20: 447 (1847).

Euterpe brevivaginata Mart., Hist. Nat. Palm. 3: 309 (1849).

Oenocarpus calaber H.Wendl., Index Palm.: 30 (1854).

Oenocarpus caracasanus H.Wendl., Index Palm.: 30 (1854).

Oenocarpus glaucus Lodd. ex H.Wendl., Index Palm.: 30 (1854).

Euterpe purpurea Engel, Linnaea 33: 669 (1865).

Euterpe antioquensis Linden, Ill. Hort. 28: 31 (1881), nom. nud.

Euterpe chaunostachys Burret, Bot. Jahrb. Syst. 63: 60 (1929).

Euterpe megalochlamys Burret, Bot. Jahrb. Syst. 63: 59 (1929). *Prestoea megalochlamys* (Burret) H.E.Moore, Gentes Herb. 9: 286 (1963).

Euterpe oocarpa Burret, Bot. Jahrb. Syst. 63: 61 (1929).

Euterpe andina Burret, Notizbl. Bot. Gart. Berlin-Dahlem 11: 322 (1932).

Euterpe aphanolepis Burret, Notizbl. Bot. Gart. Berlin-Dahlem 13: 344 (1936).

Euterpe trichoclada Burret, Notizbl. Bot. Gart. Berlin-Dahlem 13: 343 (1936). *Prestoea trichoclada* (Burret) Balslev & A.J.Hend., Principes 31: 11 (1987).

Euterpe microspadix Burret, Notizbl. Bot. Gart. Berlin-Dahlem 15: 34 (1940).

Euterpe zephyria Dugand, Revista Acad. Colomb. Ci. Exact. 8: 395 (1951).

Prestoea allenii H.E.Moore, Principes 9: 73 (1965).

var. **dasystachys** (Burret) A.J.Hend. & Galeano, Fl. Neotrop. Monogr. 72: 53 (1996).

Colombia (Santander) to NW. Venezuela. 82 VEN 83 CLM.

*Euterpe dasystachys Burret, Bot. Jahrb. Syst. 63: 62 (1929). *Prestoea dasystachys* (Burret) R.Bernal, Galeano & A.J.Hend., Taxon 38: 102 (1989).

var. **montana** (Graham) A.J.Hend. & Galeano, Fl. Neotrop. Monogr. 72: 53 (1996).

Caribbean. 81 CUB DOM HAI LEE PUE TRT WIN.

*Euterpe montana Graham, Bot. Mag. 67: t. 3874 (1841). *Prestoea montana* (Graham) G.Nicholson, Ill. Dict. Gard. 3: 216 (1886).

Oreodoxa manaele Mart., Hist. Nat. Palm. 3: 310 (1849). *Euterpe manaele* (Mart.) Griseb. & H.Wendl., Mem. Amer. Acad. Arts, n.s., 8: 530 (1862).

Acrista monticola O.F.Cook, Bull. Torrey Bot. Club 28: 557 (1901).

Euterpe pertenuis L.H.Bailey, Gentes Herb. 7: 425 (1947).

Euterpe tobagonis L.H.Bailey, Gentes Herb. 7: 423 (1947).

Prestoea allenii H.E.Moore = **Prestoea acuminata** var. **acuminata**

Prestoea asplundii H.E.Moore = **Prestoea schultzeana** (Burret) H.E.Moore

Prestoea brachyclada (Burret) R.Bernal, Galeano & A.J.Hend. = **Prestoea carderi** (W.Bull) Hook.f.

Prestoea carderi (W.Bull) Hook.f., Bot. Mag. 116: t. 7108 (1890). *Euterpe carderi* (Hook.f.) Burret, Bot. Jahrb. Syst. 63: 64 (1929).

W. South America to W. Venezuela. 82 VEN 83 CLM ECU PER.

*Geonoma carderi W.Bull, List New Pl. 1876: 9 (1876).

Euterpe brachyclada Burret, Bot. Jahrb. Syst. 63: 57 (1929). *Prestoea brachyclada* (Burret) R.Bernal, Galeano & A.J.Hend., Taxon 38: 102 (1989).

Euterpe brevicaulis Burret, Bot. Jahrb. Syst. 63: 58 (1929).

Euterpe latisecta Burret, Bot. Jahrb. Syst. 63: 55 (1929). *Prestoea latisecta* (Burret) R.Bernal, Galeano & A.J.Hend., Taxon 38: 102 (1989).

Euterpe parviflora Burret, Bot. Jahrb. Syst. 63: 54 (1929).

Euterpe simplicifrons Burret, Bot. Jahrb. Syst. 63: 53 (1929). *Prestoea simplicifrons* (Burret) A.J.Hend. & Nevers, Ann. Missouri Bot. Gard. 75: 210 (1988).

Prestoea humilis A.J.Hend. & Steyerm., Brittonia 38: 311 (1986).

Prestoea cuatrecasasii H.E.Moore = **Prestoea longipetiolata** var. **cuatrecasasii** (H.E.Moore) A.J. Hend. & Galeano

Prestoea darienensis A.J.Hend. = **Prestoea ensiformis** (Ruiz & Pav.) H.E.Moore

Prestoea dasystachys (Burret) R.Bernal, Galeano & A.J.Hend. = **Prestoea acuminata** var. **dasystachys** (Burret) A.J.Hend. & Galeano

Prestoea decurrens (H.Wendl. ex Burret) H.E.Moore, Gentes Herb. 9: 286 (1963).

Nicaragua to W. Ecuador. 80 COS NIC PAN 83 CLM ECU.

*Euterpe decurrens H.Wendl. ex Burret, Bot. Jahrb. Syst. 63: 63 (1929).

Prestoea ensiformis (Ruiz & Pav.) H.E.Moore, Gentes Herb. 9: 286 (1963).

Costa Rica to C. Peru. 80 COS PAN 83 CLM ECU PER.

Martinezia ensiformis Ruiz & Pav., Syst. Veg. Fl. Peruv. Chil.: 297 (1798). *Euterpe ensiformis* (Ruiz & Pav.) Mart., Hist. Nat. Palm. 2: 32 (1824).
Prestoea sejuncta L.H.Bailey, Gentes Herb. 6: 201 (1943).
Prestoea darienensis A.J.Hend., Brittonia 38: 266 (1986).
Prestoea integrifolia de Nevers & A.J.Hend., Ann. Missouri Bot. Gard. 75: 208 (1988).

Prestoea humilis A.J.Hend. & Steyerm. = **Prestoea carderi** (W.Bull) Hook.f.

Prestoea integrifolia de Nevers & A.J.Hend. = **Prestoea ensiformis** (Ruiz & Pav.) H.E.Moore

Prestoea latisecta (Burret) R.Bernal, Galeano & A.J.Hend. = **Prestoea carderi** (W.Bull) Hook.f.

Prestoea longipetiolata (Oerst.) H.E.Moore, Gentes Herb. 9: 286 (1963).
C. America to NW. Venezuela. 80 COS NIC PAN 82 VEN 83 CLM.
Euterpe longipetiolata Oerst., Vidensk. Meddel. Dansk Naturhist. Foren. Kjøbenhavn 1858: 31 (1858).

var. **cuatrecasasii** (H.E.Moore) A.J.Hend. & Galeano, Fl. Neotrop. Monogr. 72: 67 (1996).
NE. Colombia to NW. Venezuela. 82 VEN 83 CLM.
Prestoea cuatrecasasii H.E.Moore, Gentes Herb. 12: 34 (1980).

var. **longipetiolata**
C. America. 80 COS NIC PAN.
Euterpe brachyspatha Burret, Bot. Jahrb. Syst. 63: 56 (1929).
Malortiea simiarum Standl. & L.O.Williams, Ceiba 3: 10a (1952). *Euterpe simiarum* (Standl. & L.O. Williams) H.E.Moore, Principes 1: 145 (1957).
Euterpe williamsii Glassman, Fieldiana, Bot. 31: 5 (1964).

var. **roseospadix** (L.H.Bailey) A.J.Hend. & Galeano, Fl. Neotrop. Monogr. 72: 67 (1996).
Costa Rica to W. Panama. 80 COS PAN.
Euterpe roseospadix L.H.Bailey, Gentes Herb. 6: 201 (1943). *Prestoea roseospadix* (L.H.Bailey) H.E.Moore, Principes 9: 73 (1965).

Prestoea megalochlamys (Burret) H.E.Moore = **Prestoea acuminata** var. **acuminata**

Prestoea montana (Graham) G.Nicholson = **Prestoea acuminata** var. **montana** (Graham) A.J.Hend. & Galeano

Prestoea pubens H.E.Moore, Gentes Herb. 12: 37 (1980).
Panama to W. Colombia. 80 PAN 83 CLM.

var. **pubens**
W. Colombia. 83 CLM.

var. **semispicata** (Nevers & A.Hend.) A.J.Hend. & Galeano, Fl. Neotrop. Monogr. 72: 68 (1996).
Panama. 80 PAN.
Prestoea semispicata de Nevers & A.J.Hend., Ann. Missouri Bot. Gard. 75: 213 (1988).

Prestoea pubigera (Griseb. & H.Wendl.) Hook.f. ex B.D.Jacks., Index Kew. 2: 623 (1894).
N. Trinidad, NW. Venezuela. 81 TRT 82 VEN.
Hyospathe pubigera Griseb. & H.Wendl. in A.H.R.Grisebach, Fl. Brit. W. I.: 516 (1864).
Euterpe pubigera (Griseb. & H.Wendl.) Burret, Bot. Jahrb. Syst. 63: 53 (1929).
Prestoea trinitensis Hook.f., Bot. Mag. 116: t. 7108 (1890), nom. nov. for description in Gen. 3: 899 (1883).

Prestoea roseospadix (L.H.Bailey) H.E.Moore = **Prestoea longipetiolata** var. **roseospadix** (L.H.Bailey) A.J.Hend. & Galeano

Prestoea schultzeana (Burret) H.E.Moore, Gentes Herb. 12: 34 (1980).
C. Colombia to C. Peru. 83 CLM ECU PER.
Euterpe schultzeana Burret, Notizbl. Bot. Gart. Berlin-Dahlem 14: 326 (1939).
Prestoea asplundii H.E.Moore, Gentes Herb. 12: 33 (1980).

Prestoea sejuncta L.H.Bailey = **Prestoea ensiformis** (Ruiz & Pav.) H.E.Moore

Prestoea semispicata de Nevers & A.J.Hend. = **Prestoea pubens** var. **semispicata** (Nevers & A.Hend.) A.J.Hend. & Galeano

Prestoea simplicifolia Galeano-Garcés, Brittonia 38: 62 (1986).
Colombia (Antioquia). 83 CLM.

Prestoea simplicifrons (Burret) A.J.Hend. & Nevers = **Prestoea carderi** (W.Bull) Hook.f.

Prestoea steyermarkii H.E.Moore = **Prestoea tenuiramosa** (Dammer) H.E.Moore

Prestoea tenuiramosa (Dammer) H.E.Moore, Gentes Herb. 9: 286 (1963).
S. Venezuela to Guyana and N. Brazil. 82 GUY VEN 84 BZN.
Euterpe tenuiramosa Dammer, Notizbl. Königl. Bot. Gart. Berlin 6: 265 (1915).
Prestoea steyermarkii H.E.Moore, Principes 13: 139 (1969).

Prestoea trichoclada (Burret) Balslev & A.J.Hend. = **Prestoea acuminata** var. **acuminata**

Prestoea trinitensis Hook.f. = **Prestoea pubigera** (Griseb. & H.Wendl.) Hook.f. ex B.D.Jacks.

Pritchardia

Pritchardia Seem. & H.Wendl., Bonplandia 10: 197 (1862), nom. cons. *Eupritchardia* Kuntze, Revis. Gen. Pl. 3(2): 323 (1898), nom. illeg.
28 species, SW. & C. Pacific. 60 61 63.
Styloma O.F.Cook, J. Wash. Acad. Sci. 5: 241 (1915).

Pritchardia affinis Becc., Mem. Bernice Pauahi Bishop Mus. 8: 37 (1921).
Hawaiian Is. (Hawaii). 63 HAW.
Pritchardia affinis var. *gracilis* Becc., Mem. Bernice Pauahi Bishop Mus. 8: 40 (1921).
Pritchardia affinis var. *holaphila* Becc., Mem. Bernice Pauahi Bishop Mus. 8: 39 (1921).
Pritchardia affinis var. *rhopalocarpa* Becc., Mem. Bernice Pauahi Bishop Mus. 8: 40 (1921).

Pritchardia affinis var. *gracilis* Becc. = **Pritchardia affinis** Becc.

Pritchardia affinis var. *holaphila* Becc. = **Pritchardia affinis** Becc.

Pritchardia affinis var. *rhopalocarpa* Becc. = **Pritchardia affinis** Becc.

Pritchardia arecina Becc., Webbia 4: 224 (1913).
Styloma arecina (Becc.) O.F.Cook, J. Wash. Acad. Sci. 5: 241 (1915).
Hawaiian Is. (Maui). 63 HAW.

Pritchardia aurea Linden = **?**

Pritchardia aylmer-robinsonii H.St.John, Pacific Sci. 13: 163 (1959).
Hawaiian Is. (Niihau). 63 HAW.

Pritchardia beccariana Rock, Bull. Torrey Bot. Club 43: 386 (1916).
Hawaiian Is. (Hawaii). 63 HAW.
Pritchardia beccariana var. *giffardiana* Becc., Mem. Bernice Pauahi Bishop Mus. 8: 59 (1921).

Pritchardia beccariana var. *giffardiana* Becc. = **Pritchardia beccariana** Rock

Pritchardia brevicalyx Becc. & Rock = **Pritchardia lowreyana** Rock ex Becc.

Pritchardia donata Caum = **Pritchardia lowreyana** Rock ex Becc.

Pritchardia elliptica Rock & Caum = **?** [**63 HAW**]

Pritchardia eriophora Becc. = **Pritchardia minor** Becc.

Pritchardia eriostachya Becc. = **Pritchardia lanigera** Becc.

Pritchardia filamentosa H.Wendl. ex Franceschi = **Washingtonia filifera** (Linden ex André) H.Wendl. ex de Bary

Pritchardia filifera Linden ex André = **Washingtonia filifera** (Linden ex André) H.Wendl. ex de Bary

Pritchardia flynii Lorence & Gemmill, Novon 14: 185 (2004).
Hawaiian Is. (Kauai). 63 HAW.

Pritchardia forbesiana Rock, Mem. Bernice Pauahi Bishop Mus. 8: 52 (1921).
Hawaiian Is. (W. Maui). 63 HAW.

Pritchardia gaudichaudii (Mart.) H.Wendl. = **Pritchardia martii** (Gaudich.) H.Wendl.

Pritchardia glabrata Becc. & Rock, Mem. Bernice Pauahi Bishop Mus. 8: 42 (1921).
Hawaiian Is. (Maui). 63 HAW.

Pritchardia grandis (H.Wendl.) W.Bull = **Licuala grandis** H.Wendl.

Pritchardia grandis H.J.Veitch = **Licuala orbicularis** Becc.

Pritchardia hardyi Rock, Mem. Bernice Pauahi Bishop Mus. 8: 61 (1921).
Hawaiian Is. (Kauai). 63 HAW.
Pritchardia weissichiana Rock, Occas. Pap. Bernice Pauahi Bishop Mus. 23: 62 (1962).

Pritchardia hillebrandii Becc., Malesia 3: 292 (1890).
Washingtonia hillebrandii (Becc.) Kuntze, Revis. Gen. Pl. 2: 737 (1891). *Eupritchardia hillebrandtii* (Becc.) Kuntze, Revis. Gen. Pl. 3(2): 323 (1898). *Styloma hillebrandii* (Becc.) O.F.Cook, J. Wash. Acad. Sci. 5: 241 (1915).
Hawaiian Is. (Molokai). 63 HAW.
Pritchardia insignis Becc., Webbia 4: 219 (1913). *Styloma insignis* (Becc.) O.F.Cook, J. Wash. Acad. Sci. 5: 241 (1915).

Pritchardia insignis Becc. = **Pritchardia hillebrandii** Becc.

Pritchardia kaalae Rock, Mem. Bernice Pauahi Bishop Mus. 8: 46 (1921).
Hawaiian Is. (W. Oahu). 63 HAW.

Pritchardia kahanae Rock & Caum = **Pritchardia martii** (Gaudich.) H.Wendl.

Pritchardia kahukuensis Caum = **Pritchardia martii** (Gaudich.) H.Wendl.

Pritchardia kamapuaana Caum = **Pritchardia martii** (Gaudich.) H.Wendl.

Pritchardia lanaiensis Becc. & Rock, Mem. Bernice Pauahi Bishop Mus. 8: 41 (1921).
Hawaiian Is. (Lanai). 63 HAW.

Pritchardia lanigera Becc., Malesia 3: 298 (1890).
Washingtonia lanigera (Becc.) Kuntze, Revis. Gen. Pl. 2: 737 (1891). *Eupritchardia lanigera* (Becc.) Kuntze, Revis. Gen. Pl. 3(2): 323 (1898). *Styloma lanigera* (Becc.) O.F.Cook, J. Wash. Acad. Sci. 5: 241 (1915).
Hawaiian Is. (Hawaii). 63 HAW.
Pritchardia eriostachya Becc., Webbia 4: 232 (1913). *Styloma eriostachya* (Becc.) O.F.Cook, J. Wash. Acad. Sci. 5: 241 (1915).
Pritchardia montis-kea Rock, Mem. Bernice Pauahi Bishop Mus. 8: 65 (1921).

Pritchardia limahuliensis H.St.John, Phytologia 64: 177 (1988).
Hawaiian Is. (Kauai). 63 HAW.

Pritchardia lowreyana Rock ex Becc., Mem. Bernice Pauahi Bishop Mus. 8: 53 (1921).
Hawaiian Is. (Molokai). 63 HAW.
Pritchardia macrocarpa Linden, Ill. Hort. 26: 105 (1879). Provisional synonym.
Pritchardia brevicalyx Becc. & Rock, Mem. Bernice Pauahi Bishop Mus. 8: 56 (1921).
Pritchardia lowreyana var. *turbinata* Rock ex Becc., Mem. Bernice Pauahi Bishop Mus. 8: 55 (1921).
Pritchardia donata Caum, Occas. Pap. Bernice Pauahi Bishop Mus. 9(5): 12 (1930).

Pritchardia lowreyana var. *turbinata* Rock ex Becc. = **Pritchardia lowreyana** Rock ex Becc.

Pritchardia macdanielsii Caum = **Pritchardia martii** (Gaudich.) H.Wendl.

Pritchardia macrocarpa Linden = **Pritchardia lowreyana** Rock ex Becc.

Pritchardia maideniana Becc., Webbia 4: 213 (1913).
Styloma maideniana (Becc.) O.F.Cook, J. Wash. Acad. Sci. 5: 241 (1915).
Pacific (?). 6 +.

Pritchardia martii (Gaudich.) H.Wendl., Bonplandia 10: 199 (1862).
Hawaiian Is. (Molokai, Oahu). 63 HAW.
*Livistona martii Gaudich., Voy. Bonite, Bot.: t. 58 (1842). *Washingtonia martii* (Gaudich.) Kuntze, Revis. Gen. Pl. 2: 737 (1891). *Eupritchardia martii* (Gaudich.) Kuntze, Revis. Gen. Pl. 3(2): 323 (1898). *Styloma martii* (Gaudich.) O.F.Cook, J. Wash. Acad. Sci. 5: 241 (1915).
Livistona gaudichaudii Mart., Hist. Nat. Palm. 3(ed. 2): 242 (1849). *Pritchardia gaudichaudii* (Mart.) H.Wendl., Bonplandia 10: 199 (1862). *Washingtonia gaudichaudii* (Mart.) Kuntze, Revis. Gen. Pl. 2: 737 (1891). *Eupritchardia gaudichaudii* (Mart.) Kuntze, Revis. Gen. Pl. 3(2): 323 (1898). *Styloma gaudichaudii* (Mart.) O.F.Cook, J. Wash. Acad. Sci. 5: 241 (1915).
Livistona martiana Mart., Hist. Nat. Palm. 3(ed. 2): 242 (1849).
Pritchardia rockiana Becc., Webbia 4: 228 (1913). *Styloma rockiana* (Becc.) O.F.Cook, J. Wash. Acad. Sci. 5: 241 (1915).
Pritchardia kahanae Rock & Caum, Mem. Bernice Pauahi Bishop Mus. 8: 75 (1921).
Pritchardia martioides Rock & Caum, Mem. Bernice Pauahi Bishop Mus. 8: 76 (1921).
Pritchardia kahukuensis Caum, Occas. Pap. Bernice Pauahi Bishop Mus. 11(5): 13 (1930).

Pritchardia kamapuaana Caum, Occas. Pap. Bernice Pauahi Bishop Mus. 9(5): 10 (1930). *Eupritchardia kamapuaana* (Caum) L.H.Bailey, Gentes Herb. 4: 360 (1940).

Pritchardia macdanielsii Caum, Occas. Pap. Bernice Pauahi Bishop Mus. 9(5): 11 (1930).

Pritchardia martioides Rock & Caum = **Pritchardia martii** (Gaudich.) H.Wendl.

Pritchardia minor Becc., Webbia 3: 137 (1910). *Styloma minor* (Becc.) O.F.Cook, J. Wash. Acad. Sci. 5: 241 (1915).
Hawaiian Is. (Kauai). 63 HAW.
> *Pritchardia eriophora* Becc., Webbia 4: 235 (1913). *Styloma eriophora* (Becc.) O.F.Cook, J. Wash. Acad. Sci. 5: 241 (1915).

Pritchardia mitiaroana J.Dransf. & Y.Ehrh., Principes 39: 37 (1995).
Cook Is. (S. & SW. Mitiaro). 61 COO.

Pritchardia moensii Becc. = **Pritchardia vuylstekeana** H.Wendl.

Pritchardia montis-kea Rock = **Pritchardia lanigera** Becc.

Pritchardia munroi Rock, Mem. Bernice Pauahi Bishop Mus. 8: 62 (1921).
Hawaiian Is. (N. Molokai). 63 HAW.

Pritchardia napaliensis H.St.John, Pacific Sci. 35: 97 (1981).
Hawaiian Is. (Kauai). 63 HAW.

Pritchardia pacifica Seem. & H.Wendl., Bonplandia 10: 197 (1862). *Washingtonia pacifica* (Seem. & H.Wendl.) Kuntze, Revis. Gen. Pl. 2: 737 (1891). *Eupritchardia pacifica* (Seem. & H.Wendl.) Kuntze, Revis. Gen. Pl. 3(2): 323 (1898). *Styloma pacifica* (Seem. & H.Wendl.) O.F.Cook, J. Wash. Acad. Sci. 5: 241 (1915).
S. Pacific. 60 fij nue sam TON 61 MRQ SCI.
> *Pritchardia pacifica* var. *samoensis* Becc., Webbia 4: 206 (1913).
> *Pritchardia pacifica* var. *marquisensis* F.Br., Bernice P. Bishop Mus. Bull. 84: 118 (1931).

Pritchardia pacifica var. *marquisensis* F.Br. = **Pritchardia pacifica** Seem. & H.Wendl.

Pritchardia pacifica var. *samoensis* Becc. = **Pritchardia pacifica** Seem. & H.Wendl.

Pritchardia pericularum H.Wendl., Rev. Hort. 55: 206 (1883). *Washingtonia pericularum* (H.Wendl.) Kuntze, Revis. Gen. Pl. 2: 737 (1891). *Eupritchardia pericularum* (H.Wendl.) Kuntze, Revis. Gen. Pl. 3(2): 323 (1898). *Styloma pericularum* (H.Wendl.) O.F.Cook, J. Wash. Acad. Sci. 5: 241 (1915).
Tuamotu. 61 TUA.

Pritchardia perlmanii Gemmill, Novon 8: 18 (1998).
Hawaiian Is. (Kauai). 63 HAW.

Pritchardia remota (Kuntze) Becc., Malesia 3: 294 (1890).
Hawaiian Is. (Nihoa). 63 HAW.
> *Washingtonia remota* Kuntze, Revis. Gen. Pl. 2: 737 (1891). *Eupritchardia remota* (Kuntze) Kuntze, Revis. Gen. Pl. 3(2): 323 (1898). *Styloma remota* (Kuntze) O.F.Cook, J. Wash. Acad. Sci. 5: 241 (1915).

Pritchardia robusta (H.Wendl.) Schröt. = **Washingtonia robusta** H.Wendl.

Pritchardia rockiana Becc. = **Pritchardia martii** (Gaudich.) H.Wendl.

Pritchardia schattaueri Hodel, Principes 29: 31 (1985).
Hawaiian Is. (Hawaii). 63 HAW.

Pritchardia thurstonii F.Muell. & Drude, Gartenflora 36: 486 (1887). *Washingtonia thurstonii* (F.Muell. & Drude) Kuntze, Revis. Gen. Pl. 2: 737 (1891). *Eupritchardia thurstonii* (F.Muell. & Drude) Kuntze, Revis. Gen. Pl. 3(2): 323 (1898). *Styloma thurstonii* (F.Muell. & Drude) O.F.Cook, J. Wash. Acad. Sci. 5: 241 (1915).
E. Fiji. 60 FIJ.

Pritchardia viscosa Rock, Mem. Bernice Pauahi Bishop Mus. 8: 66 (1921).
Hawaiian Is. (Kauai). 63 HAW.

Pritchardia vuylstekeana H.Wendl., Rev. Hort. 55: 329 (1883). *Washingtonia vuylstekeana* (H.Wendl.) Kuntze, Revis. Gen. Pl. 2: 737 (1891). *Eupritchardia vluylstekeana* (H.Wendl.) Kuntze, Revis. Gen. Pl. 3(2): 323 (1898). *Styloma vuylstekeana* (H.Wendl.) O.F.Cook, J. Wash. Acad. Sci. 5: 341 (1915).
Tuamotu. 61 TUA.
> *Pritchardia moensii* Becc., Malesia 3: 300 (1890).

Pritchardia waialealeana Read, Principes 32: 135 (1988).
Hawaiian Is. (Kauai). 63 HAW.

Pritchardia weissichiana Rock = **Pritchardia hardyi** Rock

Pritchardia wrightii (Griseb. & H.Wendl. ex Voss) Becc. = **Colpothrinax wrightii** Griseb. & H.Wendl. ex Voss

Pritchardiopsis

Pritchardiopsis Becc., Webbia 3: 131 (1910).
1 species, New Caledonia. 60.

Pritchardiopsis jeanneneyi Becc., Webbia 3: 132 (1910).
S. New Caledonia. 60 NWC.

Pseudophoenix

Pseudophoenix H.Wendl. ex Sarg., Bot. Gaz. 11: 314 (1886).
4 species, S. Florida to Caribbean, SE. Mexico to Belize. 78 79 80 81.
> *Sargentia* H.Wendl. & Drude ex Salomon, Palmen: 160 (1887).
> *Cyclospathe* O.F.Cook, Mem. Torrey Bot. Club 12: 25 (1902).
> *Chamaephoenix* H.Wendl. ex Curtiss, Florida Farmer Fruit Grower 1(8): 57 (23 Feb. 1887), nom. illeg.

Pseudophoenix ekmanii Burret, Kongl. Svenska Vetenskapsakad. Handl., III, 6(7): 19 (1929).
Dominican Rep. (Barahona Pen., I. Beata). 81 DOM.

Pseudophoenix elata O.F.Cook ex Burret = **Pseudophoenix lediniana** Read

Pseudophoenix gracilis Ekman ex Burret = **Pseudophoenix sargentii** H.Wendl. ex Sarg.

Pseudophoenix insignis O.F.Cook = **Pseudophoenix vinifera** (Mart.) Becc.

Pseudophoenix lediniana Read, Gentes Herb. 10: 189 (1968).
SW. Haiti. 81 HAI.

Pseudophoenix linearis O.F.Cook = **Pseudophoenix sargentii** H.Wendl. ex Sarg.

Pseudophoenix navassana Ekman ex Burret = **Pseudophoenix sargentii** H.Wendl. ex Sarg.

Pseudophoenix saonae O.F.Cook = **Pseudophoenix sargentii** H.Wendl. ex Sarg.

Pseudophoenix sargentii H.Wendl. ex Sarg., Bot. Gaz. 11: 314 (1886). *Chamaephoenix sargentii* (H.Wendl. ex Sarg.) Curtiss, Florida Farmer Fruit Grower 1(8): 57 (1887).
S. Florida to Caribbean, SE. Mexico to Belize. 78 FLA 79 MXT 80 BLZ 81 BAH CUB DOM HAI PUE TCI WIN.
  Cyclospathe northropii O.F.Cook, Mem. Torrey Bot. Club 12: 25 (1902).
  Pseudophoenix linearis O.F.Cook, J. Wash. Acad. Sci. 13: 407 (1923).
  Pseudophoenix saonae O.F.Cook, J. Wash. Acad. Sci. 13: 406 (1923). *Pseudophoenix sargentii* subsp. *saonae* (O.F.Cook) Read, Gentes Herb. 10: 210 (1968).
  Pseudophoenix gracilis Ekman ex Burret, Kongl. Svenska Vetenskapsakad. Handl., III, 6(7): 28 (1929).
  Pseudophoenix navassana Ekman ex Burret, Kongl. Svenska Vetenskapsakad. Handl., III, 6(7): 27 (1929). *Pseudophoenix sargentii* var. *navassana* (Ekman ex Burret) Read, Gentes Herb. 10: 211 (1968).

Pseudophoenix sargentii var. *navassana* (Ekman ex Burret) Read = **Pseudophoenix sargentii** H.Wendl. ex Sarg.

Pseudophoenix sargentii subsp. *saonae* (O.F.Cook) Read = **Pseudophoenix sargentii** H.Wendl. ex Sarg.

Pseudophoenix vinifera (Mart.) Becc., Pomona Coll. J. Econ. Bot. 2(2): 268 (1912).
Haiti to SW. Dominican Rep. 81 DOM HAI.
  *Euterpe vinifera Mart., Hist. Nat. Palm. 1: t. z.II, f. 18, 19 (1831). *Cocos vinifera* (Mart.) Mart., Hist. Nat. Palm. 3: 324 (1853). *Gaussia vinifera* (Mart.) H.Wendl. in O.C.E.de Kerchove de Denterghem, Palmiers: 245 (1878). *Aeria vinifera* (Mart.) O.F.Cook, J. Wash. Acad. Sci. 13: 399 (1923).
  Pseudophoenix insignis O.F.Cook, J. Wash. Acad. Sci. 13: 400 (1923).

Pseudopinanga

Pseudopinanga Burret = **Pinanga** Blume

Pseudopinanga acuminata Burret = **?** [42 BOR] **Pinanga sp.**

Pseudopinanga albescens (Becc.) Burret = **Pinanga albescens** Becc.

Pseudopinanga anomodonta Burret = **?** [42 SUL] **Pinanga sp.**

Pseudopinanga aristata Burret = **Pinanga aristata** (Burret) J.Dransf.

Pseudopinanga barnesii (Becc.) Burret = **Pinanga maculata** Porte ex Lem.

Pseudopinanga caesia (Blume) Burret = **Pinanga caesia** Blume

Pseudopinanga insignis (Becc.) Burret = **Pinanga insignis** Becc.

Pseudopinanga kjellbergii Burret = **?** [42 SUL] **Pinanga sp.**

Pseudopinanga macrorhachis Burret = **?** [42 SUL] **Pinanga sp.**

Pseudopinanga maculata (Porte ex Lem.) Burret = **Pinanga maculata** Porte ex Lem.

Pseudopinanga multisecta Burret = **?** [42 SUL] **Pinanga sp.**

Pseudopinanga paucisecta Burret = **?** [42 BOR] **Pinanga sp.**

Pseudopinanga pilosa Burret = **Pinanga pilosa** (Burret) J.Dransf.

Pseudopinanga stylosa (Becc.) Burret = **Pinanga stylosa** Becc.

Pseudopinanga tashiroi (Hayata) Burret = **Pinanga tashiroi** Hayata

Pseudopinanga urosperma (Becc.) Burret = **Pinanga urosperma** Becc.

Ptychandra

Ptychandra Scheff. = **Heterospathe** Scheff.

Ptychandra clemensiae Burret = **Heterospathe clemensiae** (Burret) H.E.Moore

Ptychandra glabra Burret = **Heterospathe glabra** (Burret) H.E.Moore

Ptychandra glauca Scheff. = **Heterospathe glauca** (Scheff.) H.E.Moore

Ptychandra montana Burret = **Heterospathe obriensis** (Becc.) H.E.Moore

Ptychandra muelleriana Becc. = **Heterospathe muelleriana** (Becc.) Becc.

Ptychandra musschenbroekiana Becc. = **Heterospathe glauca** (Scheff.) H.E.Moore

Ptychandra obriensis Becc. = **Heterospathe obriensis** (Becc.) H.E.Moore

Ptychococcus

Ptychococcus Becc., Ann. Jard. Bot. Buitenzorg 2: 90 (1885).
2 species, Papuasia. 43.

Ptychococcus archboldianus Burret = **Ptychococcus paradoxus** (Scheff.) Becc.

Ptychococcus arecinus (Becc.) Becc. = **Ptychococcus paradoxus** (Scheff.) Becc.

Ptychococcus elatus Becc. = **Ptychococcus paradoxus** (Scheff.) Becc.

Ptychococcus guppyanus (Becc.) Burret = **Ptychococcus paradoxus** (Scheff.) Becc.

Ptychococcus kraemerianus (Becc.) Burret = **Ptychococcus paradoxus** (Scheff.) Becc.

Ptychococcus lepidotus H.E.Moore, Principes 9: 11 (1965).
New Guinea. 43 NWG.

Ptychococcus paradoxus (Scheff.) Becc., Ann. Jard. Bot. Buitenzorg 2: 96 (1885).
New Guinea. 43 NWG.
  *Drymophloeus paradoxus Scheff., Ann. Jard. Bot. Buitenzorg 1: 53 (1876).
  Ptychosperma paradoxum Scheff., Ann. Jard. Bot. Buitenzorg 1: 83 (1876).
  Ptychococcus guppyanus (Becc.) Burret, Repert. Spec. Nov. Regni Veg. 24: 262 (1928).
  Actinophloeus guppyanus Becc., Webbia 4: 264 (1914).
  Ptychococcus archboldianus Burret, J. Arnold Arbor. 20: 209 (1939).
  Ptychococcus arecinus (Becc.) Becc., Ann. Jard. Bot. Buitenzorg 2: 99 (1885).

Ptychosperma arecinum Becc., Malesia 1: 58 (1877).
Ptychococcus elatus Becc., Bot. Jahrb. Syst. 58: 451 (1923).
Ptychococcus kraemerianus (Becc.) Burret, Repert. Spec. Nov. Regni Veg. 24: 262 (1928).
*Actinophloeus kraemerianus Becc., Bot. Jahrb. Syst. 52: 30 (1914).
Ptychosperma novohibernicum Becc., Bot. Jahrb. Syst. 52: 29 (1914).

Ptychococcus schumannii (Becc.) Burret = **Brassiophoenix schumannii** (Becc.) Essig

Ptychoraphis

Ptychoraphis Becc. = **Rhopaloblaste** Scheff.

Ptychoraphis augusta (Kurz) Becc. = **Rhopaloblaste augutsa**

Ptychoraphis cagayanensis (Becc.) Becc. = **Heterospathe cagayanensis** Becc.

Ptychoraphis elmeri (Becc.) Becc. = **Heterospathe elmeri** Becc.

Ptychoraphis intermedia Becc. = **Heterospathe intermedia** (Becc.) Fernando

Ptychoraphis longiflora Ridl. = **Rhopaloblaste singaporensis** (Becc.) Hook.f.

Ptychoraphis microcarpa Becc. = **Heterospathe elmeri** Becc.

Ptychoraphis philippinensis Becc. = **Heterospathe philippinensis** (Becc.) Becc.

Ptychoraphis singaporensis (Becc.) Becc. = **Rhopaloblaste singaporensis** (Becc.) Hook.f.

Ptychosperma

Ptychosperma Labill., Mém. Cl. Sci. Math. Inst. Natl. France 9: 252 (1808 publ. 1809).
31 species, Papuasia to N. Australia and NW. Pacific. 43 50 62.
Seaforthia R.Br., Prodr.: 267 (1810).
Actinophloeus (Becc.) Becc., Ann. Jard. Bot. Buitenzorg 2: 126 (1885).
Romanowia Sander ex André, Rev. Hort. 71: 262 (1899).
Strongylocaryum Burret, Notizbl. Bot. Gart. Berlin-Dahlem 13: 95 (1936).

Ptychosperma advenum Becc. = **? [43 NWG]**

Ptychosperma album Scheff. = **Pinanga coronata** (Blume ex Mart.) Blume

Ptychosperma alexandrae F.Muell. = **Archontophoenix alexandrae** (F.Muell.) H.Wendl. & Drude

Ptychosperma ambiguum (Becc.) Becc. ex Martelli, Nuovo Giorn. Bot. Ital., n.s., 41: 709 (1935).
W. New Guinea. 43 NWG.
*Drymophloeus ambiguus Becc., Malesia 1: 42 (1877). *Actinophloeus ambiguus* (Becc.) Becc., Ann. Jard. Bot. Buitenzorg 2: 126 (1885). *Saguaster ambiguus* (Becc.) Kuntze, Revis. Gen. Pl. 2: 735 (1891).

Ptychosperma angustifolium Blume = **Drymophloeus litigiosus** (Becc.) H.E.Moore

Ptychosperma appendiculatum Blume = **Drymophloeus oliviformis** (Giseke) Mart.

Ptychosperma arecinum Becc. = **Ptychococcus paradoxus** (Scheff.) Becc.

Ptychosperma arfakianum Becc. = **Heterospathe arfakiana** (Becc.) H.E.Moore

Ptychosperma beatriceae F.Muell. = **Archontophoenix alexandrae** (F.Muell.) H.Wendl. & Drude

Ptychosperma beccarianum Warb. ex Burret = **Hydriastele microspadix** (Warb. ex K.Schum. & Lauterb.) Burret

Ptychosperma bleeseri Burret, Repert. Spec. Nov. Regni Veg. 24: 266 (1928). *Carpentaria bleeseri* (Burret) Burret, Repert. Spec. Nov. Regni Veg. 24: 268 (1928).
N. Northern Territory. 50 NTA.
Actinophloeus bleeseri Burret, Repert. Spec. Nov. Regni Veg. 24: 266 (1928).

Ptychosperma buabe Essig, Allertonia 1: 461 (1978).
Papua New Guinea. 43 NWG.

Ptychosperma burretianum Essig, Allertonia 1: 457 (1978).
New Guinea (D' Entrecasteaux Is.). 43 NWG.

Ptychosperma caesia (Blume) Miq. = **Pinanga caesia** Blume

Ptychosperma calapparia (Blume) Miq. = **Actinorhytis calapparia** (Blume) H.Wendl. & Drude ex Scheff.

Ptychosperma capitis-yorki H.Wendl. & Drude = **Ptychosperma elegans** (R.Br.) Blume

Ptychosperma caryotoides Ridl., J. Bot. 24: 358 (1886).
Papua New Guinea. 43 NWG.
Ptychosperma litigiosum Ridl., J. Bot. 24: 358 (1886), nom. illeg.
Ptychosperma ridleyi Becc., Nuovo Giorn. Bot. Ital. 20: 179 (1888).
Ptychosperma sayeri Becc., Nuovo Giorn. Bot. Ital. 20: 178 (1888).
Drymophloeus montanus K.Schum. & Lauterb., Fl. Schutzgeb. Südsee: 207 (1900). *Actinophloeus montanus* (K.Schum. & Lauterb.) Burret, Repert. Spec. Nov. Regni Veg. 24: 272 (1928). *Ptychosperma montanum* (K.Schum. & Lauterb.) Burret, Notizbl. Bot. Gart. Berlin-Dahlem 12: 595 (1935).
Ptychosperma discolor Becc., Webbia 1: 300 (1905).
Ptychosperma josephense Becc., Webbia 1: 299 (1905).
Ptychosperma polyclados Becc., Webbia 1: 301 (1905).
Ptychosperma leptocladum Burret, Notizbl. Bot. Gart. Berlin-Dahlem 13: 329 (1936).

Ptychosperma caudatum Becc. = **Pinanga rumphiana** (Mart.) J.Dransf. & Govaerts

Ptychosperma coccinea Teijsm. & Binn. = **Cyrtostachys renda** Blume

Ptychosperma cochinchinense (Blume) Miq. = **Pinanga sylvestris** (Lour.) Hodel

Ptychosperma communis (Zipp. ex Blume) Miq. = **Drymophloeus litigiosus** (Becc.) H.E.Moore

Ptychosperma coronatum (Blume ex Mart.) Miq. = **Pinanga coronata** (Blume ex Mart.) Blume

Ptychosperma costatum (Blume) Miq. = **Pinanga coronata** (Blume ex Mart.) Blume

Ptychosperma cuneatum (Burret) Burret, Notizbl. Bot. Gart. Berlin-Dahlem 12: 595 (1935).
New Guinea. 43 NWG.
*Actinophloeus cuneatus Burret, Notizbl. Bot. Gart. Berlin-Dahlem 11: 205 (1931).
Ptychosperma tenue Becc., Atti Soc. Tosc. Sci. Nat. Pisa Processi Verbali 44: 147 (1934).

Ptychosperma cunninghamianum H.Wendl. = **Archontophoenix cunninghamiana** (H.Wendl.) H.Wendl. & Drude

Ptychosperma dicksonii (Roxb.) Miq. = **Pinanga dicksonii** (Roxb.) Blume

Ptychosperma discolor Becc. = **Ptychosperma caryotoides** Ridl.

Ptychosperma distichum (Roxb.) Miq. = **Pinanga disticha** (Roxb.) H.Wendl.

Ptychosperma drudei H.Wendl. = **Archontophoenix alexandrae** (F.Muell.) H.Wendl. & Drude

Ptychosperma elegans (R.Br.) Blume, Rumphia 2: 118 (1843).
N. Northern Territory to Queensland. 50 NTA QLD.
  *Seaforthia elegans R.Br., Prodr.: 267 (1810). *Pinanga smithii* (R.Br.) Scheff., Ann. Jard. Bot. Buitenzorg 1: 154 (1876). *Saguaster elegans* (R.Br.) Kuntze, Revis. Gen. Pl. 2: 735 (1891).
  Ptychosperma seaforthii Miq., Fl. Ned. Ind. 3: 21 (1855).
  Ptychosperma capitis-yorki H.Wendl. & Drude, Linnaea 39: 217 (1875). *Saguaster capitis-yorki* (H.Wendl. & Drude) Kuntze, Revis. Gen. Pl. 2: 735 (1891). *Actinophloeus capitis-yorki* (H.Wendl. & Drude) Burret, Repert. Spec. Nov. Regni Veg. 24: 266 (1928).
  Pinanga smithii W.Hill ex Scheff., Ann. Jard. Bot. Buitenzorg 1: 154 (1876).
  Ptychosperma elegans var. *sphaerocarpum* Becc., Ann. Jard. Bot. Buitenzorg 2: 88 (1885).
  Archontophoenix jardinei F.M.Bailey, Queensland Agric. J. 2: 129 (1898). *Ptychosperma jardinei* (F.M.Bailey) F.M.Bailey, Queensland Agric. J. 23: 35 (1909).
  Ptychosperma wendlandianum Burret, Notizbl. Bot. Gart. Berlin-Dahlem 10: 205 (1927).
  Ptychosperma wendlandianum var. *sphaerocarpum* Burret, Notizbl. Bot. Gart. Berlin-Dahlem 10: 205 (1927).

Ptychosperma elegans var. *sphaerocarpum* Becc. = **Ptychosperma elegans** (R.Br.) Blume

Ptychosperma filiferum H.Wendl. = **Veitchia filifera** (H.Wendl.) H.E.Moore

Ptychosperma furcatum (Becc.) Becc. ex Martelli, Nuovo Giorn. Bot. Ital., n.s., 41: 710 (1935).
Papua New Guinea. 43 NWG.
  *Actinophloeus furcatus Becc., Webbia 1: 302 (1905).

Ptychosperma furfuraceum (Blume) Miq. = **Pinanga furfuracea** Blume

Ptychosperma gracile Labill., Mém. Cl. Sci. Math. Inst. Natl. France 9: 253 (1808 publ. 1809). *Saguaster gracilis* (Labill.) Kuntze, Revis. Gen. Pl. 2: 735 (1891).
Bismarck Arch. 43 BIS.
  Seaforthia ptychosperma Mart., Hist. Nat. Palm. 3: 182 (1838).

Ptychosperma hartmannii Becc., Webbia 1: 301 (1905).
New Guinea. 43 NWG.

Ptychosperma hentyi Essig = **Drymophloeus hentyi** (Essig) Zona

Ptychosperma hollrungii Warb. ex Burret = **Ptychosperma lauterbachii** Becc.

Ptychosperma hosinoi (Kaneh.) H.E.Moore & Fosberg = **Ponapea hosinoi** Kaneh.

Ptychosperma hospitum (Burret) Burret = **Ptychosperma macarthurii** (H.Wendl. ex H.J.Veitch) H.Wendl. ex Hook.f.

Ptychosperma inaequalis (Blume) Miq. = **Pinanga inaequalis** Blume

Ptychosperma jardinei (F.M.Bailey) F.M.Bailey = **Ptychosperma elegans** (R.Br.) Blume

Ptychosperma josephense Becc. = **Ptychosperma caryotoides** Ridl.

Ptychosperma julianettii Becc. = **Ptychosperma macarthurii** (H.Wendl. ex H.J.Veitch) H.Wendl. ex Hook.f.

Ptychosperma junghuhnii (Miq.) Miq. = **Pinanga patula** Blume

Ptychosperma kajewskii Burret = **Ptychosperma salomonense** Burret

Ptychosperma kasesa Lauterb. = **Hydriastele kasesa** (Lauterb.) Burret

Ptychosperma keiense (Becc.) Becc. ex Martelli = **Ptychosperma propinquum** (Becc.) Becc. ex Martelli

Ptychosperma kerstenianum (Sander) Burret = **Balaka seemannii** (H.Wendl.) Becc.

Ptychosperma kuhlii (Blume) Miq. = **Pinanga coronata** (Blume ex Mart.) Blume

Ptychosperma laccospadix Benth. = **Laccospadix australasicus** H.Wendl. & Drude

Ptychosperma latisectum (Blume) Miq. = **Pinanga latisecta** Blume

Ptychosperma lauterbachii Becc., Bot. Jahrb. Syst. 52: 29 (1914).
NE. New Guinea. 43 NWG.
  Actinophloeus punctulatus Becc., Bot. Jahrb. Syst. 52: 31 (1914). *Ptychosperma punctulatum* (Becc.) Becc. ex Martelli, Nuovo Giorn. Bot. Ital., n.s., 41: 709 (1934 publ. 1935).
  Ptychosperma hollrungii Warb. ex Burret, Repert. Spec. Nov. Regni Veg. 24: 265 (1928).

Ptychosperma ledermannianum (Becc.) H.E.Moore & Fosberg = **Ponapea ledermanniana** Becc.

Ptychosperma leptocladum Burret = **Ptychosperma caryotoides** Ridl.

Ptychosperma lineare (Burret) Burret, Notizbl. Bot. Gart. Berlin-Dahlem 12: 596 (1935).
SE. New Guinea. 43 NWG.
  *Actinophloeus linearis Burret, J. Arnold Arbor. 12: 268 (1931).

Ptychosperma litigiosum Ridl. = **Ptychosperma caryotoides** Ridl.

Ptychosperma litigiosum Becc. = **Drymophloeus litigiosus** (Becc.) H.E.Moore

Ptychosperma litigiosum var. *oninense* Becc. = **Drymophloeus oninensis** (Becc.) H.E.Moore

Ptychosperma macarthurii (H.Wendl. ex H.J.Veitch) H.Wendl. ex Hook.f., Rep. Progr. Condition Roy. Bot. Gard. Kew 1882: 55 (1884).
New Guinea to N. & NE. Queensland. 43 NWG 50 QLD.
  *Kentia macarthurii H.Wendl. ex H.J.Veitch, Cat. Pl. 1879: 26 (1879). *Saguaster macarthurii* (H.Wendl. ex H.J.Veitch) Kuntze, Revis. Gen. Pl. 2: 735 (1891). *Actinophloeus macarthurii* (H.Wendl. ex H.J.Veitch) Becc. ex Raderm., Ann. Jard. Bot. Buitenzorg 35: 12 (1925).
  Actinophloeus hospitus Burret, Notizbl. Bot. Gart. Berlin-Dahlem 11: 206 (1931). *Ptychosperma hospitum* (Burret) Burret, Notizbl. Bot. Gart. Berlin-Dahlem 12: 596 (1935).

Ptychosperma julianettii Becc., Atti Soc. Tosc. Sci. Nat. Pisa Processi Verbali 44: 143 (1934).

Ptychosperma macrocerum Becc., Atti Soc. Tosc. Sci. Nat. Pisa Processi Verbali 44: 142 (1934). SW. New Guinea. 43 NWG.

Ptychosperma macrospadix (Burret) Burret = **Ptychosperma microcarpum** (Burret) Burret

Ptychosperma maculatum (Porte ex Lem.) Seem. = **Pinanga maculata** Porte ex Lem.

Ptychosperma malaianum (Mart.) Miq. = **Pinanga malaiana** (Mart.) Scheff.

Ptychosperma mambare (F.M.Bailey) Becc. ex Martelli, Nuovo Giorn. Bot. Ital., n.s., 41: 709 (1934 publ. 1935). Papua New Guinea. 43 NWG.
*Drymophloeus mambare F.M.Bailey, Queensland Agric. J. 3: 202 (1899).

Ptychosperma micranthum Becc., Malesia 1: 52 (1877). *Rhopaloblaste micrantha* (Becc.) Hook.f. ex B.D.Jacks., Index Kew. 2: 713 (1895). *Heterospathe micrantha* (Becc.) H.E.Moore, Principes 14: 92 (1970). New Guinea. 43 NWG.

Ptychosperma microcarpum (Burret) Burret, Notizbl. Bot. Gart. Berlin-Dahlem 12: 596 (1935). Papua New Guinea. 43 NWG.
*Actinophloeus microcarpus Burret, J. Arnold Arbor. 12: 267 (1931).
Actinophloeus macrospadix Burret, Notizbl. Bot. Gart. Berlin-Dahlem 12: 343 (1935). *Ptychosperma macrospadix* (Burret) Burret, Notizbl. Bot. Gart. Berlin-Dahlem 12: 596 (1935).

Ptychosperma minor (Blume) Miq. = **Pinanga minor** Blume

Ptychosperma montanum (K.Schum. & Lauterb.) Burret = **Ptychosperma caryotoides** Ridl.

Ptychosperma mooreanum Essig, Principes 19: 75 (1975). SE. New Guinea. 43 NWG.

Ptychosperma multiramosum Burret = **Ptychosperma salomonense** Burret

Ptychosperma musschenbroekianum Becc. = **Heterospathe glauca** (Scheff.) H.E.Moore

Ptychosperma nenga (Blume ex Mart.) Teijsm. & Binn. = **Nenga pumila** var. **pumila**

Ptychosperma nicolai (Sander ex André) Burret, Repert. Spec. Nov. Regni Veg. 24: 263 (1928). New Guinea. 43 NWG.
*Romanovia nicolai Sander ex André, Rev. Hort. 71: 262 (1899). *Actinophloeus nicolai* (Sander ex André) Burret, Repert. Spec. Nov. Regni Veg. 24: 263 (1928).

Ptychosperma nobilis Seem. = **?**

Ptychosperma normanbyi (F.Muell.) F.Muell. = **Normanbya normanbyi** (F.Muell.) L.H.Bailey

Ptychosperma normanbyi Becc. = **Calyptrocalyx albertisianus** Becc.

Ptychosperma novohibernicum Becc. = **Ptychococcus paradoxus** (Scheff.) Becc.

Ptychosperma noxa (Blume) Miq. = **Pinanga coronata** (Blume ex Mart.) Blume

Ptychosperma pachycarpum Burret = **Ptychosperma salomonense** Burret

Ptychosperma palauense (Kaneh.) H.E.Moore & Fosberg = **Ponapea palauensis** Kaneh.,

Ptychosperma paniculatum Miq. = **Areca vestiaria** Giseke

Ptychosperma paradoxum Scheff. = **Ptychococcus paradoxus** (Scheff.) Becc.

Ptychosperma patulum (Blume) Miq. = **Pinanga patula** Blume

Ptychosperma pauciflorum H.Wendl. = **Balaka pauciflora** (H.Wendl.) H.E.Moore

Ptychosperma perbreve H.Wendl. = **Balaka seemannii** (H.Wendl.) Becc.

Ptychosperma pickeringii H.Wendl. = **Veitchia vitiensis** (H.Wendl.) H.E.Moore

Ptychosperma polyclados Becc. = **Ptychosperma caryotoides** Ridl.

Ptychosperma polystachyum Miq. = **Areca triandra** Roxb. ex Buch.-Ham.

Ptychosperma praemorsum Becc., Atti Soc. Tosc. Sci. Nat. Pisa Processi Verbali 44: 145 (1934). New Guinea. 43 NWG.

Ptychosperma propinquum (Becc.) Becc. ex Martelli, Nuovo Giorn. Bot. Ital., n.s., 41: 710 (1934 publ. 1935). New Guinea (Kep. Aru, Kep. Kai, Kep. Salawati). 42 MOL 43 NWG.
*Drymophloeus propinquus Becc., Malesia 1: 43 (1877). *Actinophloeus propinquus* (Becc.) Becc., Ann. Jard. Bot. Buitenzorg 2: 126 (1885). *Drymophloeus propinquus* var. *keiensis* Becc., Malesia 1: 43 (1877). *Ptychosperma keiense* (Becc.) Becc. ex Martelli, Nuovo Giorn. Bot. Ital., n.s., 41: 710 (1934 publ. 1935). *Saguaster propinquus* (Becc.) Kuntze, Revis. Gen. Pl. 2: 735 (1891).

Ptychosperma pullenii Essig, Allertonia 1: 462 (1978). Papua New Guinea. 43 NWG.

Ptychosperma punctulatum (Becc.) Becc. ex Martelli = **Ptychosperma lauterbachii** Becc.

Ptychosperma puniceum (Zipp. ex Blume) Miq. = **Pinanga rumphiana** (Mart.) J.Dransf. & Govaerts

Ptychosperma ramosissimum Essig, Allertonia 1: 441 (1978). New Guinea (Louisiade Arch.). 43 NWG.

Ptychosperma ridleyi Becc. = **Ptychosperma caryotoides** Ridl.

Ptychosperma rosselense Essig, Allertonia 1: 437 (1978). Papua New Guinea. 43 NWG.

Ptychosperma rumphii Blume = **Drymophloeus oliviformis** (Giseke) Mart.

Ptychosperma rupicola Thwaites = **Loxococcus rupicola** (Thwaites) H.Wendl. & Drude

Ptychosperma salicifolium (Blume) Miq. = **Pinanga salicifolia** Blume

Ptychosperma salomonense Burret, Notizbl. Bot. Gart. Berlin-Dahlem 12: 45 (1934). E. New Guinea to Solomon Is. 43 NWG SOL.
Ptychosperma kajewskii Burret, Notizbl. Bot. Gart. Berlin-Dahlem 13: 89 (1936).
Ptychosperma multiramosum Burret, Notizbl. Bot. Gart. Berlin-Dahlem 13: 93 (1936).

Ptychosperma pachycarpum Burret, Notizbl. Bot. Gart. Berlin-Dahlem 13: 91 (1936).

Strongylocaryum brassii Burret, Notizbl. Bot. Gart. Berlin-Dahlem 13: 98 (1936).

Strongylocaryum latius Burret, Notizbl. Bot. Gart. Berlin-Dahlem 13: 100 (1936).

Strongylocaryum macranthum Burret, Notizbl. Bot. Gart. Berlin-Dahlem 13: 96 (1936).

Ptychosperma sanderianum Ridl., Gard. Chron. 1898(2): 330 (1898). *Kentia sanderiana* (Ridl.) Sander, Cat. 1899: 29 (1899). *Actinophloeus sanderianus* (Ridl.) Burret, Repert. Spec. Nov. Regni Veg. 24: 270 (1928).
SE. New Guinea. 43 NWG.

Ptychosperma saxatilis (Burm.f. ex Giseke) Blume = **Areca oriziformis** var. **saxatilis**

Ptychosperma sayeri Becc. = **Ptychosperma caryotoides** Ridl.

Ptychosperma schefferi Becc. ex Martelli, Nuovo Giorn. Bot. Ital., n.s., 42: 79 (1935).
N. & C. New Guinea. 43 NWG.

Ptychosperma seaforthii Miq. = **Ptychosperma elegans** (R.Br.) Blume

Ptychosperma seemannii H.Wendl. = **Balaka seemannii** (H.Wendl.) Becc.

Ptychosperma simplicifrons Miq. = **Pinanga simplicifrons** (Miq.) Becc.

Ptychosperma singaporensis Becc. = **Rhopaloblaste singaporensis** (Becc.) Hook.f.

Ptychosperma streimannii Essig, Allertonia 1: 457 (1978).
Papua New Guinea. 43 NWG.

Ptychosperma seaforthii Miq. = **Ptychosperma elegans** (R.Br.) Blume

Ptychosperma seemannii H.Wendl. = **Balaka seemannii** (H.Wendl.) Becc.

Ptychosperma simplicifrons Miq. = **Pinanga simplicifrons** (Miq.) Becc.

Ptychosperma singaporensis Becc. = **Rhopaloblaste singaporensis** (Becc.) Hook.f.

Ptychosperma tagulense Essig, Allertonia 1: 441 (1978).
New Guinea (Louisiade Arch.). 43 NWG.

Ptychosperma tahitensis H.Wendl. = **Balaka tahitensis** (H.Wendl.) Becc.

Ptychosperma tenella H.Wendl. = **Pinanga tenella** (H.Wendl.) Scheff.

Ptychosperma tenue Becc. = **Ptychosperma cuneatum** (Burret) Burret

Ptychosperma vestiarium (Giseke) Miq. = **Areca vestiaria** Giseke

Ptychosperma vestitum Essig, Allertonia 1: 463 (1978).
Papua New Guinea. 43 NWG.

Ptychosperma vitiensis H.Wendl. = **Veitchia vitiensis** (H.Wendl.) H.E.Moore

Ptychosperma waitianum Essig, Principes 16: 39 (1972).
SE. New Guinea. 43 NWG.

Ptychosperma warleti Sander = **?**

Ptychosperma wendlandianum Burret = **Ptychosperma elegans** (R.Br.) Blume

Ptychosperma wendlandianum var. *sphaerocarpum* Burret = **Ptychosperma elegans** (R.Br.) Blume

Pyrenoglyphis

Pyrenoglyphis H.Karst. = **Bactris** Jacq. ex Scop.

Pyrenoglyphis aristata (Mart.) Burret = **Bactris fissifrons** Mart.

Pyrenoglyphis balanoidea (Oerst.) H.Karst. = **Bactris major** var. **major**

Pyrenoglyphis bicuspidata (Spruce) Burret = **Bactris acanthocarpa** var. **acanthocarpa**

Pyrenoglyphis bifida (Mart.) Burret = **Bactris bifida** Mart.

Pyrenoglyphis brongniartii (Mart.) Burret = **Bactris brongniartii** Mart.

Pyrenoglyphis chaetorhachis (Mart.) Burret = **Bactris major** var. **major**

Pyrenoglyphis chapadensis (Barb.Rodr.) Burret = **Bactris major** var. **infesta** (Mart.) Drude

Pyrenoglyphis concinna (Mart.) Burret = **Bactris concinna** Mart.

Pyrenoglyphis cruegeriana (Griseb.) H.Karst. = **Bactris major** var. **major**

Pyrenoglyphis curuena (Trail) Burret = **Bactris major** var. **infesta** (Mart.) Drude

Pyrenoglyphis exaltata (Barb.Rodr.) Burret = **Bactris major** var. **infesta** (Mart.) Drude

Pyrenoglyphis gastoniana (Barb.Rodr.) Burret = **Bactris gastoniana** Barb.Rodr.

Pyrenoglyphis gaviona (Trail) Burret = **Bactris major** var. **infesta** (Mart.) Drude

Pyrenoglyphis hoppii Burret = **Bactris hirta** var. **pectinata** (Mart.) Govaerts

Pyrenoglyphis infesta (Mart.) Burret = **Bactris major** var. **infesta** (Mart.) Drude

Pyrenoglyphis leucantha (H.Wendl.) Burret = **Bactris maraja** var. **maraja**

Pyrenoglyphis major (Jacq.) H.Karst. = **Bactris major** Jacq.

Pyrenoglyphis maraja (Mart.) Burret = **Bactris maraja** Mart.

Pyrenoglyphis mattogrossensis (Barb.Rodr.) Burret = **Bactris major** var. **infesta** (Mart.) Drude

Pyrenoglyphis microcarpa Burret = **Bactris brongniartii** Mart.

Pyrenoglyphis nemorosa (Barb.Rodr.) Burret = **Bactris major** var. **infesta** (Mart.) Drude

Pyrenoglyphis oligocarpa (Barb.Rodr.) Burret = **Bactris oligocarpa** Barb.Rodr.

Pyrenoglyphis ottostapfiana (Barb.Rodr.) Burret = **Bactris major** var. **major**

Pyrenoglyphis ovata (Oerst.) H.Karst. = **Bactris major** var. **major**

Pyrenoglyphis pallidispina (Mart.) Burret = **Bactris brongniartii** Mart.

Pyrenoglyphis piscatorum (Wedd. ex Drude) Burret = **Bactris brongniartii** Mart.

Pyrenoglyphis rivularis (Barb.Rodr.) Burret = **Bactris brongniartii** Mart.

Pyrenoglyphis socialis (Mart.) Burret = **Bactris major** var. **socialis** (Mart.) Drude

Pyrenoglyphis superior (L.H.Bailey) Burret = **Bactris major** var. **major**

Pyrenoglyphis tenera (H.Karst.) Burret = **Bactris brongniartii** Mart.

Pyrenoglyphis turbinocarpa (Barb.Rodr.) Burret = **Bactris turbinocarpa** Barb.Rodr.

Ranevea

Ranevea L.H.Bailey = **Ravenea** H.Wendl. ex C.D.Bouché

Ranevea hildebrandtii (H.Wendl. ex C.D.Bouché) L.H.Bailey = **Ravenea hildebrandtii** H.Wendl. ex C.D.Bouché

Raphia

Raphia P.Beauv., Fl. Oware 1: 75 (1806).
  20 species, Trop. & S. Africa, W. Indian Ocean, C. America to Brazil. 22 23 24 25 26 27 29 80 83 84.
  Sagus Rumph. ex Gaertn., Fruct. Sem. Pl. 1: 27 (1788), nom. illeg.

Raphia africana Otedoh, J. Nigerian Inst. Oil Palm Res. 6(22): 156 (1982).
  SE. Nigeria to Cameroon. 22 NGA 23 CMN.

Raphia angolensis Rendle = **Raphia hookeri** G.Mann & H.Wendl.

Raphia aulacolepis Burret = **Raphia taedigera** (Mart.) Mart.

Raphia australis Oberm. & Strey, Bothalia 10: 29 (1969).
  S. Mozambique to Cape Prov. 26 MOZ 27 CPP NAT.

Raphia bandamensis A.Chev. = **Raphia sudanica** A.Chev.

Raphia diasticha Burret = **Raphia vinifera** P.Beauv.

Raphia dolichocarpa Burret = **Raphia monbuttorum** var. **monbuttorum**

Raphia farinifera (Gaertn.) Hyl., Lustgården 31-32: 88 (1952).
  Trop. Africa, N. & E. Madagascar. 22 BEN BKN GAM GHA GUI IVO NGA SEN SIE TOG 23 CMN 25 KEN TAN UGA 26 ANG MLW MOZ ZAM ZIM 29 mau MDG sey.
  *Sagus farinifera Gaertn., Fruct. Sem. Pl. 2: 186 (1791).
  Sagus ruffia Jacq., Fragm. Bot.: 7 (1800). *Metroxylon ruffia* (Jacq.) Spreng., Syst. Veg. 2: 139 (1825). *Raphia ruffia* (Jacq.) Mart., Hist. Nat. Palm. 3: 217 (1838).
  Raphia pedunculata P.Beauv., Fl. Oware 1: 78 (1806). *Sagus pedunculata* (P.Beauv.) Poir. in J.B.A.M.de Lamarck, Encycl., Suppl. 5: 13 (1817).
  Raphia lyciosa Comm. ex Kunth, Enum. Pl. 3: 217 (1841).
  Raphia polymita Comm. ex Kunth, Enum. Pl. 3: 217 (1841).
  Raphia tamatavensis Sadeb., Bot. Jahrb. Syst. 36: 354 (1905).
  Raphia kirkii Engl. ex Becc., Webbia 3: 58 (1910).
  Raphia kirkii Engl. ex Becc., Agric. Colon. 4: t. II, f. 1-2 (1910).
  Raphia kirkii var. *grandis* Engl. ex Becc., Webbia 3: 64 (1910).
  Raphia kirkii var. *longicarpa* Engl. ex Becc., Webbia 3: 63 (1910).

Raphia gaertneri G.Mann & H.Wendl. = **Raphia palmapinus** subsp. **palma-pinus**

Raphia gentiliana De Wild., Miss. Ém. Laurent: 29 (1905).
  Central African Rep. to N. & C. Zaïre. 23 CAF ZAI.
  Raphia gentiliana var. *gilletii* De Wild., Miss. Ém. Laurent: 30 (1905). *Raphia gilletii* (De Wild.) Becc., Webbia 3: 105 (1910).
  Raphia sankuruensis De Wild., Bull. Jard. Bot. État 5: 145 (1916).

Raphia gentiliana var. *gilletii* De Wild. = **Raphia gentiliana** De Wild.

Raphia gigantea A.Chev. = **Raphia hookeri** G.Mann & H.Wendl.

Raphia gilletii (De Wild.) Becc. = **Raphia gentiliana** De Wild.

Raphia gossweileri Burret = **Raphia matombe** De Wild.

Raphia gracilis Becc. = **Raphia palma-pinus** subsp. **palma-pinus**

Raphia heberostris Becc. = **Raphia sudanica** A.Chev.

Raphia hookeri G.Mann & H.Wendl., Trans. Linn. Soc. London 24: 438 (1864).
  W. & WC. Trop. Africa to Angola. 22 BEN GHA GUI IVO LBR NGA SIE TOG 23 CMN EQG GAB GGI ZAI 26 ANG.
  Raphia maxima Puechuel Loesnck, Loesche Loango Exped. 3(1): 155 (1882).
  Raphia angolensis Rendle in W.P.Hiern, Cat. Afr. Pl. 2: 83 (1899).
  Raphia longirostris Becc., Webbia 3: 108 (1910).
  Raphia gigantea A.Chev., Rev. Bot. Appl. Agric. Trop. 12: 198 (1932).
  Raphia sassandrensis A.Chev., Rev. Bot. Appl. Agric. Trop. 12: 199 (1932).
  Raphia hookeri var. *planifoliola* Otedoh, J. Nigerian Inst. Oil Palm Res. 6(22): 152 (1982).
  Raphia hookeri var. *rubrifolia* Otedoh, J. Nigerian Inst. Oil Palm Res. 6(22): 153 (1982).

Raphia hookeri var. *planifoliola* Otedoh = **Raphia hookeri** G.Mann & H.Wendl.

Raphia hookeri var. *rubrifolia* Otedoh = **Raphia hookeri** G.Mann & H.Wendl.

Raphia humilis A.Chev. = **Raphia sudanica** A.Chev.

Raphia insignis Burret = **Raphia regalis** Becc.

Raphia kirkii Engl. ex Becc. = **Raphia farinifera** (Gaertn.) Hyl.

Raphia kirkii Engl. ex Becc. = **Raphia farinifera** (Gaertn.) Hyl.

Raphia kirkii var. *grandis* Engl. ex Becc. = **Raphia farinifera** (Gaertn.) Hyl.

Raphia kirkii var. *longicarpa* Engl. ex Becc. = **Raphia farinifera** (Gaertn.) Hyl.

Raphia laurentii De Wild., Miss. Ém. Laurent: 26 (1905).
  C. & S. Zaïre to Angola. 23 ZAI 26 ANG.

Raphia longiflora G.Mann & H.Wendl., Trans. Linn. Soc. London 24: 438 (1864).
  S. Nigeria to W. Zaïre. 22 GUI? NGA 23 CMN EQG GAB ZAI.

Raphia longirostris Becc. = **Raphia hookeri** G.Mann & H.Wendl.

Raphia lyciosa Comm. ex Kunth = **Raphia farinifera** (Gaertn.) Hyl.

Raphia macrocarpa Burret = **Raphia matombe** De Wild.

Raphia mambillensis Otedoh, J. Nigerian Inst. Oil Palm Res. 6(22): 163 (1982).
Nigeria to Sudan. 22 NGA 23 CAF CMN 24 SUD.

Raphia mannii Becc., Agric. Colon. 4: t. VI, f. 8-9 (1910).
S. Nigeria, Bioko. 22 NGA 23 GGI.
Raphia wendlandii Becc., Webbia 3: 81 (1910).

Raphia matombe De Wild., Bull. Jard. Bot. État 5: 144 (1916).
Cabinda to S. Zaïre. 23 CAB ZAI 26 ANG?
Raphia gossweileri Burret, Notizbl. Bot. Gart. Berlin-Dahlem 12: 305 (1935).
Raphia macrocarpa Burret, Notizbl. Bot. Gart. Berlin-Dahlem 15: 743 (1942).

Raphia maxima Puechuel Loesnck = **Raphia hookeri** G.Mann & H.Wendl.

Raphia monbuttorum Drude, Bot. Jahrb. Syst. 21: 111 (1895).
S. Nigeria to WC. Trop. Africa and S. Sudan. 22 NGA 23 CAF CMN ZAI 24 CHA SUD.

var. **monbuttorum**
S. Nigeria to WC. Trop. Africa and S. Sudan. 22 NGA 23 CAF CMN ZAI 24 CHA SUD.
Raphia dolichocarpa Burret, Notizbl. Bot. Gart. Berlin-Dahlem 15: 741 (1942).
Raphia pycnosticha Burret, Notizbl. Bot. Gart. Berlin-Dahlem 15: 740 (1942).

var. **mortehanii** (De Wild.) Otedoh, J. Nigerian Inst. Oil Palm Res. 6(22): 159 (1982).
Zaïre. 23 ZAI.
Raphia mortehanii De Wild., Bull. Jard. Bot. État 5: 145 (1916).

Raphia monbuttorum var. *macrocarpa* Robyns & Tournay = **Raphia ruwenzorica** Otedoh

Raphia mortehanii De Wild. = **Raphia monbuttorum** var. **mortehanii** (De Wild.) Otedoh

Raphia nicaraguensis Oerst. = **Raphia taedigera** (Mart.) Mart.

Raphia palma-pinus (Gaertn.) Hutch. in J.Hutchinson & J.M.Dalziel, Fl. W. Trop. Afr. 2: 387 (1936).
W. & WC. Trop. Africa. 22 GAM GHA GNB GUI IVO LBR SEN SIE 23 CAB CMN.
Sagus palma-pinus Gaertn., Fruct. Sem. Pl. 1: 27 (1788).

subsp. **nodostachys** Otedoh, J. Nigerian Inst. Oil Palm Res. 2(66): 159 (1982).
W. Trop. Africa. 22 IVO LBR SIE.

subsp. **palma-pinus**
W. & WC. Trop. Africa. 22 GAM GHA GNB GUI IVO LBR SEN SIE 23 CAB CMN.
Raphia gaertneri G.Mann & H.Wendl., Trans. Linn. Soc. London 24: 437 (1864).
Raphia gracilis Becc., Webbia 3: 92 (1910).

Raphia pedunculata P.Beauv. = **Raphia farinifera** (Gaertn.) Hyl.

Raphia polymita Comm. ex Kunth = **Raphia farinifera** (Gaertn.) Hyl.

Raphia pseudotextilis Burret = **Raphia textilis** Welw.

Raphia pycnosticha Burret = **Raphia monbuttorum** var. **monbuttorum**

Raphia regalis Becc., Webbia 3: 125 (1910).
S. Nigeria to Angola. 22 NGA 23 CAB CMN GAB ZAI 26 ANG.

Raphia insignis Burret, Notizbl. Bot. Gart. Berlin-Dahlem 15: 744 (1942).

Raphia rostrata Burret, Notizbl. Bot. Gart. Berlin-Dahlem 12: 307 (1935).
WC. Trop. Africa. 23 CAB ZAI.

Raphia ruffia (Jacq.) Mart. = **Raphia farinifera** (Gaertn.) Hyl.

Raphia ruwenzorica Otedoh, J. Nigerian Inst. Oil Palm Res. 6(22): 148 (1982).
E. Zaïre to Burundi. 23 BUR RWA ZAI.
Raphia monbuttorum var. *macrocarpa* Robyns & Tournay, Bull. Jard. Bot. État 25: 250 (1955).

Raphia sankuruensis De Wild. = **Raphia gentiliana** De Wild.

Raphia sassandrensis A.Chev. = **Raphia hookeri** G.Mann & H.Wendl.

Raphia sese De Wild., Miss. Ém. Laurent: 28 (1905).
Zaïre. 23 ZAI.

Raphia sudanica A.Chev., Bull. Soc. Bot. France 55(8): 95 (1908).
W. Trop. Africa to Cameroon. 22 BEN BKN GAM GHA GUI IVO MLI NGA NGR SEN SIE TOG 23 CMN.
Raphia heberostris Becc., Webbia 3: 96 (1910).
Raphia bandamensis A.Chev., Rev. Bot. Appl. Agric. Trop. 12: 205 (1932).
Raphia humilis A.Chev., Rev. Bot. Appl. Agric. Trop. 12: 204 (1932).

Raphia taedigera (Mart.) Mart., Hist. Nat. Palm. 3: 217 (1838).
Nigeria to Cameroon, C. America to Colombia, Brazil (Pará). 22 NGA 23 CMN 80 COS NIC PAN 83 CLM 84 BZN.
Sagus taedigera Mart., Hist. Nat. Palm. 2: 54 (1824).
Metroxylon taedigerum (Mart.) Spreng., Syst. Veg. 2: 139 (1825).
Raphia nicaraguensis Oerst., Vidensk. Meddel. Dansk Naturhist. Foren. Kjøbenhavn 1858: 52 (1858).
Raphia aulacolepis Burret, Notizbl. Bot. Gart. Berlin-Dahlem 15: 742 (1942).

Raphia tamatavensis Sadeb. = **Raphia farinifera** (Gaertn.) Hyl.

Raphia textilis Welw., Apont.: 584 (1858 publ. 1859).
WC. Trop. Africa to Angola. 23 CAB GAB ZAI 26 ANG.
Metroxylon textile Welw., Apont.: 584 (1858 publ. 1859).
Raphia welwitschii H.Wendl., Trans. Linn. Soc. London 24: 439 (1864).
Raphia pseudotextilis Burret, Notizbl. Bot. Gart. Berlin-Dahlem 15: 737 (1942).

Raphia vinifera P.Beauv., Fl. Oware 1: 77 (1806). *Sagus vinifera* (P.Beauv.) Pers., Syn. Pl. 2: 562 (1807). *Sagus raphia* Poir. in J.B.A.M.de Lamarck, Encycl., Suppl. 6: 395 (1817), nom. illeg. *Metroxylon viniferum* (P.Beauv.) Spreng., Syst. Veg. 2: 139 (1825).
W. & WC. Trop. Africa. 22 BEN GHA NGA 23 CMN GAB GGI ZAI.
Raphia diasticha Burret, Notizbl. Bot. Gart. Berlin-Dahlem 15: 739 (1942).
Raphia vinifera var. *nigerica* Otedoh, J. Nigerian Inst. Oil Palm Res. 6(22): 161 (1982).

Raphia vinifera var. *nigerica* Otedoh = **Raphia vinifera** P.Beauv.

Raphia welwitschii H.Wendl. = **Raphia textilis** Welw.

Raphia wendlandii Becc. = **Raphia mannii** Becc.

Rathea

Rathea H.Karst. = **Synechanthus** H.Wendl.

Rathea fibrosa (H.Wendl.) H.Karst. = **Synechanthus fibrosus** (H.Wendl.) H.Wendl.

Ravenea

Ravenea H.Wendl. ex C.D.Bouché, Monatsschr. Vereines Beförd. Gartenbaues Königl. Preuss. Staaten 21: 324 (1878). *Ranevea* L.H.Bailey, Cycl. Amer. Hort.: 1497 (1902), nom. illeg.
17 species, Comoros, Madagascar. 29.
Louvelia Jum. & H.Perrier, Compt. Rend. Hebd. Séances Acad. Sci. 155: 411 (1912).

Ravenea albicans (Jum.) Beentje, Kew Bull. 49: 663 (1994).
NE. Madagascar. 29 MDG.
*Louvelia albicans Jum., Ann. Inst. Bot.-Géol. Colon. Marseille, V, 1(1): 5 (1933).

Ravenea amara Jum. = **Ravenea sambiranensis** Jum. & H.Perrier

Ravenea dransfieldii Beentje, Kew Bull. 49: 656 (1994).
E. Madagascar. 29 MDG.

Ravenea glauca Jum. & H.Perrier, Ann. Inst. Bot.-Géol. Colon. Marseille, III, 1: 56 (1913).
SC. Madagascar. 29 MDG.

Ravenea hildebrandtii H.Wendl. ex C.D.Bouché, Monatsschr. Vereines Beförd. Gartenbaues Königl. Preuss. Staaten 21: 324 (1878). *Ranevea hildebrandtii* (H.Wendl. ex C.D.Bouché) L.H.Bailey, Cycl. Amer. Hort.: 1497 (1902).
Comoros. 29 COM.

Ravenea julietiae Beentje, Kew Bull. 49: 646 (1994).
E. Madagascar. 29 MDG.

Ravenea krociana Beentje, Kew Bull. 49: 636 (1994).
SE. Madagascar. 29 MDG.

Ravenea lakatra (Jum.) Beentje, Kew Bull. 49: 662 (1994).
E. Madagascar. 29 MDG.
*Louvelia lakatra Jum., Ann. Inst. Bot.-Géol. Colon. Marseille, IV, 5(1): 50 (1927).

Ravenea latisecta Jum., Ann. Inst. Bot.-Géol. Colon. Marseille, IV, 5(1): 35 (1927).
EC. Madagascar (Andasibe reg.). 29 MDG.

Ravenea louvelii Beentje, Kew Bull. 49: 664 (1994).
EC. Madagascar (Andasibe reg.). 29 MDG.
*Louvelia madagascariensis Jum. & H.Perrier, Compt. Rend. Hebd. Séances Acad. Sci. 155: 411 (1912).

Ravenea madagascariensis Becc., Bot. Jahrb. Syst. 38(87): 40 (1906).
C. & SE. Madagascar. 29 MDG.
Ravenea madagascariensis var. *monticola* Jum. & H.Perrier, in Fl. Madag. 30: 179 (1945).

Ravenea madagascariensis var. *monticola* Jum. & H.Perrier = **Ravenea madagascariensis** Becc.

Ravenea moorei J.Dransf. & N.W.Uhl, Principes 30: 159 (1986).
Comoros (Njazidja). 29 COM.

Ravenea musicalis Beentje, Principes 37: 199 (1993).
SE. Madagascar. 29 MDG.

Ravenea nana Beentje, Kew Bull. 49: 665 (1994).
N. & SE. Madagascar. 29 MDG.

Ravenea rivularis Jum. & H.Perrier, Ann. Inst. Bot.-Géol. Colon. Marseille, III, 1: 54 (1913).
SW. Madagascar. 29 MDG.

Ravenea robustior Jum. & H.Perrier, Ann. Inst. Bot.-Géol. Colon. Marseille, III, 1: 49 (1913).
N. & E. Madagascar. 29 MDG.
Ravenea robustior var. *kouna* Jum. & H.Perrier, Ann. Inst. Bot.-Géol. Colon. Marseille, IV, 5(1): 45 (1927).

Ravenea robustior var. *kouna* Jum. & H.Perrier = **Ravenea robustior** Jum. & H.Perrier

Ravenea sambiranensis Jum. & H.Perrier, Ann. Inst. Bot.-Géol. Colon. Marseille, III, 1: 50 (1913).
Madagascar. 29 MDG.
Ravenea amara Jum., Ann. Inst. Bot.-Géol. Colon. Marseille, IV, 5(1): 33 (1927).

Ravenea xerophila Jum., Ann. Inst. Bot.-Géol. Colon. Marseille, V, 1(1): 28 (1933).
S. Madagascar. 29 MDG.

Regelia

Regelia H.Wendl. = **Verschaffeltia** H.Wendl.

Regelia magnifica H.Wendl. = **Verschaffeltia splendida** H.Wendl.

Regelia majestica auct. = **Verschaffeltia splendida** H.Wendl.

Regelia princeps Balf.f. = **Verschaffeltia splendida** H.Wendl.

Rehderophoenix

Rehderophoenix Burret = **Drymophloeus** Zipp.

Rehderophoenix pachyclada Burret = **Drymophloeus pachycladus** (Burret) H.E.Moore

Rehderophoenix subdisticha H.E.Moore = **Drymophloeus subdistichus** (H.E.Moore) H.E.Moore

Reineckia

Reineckia H.Karst. = **Synechanthus** H.Wendl.

Reineckia triandra H.Karst. = **Synechanthus warscewiczianus** H.Wendl.

Reinhardtia

Reinhardtia Liebm., Overs. Kongel. Danske Vidensk. Selsk. Forh. Medlemmers Arbeider 1845: 9 (1846).
6 species, S. Mexico to Colombia, S. Hispaniola. 79 80 81 83.
Malortiea H.Wendl., Allg. Gartenzeitung 21: 25 (1853).

Reinhardtia elegans Liebm., Overs. Kongel. Danske Vidensk. Selsk. Forh. Medlemmers Arbeider 1845: 9 (1846).
Mexico (Oaxaca, Chiapas) to Honduras. 79 MXS MXT 80 HON.
Reinhardtia spinigera L.H.Bailey, Gentes Herb. 8: 191 (1949). Provisional synonym.

Reinhardtia gracilior Burret = **Reinhardtia gracilis** var. **gracilior** (Burret) H.E.Moore

Reinhardtia gracilis (H.Wendl.) Burret, Notizbl. Bot. Gart. Berlin-Dahlem 11: 554 (1932).
Mexico to C. America. 79 MXG MXS MXT 80 BLZ GUA HON NIC.
*Malortiea gracilis H.Wendl., Allg. Gartenzeitung 21: 26 (1853).

var. **gracilior** (Burret) H.E.Moore, Principes 1: 140 (1957).
Mexico (Veracruz, Oaxaca, Chiapas) to C. America. 79 MXG MXS MXT 80 BLZ HON.
Reinhardtia gracilior Burret, Notizbl. Bot. Gart. Berlin-Dahlem 11: 555 (1943).

var. **gracilis**
C. America. 80 BLZ GUA HON NIC.

var. **rostrata** (Burret) H.E.Moore, Principes 1: 140 (1957).
C. America. 80 COS NIC.
Reinhardtia rostrata Burret, Ann. Naturhist. Mus. Wien 46: 228 (1933). *Malortiea rostrata* (Burret) L.H.Bailey, Gentes Herb. 6: 260 (1943).

var. **tenuissima** H.E.Moore, Principes 1: 140 (1957).
Mexico (Oaxaca). 79 MXS.

Reinhardtia koschnyana (H.Wendl. & Dammer) Burret, Notizbl. Bot. Gart. Berlin-Dahlem 11: 554 (1932).
C. America to NW. Colombia. 80 COS HON NIC PAN 83 CLM.
Malortiea koschnyana H.Wendl. & Dammer, Gard. Chron., III, 29: 341 (1901).
Malortiea pumila Dugand, Revista Acad. Colomb. Ci. Exact. 7: 515 (1950).

Reinhardtia latisecta (H.Wendl.) Burret, Notizbl. Bot. Gart. Berlin-Dahlem 11: 554 (1932).
C. America. 80 BLZ COS HON NIC.
Malortiea latisecta H.Wendl., Allg. Gartenzeitung 21: 146 (1853).

Reinhardtia paiewonskiana Read, Zanoni & M.M.Mejía, Brittonia 39: 20 (1987).
SW. Dominican Rep. 81 DOM.

Reinhardtia rostrata Burret = **Reinhardtia gracilis** var. **rostrata** (Burret) H.E.Moore

Reinhardtia simplex (H.Wendl.) Burret, Notizbl. Bot. Gart. Berlin-Dahlem 11: 554 (1932).
C. America to NW. Colombia. 80 COS HON NIC PAN 83 CLM.
Malortiea simplex H.Wendl., Bot. Zeitung (Berlin) 17: 5 (1859).

Reinhardtia spinigera L.H.Bailey = **Reinhardtia elegans** Liebm.

Retispatha

Retispatha J.Dransf., Kew Bull. 34: 529 (1980).
1 species, Borneo. 42.

Retispatha dumetosa J.Dransf., Kew Bull. 34: 531 (1980).
Borneo. 42 BOR.

Rhapidophyllum

Rhapidophyllum H.Wendl. & Drude, Bot. Zeitung (Berlin) 34: 803 (1876).
1 species, SE. U.S.A. 78.

Rhapidophyllum hystrix (Frazer ex Thouin) H.Wendl. & Drude, Bot. Zeitung (Berlin) 34: 803 (1976).
SE. U.S.A. 78 FLA GEO MSI NCA.
Corypha hystrix Frazer ex Thouin, Ann. Mus. Natl. Hist. Nat. 2: 252 (1803). *Chamaerops hystrix* (Frazer ex Thouin) Pursh, Fl. Amer. Sept. 1: 240 (1813). *Sabal hystrix* (Frazer ex Thouin) Nutt., Amer. J. Sci. Arts 5: 293 (1822).
Rhapis caroliniana Kunth, Enum. Pl. 3: 246 (1841).

Rhapis

Rhapis L.f. ex Aiton, Hort. Kew. 3: 473 (1789).
8 species, S. China to Indo-China. 36 38 41 42.

Rhapis acaulis (Michx.) Walter ex Willd. = **Sabal minor** (Jacq.) Pers.

Rhapis arundinacea Aiton = **Sabal minor** (Jacq.) Pers.

Rhapis aspera W.Baxter = **Rhapis excelsa** (Thunb.) Henry

Rhapis caroliniana Kunth = **Rhapidophyllum hystrix** (Frazer ex Thouin) H.Wendl. & Drude

Rhapis cochinchinensis (Lour.) Mart. = **Livistona saribus** (Lour.) Merr. ex A.Chev.

Rhapis cordata W.Baxter = **Rhapis excelsa** (Thunb.) Henry

Rhapis divaricata Gagnep. = **Rhapis excelsa** (Thunb.) Henry

Rhapis excelsa (Thunb.) Henry, J. Arnold Arbor. 11: 153 (1930).
S. China to N. Vietnam. 36 CHC CHH CHS (38) jap nns 41 VIE.
Chamaerops excelsa Thunb. in J.A.Murray, Syst. Veg. ed. 14: 984 (1784). *Rhapis flabelliformis* L'Hér. ex Aiton, Hort. Kew. 3: 473 (1789), nom. illeg. *Trachycarpus excelsus* (Thunb.) H.Wendl., Bull. Soc. Bot. France 8: 429 (1861).
Rhapis aspera W.Baxter in J.C.Loudon, Hort. Brit., Suppl. 3: 624 (1839), nom. nud.
Rhapis cordata W.Baxter in J.C.Loudon, Hort. Brit., Suppl. 3: 624 (1839), nom. nud.
Rhapis major Blume, Rumphia 2: 55 (1839).
Chamaerops kwanwortsik Siebold ex H.Wendl., Index Palm.: 34 (1854), nom. nud.
Rhapis kwamwonzick Siebold ex Linden, Ill. Hort. 34: 39 (1887).
Rhapis divaricata Gagnep., Notul. Syst. (Paris) 6: 158 (1937).

Rhapis filiformis Burret = **Guihaia grossifibrosa** (Gagnep.) J.Dransf., S.K.Lee & F.N.Wei

Rhapis flabelliformis L'Hér. ex Aiton = **Rhapis excelsa** (Thunb.) Henry

Rhapis gracilis Burret, Notizbl. Bot. Gart. Berlin-Dahlem 10: 883 (1930).
Laos to SE. China. 36 CHH CHS 41 LAO.

Rhapis grossifibrosa Gagnep. = **Guihaia grossifibrosa** (Gagnep.) J.Dransf., S.K.Lee & F.N.Wei

Rhapis humilis Blume, Rumphia 2: 54 (1839).
S. China, Japan (S. Kyushu). 36 CHC CHS 38 JAP (42) jaw.
Chamaerops excelsa var. *humilior* Thunb. in J.A.Murray, Syst. Veg. ed. 14: 984 (1784).
Licuala waraguh Blume in J.J.Roemer & J.A.Schultes, Syst. Veg. 7: 1305 (1830), nom. rejic. prop.
Licuala wixu Blume in J.J.Roemer & J.A.Schultes, Syst. Veg. 7: 1305 (1830), nom. rejic. prop.
Rhapis javanica Blume, Rumphia 2: 56 (1839).
Chamaerops sirotsik H.Wendl., Index Palm.: 17 (1854).
Rhapis sirotsik H.Wendl. in O.C.E.de Kerchove de Denterghem, Palmiers: 255 (1878), nom. nud.

Rhapis javanica Blume = **Rhapis humilis** Blume

Rhapis kwamwonzick Siebold ex Linden = **Rhapis excelsa** (Thunb.) Henry

Rhapis laosensis Becc., Webbia 3: 225 (1910).
E. Thailand to Vietnam. 41 LAO THA VIE.

Rhapis macrantha Gagnep., Notul. Syst. (Paris) 6: 160 (1937).

Rhapis macrantha Gagnep. = **Rhapis laosensis** Becc.

Rhapis major Blume = **Rhapis excelsa** (Thunb.) Henry

Rhapis micrantha Becc., Webbia 3: 220 (1910).
Laos to NW. Vietnam. 41 LAO VIE.

Rhapis multifida Burret, Notizbl. Bot. Gart. Berlin-Dahlem 13: 588 (1937).
China (Yunnan to Guangxi). 36 CHC CHS.

Rhapis robusta Burret, Notizbl. Bot. Gart. Berlin-Dahlem 13: 587 (1937).
China (Guangxi). 36 CHS.

Rhapis siamensis Hodel = **Rhapis subtilis** Becc.

Rhapis sirotsik H.Wendl. = **Rhapis humilis** Blume

Rhapis subtilis Becc., Webbia 3: 227 (1910).
Indo-China to Sumatera. 41 CBD LAO THA 42 SUM.
Rhapis siamensis Hodel, Palm J. 136: 19 (1997).

Rhopaloblaste

Rhopaloblaste Scheff., Ann. Jard. Bot. Buitenzorg 1: 137 (1876).
6 species, Nicobar Is. to Papuasia. 41 42 43.
Ptychoraphis Becc., Ann. Jard. Bot. Buitenzorg 2: 90 (1885).

Rhopaloblaste augusta (Kurz) H.E.Moore, Principes 14: 79 (1970).
Nicobar Is. 41 NCB.
*Areca augusta Kurz, J. Bot. 13: 331 (1875).
Ptychoraphis augusta (Kurz) Becc., Ann. Jard. Bot. Buitenzorg 2: 90 (1885).

Rhopaloblaste brassii H.E.Moore = **Rhopaloblaste ledermanniana** Becc.

Rhopaloblaste ceramica (Miq.) Burret, Repert. Spec. Nov. Regni Veg. 24: 288 (1928).
Maluku to New Guinea. 42 MOL 43 NWG.
*Bentinckia ceramica Miq., Verh. Kon. Ned. Akad. Wetensch., Afd. Natuurk. 11: 8 (1868). *Cyrtostachys ceramica* (Miq.) H.Wendl. in O.C.E.de Kerchove de Denterghem, Palmiers: 242 (1878).
Rhopaloblaste hexandra Scheff., Ann. Jard. Bot. Buitenzorg 1: 156 (1876).
Rhopaloblaste micrantha Burret, Notizbl. Bot. Gart. Berlin-Dahlem 15: 10 (1940), nom. illeg.
Rhopaloblaste dyscrita H.E.Moore, Principes 14: 83 (1970).

Rhopaloblaste dyscrita = **Rhopaloblaste ceramica** (Miq.) Burret

Rhopaloblaste elegans H.E.Moore, Principes 11: 94 (1966).
Solomon Is. 43 SOL.

Rhopaloblaste elmeri (Becc.) Becc. = **Heterospathe elmeri** Becc.

Rhopaloblaste gideonii R.Banka, Kew Bull. 59: 56 (2004).
Bismarck Arch. (New Ireland). 43 BIS.

Rhopaloblaste hexandra Scheff. = **Rhopaloblaste ceramica** (Miq.) Burret

Rhopaloblaste intermedia (Becc.) Becc. = **Heterospathe intermedia** (Becc.) Fernando

Rhopaloblaste ledermanniana Becc., Bot. Jahrb. Syst. 58: 451 (1923).
New Guinea. 43 NWG.

Rhopaloblaste brassii H.E.Moore, Principes 14: 81 (1970).

Rhopaloblaste macgregorii Becc. = **Heterospathe macgregorii** (Becc.) H.E.Moore

Rhopaloblaste micrantha Burret = **Rhopaloblaste dyscrita** H.E.Moore

Rhopaloblaste micrantha (Becc.) Hook.f. ex B.D.Jacks. = **Heterospathe micrantha**

Rhopaloblaste microcarpa (Becc.) Becc. = **Heterospathe elmeri** Becc.

Rhopaloblaste princeps W.Bull = **?**

Rhopaloblaste singaporensis (Becc.) Hook.f. in G.Bentham & J.D.Hooker, Gen. Pl. 3: 892 (1883).
Pen. Malaysia. 42 MLY.
*Ptychosperma singaporensis Becc., Malesia 1: 61 (1877). *Drymophloeus singaporensis* (Becc.) Hook.f., Rep. Progr. Condition Roy. Bot. Gard. Kew 1882: 55 (1884). *Ptychoraphis singaporensis* (Becc.) Becc., Ann. Jard. Bot. Buitenzorg 2: 90 (1885).
Ptychoraphis longiflora Ridl., J. Straits Branch Roy. Asiat. Soc. 41: 39 (1904).

Rhopalostylis

Rhopalostylis H.Wendl. & Drude, Linnaea 39: 180 (1875). *Eora* O.F.Cook, J. Heredity 18: 409 (1927).
2 species, Norfolk I., New Zealand. 50 51.

Rhopalostylis baueri (Hook.f. ex Lem.) H.Wendl. & Drude, Bot. Zeitung (Berlin) 35: 638 (1877).
Norfolk I., Kermadec Is. 50 NFK 51 KER.
*Areca baueri Hook.f. ex Lem., Ill. Hort. 15: 575 (1868). *Kentia baueri* (Hook.f. ex Lem.) Seem., Fl. Vit.: 269 (1868). *Eora baueri* (Hook.f. ex Lem.) O.F.Cook, J. Heredity 18: 409 (1927).

var. **baueri**
Norfolk I. 50 NFK.
Seaforthia robusta H.Wendl. in O.C.E.de Kerchove de Denterghem, Palmiers: 257 (1878), nom. inval.

var. **cheesemanii** (Becc. ex Cheeseman) Sykes, Bull. New Zealand Dept. Sci. Industr. Res. 119: 184 (1977).
Kermadec Is. (Raoul I.). 51 KER.
*Rhopalostylis cheesemanii Becc. ex Cheeseman, Trans. & Proc. New Zealand Inst. 48: 215 (1916). *Eora cheesemanii* (Becc. ex Cheeseman) O.F.Cook, J. Heredity 18: 409 (1927). *Eora ultima* O.F.Cook, J. Heredity 18: 409 (1927).

Rhopalostylis cheesemanii Becc. ex Cheeseman = **Rhopalostylis baueri** var. **cheesemanii** (Becc. ex Cheeseman) Sykes

Rhopalostylis sapida (Sol. ex G.Forst.) H.Wendl. & Drude in O.C.E.de Kerchove de Denterghem, Palmiers: 255 (1878).
N. & C. New Zealand, Chatham Is. 51 CTM NZN NZS.
*Areca sapida Sol. ex G.Forst., Pl. Esc.: 66 (1786). *Kentia sapida* (Sol. ex G.Forst.) Mart., Hist. Nat. Palm. 3: 312 (1849). *Eora sapida* (Sol. ex G.Forst.) O.F.Cook, J. Heredity 18: 409 (1927).
Areca banksii A.Cunn. ex Kunth, Enum. Pl. 3: 185 (1841), pro syn.

Rhynchocarpa

Rhynchocarpa Becc. = **Burretiokentia** Pic.Serm.

Rhynchocarpa pilosa Burret = **Heterospathe pilosa** (Burret) Burret

Rhynchocarpa vieillardii (Brongn. & Gris) Becc. = **Burretiokentia vieillardii** (Brongn. & Gris) Pic.Serm.

Rhyticocos

Rhyticocos Becc. = **Syagrus** Mart.

Rhyticocos amara (Jacq.) Becc. = **Syagrus amara** (Jacq.) Mart.

Roebelia

Roebelia Engel = **Geonoma** Willd.

Roebelia solitaria Engel = **Geonoma weberbaueri** Dammer ex Burret

Romanowia

Romanowia Sander ex André = **Ptychosperma** Labill.

Rooseveltia

Rooseveltia O.F.Cook = **Euterpe** Mart.

Rooseveltia frankliniana O.F.Cook = **Euterpe precatoria** var. **longivaginata** (Mart.) A.J.Hend.

Roscheria

Roscheria H.Wendl. ex Balf.f. in J.G.Baker, Fl. Mauritius: 386 (1877).
1 species, Seychelles. 29.

Roscheria melanochaetes (H.Wendl.) H.Wendl. ex Balf.f. in J.G.Baker, Fl. Mauritius: 387 (1877).
Seychelles. 29 SEY.
  *Verschaffeltia melanochaetes H.Wendl., Ill. Hort. 18: 54 (1871).
  Phoenicophorium viridifolium H.Wendl. in O.C.E.de Kerchove de Denterghem, Palmiers: 252 (1878).

Rotang

Rotang Adans. = **Calamus** L.

Rotang accedens (Blume) Baill. = **Daemonorops rubra** (Reinw. ex Mart.) Blume

Rotang albus (Pers.) Baill. = **Calamus albus** Pers.

Rotang asperrimus (Blume) Baill. = **Calamus asperrimus** Blume

Rotang barbatus (Zipp. ex Blume) Baill. = **Calamus barbatus** Zipp. ex Blume

Rotang blancoi (Kunth) Baill. = **Calamus usitatus** Blanco

Rotang caesius (Blume) Baill. = **Calamus caesius** Blume

Rotang cawa (Blume) Baill. = **Calamus cawa** Blume

Rotang crinitus (Blume) Baill. = **Daemonorops crinita** Blume

Rotang equestris (Willd.) Baill. = **Calamus equestris** Willd.

Rotang graminosus (Blume) Baill. = **Calamus graminosus** Blume

Rotang heteroideus (Blume) Baill. = **Calamus heteroideus** Blume

Rotang latispinus (Miq.) Baill. = **Calamus latispinus** Miq.

Rotang linnaei Baill. = **Calamus rotang** L.

Rotang longipes (Griff.) Baill. = **Daemonorops longipes** (Griff.) Mart.

Rotang manan (Miq.) Baill. = **Calamus manan** Miq.

Rotang maximus Baill. = **Plectocomia elongata** Mart. & Blume

Rotang melanoloma (Mart.) Baill. = **Calamus melanoloma** Mart.

Rotang micracanthus (Griff.) Baill. = **Daemonorops micracantha** (Griff.) Becc.

Rotang niger (Willd.) Baill. = **Daemonorops nigra** (Willd.) Blume

Rotang ornatus (Blume) Baill. = **Calamus ornatus** Blume

Rotang palembanicus (Blume) Baill. = **Daemonorops palembanica** Blume

Rotang periacanthus (Miq.) Baill. = **Daemonorops periacantha** Miq.

Rotang petraeus (Lour.) Baill. = **Calamus petraeus**

Rotang pisicarpus (Blume) Baill. = **Calamus pisicarpus** Blume

Rotang rhomboideus (Blume) Baill. = **Calamus rhomboideus** Blume

Rotang royleanus (Griff.) Baill. = **Calamus tenuis** Roxb.

Rotang rudentum (Lour.) Baill. = **Calamus rudentum** Lour.

Rotang scandens (Blume) Baill. = **Daemonorops scandens ?**

Rotang scipionum (Lour.) Baill. = **Calamus scipionum** Lour.

Rotang spectabilis (Blume) Baill. = **Calamus spectabilis** Blume

Rotang verus (Lour.) Baill. = **Calamus verus**

Rotang viminalis (Willd.) Baill. = **Calamus viminalis** Willd.

Rotanga

Rotanga Boehm. = **Calamus** L.

Rotanga calamus Crantz = **Calamus rotang** L.

Roystonea

Roystonea O.F.Cook, Science, II, 12: 479 (1900).
10 species, S. Florida, Caribbean, Mexico to Venezuela. 78 79 80 81 82 83.
  Gorgasia O.F.Cook, Natl. Hort. Mag. 18: 112 (1939), no latin descr.

Roystonea altissima (Mill.) H.E.Moore, Gentes Herb. 9: 239 (1963).
Jamaica. 81 JAM.
  *Palma altissima Mill., Gard. Dict. ed. 6: 4 (1752), nom. inval.
  Roystonea jamaicana L.H.Bailey, Gentes Herb. 3: 384 (1935).

Roystonea aitia O.F.Cook = **?** [81 HAI]

Roystonea borinquena O.F.Cook, Bull. Torrey Bot. Club 28: 552 (1901). *Oreodoxa borinquena* (O.F.Cook) Reasoner ex L.H.Bailey, Stand. Cycl. Hort. 4: 2405 (1917 publ. 1916).
Hispaniola to Virgin Is. 81 DOM HAI LEE PUE.
  Roystonea hispaniolana L.H.Bailey, Gentes Herb. 4: 268 (1939).

Roystonea peregrina L.H.Bailey, Gentes Herb. 8: 127 (1949).

Roystonea caribaea (Spreng.) P.Wilson = **Roystonea oleracea** (Jacq.) O.F.Cook

Roystonea dunlapiana P.H.Allen, Ceiba 3: 15 (1952).
SE. Mexico to C. America. 79 MXT 80 HON NIC.

Roystonea elata (W.Bartram) F.Harper = **Roystonea regia** (Kunth) O.F.Cook

Roystonea floridana O.F.Cook = **Roystonea regia** (Kunth) O.F.Cook

Roystonea hispaniolana L.H.Bailey = **Roystonea borinquena** O.F.Cook

Roystonea jamaicana L.H.Bailey = **Roystonea altissima** (Mill.) H.E.Moore

Roystonea jenmanii (C.H.Wright) Burret = **Roystonea regia** (Kunth) O.F.Cook

Roystonea lenis Léon, Mem. Soc. Cub. Hist. Nat. "Felipe Poey" 17: 8 (1943).
E. Cuba. 81 CUB.
　　Roystonea regia var. *pinguis* L.H.Bailey, Gentes Herb. 3: 378 (1935).

Roystonea maisiana (L.H.Bailey) Zona, Fl. Neotrop. Monogr. 71: 22 (1996).
E. Cuba. 81 CUB.
　　Roystonea regia var. *maisiana* L.H.Bailey, Gentes Herb. 3: 376 (1935).

Roystonea oleracea (Jacq.) O.F.Cook, Bull. Torrey Bot. Club 28: 554 (1901).
Lesser Antilles to NE. Colombia. 81 LEE TRT WIN 82 guy VEN 83 CLM.
　　Areca oleracea Jacq., Select. Stirp. Amer. Hist.: 278 (1763). *Oreodoxa oleracea* (Jacq.) Mart., Hist. Nat. Palm. 3: 166 (1838). *Kentia oleracea* (Jacq.) Seem. ex H.Wendl. in O.C.E.de Kerchove de Denterghem, Palmiers: 248 (1878). *Gorgasia oleracea* (Jacq.) O.F.Cook, Natl. Hort. Mag. 18: 114 (1939), nom. inval.
　　Euterpe caribaea Spreng., Syst. Veg. 2: 140 (1825). *Oreodoxa caribaea* (Spreng.) Dammer & Urb. in I.Urban, Symb. Antill. 4: 129 (1903). *Roystonea caribaea* (Spreng.) P.Wilson, Bull. New York Bot. Gard. 8: 385 (1917).
　　Oreodoxa regia var. *jenmanii* Waby, J. Board Agric. Brit. Guiana 12: 51 (1919). *Roystonea oleracea* var. *excelsior* L.H.Bailey, Gentes Herb. 3: 369 (1935), nom. illeg. *Roystonea oleracea* var. *jenmanii* (Waby) Zona, Fl. Neotrop. Monogr. 71: 24 (1996).
　　Roystonea venezuelana L.H.Bailey, Gentes Herb. 8: 124 (1949).

Roystonea oleracea var. *excelsior* L.H.Bailey = **Roystonea oleracea** (Jacq.) O.F.Cook

Roystonea oleracea var. *jenmanii* (Waby) Zona = **Roystonea oleracea** (Jacq.) O.F.Cook

Roystonea peregrina L.H.Bailey = **Roystonea borinquena** O.F.Cook

Roystonea princeps (Becc.) Burret, Bot. Jahrb. Syst. 63: 76 (1929).
SW. Jamaica. 81 JAM.
　　Oreodoxa princeps Becc., Pomona Coll. J. Econ. Bot. 2: 266 (1912).

Roystonea regia (Kunth) O.F.Cook, Science, II, 12: 479 (1900).
S. Florida, NC. & SE. Mexico to C. America, Caribbean. 78 FLA 79 MXG MXT 80 BLZ HON 81 BAH CAY CUB.

Palma elata W.Bartram, Travels Carolina: iv, 115 (1791), nom. rejic. *Roystonea elata* (W.Bartram) F.Harper, Proc. Biol. Soc. Wash. 59: 29 (1946).
　　Oreodoxa regia Kunth in F.W.H.von Humboldt, A.J.A.Bonpland & C.S.Kunth, Nov. Gen. Sp. 1: 305 (1816), nom. cons. *Oenocarpus regius* (Kunth) Spreng., Syst. Veg. 2: 140 (1825).
　　Roystonea floridana O.F.Cook, Bull. Torrey Bot. Club 28: 554 (1901).
　　Euterpe jenmanii C.H.Wright, Bull. Misc. Inform. Kew 1906: 203 (1906). *Roystonea jenmanii* (C.H.Wright) Burret, Bot. Jahrb. Syst. 63: 76 (1929).
　　Euterpe ventricosa C.H.Wright, Bull. Misc. Inform. Kew 1906: 203 (1906). *Roystonea ventricosa* (C.H. Wright) L.H.Bailey, Gentes Herb. 8: 113 (1949).

Roystonea regia var. *maisiana* L.H.Bailey = **Roystonea maisiana** (L.H.Bailey) Zona

Roystonea regia var. *pinguis* L.H.Bailey = **Roystonea lenis** Léon

Roystonea stellata Léon, Mem. Soc. Cub. Hist. Nat. "Felipe Poey" 17: 11 (1943).
E. Cuba (Maisí reg.). 81 CUB.

Roystonea venezuelana L.H.Bailey = **Roystonea oleracea** (Jacq.) O.F.Cook

Roystonea ventricosa (C.H.Wright) L.H.Bailey = **Roystonea regia** (Kunth) O.F.Cook

Roystonea violacea Léon, Mem. Soc. Cub. Hist. Nat. "Felipe Poey" 17: 10 (1943).
E. Cuba (Maisí reg.). 81 CUB.

Sabal

Sabal Adans., Fam. Pl. 2: 495 (1763).
16 species, S. Oklahoma to Venezuala, Caribbean. 74 77 78 79 80 81 82 83.
　　Inodes O.F.Cook, Bull. Torrey Bot. Club 28: 529 (1901).

Sabal adansonii Guerns. = **Sabal minor** (Jacq.) Pers.

Sabal adansonii var. *megacarpa* Chapm. = **Sabal etonia** Swingle ex Nash

Sabal adiantina Raf. = **Sabal minor** (Jacq.) Pers.

Sabal allenii L.H.Bailey = **Sabal mauritiiformis** (H.Karst.) Griseb. & H.Wendl.

Sabal bahamensis (Becc.) L.H.Bailey = **Sabal palmetto** (Walter) Lodd. ex Schult. & Schult.f.

Sabal beccariana L.H.Bailey = **Sabal bermudana** L.H.Bailey

Sabal bermudana L.H.Bailey, Gentes Herb. 3: 326 (1934).
Bermuda. 81 BER.
　　Sabal princeps Becc., Webbia 2: 59 (1908), nom. illeg. *Sabal beccariana* L.H.Bailey, Gentes Herb. 4: 387 (1940). *Inodes princeps* (Becc.) Cif. & Giacom., Nomencl. Fl. Ital. 1: 72 (1950).

Sabal blackburniana Glazebr. = **Sabal palmetto** (Walter) Lodd. ex Schult. & Schult.f.

Sabal caroliniana Poir. = **Sabal minor** (Jacq.) Pers.

Sabal causiarum (O.F.Cook) Becc., Webbia 2: 71 (1908).
S. Hispaniola to British Virgin Is. 81 DOM HAI LEE NLA? PUE.
　　Inodes causiarum O.F.Cook, Bull. Torrey Bot. Club 28: 531 (1901).
　　Inodes glauca Dammer in I.Urban, Symb. Antill. 4: 127 (1903).

Sabal haitensis Becc. ex Martelli, Ann. Roy. Bot. Gard. (Calcutta) 13: 293 (1931).

Sabal questeliana L.H.Bailey, Gentes Herb. 6: 422 (1944).

Sabal coerulescens auct. = **Sabal mauritiiformis** (H.Karst.) Griseb. & H.Wendl.

Sabal deeringiana Small = **Sabal minor** (Jacq.) Pers.

Sabal domingensis Becc., Webbia 2: 49 (1908).
E. Cuba to NW. & C. Hispaniola. 81 CUB DOM HAI.
Sabal neglecta Becc., Webbia 2: 40 (1908).

Sabal dugesii S.Watson ex L.H.Bailey = **Sabal pumos** (Kunth) Burret

Sabal elata Mart. = **?**

Sabal etonia Swingle ex Nash, Bull. Torrey Bot. Club 23: 99 (1896).
C. & SE. Florida. 78 FLA.
Sabal adansonii var. *megacarpa* Chapm., Fl. South. U.S., ed. 3: 462 (1897). *Sabal megacarpa* (Chapm.) Small, Fl. S.E. U.S.: 223 (1903).

Sabal exul (O.F.Cook) L.H.Bailey = **Sabal mexicana** Mart.

Sabal floribunda Katzenstein = **Sabal minor** (Jacq.) Pers.

Sabal florida Becc. = **Sabal maritima** (Kunth) Burret

Sabal ghiesbrechtii Pfister = **?**

Sabal glabra (Mill.) Sarg. = **Sabal minor** (Jacq.) Pers.

Sabal glaucescens Lodd. ex H.E.Moore = **Sabal mauritiiformis** (H.Karst.) Griseb. & H.Wendl.

Sabal gretherae H.J.Quero, Principes 35: 219 (1991).
Mexico (Quintana Roo). 79 MXT.

Sabal guatemalensis Becc., Webbia 2: 68 (1908).
S. Mexico to Guatemala. 79 MXS MXT 80 GUA.

Sabal haitensis Becc. ex Martelli = **Sabal causiarum** (O.F.Cook) Becc.

Sabal havanensis Lodd. ex Mart. = **?**

Sabal hoogendorpii (Teijsm. & Binn. ex Miq.) L.H.Bailey = **Livistona saribus** (Lour.) Merr. ex A.Chev.

Sabal hystrix (Frazer ex Thouin) Nutt. = **Rhapidophyllum hystrix** (Frazer ex Thouin) H.Wendl. & Drude

Sabal jamaicensis Becc. = **Sabal maritima** (Kunth) Burret

Sabal jamesiana Small = **Sabal palmetto** (Walter) Lodd. ex Schult. & Schult.f.

Sabal louisiana (Darby) Bomhard = **Sabal minor** (Jacq.) Pers.

Sabal maritima (Kunth) Burret, Repert. Spec. Nov. Regni Veg. 32: 101 (1933).
Cuba, Jamaica. 81 CUB JAM.
Corypha maritima Kunth in F.W.H.von Humboldt, A.J.A.Bonpland & C.S.Kunth, Nov. Gen. Sp. 1: 298 (1816). *Copernicia maritima* (Kunth) Mart., Hist. Nat. Palm. 3: 243 (1849).
Sabal taurina Mart., Hist. Nat. Palm. 3: 320 (1853). Provisional synonym.
Sabal florida Becc., Webbia 2: 46 (1908).
Sabal jamaicensis Becc., Repert. Spec. Nov. Regni Veg. 6: 94 (1908).

Sabal mauritiiformis (H.Karst.) Griseb. & H.Wendl. in A.H.R.Grisebach, Fl. Brit. W. I.: 514 (1864).
NC. & S. Mexico to Venezuela, Trinidad. 79 MXG MXS MXT 80 BLZ COS GUA HON PAN 81 TRT 82 VEN 83 CLM.
*Trithrinax mauritiiformis H.Karst., Linnaea 28: 244 (1856).
Sabal coerulescens auct., Gard. Chron. 1875(1): 589 (1875). Provisional synonym.
Sabal nematoclada Burret, Repert. Spec. Nov. Regni Veg. 48: 256 (1940).
Sabal allenii L.H.Bailey, Gentes Herb. 6: 200 (1943).
Sabal morrisiana Bartlett ex L.H.Bailey, Gentes Herb. 6: 412 (1944).
Sabal glaucescens Lodd. ex H.E.Moore, Gentes Herb. 9: 287 (1963).

Sabal mayara Bartlett = **Sabal yapa** C.Wright ex Becc.

Sabal megacarpa (Chapm.) Small = **Sabal etonia** Swingle ex Nash

Sabal mexicana Mart., Hist. Nat. Palm. 3: 246 (1838).
Inodes mexicana (Mart.) Standl., Contr. U. S. Natl. Herb. 13: 71 (1920).
S. Texas to C. America. 77 TEX 79 MXE MXG MXS MXT 80 ELS HON NIC.
Inodes texana O.F.Cook, Bull. Torrey Bot. Club 28: 534 (1901). *Sabal texana* (O.F.Cook) Becc., Webbia 2: 73 (1908).
Inodes exul O.F.Cook, U.S.D.A. Bur. Pl. Industr. Circ. 113: 14 (1913). *Sabal exul* (O.F.Cook) L.H.Bailey, Rhodora 18: 155 (1916).
Erythea loretensis M.E.Jones, Contr. W. Bot. 18: 29 (1933).

Sabal miamiensis Zona, Brittonia 37: 366 (1985).
SE. Florida. 78 FLA.

Sabal minima Nutt. = **Sabal minor** (Jacq.) Pers.

Sabal minor (Jacq.) Pers., Syn. Pl. 1: 399 (1805).
S. Oklahoma to SE. U.S.A. 74 OKL 77 TEX 78 ALA ARK FLA LOU MSI NCA SCA.
Chamaerops glabra Mill., Gard. Dict. ed. 8: 2 (1768). Provisional synonym. *Sabal glabra* (Mill.) Sarg., Silva 10: 38 (1896).
*Corypha minor Jacq., Hort. Bot. Vindob. 3: 8 (1776). *Sabal adansonii* Guerns., Bull. Soc. Philom. Paris 3: 206 (1803), nom. illeg.
Corypha pumila Walter, Fl. Carol.: 119 (1788). *Sabal pumila* (Walter) Elliott, Sketch Bot. S. Carolina 1: 430 (1817).
Rhapis arundinacea Aiton, Hort. Kew. 3: 474 (1789). *Chamaerops arundinacea* (Aiton) Sm. in A.Rees, Cycl. 7: 3 (1806).
Chamaerops acaulis Michx., Fl. Bor.-Amer. 2: 207 (1803). *Rhapis acaulis* (Michx.) Walter ex Willd., Sp. Pl. 4: 1093 (1806).
Sabal caroliniana Poir. in J.B.A.M.de Lamarck, Encycl. 6: 356 (1804), nom. nud.
Chamaerops louisiana Darby, Geogr. Descr. Louisiana: 194 (1816). *Sabal louisiana* (Darby) Bomhard, J. Wash. Acad. Sci. 25: 44 (1935).
Sabal adiantina Raf., Fl. Ludov.: 17 (1817).
Sabal minima Nutt., Amer. J. Sci. Arts 5: 293 (1822). *Brahea minima* (Nutt.) H.Wendl. in O.C.E.de Kerchove de Denterghem, Palmiers: 235 (1878).
Chamaerops sabaloides Baldwin ex Darl., Reliq. Baldw.: 334 (1843).
Sabal deeringiana Small, Torreya 26: 34 (1926).
Sabal floribunda Katzenstein, Cat. 1934: (1934).

Sabal morrisiana Bartlett ex L.H.Bailey = **Sabal mauritiiformis** (H.Karst.) Griseb. & H.Wendl.

Sabal neglecta Becc. = **Sabal domingensis** Becc.

Sabal nematoclada Burret = **Sabal mauritiiformis** (H.Karst.) Griseb. & H.Wendl.

Sabal oleracea Mart. = **?**

Sabal palmetto (Walter) Lodd. ex Schult. & Schult.f., Syst. Veg. 7: 1487 (1830).
SE. U.S.A. to W. Cuba. 78 FLA GEO NCA SCA 81 BAH CUB.
*Corypha palmetto Walter, Fl. Carol.: 119 (1788).
Chamaerops palmetto (Walter) Michx., Fl. Bor.-Amer. 1: 206 (1803). *Inodes palmetto* (Walter) O.F.Cook, Bull. Torrey Bot. Club 28: 532 (1901).
Corypha umbraculifera Jacq., Fragm. Bot.: 12 (1800), nom. illeg. *Sabal umbraculifera* Mart., Hist. Nat. Palm. 3: 245 (1838). Provisional synonym.
Sabal blackburniana Glazebr., Gard. Mag. (London) 5: 52 (1829). Provisional synonym.
Inodes schwarzii O.F.Cook, Bull. Torrey Bot. Club 28: 532 (1901). *Sabal schwarzii* (O.F.Cook) Becc., Webbia 2: 39 (1908).
Sabal palmetto var. *bahamensis* Becc., Webbia 2: 38 (1907). *Sabal bahamensis* (Becc.) L.H.Bailey, Gentes Herb. 6: 417 (1944).
Sabal parviflora Becc., Webbia 2: 43 (1908).
Sabal jamesiana Small, J. New York Bot. Gard. 28: 182 (1927).
Sabal viatoris L.H.Bailey, Gentes Herb. 6: 403 (1944).

Sabal palmetto var. *bahamensis* Becc. = **Sabal palmetto** (Walter) Lodd. ex Schult. & Schult.f.

Sabal parviflora Becc. = **Sabal palmetto** (Walter) Lodd. ex Schult. & Schult.f.

Sabal peregrina L.H.Bailey = **Sabal yapa** C.Wright ex Becc.

Sabal princeps Becc. = **Sabal bermudana** L.H.Bailey

Sabal princeps Knuth = **?**

Sabal pumila (Walter) Elliott = **Sabal minor** (Jacq.) Pers.

Sabal pumos (Kunth) Burret, Repert. Spec. Nov. Regni Veg. 32: 101 (1933).
SW. & WC. Mexico. 79 MXC MXE MXS.
*Corypha pumos Kunth in F.W.H.von Humboldt, A.J.A.Bonpland & C.S.Kunth, Nov. Gen. Sp. 1: 298 (1816). *Copernicia pumos* (Kunth) Mart., Hist. Nat. Palm. 3: 243 (1849).
Sabal dugesii S.Watson ex L.H.Bailey, Gentes Herb. 3: 335 (1934).

Sabal questeliana L.H.Bailey = **Sabal causiarum** (O.F.Cook) Becc.

Sabal rosei (O.F.Cook) Becc., Webbia 2: 83 (1908).
Mexico (Sinaloa to Jalisco). 79 MXS MXN.
*Inodes rosei O.F.Cook, Bull. Torrey Bot. Club 28: 534 (1901). *Sabal uresana* var. *rosei* (O.F.Cook) I.M. Johnst., Proc. Calif. Acad. Sci., IV, 12: 995 (1924).

Sabal schwarzii (O.F.Cook) Becc. = **Sabal palmetto** (Walter) Lodd. ex Schult. & Schult.f.

Sabal serrulata (Michx.) Schult.f. = **Serenoa repens** (W.Bartram) Small

Sabal speciosa L.H.Bailey = **?**

Sabal taurina Mart. = **Sabal maritima** (Kunth) Burret

Sabal texana (O.F.Cook) Becc. = **Sabal mexicana** Mart.

Sabal umbraculifera Mart. = **Sabal palmetto** (Walter) Lodd. ex Schult. & Schult.f.

Sabal uresana Trel., Rep. (Annual) Missouri Bot. Gard. 12: 80 (1900). *Inodes uresana* (Trel.) O.F.Cook, Bull.

Torrey Bot. Club 28: 534 (1901).
Mexico (Chihuahua, Sonora). 79 MXE MXN.

Sabal uresana var. *rosei* (O.F.Cook) I.M.Johnst. = **Sabal rosei** (O.F.Cook) Becc.

Sabal viatoris L.H.Bailey = **Sabal palmetto** (Walter) Lodd. ex Schult. & Schult.f.

Sabal yapa C.Wright ex Becc., Webbia 2: 64 (1907).
Inodes yapa (C.Wright ex Becc.) Standl., Publ. Field Columbian Mus., Bot. Ser. 3: 219 (1930).
SE. Mexico to Belize, W. Cuba. 79 MXT 80 BLZ 81 CUB.
Sabal mayara Bartlett, Publ. Carnegie Inst. Wash. 461: 35 (1935).
Sabal peregrina L.H.Bailey, Gentes Herb. 6: 400 (1944).
Sabal yucatanica L.H.Bailey, Gentes Herb. 6: 418 (1944).

Sabal yucatanica L.H.Bailey = **Sabal yapa** C.Wright ex Becc.

Saguaster

Saguaster Kuntze = **Drymophloeus** Zipp.

Saguaster ambiguus (Becc.) Kuntze = **Ptychosperma ambiguum** (Becc.) Becc. ex Martelli

Saguaster angustifolius (Blume) Kuntze = **Ptychosperma angustifolium**

Saguaster appendiculatus (Blume) Kuntze = **Drymophloeus oliviformis** (Giseke) Mart.

Saguaster bifidus (Becc.) Kuntze = **Drymophloeus oliviformis** (Giseke) Mart.

Saguaster capitis-yorki (H.Wendl. & Drude) Kuntze = **Ptychosperma elegans** (R.Br.) Blume

Saguaster drudei (H.Wendl.) Kuntze = **Archontophoenix alexandrae** (F.Muell.) H.Wendl. & Drude

Saguaster elegans (R.Br.) Kuntze = **Ptychosperma elegans** (R.Br.) Blume

Saguaster gracilis (Labill.) Kuntze = **Ptychosperma gracile** Labill.

Saguaster leprosus (Becc.) Kuntze = **Drymophloeus oliviformis** (Giseke) Mart.

Saguaster macarthurii (H.Wendl. ex H.J.Veitch) Kuntze = **Ptychosperma macarthurii** (H.Wendl. ex H.J.Veitch) H.Wendl. ex Hook.f.

Saguaster normanbyi (F.Muell.) Kuntze = **Ptychosperma normanbyi**

Saguaster oliviformis (Giseke) Kuntze = **Drymophloeus oliviformis** (Giseke) Mart.

Saguaster oninensis (Becc.) Kuntze = **Drymophloeus oninensis** (Becc.) H.E.Moore

Saguaster pauciflora (H.Wendl.) Kuntze = **Balaka pauciflora** (H.Wendl.) H.E.Moore

Saguaster perbrevis (H.Wendl.) Kuntze = **Balaka seemannii** (H.Wendl.) Becc.

Saguaster pickeringii (H.Wendl.) Kuntze = **Veitchia vitiensis** (H.Wendl.) H.E.Moore

Saguaster propinquus (Becc.) Kuntze = **Ptychosperma propinquum** (Becc.) Becc. ex Martelli

Saguaster puniceus (Zipp. ex Blume) Kuntze = **Pinanga rumphiana** (Mart.) J.Dransf. & Govaerts

Saguaster saxatilis (Burm.f. ex Giseke) Kuntze = **Areca oryziformis** var. **saxatilis**

Saguaster seemannii (H.Wendl.) Kuntze = **Balaka seemannii** (H.Wendl.) Becc.

Saguaster tahitensis (H.Wendl.) Kuntze = **Balaka tahitensis** (H.Wendl.) Becc.

Saguaster vitiensis (H.Wendl.) Kuntze = **Veitchia vitiensis** (H.Wendl.) H.E.Moore

Saguerus

Saguerus Steck = **Arenga** Labill. ex DC.

Saguerus americanus H.Wendl. = **Mauritia flexuosa** L.f.

Saguerus australasicus H.Wendl. & Drude = **Arenga australasica** (H.Wendl. & Drude) T.S.Blake ex H.E.Moore

Saguerus brevipes (Becc.) Kuntze = **Arenga brevipes** Becc.

Saguerus gamuto Houtt. = **Arenga pinnata** (Wurmb) Merr.

Saguerus langbak Blume = **Arenga obtusifola**

Saguerus mindorensis (Becc.) O.F.Cook = **Arenga mindorensis** Becc.

Saguerus pinnatus Wurmb = **Arenga pinnata** (Wurmb) Merr.

Saguerus rumphii (Corrêa) Roxb. ex Ainslie = **Arenga pinnata** (Wurmb) Merr.

Saguerus saccharifer (Labill. ex DC.) Blume = **Arenga pinnata** (Wurmb) Merr.

Saguerus undulatifolius (Becc.) Kuntze = **Arenga undulatifolia** Becc.

Saguerus westerhoutii (Griff.) H.Wendl. & Drude = **Arenga westerhoutii** Griff.

Saguerus wightii (Griff.) H.Wendl. & Drude = **Arenga wightii** Griff.

Sagus

Sagus Steck = **Metroxylon** Rottb.

Sagus Rumph. ex Gaertn. = **Raphia** P.Beauv.

Sagus americana Poir. = **Metroxylon sagu** Rottb.

Sagus amicarum H.Wendl. = **Metroxylon amicarum** (H.Wendl.) Hook.f.

Sagus elata (Mart.) Reinw. ex Blume = **Pigafetta elata** (Mart.) H.Wendl.

Sagus farinifera Gaertn. = **Raphia farinifera** (Gaertn.) Hyl.

Sagus filaris Giseke = **Pigafetta filaris** (Giseke) Becc.

Sagus genuina Giseke = **Metroxylon sagu** Rottb.

Sagus genuina Giseke = **Metroxylon sagu** Rottb.

Sagus genuina Giseke = **Metroxylon sagu** Rottb.

Sagus genuina Giseke = **Metroxylon sagu** Rottb.

Sagus gomutus (Lour.) Perr. = **Arenga pinnata** (Wurmb) Merr.

Sagus hospita Kerch. = **?**

Sagus inermis Roxb. = **Metroxylon sagu** Rottb.

Sagus koenigii Griff. = **Metroxylon sagu** Rottb.

Sagus laevis Jack = **Metroxylon sagu** Rottb.

Sagus longispina (Giseke) Blume = **Metroxylon sagu** Rottb.

Sagus micracantha (Mart.) Blume = **Metroxylon sagu** Rottb.

Sagus microcarpa Zipp. ex Hall = **Pigafetta filaris** (Giseke) Becc.

Sagus microsperma Zipp. ex Hall = **Pigafetta filaris** (Giseke) Becc.

Sagus palma-pinus Gaertn. = **Raphia palma-pinus** (Gaertn.) Hutch.

Sagus pedunculata (P.Beauv.) Poir. = **Raphia faranifera**

Sagus raphia Poir. = **Raphia vinifera** P.Beauv.

Sagus ruffia Jacq. = **Raphia farinifera** (Gaertn.) Hyl.

Sagus rumphii Willd. = **Metroxylon sagu** Rottb.

Sagus sagu (Rottb.) H.Karst. = **Metroxylon sagu** Rottb.

Sagus spinosa Roxb. = **Metroxylon sagu** Rottb.

Sagus sylvestris (Giseke) Blume = **Metroxylon sagu** Rottb.

Sagus taedigera Mart. = **Raphia taedigera** (Mart.) Mart.

Sagus vinifera (P.Beauv.) Pers. = **Raphia vinifera** P.Beauv.

Sagus vitiensis (H.Wendl.) H.Wendl. = **Metroxylon vitiense** (H.Wendl.) Hook.f.

Salacca

Salacca Reinw., Syll. Ratisb. 2: 3 (1828).
20 species, Assam to SC. China and W. Malesia. 36 40 41 42.
Salakka Reinw. ex Blume, Catalogus: 112 (1823), orth. var.
Zalacca Rumph. ex Blume in J.J.Roemer & J.A. Schultes, Syst. Veg. 7: 1333 (1830), orth. var.
Lophospatha Burret, Notizbl. Bot. Gart. Berlin-Dahlem 15: 752 (1942).

Salacca affinis Griff., Calcutta J. Nat. Hist. 5: 9 (1845). W. Malesia. 42 BOR MLY SUM.
Salacca borneensis Becc., Malesia 3: 68 (1886).
Salacca affinis var. *borneensis* (Becc.) Furtado, Gard. Bull. Singapore 12: 399 (1949).
Salacca dubia Becc., Malesia 3: 68 (1886).

Salacca affinis var. *borneensis* (Becc.) Furtado = **Salacca affinis** Griff.

Salacca beccarii Hook.f. = **Salacca wallichiana** Mart.

Salacca blumeana Mart. = **Salacca zalacca** (Gaertn.) Voss

Salacca borneensis Becc. = **Salacca affinis** Griff.

Salacca clemensiana Becc., Philipp. J. Sci., C 4: 618 (1909).
Philippines to N. Borneo. 42 BOR 42 PHI.

Salacca conferta Griff. = **Eleiodoxa conferta** (Griff.) Burret

Salacca dolicholepis Burret, Notizbl. Bot. Gart. Berlin-Dahlem 15: 731 (1942).
Borneo (Sabah). 42 BOR.

Salacca dransfieldiana Mogea, Reinwardtia 9: 463 (1980).
Borneo (Kalimantan). 42 BOR.

Salacca dubia Becc. = **Salacca affinis** Griff.

Salacca edulis Reinw. = **Salacca zalacca** (Gaertn.) Voss

Salacca edulis var. *amboinensis* Becc. = **Salacca zalacca** (Gaertn.) Voss

Salacca flabellata Furtado, Gard. Bull. Singapore 12: 387 (1949).
Pen. Malaysia. 42 MLY.

Salacca glabrescens Griff., Calcutta J. Nat. Hist. 5: 14 (1845).
Pen. Thailand to Pen. Malaysia. 41 THA 42 MLY.

Salacca graciliflora Mogea, Fed. Mus. J. (Kuala Lumpur) 29: 6 (1984).
Pen. Malaysia. 42 MLY.

Salacca lophospatha J.Dransf. & Mogea, Principes 25: 180 (1981).
Borneo. 42 BOR.
*Lophospatha borneensis Burret, Notizbl. Bot. Gart. Berlin-Dahlem 15: 753 (1942).

Salacca macrostachya Griff. = **Salacca wallichiana** Mart.

Salacca magnifica Mogea, Reinwardtia 9: 468 (1980).
Borneo (Sarawak). 42 BOR.

Salacca minuta Mogea, Fed. Mus. J. (Kuala Lumpur) 29: 11 (1984).
Pen. Malaysia. 42 MLY.

Salacca multiflora Mogea, Fed. Mus. J. (Kuala Lumpur) 29: 13 (1984).
Pen. Malaysia. 42 MLY.

Salacca nitida W.Bull = ? [22]

Salacca ramosiana Mogea, Principes 30: 161 (1986).
Philippines to N. Borneo. 42 BOR PHI.

Salacca rumphii Wall. = **Salacca zalacca** (Gaertn.) Voss

Salacca rupicola J.Dransf., Bot. J. Linn. Soc. 81: 36 (1980).
Borneo (Sarawak). 42 BOR.

Salacca sarawakensis Mogea, Reinwardtia 9: 473 (1980).
Borneo (Sarawak). 42 BOR.

Salacca scortechinii Becc. = **Eleiodoxa conferta** (Griff.) Burret

Salacca secunda Griff., Calcutta J. Nat. Hist. 5: 12 (1845).
Assam to SC. China. 36 CHC 40 ASS 41 THA MYA.

Salacca stolonifera Hodel, Palm J. 134: 35 (1997).
Pen. Thailand. 41 THA.

Salacca sumatrana Becc., Ann. Roy. Bot. Gard. (Calcutta) 12(2): 80 (1918).
Sumatera. 42 SUM.

Salacca vermicularis Becc., Malesia 3: 66 (1886).
Borneo. 42 BOR.

Salacca wallichiana Mart., Hist. Nat. Palm. 3: 201 (1838).
Indo-China to Sumatera. 41 MYA THA VIE 42 MLY SUM.
Calamus zalacca Roxb., Fl. Ind. ed. 1832, 3: 773 (1832).
Salacca macrostachya Griff., Calcutta J. Nat. Hist. 5: 13 (1845).
Salacca beccarii Hook.f., Fl. Brit. India 6: 474 (1893).

Salacca zalacca (Gaertn.) Voss, Vilm. Blumengärtn. ed. 3, 1: 1152 (1895).
S. Sumatera to SW. Jawa, widely cultivated elsewere. 42 bor JAW lsi mly mol sul SUM.
*Calamus zalacca Gaertn., Fruct. Sem. Pl. 2: 267 (1791). *Salacca rumphii* Wall., Pl. Asiat. Rar. 3: t. 223 (1831), nom. illeg.
Salacca edulis Reinw., Syll. Ratisb. 2: 3 (1828).
Salacca blumeana Mart., Hist. Nat. Palm. 3: 202 (1838).
Calamus salakka Willd. ex Steud., Nomencl. Bot., ed. 2, 1: 252 (1840).

Salacca edulis var. *amboinensis* Becc., Ann. Roy. Bot. Gard. (Calcutta) 12(2): 74 (1918). *Salacca zalacca* var. *amboinensis* (Becc.) Mogea, Principes 26: 71 (1982).

Salacca zalacca var. *amboinensis* (Becc.) Mogea = **Salacca zalacca** (Gaertn.) Voss

Salakka

Salakka Reinw. ex Blume = **Salacca** Reinw.

Sargentia

Sargentia H.Wendl. & Drude ex Salomon = **Pseudophoenix** H.Wendl. ex Sarg.

Saribus

Saribus Blume = **Livistona** R.Br.

Saribus chinensis (Jacq.) Blume = **Livistona chinensis** (Jacq.) R.Br. ex Mart.

Saribus cochinchinensis (Lour.) Blume = **Livistona saribus** (Lour.) Merr. ex A.Chev.

Saribus hasseltii Hassk. = **Livistona saribus** (Lour.) Merr. ex A.Chev.

Saribus hoogendorpii (Teijsm. & Binn. ex Miq.) Kuntze = **Livistona saribus** (Lour.) Merr. ex A.Chev.

Saribus humilis (R.Br.) Kuntze = **Livistona humilis** R.Br.

Saribus inermis (R.Br.) Kuntze = **Livistona inermis** R.Br.

Saribus jenkensii (Griff.) Kuntze = **Livistona jenkinsiana** Griff.

Saribus kingianus (Becc.) Kuntze = **Pholidocarpus kingianus** (Becc.) Ridl.

Saribus mariae (F.Muell.) Kuntze = **Livistona mariae** F.Muell.

Saribus oliviformis Hassk. = **Livistona chinensis** var. **chinensis**

Saribus papuanus (Becc.) Kuntze = **Livistona papuana** Becc.

Saribus rotundifolius (Lam.) Blume = **Livistona rotundifolia** (Lam.) Mart.

Saribus speciosus (Kurz) Kuntze = **Livistona speciosa** Kurz

Saribus subglobosus Hassk. = **Livistona chinensis** var. **subglobosa** (Hassk.) Becc.

Sarinia

Sarinia O.F.Cook = **Attalea** Kunth

Sarinia funifera (Mart.) O.F.Cook = **Attalea funifera** Mart.

Satakentia

Satakentia H.E.Moore, Principes 13: 5 (1969).
1 species, Nansei-shoto. 38.

Satakentia liukiuensis (Hatus.) H.E.Moore, Principes 13: 5 (1969).
Nansei-shoto (Ishigaki, Iriomote). 38 NNS.
*Gulubia liukiuensis Hatus., Mem. Fac. Agric. Kagoshima Univ. 5(1): 39 (1964).

Satranala

Satranala J.Dransf. & Beentje, Kew Bull. 50: 87 (1995). 1 species, Madagascar. 29.

Satranala decussilvae Beentje & J.Dransf., Kew Bull. 50: 89 (1995). NE. Madagascar. 29 MDG.

Scheelea

Scheelea H.Karst. = **Attalea** Kunth

Scheelea amylacea Barb.Rodr. = **Attalea amylacea** (Barb.Rodr.) Zona

Scheelea anisitsiana Barb.Rodr. = **Attalea anisitsiana** (Barb.Rodr.) Zona

Scheelea attaleoides H.Karst. = **Attalea insignis** (Mart.) Drude

Scheelea bassleriana Burret = **Attalea bassleriana** (Burret) Zona

Scheelea blepharopus (Mart.) Burret = **Attalea blepharopus** Mart.

Scheelea brachyclada Burret = **Attalea bassleriana** (Burret) Zona

Scheelea butyracea (Mutis ex L.f.) H.Karst. ex H.Wendl. = **Attalea butyracea** (Mutis ex L.f.) Wess.Boer

Scheelea camopiensis Glassman = **Attalea camopiensis** (Glassman) Zona

Scheelea cephalotes (Poepp. ex Mart.) H.Karst. = **Attalea cephalotes**

Scheelea corumbaensis (Barb.Rodr.) Barb.Rodr. = **Attalea phalerata** Mart. ex Spreng.

Scheelea costaricensis Burret = **Attalea rostrata** Oerst.

Scheelea cubensis Burret = **? [81 CUB]**

Scheelea curvifrons L.H.Bailey = **Attalea osmantha** (Barb.Rodr.) Wess.Boer

Scheelea degranvillei Glassman = **Attalea degranvillei** (Glassman) Zona

Scheelea dryanderae Burret = **Attalea butyracea** (Mutis ex L.f.) Wess.Boer

Scheelea dubia (Mart.) Burret = **Attalea dubia** (Mart.) Burret

Scheelea excelsa Barb.Rodr. = **Attalea butyracea** (Mutis ex L.f.) Wess.Boer

Scheelea excelsa H.Karst. = **Attalea butyracea** (Mutis ex L.f.) Wess.Boer

Scheelea fairchildensis Glassman = **Attalea fairchildensis** (Glassman) Zona

Scheelea goeldiana (Huber) Burret = **Attalea insignis** (Mart.) Drude

Scheelea gomphococca (Mart.) Burret = **Attalea gomphococca**

Scheelea guianensis Glassman = **Attalea guianensis** (Glassman) Zona

Scheelea huebneri Burret = **Attalea huebneri** (Burret) Zona

Scheelea humboldtiana (Spruce) Burret = **Attalea butyracea** (Mutis ex L.f.) Wess.Boer

Scheelea imperialis auct. = **? [83 CLM]**

Scheelea insignis (Mart.) H.Karst. = **Attalea insignis** (Mart.) Drude

Scheelea kewensis Hook.f. = **Attalea kewensis** (Hook.f.) Zona

Scheelea lauromulleriana Barb.Rodr. = **Attalea lauromulleriana**

Scheelea leandroana Barb.Rodr. = **Attalea leandroana** (Barb.Rodr.) Zona

Scheelea liebmannii Becc. = **Attalea liebmannii** (Becc.) Zona

Scheelea lundellii Bartlett = **Attalea lundellii** (Bartlett) Zona

Scheelea macrocarpa H.Karst. = **Attalea macrocarpa** (H.Karst.) Linden

Scheelea macrolepis Burret = **Attalea macrolepis** (Burret) Wess.Boer

Scheelea magdalenica Dugand = **Attalea magdalenica** (Dugand) Zona

Scheelea maracaibensis (Mart.) Burret = **Attalea maracaibensis** Mart.

Scheelea maripa (Aubl.) H.Wendl. = **Attalea maripa** (Aubl.) Mart.

Scheelea maripensis Glassman = **Attalea maripensis** (Glassman) Zona

Scheelea martiana Burret = **Attalea phalerata** Mart. ex Spreng.

Scheelea microspadix Burret = **Attalea phalerata** Mart. ex Spreng.

Scheelea moorei Glassman = **Attalea moorei** (Glassman) Zona

Scheelea osmantha Barb.Rodr. = **Attalea osmantha** (Barb.Rodr.) Wess.Boer

Scheelea parviflora (Barb.Rodr.) Barb.Rodr. = **Attalea anisitsiana** (Barb.Rodr.) Zona

Scheelea passargei Burret = **Attalea macrocarpa** (H.Karst.) Linden

Scheelea phalerata (Mart. ex Spreng.) Burret = **Attalea phalerata** Mart. ex Spreng.

Scheelea plowmanii Glassman = **Attalea plowmanii** (Glassman) Zona

Scheelea preussii Burret = **Attalea rostrata** Oerst.

Scheelea princeps (Mart.) H.Karst. = **Attalea princeps** Mart.

Scheelea princeps var. *corumbaensis* Barb.Rodr. = **Attalea phalerata** Mart. ex Spreng.

Scheelea quadrisperma Barb.Rodr. = **Attalea anisitsiana** (Barb.Rodr.) Zona

Scheelea quadrisulcata Barb.Rodr. = **Attalea anisitsiana** (Barb.Rodr.) Zona

Scheelea regia H.Karst. = **Attalea butyracea** (Mutis ex L.f.) Wess.Boer

Scheelea rostrata (Oerst.) Burret = **Attalea rostrata** Oerst.

Scheelea salazarii Glassman = **Attalea salazarii** (Glassman) Zona

Scheelea stenorhyncha Burret = **Attalea bassleriana** (Burret) Zona

Scheelea tessmannii Burret = **Attalea peruviana** Zona

Scheelea tetrasticha (Drude) Burret = **Attalea maripa** (Aubl.) Mart.

Scheelea unguis G.Nicholson = **?**

Scheelea urbaniana Burret = **Attalea osmantha** (Barb. Rodr.) Wess.Boer

Scheelea wallisii (Huber) Burret = **Attalea butyracea** (Mutis ex L.f.) Wess.Boer

Scheelea weberbaueri Burret = **Attalea weberbaueri** (Burret) Zona

Scheelea wesselsboeri Glassman = **Attalea wesselsboeri** (Glassman) Zona

Scheelea zonensis L.H.Bailey = **Attalea rostrata** Oerst.

Schippia

Schippia Burret, Notizbl. Bot. Gart. Berlin-Dahlem 11: 867 (1933).
1 species, C. America. 80.

Schippia concolor Burret, Notizbl. Bot. Gart. Berlin-Dahlem 11: 868 (1933).
Belize to N. Guatemala. 80 BLZ GUA.

Schizospatha

Schizospatha Furtado = **Calamus** L.

Schizospatha setigera (Burret) Furtado = **Calamus anomalus** Burret

Schunda-pana

Schunda-Pana Adans. = **Caryota** L.

Sclerosperma

Sclerosperma G.Mann & H.Wendl., Trans. Linn. Soc. London 24: 427 (1864).
3 species, Ghana to Angola. 22 23 25.

Sclerosperma dubium Becc., Atti Soc. Tosc. Sci. Nat. Pisa Processi Verbali 44: 176 (1934).
Equatorial Guinea. 23 EQG.

Sclerosperma mannii H.Wendl., Trans. Linn. Soc. London 24: 427 (1964).
Ghana to Angola. 22 GHA NGA TOG 23 CMN GAB 26 ANG.

Sclerosperma walkeri A.Chev., Rev. Bot. Appl. Agric. Trop. 11: 237 (1931).
Gabon. 23 GAB.

Seaforthia

Seaforthia R.Br. = **Ptychosperma** Labill.

Seaforthia angustifolia Mart. = **?** [43 NWG]

Seaforthia appendiculata (Blume) Juss. ex Kunth = **Drymophloeus oliviformis** (Giseke) Mart.

Seaforthia blumei Juss. ex Kunth = **Drymophloeus oliviformis** (Giseke) Mart.

Seaforthia caesia (Blume) Mart. = **Pinanga caesia** Blume

Seaforthia calapparia (Blume) Mart. = **Actinorhytis calapparia** (Blume) H.Wendl. & Drude ex Scheff.

Seaforthia cochinchinensis (Blume) Mart. = **Pinanga sylvestris** (Lour.) Hodel

Seaforthia communis (Zipp. ex Blume) Mart. = **Drymophloeus litigiosus** (Becc.) H.E.Moore

Seaforthia coronata (Blume ex Mart.) Mart. = **Pinanga coronata** (Blume ex Mart.) Blume

Seaforthia costata (Blume) Mart. = **Pinanga coronata** (Blume ex Mart.) Blume

Seaforthia dicksonii (Roxb.) Mart. = **Pinanga dicksonii** (Roxb.) Blume

Seaforthia disticha (Roxb.) Mart. = **Pinanga disticha** (Roxb.) H.Wendl.

Seaforthia elegans Hook. = **Archontophoenix cunninghamiana** (H.Wendl.) H.Wendl. & Drude

Seaforthia elegans R.Br. = **Ptychosperma elegans** (R.Br.) Blume

Seaforthia furfuracea (Blume) Mart. = **Pinanga furfuracea** Blume

Seaforthia gracilis (Blume) Mart. = **Pinanga gracilis** Blume

Seaforthia inaequalis (Blume) Mart. = **Pinanga inaequalis** Blume

Seaforthia jaculatoria Mart. = **Drymophloeus oliviformis** (Giseke) Mart.

Seaforthia kuhlii (Blume) Mart. = **Pinanga coronata** (Blume ex Mart.) Blume

Seaforthia latisecta (Blume) Mart. = **Pinanga latisecta** Blume

Seaforthia malaiana Mart. = **Pinanga malaiana** (Mart.) Scheff.

Seaforthia minor (Blume) Mart. = **Pinanga minor** Blume

Seaforthia montana Blume ex Mart. = **Pinanga coronata** (Blume ex Mart.) Blume

Seaforthia oliviformis (Giseke) Mart. = **Drymophloeus oliviformis** (Giseke) Mart.

Seaforthia oriziformis (Gaertn.) Mart. = **Pinanga globulifera** (Lam.) Blume

Seaforthia patula (Blume) Mart. = **Pinanga patula** Blume

Seaforthia ptychosperma Mart. = **Ptychosperma gracile** Labill.

Seaforthia reinwardtiana Mart. = **Pinanga coronata** (Blume ex Mart.) Blume

Seaforthia robusta H.Wendl. = **Rhopalostylis baueri** var. **baueri**

Seaforthia rumphiana Mart. = **Pinanga rumphiana** (Mart.) J.Dransf. & Govaerts

Seaforthia salicifolia (Blume) Mart. = **Pinanga salicifolia** Blume

Seaforthia saxatilis (Burm.f. ex Giseke) Blume ex Mart. = **Areca oryziformis** var. **saxatilis**

Seaforthia sylvestris (Lour.) Blume ex Mart. = **Pinanga sylvestris** (Lour.) Hodel

Seaforthia veitchii auct. = **?** [50]

Seaforthia vestiaria (Giseke) Mart. = **Areca vestiaria** Giseke

Serenoa

Serenoa Hook.f. in G.Bentham & J.D.Hooker, Gen. Pl. 3: 1228 (1883).
1 species, SE. U.S.A. 78.
Diglossophyllum H.Wendl. ex Salomon, Palmen: 155 (1887).

Serenoa arborescens Sarg. = **Acoelorrhaphe wrightii** (Griseb. & H.Wendl.) H.Wendl. ex Becc.

Serenoa repens (W.Bartram) Small, J. New York Bot. Gard. 27: 193 (1926).
SE. U.S.A. 78 ALA FLA GEO LOU MSI SCA.

Corypha obliqua W.Bartram, Travels Carolina: 61 (1791).

**Corypha repens* W.Bartram, Travels Carolina: 61 (1791).

Chamaerops serrulata Michx., Fl. Bor.-Amer. 1: 239 (1803). *Sabal serrulata* (Michx.) Schult.f., Syst. Veg. 7: 1486 (1830). *Brahea serrulata* (Michx.) H.Wendl. in O.C.E.de Kerchove de Denterghem, Palmiers: 235 (1878). *Diglossophyllum serrulatum* (Michx.) H.Wendl. ex Salomon, Palmen: 155 (1887). *Serenoa serrulata* (Michx.) Hook.f. ex B.D.Jacks., Index Kew. 2: 884 (1895).

Serenoa serrulata (Michx.) Hook.f. ex B.D.Jacks. = **Serenoa repens** (W.Bartram) Small

Simpsonia

Simpsonia O.F.Cook = **Thrinax** L.f. ex Sw.

Simpsonia microcarpa (Sarg.) O.F.Cook = **Thrinax morrisii** H.Wendl.

Sindroa

Sindroa Jum. = **Orania** Zipp.

Sindroa longisquama Jum. = **Orania longisquama** (Jum.) J.Dransf. & N.W.Uhl

Siphokentia

Siphokentia Burret = **Hydriastele** H.Wendl. & Drude

Siphokentia beguinii Burret = **Hydriastele beguinii** (Burret) W.J.Baker & Loo

Siphokentia dransfieldii Hambali & al. = **Hydriastele dransfieldii** (Hambali & al.) W.J.Baker & Loo

Siphokentia pachypus Burret = **Hydriastele beguinii** (Burret) W.J.Baker & Loo

Slackia

Slackia Griff. = **Iguanura** Blume

Slackia geonomiformis (Mart.) Griff. = **Iguanura geonomiformis** Mart.

Slackia insignis Griff. = **Iguanura wallichiana** var. **wallichiana**

Socratea

Socratea H.Karst., Linnaea 28: 263 (1856).
5 species, C. & S. Trop. America. 80 82 83 84.
Metasocratea Dugand, Revista Acad. Colomb. Ci. Exact. 8: 389 (1951).

Socratea albolineata Steyerm. = **Socratea exorrhiza** (Mart.) H.Wendl.

Socratea altissima Burret = **? [82 VEN]**

Socratea durissima (Oerst.) H.Wendl. = **Socratea exorrhiza** (Mart.) H.Wendl.

Socratea elegans H.Karst. = **Socratea exorrhiza** (Mart.) H.Wendl.

Socratea exorrhiza (Mart.) H.Wendl., Bonplandia 8: 103 (1860).
C. & S. Trop. America. 80 COS NIC PAN 82 FRG GUY SUR VEN 83 BOL CLM ECU PER 84 BZN.
**Iriartea exorrhiza* Mart., Hist. Nat. Palm. 2: 36 (1824).
Iriartea orbignyana Mart., Hist. Nat. Palm. 3: 189 (1838). *Socratea orbignyana* (Mart.) H.Karst., Linnaea 28: 264 (1856).

Socratea elegans H.Karst., Linnaea 28: 264 (1856).

Iriartea durissima Oerst., Vidensk. Meddel. Dansk Naturhist. Foren. Kjøbenhavn 1858: 30 (1858). *Socratea durissima* (Oerst.) H.Wendl., Bonplandia 8: 103 (1860).

Iriartea philonotia Barb.Rodr., Enum. Palm. Nov.: 13 (1875). *Socratea philonotia* (Barb.Rodr.) Hook.f. in G.Bentham & J.D.Hooker, Gen. Pl. 3: 900 (1883).

Socratea macrochlamys Burret, Notizbl. Bot. Gart. Berlin-Dahlem 10: 918 (1930).

Socratea hoppii Burret, Notizbl. Bot. Gart. Berlin-Dahlem 11: 232 (1931).

Socratea gracilis Burret, Notizbl. Bot. Gart. Berlin-Dahlem 15: 1 (1940).

Socratea albolineata Steyerm., Fieldiana, Bot. 28(1): 9 (1951).

Socratea forgetiana (auct.) L.H.Bailey = **Ceratolobus forgetianus**

Socratea fusca H.Karst. = **Dictyocaryum fuscum** (H.Wendl.) H.Wendl.

Socratea gracilis Burret = **Socratea exorrhiza** (Mart.) H.Wendl.

Socratea hecatonandra (Dugand) R.Bernal, Kew Bull. 41: 152 (1986).
W. Colombia to NW. Ecuador. 83 CLM ECU.
**Metasocratea hecatonandra* Dugand, Revista Acad. Colomb. Ci. Exact. 8: 389 (1951).

Socratea hoppii Burret = **Socratea exorrhiza** (Mart.) H.Wendl.

Socratea macrochlamys Burret = **Socratea exorrhiza** (Mart.) H.Wendl.

Socratea montana R.Bernal & A.J.Hend., Brittonia 38: 55 (1986).
W. Colombia to NW. Ecuador. 83 CLM ECU.

Socratea orbignyana (Mart.) H.Karst. = **Socratea exorrhiza** (Mart.) H.Wendl.

Socratea philonotia (Barb.Rodr.) Hook.f. = **Socratea exorrhiza** (Mart.) H.Wendl.

Socratea rostrata Burret, Notizbl. Bot. Gart. Berlin-Dahlem 15: 31 (1940).
S. Colombia to N. Peru. 83 CLM ECU PER.

Socratea salazarii H.E.Moore, Principes 7: 112 (1963).
Peru to Northern Bolivia, Brazil (Acre). 83 BOL PER 84 BZN.

Solfia

Solfia Rech., Repert. Spec. Nov. Regni Veg. 4: 232 (1907).
1 species, Samoa. 60.

Solfia samoensis Rech., Repert. Spec. Nov. Regni Veg. 4: 233 (1907). *Drymophloeus samoensis* (Rech.) Becc. ex Martelli, Nuovo Giorn. Bot. Ital., n.s., 42: 44 (1935).
Samoa. 60 SAM.

Solfia whitmeeana (Becc.) Burret = **Drymophloeus whitmeeanus** Becc.

Sommieria

Sommieria Becc., Malesia 1: 66 (1877).
1 species, W. & C. New Guinea. 32.

Sommieria affinis Becc. = **Sommieria leucophylla** Becc.

Sommieria elegans Becc. = **Sommieria leucophylla** Becc.

Sommieria leucophylla Becc., Malesia 1: 67 (1877). W. & C. New Guinea. 43 NWG.

Sommieria elegans Becc., Malesia 1: 68 (1877).

Sommieria affinis Becc., Bot. Jahrb. Syst. 52: 37 (1914).

Spathoscaphe

Spathoscaphe Oerst. = **Chamaedorea** Willd.

Spathoscaphe arenbergiana (H.Wendl.) Oerst. = **Chamaedorea arenbergiana** H.Wendl.

Stachyophorbe

Stachyophorbe (Liebm.) Liebm. ex Klotzsch = **Chamaedorea** Willd.

Stachyophorbe cataractarum (Mart.) Liebm. ex Klotzsch = **Chamaedorea cataractarum** Mart.

Stachyophorbe deckeriana Klotzsch = **Chamaedorea deckeriana** (Klotzsch) Hemsl.

Stachyophorbe filipes O.F.Cook = **Chamaedorea oreophila** Mart.

Stachyophorbe montana Liebm. ex Oerst. = **Chamaedorea oreophila** Mart.

Stachyophorbe oreophila (Mart.) O.F.Cook = **Chamaedorea oreophila** Mart.

Stachyophorbe pygmaea (H.Wendl.) Oerst. = **Chamaedorea pygmaea** H.Wendl.

Stephanostachys

Stephanostachys Klotzsch ex Oerst. = **Chamaedorea** Willd.

Stephanostachys casperiana (Klotzsch) Oerst. = **Chamaedorea casperiana**

Stephanostachys martiana (H.Wendl.) Oerst. = **Chamaedorea cataractarum** Mart.

Stephanostachys tepejilote (Liebm.) Oerst. = **Chamaedorea tepejilote** Liebm.

Stephanostachys wendlandiana Oerst. = **Chamaedorea tepejilote** Liebm.

Stevensonia

Stevensonia Duncan ex Balf.f. = **Phoenicophorium** H.Wendl.

Stevensonia borsigiana (K.Koch) L.H.Bailey = **Phoenicophorium borsigianum** (K.Koch) Stuntz

Stevensonia grandifolia Duncan ex Balf.f. = **Phoenicophorium borsigianum** (K.Koch) Stuntz

Stevensonia viridifolia Duncan = **Verschaffeltia splendida** H.Wendl.

Strongylocaryum

Strongylocaryum Burret = **Ptychosperma** Labill.

Strongylocaryum brassii Burret = **Ptychosperma salomonense** Burret

Strongylocaryum latius Burret = **Ptychosperma salomonense** Burret

Strongylocaryum macranthum Burret = **Ptychosperma salomonense** Burret

Styloma

Styloma O.F.Cook = **Pritchardia** Seem. & H.Wendl.

Styloma arecina (Becc.) O.F.Cook = **Pritchardia arecina** Becc.

Styloma eriophora (Becc.) O.F.Cook = **Pritchardia minor** Becc.

Styloma eriostachya (Becc.) O.F.Cook = **Pritchardia lanigera** Becc.

Styloma gaudichaudii (Mart.) O.F.Cook = **Pritchardia martii** (Gaudich.) H.Wendl.

Styloma hillebrandii (Becc.) O.F.Cook = **Pritchardia hillebrandii** Becc.

Styloma insignis (Becc.) O.F.Cook = **Pritchardia hillebrandii** Becc.

Styloma lanigera (Becc.) O.F.Cook = **Pritchardia lanigera** Becc.

Styloma maideniana (Becc.) O.F.Cook = **Pritchardia maideniana** Becc.

Styloma martii (Gaudich.) O.F.Cook = **Pritchardia martii** (Gaudich.) H.Wendl.

Styloma minor (Becc.) O.F.Cook = **Pritchardia minor** Becc.

Styloma pacifica (Seem. & H.Wendl.) O.F.Cook = **Pritchardia pacifica** Seem. & H.Wendl.

Styloma pericularum (H.Wendl.) O.F.Cook = **Pritchardia pericularum** H.Wendl.

Styloma remota (Kuntze) O.F.Cook = **Pritchardia remota** (Kuntze) Becc.

Styloma rockiana (Becc.) O.F.Cook = **Pritchardia martii** (Gaudich.) H.Wendl.

Styloma thurstonii (F.Muell. & Drude) O.F.Cook = **Pritchardia thurstonii** F.Muell. & Drude

Styloma vuylstekeana (H.Wendl.) O.F.Cook = **Pritchardia vuylstekeana** H.Wendl.

Sublimia

Sublimia Comm. ex Mart. = **Hyophorbe** Gaertn.

Sublimia aevidaps Comm. ex Mart. = **Hyophorbe amaricaulis** Mart.

Sublimia amaricaulis Comm. ex Mart. = **Hyophorbe amaricaulis** Mart.

Sublimia areca Comm. ex Mart. = **Areca catechu** L.

Sublimia centennina Comm. ex Mart. = **Acanthophoenix rubra** (Bory) H.Wendl.

Sublimia palmicaulis Comm. ex Mart. = **Dictyosperma album** var. **album**

Sublimia vilicaulis Comm. ex Mart. = **Hyophorbe indica** Gaertn.

Syagrus

Syagrus Mart., Hist. Nat. Palm. 2: 129 (1826). 31 species, Lesser Antilles to S. Trop. America. 81 82 83 84 85.

Langsdorffia Raddi, Mem. Mat. Fis. Soc. Ital. 18: 347 (1820).

Platenia H.Karst., Linnaea 28: 250 (1856).

Rhyticocos Becc., Malpighia 1: 350 (1886).

Barbosa Becc., Malpighia 1: 339 (1887).

Arikuryroba Barb.Rodr., Pl. Jard. Rio de Janeiro 1: 5 (1891).

Arecastrum (Druce) Becc., Agric. Colon. 10: 446 (1916).

Arikury Becc., Agric. Colon. 10: 445 (1916).

Chrysallidosperma H.E.Moore, Principes 7: 109 (1963).

Syagrus acaulis (Drude) Becc. = **Syagrus comosa** (Mart.) Mart.

Syagrus allenii Glassman = **Syagrus orinocensis** (Spruce) Burret

Syagrus amadelpha (Barb.Rodr.) Frambach ex Dahlgren = **Butia paraguayensis** (Barb.Rodr.) L.H.Bailey

Syagrus amara (Jacq.) Mart. in A.D.d'Orbigny, Voy. Amér. Mér. 7(3): 132 (1847).
Lesser Antilles. 81 LEE WIN.
*Cocos amara Jacq., Select. Stirp. Amer. Hist.: 277 (1763). *Rhyticocos amara* (Jacq.) Becc., Malpighia 1: 353 (1886). *Calappa amara* (Jacq.) Kuntze, Revis. Gen. Pl. 2: 982 (1891).

Syagrus apaensis (Barb.Rodr.) Becc. = **Syagrus campylospatha** (Barb.Rodr.) Becc.

Syagrus archeri Glassman = **Butia archeri** (Glassman) Glassman

Syagrus arenicola (Barb.Rodr.) Frambach = **Butia paraguayensis** (Barb.Rodr.) L.H.Bailey

Syagrus argentea (Engel) Becc. = **Syagrus sancona** (Kunth) H.Karst.

Syagrus botryophora (Mart.) Mart. in A.D.d'Orbigny, Voy. Amér. Mér. 7(3): 133 (1847).
E. Brazil (S. Sergipe to N. Espírito Santo). 84 BZE BZL.
*Cocos botryophora Mart., Hist. Nat. Palm. 2: 118 (1826). *Calappa botryophora* (Mart.) Kuntze, Revis. Gen. Pl. 2: 982 (1891).

Syagrus brachyrhyncha Burret = **Syagrus cocoides** Mart.

Syagrus campicola (Barb.Rodr.) Becc. = **Butia campicola** (Barb.Rodr.) Noblick

Syagrus × campos-portoana (Bondar) Glassman, Rhodora 65: 260 (1963). S. coronata × S. romanzoffiana.
NE. Brazil. 84 BZE.
*Cocos × campos-portoana Bondar, Publ. Field Mus. Nat. Hist., Bot. Ser. 22: 460 (1942). *Arecastrum × campos-portoanum* (Bondar) A.D.Hawkes, Arq. Bot. Estado São Paulo, n.s., f.m., 2: 175 (1952).

Syagrus campylospatha (Barb.Rodr.) Becc., Agric. Colon. 10: 465 (1916).
E. Paraguay. 85 PAR.
*Cocos campylospatha Barb.Rodr., Palm. Hassler.: 9 (1900).
Cocos apaensis Barb.Rodr., Sert. Palm. Brasil. 1: 100 (1903). *Syagrus apaensis* (Barb.Rodr.) Becc., Agric. Colon. 10: 465 (1916).
Cocos hassleriana Barb.Rodr., Sert. Palm. Brasil. 1: 101 (1903). *Syagrus hassleriana* (Barb.Rodr.) Becc., Agric. Colon. 10: 466 (1916).

Syagrus capitata (Mart.) Glassman = **Butia capitata** (Mart.) Becc.

Syagrus cardenasii Glassman, Fieldiana, Bot. 31: 238 (1967).
SC. Bolivia. 83 BOL.

Syagrus catechucarpa (Barb.Rodr.) Becc. = **Syagrus picrophylla** Barb.Rodr.

Syagrus chavesiana (Barb.Rodr. ex Becc.) Barb.Rodr. = **Syagrus inajai** (Spruce) Becc.

Syagrus chiragua (H.Karst.) H.Wendl. = **Syagrus sancona** (Kunth) H.Karst.

Syagrus chloroleuca (Barb.Rodr.) Burret = **Cocos chloroleuca**

Syagrus cocoides Mart., Hist. Nat. Palm. 2: 130 (1826).
Guyana to Brazil. 82 GUY 84 BZC BZE BZN.
Cocos syagrus Drude in C.F.P.von Martius & auct. suc. (eds.), Fl. Bras. 3(2): 406 (1881).
Cocos weddellii Drude in C.F.P.von Martius & auct. suc. (eds.), Fl. Bras. 3(2): 411 (1881). *Calappa weddellii* (Drude) Kuntze, Revis. Gen. Pl. 2: 982 (1891).
Cocos drudei Becc., Malpighia 1: 445 (1887). *Syagrus drudei* (Becc.) Becc., Agric. Colon. 10: 466 (1916).
Calappa cocoides Kuntze, Revis. Gen. Pl. 2: 982 (1891).
Syagrus brachyrhyncha Burret, Notizbl. Bot. Gart. Berlin-Dahlem 13: 686 (1937).

Syagrus comosa (Mart.) Mart. in A.D.d'Orbigny, Voy. Amér. Mér. 7(3): 134 (1847).
Brazil. 84 BZC BZE BZL BZN.
*Cocos comosa Mart., Hist. Nat. Palm. 2: 121 (1826). *Calappa comosa* (Mart.) Kuntze, Revis. Gen. Pl. 2: 982 (1891).
Cocos plumosa Lodd. ex Loudon, Hort. Brit.: 381 (1830).
Cocos acaulis Drude in C.F.P.von Martius & auct. suc. (eds.), Fl. Bras. 3(2): 426 (1881). *Calappa acaulis* (Drude) Kuntze, Revis. Gen. Pl. 2: 982 (1891). *Syagrus acaulis* (Drude) Becc., Agric. Colon. 10: 465 (1916).
Cocos acaulis var. *glabra* Drude in C.F.P.von Martius & auct. suc. (eds.), Fl. Bras. 3(2): 426 (1881).

Syagrus coronata (Mart.) Becc., Agric. Colon. 10: 466 (1916).
NE. Brazil (to N. Minas Gerais). 84 BZE BZL.
*Cocos coronata Mart., Hist. Nat. Palm. 2: 115 (1826). *Calappa coronata* (Mart.) Kuntze, Revis. Gen. Pl. 2: 982 (1891).
Cocos quinquefaria Barb.Rodr., Palm. Hassler.: 13 (1900). *Syagrus quinquefaria* (Barb.Rodr.) Becc., Agric. Colon. 10: 467 (1916).
Glaziova treubiana Becc., Ann. Jard. Bot. Buitenzorg, Suppl. 3: 791 (1910). *Syagrus treubiana* (Becc.) Becc., Agric. Colon. 10: 468 (1916).

Syagrus × costae Glassman, Fieldiana, Bot. 32: 244 (1970). S. coronata × S. oleracea.
NE. Brazil. 84 BZE.

Syagrus drudei (Becc.) Becc. = **Syagrus cocoides** Mart.

Syagrus duartei Glassman, Fieldiana, Bot. 31: 289 (1968).
Brazil (Minas Gerais: Serra do Cipó). 84 BZL.

Syagrus dyeriana (Barb.Rodr.) Becc. = **Butia paraguayensis** (Barb.Rodr.) L.H.Bailey

Syagrus ecuadorensis Becc. = **Syagrus sancona** (Kunth) H.Karst.

Syagrus edulis (Barb.Rodr.) Frambach ex Dahlgren = **Cocos edulis**

Syagrus eriospatha (Mart. ex Drude) Glassman = **Butia eriospatha** (Mart. ex Drude) Becc.

Syagrus × fairchildensis Glassman = **Butia capitata** × **Syagrus romanzoffiana**

Syagrus flexuosa (Mart.) Becc., Agric. Colon. 10: 466 (1916).
WC. & E. Brazil. 84 BZC BZE BZL.
Cocos campestris Mart., Hist. Nat. Palm. 2: 121 (1826). *Calappa campestris* (Mart.) Kuntze, Revis. Gen. Pl. 2: 982 (1891).
*Cocos flexuosa Mart., Hist. Nat. Palm. 2: 120 (1826). *Calappa flexuosa* (Mart.) Kuntze, Revis. Gen. Pl. 2: 982 (1891).

Cocos urbaniana Dammer, Bot. Jahrb. Syst. 31(70): 22 (1902). *Syagrus urbaniana* (Dammer) Becc., Agric. Colon. 10: 468 (1916).

Syagrus getuliana (Bondar) Glassman = **Syagrus macrocarpa** Barb.Rodr.

Syagrus glaucescens Glaz. ex Becc., Agric. Colon. 10: 470 (1916).
Brazil (Minas Gerais: Serra da Diamantina). 84 BZL.

Syagrus glazioviana (Dammer) Becc. = **Syagrus petraea** (Mart.) Becc.

Syagrus gomesii Glassman = **Syagrus oleracea** (Mart.) Becc.

Syagrus graminifolia (Drude) Becc., Agric. Colon. 10: 466 (1916).
Brazil (S. Goiás, Mato Grosso do Sul) to E. Paraguay. 84 BZC 85 PAR.
   *Cocos graminifolia Drude in C.F.P.von Martius & auct. suc. (eds.), Fl. Bras. 3(2): 415 (1881). *Calappa graminifolia* (Drude) Kuntze, Revis. Gen. Pl. 2: 982 (1891).
   Cocos lilliputiana Barb.Rodr., Palm. Hassler.: 5 (1900). *Syagrus lilliputiana* (Barb.Rodr.) Becc., Agric. Colon. 10: 467 (1916).

Syagrus harleyi Glassman, Phytologia 39: 401 (1978).
Brazil (Bahia). 84 BZE.

Syagrus hassleriana (Barb.Rodr.) Becc. = **Syagrus campylospatha** (Barb.Rodr.) Becc.

Syagrus hatschbachii Glassman = **Butia microspadix** Burret

Syagrus hoehnei Burret = **Lytocaryum hoehnei** (Burret) Toledo

Syagrus inajai (Spruce) Becc., Agric. Colon. 10: 467 (1916).
N. South America to N. & NE. Brazil. 82 FRG GUY SUR 84 BZE BZN.
   *Maximiliana inajai Spruce, J. Linn. Soc., Bot. 11: 163 (1869). *Cocos inajai* (Spruce) Trail, J. Bot. 15: 79 (1877).
   Cocos aequatorialis Barb.Rodr., Enum. Palm. Nov.: 38 (1875).
   Cocos speciosa Barb.Rodr., Enum. Palm. Nov.: 38 (1875). *Calappa speciosa* (Barb.Rodr.) Kuntze, Revis. Gen. Pl. 2: 982 (1891).
   Cocos chavesiana Barb.Rodr. ex Becc., Malpighia 1: 445 (1887). *Syagrus chavesiana* (Barb.Rodr. ex Becc.) Barb.Rodr., Vellosia 1: 52 (1888).
   Cocos equatorialis Barb.Rodr., Palm. Hassler.: 38 (1900).

Syagrus insignis (Drude) Becc. = **Lytocaryum weddellianum** (H.Wendl.) Toledo

Syagrus leptospatha Burret, Notizbl. Bot. Gart. Berlin-Dahlem 15: 105 (1940).
Brazil (Mato Grosso). 84 BZC.

Syagrus lilliputiana (Barb.Rodr.) Becc. = **Syagrus graminifolia** (Drude) Becc.

Syagrus loefgrenii Glassman = **Syagrus petraea** (Mart.) Becc.

Syagrus macrocarpa Barb.Rodr., Prot.-App. Enum. Palm. Nov.: 46 (1879). *Cocos marocarpa* (Barb.Rodr.) Barb.Rodr., Palmiers: 26 (1882).
Brazil (S. Espírito Santo, SE. Minas Gerais). 84 BZL.
   Cocos procopiana Glaz. ex Drude in C.F.P.von Martius & auct. suc. (eds.), Fl. Bras. 3(2): 412 (1881). *Calappa procopiana* (Glaz. ex Drude) Kuntze, Revis. Gen. Pl. 2: 982 (1891).

Cocos getuliana Bondar, Bol. Inst. Centr. Fomento Econ. Bahia 9: 35 (1941). *Barbosa getuliana* (Bondar) A.D.Hawkes, Arq. Bot. Estado São Paulo, n.s., f.m., 2: 177 (1952). *Syagrus getuliana* (Bondar) Glassman, Rhodora 65: 260 (1963).

Syagrus × mataforme (Bondar) A.D.Hawkes, Arq. Bot. Estado São Paulo, n.s., f.m., 2: 178 (1952). S. coronata × S. vagans.
NE. Brazil. 84 BZE.
   *Cocos × mataforme Bondar, Publ. Field Mus. Nat. Hist., Bot. Ser. 22: 459 (1942).

Syagrus mendanhensis Glassman = **Syagrus pleioclada** Burret

Syagrus microphylla Burret, Repert. Spec. Nov. Regni Veg. 32: 111 (1933).
Brazil (Bahia). 84 BZE.

Syagrus mikaniana (Mart.) Mart. = **Syagrus pseudococos** (Raddi) Glassman

Syagrus × nabonnandii (Prosch.) Demoly = **? Butyagrus nabonnandii**

Syagrus oleracea (Mart.) Becc., Agric. Colon. 10: 467 (1916).
E. Brazil to E. Paraguay. 84 BZE BZL 85 PAR.
   *Cocos oleracea Mart., Hist. Nat. Palm. 2: 117 (1826). *Calappa oleracea* (Mart.) Kuntze, Revis. Gen. Pl. 2: 982 (1891).
   Cocos oleracea var. *platyphylla* Drude in C.F.P.von Martius & auct. suc. (eds.), Fl. Bras. 3(2): 417 (1881).
   Cocos picrophylla Barb.Rodr. ex Becc., Malpighia 1: 448 (1889).
   Syagrus gomesii Glassman, Fieldiana, Bot. 32: 22 (1968).

Syagrus orinocensis (Spruce) Burret, Notizbl. Bot. Gart. Berlin-Dahlem 13: 695 (1937).
Colombia to Venezuela. 82 VEN 83 CLM.
   *Cocos orinocensis Spruce, J. Linn. Soc., Bot. 11: 161 (1871). *Calappa orinocensis* (Spruce) Kuntze, Revis. Gen. Pl. 2: 982 (1891). *Maximiliana orinocensis* (Spruce) Speg., Physis (Buenos Aires) 3: 170 (1917).
   Syagrus stenopetala Burret, Notizbl. Bot. Gart. Berlin-Dahlem 11: 322 (1932).
   Syagrus allenii Glassman, Fieldiana, Bot. 31: 285 (1968).

Syagrus paraguayensis (Barb.Rodr.) Glassman = **Butia paraguayensis** (Barb.Rodr.) L.H.Bailey

Syagrus petraea (Mart.) Becc., Agric. Colon. 10: 467 (1916).
Brazil to E. Bolivia and E. Paraguay. 83 BOL 84 BZC BZE BZL BZN 85 PAR.
   *Cocos petraea Mart. in A.D.d'Orbigny, Voy. Amér. Mér. 7(3): 100 (1844). *Calappa petraea* (Mart.) Kuntze, Revis. Gen. Pl. 2: 982 (1891).
   Cocos rupestris Barb.Rodr., Prot.-App. Enum. Palm. Nov.: 45 (1879).
   Cocos glazioviana Dammer, Bot. Jahrb. Syst. 31(70): 21 (1902). *Syagrus glazioviana* (Dammer) Becc., Agric. Colon. 10: 466 (1916).
   Syagrus loefgrenii Glassman, Fieldiana, Bot. 31: 240 (1967).
   Syagrus rachidii Glassman, Fieldiana, Bot. 31: 245 (1967).

Syagrus picrophylla Barb.Rodr., Prot.-App. Enum. Palm. Nov.: 45 (1879).
Brazil (S. Bahia to Rio de Janeiro). 84 BZE BZL.

Cocos catechucarpa Barb.Rodr., Contr. Jard. Bot. Rio de Janeiro 1: 41 (1901). *Syagrus catechucarpa* (Barb.Rodr.) Becc., Agric. Colon. 10: 465 (1916).

Syagrus pleioclada Burret, Repert. Spec. Nov. Regni Veg. 32: 110 (1933).
Brazil (Minas Gerais, São Paulo). 84 BZL.
Syagrus mendanhensis Glassman, Fieldiana, Bot. 31: 298 (1968).

Syagrus pseudococos (Raddi) Glassman, Fieldiana, Bot. 32: 233 (1970).
SE. Brazil. 84 BZL.
Langsdorffia pseudococos Raddi, Mem. Mat. Fis. Soc. Ital. 18: 347 (1820).
Cocos mikaniana Mart., Hist. Nat. Palm. 2: 128 (1826). *Syagrus mikaniana* (Mart.) Mart. in A.D.d'Orbigny, Voy. Amér. Mér. 7(3): 133 (1847). *Calappa mikaniana* (Mart.) Kuntze, Revis. Gen. Pl. 2: 982 (1891).
Barbosa pseudococos Becc., Malpighia 1: 352 (1887).

Syagrus purusana (Huber) Burret = **Cocos purusana**

Syagrus quinquefaria (Barb.Rodr.) Becc. = **Syagrus coronata** (Mart.) Becc.

Syagrus rachidii Glassman = **Syagrus petraea** (Mart.) Becc.

Syagrus romanzoffiana (Cham.) Glassman, Fieldiana, Bot. 31: 382 (1968).
Brazil to NE. Argentina. 83 BOL? 84 BZC BZE BZL BZS 85 AGE PAR URU.
Cocos romanzoffiana Cham., Choris Voy. Pittor. (Chili): 5, t. 6 (1822). *Calappa romanzoffiana* (Cham.) Kuntze, Revis. Gen. Pl. 2: 982 (1891). *Arecastrum romanzoffianum* (Cham.) Becc., Agric. Colon. 10: 455 (1916).
Cocos australis Mart. in A.D.d'Orbigny, Voy. Amér. Mér. 7(3): 95 (1844). *Calappa australis* (Mart.) Kuntze, Revis. Gen. Pl. 2: 982 (1891).
Cocos plumosa Hook.f., Bot. Mag. 86: t. 5180 (1860). *Calappa plumosa* (Hook.f.) Kuntze, Revis. Gen. Pl. 2: 982 (1891).
Cocos datil Drude & Griseb., Abh. Königl. Ges. Wiss. Göttingen 24: 283 (1879). *Calappa datil* (Drude & Griseb.) Kuntze, Revis. Gen. Pl. 2: 982 (1891).
Cocos geriba Barb.Rodr., Prot.-App. Enum. Palm. Nov.: 43 (1879).
Cocos acrocomioides Drude in C.F.P.von Martius & auct. suc. (eds.), Fl. Bras. 3(2): 409 (1881). *Calappa acrocomioides* (Drude) Kuntze, Revis. Gen. Pl. 2: 982 (1891).
Cocos martiana Drude & Glaz. in C.F.P.von Martius & auct. suc. (eds.), Fl. Bras. 3(2): 418 (1881). *Calappa martiana* (Drude & Glaz.) Kuntze, Revis. Gen. Pl. 2: 982 (1891).
Cocos australis Drude & Brandt, Gartenflora 38: 451 (1889).
Cocos arechavaletana Barb.Rodr., Contr. Jard. Bot. Rio de Janeiro 1: 43 (1901).

Syagrus ruschiana (Bondar) Glassman, Rhodora 65: 261 (1963).
Brazil (Espírito Santo to E. Minas Gerais). 84 BZL.
Cocos ruschiana Bondar, Bol. Inst. Centr. Fomento Econ. Bahia 9: 45 (1941). *Arikuryroba ruschiana* (Bondar) Toledo, Arq. Bot. Estado São Paulo, n.s., f.m., 2: 6 (1944).

Syagrus sancona (Kunth) H.Karst., Linnaea 28: 247 (1856).
W. South America to W. Venezuela and N. Brazil. 82 VEN 83 BOL CLM ECU PER 84 BZN.

Oreodoxa sancona Kunth in F.W.H.von Humboldt, A.J.A.Bonpland & C.S.Kunth, Nov. Gen. Sp. 1: 304 (1816). *Oenocarpus sancona* (Kunth) Spreng., Syst. Veg. 2: 140 (1825). *Palma sancona* (Kunth) Kunth, Enum. Pl. 3: 182 (1841). *Cocos sancona* (Kunth) Hook.f., Rep. Progr. Condition Roy. Bot. Gard. Kew 1882: 72 (1884). *Calappa sancona* (Kunth) Kuntze, Revis. Gen. Pl. 2: 982 (1891).
Platenia chiragua H.Karst., Linnaea 28: 250 (1856). *Syagrus chiragua* (H.Karst.) H.Wendl. in O.C.E.de Kerchove de Denterghem, Palmiers: 257 (1878). *Cocos chiragua* (H.Karst.) Becc., Malpighia 1: 446 (1887).
Cocos argentea Engel, Linnaea 33: 690 (1865). *Syagrus argentea* (Engel) Becc., Agric. Colon. 10: 465 (1916). *Butia argentea* (Engel) Nehrl., Amer. Eagle 24(17): 1 (5 Sept. 1929).
Syagrus ecuadorensis Becc., Agric. Colon. 10: 469 (1916).
Syagrus tessmannii Burret, Repert. Spec. Nov. Regni Veg. 32: 106 (1933).

Syagrus sapida (Barb.Rodr.) Becc. = **Cocos sapida**

Syagrus schizophylla (Mart.) Glassman, Fieldiana, Bot. 31: 386 (1968).
NE. Brazil (to N. Espírito Santo). 84 BZE BZL.
Cocos aricui Wied-Neuw., Reise Bras. 1: 272 (1820), nom. inval.
Cocos schizophylla Mart., Hist. Nat. Palm. 2: 119 (1826). *Calappa schizophylla* (Mart.) Kuntze, Revis. Gen. Pl. 2: 982 (1891). *Arikury schizophylla* (Mart.) Becc., Agric. Colon. 10: 445 (1916). *Arikuryroba schizophylla* (Mart.) L.H.Bailey, Gentes Herb. 2: 196 (1930).
Arikuryroba capanemae Barb.Rodr., Pl. Jard. Rio de Janeiro 1: 5 (1891). *Cocos arikuryroba* Barb.Rodr., Palm. Matogross. Nov.: 25 (1898). *Cocos capanemae* (Barb.Rodr.) Drude in H.G.A.Engler & K.A.E.Prantl (eds.), Nat. Pflanzenfam., Nachtr. 2: 56 (1900).

Syagrus smithii (H.E.Moore) Glassman, Fieldiana, Bot. 31: 231 (1970).
SE. Colombia to N. Peru and N. Brazil. 83 CLM PER 84 BZN.
Chrysallidosperma smithii H.E.Moore, Principes 7: 110 (1963).

Syagrus stenopetala Burret = **Syagrus orinocensis** (Spruce) Burret

Syagrus stratincola Wess.Boer, in Fl. Suriname 5: 170 (1965).
Guianas. 82 FRG SUR 84 BZN?

Syagrus × teixeiriana Glassman = **?** [84 BZL]

Syagrus tessmannii Burret = **Syagrus sancona** (Kunth) H.Karst.

Syagrus × tostana (Bondar) Glassman, Rhodora 65: 261 (1963). S. coronata × S. schizophylla.
NE. Brazil. 84 BZE.
Cocos × tostana Bondar, Publ. Field Mus. Nat. Hist., Bot. Ser. 22: 458 (1942). *Arikuryroba × tostana* (Bondar) A.D.Hawkes, Arq. Bot. Estado São Paulo, n.s., f.m., 2: 175 (1952).

Syagrus treubiana (Becc.) Becc. = **Syagrus coronata** (Mart.) Becc.

Syagrus urbaniana (Dammer) Becc. = **Syagrus flexuosa** (Mart.) Becc.

Syagrus vagans (Bondar) A.D.Hawkes, Arq. Bot. Estado São Paulo, n.s., f.m., 2: 178 (1952).

Brazil (Bahia to NE. Minas Gerais). 84 BZE BZL.
Cocos vagans Bondar, Publ. Field Mus. Nat. Hist.,
Bot. Ser. 22: 457 (1942).

Syagrus wallisii Linden = ? [83 CLM]

Syagrus weddelliana (H.Wendl.) Becc. = **Lytocaryum weddellianum** (H.Wendl.) Toledo

Syagrus werdermannii Burret, Repert. Spec. Nov.
Regni Veg. 32: 109 (1933).
Brazil (S. Bahia: N. Serra do Espinhaço). 84 BZE.

Syagrus wildemaniana (Barb.Rodr.) Frambach ex
Dahlgren = **Butia paraguayensis** (Barb.Rodr.)
L.H.Bailey

Syagrus yatay (Mart.) Glassman = **Butia yatay** (Mart.)
Becc.

Syagrus yungasensis M.Moraes, Novon 6: 89 (1996).
Bolivia. 83 BOL.

Symphyogyne

Symphyogyne Burret = **Maxburretia** Furtado

Symphyogyne gracilis Burret = **Maxburretia gracilis**
(Burret) J.Dransf.

Symphyogyne rupicola (Ridl.) Burret = **Maxburretia
rupicola** (Ridl.) Furtado

Synechanthus

Synechanthus H.Wendl., Bot. Zeitung (Berlin) 16: 145
(1858).
2 species, Mexico to Ecuador. 79 80 83.

Synechanthus angustifolius H.Wendl. = **Synechanthus
warscewiczianus** H.Wendl.

Synechanthus ecuadorensis Burret = **Synechanthus
warscewiczianus** H.Wendl.

Synechanthus fibrosus (H.Wendl.) H.Wendl., Bot.
Zeitung (Berlin) 16: 145 (1858).
Mexico (Veracruz, Oaxaca, Chiapas) to C. America.
79 MXG MXS MXT 80 BLZ COS GUA HON
NIC.
Chamaedorea fibrosa H.Wendl., Index Palm.: 57
(1854). *Rathea fibrosa* (H.Wendl.) H.Karst.,
Wochenschr. Gärtnerei Pflanzenk. 1: 377 (1858).
Collinia fibrosa (H.Wendl.) Oerst., Vidensk.
Meddel. Dansk Naturhist. Foren. Kjøbenhavn 1858:
5 (1858).
Synechanthus mexicanus L.H.Bailey ex H.E.Moore,
Gentes Herb. 8: 199 (1949).

Synechanthus mexicanus L.H.Bailey ex H.E.Moore =
Synechanthus fibrosus (H.Wendl.) H.Wendl.

Synechanthus panamensis H.E.Moore = **Synechanthus
warscewiczianus** H.Wendl.

Synechanthus warscewiczianus H.Wendl., Bot.
Zeitung (Berlin) 16: 145 (1858).
C. America to W. Ecuador. 80 COS NIC PAN 83 CLM
ECU.
Reineckia triandra H.Karst., Wochenschr. Gärtnerei
Pflanzenk. 1: 349 (1858).
Synechanthus angustifolius H.Wendl., Wochenschr.
Gärtnerei Pflanzenk. 2: 15 (1859).
Synechanthus ecuadorensis Burret, Notizbl. Bot. Gart.
Berlin-Dahlem 13: 339 (1936).
Synechanthus panamensis H.E.Moore, Gentes Herb. 8:
201 (1949).

Taenianthera

Taenianthera Burret = **Geonoma** Willd.

Taenianthera acaulis (Mart.) Burret = **Geonoma
macrostachys** var. **acaulis** (Mart.) Andrew Hend.

Taenianthera camana (Trail) Burret = **Geonoma
camana** Trail

Taenianthera dammeri (Huber) Burret = **Geonoma
macrostachys** var. **poiteauana** (Kunth) A.J.Hend.

Taenianthera gracilis Burret = **Geonoma macrostachys**
var. **acaulis** (Mart.) Andrew Hend.

Taenianthera lagesiana (Dammer) Burret = **Geonoma
camana** Trail

Taenianthera lakoi Burret = **Geonoma macrostachys**
var. **poiteauana** (Kunth) A.J.Hend.

Taenianthera macrostachys (Mart.) Burret = **Geonoma
macrostachys** Mart.

Taenianthera minor Burret = **Geonoma macrostachys**
var. **acaulis** (Mart.) Andrew Hend.

Taenianthera multisecta Burret = **Geonoma jussieuana**
Mart.

Taenianthera oligosticha Burret = **Geonoma
macrostachys** var. **acaulis** (Mart.) Andrew Hend.

Taenianthera tamandua (Trail) Burret = **Geonoma
macrostachys** var. **macrostachys**

Taenianthera tapajotensis (Trail) Burret = **Geonoma
macrostachys** var. **acaulis** (Mart.) Andrew Hend.

Taenianthera weberbaueri Burret = **Geonoma
jussieuana** Mart.

Taliera

Taliera Mart. = **Corypha** L.

Taliera bengalensis Spreng. = **Corypha taliera** Roxb.

Taliera elata (Roxb.) Wall. = **Corypha utan** Lam.

Taliera gembanga Blume = **Corypha utan** Lam.

Taliera sylvestris Blume = **Corypha utan** Lam.

Taliera tali Mart. ex Blume = **Corypha taliera** Roxb.

Taveunia

Taveunia Burret = **Cyphosperma** H.Wendl. ex Hook.f.

Taveunia tanga H.E.Moore = **Cyphosperma tanga**
(H.E.Moore) H.E.Moore

Taveunia trichospadix Burret = **Cyphosperma
trichospadix** (Burret) H.E.Moore

Tectiphiala

Tectiphiala H.E.Moore, Gentes Herb. 11: 285 (1978).

Tectiphiala ferox H.E.Moore, Gentes Herb. 11: 285
(1978).
1 species, SC. Mauritius. 29 MAU.

Temenia

Temenia O.F.Cook = **Attalea** Kunth

Temenia regia (Mart.) O.F.Cook = **Attalea maripa**
(Aubl.) Mart.

Tessmanniodoxa

Tessmanniodoxa Burret = **Chelyocarpus** Dammer

Tessmanniodoxa chuco (Mart.) Burret = **Chelyocarpus
chuco** (Mart.) H.E.Moore

Tessmanniodoxa dianeura (Burret) Burret = **Chelyocarpus dianeurus** (Burret) H.E.Moore

Tessmanniophoenix

Tessmanniophoenix Burret = **Chelyocarpus** Dammer

Tessmanniophoenix chuco (Mart.) Burret = **Chelyocarpus chuco** (Mart.) H.E.Moore

Tessmanniophoenix dianeura Burret = **Chelyocarpus dianeurus** (Burret) H.E.Moore

Tessmanniophoenix longibracteata Burret = **Chelyocarpus ulei** Dammer

Tessmanniophoenix wallisii (H.Wendl.) Burret = **Acanthorrhiza wallisii**

Teysmannia

Teysmannia Rchb.f. & Zoll. = **Johannesteijsmannia** H.E.Moore

Teysmannia altifrons Rchb.f. & Zoll. = **Johannesteijsmannia altifrons** (Rchb.f. & Zoll.) H.E.Moore

Thrinax

Thrinax L.f. ex Sw., Prodr.: 57 (1788).
5 species, S. Florida, SE. Mexico to Honduras, Caribbean. 78 79 80 81.
Porothrinax H.Wendl. ex Griseb., Cat. Pl. Cub.: 221 (1866).
Simpsonia O.F.Cook, Science, n.s., 85: 333 (1937).

Thrinax acuminata Griseb. & H.Wendl. ex Sarg. = **Coccothrinax miraguama** subsp. **miraguama**

Thrinax altissima N.Taylor = **Coccothrinax argentata** (Jacq.) L.H.Bailey

Thrinax arborea Hook.f. = **?**

Thrinax argentea Lodd. ex Schult. & Schult.f. = **Coccothrinax argentea** (Lodd. ex Schult. & Schult.f.) Sarg. ex Becc.

Thrinax aurantia Fulchir. ex Schult. & Schult.f. = **?**

Thrinax aurata Kunth = **Thrinax radiata** Lodd. ex Schult. & Schult.f.

Thrinax bahamensis O.F.Cook = **Thrinax morrisii** H.Wendl.

Thrinax barbadensis Lodd. ex Mart. = **Coccothrinax barbadensis** (Lodd. ex Mart.) Becc.

Thrinax brasiliensis (Mart.) Mart. = **Trithrinax brasiliensis** Mart.

Thrinax chuco Mart. = **Chelyocarpus chuco** (Mart.) H.E.Moore

Thrinax compacta (Griseb. & H.Wendl.) Borhidi & O.Muñiz = **Hemithrinax compacta** (Griseb. & H.Wendl.) Hook.f.

Thrinax crinita Griseb. & H.Wendl. ex C.H.Wright = **Coccothrinax crinita** (Griseb. & H.Wendl. ex C.H.Wright) Becc.

Thrinax drudei Becc. = **Thrinax morrisii** H.Wendl.

Thrinax ekmaniana (Burret) Borhidi & O.Muñiz, Acta Bot. Hung. 31: 227 (1985).
C. Cuba (Las Villas). 81 CUB.
Hemithrinax ekmaniana Burret, Kongl. Svenska Vetenskapsakad. Handl., III, 6(7): 9 (1929).

Thrinax ekmanii Burret = **Thrinax morrisii** H.Wendl.

Thrinax elegans Schult. & Schult.f. = **Thrinax radiata** Lodd. ex Schult. & Schult.f.

Thrinax elegantissima Hook.f. = **Thrinax radiata** Lodd. ex Schult. & Schult.f.

Thrinax excelsa Lodd. ex Mart., Hist. Nat. Palm. 3: 320 (1853).
Jamaica (John Crow Mts.). 81 JAM.
Thrinax rex Britton & Harris, Bull. Torrey Bot. Club 37: 352 (1910).

Thrinax ferruginea Lodd. ex Mart. = **?** [81 JAM]

Thrinax floridana Sarg. = **Thrinax radiata** Lodd. ex Schult. & Schult.f.

Thrinax garberi Chapm. = **Coccothrinax argentata** (Jacq.) L.H.Bailey

Thrinax gracilis Schult. & Schult.f. = **Thrinax radiata** Lodd. ex Schult. & Schult.f.

Thrinax graminifolia H.Wendl. = **Coccothrinax argentea** (Lodd. ex Schult. & Schult.f.) Sarg. ex Becc.

Thrinax grandis Kerch. = **?** [81 CUB]

Thrinax harrisiana Becc. = **Thrinax parviflora** subsp. **parviflora**

Thrinax havanensis auct. = **Thrinax morrisii** H.Wendl.

Thrinax keyensis Sarg. = **Thrinax morrisii** H.Wendl.

Thrinax longistyla Becc. = **Coccothrinax argentea** (Lodd. ex Schult. & Schult.f.) Sarg. ex Becc.

Thrinax maritima Lodd. ex Mart. = **Thrinax radiata** Lodd. ex Schult. & Schult.f.

Thrinax martii Griseb. = **Thrinax radiata** Lodd. ex Schult. & Schult.f.

Thrinax mexicana Lodd. ex Mart. = **Thrinax radiata** Lodd. ex Schult. & Schult.f.

Thrinax microcarpa Sarg. = **Thrinax morrisii** H.Wendl.

Thrinax miraguama (Kunth) Mart. = **Coccothrinax miraguama** (Kunth) Becc.

Thrinax montana Lodd. ex Mart. = **Thrinax radiata** Lodd. ex Schult. & Schult.f.

Thrinax morrisii H.Wendl., Gard. Chron. 1891(1): 700 (1891).
Florida Keys to Caribbean. 78 FLA 81 BAH CUB HAI LEE PUE.
Thrinax havanensis auct., Gard. Chron. 1870: 37 (1870), nom. nud.
Thrinax microcarpa Sarg., Gard. & Forest 9: 162 (1896). *Simpsonia microcarpa* (Sarg.) O.F.Cook, Science, n.s., 85: 333 (1937).
Thrinax keyensis Sarg., Bot. Gaz. 27: 86 (1899).
Thrinax ponceana O.F.Cook, Bull. Torrey Bot. Club 28: 536 (1901).
Thrinax praeceps O.F.Cook, Bull. Torrey Bot. Club 28: 536 (1901).
Thrinax bahamensis O.F.Cook, Mem. Torrey Bot. Club 12: 20 (1902).
Thrinax drudei Becc., Webbia 2: 269 (1908).
Thrinax punctulata Becc., Webbia 2: 280 (1908).
Thrinax ekmanii Burret, Kongl. Svenska Vetenskapsakad. Handl., III, 6(7): 27 (1929).

Thrinax multiflora Mart. = **Coccothrinax argentea** (Lodd. ex Schult. & Schult.f.) Sarg. ex Becc.

Thrinax parviflora Maycock = **Coccothrinax barbadensis** (Lodd. ex Mart.) Becc.

Thrinax parviflora Sw., Prodr.: 57 (1788).
Jamaica. 81 JAM.

subsp. **parviflora**
Jamaica. 81 JAM.
Corypha palmacea Steud., Nomencl. Bot. 1: 229 (1821).
Thrinax harrisiana Becc., Repert. Spec. Nov. Regni Veg. 6: 94 (1908).
Thrinax tessellata Becc., Webbia 2: 271 (1908).

subsp. **puberula** Read, Smithsonian Contr. Bot. 19: 76 (1975).
Jamaica. 81 JAM.

Thrinax ponceana O.F.Cook = **Thrinax morrisii** H.Wendl.

Thrinax praeceps O.F.Cook = **Thrinax morrisii** H.Wendl.

Thrinax pumilio Lodd. ex Schult. & Schult.f. = **?**

Thrinax punctulata Becc. = **Thrinax morrisii** H.Wendl.

Thrinax radiata Lodd. ex Schult. & Schult.f., Syst. Veg. 7(2): 1301 (1830). *Coccothrinax radiata* (Lodd. ex Schult. & Schult.f.) Sarg., Just's Bot. Jahresber. 27(1): 469 (1901).
S. Florida, SE. Mexico to Honduras, N. Caribbean. 78 FLA 79 MXT 80 BLZ HON 81 BAH CAY CUB DOM HAI JAM.
Thrinax elegans Schult. & Schult.f., Syst. Veg. 7: 1301 (1830).
Thrinax gracilis Schult. & Schult.f., Syst. Veg. 7: 1301 (1830).
Thrinax aurata Kunth, Enum. Pl. 3: 254 (1841).
Thrinax maritima Lodd. ex Mart., Hist. Nat. Palm. 3: 320 (1853).
Thrinax mexicana Lodd. ex Mart., Hist. Nat. Palm. 3: 320 (1853).
Thrinax montana Lodd. ex Mart., Hist. Nat. Palm. 3: 320 (1853).
Porothrinax pumilio H.Wendl. ex Griseb., Cat. Pl. Cub.: 221 (1866).
Thrinax martii Griseb., Cat. Pl. Cub.: 221 (1866). *Coccothrinax martii* (Griseb.) Becc., Webbia 2: 305 (1908).
Thrinax elegantissima Hook.f., Rep. Progr. Condition Roy. Bot. Gard. Kew 1882: 66 (1884).
Thrinax floridana Sarg., Bot. Gaz. 27: 84 (1899).
Thrinax wendlandiana Becc., Webbia 2: 265 (1908).

Thrinax rex Britton & Harris = **Thrinax excelsa** Lodd. ex Mart.

Thrinax rigida Griseb. & H.Wendl. = **Coccothrinax rigida** (Griseb. & H.Wendl.) Becc.

Thrinax rivularis (Léon) Borhidi & O.Muñiz = **Hemithrinax rivularis** Léon

Thrinax rivularis var. *savannarum* (Léon) Borhidi & O.Muñiz = **Hemithrinax rivularis** var. **savannarum** (Léon) O.Muñiz

Thrinax stellata Lodd. ex Mart. = **Coccothrinax miraguama** subsp. **miraguama**

Thrinax tessellata Becc. = **Thrinax parviflora** subsp. **parviflora**

Thrinax tunica Hook.f. = **Brahea dulcis** (Kunth) Mart.

Thrinax wendlandiana Becc. = **Thrinax radiata** Lodd. ex Schult. & Schult.f.

Thrinax yuraguana A.Rich. = **Coccothrinax yuraguana** (A.Rich.) Léon

Thrincoma

Thrincoma O.F.Cook = **Coccothrinax** Sarg.

Thrincoma alta O.F.Cook = **Coccothrinax barbadensis** (Lodd. ex Mart.) Becc.

Thringis

Thringis O.F.Cook = **Coccothrinax** Sarg.

Thringis latifrons O.F.Cook = **Coccothrinax barbadensis** (Lodd. ex Mart.) Becc.

Thringis laxa O.F.Cook = **Coccothrinax barbadensis** (Lodd. ex Mart.) Becc.

Thuessinkia

Thuessinkia Korth. ex Miq. = **Caryota** L.

Thuessinkia speciosa Korth. = **Caryota mitis** Lour.

Tilmia

Tilmia O.F.Cook = **Aiphanes** Willd.

Tilmia caryotifolia (Kunth) O.F.Cook = **Aiphanes horrida** (Jacq.) Burret

Tilmia disticha (Linden) O.F.Cook = **Martinezia disticha**

Toxophoenix

Toxophoenix Schott = **Astrocaryum** G.Mey.

Toxophoenix aculeatissima Schott = **Astrocaryum aculeatissimum** (Schott) Burret

Trachycarpus

Trachycarpus H.Wendl., Bull. Soc. Bot. France 8: 429 (1861).
8 species, Himalaya to SC. China. 36 (38) 40 41.

Trachycarpus argyratus S.K.Lee & F.N.Wei = **Guihaia argyrata** (S.K.Lee & F.N.Wei) S.K.Lee, F.N.Wei & J.Dransf.

Trachycarpus caespitosus Becc. = **Trachycarpus fortunei** (Hook.) H.Wendl.

Trachycarpus dracocephalus Ching & Y.C.Hsu = **Trachycarpus nanus** Becc.

Trachycarpus excelsus (Thunb.) H.Wendl. = **Rhapis excelsa** (Thunb.) Henry

Trachycarpus fortunei (Hook.) H.Wendl., Bull. Soc. Bot. France 8: 429 (1861).
N. Myanmar to C. China. 36 CHC (38) jap 41 MYA.
Chamaerops fortunei Hook., Bot. Mag. 86: t. 5221 (1860).
Trachycarpus caespitosus Becc., Bull. Soc. Tosc. Ortic., III, 20: 164 (1915).
Trachycarpus wagnerianus Becc., Webbia 5: 70 (1921).

Trachycarpus geminisectus Spanner & al., Palms 47: 146 (2003).
Vietnam. 41 VIE.

Trachycarpus griffithii (Lodd. ex Verl.) auct. = **Trachycarpus martianus** (Wall. ex Mart.) H.Wendl.

Trachycarpus khasyanus (Griff.) H.Wendl. = **Trachycarpus martianus** (Wall. ex Mart.) H.Wendl.

Trachycarpus latisectus Spanner, H.J.Noltie & M.Gibbons, Edinburgh J. Bot. 54: 257 (1997).
Darjiling. 40 EHM.

Trachycarpus martianus (Wall. ex Mart.) H.Wendl., Bull. Soc. Bot. France 8: 429 (1861).
C. Himalaya to SC. China. 36 CHC 40 ASS EHM NEP 41 MYA.
Chamaerops nepalensis Lodd. ex Schult. & Schult.f., Syst. Veg. 7: 1489 (1830), nom. nud.

Chamaerops martiana Wall. ex Mart. in N.Wallich, Pl. Asiat. Rar. 3: 5 (1831).

Chamaerops khasyana Griff., Calcutta J. Nat. Hist. 5: 341 (1845). *Trachycarpus khasyanus* (Griff.) H.Wendl., Bull. Soc. Bot. France 8: 429 (1861).

Chamaerops tomentosa C.Morren, Ann. Soc. Roy. Agric. Gand 1: 488 (1845).

Chamaerops griffithii Lodd. ex Verl., Rev. Hort. 42: t. 276 (1870). *Trachycarpus griffithii* (Lodd. ex Verl.) auct., Rev. Hort. 51: 212 (1879).

Trachycarpus nanus Becc., Webbia 3: 187 (1910). *Chamaerops nana* (Becc.) Chabaud, Palmiers: 60 (1915).
China (Yunnan). 36 CHC.
Trachycarpus dracocephalus Ching & Y.C.Hsu, Acta Phytotax. Sin. 3: 367 (1955).

Trachycarpus oreophilus Gibbons & Spanner, Principes 41: 205 (1997).
N. Thailand. 41 THA.

Trachycarpus princeps Gibbons, Spanner & San Y.Chen, Principes 39: 73 (1995).
China (Yunnan). 36 CHC.

Trachycarpus takil Becc., Webbia 1: 52 (1905).
WC. Himalaya. 40 NEP? WHM.

Trachycarpus wagnerianus Becc. = **Trachycarpus fortunei** (Hook.) H.Wendl.

Trichodypsis

Trichodypsis Baill. = **Dypsis** Noronha ex Mart.

Trichodypsis glabrescens Becc. = **Dypsis glabrescens** (Becc.) Becc.

Trichodypsis hildebrandtii Baill. = **Dypsis hildebrandtii** (Baill.) Becc.

Trichodypsis mocquerysiana Becc. = **Dypsis mocquerysiana** (Becc.) Becc.

Trithrinax

Trithrinax Mart., Hist. Nat. Palm. 2: 149 (1837).
3 species, Bolivia to N. Argentina and Brazil. 83 84 85.
Diodosperma H.Wendl., Bot. Zeitung (Berlin) 36: 118 (1878).
Chamaethrinax H.Wendl. ex R.Pfister, Beitr. Vergl. Anat. Sabaleenblatter: 19 (1891), nom. inval.

Trithrinax acanthocoma Drude = **Trithrinax brasiliensis** Mart.

Trithrinax aculeata Liebm. ex Mart. = **Cryosophila nana** (Kunth) Blume

Trithrinax biflabellata Barb.Rodr. = **Trithrinax schizophylla** Drude

Trithrinax brasiliensis Mart., Hist. Nat. Palm. 2: 150 (1837). *Thrinax brasiliensis* (Mart.) Mart., Hist. Nat. Palm. 3: 320 (1853).
S. Brazil. 84 BZS.
Trithrinax acanthocoma Drude, Gartenflora 27: 361 (1878).

Trithrinax campestris (Burmeist.) Drude & Griseb., Abh. Königl. Ges. Wiss. Göttingen 24: 283 (1879).
NC. & NE. Argentina to W. Uruguay. 85 AGE AGW URU.
Copernicia campestris Burmeist., Reise La Plata Staaten 2: 48 (1861).
Chamaethrinax hookeriana H.Wendl. ex R.Pfister, Beitr. Vergl. Anat. Sabaleenblatter: 46 (1891), nom. inval.

Trithrinax chuco (Mart.) Walp. = **Chelyocarpus chuco** (Mart.) H.E.Moore

Trithrinax compacta Griseb. & H.Wendl. = **Hemithrinax compacta** (Griseb. & H.Wendl.) Hook.f.

Trithrinax mauritiiformis H.Karst. = **Sabal mauritiiformis** (H.Karst.) Griseb. & H.Wendl.

Trithrinax schizophylla Drude in C.F.P.von Martius & auct. suc. (eds.), Fl. Bras. 3(2): 551 (1882).
WC. Brazil to NW. & NC. Argentina. 83 BOL 84 BZC 85 AGE AGW PAR.
Diodosperma burity H.Wendl., Bot. Zeitung (Berlin) 36: 118 (1878). Provisonal synonym.
Trithrinax biflabellata Barb.Rodr., Palm. Paraguay.: 2 (1899).

Tuerckheimia

Tuerckheimia Dammer = **Chamaedorea** Willd.

Tuerckheimia ascendens Dammer = **Chamaedorea adscendens** (Dammer) Burret

Vadia

Vadia O.F.Cook = **Chamaedorea** Willd.

Vadia atrovirens (Mart.) O.F.Cook = **Chamaedorea atrovirens** Mart.

Vadia jotolana O.F.Cook = **Chamaedorea cataractarum** Mart.

Veillonia

Veillonia H.E.Moore, Gentes Herb. 11: 299 (1978).
1 species, New Caledonia. 60.

Veillonia alba H.E.Moore, Gentes Herb. 11: 301 (1978).
NE. New Caledonia. 60 NWC.

Veitchia

Veitchia H.Wendl. in B.Seemann, Fl. Vit.: 270 (1868).
8 species, SW. Pacific. 60.
Vitiphoenix Becc., Ann. Jard. Bot. Buitenzorg 2: 91 (1885).
Kajewskia Guillaumin, J. Arnold Arbor. 13: 113 (1932).

Veitchia arecina Becc., Webbia 5: 78 (1921).
Vanuatu. 60 VAN.
Veitchia hookeriana Becc., Webbia 5: 77 (1921).
Veitchia macdanielsii H.E.Moore, Gentes Herb. 8: 496 (1957).
Veitchia montgomeryana H.E.Moore, Gentes Herb. 8: 492 (1957).

Veitchia canterburyana (C.Moore & F.Muell.) H.Wendl. = **Hedyscepe canterburyana** (C.Moore & F.Muell.) H.Wendl. & Drude

Veitchia filifera (H.Wendl.) H.E.Moore, Gentes Herb. 8: 533 (1957).
Fiji (Vanua Levu, S. Taveuni). 60 FIJ.
Ptychosperma filiferum H.Wendl., Bonplandia 10: 195 (1862). *Drymophloeus filiferus* (H.Wendl.) Scheff., Ann. Jard. Bot. Buitenzorg 1: 158 (1876). *Vitiphoenix filifera* (H.Wendl.) Becc., Ann. Jard. Bot. Buitenzorg 2: 91 (1885).
Vitiphoenix petiolata Burret, Occas. Pap. Bernice Pauahi Bishop Mus. 11(4): 8 (1935). *Veitchia petiolata* (Burret) H.E.Moore, Gentes Herb. 8: 520 (1957).
Vitiphoenix sessilifolia Burret, Occas. Pap. Bernice Pauahi Bishop Mus. 11(4): 9 (1935). *Veitchia*

sessilifolia (Burret) H.E.Moore, Gentes Herb. 8: 521 (1957).

Vitiphoenix pedionoma A.C.Sm., J. Arnold Arbor. 31: 145 (1950). *Veitchia pedionoma* (A.C.Sm.) H.E. Moore, Gentes Herb. 8: 524 (1957).

Veitchia hookeriana Becc. = **Veitchia arecina** Becc.

Veitchia joannis H.Wendl. in B.Seemann, Fl. Vit.: 271 (1868). *Kentia joannis* (H.Wendl.) F.Muell., Fragm. 7: 101 (1870).
Fiji. 60 FIJ ton.

Veitchia macdanielsii H.E.Moore = **Veitchia arecina** Becc.

Veitchia merrillii (Becc.) H.E.Moore = **Adonidia merrillii** (Becc.) Becc.

Veitchia metiti Becc., Webbia 5: 77 (1921).
N. Vanuatu (Vanua Lava, Uréparapara). 60 VAN.

Veitchia montgomeryana H.E.Moore = **Veitchia arecina** Becc.

Veitchia pedionoma (A.C.Sm.) H.E.Moore = **Veitchia filifera** (H.Wendl.) H.E.Moore

Veitchia petiolata (Burret) H.E.Moore = **Veitchia filifera** (H.Wendl.) H.E.Moore

Veitchia pickeringii (H.Wendl.) H.E.Moore = **Veitchia vitiensis** (H.Wendl.) H.E.Moore

Veitchia sessilifolia (Burret) H.E.Moore = **Veitchia filifera** (H.Wendl.) H.E.Moore

Veitchia simulans H.E.Moore, Gentes Herb. 8: 506 (1957).
Fiji (Taveuni). 60 FIJ.

Veitchia smithii (Burret) H.E.Moore = **Veitchia vitiensis** (H.Wendl.) H.E.Moore

Veitchia spiralis H.Wendl. in B.Seemann, Fl. Vit.: 270 (1868).
Vanuatu (Anatom, Tanna). 60 VAN.
Kajewskia aneityensis Guillaumin, J. Arnold Arbor. 13: 113 (1932).

Veitchia storckii H.Wendl. = **Neoveitchia storckii** (H.Wendl.) Becc.

Veitchia subglobosa H.Wendl. = ? [60 FIJ]

Veitchia vitiensis (H.Wendl.) H.E.Moore, Gentes Herb. 8: 514 (1957).
Fiji. 60 FIJ.
*Ptychosperma vitiensis H.Wendl., Bonplandia 9: 260 (1861). *Saguaster vitiensis* (H.Wendl.) Kuntze, Revis. Gen. Pl. 2: 735 (1891). *Vitiphoenix vitiensis* (H.Wendl.) Burret, Repert. Spec. Nov. Regni Veg. 24: 271 (1928).
Ptychosperma pickeringii H.Wendl., Bonplandia 10: 194 (1862). *Saguaster pickeringii* (H.Wendl.) Kuntze, Revis. Gen. Pl. 2: 735 (1891). *Vitiphoenix pickeringii* (H.Wendl.) Burret, Repert. Spec. Nov. Regni Veg. 24: 270 (1928). *Veitchia pickeringii* (H.Wendl.) H.E.Moore, Gentes Herb. 8: 534 (1957).
Vitiphoenix smithii Burret, Occas. Pap. Bernice Pauahi Bishop Mus. 11(4): 7 (1935). *Veitchia smithii* (Burret) H.E.Moore, Gentes Herb. 8: 519 (1957).
Balaka spectabilis Burret, Bernice P. Bishop Mus. Bull. 141: 13 (1936).
Veitchia vitiensis var. *microcarpa* H.E.Moore, Gentes Herb. 8: 518 (1957).
Veitchia vitiensis var. *parhamiorum* H.E.Moore, Gentes Herb. 8: 518 (1957).

Veitchia vitiensis var. *microcarpa* H.E.Moore = **Veitchia vitiensis** (H.Wendl.) H.E.Moore

Veitchia vitiensis var. *parhamiorum* H.E.Moore = **Veitchia vitiensis** (H.Wendl.) H.E.Moore

Veitchia winin H.E.Moore, Gentes Herb. 8: 499 (1957).
Vanuatu. 60 VAN.

Verschaffeltia

Verschaffeltia H.Wendl., Ill. Hort. 12(Misc.): 5 (1865).
1 species, Seychelles. 29.
Regelia H.Wendl., Ill. Hort. 12(Misc.): 6 (1865), nom. illeg.

Verschaffeltia melanochaetes H.Wendl. = **Roscheria melanochaetes** (H.Wendl.) H.Wendl. ex Balf.f.

Verschaffeltia splendida H.Wendl., Ill. Hort. 12(Misc.): 6 (1865).
Seychelles. 29 SEY.
Regelia magnifica H.Wendl., Ill. Hort. 12(Misc.): 6 (1865).
Regelia majestica auct., Gard. Chron. 1865: 292 (1865).
Stevensonia viridifolia Duncan, Gard. Chron. 1870: 697 (1870).
Regelia princeps Balf.f. in J.G.Baker, Fl. Mauritius: 388 (1877).

Vitiphoenix

Vitiphoenix Becc. = **Veitchia** H.Wendl.

Vitiphoenix filifera (H.Wendl.) Becc. = **Veitchia filifera** (H.Wendl.) H.E.Moore

Vitiphoenix minuta (Rech.) Burret = **Balaka minuta** Burret

Vitiphoenix pauciflora (H.Wendl.) Burret = **Balaka pauciflora** (H.Wendl.) H.E.Moore

Vitiphoenix pedionoma A.C.Sm. = **Veitchia filifera** (H.Wendl.) H.E.Moore

Vitiphoenix petiolata Burret = **Veitchia filifera** (H.Wendl.) H.E.Moore

Vitiphoenix pickeringii (H.Wendl.) Burret = **Veitchjia pikeringii**

Vitiphoenix polyclada Burret = **Balaka tahitensis** (H.Wendl.) Becc.

Vitiphoenix samoensis (Becc.) Burret = **Balaka samoensis** Becc.

Vitiphoenix seemannii Becc. ex Martelli = **Balaka seemannii** (H.Wendl.) Becc.

Vitiphoenix sessilifolia Burret = **Veitchia filifera** (H.Wendl.) H.E.Moore

Vitiphoenix smithii Burret = **Veitchia vitiensis** (H.Wendl.) H.E.Moore

Vitiphoenix vitiensis (H.Wendl.) Burret = **Veitchia vitiensis** (H.Wendl.) H.E.Moore

Vitiphoenix withmeeana (Becc.) Burret = **Drymophloeus withmeeanus**

Voanioala

Voanioala J.Dransf., Kew Bull. 44: 192 (1989).
1 species, Madagascar. 29.

Voanioala gerardii J.Dransf., Kew Bull. 44: 195 (1989).
NE. Madagascar. 29 MDG.

Vonitra

Vonitra Becc. = **Dypsis** Noronha ex Mart.

Vonitra crinita Jum. & H.Perrier = **Dypsis crinita** (Jum. & H.Perrier) Beentje & J.Dransf.

Vonitra fibrosa (C.H.Wright) Becc. = **Dypsis fibrosa** (C.H.Wright) Beentje & J.Dransf.

Vonitra loucoubensis (Baill.) Jum. = **Dypsis nossibensis** (Becc.) Beentje & J.Dransf.

Vonitra nossibensis (Becc.) H.Perrier = **Dypsis nossibensis** (Becc.) Beentje & J.Dransf.

Vonitra thouarsiana (Baill.) Becc. = **Dypsis thouarsiana** Baill.

Vonitra utilis Jum. = **Dypsis utilis** (Jum.) Beentje & J.Dransf.

Vouay

Vouay Aubl. = **Geonoma** Willd.

Wallichia

Wallichia Roxb., Pl. Coromandel 3: 91 (1820).
10 species, Himalaya to S. China. 36 40 41.
Harina Buch.-Ham., Mem. Wern. Nat. Hist. Soc. 5: 317 (1826).
Wrightea Roxb., Fl. Ind. ed. 1832, 3: 621 (1832).
Asraoa J.Joseph, Bull. Bot. Surv. India 14: 144 (1972 publ. 1975).

Wallichia caryotoides Roxb., Pl. Coromandel 3: 91 (1820). *Harina caryotoides* (Roxb.) Buch.-Ham., Mem. Wern. Nat. Hist. Soc. 5: 317 (1826). *Wrightea caryotoides* (Roxb.) Roxb., Fl. Ind. ed. 1832, 3: 621 (1832).
E. Himalaya to China (Yunnan). 36 CHC 40 BAN EHM 41 MYA.
Harina wallichia Steud. ex Saloman, Palmen: 127 (1877).

Wallichia caudata (Lour.) Mart. = **Arenga caudata** (Lour.) H.E.Moore

Wallichia chinensis Burret, Notizbl. Bot. Gart. Berlin-Dahlem 13: 602 (1937).
S. China to Hainan. 36 CHC CHH CHS.

Wallichia densiflora Mart., Hist. Nat. Palm. 3: 190 (1845). *Harina densiflora* (Mart.) Walp., Ann. Bot. Syst. 3: 1032 (1853).
Himalaya to China (Yunnan). 36 CHC 40 ASS EHM NEP WHM 41 MYA.
Wallichia oblongifolia Griff., Calcutta J. Nat. Hist. 5: 486 (1845). *Harina oblongifolia* (Griff.) Griff., Palms Brit. E. Ind.: 175 (1850).

Wallichia disticha T.Anderson, J. Linn. Soc., Bot. 11: 6 (1871). *Didymosperma distichum* (T.Anderson) Hook.f., Rep. Progr. Condition Roy. Bot. Gard. Kew 1882: 61 (1884).
E. Himalaya to China (Yunnan). 36 CHC 40 ASS BAN EHM 41 MYA THA.
Wallichia yomae Kurz, Forest Fl. Burma 2: 533 (1877).

Wallichia gracilis Becc., Webbia 3: 211 (1910).
N. Vietnam. 41 VIE.

Wallichia horsfieldii Blume = **Arenga porphyrocarpa** (Blume) H.E.Moore

Wallichia marianneae Hodel, Palm J. 137: 8 (1997).
SW. & Pen. Thailand. 41 THA.

Wallichia marianniae Hodel, Palm J. 137: 8 (1997).
Thailand. 41 THA.

Wallichia mooreana S.K.Basu, Taiwania 28: 146 (1983).
China (Yunnan). 36 CHC.

Wallichia nana Griff. = **Arenga nana** (Griff.) H.E.Moore

Wallichia oblongifolia Griff. = **Wallichia densiflora** Mart.

Wallichia orania Blume = **Arenga porphyrocarpa** (Blume) H.E.Moore

Wallichia porphyrocarpa (Blume) Mart. = **Arenga porphyrocarpa** (Blume) H.E.Moore

Wallichia reinwardtiana Miq. = **Arenga porphyrocarpa** (Blume) H.E.Moore

Wallichia siamensis Becc., Atti Soc. Tosc. Sci. Nat. Pisa Processi Verbali 44: 175 (1934).
China (Yunnan) to N. Thailand. 36 CHC 41 THA.

Wallichia tremula (Blanco) Mart. = **Arenga tremula** (Blanco) Becc.

Wallichia triandra (J.Joseph) S.K.Basu, Principes 20: 120 (1976).
Assam. 40 ASS.
*Asraoa triandra J.Joseph, Bull. Bot. Surv. India 14: 144 (1972 publ. 1975).

Wallichia yomae Kurz = **Wallichia disticha** T.Anderson

Wallichia zebrina B.S.Williams = **?**

Washingtonia

Washingtonia H.Wendl., Bot. Zeitung (Berlin) 37: 68 (1879), nom. cons. *Neowashingtonia* Sudw., U.S.D.A. Div. Forest. Bull. 14: 105 (1897), nom. illeg.
2 species, SW. U.S.A. to NW. Mexico. 76 79.

Washingtonia filamentosa (H.Wendl. ex Franceschi) Kuntze = **Washingtonia filifera** (Linden ex André) H.Wendl. ex de Bary

Washingtonia filifera (Linden ex André) H.Wendl. ex de Bary, Bot. Zeitung (Berlin) 37: LXI (1879).
S. California, W. Arizona, Mexico (NE. Baja California). 76 ARI CAL nev 79 MXN.
Brahea dulcis J.G.Cooper, Rep. (Annual) Board Regents Smithsonian Inst. 1860: 442 (1860), nom. illeg.
*Pritchardia filifera Linden ex André, Ill. Hort. 21: 28 (1874). *Brahea filifera* (Linden ex André) W.Watson, Bull. Misc. Inform. Kew 1889: 296 (1889). *Neowashingtonia filifera* (Linden ex André) Sudw., For. Trees Pacif. Slope: 199 (1908).
Brahea filamentosa S.Watson, Proc. Amer. Acad. Arts 11: 147 (1876).
Pritchardia filamentosa H.Wendl. ex Franceschi, Boll. Reale Soc. Tosc. Ortic. 1: 116 (1876). *Washingtonia filamentosa* (H.Wendl. ex Franceschi) Kuntze, Revis. Gen. Pl. 2: 737 (1891). *Livistona filamentosa* (H.Wendl. ex Franceschi) Pfister, Beitr. Vergl. Anat. Sabaleenblatter: 25 (1892). *Neowashingtonia filamentosa* (H.Wendl. ex Franceschi) Sudw., U.S.D.A. Div. Forest. Bull. 14: 105 (1897).
Washingtonia filifera var. *microsperma* Becc., Webbia 2: 191 (1907).

Washingtonia filifera var. *microsperma* Becc. = **Washingtonia filifera** (Linden ex André) H.Wendl. ex de Bary

Washingtonia filifera var. *robusta* (H.Wendl.) Parish = **Washingtonia robusta** H.Wendl.

Washingtonia gaudichaudii (Mart.) Kuntze = **Pritchardia martii** (Gaudich.) H.Wendl.

Washingtonia gracilis Parish = **Washingtonia robusta** H.Wendl.

Washingtonia hillebrandii (Becc.) Kuntze = **Pritchardia hillebrandii** Becc.

Washingtonia lanigera (Becc.) Kuntze = **Pritchardia lanigera** Becc.

Washingtonia martii (Gaudich.) Kuntze = **Pritchardia martii** (Gaudich.) H.Wendl.

Washingtonia pacifica (Seem. & H.Wendl.) Kuntze = **Pritchardia pacifica** Seem. & H.Wendl.

Washingtonia pericularum (H.Wendl.) Kuntze = **Pritchardia pericularum** H.Wendl.

Washingtonia remota Kuntze = **Pritchardia remota** (Kuntze) Becc.

Washingtonia robusta H.Wendl., Berliner Allg. Gartenzeitung 2: 198 (1883). *Neowashingtonia robusta* (H.Wendl.) A.Heller, Cat. N. Amer. Pl.: 3 (1898). *Washingtonia filifera* var. *robusta* (H.Wendl.) Parish, Bot. Gaz. 44: 420 (1907). *Pritchardia robusta* (H.Wendl.) Schröt., Schweiz. Gartenbau 7: 8 (1931).
Mexico (C. & S. Baja California, W. Sonora). (76) cal (78) fla 79 MXN.
> *Washingtonia sonorae* S.Watson, Proc. Amer. Acad. Arts 24: 79 (1889). *Neowashingtonia sonorae* (S.Watson) Rose, Contr. U. S. Natl. Herb. 5: 255 (1897).
> *Brahea robusta* Voss, Vilm. Blumengärtn. ed. 3, 1: 1149 (1895), pro syn.
> *Washingtonia gracilis* Parish, Bot. Gaz. 44: 420 (1907).

Washingtonia sonorae S.Watson = **Washingtonia robusta** H.Wendl.

Washingtonia thurstonii (F.Muell. & Drude) Kuntze = **Pritchardia thurstonii** F.Muell. & Drude

Washingtonia vuylstekeana (H.Wendl.) Kuntze = **Pritchardia vuylstekeana** H.Wendl.

Welfia

Welfia H.Wendl., Gartenflora 18: 242 (1869).
1 species, C. America to Ecuador. 80 82.

Welfia georgii H.Wendl. = **Welfia regia** H.Wendl.

Welfia microcarpa Burret = **Welfia regia** H.Wendl.

Welfia regia H.Wendl., Ill. Hort. 18: 93 (1871).
C. America to W. Ecuador. 80 COS HON NIC PAN 83 CLM ECU.
Welfia georgii H.Wendl., Ill. Hort. 18: 94 (1871).
Welfia microcarpa Burret, Bot. Jahrb. Syst. 63: 129 (1930).

Wendlandiella

Wendlandiella Dammer, Bot. Jahrb. Syst. 36(80): 31 (1905).
1 species, Peru to Bolivia and N. Brazil. 83 84.

Wendlandiella gracilis Dammer, Bot. Jahrb. Syst. 36(80): 32 (1905).
Brazil (Acre) to Bolivia. 83 BOL PER 84 BZN.

var. **gracilis**
Brazil (Acre) to N. Peru. 83 PER 84 BZN.

var. **polyclada** (Burret) A.J.Hend., Palms Amazon: 88 (1995).
N. Peru. 83 PER.
> *Wendlandiella polyclada* Burret, Notizbl. Bot. Gart. Berlin-Dahlem 11: 203 (1931).

var. **simplicifrons** (Burret) A.J.Hend., Palms Amazon: 88 (1995).
Peru to Bolivia. 83 BOL PER.
> *Wendlandiella simplicifrons* Burret, Notizbl. Bot. Gart. Berlin-Dahlem 11: 316 (1932).

Wendlandiella polyclada Burret = **Wendlandiella gracilis** var. **polyclada** (Burret) A.J.Hend.

Wendlandiella simplicifrons Burret = **Wendlandiella gracilis** var. **simplicifrons** (Burret) A.J.Hend.

Wettinella

Wettinella O.F.Cook & Doyle = **Wettinia** Poepp. ex Endl.

Wettinella maynensis (Spruce) O.F.Cook & Doyle = **Wettinia maynensis** Spruce

Wettinella quinaria O.F.Cook & Doyle = **Wettinia quinaria** (O.F.Cook & Doyle) Burret

Wettinia

Wettinia Poepp. ex Endl., Gen. Pl.: 243 (1837).
21 species, Panama to S. Trop. America. 80 82 83 84.
> *Catoblastus* H.Wendl., Bonplandia 8: 104 (1860).
> *Acrostigma* O.F.Cook & Doyle, Contr. U. S. Natl. Herb. 16: 228 (1913).
> *Catostigma* O.F.Cook & Doyle, Contr. U. S. Natl. Herb. 16: 230 (1913).
> *Wettinella* O.F.Cook & Doyle, Contr. U. S. Natl. Herb. 16: 235 (1913).
> *Wettiniicarpus* Burret, Notizbl. Bot. Gart. Berlin-Dahlem 10: 937 (1930).

Wettinia aequalis (O.F.Cook & Doyle) R.Bernal, Caldasia 17: 368 (1995).
Panama to W. Ecuador. 80 PAN 83 CLM ECU.
> *Acrostigma aequale* O.F.Cook & Doyle, Contr. U. S. Natl. Herb. 16: 228 (1913). *Catostigma aequale* (O.F.Cook & Doyle) Burret, Notizbl. Bot. Gart. Berlin-Dahlem 10: 934 (1930). *Catoblastus aequalis* (O.F.Cook & Doyle) Burret, Notizbl. Bot. Gart. Berlin-Dahlem 10: 935 (1930).
> *Catoblastus velutinus* Burret, Notizbl. Bot. Gart. Berlin-Dahlem 10: 929 (1930), nom. alternativ.

Wettinia aequatorialis R.Bernal, Caldasia 17: 369 (1995).
SE. Ecuador. 83 ECU.

Wettinia anomala (Burret) R.Bernal, Caldasia 17: 368 (1995).
S. Colombia to N. Ecuador. 83 CLM ECU.
> *Catostigma anomalum* Burret, Notizbl. Bot. Gart. Berlin-Dahlem 11: 4 (1930). *Catoblastus anomalus* (Burret) Burret, Notizbl. Bot. Gart. Berlin-Dahlem 11: 5 (1939).

Wettinia augusta Poepp. & Endl., Nov. Gen. Sp. Pl. 2: 39 (1838).
SE. Colombia to N. Bolivia. 83 BOL CLM PER 84 BZN.
> *Wettinia poeppigii* Kunth, Enum. Pl. 3: 109 (1841).
> *Wettinia weberbaueri* Burret, Notizbl. Bot. Gart. Berlin-Dahlem 10: 939 (1930).

Wettinia castanea H.E.Moore & J.Dransf., Notes Roy. Bot. Gard. Edinburgh 36: 263 (1978).
NW. Colombia. 83 CLM.

Wettinia cladospadix (Dugand) H.E.Moore & J.Dransf. = **Wettinia fascicularis** (Burret) H.E.Moore & J.Dransf.

Wettinia disticha (R.Bernal) R.Bernal, Caldasia 17: 368 (1995).
W. Colombia. 83 CLM.
 *Catoblastus distichus R.Bernal, Principes 30: 38 (1986).

Wettinia drudei (O.F.Cook & Doyle) A.J.Hend., Palms Amazon: 98 (1995).
Colombia to N. Peru and N. Brazil. 83 CLM ECU PER 84 BZN.
 *Catoblastus drudei O.F.Cook & Doyle, Contr. U. S. Natl. Herb. 16: 233 (1913). *Catostigma drudei* (O.F.Cook & Doyle) Burret, Notizbl. Bot. Gart. Berlin-Dahlem 10: 924 (1930).

Wettinia fascicularis (Burret) H.E.Moore & J.Dransf., Notes Roy. Bot. Gard. Edinburgh 36: 264 (1978).
Colombia to N. Ecuador. 83 CLM ECU.
 *Wettiniicarpus fascicularis Burret, Notizbl. Bot. Gart. Berlin-Dahlem 10: 938 (1930).
 Wettiniicarpus cladospadix Dugand, Caldasia 7: 138 (1955). *Wettinia cladospadix* (Dugand) H.E.Moore & J.Dransf., Notes Roy. Bot. Gard. Edinburgh 36: 264 (1978).

Wettinia hirsuta Burret, Notizbl. Bot. Gart. Berlin-Dahlem 10: 941 (1930).
Colombia. 83 CLM.

Wettinia illaqueans Spruce = **Wettinia maynensis** Spruce

Wettinia kalbreyeri (Burret) R.Bernal, Caldasia 17: 368 (1995).
Colombia to Ecuador. 83 CLM ECU.
 Catoblastus kalbreyeri Burret, Notizbl. Bot. Gart. Berlin-Dahlem 10: 936 (1930), nom. alternativ.
 Catoblastus megalocarpus Burret, Notizbl. Bot. Gart. Berlin-Dahlem 10: 935 (1930), nom. alternativ.
 Catoblastus microcaryus Burret, Notizbl. Bot. Gart. Berlin-Dahlem 10: 937 (1930), nom. alternativ.
 Catoblastus sphaerocarpus Burret, Notizbl. Bot. Gart. Berlin-Dahlem 10: 937 (1930), nom. alternativ.
 *Catostigma kalbreyeri Burret, Notizbl. Bot. Gart. Berlin-Dahlem 10: 935 (1930).
 Catostigma megalocarpum Burret, Notizbl. Bot. Gart. Berlin-Dahlem 10: 935 (1930), nom. alternativ.
 Catostigma microcaryum Burret, Notizbl. Bot. Gart. Berlin-Dahlem 10: 937 (1930).
 Catostigma sphaerocarpum Burret, Notizbl. Bot. Gart. Berlin-Dahlem 10: 936 (1930).
 Catostigma sphaerocarpum var. *microcaryum* Burret, Notizbl. Bot. Gart. Berlin-Dahlem 10: 937 (1930).
 Catostigma inconstans Dugand, Caldasia 2: 392 (1944). *Catoblastus inconstans* (Dugand) Glassman, Phanerog. Monogr. 6: 63 (1972).

Wettinia lanata R.Bernal, Caldasia 17: 371 (1995).
W. Colombia. 83 CLM.

Wettinia longipetala A.H.Gentry, Ann. Missouri Bot. Gard. 73: 160 (1986).
Peru (Pasco). 83 PER.

Wettinia maynensis Spruce, J. Proc. Linn. Soc., Bot. 3: 191 (1859). *Catoblastus maynensis* (Spruce) Drude in C.F.P.von Martius & auct. suc. (eds.), Fl. Bras. 3(2): 544 (1882). *Wettinella maynensis* (Spruce) O.F.Cook & Doyle, Contr. U. S. Natl. Herb. 16: 237 (1913).
S. Colombia to Peru. 83 CLM ECU PER.
 Wettinia illaqueans Spruce, J. Proc. Linn. Soc., Bot. 3: 191 (1859).

Wettinia mesocarpa (Burret) Wess.Boer = **Wettinia praemorsa** (Willd.) Wess.Boer

Wettinia microcarpa (Burret) R.Bernal, Caldasia 17: 368 (1995).
NE. Colombia (Norte de Santander). 83 CLM.
 *Catoblastus microcarpus Burret, Notizbl. Bot. Gart. Berlin-Dahlem 10: 931 (1930).

Wettinia minima R.Bernal, Caldasia 17: 373 (1995).
Ecuador (Cordillera de Cutucú). 83 ECU.

Wettinia oxycarpa Galeano-Garcés & R.Bernal, Caldasia 13: 695 (1983).
W. Colombia to NW. Ecuador. 83 CLM ECU.

Wettinia panamensis R.Bernal, Caldasia 17: 373 (1995).
Panama. 80 PAN.

Wettinia poeppigii Kunth = **Wettinia augusta** Poepp. & Endl.

Wettinia praemorsa (Willd.) Wess.Boer, Pittieria 17: 185 (1988).
Colombia to Venezuela. 82 VEN 83 CLM.
 *Oreodoxa praemorsa Willd., Mém. Acad. Roy. Sci. Hist. (Berlin) 1804: 36 (1807). *Iriartea praemorsa* (Willd.) Klotzsch, Linnaea 20: 448 (1847). *Catoblastus praemorsus* (Willd.) H.Wendl., Bonplandia 8: 104 (1860).
 Iriartea pubescens H.Karst., Linnaea 28: 262 (1856). *Catoblastus pubescens* (H.Karst.) H.Wendl., Bonplandia 8: 104 (1860).
 Catoblastus engelii H.Wendl. ex Burret, Notizbl. Bot. Gart. Berlin-Dahlem 10: 928 (1930), nom. alternativ.
 Catoblastus mesocarpus Burret, Notizbl. Bot. Gart. Berlin-Dahlem 10: 930 (1930), nom. alternativ. *Wettinia mesocarpa* (Burret) Wess.Boer, Pittieria 17: 187 (1988).
 Catoblastus andinus Dugand, Caldasia 1(1): 15 (1940).
 Catoblastus cuatrecasasii Dugand, Caldasia 1(1): 17 (1940).

Wettinia quinaria (O.F.Cook & Doyle) Burret, Notizbl. Bot. Gart. Berlin-Dahlem 10: 942 (1930).
W. Colombia to NW. Ecuador. 83 CLM ECU.
 *Wettinella quinaria O.F.Cook & Doyle, Contr. U. S. Natl. Herb. 16: 236 (1913).
 Wettinia utilis Little, Phytologia 19: 251 (1970).

Wettinia radiata (O.F.Cook & Doyle) R.Bernal, Caldasia 17: 368 (1995).
E. Panama to W. Colombia. 80 PAN 83 CLM.
 *Catostigma radiatum O.F.Cook & Doyle, Contr. U. S. Natl. Herb. 16: 231 (1913). *Catoblastus radiatus* (O.F.Cook & Doyle) Burret, Notizbl. Bot. Gart. Berlin-Dahlem 10: 934 (1930).
 Catoblastus dryanderae Burret, Notizbl. Bot. Gart. Berlin-Dahlem 11: 861 (1933), nom. alternativ.
 Catostigma dryanderae Burret, Notizbl. Bot. Gart. Berlin-Dahlem 11: 860 (1933).

Wettinia utilis Little = **Wettinia quinaria** (O.F.Cook & Doyle) Burret

Wettinia verruculosa H.E.Moore, Principes 26: 42 (1982).
SW. Colombia (Nariño) to NW. Ecuador (Carchi). 83 CLM ECU.

Wettinia weberbaueri Burret = **Wettinia augusta** Poepp. & Endl.

Wettiniicarpus

Wettiniicarpus Burret = **Wettinia** Poepp. ex Endl.

Wettiniicarpus cladospadix Dugand = **Wettinia fascicularis** (Burret) H.E.Moore & J.Dransf.

Wettiniicarpus fascicularis Burret = **Wettinia fascicularis** (Burret) H.E.Moore & J.Dransf.

Wissmannia

Wissmannia Burret = **Livistona** R.Br.

Wissmannia carinensis (Chiov.) Burret = **Livistona carinensis** (Chiov.) J.Dransf. & N.W.Uhl

Wodyetia

Wodyetia A.K.Irvine, Principes 27: 161 (1983).
1 species, NE. Australia. 50.

Wodyetia bifurcata A.K.Irvine, Principes 27: 163 (1983).
N. Queensland (Melville Range). 50 QLD.

Woodsonia

Woodsonia L.H.Bailey = **Neonicholsonia** Dammer

Woodsonia scheryi L.H.Bailey = **Neonicholsonia watsonii** Dammer

Wrightea

Wrightea Roxb. = **Wallichia** Roxb.

Wrightea caryotoides (Roxb.) Roxb. = **Wallichia caryotoides** Roxb.

Yarina

Yarina O.F.Cook = **Phytelephas** Ruiz & Pav.

Yarina microcarpa (Ruiz & Pav.) O.F.Cook = **Phytelephas microcarpa**

Ynesa

Ynesa O.F.Cook = **Attalea** Kunth

Ynesa colenda O.F.Cook = **Attalea colenda** (O.F.Cook) Balslev & A.J.Hend.

Yuyba

Yuyba L.H.Bailey = **Bactris** Jacq. ex Scop.

Yuyba dakamana L.H.Bailey = **Bactris simplicifrons** Mart.

Yuyba essequiboensis L.H.Bailey = **Bactris simplicifrons** Mart.

Yuyba gleasonii L.H.Bailey = **Bactris simplicifrons** Mart.

Yuyba maguirei L.H.Bailey = **Bactris simplicifrons** Mart.

Yuyba paula (L.H.Bailey) L.H.Bailey = **Bactris hondurensis** Standl.

Yuyba schultesii L.H.Bailey = **Bactris schultesii** (L.H. Bailey) Glassman

Yuyba simplicifrons (Mart.) L.H.Bailey = **Bactris simplicifrons** Mart.

Yuyba stahelii L.H.Bailey = **Bactris simplicifrons** Mart.

Yuyba trinitensis L.H.Bailey = **Bactris simplicifrons** Mart.

Zalacca

Zalacca Rumph. ex Blume = **Salacca** Reinw.

Zalaccella

Zalaccella Becc. = **Calamus** L.

Zalaccella harmandii (Pierre ex Becc.) Becc. = **Calamus harmandii** Pierre ex Becc.

Zelonops

Zelonops Raf. = **Phoenix** L.

Zelonops pusilla (Gaertn.) Raf. = **Phoenix pusilla** Gaertn.

Zombia

Zombia L.H.Bailey, Gentes Herb. 4: 240 (1939).
1 species, Hispaniola. 81.
Oothrinax (Becc.) O.F.Cook, Natl. Hort. Mag. 20: 21 (1941).

Zombia antillarum (Desc.) L.H.Bailey, Gentes Herb. 4: 242 (1939).
Hispaniola. 81 DOM HAI.
*Chamaerops antillarum Desc., Fl. Méd. Antilles 1: t. 28 (1821).
Coccothrinax anomala Becc., Repert. Spec. Nov. Regni Veg. 6: 95 (1908). *Oothrinax anomala* (Becc.) O.F.Cook, Natl. Hort. Mag. 10: 21 (1941).

Unplaced names

Acanthophoenix grandis auct., Ill. Hort. 1895: 185 (1895) = ?

Acanthorrhiza wallisii H.Wendl., Gartenflora 28: 163 (1879) = ?

Aiphanes leiospatha Burret, Notizbl. Bot. Gart. Berlin-Dahlem 11: 571 (1932) = ? [83 CLM]

Areca angulosa Giseke, Prael. Ord. Nat. Pl.: 80 (1792) = ? [40 IND]

Areca cornuta Giseke, Prael. Ord. Nat. Pl.: 81 (1792) = ? [40 IND]

Areca cuneifolia Stokes, Bot. Mat. Med. 2: 318 (1812) = ? Drymophloeus sp

Areca glandiformis Lam., Encycl. 1: 241 (1783) = ? [42 MOL]

Areca ilsemannii André, Rev. Hort. 70: 261 (1898) = ?

Areca lansiformis Giseke, Prael. Ord. Nat. Pl.: 81 (1792) = ? [40 IND]

Areca madagascariensis Mart., Hist. Nat. Palm. 3: 179 (1838) = ? [29 MDG]

Areca micholitzii Sander, Cat. 1895: 46 (1895) = ? [43 NWG]

Areca oriziformis var. **saxatilis** Burm.f. ex Giseke, Prael. Ord. Nat. Pl.: 76 (1792) = ? [42 MOL] Pinanga ?

Areca oviformis Giseke, Prael. Ord. Nat. Pl.: 81 (1792) = ? [40 IND]

Areca purpurea auct., Ill. Hort. 24: t. 298 (1877) = ? [29 MAU]

Arenga bonnetti Hook.f., Rep. Progr. Condition Roy. Bot. Gard. Kew 1882: 61 (1884) = ? [40 IND]

Arenga javanica H.Wendl. in O.C.E.de Kerchove de Denterghem, Palmiers: 232 (1878) = ? [42 JAW] Korthalsia sp.

Arenga manillensis (H.Lodd) H.Wendl., Index Palm.: 3 (1854) = ? [42 PHI]

*Saguerus manillensis H.Lodd

Astrocaryum arenarium Barb.Rodr., Palm. Mattogross.: 53 (1898) = ? [84 BZC]

Astrocaryum argenteum auct., Gard. Chron. 1875(1): 491 (1875) = ?

Astrocaryum burity Barb.Rodr., Sert. Palm. Brasil. 2: 74 (1903) = ? [84]

Astrocaryum echinatum Barb.Rodr., Palm. Mattogross.: 51 (1898) = ? [84 BZC]

Astrocaryum filare W.Bull ex Hook.f., Rep. Progr. Condition Roy. Bot. Gard. Kew 1882: 71 (1884) = ?

Astrocaryum flexuosum H.Wendl. in O.C.E.de Kerchove de Denterghem, Palmiers: 232 (1878) = ?

Astrocaryum iriartoides Willis ex Regel., Gartenflora 29: 230 (1880) = ?

Astrocaryum kewense Barb.Rodr., Sert. Palm. Brasil. 2: 70 (1903) = ? [84]

Astrocaryum panamense Linden, Ill. Hort. 38: 15 (1881) = ? [83 CLM]

Astrocaryum pumilum H.Wendl. in O.C.E.de Kerchove de Denterghem, Palmiers: 232 (1878) = ?

Astrocaryum pygmaeum Drude in C.F.P.von Martius & auct. suc. (eds.), Fl. Bras. 3(2): 385 (1881) = ? [84]

Astrocaryum sclerophyllum Drude in C.F.P.von Martius & auct. suc. (eds.), Fl. Bras. 3(2): 377 (1881) = ? [84]

Astrocaryum tenuifolium Linden, Ill. Hort. 28: 15 (1881) = ? [84]

Astrocaryum warszewiczii H.Karst., Wochenschr. Gärtnerei Pflanzenk. 1: 297 (1858) = ?

Astrocaryum weddellii Drude in C.F.P.von Martius & auct. suc. (eds.), Fl. Bras. 3(2): 383 (1881) = ? [84]

Attalea boehmii Drude in H.G.A.Engler & K.A.E.Prantl (eds.), Nat. Pflanzenfam., Nachtr. 1: 59 (1897) = ? [2]

Attalea ceraensis Barb.Rodr., Pl. Jard. Rio de Janeiro 6: 22 (1898) = ? [84]

Attalea coronata Lodd. ex H.Wendl., Index Palm.: 5 (1854) = ?

Attalea grandis H.Wendl. in O.C.E.de Kerchove de Denterghem, Palmiers: 233 (1878) = ?

Attalea limbata Seem. ex H.Wendl. in O.C.E.de Kerchove de Denterghem, Palmiers: 233 (1878) = ?

Attalea magdalenae Linden, Ill. Hort. 28: 15 (1881) = ? [83 CLM]

Attalea manaca Linden, Ill. Hort. 28: 15 (1881) = ? [83 CLM]

Attalea purpurea Linden, Ill. Hort. 28: 16 (1881) = ? [83 CLM]

Attalea puruensis Linden, Ill. Hort. 28: 16 (1881) = ? [84 BZN]

Attalea rossii Lodd. ex Loudon, Hort. Brit.: 387 (1830) = ? [84]

Attalea spinosa Meyen, Observ. Bot. 1: 469 (1843) = ? [83 PER]

Attalea tiasse Linden, Ill. Hort. 28: 16 (1881) = ? [84]

Attalea venatorum Mart., Hist. Nat. Palm. 3: 325 (1853) = ?

Avoira conanam Giseke, Prael. Ord. Nat. Pl.: 38 (1792) = ? [82 FRG]

Avoira nodosa Giseke, Prael. Ord. Nat. Pl.: 38 (1792) = ? [82 FRG]

Avoira scandens Giseke, Prael. Ord. Nat. Pl.: 38 (1792) = ? [82 FRG]

Avoira sylvestris Giseke, Prael. Ord. Nat. Pl.: 38 (1792) = ? [82 FRG]

Avoira uliginosa Giseke, Prael. Ord. Nat. Pl.: 38 (1792) = ? [82 FRG]

Avoira vulgaris Giseke, Prael. Ord. Nat. Pl.: 38 (1792) = ? [82 FRG] Astrocaryum sp.

Bactris confluens Linden & H.Wendl., Linnaea 28: 347 (1856) = ? [82 VEN]

Bactris megistocarpa Burret, Notizbl. Bot. Gart. Berlin-Dahlem 12: 623 (1935) = ? [84 BZN] Astrocaryum ?

Bactris vexans Burret, Notizbl. Bot. Gart. Berlin-Dahlem 14: 266 (1938) = ? [84 BZN]

Brahea conduplicala Linden, Ill. Hort. 28: 16 (1881) = ?

× **Butiarecastrum** Prosch., Rev. Hort. 93: 290 (1921) = ? Butyagrus

× **Butiarecastrum nabonnandii** Prosch., Rev. Hort. 93: 290 (1921) = × Butyagrus nabonnandii

× **Butyagrus** Vorster, Taxon 39: 662 (1990) = Butia ? Syagrus

× **Butyagrus nabonnandii** (Prosch.) Vorster, Taxon 39: 662 (1990) = Butia capitata × Syagrus romzanoffiana

*× Butiarecastrum nabonnandii Prosch.

Calamus cochinchinensis Hook.f., Rep. Progr. Condition Roy. Bot. Gard. Kew 1882: 67 (1884), nom. inval. = ? [41 VIE]

Calamus elegans H.Wendl. in O.C.E.de Kerchove de Denterghem, Palmiers: 236 (1878) = ? [41 VIE]

Calamus extensus Roxb., Fl. Ind. ed. 1832, 3: 777 (1832) = ? [40 BAN]

Calamus maximus Reinw. ex de Vriese, Hort. Spaarn-Berg.: 35 (1839), nom. illeg. = ?

Calamus maximus Blanco, Fl. Filip.: 266 (1837), nom. illeg. = ? [42 PHI]

Calamus nicolai Hook.f., Rep. Progr. Condition Roy. Bot. Gard. Kew 1882: 67 (1884), nom. inval. = ?

Calamus petraeus Lour., Fl. Cochinch.: 209 (1790) = ? [41 VIE] Plectocomia ?

Calamus robustus L.Linden & Rodigas, Ill. Hort. 1893: 19 (1893) = ? [42 BOR]

Calamus rubiginosus Ridl., Agric. Bull. Straits Fed. Malay States, II, 2: 7 (1906), nom. nud. = ? [42 BOR]

Calamus scipionum Lam., Encycl. 6: 304 (1804), nom. illeg. = ?

Calamus trinervis W.Watson, Gard. Chron., n.s., 1884: 729 (1884) = ?

Calamus verus Lour., Fl. Cochinch.: 210 (1790) = ? [41 VIE] Daemonorops sp.

Calamus viminalis Reinw. ex Mart., Hist. Nat. Palm. 3: 335 (1853), nom. illeg. = ?

Calyptrogyne elata H.Wendl. in ? = ?

Caryota arenga Mezieres ex Desjardins, Rapp. Annuel Trav. Soc. Hist. Nat. Ile Maurice 6: 28 (1835) = ?

Caryota javanica Osbeck, Dagb. Ostind. Resa: 270 (1757) = Korthalsia sp.

Caryota kiriwongensis Hodel, Palm J. 139: 53 (1998), without diagnostic latin descr. = ? [41 THA]

Caryota majestica Linden, Ill. Hort. 28: 16 (1881) = ? [42 PHI]

Caryota princeps Voigt, Syll. Ratisb. 2: 51 (1828) = ?

Ceratolobus forgetiana auct., Gard. Chron., III, 51(Suppl.): xv (1912) = ?

Ceratolobus micholtziana auct., Gard. Chron., III, 23: 251 (1898) = Plectocomia sp. ?

Chamaedorea andreana Linden, Ill. Hort. 28: 16 (1881) = ? [83 CLM]

Chamaedorea glauca Linden, Ill. Hort. 28: 16 (1881) = ? [8]

Chamaedorea × katzeri Loebner, Gartenwelt 13: 159 (1909) = C. ernesti-augustii × C. sartorii

Chamaedorea × romana Guillaumin, J. Soc. Natl. Hort. France, IV, 24: 243 (1923) = Chamaedorea × katzeri

Chamaedorea wallisii Linden, Ill. Hort. 28: 16 (1881) = ? [8]

Chamaedorea wendlandii H.Wendl. in O.C.E.de Kerchove de Denterghem, Palmiers: 240 (1878) = ?

Chamaerops ghiesbreghtii H.Wendl. in O.C.E.de Kerchove de Denterghem, Palmiers: 240 (1878) = Sabal ?

Chamaerops macrocarpa Linden, Cat. Gén. 87: 87 (1871) = ?

Chamaerops mitis J.Mey., Mém. Acad. Roy. Sci. Hist. (Berlin) 1769: 29 (1769) = ?

Coccothrinax baileyana O.F.Cook, Natl. Hort. Mag. 20: 30 (1941) = ?

Cocos barbosii Barb.Rodr., Sert. Palm. Brasil. 1: 86 (1903) = ?

Cocos chloroleuca Barb.Rodr., Contr. Jard. Bot. Rio de Janeiro 6: 135 (1907) = ?

Cocos cogniauxiana Barb.Rodr., Sert. Palm. Brasil. 1: 102 (1903) = ?

Cocos edulis Barb.Rodr., Contr. Jard. Bot. Rio de Janeiro 4: 105 (1907) = ?

Cocos elegantissima H.Wendl. in O.C.E.de Kerchove de Denterghem, Palmiers: 241 (1878) = ?

Cocos gaertneri W.Watson, Gard. Chron., n.s., 1885(1): 439 (1885) = ?

Cocos iagua Sessé & Moç., Fl. Mexic., ed. 2: 240 (1894) = ?

Cocos lapidea Gaertn., Fruct. Sem. Pl. 1: 16 (1788) = ? [84]

Cocos mamillaris Blanco, Fl. Filip.: 722 (1837) = ?

Cocos naja Arruda ex Kunth, Enum. Pl. 3: 288 (1841) = ?

Cocos nolaia-assu Wied-Neuw., Reise Bras. 1: 271 (1820) = ?

Cocos orbignyana Becc., Malpighia 2: 147 (1888) = ?

Cocos pityrophylla Mart. in A.D.d'Orbigny, Voy. Amér. Mér. 7(3): 99 (1844) = ?

Cocos purusana Huber, Bull. Herb. Boissier, II, 6: 271 (1906) = ? [84 BZN]

Cocos romanzoffanopulposa Barb.Rodr., Sert. Palm. Brasil. 1: 116 (1903) = ?

Cocos sapida Barb.Rodr., Palm. Paraguay.: 12 (1899) = ?

Cocos ventricosa Arruda in H.Koster, Trav. Brazil: 485 (1816) = ?

Cocos virgata A.Usteri, Guia Bot. Praca Rep. e Jard. Luz: 13 (1919) = ?

Cocos yurumaguas H.Wendl. in O.C.E.de Kerchove de Denterghem, Palmiers: 241 (1878) = ?

Copernicia robusta Linden, Ill. Hort. 28: 16 (1881) = Licuala ?

Corypha glaucescens Lodd. ex Loudon, Hort. Brit.: 125 (1830) = Sabal sp.

Cyphokentia heanei auct., Rev. Hort. 58: 230 (1886) = ?

Daemonorops plumosus W.Bull, Gard. Chron. 1870: 1086, f. 206 (1870) = ?

Deckeria elegans Linden, Ill. Hort. 28: 16 (1881) = ?

Deckeria nobilis H.Wendl. ex Seem. in ? = ?

Desmoncus andicola Pasq., Cat. Orto Bot. Napoli: 36 (1867) = ? [83]

Desmoncus granatensis W.Bull, Cat. 1876: 142 (1876) = ? [83 CLM]

Desmoncus grandifolius Linden, Ill. Hort. 28: 16 (1881) = ? [83 CLM]

Desmoncus intermedius Mart. ex H.Wendl. in O.C.E.de Kerchove de Denerghem, Palmiers: 243 (1878) = ?

Desmoncus latifrons W.Bull in ? = ?

Desmoncus panamensis Linden, Ill. Hort. 28: 16 (1881) = ?

Desmoncus wallisii Linden, Ill. Hort. 28: 16 (1881) = ? [84]

Dictyocaryum glaucescens Linden, Ill. Hort. 28: 31 (1881) = ?

Dictyocaryum wallisii H.Wendl. in O.C.E.de Kerchove de Denerghem, Palmiers: 243 (1878) = ?

Didymosperma gracile Hook.f., Fl. Brit. India 6: 420 (1892) = ? [40 ASS]

Didymosperma humile K.Schum. & Lauterb., Fl. Schutzgeb. Südsee: 204 (1900) = ? [43 NWG]

Diplothemium henryanum F.Br., Bernice P. Bishop Mus. Bull. 84: 128 (1931) = ? [61 MRQ]

Drymophloeus mooreanus auct., Gard. Chron. 1903: t. 266 (1903) = ?

Dypsis humbertii (Jum.) Beentje & J.Dransf., Palms Madagascar: 239 (1995), nom. illeg. = ? [29 MDG]

*Neophloga humbertii Jum.

Dypsis madagascariensis (Becc.) Beentje & J.Dransf., Palms Madagascar: 185 (1995), nom. illeg. = ? [29 MDG]

*Chrysalidocarpus madagascariensis Becc.

Elaeis pernambucana Lodd. ex G.Don in J.C.Loudon, Hort. Brit.: 399 (1830) = ? [84]

Elaeis spectabilis Lodd. ex Sweet, Hort. Brit., ed. 3: 716 (1839) = ? [40]

Euterpe disticha H.Wendl. ex Linden, Cat. Gén. 23: (1868) = ? [83 CLM]

Euterpe elegans Linden, Ill. Hort. 28: 31 (1881) = ? [83 CLM]

Euterpe gracilis Linden, Cat. Gén. 17: (1862) = ? [84]

Euterpe puruensis Linden, Ill. Hort. 28: 32 (1881) = ? [84]

Gaussia ghiesbreghtii H.Wendl. in O.C.E.de Kerchove de Denterghem, Palmiers: 245 (1878) = ? [81]

Geonoma amazonica H.Wendl. in O.C.E.de Kerchove de Denterghem, Palmiers: 245 (1878) = ? [84]

Geonoma bluntii auct., Gard. Chron., n.s., 15: 766 (1881) = ?

Geonoma caudescens H.Wendl. ex Drude in C.F.P.von Martius & auct. suc. (eds.), Fl. Bras. 3(2): 504 (1882) = ? [84]

Geonoma chiriquensis Linden ex Hook.f., Rep. Progr. Condition Roy. Bot. Gard. Kew 1882: 60 (1884) = ?

Geonoma congestissima Burret, Bot. Jahrb. Syst. 63: 224 (1930) = ? [83 PER]

Geonoma decora L.Linden & Rodigas, Ill. Hort. 41: 364 (1894) = ? [84]

Geonoma frigida Linden, Ill. Hort. 28: 31 (1881) = ? [83 CLM]

Geonoma gracilipes Dammer ex Burret, Bot. Jahrb. Syst. 63: 173 (1930), nom. provis. = ? [83 PER]

Geonoma herbstii auct., Garden (London 1871-1927) 35: 463 (1889) = ?

Geonoma hoppii Burret, Notizbl. Bot. Gart. Berlin-Dahlem 11: 235 (1931) = ? [83 ECU]

Geonoma imperialis Linden, Ill. Hort. 27: 31 (1881) = ? [83 CLM]

Geonoma insignis Burret, Notizbl. Bot. Gart. Berlin-Dahlem 15: 28 (1940) = ? [83 ECU]

Geonoma iraze Linden, Ill. Hort. 28: 31 (1881) = ? [82 VEN]

Geonoma lacerata auct., Fl. Mag. (London) 8: t. 446 (1869) = ? [80]

Geonoma macrophylla Burret, Notizbl. Bot. Gart. Berlin-Dahlem 15: 27 (1940) = ? [83 CLM]

Geonoma princeps Linden, Ill. Hort. 28: 31 (1881) = ? [83 PER]

Geonoma pulchella H.Wendl. ex Linden, Ill. Hort. 28: 31 (1881) = ? [83 CLM]

Geonoma riedeliana H.Wendl. in O.C.E.de Kerchove de Denterghem, Palmiers: 245 (1878) = ? [84]

Geonoma seemannii auct., Fl. Mag. (London) 8: t. 428 (1869) = ? [80]

Geonoma stenothyrsa Burret, Bot. Jahrb. Syst. 63: 197 (1930) = ? [83 CLM]

Geonoma tenuifolia auct., Ill. Hort. 42: 186 (1895) = ? [83 PER]

Geonoma trichostachys Burret, Notizbl. Bot. Gart. Berlin-Dahlem 11: 862 (1933) = ? [83 CLM]

Geonoma ventricosa Engel, Linnaea 33: 688 (1865) = ? [83 CLM]

Geonoma verdugo Linden, Ill. Hort. 28: 31 (1881) = ? [83 CLM]

Geonoma wendlandii auct., Gard. Chron., n.s., 9: 440 (1878) = ?

Geonoma zamorensis Linden ex H.Wendl. in O.C.E.de Kerchove de Denterghem, Palmiers: 246 (1878) = ? [83 ECU]

Haplophloga comorensis Baill., Bull. Mens. Soc. Linn. Paris 2: 1171 (1894) = ? [29 COM]

Hyospathe chiriquensis Linden ex H.Wendl. in O.C.E.de Kerchove de Denterghem, Palmiers: 247 (1878) = ? [80 COS]

Hyospathe elata Hook.f., Rep. Progr. Condition Roy. Bot. Gard. Kew 1882: 68 (1884) = ? [42 SUL]

Hyphaene cuciphera Pers., Syn. Pl. 2: 623 (1807) = ? [2]

Hyphaene violascens Becc., Bot. Jahrb. Syst. 38(87): 8 (1906) = ?

Iguanura speranskyana Bois, J. Hort. Soc. France 1899: 665 (1899) = Geonoma ? [84]

Inodes vestita O.F.Cook, Bull. Torrey Bot. Club 28: 533 (1901) = ? [81]

Iriartea affinis H.Karst. ex Linden, Ill. Hort. 28: 31 (1881) = ? [83 CLM]

Iriartea costata Linden, Ill. Hort. 28: 31 (1881) = ? [83 CLM]

Iriartea glaucescens Linden, Ill. Hort. 28: 31 (1881) = ? [83 CLM]

Iriartea pygmaea Linden, Ill. Hort. 28: 31 (1881), nom. nud. = ? [84]

Iriartea xanthorhiza Klotzsch ex Linden, Ill. Hort. 28: 31 (1881) = ? [82 VEN]

Iriartea zamorensis Linden, Ill. Hort. 28: 31 (1881) = ? [83 ECU]

× **Jubautia** Demoly, J. Bot. Soc. Bot. France 18-19: 189 (2002 publ. 2003) = Butia × Jubaea

Kentia rubra A.Usteri, Guia Bot. Praca Rep. e Jard. Luz: 14 (1919) = ?

Kentia rupicola Linden, Gard. Chron., n.s., 9: 440 (1878) = ?

Klopstockia interrupta H.Karst., Linnaea 28: 252 (1856) = ? [83 CLM]

Klopstockia utilis H.Karst., Linnaea 28: 252 (1856) = ? [83 CLM]

Licuala hospita Burret, Notizbl. Bot. Gart. Berlin-Dahlem 15: 332 (1941) = ? [42 MLY]

Licuala kersteniana André, Rev. Hort. 67: 249 (1895) = ?

Licuala kirsteniana André, Ill. Hort. 42: 189 (1895) = ?

Linospadix leopoldii Sander, Gard. Chron. 1903(1, Suppl. Apr.): 25 (1903), nom. inval. = ?

Livistona bissula Mart., Hist. Nat. Palm. 3: 329 (1853) = ? [42 SUL] Licuala sp.

Livistona enervis auct., Wiener Ill. Gart.-Zeitung 16: 346 (1891) = ?

Livistona macrophylla Roster, Bull. Soc. Tosc. Ortic. 29: 82 (1904) = ?

Livistona moluccana H.Wendl. in O.C.E.de Kerchove de Denterghem, Palmiers: 250 (1878) = ? [42]

Malortiea lacerata H.Wendl. in O.C.E.de Kerchove de Denterghem, Palmiers: 250 (1878) = ?

Martinezia abrupta Ruiz & Pav., Fl. Peruv. Prodr.: 148 (1794) = ? [83 PER]

Martinezia disticha Linden, Cat. Gén. 93: 32 (1875) = ? [83 CLM]

Martinezia leucophoeus auct., Gard. Chron. 1875(1): 589 (1875) = ? [83 CLM]

Martinezia roezlii auct., Gard. Chron. 1876(1): 735 (1876) = ? [83 CLM]

Mauritia piritu Linden, Ill. Hort. 28: 32 (1881) = ? [84]

Maximiliana argentinensis Speg., Physis (Buenos Aires) 3: 169 (1917) = ? [85 AG]

Maximiliana venatorum H.Wendl. in O.C.E.de Kerchove de Denterghem, Palmiers: 251 (1878) = ? [83 PER]

Micronoma H.Wendl. ex Benth. & Hook.f., Gen. Pl. 3: 882 (1883) = ?

× **Microphoenix** Naudin ex Carr., Rev. Hort. 57: 513 (1885), nom. inval. = Chamaedorea × Phoenix

× **Microphoenix decipiens** Naudin ex Carr., Rev. Hort. 57: 513 (1885) = Chamaedorea humilis × Phoenix dactylifera

× **Microphoenix sahuti** Carrière, Rev. Hort. 57: 513 (1885) = Chamaedorea humilis × Phoenix dactylifera × Trachycarpus fortunei

Nunnezharia demaniana Kuntze, Revis. Gen. Pl. 2: 731 (1891), nom. inval. = ?

Nunnezharia eburnea Kuntze, Revis. Gen. Pl. 2: 731 (1891), nom. inval. = ?

Nunnezharia liboniana Kuntze, Revis. Gen. Pl. 2: 731 (1891), nom. inval. = ?

Nunnezharia polita Kuntze, Revis. Gen. Pl. 2: 731 (1891), nom. inval. = ?

Oenocarpus altissimus Klotzsch ex H.Wendl., Index Palm.: 30 (1854) = ? [82 VEN]

Oenocarpus bolivianus H.Wendl., Index Palm.: 30 (1854) = ? [83 BOL]

Oenocarpus chiragua H.Wendl., Index Palm.: 30 (1854) = ? [82 VEN]

Oenocarpus cubarro H.Wendl., Index Palm.: 30 (1854) = ?

Oenocarpus edulis W.Watson, Gard. Chron. 1887(2): 157 (1887) = ?

Oenocarpus iriartoides H.Karst. & Triana in J.J.Triana, Nuev. Jen. Esp.: 15 (1855) = ? [83 CLM]

Oenocarpus pulchellus Linden, Ill. Hort. 28: 32 (1881) = ? [83 CLM]

Orbignya excelsa Barb.Rodr., Pl. Jard. Rio de Janeiro 1: 32 (1891) = ? [84]

Oreodoxa granatensis W.Watson, Gard. Chron. 1887(2): 304 (1887) = ? [83 CLM]

Palma amboinensis Garsault, Fig. Pl. Méd.: t. 19 (1764) = ? [42 MOL] Daemonorops ?

Palma avoira Aubl., Hist. Pl. Guiane, Suppl.: 95 (1775) = ? [82 FRG]

Palma bache Aubl., Hist. Pl. Guiane, Suppl.: 103 (1775) = ? [82 FRG]

Palma coman Aubl., Hist. Pl. Guiane, Suppl.: 102 (1775) = ? [82 FRG]

Palma mocaia Aubl., Hist. Pl. Guiane 2: 976 (1775) = ? [82 FRG]

Palma paripou Aubl., Hist. Pl. Guiane 2: 974 (1775) = ? [82 FRG]

Palma patavoua Aubl., Hist. Pl. Guiane, Suppl.: 102 (1775) = ? [82 FRG]

Palma pinao Aubl., Hist. Pl. Guiane 2: 974 (1775) = ? [82 FRG] Calyptrogyne sp.

Palma zagueneti Aubl., Hist. Pl. Guiane 2: 976 (1775) = ? [82 FRG]

Palmijuncus maximus Kuntze, Revis. Gen. Pl. 2: 733 (1891) = ?

*Calamus maximus Blanco

Paripon Voigt, Syll. Ratisb. 2: 51 (1828) = ?

Paripon palmiri Voigt, Syll. Ratisb. 2: 51 (1828) = ?

Phloga microphoenix Baill., Bull. Mens. Soc. Linn. Paris 2: 1185 (1895) = ? [29 MDG]

Phytelephas bonplandiana Gaudich., Voy. Bonite, Bot.: t. 30 (1841) = ?

Phytelephas endlicheriana Gaudich., Voy. Bonite, Bot.: t. 30 (1841) = ?

Phytelephas humboldtiana Gaudich., Voy. Bonite, Bot.: t. 20 (1841) = ?

Phytelephas kunthiana Gaudich., Voy. Bonite, Bot.: t. 30 (1841) = ?

Phytelephas orbignyana Gaudich., Voy. Bonite, Bot.: t. 29 (1841) = ?

Phytelephas pavonii Gaudich., Voy. Bonite, Bot.: t. 29 (1841) = ?

Phytelephas persooniana Gaudich., Voy. Bonite, Bot.: t. 30 (1841) = ?

Phytelephas poeppigii Gaudich., Voy. Bonite, Bot.: t. 16 (1841) = ?

Phytelephas ruizii Gaudich., Voy. Bonite, Bot.: t. 15 (1841) = ?

Phytelephas willdenowiana Gaudich., Voy. Bonite, Bot.: t. 30 (1841) = ?

Pinanga lepida W.Bull, Gard. Chron. 1888(2): 273 (1888) = ? [40 IND]

Pinanga sanderiana W.Bull, Cat. 1885: 15 (1885) = ? [42 MLY]

Pinanga spectabilis W.Bull ex Becc., Malesia 3: 115 (1886) = ?

Pinanga vonmohlii Linden ex W.Watson, Gard. Chron., III, 13: 260 (1893) = ?

Pritchardia aurea Linden, Rev. Hort. 50: 186 (1878) = ?

Pritchardia elliptica Rock & Caum, Occas. Pap. Bernice Pauahi Bishop Mus. 9(5): 14 (1930) = ? [63 HAW]

Pseudopinanga acuminata Burret, Notizbl. Bot. Gart. Berlin-Dahlem 15: 210 (1940) = ? [42 BOR] Pinanga sp.

Pseudopinanga anomodonta Burret, Notizbl. Bot. Gart. Berlin-Dahlem 13: 196 (1936) = ? [42 SUL] Pinanga sp.

Pseudopinanga kjellbergii Burret, Notizbl. Bot. Gart. Berlin-Dahlem 13: 194 (1936) = ? [42 SUL] Pinanga sp.

Pseudopinanga macrorhachis Burret, Notizbl. Bot. Gart. Berlin-Dahlem 13: 194 (1936) = ? [42 SUL] Pinanga sp.

Pseudopinanga multisecta Burret, Notizbl. Bot. Gart. Berlin-Dahlem 13: 197 (1936) = ? [42 SUL] Pinanga sp.

Pseudopinanga paucisecta Burret, Notizbl. Bot. Gart. Berlin-Dahlem 13: 192 (1936) = ? [42 BOR] Pinanga sp.

Ptychorhaphis siebertiana auct., Gard. Chron., III, 43: 257 (1908) = ?

Ptychosperma advenum Becc., Atti Soc. Tosc. Sci. Nat. Pisa Processi Verbali 44: 141 (1934) = ? [43 NWG]

Ptychosperma nobilis Seem., Gard. Chron. 1870: 697 (1870) = ?

Ptychosperma warleti Sander, Gard. Chron. 1898(1): 242 (1898) = ?

Rhopaloblaste princeps W.Bull, Gard. Chron., n.s., 13: 759 (1880) = ?

Rotang scandens (Blume) Baill., Hist. Pl. 13: 300 (1895) = Daemonorops scandens

*Daemonorops scandens Blume

Roystonea aitia O.F.Cook, Natl. Hort. Mag. 18: 106 (1939), no latin descr. = ? [81 HAI]

Sabal elata Mart., Hist. Nat. Palm. 3: 320 (1853) = ?

Sabal ghiesbrechtii Pfister, Beitr. Vergl. Anat. Sabaleenblatter: 41 (1892) = ?

Sabal havanensis Lodd. ex Mart., Hist. Nat. Palm. 3: 320 (1853) = ?

Sabal oleracea Mart., Hist. Nat. Palm. 3: 348 (1853) = ?

Sabal princeps Knuth, Handb. Blutenb. 3: 65 (1904) = ?

Sabal speciosa L.H.Bailey, Gentes Herb. 3: 303 (1934) = ?

Sagus blackallii W.Hill, Catalog. Pl. Brisbane Bot. Gard. 21 (1875) = ?

Sagus hospita Kerch., Palmiers: 162 (1878) = ?

Salacca dubia Becc., Malesia 3: 68 (1886) = Salacca affinis

Salacca nitida W.Bull, Cat. 1886: 9 (1886) = ? [22]

Scheelea cubensis Burret, Notizbl. Bot. Gart. Berlin-Dahlem 10: 671 (1929) = ? [81 CUB]

Scheelea imperialis auct., Gard. Chron. 1875(1): 589 (1875) = ? [83 CLM]

Scheelea unguis G.Nicholson, Ill. Dict. Gard. 3: 385 (1886) = ?

Seaforthia angustifolia Mart., Hist. Nat. Palm. 3: 314 (1849) = ? [43 NWG]

Seaforthia veitchii auct., Gard. Chron. 1870: 37 (1870) = ? [50]

Socratea altissima Burret, Notizbl. Bot. Gart. Berlin-Dahlem 10: 925 (1930), nom. provis. = ? [82 VEN]

Syagrus × nabonnandii (Prosch.) Demoly, Bull. Assoc. Parcs Bot. France 12: 30 (1989) = × Butyagrus nabonnandii

*× Butiarecastrum nabonnandii Prosch.

Syagrus × teixeiriana Glassman, Fieldiana, Bot. 32: 27 (1968) = ? [84 BZL]

Syagrus wallisii Linden, Cat. Gén. 16: (1861) = ? [83 CLM]

Thrinax arborea Hook.f., Rep. Progr. Condition Roy. Bot. Gard. Kew 1882: 64 (1884) = ?

Thrinax aurantia Fulchir. ex Schult. & Schult.f., Syst. Veg. 7: 1301 (1830) = ?

Thrinax ferruginea Lodd. ex Mart., Hist. Nat. Palm. 3: 320 (1853) = ? [81 JAM]

Thrinax grandis Kerch., Palmiers: 258 (1878) = ? [81 CUB]

Thrinax pumilio Lodd. ex Schult. & Schult.f., Syst. Veg. 7: 1301 (1830) = ?

Veitchia subglobosa H.Wendl. in B.Seemann, Fl. Vit.: 272 (1868) = ? [60 FIJ]

Wallichia zebrina B.S.Williams, Gard. Chron., n.s., 13: 759 (1880) = ?

Printed in the United Kingdom
by Lightning Source UK Ltd.
103918UKS00002B/1-60